AF778325

Isaac N. Gibra

Stevens Institute of Technology

PROBABILITY AND STATISTICAL INFERENCE FOR SCIENTISTS AND ENGINEERS

Prentice-Hall, Inc.

Englewood Cliffs, New Jersey

© 1973 by Prentice-Hall, Inc.
Englewood Cliffs, N.J.

All rights reserved. No part of this book may be reproduced in any form or by any means without permission in writing from the publisher.

10 9 8 7 6 5 4 3 2 1

ISBN: 0-13-711622-5

Library of Congress Catalog Card No. 78-107395

Printed in the United States of America

PRENTICE-HALL INTERNATIONAL, INC., *London*
PRENTICE-HALL OF AUSTRALIA, PTY. LTD., *Sydney*
PRENTICE-HALL OF CANADA, LTD., *Toronto*
PRENTICE-HALL OF INDIA PRIVATE LIMITED, *New Delhi*
PRENTICE-HALL OF JAPAN, INC., *Tokyo*

To the memory of
Esther C. Gibra

CONTENTS

Part **II**

STATISTICAL INFERENCE AND APPLICATIONS

APPENDIX TABLES 550

PREFACE

In recent years modern techniques of probability theory have been applied to a wide variety of fields. The differentiated and quantitative concept of probability provided the structure of mathematical statistics. Probability and statistical techniques are finding increased use in the fields of engineering, science and industry. The broad application of probability and statistics demands that greater attention be paid to the technical training of scientists and engineers.

Although there are many excellent textbooks on probability and statistics, few are directed toward engineering students who have an elementary knowledge of differential and integral calculus. Many capable engineering students have experienced difficulty in achieving an appropriate background in both disciplines. It seems that this difficulty is due to a rigorous, sophisticated and over-complicated approach to the subject.

The author's prime purpose has been to provide a text where the exposition has been attuned to possible self-study and a minimum of mathematical requirement. Thus, the book can serve well as a text for an in-plant training program and also as a reference for practicing engineers and research workers.

This book is divided into two parts:

Part I provides a fundamental one-semester course in probability theory, whereas Part II provides a one-semester course in statistical inference. A balanced selection from both parts will make a solid one-semester course in probability and statistical inference. More material is covered than could normally be studied in just a single semester, and this makes it possible for the instructor to have some latitude in the choice of topics to be covered and the depth to which any given topic will be studied. The textual material is designed to provide a sound basis for advanced courses in both disciplines. However, should the reader become interested in further explication of a certain topic he can easily pursue that subject through the references provided by the extensive bibliography given at the end of each chapter. Detailed solutions of numerous illustrative examples are given to clarify the concepts that are established. Many problems of varying levels of difficulty are included at the end of each chapter to enable the reader to check his understanding of the concepts introduced. The reader is also provided with answers to selected exercises to enable him to test his achievement.

I am indebted to the publishers and authors who have given permission to reproduce their charts and tables. Specific acknowledgements are made throughout the text.

With deep gratitude I wish to acknowledge the moral encouragement, vital contribution, and constructive suggestions made by Dr. S. B. Littauer, Professor of Operations Research, Columbia University.

I express my appreciation to S. Parris who carefully typed much of the manuscript. Errors may escape detection in the most carefully and diligently checked manuscript; for these I must assume final responsibility. Comments and suggestions for improvement of the text will be greatly appreciated.

ISAAC N. GIBRA

Part I

PROBABILITY

1

ELEMENTS OF PROBABILITY THEORY

1.1 Historical Introduction to Probability Theory

The theory of *probability* was originated by two famous French mathematicians, Blaise Pascal and Pierre de Fermat, in 1654. Antoine Gombaud, *Chevalier de Mere* (a French nobleman with deep interest in gambling), evoked Pascal's attention to an apparent contradiction concerning a popular dice game. This gambling problem and others posed by the French nobleman motivated Pascal and Fermat to formulate the fundamental principles of probability theory. No general theory of probability was developed before 1654, though a few problems on games, of chance had been solved by some Italian mathematicians in the fifteenth and sixteenth centuries.

In 1657 the Dutch Scientist Christian Huygens published the first book on probability that dealt primarily with problems associated with gambling. Probability theory soon became popular, and the subject developed rapidly during the eighteenth century. Among the major contributors during that period were J. Bernoulli, A. de Moivre, and P. de Laplace. The basic mathematical techniques of the calculus of probability were introduced by Laplace in his book, *Théorie Analytique des Probabilitiés*. Since Laplace's time many scientists have contributed to the theory, among the most important being Chebyshev, Markov, von Mises, and Kolmogorov.

In the mideighteenth century the theory was applied to errors of measurement, actuarial mathematics, and statistical mechanics.

In recent decades the growth of probability theory has been felt by the variety of its applications. The most important feature of its growth has been in the field of mathematical statistics such as statistical inference and decision theory. Other applications occur in such different fields as genetics, psychology, economics, and the physical, political, and engineering sciences.

In recent years the differentiated and quantitative concept of probability provided the structure of modern techniques such as operations research, game theory, information theory, queueing theory, statistical quality control, and reliability.

1.2 Some Basic Notions of Set Theory

1.2.1 Sets

The terminology and notation of *sets* are extensively applied in the formulation of modern probability theory.

Definition 1.1. A *set* is merely an aggregate or collection of objects viewed as a single entity.

Sets usually are denoted by capital letters: $A, B, C, \ldots, Y, Z$; *elements* are designated by small letters: $a, b, c, \ldots, y, z$. If a is an element and A a set, then we write $a \in A$ as an abbreviation for "a is an element of A" and $a \notin A$ for "a is not an element of A."

Example 1.1 The set whose elements are the integers 1, 3, 5, and 7 is a finite set with four elements. If we denote this set by X, then it can be written using the roster notation $X = \{1, 3, 5, 7\}$. The statements "$1 \in X$," "$3 \in X$," "$5 \in X$," and "$4 \notin X$" are true.

Two sets X and Y are said to be *equal* (or identical) if and only if they consist of exactly the *same* elements, in which case we write $X = Y$. If one of the sets has an element *not* in the other, they are *unequal* and we write $X \neq Y$.

Example 1.2 The two sets $\{2, 4, 6\}$ and $\{6, 2, 4\}$ are equal since they have exactly the same three elements. Thus, when the roster notation is used to describe a set, the order in which we list the elements is immaterial.

1.2.2 Subsets

From a given set we may form new sets, called *subsets* of the given set. We say a set X is a subset of set Y if each element of X is also an element of Y, written $X \subseteq Y$. The relation $\subseteq$ is referred to as *set inclusion*. The definition of subset implies that a *set is a subset of itself*, i.e., $X \subseteq X$ is always true. The set inclusion is a *transitive* relation, because if $X \subseteq Y$ and $Y \subseteq Z$, it follows that $X \subseteq Z$. The basic principle of set equality in the preceding section was formulated in terms of the subset relation. In other words,

$$X = Y \quad \text{if and only if} \quad X \subseteq Y \quad \text{and} \quad Y \subseteq X$$

When a set contains no elements, it is called an *empty or void or null set* and is denoted by the symbol $\varnothing$.

Example 1.3 Let Y be the set of 36 outcomes of the experiment in which two perfect dice are rolled. Accordingly we have 2^{36} subsets of Y. Suppose, now, we define subset X as the subset containing the outcomes for which the sum of spots on the two dice is 4; then we can write

$$X = \{(1, 3), (2, 2), (3, 1)\}$$

1.2.3 The Algebra of Sets

In this section we consider the relations that can exist among sets and the operations that can be performed on sets.

We now define the three basic operations on sets:

Definition 1.2. $A \cup B$ (*A union B*) is the set of elements that belong to at *least* one of the sets A and B, i.e., to A or B.

Definition 1.3. $A \cap B$ (*A intersection B*) is the set of elements *common* to both A and B

Definition 1.4. A' (*complement* of A) is the set of elements of a *universal set* (A and B are subsets of the universal set) that do *not* belong to A.

When considering sets and operations on sets, it is helpful to represent sets diagrammatically. Such diagrams, called *Venn diagrams*, can be drawn to illustrate the relations among subsets. A rectangle is usually used to represent the universal set, points to designate elements of the set, and circles or ellipses to depict subsets of the universal set. Figure 1.1 shows some of these relations. In order to consider another application of Venn diagrams, we include the following definition:

Definition 1.5. Two sets are said to be *disjoint* or *mutually exclusive* if they have no elements in common, i.e., if $A \cap B = \varnothing$.

Operations on sets have many of the algebraic properties of ordinary addition and multiplication of numbers. Among the important algebraic properties are the following relations, which hold for any subsets A, B, and C of a universal set U:

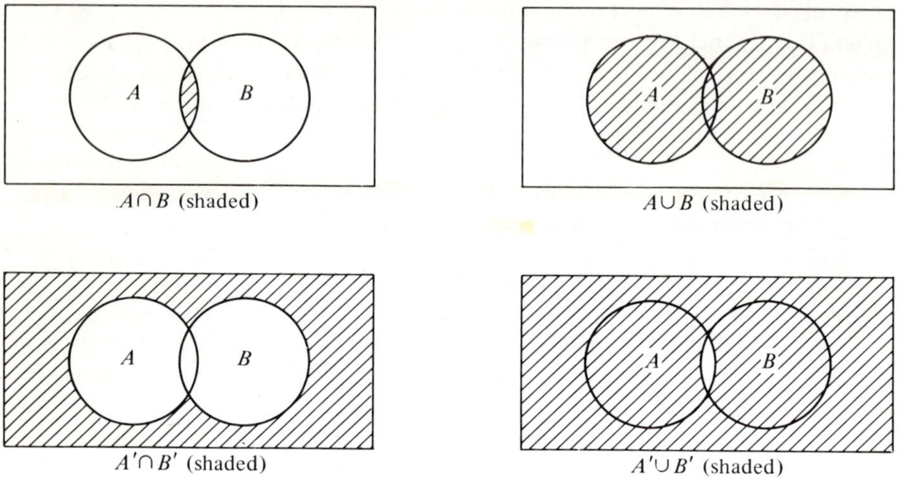

Fig. 1.1 Venn diagrams.

Commutative Law:		
(1.1)	$A \cup B = B \cup A$	$A \cap B = B \cap A$
Associative Law:		
(1.2)	$A \cup (B \cup C) = (A \cup B) \cup C$	$A \cap (B \cap C) = (A \cap B) \cap C$
Distributive Law:		
(1.3)	$A \cup (B \cap C)$ $= (A \cup B) \cap (A \cup C)$	$A \cap (B \cup C)$ $= (A \cap B) \cup (A \cap C)$
Idempotency Law:		
(1.4)	$A \cup A = A$	$A \cap A = A$
De Morgan's Law:		
(1.5)	$(A \cup B)' = A' \cap B'$	$(A \cap B)' = A' \cup B'$
Complementary Law:		
(1.6)	$A \cup A' = u$	$A \cap A' = \varnothing$ (null set)
Identity Law:		
(1.7)	$A \cup \varnothing = A$	$A \cap \varnothing = \varnothing$

Example 1.4 Prove the following relations using the Venn diagram:

(a) $(A \cup B) - B = A \cap B'$

(b) $A \cap (A \cup B) = A$

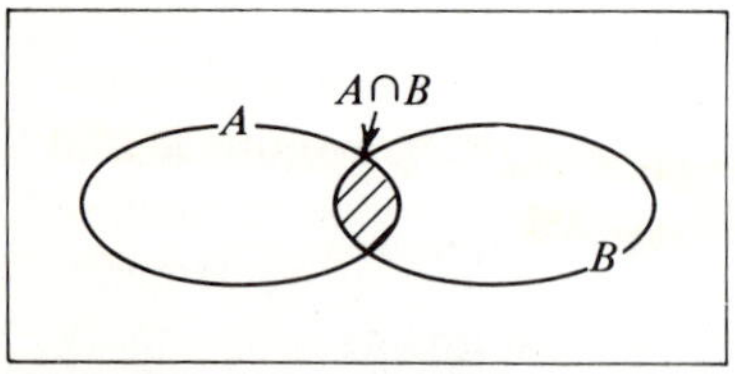

Fig. 1.2 Venn diagram representing $A \cap B$.

Solution. From the Venn diagram of Figure 1.2 we have

(a) $A \cap B' = A - A \cap B$
$\qquad = (A \cup B) - B$

(b) $A \cap (A \cup B) = A \cap \{A + B - (A \cap B)\}$
$\qquad = A + (A \cap B) - (A \cap B)$
$\qquad = A$

1.3 Sample Space

Probability theory is fundamentally concerned with specific properties of random phenomena. To make meaningful statements about those properties, one is required to perform a *real or conceptual experiment* and to assign meaningful interpretation to its possible outcomes.

Before formulating these basic conventions, we proceed to discuss a few typical examples.

Example 1.5 Toss a coin twice. Let us agree to regard "head" and "tail" as the only possible outcomes. If we denote these outcomes by H and T respectively, the possible outcomes of the experiment are

$$\{(H, T), (H, H), (T, H), (T, T)\}$$

Example 1.6 Draw three items at random from a lot of material. Inspect each item and classify it as good or bad, denoted by G and B, respectively. The possible outcomes are

$$\{(G, G, G), (G, G, B), (G, B, G), (B, G, G), (B, B, G), (B, G, B), (G, B, B), (B, B, B)\}$$

Example 1.7 Roll two fair dice. Let us assume that the sum of the numbers on the dice is of interest. Then the possible outcomes of the experiment are

$$\{2, 3, 4, 5, 6, 7, 8, 9, 10, 11, 12\}$$

Example 1.8 Toss a coin until it falls tails for the first time. Let ω denote this outcome. We may obtain an infinite sequence of heads until the first tail is obtained. If a tail is obtained, we specify the outcome by recording the number of the toss that produced the first tail. The set of the possible outcomes is

$$\{\omega, 1, 2, 3, 4, \ldots\}$$

In the preceding examples the experiment is specified by a *prescribed action* as tossing a coin twice, drawing three items from a lot of material, rolling two fair dice, etc. Thus an experiment is defined as any *clearly specified procedure*, whereas a *trial* is a single performance of the experiment. Each possible result of an experiment that is of interest is called an *outcome*.

Definition 1.6. The totality of *outcomes* associated with a real or conceptual experiment is called the *sample space* (denoted by S). Therefore S is a set of outcomes. An outcome is sometimes referred to as a *point* or an *element* in the sample space. Stated differently, a sample space S is a set of elements or points each of which denotes an outcome of the experiment.

A sample space can be either a *finite or infinite set*. The sample spaces of

Examples 1.5, 1.6, and 1.7 are finite, whereas the sample space of Example 1.8 is infinite. Most engineering problems are restricted to finite sample spaces.

As our discussion indicates, once the experiment is defined, we can determine the associated sample space. Then we may be interested in the *occurrence* of a variety of *events* associated with the experiment. To illustrate this notion, let us consider again Example 1.5. The sample space, in brace notation, can be written as

$$S = \{(\mathrm{H, T}), (\mathrm{H, H}), (\mathrm{T, H}), (\mathrm{T, T})\}$$

Suppose now we are interested in the event (denoted by E) of heads occurring on the first toss. This event consists of two elements and can be written as

$$E = \{(\mathrm{H, T}), (\mathrm{H, H})\}$$

Another event of interest may be heads occurring on the first and second toss. This event is

$$E = \{(\mathrm{H, H})\}$$

In the light of this discussion an *event* is defined as a *set of outcomes*. An event E is said to have *occurred* if the outcome of the experiment *corresponds* to an element of subset E.

It is noteworthy that there are many possible events associated with a real or conceptual experiment in whose occurrence we may have some interest. For example, the subset consisting of *no outcomes* is considered to be an event. It is called the *null event* (denoted by $\varnothing$). If subset E consists of *all* possible outcomes of the experiment, then E describes the sample space and accordingly the sample space is considered to be an event.

Suppose the sample space S consists of n elements and each element is considered a subset (denoted by Event E_j); now if we assign a number (denoted by P) to E_j, then we say that $P\{E_j\}$ is the *probability of event* E_j. In other words probability is a specific kind of *measure* defined on a specific type of Boolean algebra of sets. This probability measure *must* satisfy the following conditions:

$$(1.8)\quad \begin{cases} (1)\quad 0 \le P\{E_j\} \le 1 \qquad (j = 1, 2, \ldots, n) \\ (2)\quad P\{\varnothing\} = 0 \\ (3)\quad P\{S\} = \sum_{j=1}^{n} P\{E_j\} = 1 \\ (4)\quad P\{E_1 \cup E_2\} = P\{E_1\} + P\{E_2\} - P\{E_1 \cap E_2\} \end{cases}$$

Two approaches to probability of the occurrence of an event are usually known as the *classical theory* and the *relative frequency theory*. The former is founded upon the indefinable concept of equally likely events, expressed as

$$(1.9)\qquad P\{E_1\} = P\{E_2\} = \cdots = P\{E_n\} = \frac{1}{n}$$

The latter is founded upon an observational concept and is defined operationally as

(1.10) $$P\{E_j\} = \lim_{k \to \infty} \frac{k\{E_j\}}{k}$$

where k is the number of times an experiment is repeated and $k\{E_j\}$ is the number of times the event E_j has occurred during the k trials.

Example 1.9 A fair coin is tossed three times.

(a) Define the sample space.
(b) Find the probability of event E_1 that at least one head occurs.
(c) Find the probability of event E_2 that at least two tails occur.
(d) Find $P\{E_1 \cap E_2\}$.

Solution

(a) The sample space is

$$S = \{(T, H, T), (T, H, H), (T, T, T), (T, T, H), (H, T, T), (H, T, H), (H, H, T), (H, H, H)\}$$

(b) We assign probability $\frac{1}{8}$ to each point in the sample space:

$$E_1 = \{(T, H, T) \cup (T, H, H) \cup (T, T, H) \cup (H, T, T) \cup (H, T, H) \cup (H, H, T) \cup (H, H, H)\}$$

Hence

$$P\{E_1\} = \tfrac{1}{8} + \tfrac{1}{8} + \tfrac{1}{8} + \tfrac{1}{8} + \tfrac{1}{8} + \tfrac{1}{8} + \tfrac{1}{8} = \tfrac{7}{8}$$

or simply

$$\begin{aligned} P\{E_1\} &= 1\text{—probability that no tails occur} \\ &= 1\text{—}P\{(T, T, T)\} \\ &= 1\text{—}\tfrac{1}{8} = \tfrac{7}{8} \end{aligned}$$

(c) $E_2 = \{(T, H, T) \cup (T, T, T) \cup (T, T, H) \cup (H, T, T)\}$

Hence

$$P\{E_2\} = \tfrac{1}{8} + \tfrac{1}{8} + \tfrac{1}{8} + \tfrac{1}{8} = \tfrac{1}{2}$$

(d) $$\begin{aligned} P\{E_1 \cap E_2\} &= P\{T, H, T\} + P\{T, T, H\} + P\{H, T, T\} \\ &= \tfrac{1}{8} + \tfrac{1}{8} + \tfrac{1}{8} = \tfrac{3}{8} \end{aligned}$$

1.4 Fundamental Probability Theorems

Theorem 1.1. *Addition Rule.*

(1.11) $$P\{A + B\} = P\{A\} + P\{B\} - P\{A \cap B\}$$

In words, *the probability that at least one of the events A* and *B occurs is equal to the sum of their individual probabilities minus the probability of their simultaneous occurrence.*

The above theorem can be extended by induction to any finite number of events. To illustrate, let A, B, and C be any three events; then we write

$$P\{A + B + C\} = P\{A\} + P\{B\} + P\{C\} - P\{A \cap B\} - P\{A \cap C\} - P\{B \cap C\} + P\{A \cap B \cap C\} \tag{1.12}$$

If these events are mutually exclusive, then

$$P\{A + B + C\} = P\{A\} + P\{B\} + P\{C\}$$

a result which can be extended by induction to any finite number of mutually exclusive events. In words, *the probability of the occurrence of at least one of N mutually exclusive events is the sum of their individual probabilities.*

Example 1.10 A die is thrown once. What is the probability that the number of points is either odd or a multiple of two?

Solution

$$S = \{1, 2, 3, 4, 5, 6\}$$

The sample space consists of six elements. According to the classical theory, we assign the probability $\frac{1}{6}$ to each element. The event odd (denoted by A) is the subset $\{1, 3, 5\}$; the event "a multiple of 2" (denoted by B) is the subset $\{2, 4, 6\}$. The required probability is given by

$$\begin{aligned} P\{(A + B)\} &= P\{A\} + P\{B\} - P\{A \cap B\} \\ &= \tfrac{3}{6} + \tfrac{3}{6} - P\{\varnothing\} \\ &= 1 \end{aligned}$$

Theorem 1.2. *Complementary Rule*. Let A and A' *be any complementary events.*
Then

$$P\{A\} + P\{A'\} = 1 \tag{1.13}$$

In words, *the probability that A does occur plus the probability that A does not occur is equal to unity.*

Theorem 1.3. *Subevent Rule. Let A and B be the events on a probability sample space S such that $A \subseteq B$. Then*

$$P\{A\} \leq P\{B\} \tag{1.14}$$

In words, *if A implies B, then the probability of A cannot exceed the probability of B.* From the preceding rule we may derive the fact that the probability of any event is a *real number* in the range of 0 to 1.

Theorem 1.4. *Conditional Probability Rule*. We will consider a typical example to illustrate the notion of conditional probability. Suppose a population of N people includes N_S smokers and N_F females. Let S represent the event that a person chosen at random is a smoker and F represent the event that a person chosen at random is a female. We are interested to find the probability that the person chosen is a smoker, under the assumption that the person chosen is female. In other words, given two events S and F, by the *conditional probability* of the event S given the event F (denoted by $P\{S \mid F\}$), we mean that S will occur provided that F has occurred. Then, in terms of frequency definition, we write

$$P\{S\} = \frac{N_S}{N}$$

$$P\{F\} = \frac{N_F}{N}$$

Suppose now we consider the female subpopulation to find the probability that a female chosen at random is a smoker. Obviously this probability is

$$\frac{N_{SF}}{N_F}$$

where N_{SF} is the number of female smokers. We now introduce the formal definition of *conditional probability* given by

$$P\{S \mid F\} = \frac{N_{SF}}{N_F} = \frac{P\{S \cap F\}}{P\{F\}} \qquad P\{F\} > 0 \tag{1.15}$$

Conditional probabilities remain *undefined* when the hypothesis has *zero probability*.

Example 1.11 A die is thrown and the result is known to be an even number. What is the probability that this number is divisible by 1.5?

Solution. The sample space is given by

$$S = \{1, 2, 3, 4, 5, 6\}$$

We assign point probabilities $\frac{1}{6}$ to each element of S. The event even numbers (denoted by B) is the subset $\{2, 4, 6\}$. The event divisible by 1.5 (denoted by A) is the subset $\{3, 6\}$. Therefore, by Eq. (1.15), we have

$$P\{A \mid B\} = \frac{P\{A \cap B\}}{P\{B\}}$$

$$= \frac{1/6}{3/6} = \frac{1}{3}$$

Example 1.12 The probability of having various numbers of children in a family in a certain community is given by

Number of children:	0	1	2	3
Probability:	.20	.50	.25	.05

Furthermore, assume a 50-50 chance that a child born will be a boy or girl:

(a) Find the probability that the family has only one boy.

(b) What is the probability that there are two children in the family given the family has one boy?

Solution

(a) $P\{1 \text{ boy}\} = \sum_{j=0}^{3} P\{1 \text{ boy} \mid j\}P\{j\}$

where j is the number of children in a family, then

$$P\{1 \text{ boy}\} = 0 + (.50)^2 + (.50 \times .25) + (.375 \times .05)$$
$$= .39375$$

(b) Let A denote the event that there are two children in the family and B the event that the family has one boy. Then, by Eq. (1.15),

$$P\{A \cap B\} = P\{B \mid A\}P\{A\}$$

However,

$$P\{B \mid A\} = P\{(\text{boy, girl}), (\text{girl, boy})\} = .25 + .25 = .50$$

and

$$P\{A \cap B\} = (.50)(.25) = .125$$

Hence

$$P\{A \mid B\} = \frac{P\{A \cap B\}}{P\{B\}} = \frac{.25 \times .50}{.39375}$$
$$= .3174$$

Theorem 1.5. *Compound Probability Rule.* The *probability of the simultaneous occurrence of two or more events is called the compound probability or, synonymously, the joint probability of the events.* Let us consider the simultaneous occurrence of two events A and B. According to Theorem 1.4, we can write

$$P\{A \cap B\} = P\{A \mid B\}P\{B\}$$

However, $P\{A \cap B\}$ is equivalent to $P\{B \cap A\}$; therefore, we have

$$P\{A \cap B\} = P\{A \mid B\}P\{B\} = P\{B \mid A\}P\{A\} \tag{1.16}$$

This means that we can express the *joint probability* of any two events in two (2!) ways. By the same line of argument, we can prove that the joint probability of any three events can be expressed by six (3!) equivalent expressions.

In general it can be stated that *any finite number N of events A, B, C, . . . , N can be expressed by N*! equivalent expressions. In symbols, one of these is

$$P\{A \cap B \cap C, \ldots, M \cap N\} = P\{A\}P\{B \mid A\}P\{C \mid A \cap B\} \ldots P\{N \mid A \cap B \cap C, \ldots, \cap M\} \qquad (N \geq 2)$$

The above expression is known as the *General Law of Compound Probability.*

An important case related to conditional probability arises when the events are *independent.* What do we mean by *independence*? *Two events A and B are said to be (statistically) independent when and only when neither affects the probability of the occurrence of the other.* This means that the probability that event A occurs given event B has occurred is equivalent to the probability of occurrence of event A. In symbols,

$$P\{A \mid B\} = P\{A\} \qquad (\text{if } P\{A\} > 0) \tag{1.17}$$

Similarly,

$$P\{B \mid A\} = P\{B\} \qquad (\text{if } P\{B\} > 0) \tag{1.18}$$

Hence we can write

$$P\{A \mid B\} = \frac{P\{A \cap B\}}{P\{B\}} = P\{A\} \tag{1.19}$$

or

$$P\{A \cap B\} = P\{A\}P\{B\}$$

This rule can be extended to any finite number N of events $A, B, C, \ldots, N$. In symbols,

$$P\{A \cap B \cap C, \ldots, \cap N\} = P\{A\}P\{B\}P\{C\}, \ldots, P\{N\} \tag{1.20}$$

The above rule is known as *the multiplication rule. It is worthy to note that mutually exclusive events are not independent unless one of them is zero.* This is true since, by definition, the joint probability of mutually exclusive events is necessarily equal to zero.

Example 1.13 A fair die is thrown twice. Let A, B, and C denote the following events:

A: first toss is odd
B: second toss is even
C: total number of spots is equal to seven

(a) Describe the sample space.
(b) Compute $P\{A\}$, $P\{B\}$, and $P\{C\}$.
(c) Show that A, B, and C are independent in pairs.
(d) Show that A, B, and C are not independent.

Solution. The sample space is given by

(a)

$$S = \begin{Bmatrix} (1, 1), (1, 2), (1, 3), (1, 4), (1, 5), (1, 6) \\ (2, 1), (2, 2), (2, 3), (2, 4), (2, 5), (2, 6) \\ (3, 1), (3, 2), (3, 3), (3, 4), (3, 5), (3, 6) \\ (4, 1), (4, 2), (4, 3), (4, 4), (4, 5), (4, 6) \\ (5, 1), (5, 2), (5, 3), (5, 4), (5, 5), (5, 6) \\ (6, 1), (6, 2), (6, 3), (6, 4), (6, 5), (6, 6) \end{Bmatrix}$$

(b) Event A is given by

$$A = \{\text{element of first, third, and fifth rows}\}$$

Hence

$$P\{A\} = \tfrac{1}{2}$$

Similarly,

$$B = \{\text{elements of second, fourth and sixth columns}\}$$

and

$$P\{B\} = \tfrac{1}{2}$$

Finally,

$$C = \{(1, 6), (2, 5), (3, 4), (4, 3), (5, 2), (6, 1)\}$$

and

$$P\{C\} = \tfrac{1}{6}$$

(c) $\{A \cap B\} = \{(1, 2), (1, 4), (1, 6), (3, 2), (3, 4), (3, 6), (5, 2), (5, 4), (5, 6)\}$

Hence

$$P\{A \cap B\} = \tfrac{1}{4}$$

However,

$$P\{A\}P\{B\} = \tfrac{1}{2} \times \tfrac{1}{2} = \tfrac{1}{4}$$

Therefore, we have

$$P\{A \cap B\} = P\{A\}P\{B\}$$

We conclude that events A and B are independent.
Similarly,

$$\{A \cap C\} = \{(1, 6), (3, 4), (5, 2)\}$$

Hence

$$P\{A \cap C\} = \tfrac{1}{12}$$
$$= P\{A\}P\{C\} = \tfrac{1}{2} \times \tfrac{1}{6} = \tfrac{1}{12}$$

Finally,

$$\{B \cap C\} = \{(5, 2), (3, 4), (1, 6)\}$$

Hence

$$P\{B \cap C\} = \tfrac{1}{12}$$
$$= P\{B\}P\{C\} = \tfrac{1}{2} \times \tfrac{1}{6} = \tfrac{1}{12}$$

(d) $\{A \cap B \cap C\} = \{(1, 6), (3, 4), (5, 2)\}$

$P\{A \cap B \cap C\} = \tfrac{1}{12}$

$P\{A\}P\{B\}\{C\} = \tfrac{1}{2} \times \tfrac{1}{2} \times \tfrac{1}{6} = \tfrac{1}{24}$

Hence, we conclude that events A, B, and C are not independent.

Here it is important to note the significance of the notion of independence. Referring to the dice-rolling experiment in Example 1.13, we note that the specified events are *independent pairwise*, but given any two, the third event is *not independent*. This fact was proved by showing that Eq. (1.20) holds for any set of two events but does not hold for the set of three events. Sometimes one may have the intuitive feeling that a set of events is either dependent or independent. For instance, intuitively one might feel that events A and B are independent because the outcome of the experiment on the first toss does not affect the outcome associated with the second toss. In other words, event A consists of outcomes, $\{1, 3, 5\}$, the occurrences of which are determined by the first toss. Similarly, the occurrences of the outcomes associated with event B are determined by the second toss. Since each toss is physically independent of the other, events A and B are independent. Now consider the set $\{A \cap B \cap C\}$. Even C consists of outcomes the occurrences of which are determined by both tosses. It follows that events A, B, and C are not independent. From this intuitive concept we make the following statement: *If an experiment consists of N physically independent successive trials and a set of events* $E_1, E_2, \ldots, E_N$ *is defined such that* E_j *is determined by the* jth *trial* ($j = 1, 2, \ldots, N$), *then the events so defined are independent.*

Example 1.14 A batch of material consists of 120 items; 40 items are oversized, 20 are undersized, and 60 conform to specification. Two items are drawn at random one after the other (without replacement). What is the probability that both items will be of the same quality?

Solution. Let o, u, and g designate oversized, undersized, and conforming to specification items, respectively. This problem can be considered as a *compound experiment*. As a rule a compound experiment of n trials will always have as sample space the set S of ordered n-tuples denoting possible outcomes of the experiment. In symbols,

$$S = \begin{Bmatrix} (o, o) & (u, g) & (g, g) \\ (o, u) & (u, u) & (g, u) \\ (o, g) & (u, o) & (g, o) \end{Bmatrix}$$

In this case the elements of the sample space do not have equal probabilities. Let us evaluate the probability of each element in the sample space. Under the assumption that each item has the same probability of being selected, then the first item drawn can either be o, u, or g. These probabilities are

$$P\{o\} = \tfrac{1}{3} \qquad P\{u\} = \tfrac{1}{6} \qquad P\{g\} = \tfrac{1}{2}$$

Now, what is the probability that the second draw yields an oversized item given that the first item chosen is oversized? Since these two events can be considered as independent, this conditional probability is given by

$$\{o, o\} = \tfrac{1}{3} \times \tfrac{39}{119} = \tfrac{13}{119}$$

Similarly, we can calculate the conditional probabilities of each element of the sample space. These probabilities are listed in Table 1.1. We note that each probability is positive and that their sum is 1.

Now the event "both items selected will be of the same quality" (denoted by E) is the set given by

$$E = \{o, o\} \cup \{u, u\} \cup \{g, g\}$$

TABLE 1.1

Element	Probability
$\{o, o\}$	$\frac{1}{3} \times \frac{39}{119} = \frac{13}{119}$
$\{o, u\}$	$\frac{1}{3} \times \frac{20}{119} = \frac{20}{357}$
$\{o, g\}$	$\frac{1}{3} \times \frac{60}{119} = \frac{20}{119}$
$\{u, g\}$	$\frac{1}{6} \times \frac{60}{119} = \frac{10}{119}$
$\{u, u\}$	$\frac{1}{6} \times \frac{19}{119} = \frac{19}{714}$
$\{u, o\}$	$\frac{1}{6} \times \frac{40}{119} = \frac{40}{714}$
$\{g, g\}$	$\frac{1}{2} \times \frac{59}{119} = \frac{59}{238}$
$\{g, u\}$	$\frac{1}{2} \times \frac{20}{119} = \frac{10}{119}$
$\{g, o\}$	$\frac{1}{2} \times \frac{40}{119} = \frac{20}{119}$

Hence

$$\begin{aligned} P\{E\} &= P\{o, o\} + P\{u, u\} + P\{g, g\} \\ &= \frac{13}{119} + \frac{19}{714} + \frac{59}{238} = \frac{274}{714} \cong .383 \end{aligned}$$

Example 1.15 Suppose that the probability of successful launching of a special satellite is $\frac{1}{2}$. Suppose further that each trial is physically independent of preceding trials. Determine the number of trials required to make the probability of a successful launching at least .99.

Solution. Let N denote the required number of trials. The probability of successful launching on the ith trial is given by

$$\left(\frac{1}{2}\right)^{i-1} \times \frac{1}{2}$$

Hence

$$\frac{1}{2} \sum_{i=1}^{N} \left(\frac{1}{2}\right)^{i-1} \geq .99$$

and accordingly $N = 7$ trials.

Example 1.16 The distribution of the number of absent operators from departments A and B per week is as follows:

Number of absentees:	1	2	3	4	5	6	7	8	9	10
Department A:	.02	.05	.10	.25	.23	.11	.09	.09	.05	.01
Department B:	.03	.07	.10	.10	.40	.10	.10	.07	.02	.01

Find the following probabilities:
(a) Plant A has at least four absentees per week
(b) Absentees from A are less than two and from B more than eight.
(c) The absenteeism from A is more than from B.
(d) The absenteeism from A is greater than from B, given two operators are absent from B.

Solution. Let N denote the number of absent operators: then

(a) $$P\{N_A \geq 4\} = P\{N_A = 4\} + P\{N_A = 5\} + \cdots + P\{N_A = 10\} = .83$$

(b) $$P\{N_A < 2) \cap (N_B > 8)\} = P\{N_A < 2\}P\{N_B > 8\} = .02 \times .03 = .0006$$

(c) $$P\{N_A > N_B\} = \sum_{j=1}^{10} P\{N_A > j\}P\{N_B = j\} = .98 \times .03 + .93 \times .07 + \cdots + .01 \times .02 = .4189$$

(d) $$P\{(N_A > N_B) \mid N_B = 2\} = \frac{P\{N_A > N_B\} \cap P\{N_B = 2\}}{P\{N_B = 2\}} = P\{N_A > N_B\} = P\{N_A > 2\} = .93$$

Theorem 1.6. *Bayes' Rule:* If $E_1, E_2, \ldots, E_n$ *are mutually exclusive events of which one event must occur, i.e.,* $\sum_{i=1}^{n} P\{E_i\} = 1$, *then*

$$P\{E_r \mid A\} = \frac{P\{E_r\}P\{A \mid E_r\}}{\sum_{i=1}^{n} P\{A \mid E_i\}P\{E_i\}} \tag{1.21}$$

where $i = 1, 2, \ldots, n$, $r = 1, 2, \ldots, n$, and A is any event, which is a subset of the sample space. Bayes' rule can be derived from the basic laws of probability.

Proof. Suppose the sample space is the set of mutually exclusive events. Symbolically,

$$S = \{E_1, E_2, \ldots, E_n\}$$

In words, the sample space is partitioned into n mutually exclusive events. Such a set is called the "partition of the sample space" and must satisfy the following conditions:

$$\begin{cases} (1)\ E_i \subseteq S \quad (i = 1, 2, \ldots, n) \\ (2)\ E_i \cap E_r = \varnothing \quad (i = 1, 2, \ldots, n,\ r = 1, \ldots, n,\ i \neq r) \\ (3)\ E_1 \cup E_2 \cup \cdots \cup E_n = S \end{cases} \tag{1.22}$$

Let A be any event that is a subset of S, thus we can write

$$\begin{aligned} A &= A \cap S \\ &= A \cap \{E_1, E_2, \ldots, E_n\} \\ &= A \cap \{E_1 \cup E_2, \ldots, \cup E_n\} \\ &= (A \cap E_1) \cup (A \cap E_2) \cup \cdots \cup (A \cap E_n) \end{aligned}$$

Since $E_1, E_2, \ldots, E_n$ are mutually exclusive events by Theorem 1.1, we write

$$P\{A\} = P\{A \cap E_1\} + P\{A \cap E_2\} + \cdots + P\{A \cap E_n\}$$

Applying the conditional probability rule, this becomes

$$= P\{A \mid E_1\}P\{E_1\} + P\{A \mid E_2\}P\{E_2\} + \cdots + P\{A \mid E_n\}P\{E_n\}$$

$$= \sum_{i=1}^{n} P\{A \mid E_i\}P\{E_i\}$$

However,

$$P\{E_r \cap A\} = P\{E_r \mid A\}P\{A\} = P\{A \mid E_r\}P\{E_r\}$$

Therefore

$$\begin{aligned} P\{E_r \mid A\} &= \frac{P\{A \mid E_r\}P\{E_r\}}{P\{A\}} \\ &= \frac{P\{A \mid E_r\}P\{E_r\}}{\sum_{i=1}^{n} P\{A \mid E_i\}P\{E_i\}} \end{aligned}$$

The following two examples will illustrate this rule:

Example 1.17 Suppose the total inventory of a firm is a collection of lots from four different suppliers, A, B, C, and D, as indicated below:

Supplier	*Percent of Inventory*
A	60
B	20
C	5
D	15

Furthermore, suppose that past records indicate that lots from suppliers A, B, C, and D are 5, 3, 2, and 8% defective, respectively. Find the probability that a defective item selected at random is from supplier B.

Solution. Applying the rule of Bayes we obtain

$$\begin{aligned} P\{\text{B} \mid \text{defective}\} &= \frac{(.20)(.03)}{(.60)(.05) + (.20)(.03) + (.05)(.02) + (.15)(.08)} \\ &= .122 \end{aligned}$$

Example 1.18 Suppose three urns, A, B, and C, contain white, red, and black balls as tabulated below:

Urn	*White*	*Red*	*Black*
A	10	5	10
B	8	8	4
C	5	5	5

Two balls are drawn (without replacement) at random and are found to be white and black. What is the probability that the two balls come from urn C?

Solution. Since the urn is chosen at random,

$$P\{A\} = P\{B\} = P\{C\}$$

Let $\{(w, b), (b, w)\}$ denote the event "two balls selected are white and black." We can immediately calculate the following conditional probabilities:

$$P\{(w, b), (b, w) \mid \mathrm{A}\} = \tfrac{10}{25} \times \tfrac{10}{24} + \tfrac{10}{25} \times \tfrac{10}{24} = \tfrac{1}{3}$$

$$P\{(w, b), (b, w) \mid \mathrm{B}\} = \tfrac{8}{20} \times \tfrac{4}{19} + \tfrac{4}{20} \times \tfrac{8}{19} = \tfrac{8}{45}$$

$$P\{(w, b), (b, w) \mid \mathrm{C}\} = \tfrac{5}{15} \times \tfrac{5}{14} + \tfrac{5}{15} \times \tfrac{5}{14} = \tfrac{5}{21}$$

Applying the rule of Bayes we obtain

$$P\{C \mid (w, b), (b, w)\} = \frac{(\frac{1}{3})(\frac{5}{21})}{(\frac{1}{3})(\frac{1}{3}) + (\frac{1}{3})(\frac{8}{45}) + \frac{1}{3}(\frac{5}{21})}$$

$$= .317$$

1.5 Combinatorial Analysis

Computational procedures of probabilities related to finite sample spaces with comparatively few elements seem not too complicated. In real life many interesting and important experiments yield sample spaces with large numbers of possible outcomes (elements); thereby the computational procedures involve tedious counting. In order to overcome this difficulty we need to derive some *combinatorial formulas* to provide a short and convenient way of facilitating the required counting. The formulas to be developed will be especially helpful in computing probabilities for events with finite sample spaces and equiprobable outcomes. We will develop a few of these techniques in this section.

1.5.1 Multiplets

A simple example will help to clarify this notion:

Example 1.19 Suppose a TV manufacturer produces color and black and white TV sets. Furthermore, suppose that each set is classified according to screen size 19, 23, and 25 in. How many models does he produce in all? Obviously the answer is six.

Suppose now that each set can be classified as portable and stationary; then we have $2 \times 3 \times 2 = 12$ models in all. Accordingly, we state the following rule: *Given r sets* ($r = 1, 2, \ldots, r$) *and each set contains* $n_1, n_2, \ldots, n_r$ *elements, it is possible to form* $n_1 \times n_2 \times \cdots \times n_r$ *ordered r-tuplets containing one element of each kind.*

1.5.2 Ordered Samples

Samples can be drawn from a finite population according to two procedures. First, sampling *with replacement* means the same element can be selected *more* than once. Second, sampling *without replacement* means an element once selected is *removed* from the population.

Example 1.20 Consider a simple experiment of sampling with replacement from a finite population of five elements labeled A, B, C, D, and E. Suppose elements, A, B, and C have common characteristics whereas elements D and E have another common characteristic. In other words, elements A, B, and C may be classified as good (denoted by g) whereas elements D and E classified as bad (denoted by b). Then, if one sample is drawn at random, our sample space is given by

$$S = \{A_g, B_g, C_g, D_b, E_b\}$$

If two samples are drawn, the number of possible samples is the sum of the following outcomes:

$$\begin{array}{ccccc}
(A_g, B_g) & (B_g, A_g) & (C_g, A_g) & (D_b, A_g) & (E_b, A_g) \\
(A_g, C_g) & (B_g, C_g) & (C_g, B_g) & (D_b, B_g) & (E_b, B_g) \\
(A_g, D_b) & (B_g, D_b) & (C_g, D_b) & (D_b, C_g) & (E_b, C_g) \\
(A_g, E_b) & (B_g, E_b) & (C_g, E_b) & (D_b, E_b) & (E_b, D_b) \\
(A_g, A_g) & (B_g, B_g) & (C_g, C_g) & (D_b, D_b) & (E_b, E_b)
\end{array}$$

and is equal to 25. In symbols, the number of possible samples is N^n, where N is the population size and n is the sample size.

Consider now the same experiment where sampling is without replacement. The number of possible samples is the sum of the outcomes listed in the four first rows above. We exclude the elements of the fifth row simply because when A_g is drawn and not replaced, the second sample cannot be A_g and so on. Therefore we have 20 possible samples in all. Symbolically, the number of possible samples is given by

$$N(N-1)\cdots(N-n+1) \qquad n \leq N$$

It is of interest to calculate the probability that both items selected are bad. This probability is

$$P\{(D_b, D_b), (D_b, E_b), (E_b, D_b), (E_b, E_b)\} = \tfrac{4}{25} \quad \text{if sampling with replacement}$$

or

$$P\{(D_b, E_b), (E_b, D_b)\} = \tfrac{1}{10} \quad \text{if sampling is without replacement}$$

1.5.3 Permutations

We will consider again the case of sampling without replacement in Example 1.20. We can state this problem in a different way: given a set of five *distinct objects*, what is the number of *ordered pairs*? We found the number of ordered pairs to be 20 by simply counting the elements of the sample space. The same result can be obtained in a different way. For the first object selected there are five choices. When the first choice has been made, the element selected is removed from the sample space and there remain four ways to select the second element. This means we have two sets; the first contains five elements and the second contains four elements. Therefore, by the rule of multiplets the number of ordered pairs that can be formed by these choices is

$$N(N - 1) = 5 \times 4 = 20$$

In general we can state that the number of order n-tuples from a set of N elements is

$$N(n - 1)(N - 2) \cdots (N - n + 1) \tag{1.23}$$

The above expression is known as the number of *permutations* of n elements chosen from a set of N elements and is denoted by $P(N, n)$. Multiplying and dividing the above expression by $(N - n)!$ we obtain

$$P(N, n) = \frac{N(N - 1)(N - 2) \cdots (N - n + 1)(N - n)!}{(N - n)!}$$

$$= \frac{N!}{(N - n)!} \qquad n \leq N \tag{1.24}$$

When $n = N$, Eq. (1.24) becomes

$$P(N, n) = \frac{N!}{0!} = N! \quad \text{(0! by definition is equal to 1)} \tag{1.25}$$

Example 1.21 A product undergoes five operations that can be performed if any sequence. In how many different ways can it be produced?

Solution. The number of different ways is

$$N! = 5! = 120$$

1.5.4 Combinations

Suppose we have N distinct objects and we must select n of them. In how many ways can this be done? Furthermore, suppose that the order in which the choices are made is irrelevant. We call a group of n elements (without regard to order) chosen at random from N elements a *combination* of n elements from N and denote if by $C(N, n)$.

To answer this question, we will consider the following example:

Example 1.22 An urn contains one white, one black, one red, and one green sphere, all of equal size. Three spheres are drawn at random. In how many ways can this be done? In other words, we want to find the number of subsets with exactly three elements from a set of four elements, provided that the order in which the choices are made is irrelevant.

Solution. The number of permutations of three objects from a set of four objects is (order is relevant)

$$P(4, 3) = \frac{4!}{(4-3)!} = 24$$

Since the order of elements within a subset is irrelevant, by how much can we reduce the number of subsets? Consider a subset with three colors: green (g), red (r), and white (w). The number of permutations of these three colors is 3!. In symbols,

$$\{g, r, w\}, \{g, w, r\}, \{r, g, w\}, \{r, w, g\}, \{w, g, r\}, \{w, r, g\}$$

This means that each subset is counted 3! times. *In general we state that the permutations of n objects from a set of N counts an n element subset exactly n! times.* Therefore we have the following relationship:

$$C(N, n) = \frac{P(N, n)}{n!} = \frac{N!}{n!(N-n)!}$$

Hence

$$C(4, 3) = \frac{4!}{3!(4-3)!} = 4$$

The standard notation for $C(N, n)$ is $\binom{N}{n}$ and is known as the *binomial coefficient*. Note also that $\binom{N}{n} = \binom{N}{N-n}$.

Example 1.23 How many subsets can be formed from a set of N elements?

Solution. The number of subsets is

$$\binom{N}{0} + \binom{N}{1} + \binom{N}{2} + \cdots + \binom{N}{N} = \sum_{n=0}^{N} \binom{N}{n}$$

(the first term is the null event, $\varnothing$, whereas the last term is the sample space S) By Newton's binomial theorem we have

$$\sum_{n=0}^{N} \binom{N}{n} = (1 + 1)^N = 2^N$$

A further extension of the combination rule is the following rule: *The number of ways in which N distinct objects can be divided into n_i objects, such that $\sum_{i=1}^{k} n_i = N$, is given by*

$$\frac{N!}{n_1!n_2! \cdots n_k!} \tag{1.26}$$

usually called the multinomial coefficients. The following example will prove this rule.

Example 1.24 In how may different ways can we divide a deck of playing cards (52 cards) among four players such that each player gets 13 cards? Here the set of distinct objects is a deck of cards ($N = 52$). Each player gets 13 cards ($n_1 = n_2 = n_3 = n_4 = 13$.)

Let the first player choose his 13 cards at random from the full deck. Therefore the first player can choose his cards in $\binom{52}{13}$ ways. Now let the second player choose his cards from the remaining 39 cards; this can be done in $\binom{39}{13}$ ways. Similarly, the third player can choose his cards in $\binom{26}{13}$ ways. Finally, the fourth player will have one choice, namely, getting the remaining 13 cards. Therefore the number of different ways a deck can be divided equally among four players is

$$\binom{52}{13}\binom{39}{13}\binom{26}{13} \tag{1.27}$$

In symbols this is equivalent to

$$\binom{N}{n_1}\binom{N-n_1}{n_2}\binom{N-n_1-n_2}{n_3}$$

$$= \frac{N!}{n_1!(N-n_1)!} \cdot \frac{(N-n_1)!}{n_2!(N-n_1-n_2)!} \cdot \frac{(N-n_1-n_2)!}{n_3!(N-n_1-n_2-n_3)!}$$

$$= \frac{N!}{n_1!n_2!n_3!(N-n_1-n_2-n_3)!} \tag{1.28}$$

Since $N = n_1 + n_2 + n_3 + n_4$, Eq. (1.28) becomes

$$\frac{N!}{n_1!n_2!n_3!n_4!} \tag{1.29}$$

which proves the rule.

Example 1.25 An urn contains 20 balls of equal size: 5 red, 7 green, 2 black, and 6 white. Four balls are drawn at random (without replacement). What is the probability that all of them are of different colors?

Solution. Consider each color as a subpopulation. The choice of one of the red balls can be done in $\binom{5}{1}$ ways. Then one green ball can be chosen in $\binom{7}{1}$ ways. Similarly, one black ball can be chosen in $\binom{2}{1}$ ways. Finally, the white ball can be chosen in $\binom{6}{1}$ ways. Since the four balls can be chosen in $\binom{20}{4}$ ways, the required probability is therefore

$$\frac{\binom{5}{1}\binom{7}{1}\binom{2}{1}\binom{6}{1}}{\binom{20}{4}} = \frac{28}{323}$$

1.5.5 Stirling's Formula

The calculation of $N!$ (N factorial) for large values of N is rather a tedious job. Stirling's formula can be used to obtain an excellent numerical approximation:

$$N! \sim \sqrt{2\pi N}\left(\frac{N}{e}\right)^N \tag{1.30}$$

where e is the base of the natural logarithms ($e = 2.71828\ldots$). The symbol $\sim$ means that the ratio of the two sides tends to unity as N tends to infinity. The interested reader will find the proof of this formula in Reference [5], page 50. For $N > 10$, Stirling's formula gives the value of $N!$ with an accuracy of better than .8%. For $N = 100$ the error is .08%.

Example 1.26 A fair coin is tossed 50 times. What is the probability that the number of heads is exactly equal to 25?

Solution. We define $S_1 = \{\text{T, H}\}$, as the sample space for a single toss. For 50 tosses the sample space is given by the Cartesian product

$$S = S_1 \times S_1 \times \cdots \times S_1 \quad \text{50 times}$$

Since these tosses are independent the probability that any permutation of S resulting is 25 heads is

$$\left(\frac{1}{2}\right)^{25}\left(\frac{1}{2}\right)^{25}$$

The number of 50-tuples containing exactly 25 heads is $\binom{50}{25}$; therefore the required probability is

$$\binom{50}{25}\left(\frac{1}{2}\right)^{25}\left(\frac{1}{2}\right)^{25} \tag{1.31}$$

(known as *binomial probability law*)

$$= \frac{50!}{25!25!}\left(\frac{1}{2}\right)^{50}$$

Using Stirling's approximation Eq. (1.31) becomes

$$\frac{\sqrt{2\pi(50)}(50/e)^{50}(1/2)^{50}}{2\pi(25)(25/e)^{50}} = \frac{1}{\sqrt{25\pi}} = .11298$$

Example 1.27 A group of 100 boys and 100 girls is divided into two equal groups. Find the probability p that each group will be equally divided into boys and girls.

Solution

$$p = \frac{\binom{100}{50}\binom{100}{50}}{\binom{200}{100}}$$

$$= \frac{100!}{50!50!} \cdot \frac{100!}{50!50!} \cdot \frac{100!100!}{200!}$$

Using Stirling's approximation,

$$p \cong \left[\frac{\sqrt{200\pi}(100/e)^{100}}{\sqrt{100\pi}(50/e)^{50}}\right]^4 \left[\frac{e^{200}}{\sqrt{400\pi}(200)^{200}}\right]$$

$$\cong \frac{1}{5\sqrt{\pi}} = .11298$$

PROBLEMS

1.1. An experiment consists of five tosses with a fair coin. Give a verbal description of the complement of each of the following events:
(a) at least three heads
(b) at most three heads
(c) no heads.

1.2. Let A, B, and C be any subsets of a universal set U. Determine whether the following statements are false or correct:
(a) If A, B are exclusive and B, C are exclusive, then A, C are exclusive.
(b) If A, B are exclusive and A, C are exclusive, then A (B or C) are exclusive.
(c) If A, (B or C) are exclusive, then A, B are exclusive.

1.3. Let $U = \{1, 2, 3, 4, 5, 6, 7\}$ be the universal set, and consider the subsets of U given by

$$A = \{1, 2, 3\} \qquad B = \{2, 4, 6\} \qquad C = \{1, 3, 5, 7\}$$

List the elements of the following sets:
(a) A', B', and C'
(b) $A \cup B$
(c) $A \cap B$
(d) $B \cap C$.

1.4. Let A, B, and C be three arbitrary events. Find expressions for the events that of A, B, and C
(a) only A occurs
(b) all three events occur
(c) none occurs
(d) at least one occurs
(e) not more than two occur
(f) at least two occur.

1.5. Verify the following relations:
(a) $A \cup B = (A - A \cap B) \cup B$
(b) $(A \cup B)' = A' \cap B'$
(c) $(A \cap C) \cup (B \cap C) = (A \cup B)C$.

1.6. Prove that

$$P\{A \cup B \cup C\} = P\{A\} + P\{B\} + P\{C\} - P\{A \cap B\} - P\{A \cap C\} - P\{B \cap C\} + P\{A \cap B \cap C\}$$

1.7. Let A, B, and C be subsets of the sample space S defined by

$$S = \{1, 2, 3, 4, 5, 6\}$$
$$A = \{1, 3, 5\}$$
$$B = \{1, 2, 3\}$$
$$C = \{2, 4, 6\}$$

Find the events
(a) $A \cap (B \cap C)$
(b) $A \cup B \cup C$
(c) $A \cup A'$
(d) $(A \cup B)$.

1.8. Two dice are thrown. Let A be the event that the sum of the faces is 5 and B the event of at least one ace. Describe the events $A \cap B$, $A \cup B$, and $A \cap B'$.

1.9. In the preceding problem find the probabilities of the events assuming that all sample points have equal probabilities.

1.10. A fair coin is tossed until for the first time heads occur. Find the probability that the experiment ends on or before the sixth toss.

1.11. Two coins are tossed.
(a) Describe the sample space.
(b) What is the probability that no heads will occur?
(c) What is the probability that a head and a tail will occur?

1.12. An urn contains three green spheres and two blue ones. What is the probability that two spheres drawn from the urn will both be green?

1.13. An optical device consists of three different lenses denoted by L_1, L_2, and L_3. Let the probabilities that lenses are defective be $P_D(L_1) = .05$, $P_D(L_2) = .01$, and $P_D(L_3) = .04$. Find the following probabilities:
(a) The device has no defective lens.
(b) The device has three defective lenses.
(c) At most one lens is defective.

1.14. Three fair coins are tossed simultaneously.
(a) What is the probability of obtaining three tails?
(b) What is the probability that at least two tails will appear?

1.15. Three defective items are known to be in a container of 40 items. Five items are selected at random without replacement.
(a) What is the probability that the sample will contain no defectives?
(b) What is the probability that the sample will contain exactly two defectives?

1.16. An integer is chosen at random from the first 200 positive integers. What is the probability that the integer chosen is divisible by 4 or 6?

1.17. A consumer survey investigating the ownership of television sets and cars for a sample of 1000 families in a certain community finds the following results:

	TV	*No TV*
Car	115	245
No car	380	260

What is the probability that a family has
(a) a TV set
(b) a car?
(c) both?
(d) either a TV set or a car or both?

1.18. A card is chosen at random from a standard deck. What is the probability that the chosen card is a spade, an honor card, or a deuce?

1.19. An integer is chosen at random from the first 200 positive integers. What is the probability that the integer chosen is divisible by 4 or 6 or 8?

1.20. A fair coin is tossed repeatedly. What is the minimum number of independent tosses so that the probability of obtaining at least one tail is more than or equal to 0.96?

1.21. A person repeatedly casts a pair of dice. What is the probability of throwing a "five" before a "nine"?

1.22. A fair coin is tossed three successive times. What is the probability of obtaining at least two heads given that the first toss resulted in a head?

1.23. Assume a 50–50 chance that a child born will be a boy or a girl. A family has two children. What is the probability that both children are girls given that
(a) the older child is a girl?
(b) at least one of the children is a girl?

1.24. A box contains a dozen ball-point pens (four red, six green, and two blue). Two pens are drawn without replacement. What is the probability that both pens will be of the same color?

1.25. From a standard deck five individual drawings are made with replacement. What is the probability of drawing one diamond, two spades, one heart, and one club?

1.26. A young lady keeps her nail polish in two drawers, of which one contains one red and one gray nail polish and the other three red and one orange nail polish. A drawer is chosen at random and from it a nail polish is drawn at random. What is the probability of getting a red nail polish?

1.27. A fair die is thrown twice. What is the probability of getting at least four points on each throw?

1.28. What is the probability that a poker hand is a diamond flush, given that it consists only of black cards?

1.29. The distribution of the number of accidents in plant I and plant II per month is as follows:

Number of accidents:	1	2	3	4	5	6	7	8	9	10
Plant I:	.08	.02	.10	.05	.21	.32	.11	.05	.04	.02
Plant II:	.06	.04	.13	.18	.14	.16	.11	.08	.07	.03

Find the following probabilities:
(a) Plant I has a record of at least six accidents per month.
(b) The number of accidents in plant I is less than four and in plant II is more than seven.
(c) The number of accidents in plant I is more than in plant II.
(d) The number of accidents in plant I is greater than in plant II, given that two accidents have occurred in plant II.

1.30. The following data represent the results of inspection of ball bearings supplied by three different companies:

Parts	*Company A*	*Company B*	*Company C*	*Total*
Satisfactory	3,000	2,500	1,500	7,000
Oversized	880	100	120	1,100
Undersized	200	150	250	600

Find the probability that
(a) a defective part is oversized
(b) Company A supplied a defective part
(c) a part is both defective and from Company C.

1.31. Three urns contain colored spheres as specified below:

Urn	*Red*	*Green*	*Yellow*
1	5	8	3
2	6	4	2
3	2	4	2

One urn is chosen at random and a sphere is withdrawn. If the chosen sphere is red, what is the probability that it came from urn 3?

1.32. A factory has three machines that produce plastic gears. The output of machine 1 is 1.2 times the output of machines 2 and 3. The probability of producing a defective gear by machines 1, 2, and 3 are .04, .02, and .06, respectively. The output of the three machines is fed into a stockpile. A gear chosen at random from the stockpile is found to be defective. What is the probability that it was produced by machine 3?

1.33. A certain diagnostic test gives the following positive reactions. It indicates a positive reaction to about 85% of the persons having the disease and to about 5% of the persons who do not. From a large population of which only 1% have the disease, one person is selected at random, takes the diagnostic test, and gets a positive reaction. What is the probability that the person has the disease?

1.34. Three fair dice are thrown simultaneously.
(a) What is the number of possible outcomes (order is relevant)?
(b) What is the probability that all three dice will show the same value?

1.35. An examination consists of five questions. How many different orders of answering can a student use?

1.36. A box contains 12 electronic tubes of which two are defective.
(a) In how many ways can a sample of three tubes be taken?
(b) What is the probability of getting a sample from the box with exactly one defective tube and two good tubes?

1.37. In a straight poker hand what is the probability of getting a full house (three of one kind and two of another kind)?

1.38. An electric switch consists of three components. A switch is inoperable if any of the three components is defective. Suppose that the probabilities that the first, second, and third components will be defective are $\frac{1}{4}$, $\frac{1}{5}$, and $\frac{1}{3}$ respectively. If a sample of five switches is chosen and tested, what is the probability that at least four switches are satisfactory?

1.39. A transmission shaft is assembled by using either a set of three, four, or five gears. If there are available six different gears, how many different assemblies can be obtained?

1.40. A fair coin is tossed 100 times. Using Stirling's approximation find the probability of obtaining exactly 50 heads.

SUGGESTED REFERENCES

[1] Bizley, M. T. L., *Probability: An Intermediate Textbook*, Cambridge University Press, New York, 1957.

[2] Breuer, J., *Introduction to the Theory of Sets*, trans. H. F. Fehr: Prentice-Hall, Inc., Englewood Cliffs, N.J., 1958.

[3] Chernoff, H., and L. E. Moses, *Elementary Decision Theory*, John Wiley & Sons, Inc., New York, 1959.

[4] Cramér, H., *The Elements of Probability Theory and Some of Its Applications*, John Wiley & Sons, Inc., New York, 1955.

[5] Feller, W., *An Introduction to Probability Theory and Its Applications*, 2nd ed., John Wiley & Sons, Inc., New York, 1957.

[6] Kemeny, J. G., H. Mirkil, and J. L. Thompson, *Finite Mathematical Structures*, Prentice-Hall, Inc., Englewood Cliffs, N.J., 1959.

[7] Kolmogoroff, A. N., *Foundations of the Theory of Probability*, 2nd ed, Chelsea Publishing Co., New York, 1950.

[8] May, K. O., *Elements of Modern Mathematics*, Addison-Wesley Publishing Co., Reading, Mass., 1959.

[9] Munroe, M. E., *Theory of Probability*, McGraw-Hill Book Company, New York, 1951.

[10] Neyman, J., *First Course in Probability and Statistics*, Holt, Rinehart & Winston, Inc., New York, 1950.

[11] Parzen, E., *Modern Probability Theory and Its Applications*, John Wiley & Sons, Inc., New York, 1960.

[12] Schlaifer, R., *Probability and Statistics for Business Decisions*, McGraw-Hill Book Company, New York, 1959.

[13] Todhunter, I., *A History of the Mathematical Theory of Probability*, Chelsea Publishing Co., New York, 1959.

[14] Uspensky, J. V., *Introduction to Mathematical Probability*, McGraw-Hill Book Company, New York, 1937.

2

PROBABILITY FUNCTIONS

2.1 Random Variables

In previous sections we developed some basic probability concepts in terms of *nonnumerical outcomes*, such as "heads or tails," "good or bad," and "red or white." In many experiments we are interested in *numerical values* associated with the outcomes of the experiment. To illustrate: A coin is tossed n times, and we may be interested to know the number of tails; two dice are rolled, and we may be interested to know only the sum of the points. If a sample of n items is drawn at random, we may be interested to know the number of defective items. In all these examples we have associated a *single real number* with each outcome of the experiment. Whenever this is done, a *function* is thereby defined.

Definition 2.1. A *function* is a rule that assigns to each outcome of an experiment (real or conceptual) a *single* real number; a function is denoted by f. The set of possible outcomes is called *domain* of the function. The set of all possible values of a function is called the *range* of the function.

Example 2.1

$$f(x) = x^2 \quad \text{where } x = 1, 2, \ldots, 10$$

The domain of this function is the set of positive integers from 1 to 10. The range of this function is the set {1, 4, 9, 16, 25, 36, 49, 64, 81, 100}. The numbers 1, 4, . . . , 100 are the elements of the range.

Definition 2.2. A *function* whose domain is a sample space and whose range is some set of real numbers is called a *random variable* or *variate*. In other words, a random variable is a *numerically valued function* defined over the sample space.

A random variable is *two dimensional* whenever the range is a collection of *ordered pairs*. More generally, an n-dimensional random variable is a function whose range is a collection of *n-tuples* of real numbers.

Random variables may be classified as follows:

1. Discrete random variables are those that take on a finite or denumerably infinite number of distinct values.
2. Continuous random variables are those that take on a continuum of values within the given range.

Random variables are usually denoted by capital latin letters say X, Y, and Z. A typical outcome is usually denoted by ω. Thus $X(\omega)$ denotes that real number which the random variable X associates with the outcome ω.

A few examples of discrete and continuous random variables are as follows:

1. Discrete:
   - (a) The daily demand for a specific newspaper.
   - (b) The daily number of absentees from an office.
   - (c) The number of pinholes per aluminum sheet.
   - (d) The number of customers arriving at a counter per hour.
2. Continuous:
   - (a) The waiting time of a commuter for a train in a certain subway.
   - (b) The daily consumption of fuel in a plant.
   - (c) The amount of rain that falls per season on a certain territory.
   - (d) The dimension of a part produced by a certain machine.

Example 2.2 A newspaper boy buys a paper for \$.04 and sells it for \$.07. Unsold papers cannot be returned. Suppose the daily demand for the paper is a random variable (Y; unknown). Describe the random variable G that denotes his daily profit if the newsboy orders quantity X.

Solution

$$G(Y) = \begin{cases} \$.03X & Y \geq X \\ \$.03Y - \$.04(X - Y) & Y < X \end{cases}$$

where $Y = 0, 1, 2, \ldots$.

Example 2.3 A die is rolled twice. Describe the random variable X that denotes the sum of the points.

Solution

$$X(\omega) = \omega \qquad \omega = 2, 3, \ldots, 12$$

Example 2.4 A sample of five items is drawn at random from a lot. Describe the random variable X that denotes the number of defective items in the sample.

Solution

$$X(\omega) = \omega \qquad \omega = 0, 1, 2, 3, 4, 5$$

Let the random variable Y denote the number of nondefective items in the sample. Define the random variable Z as the absolute difference between X and Y, i.e., $Z = |X - Y|$. Describe the random variable Z:

$$Z(\omega) = \omega \qquad \omega = 1, 3, 5$$

2.2 Distribution Functions

We are very often interested in computing the *probability of events associated with a given random variable*. Let X denote a random variable defined on a sample space S. Then, if we define event E_a as the set of all outcomes (ω) such that $X_\omega \leq a$, we can derive the probability of event E_a for any real number a. In symbols,

$$E_a = \{\omega \,|\, X(\omega) \leq a\}$$

or a more convenient notation:

$$E_a = \{X \leq a\}$$

and

$$P\{E_a\} = P\{X \leq a\}$$

If we denote the probability of this event by a new function F called the *cumulative distribution function* or just the *distribution function* (c.d.f.) of the random variable X, we can write

$$F_X(a) = P\{X \leq a\} \qquad (-\infty < a < \infty)$$

In words, $F_X(a)$ *denotes the particular value taken by the distribution function of the random variable X at a.* The distribution function has the set of *all real numbers* as its domain.

Example 2.5 A fair die is rolled once. Find the probability that the number of points on the upturned face is less than or equal to three.

Solution. We define the sample space

$$S = \{1, 2, 3, 4, 5, 6\}$$

If X is the random variable whose value is less than or equal to three, then the event $X \leq 3$ is the subset $\{1, 2, 3\}$; therefore

$$F_X(3) = P\{X \le 3\} = P\{1, 2, 3\}$$
$$= P\{1\} + P\{2\} + P\{3\}$$
$$= \tfrac{1}{2}$$

Similarly, we can compute $F_X(1)$, $F_X(2)$, . . . , $F_X(6)$. The graph of the distribution function is given in Fig. 2.1. The distribution function has the following properties:

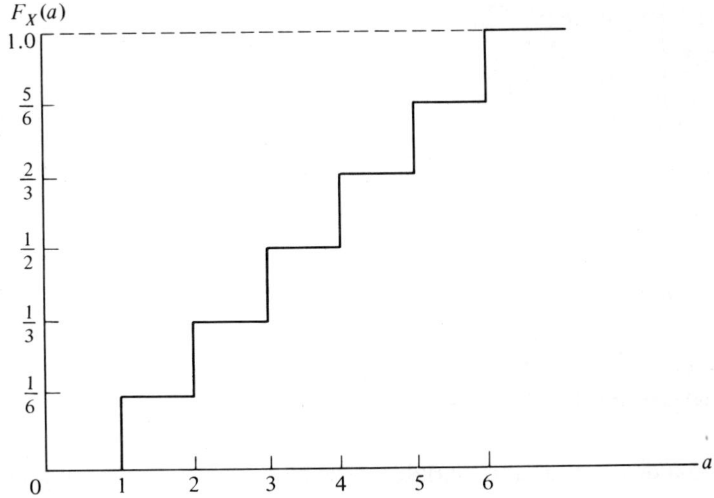

Fig. 2.1 Cumulative distribution of the discrete variate of Example 2.5.

$$\text{(2.1)} \qquad \begin{cases} \text{(1) It is a nondecreasing function} \\ \text{(2) } F_X(-\infty) = 0 \\ \text{(3) } F_X(+\infty) = 1 \end{cases}$$

Theorem 2.1. *Let F_X denote the distribution function of random variable X. Then, if a and b are any real numbers such that $b > a$, we have*

$$\text{(2.2)} \qquad F_X(b) - F_X(a) = P\{a < X \le b\}$$

Proof. We have

$$E_{a,b} = \{a < X \le b\}$$
$$E_a = \{X \le a\}$$

Therefore

$$F_X(b) = P\{X \le a\} + P\{a < X \le b\}$$
$$= F_X(a) + P\{a < X \le b\}$$

and

$$F_X(b) - F_X(a) = P\{a < X \le b\}$$

Example 2.6 The distribution function of a random variable X is given as follows:

$$F_X(a) = \begin{cases} 0 & a < 0 \\ \frac{1}{4} & 0 \leq a < 1 \\ \frac{2}{3} & 1 \leq a < 2 \\ 1 & a \geq 2 \end{cases}$$

Compute the following:
(a) $P\{X \leq 1\}$
(b) $P\{0 < X \leq 2\}$
(c) $P\{1 < X \leq 2\}$
(d) $P\{1 \leq X \leq 2\}$.

Solution
(a) $P\{X \leq 1\} = F_X(1) = \frac{2}{3}$
(b) $P\{0 < X \leq 2\} = F_X(2) - F_X(0) = 1 - \frac{1}{4} = \frac{3}{4}$
(c) $P\{1 < X \leq 2\} = F_X(2) - F_X(1) = 1 - \frac{2}{3} = \frac{1}{3}$
(d) $P\{1 \leq X \leq 2\} = F_X(2) - F_X(0) = 1 - \frac{1}{4} = \frac{3}{4}$

2.3 Probability Functions

When a discrete random variable X takes on any given one of its possible finite values, say k, then the probability that X is equal to k is called the *probability function* of the random variable X and is denoted by $P_X(k)$, where

$$P_X(k) = P\{X = k\}$$

The probability function P_X has the following properties:

(2.3)
(a) $P_X(k) > 0 \quad \text{for } k = 1, 2, \ldots, N$
In words, *the probability function is positive for a finite or countably infinite distinct set of values of X.*
(b) $$\sum_{k=1}^{N} P_X(k) = 1$$

The cumulative distribution of a discrete random variable can be expressed as

$$F_X(a) = \sum_{k \leq a} P_X(k) \qquad (-\infty < a < \infty) \tag{2.4}$$

Example 2.7 A sample of 3 items is drawn without replacement from a batch of 15 items, of which two are defective. Let X denote the number of defective items in the sample.
(a) Determine the probability function of X.
(b) Sketch the graph of the probability function.
(c) Determine the distribution function.

Solution

(a)
$$P_X(0) = (\tfrac{13}{15})(\tfrac{12}{14})(\tfrac{11}{13}) = \tfrac{22}{35}$$
$$P_X(1) = P\{G, B, G\} + P\{G, G, B\} + P\{B, G, G\}$$

where G and B denote good and bad items, respectively. Therefore

$$P_X(1) = (\tfrac{13}{15})(\tfrac{2}{14})(\tfrac{12}{13}) + (\tfrac{13}{15})(\tfrac{12}{14})(\tfrac{2}{13}) + (\tfrac{2}{15})(\tfrac{13}{14})(\tfrac{12}{13}) = \tfrac{12}{35}$$

and

$$P_X(2) = P\{G, B, B\} + P\{B, B, G\} + P\{B, G, B\}$$
$$= (\tfrac{13}{15})(\tfrac{2}{14})(\tfrac{1}{13}) + (\tfrac{2}{15})(\tfrac{1}{14})(\tfrac{13}{13}) + (\tfrac{2}{15})(\tfrac{13}{14})(\tfrac{1}{13}) = \tfrac{1}{35}$$

Hence the probability function of X is given by

$$P_X(k) = \begin{cases} \frac{22}{35} & k = 0 \\ \frac{12}{35} & k = 1 \\ \frac{1}{35} & k = 2 \end{cases}$$

(b) The graph of the probability function for the discrete random variable shown in Fig. 2.2 is called the *line diagram.*

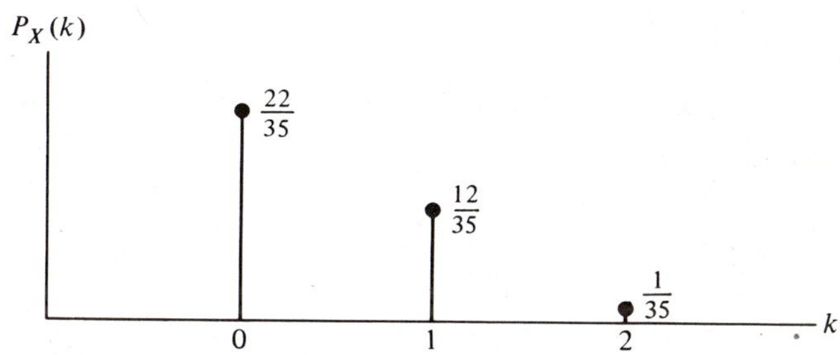

Fig. 2.2 Graph of the probability function of the discrete variate of Example 2.7.

(c) The c.d.f. of X has the form

$$F_X(k) = \begin{cases} 0 & k < 0 \\ \frac{22}{35} & 0 \le k < 1 \\ \frac{34}{35} & 1 \le k < 2 \\ 1 & k \ge 2 \end{cases}$$

2.4 Density Functions

Let X denote a continuous random variable, and suppose that its distribution function F_X is continuous everywhere on the real axis. Assume now that X can take on any real value between t and $t + \Delta$ where Δ is an arbitrary small positive number. Then we can write

$$t < X \leq t + \Delta \subseteq t \leq X \leq t + \Delta$$

Therefore

$$P\{t < X \leq t + \Delta\} \leq P\{t \leq X \leq t + \Delta\}$$

However,

$$P\{t < X \leq t + \Delta\} = F_X(t + \Delta) - F_X(t)$$

So that

$$F_X(t + \Delta) - F_X(t) \leq P\{t \leq X \leq t + \Delta\}$$

Since F_X is assumed to be continuous, the limit of the left-hand side of the above inequality as $\Delta \longrightarrow 0$ is zero, whereas the limit of the right-hand side as $\Delta \longrightarrow 0$ is $P\{X = t\}$. Hence we conclude that *whenever the distribution function of a random variable is continuous, the following condition is satisfied:*

$$P\{X = t\} = 0 \qquad \text{for all } t$$

This also means that for a continuous distribution we have

$$P\{a \leq X \leq b\} = P\{a < X \leq b\} = P\{a \leq X < b\} = P\{a < X < b\}$$

Now if $\frac{d}{dt}[F_X(t)]$ exists and is continuous in the interval (a, b), we can write

$$F_X(b) - F_X(a) = \int_a^b f_X(t)dt \qquad \text{for all } a \leq b$$

where

$$f_X(t) = \frac{d}{dt}F_X(t)$$

or

$$dF_X(t) = f_X(t)dt$$

f_X is called the *probability density function* (p.d.f.) of the random variable X and must satisfy the following conditions:

$$(2.5) \qquad \begin{cases} (1) & f_X(t) \geq 0 \qquad -\infty \leq t \leq \infty \\ (2) & \displaystyle\int_{-\infty}^{\infty} f_X(t)dt = 1 \end{cases}$$

The density function f for a continuous distribution plays the same role as the probability function P for a discrete distribution. However, in the discrete case $P_X(t)$ is the probability that $X = t$, but in the continuous case $f_X(t)$ is not the probability that $X = t$. In fact this probability is equal to zero because the distribution function is continuous for all t.

Let us elaborate further on this statement by considering an illustrative example. Suppose that the diameter of a sphere is measured by a vernier caliper. Although the diameter is assumed to be a continuous random variable, the recorded measurements (rounded to the nearest one-thousandth inch because of the discreteness of the measuring instrument) seem to come from a discrete sample space. Thus individual measurements will appear on the scale as isolated points separated by equal gaps of .001 in. each. Since it is desirable that the data fill a continuous scale, each individual measurement is considered as the midpoint of a cell. For example, if the recorded measurement is .364 in., we really mean that the diameter lies in the interval (.3635, .3645) and not that it is exactly .364 in. Thus the $P\{X=t\}=0$ does not imply that the event $\{X=t\}$ is an impossible event; it implies only that this event is very unlikely to occur. Accordingly, $f_X(t)$ cannot be interpreted as the probability that random variable X lies in the infinitesimal interval $[t-(\Delta t/2), t+(\Delta t/2)]$.

$f_X(t)$ is merely a measure of the frequency with which variate X will tend to assume a value in the interval Δt. $f_X(t)$ provides a probability measure if integrated between any two constants within the range of variate X. Thus, the probability that variate X lies between a and b ($b > a$) is $\int_a^b f_X(t)dt$.

Example 2.8 The demand for a certain commodity is a random variable specified by the probability density function

$$f_D(t) = \begin{cases} kt & 0 \leq t < 10 \\ k(20-t) & 10 \leq t < 20 \\ 0 & \text{elsewhere} \end{cases}$$

(a) Find k such that f_D is a density function.
(b) Sketch the graph of the probability density function.
(c) Sketch the graph of the distribution function.
Compute the following:
(d) $P\{D \geq 10\}$
(e) $P\{15 \leq D \leq 20\}$.

Solution

(a) f_D is a density function if $\int_{-\infty}^{\infty} f_D(t)dt = 1$. Since $f_D(t) \geq 0$,

$$\int_0^{10} kt\,dt + \int_{10}^{20} k(20-t)dt = 1$$

whence

$$100k = 1$$

or

$$k = .01$$

(b) The density function of demand:

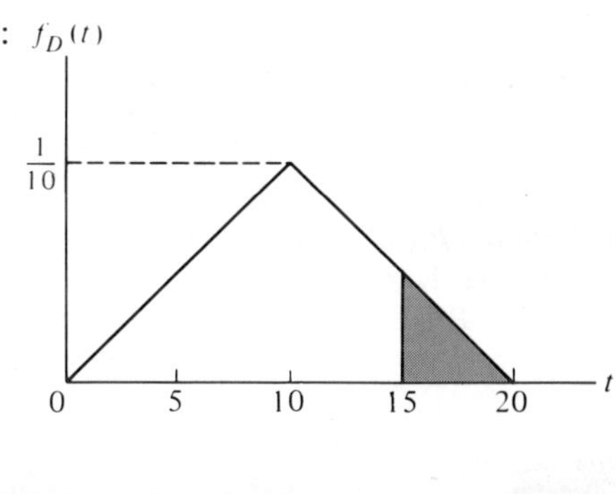

(c) The distribution function of demand:

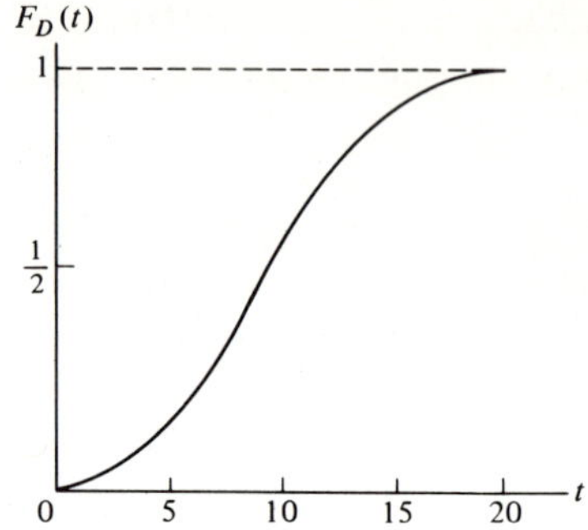

(d) $P\{D \geq 10\} = \int_{10}^{20} k(20 - t)dt = \frac{1}{2}$

(e) $P\{15 \leq D \leq 20\} = \int_{15}^{20} k(20 - t)dt = .125$

which is represented by the shaded area in (b).

Example 2.9 The random variable X has a continuous distribution function F given by

$$F_X(b) = 1 - e^{-.01b} \qquad b \geq 0$$

(a) Find f_X, the p.d.f. of X.
(b) Compute $P\{X > 100\}$.
(c) If a new random variable is given by $G(X) = 2X + 5$ find F_G.

Solution

(a)
$$f_X(t) = \frac{d}{db}F_X(b)\,|_{b=t}$$
$$= .01e^{-.01t} \qquad t \geq 0$$

(b)
$$P\{X > 100\} = \int_{100}^{\infty} .01e^{-.01t}dt$$
$$= e^{-1} = .368$$

(c)
$$F_G(a) = P\{G \leq a\} = P\{2X + 5 \leq a\}$$
$$= P\left\{X \leq \frac{a-5}{2}\right\} = F_X\left(\frac{a-5}{2}\right)$$

Therefore

$$F_G(a) = \begin{cases} 0 & a < 5 \\ 1 - e^{-.005(a-5)} & a \geq 5 \end{cases}$$

Example 2.10 A random variable X has the density function

$$f_X(b) = b + \tfrac{1}{2} \qquad 0 \leq b \leq 1$$
$$= 0 \qquad \text{otherwise}$$

(a) Determine F_X.
(b) Determine the density function of the random variable $G(X) = X^2$.
(c) Compute $P\{G > .36\}$.

Solution

(a) The cumulative distribution function of X is

$$F_X(t) = \int_0^t (b + \tfrac{1}{2})db = \tfrac{1}{2}(t + t^2)$$

hence

$$F_X(t) = \begin{cases} 0 & t < 0 \\ \tfrac{1}{2}(t + t^2) & 0 \le t \le 1 \\ 1 & t > 1 \end{cases}$$

(b)
$$F_X(a) = P\{G \le a\} = P\{X^2 \le a\}$$
$$= P\{-\sqrt{a} \le X \le \sqrt{a}\}$$

Since G is never negative,

$F_G(a) = P\{X \le \sqrt{a}\} = F_X(\sqrt{a})$ and is described by

$$F_G(a) = \begin{cases} 0 & a < 0 \\ \tfrac{1}{2}(a + \sqrt{a}) & 0 \le a \le 1 \\ 1 & a > 1 \end{cases}$$

Hence, the p.d.f. of G is given by

$$f_G(u) = \frac{1}{2}\left(1 + \frac{1}{2\sqrt{u}}\right) \qquad 0 \le u \le 1$$
$$= 0 \qquad \text{otherwise}$$

(c)
$$\begin{aligned} P\{G > .36\} &= 1 - P\{G \le .36\} \\ &= 1 - F_G(.36) \\ &= 1 - \tfrac{1}{2}(.36 + \sqrt{.36}) \\ &= .52 \end{aligned}$$

2.5 Empirical Frequency Distributions

Suppose that a set of data is obtained from performing an experiment a number of times. As an illustration, let us consider the following 100 measurements of the resistance in ohms of a certain type of resistor:

1.001	.991	1.004	.935	1.001	1.003	1.009	1.002	1.018	1.015
1.002	1.002	.993	.993	1.022	.943	1.016	1.048	1.018	1.016
1.050	.992	.954	1.013	.996	.995	1.052	1.017	.982	1.017
.980	1.003	1.004	.971	.963	.994	1.047	.965	.981	.993
.983	1.021	.987	1.006	.987	1.004	.991	.990	1.028	1.022
1.020	1.065	1.022	1.061	1.001	1.046	.978	.991	1.053	.993
.990	1.021	.976	1.023	1.024	.988	1.016	1.017	.996	1.003
1.010	1.041	1.012	1.042	1.013	1.025	1.007	1.003	1.002	1.054
.990	1.011	.979	1.005	1.014	1.015	1.026	1.027	1.005	1.008
1.030	1.036	1.030	1.033	1.039	1.038	1.037	1.036	1.032	1.031

The resistance is assumed to be a continuous random variable whose probability density function is unknown. The hundred measurements constitute a sample from the population. Our task is to determine the empirical frequency distribution of the random variable (resistance) giving rise to the data. The first step is to determine the number of *classes and boundaries* of each class. We note that the largest observation is 1.065, the smallest is .935, and the range is .130. A rough working rule is to use 10–20 classes. If we use fewer than 10 classes too much accuracy is lost, whereas using more than 15 classes entails unnecessarily laborious computations. Thus, in this problem it is reasonable to classify the observations into 14 classes. Class intervals should be equal. Since the random variable under study is continuous, although rounded to the nearest one-thousandth, it is desirable to have the classes into which we group the data fill a continuous scale. This means that class boundaries would be overlapping. Accordingly, class boundaries should be chosen halfway between two possible observations. This choice eliminates ambiguities even though the boundaries overlap. We now tally the 100 observations as shown below:

Class Boundaries	*Tally*	*Class Midpoints*	*Observed Frequency*
.9295– .9395	/	.9345	1
.9395– .9495	/	.9445	1
.9495– .9595	/	.9545	1
.9595– .9695	//	.9645	2
.9695– .9795	////	.9745	4
.9795– .9895	~~////~~ //	.9845	7
.9895– .9995	~~////~~ ~~////~~ ~~////~~	.9945	15
.9995–1.0095	~~////~~ ~~////~~ ~~////~~ ~~////~~	1.0045	20
1.0095–1.0195	~~////~~ ~~////~~ ~~////~~ /	1.0145	16
1.0195–1.0295	~~////~~ ~~////~~ //	1.0245	12
1.0295–1.0395	~~////~~ ~~////~~	1.0345	10
1.0395–1.0495	~~////~~	1.0445	5
1.0495–1.0595	////	1.0545	4
1.0595–1.0695	//	1.0645	2

Note that *class midpoint* is sometimes called *class mark*. It represents all measurements in a given class interval. The above tabulation shows the frequency of occurrence of the values of the resistance in specified classes. Such tabulation is known as a *grouped frequency distribution* of a set of observations. Now let us represent graphically the frequency distribution of our random variable. In practice, three common ways of graphic representation are used. Of these, the *frequency histogram* of Fig. 2.3(a) is extensively used. In this graph the heights of the rectangles represent the observed class frequencies and the sides of the rectangles represent the upper and lower class boundaries. Figure 2.3(b), the *frequency bar chart*, uses bars centered on the class mark; the heights of the bars represent the observed class frequencies.

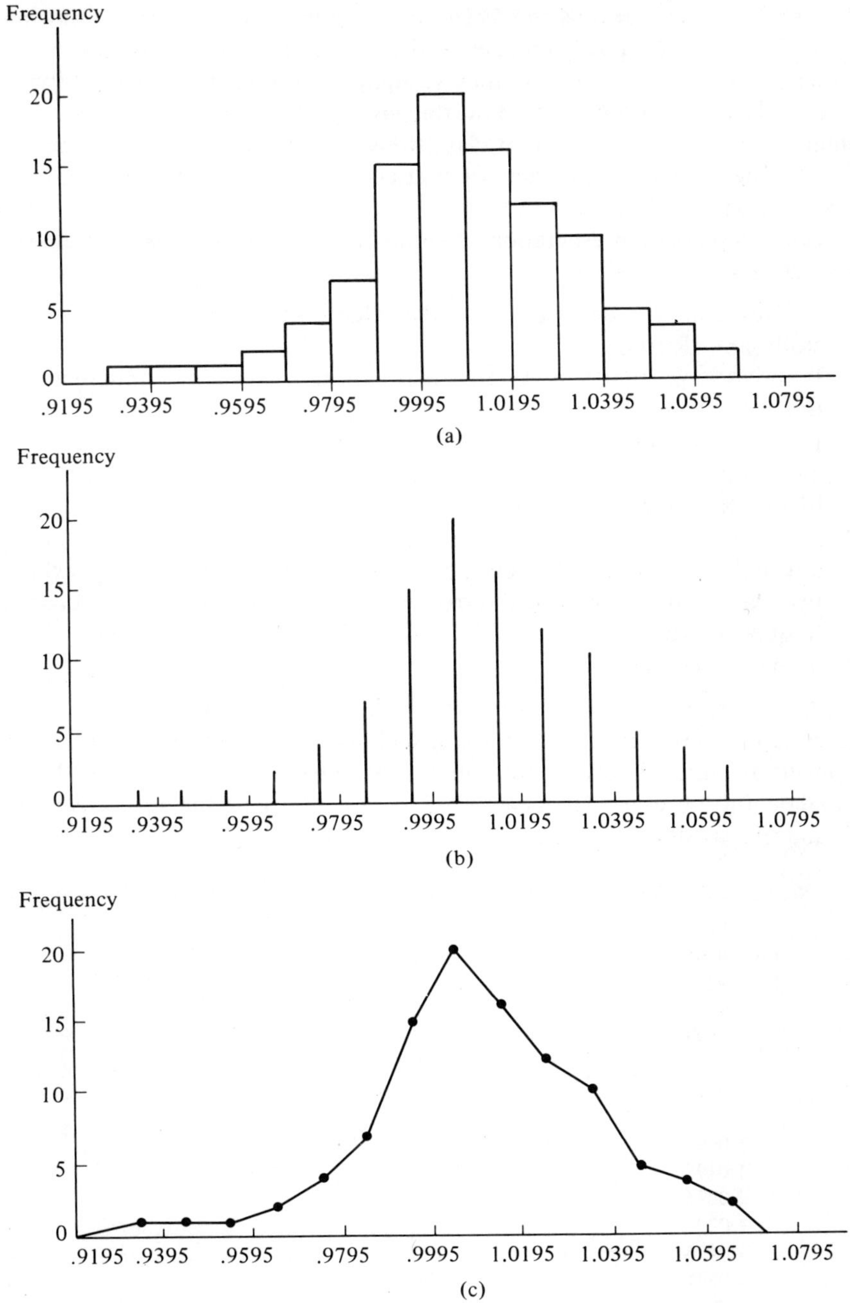

Fig. 2.3 (a) Histogram; (b) frequency bar chart; (c) frequency polygon.

Figure 2.3(c) is known as the *frequency polygon*. In this graph we first plot small circles (each circle represents the class mark and its corresponding observed frequency) and then draw straight line segments to connect successive circles. It should be noted that the *observed frequency* for any class is the number of observations in that class, whereas the *relative frequency* for any class is the observed frequency for that class divided by the total number of observations.

The graphical representation of empirical distributions has the following advantages:

1. It gives a fairly good idea of how the values of the random variable under study are distributed.
2. It shows important features that may not be immediately apparent from the data themselves.
3. It is a useful tool for deciding on an appropriate type of model (theoretical distribution) to represent the data before one can perform statistical hypothesis testing for goodness of fit.
4. It provides a basis for taking corrective action when irregularities and unusual features exist. For example, a frequency distribution that exhibits two classes with high observed frequencies separated by classes with lower frequencies (bimodal) may indicate that the process has sustained a shift in the process mean.

Cumulative frequency distributions are very useful for representing essentially discrete random variables, such as the number of defective components in an assembly or the number of defective items in a lot. Let us tabulate the observed frequencies of values less than the respective class boundaries as shown below:

Resistance Less Than	*Number of Observations*	*Empirical Probability*
.9295	0	0
.9395	1	.01
.9495	2	.02
.9595	3	.03
.9695	5	.05
.9795	9	.09
.9895	16	.16
.9995	31	.31
1.0095	51	.51
1.0195	67	.67
1.0295	79	.79
1.0395	89	.89
1.0495	94	.94
1.0595	98	.98
1.0695	100	1.00

Figure 2.4 shows the cumulative "less than" distribution of the resistance.

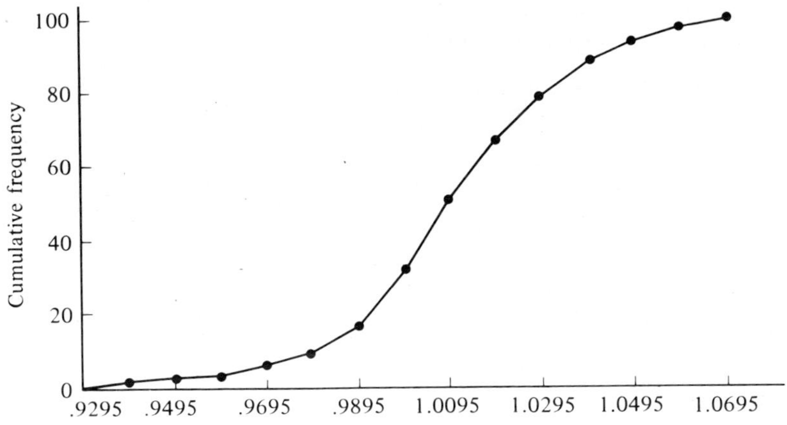

Fig. 2.4 Ogive.

This type of graph is sometimes called an *ogive*. If the relative frequency is plotted on normal probability graph paper the ogive will be a straight line for a normally distributed random variable. The normal probability graph paper is a useful device for checking whether the observations come from a normally distributed population. Needless to say, such a device is approximate. One usually rejects normality when remarkable departure from linearity is quite evident.

2.6 The Monte Carlo Technique

Simulation, in its broadest sense, means representation of reality. In recent years the widespread use of *simulation techniques* has been felt in different fields such as business, economics, social, physical, and engineering sciences. The use of high-speed digital computers has increased the power of these techniques.

The Monte Carlo technique is a device for modeling and simulating processes that involve *chance variables*. The Monte Carlo technique is resorted to under the following circumstances:

1. The systems being studied cannot be analyzed by using direct and formal analytical methods.
2. Analytical methods are complex, time-consuming, and costly.
3. Direct experimentation cannot be performed.
4. Analytical solutions are beyond the mathematical training of operations analysts.

The ingredients of the Monte Carlo technique are

1. A random variable and its probability distribution.
2. A sequence of random numbers.

Suppose that X is a random variable that describes the demand per day for a certain commodity. Its probability function is given by

$$P_X(k) = \begin{cases} .10 & k = 0 \\ .20 & k = 1 \\ .40 & k = 2 \\ .30 & k = 3 \end{cases} \tag{2.6}$$

Then its cumulative distribution function has the form

$$F_X(k) = \begin{cases} 0 & k < 0 \\ .1 & 0 \leq k < 1 \\ .30 & 1 \leq k < 2 \\ .70 & 2 \leq k < 3 \\ 1.00 & k \geq 3 \end{cases} \tag{2.7}$$

Our task is to generate values of demand (k) by chance such that the relative frequency of each value of k will be equal to its probability, given by Eq. (2.6). To accomplish this task we need a sequence of *random numbers.* A sequence of such numbers is given in Table A in the Appendix. In practice, the Monte Carlo simulation is generally very complex. In most cases high-speed digital computers are required. When a computer is used in simulation, it is not advisable to fill the computer's memory with a large number of random digits. Consequently, programs have been developed to generate predictable and reproducible random numbers. These random numbers are called *pseudorandom numbers.* There are many techniques for generating pseudorandom numbers. Two of the more commonly used are the *midsquare* and *congruential techniques.* We shall briefly describe the midsquare technique. The interested reader may consult References [7], [11], and [12] for a detailed discussion of testing and generating random numbers. The midsquare procedure consists of taking a four-digit number (preferably selected at random), squaring it, taking the four digits starting at the third from the left, recording them, squaring them, and so on. For example, if we select the four-digit number 1653, by squaring we obtain 02732409. (Note a zero is added on the left to make the number of digits equal to 8.) The set of four digits starting at the third from the left is 7324. Squaring 7324 yields 53640976, and 6409 is considered as the next set of four random digits. Proceeding in this way we obtain the sequences 1653, 7324, 6409, . . . , The "middle digits" thus generated can be looked upon as values of a random variable having the discrete uniform distribution. That is,

$$P_X(k) = \tfrac{1}{10} \qquad k = 0, 1, \ldots, 9$$

The first step in the simulation procedure is to partition the set of random numbers into a number of subsets. Since the random variable under study (the demand) has four outcomes, we will partition the set of random numbers into four subsets such that each value of the random variable would occur with relative frequency equal to its prior probability. Suppose we decided to use two-digit numbers; then the subset $\{00, 01, \ldots, 09\}$ would represent the outcome $k = 0$. It does not matter which numbers are used in each subset so long as no number is ever used more than once and so long as each subset provides the correct probability measure. Proceeding in this way, the entire assignment would be

k	*Random Numbers*
0	00–09
1	10–29
2	30–69
3	70–99

Now if we use, say, the random numbers shown on page 551, starting from the top of the fifth column, we read: 06, 53, 56, 58, 41, 38, 78, ... (Random numbers may be read from the table in any order or sequence with the restriction they not be chosen on the basis of the numbers themselves.) This sequence is interpreted as k having the respective value 0, 2, 2, 2, 2, 3, ...

The following two examples will illustrate the Monte Carlo method of analysis:

Example 2.11 Suppose that you have the following probability distribution of bus arrivals at a terminal station:

Buses per 10-Min Interval	*Probability*
1	.40
2	.50
3	.10

Take a sample of 200 hr and estimate the probability distribution of bus arrivals on an hourly basis.

Solution. Although the analytical solution of this problem is straightforward, it will be instructive to apply the Monte Carlo technique to estimate the required distribution. We shall assign two-digit numbers to each value of bus arrivals as follows:

Arrivals per 10-Min Interval	*Number Assigned*
1	00–39
2	40–89
3	90–99

Let us use the random numbers page 552, starting from the top of Column 1. Thus we read the random numbers

44, 48, 61, 38, 40, 29, 45, 72, 35, 62, 57, 97, . . .

which are interpreted, respectively, as 2, 2, 2, 1, 2, 1, 2, 2, 1, 2, 2, 3, . . . arrivals within 10-min intervals. The total number of arrivals in six intervals is the number of arrivals in 1 hr. Thus the number of arrivals in the first hour is 10, and in the second hour it is 12. Continuing in this way we can quickly evaluate the number of arrivals per hour for a sample of 200 hr. Our simulation procedure yields the following results:

Number of Arrivals per Hr.	*Frequency*	*Relative Frequency*
7	11	.055
8	20	.100
9	35	.175
10	51	.255
11	42	.210
12	21	.105
13	17	.085
14	2	.010
15	1	.005
	200	1.000

It is of interest to compare the simulated distribution with the analytical one. The analytical solution of this example yields the following probability distribution:

Number of Arrivals per Hr.	*Probability*
6	.004096
7	.030720
8	.102144
9	.198400
10	.249840
11	.214200
12	.127905
13	.053550
14	.015615
15	.003100
16	.000399
17	.000030
18	.000001
	1.000000

It can readily be seen that the simulated distribution is not identical to the theoretical distribution; however, it is close indeed. The accuracy associated with the experimental method can be improved by increasing the sample size. On the other hand, better estimates can be obtained with relatively small sample sizes by using variance reducing techniques. The interested reader may consult the suggested references for further explication of this topic.

Example 2.12 A product is assembled by Operator A and then moved by a belt conveyor to Operator B, who performs functional inspection. Assembly time in minutes is a random variable having the p.d.f. (normal with $\mu = 20$, $\sigma^2 = 4$)

$$f_X(t) = \frac{1}{2\sqrt{2\pi}} \exp\left[-\frac{1}{2}\left(\frac{t-20}{2}\right)^2\right] \qquad -\infty < t < \infty$$

Suppose that inspection time in minutes is uniformly distributed. That is,

$$\begin{aligned} f_Y(b) &= .2 \qquad 17 < b < 22 \\ &= 0 \qquad \text{otherwise} \end{aligned}$$

Using the Monte Carlo technique, estimate the mean expected number of assemblies produced in an 8-hr day. Assume that handling time is negligible.

Solution. This problem assumes that the reader is acquainted with the normal and uniform distributions. Both distributions are described in Chapter 6. We note that in this problem we will simulate observations from two continuous distributions. The first step is to construct the cumulative distribution function of the variate being studied. The second step is to generate a random decimal number between 0 and 1. The final step is finding the value of the random variable by solving $F_X(t) =$ random number. In our example, assembly time is a random variable having the normal distribution. Its c.d.f. is shown in Fig. 2.5.

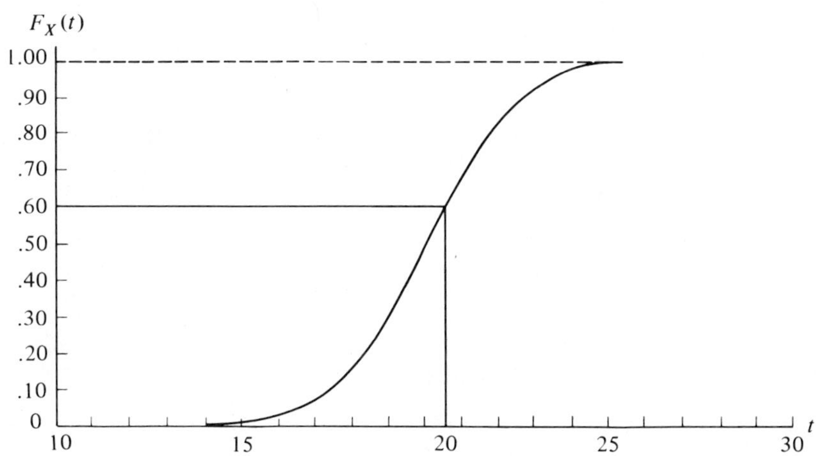

Fig. 2.5 Random selection from normal distribution.

Suppose now that random number .61 is read from Table A in the Appendix. At .61 a horizontal line is drawn to intersect the c.d.f. curve. The abscissa of the point of intersection gives the value t, which is 20.56. Repeating this procedure we can then obtain a random selection of assembly times. More conveniently, the random selection of assembly times can be obtained from the cumulative normal distribution table (Table D in the Appendix). To illustrate this procedure, if $F_X(t) = .61$, then from Table D we have

$$.28 = \frac{t-\mu}{\sigma} = \frac{t-20}{2} = z \tag{2.8}$$

where Z is the standardized normal variate. Solving (2.8) we find that $t = 20.56$.

Similarly, from the cumulative distribution of inspection time, we can obtain a random selection of inspection times. For example, suppose that random number .55 is read from Table A; then

$$F_Y(b) = .55 = \int_{17}^{b} .2\,dt = .2(b-17) \tag{2.9}$$

TABLE 2.1

Operator A					*Operator B*				
Random Number	*z*	*Assembly Time*	*Start*	*Finish*	*Random Number*	*Inspection Time*	*Start*	*Finish*	*Assembly Number*
80	.84	21.68	0	21.68	17	17.85	21.68	39.53	1
68	.47	20.94	21.68	42.62	92	21.60	42.62	64.22	2
30	−.52	18.96	42.62	61.58	89	21.45	64.22	85.67	3
67	.44	20.88	61.58	82.46	79	20.95	85.67	106.62	4
70	.52	21.04	82.46	103.50	76	20.80	106.62	127.42	5
21	−.81	18.38	103.50	121.88	02	17.10	127.42	144.52	6
62	.31	20.62	121.88	142.50	12	17.60	144.52	162.12	7
01	−2.33	15.34	142.50	157.84	01	17.05	162.12	179.17	8
79	.81	21.62	157.84	179.46	44	19.20	179.46	198.66	9
75	.67	21.34	179.46	200.80	18	17.90	200.80	218.70	10
18	−.92	18.16	200.80	218.96	33	18.65	218.96	237.61	11
53	.08	20.16	218.96	239.12	58	19.90	239.12	259.02	12
29	−.55	18.90	239.12	258.02	85	21.25	259.02	280.27	13
65	.39	20.78	258.02	278.80	93	21.65	280.27	301.92	14
19	−.88	18.24	278.80	297.04	39	18.95	301.92	320.87	15
85	1.04	21.08	297.04	318.12	72	20.60	320.87	341.47	16
68	.47	20.94	318.12	339.06	05	17.25	341.47	358.72	17
11	−1.23	17.54	339.06	356.60	12	17.60	358.72	376.32	18
62	.31	20.62	356.60	377.22	93	21.65	377.22	398.87	19
56	.15	20.30	377.22	397.52	40	19.00	398.87	417.87	20
63	.33	20.66	397.52	418.38	18	17.90	418.38	436.28	21
64	.36	20.72	418.38	439.10	84	21.20	439.10	460.30	22
39	−.28	19.44	439.10	458.54	99	21.95	460.30	482.25	23
34	−.41	19.18	458.54	477.72	41	19.05	482.25	501.30	24

whence $b = 19.75$. The results of the simulated experiment in tabular form are shown in Table 2.1.

Hence, the estimate of the mean expected number of assemblies per day is 23 approximately. Obviously this estimate is derived from a sample of 8 hours. It is absolutely inappropriate to assume that this estimate is sufficiently precise. However, statistical analysis can be conducted to determine the sample size to achieve a desired confidence level for estimates.

PROBLEMS

2.1. Suppose that sales revenue f in dollars per day from the sale of a certain product is given by the function

$$f: f(x) = 10x \qquad x = 0, 1, \ldots, 5$$

Find the domain and the range of f.

2.2. The production cost of a certain extruded aluminum tubing in dollars per foot can be described by the function

$$f(L) = \begin{cases} 2 & L < 1000 \\ 1.8 & 1000 \leq L \leq 2000 \\ 1.7 & L \geq 2000 \end{cases}$$

where L denotes extruded length in feet. If 3,000 ft are produced, what is the total production cost?

2.3. A function f is given by

$$f(x) = \begin{cases} 10 & 0 < x \leq 10 \\ 2x - 10 & 10 < x \leq 15 \\ 20 & 15 < x \leq 20 \end{cases}$$

Compute the following:

(a) $\int_0^{12} f(x)\,dx$

(b) $\int_{12}^{18} f(x)\,dx$.

2.4. A merchant in a small town stocks X Christmas trees. The demand Y for Christmas trees is a random variable. The merchant pays \$3 for each tree, and he sells it for \$8. Trees left over at the end of the season can be sold for firewood at \$.75 each. Describe the random variable denoting the profit of the merchant.

2.5. A factory has two departments (A and B). Let X be a random variable describing the number of accidents per week in department A and Y be a random variable describing the number of accidents per week in department B. Assume that the range of X is $\{0, 1, 2\}$ and that of Y is $\{0, 1, 2, 3, 4\}$. Describe the random variable denoting the number of accidents per week in the factory.

2.6. A die is rolled three times. Describe the random variable X that denotes the sum of the points.

2.7. Suppose that the number of typographical errors per typewritten page is a random variable that takes on the values {0, 1, 2}. Describe the random variable denoting the number of typographical errors in a report of five typewritten pages.

2.8. The daily demand for gasoline in gallons at a certain gas station is a random variable. Its density function is given by

$$\begin{aligned} f_D(a) &= k \qquad 3{,}000 < a < 5{,}000 \\ &= 0 \qquad \text{otherwise} \end{aligned}$$

(a) Find k such that f_D is a density function.
(b) Describe the c.d.f. of D.

2.9. The random variable X has the p.d.f. given by

$$f_X(a) = \begin{cases} 1 & .5 \le a \le 1 \\ k & 1 < a \le 2 \\ 0 & \text{otherwise} \end{cases}$$

(a) Determine k such that f_X is a density function.
(b) Find F_X, the c.d.f. of X.
(c) Compute $P\{X > 1.5\}$.

2.10. The p.d.f. of random variable X is

$$f_X(a) = .01e^{-.01a} \qquad 0 \le a \le \infty$$

(a) Find F_X, the c.d.f. of X.
Compute the following:
(b) $P\{X > 120\}$
(c) $P\{80 \le X \le 100\}$.

2.11. The random variable X has the p.d.f. given by

$$\begin{aligned} f_X(t) &= \frac{100}{t^2} \qquad t \ge 100 \\ &= 0 \qquad \text{otherwise} \end{aligned}$$

(a) Describe F_X, the c.d.f. of X.
(b) Compute $P\{X > 200\}$.

2.12. Given

$$P\{X > a\} = e^{-\lambda a}(\lambda a + 1) \qquad \lambda > 0 \qquad a \ge 0$$

find
(a) F_X, the c.d.f. of X
(b) f_X, the p.d.f. of X
(c) $P\left\{X > \frac{1}{\lambda}\right\}$.

2.13. Suppose that the p.d.f. of X is

$$f_X(a) = \begin{cases} a & 0 \le a \le 1 \\ 2 - a & 1 < a \le 2 \\ 0 & \text{otherwise} \end{cases}$$

(a) Describe F_X, the c.d.f. of X.
(b) Compute $P\{.4 \le X \le 1.6\}$.

2.14. The variate X is distributed as follows:

$$f_X(a) = ka^4e^{-a} \qquad a \geq 0$$
$$= 0 \qquad \text{otherwise}$$

Determine the value of k such that f_X is a density function.

2.15. A measurable characteristic is distributed as follows:

$$f_X(t) = 5 \qquad 2.65 < t < 2.85$$
$$= 0 \qquad \text{otherwise}$$

(a) Prove that the above is a p.d.f.
(b) Find F_X, the c.d.f. of X.
(c) Compute $P\{X \geq 2.75\}$.
(d) If $P\{X < k\} = 4P\{X > k\}$, what is the value of k?

2.16. The density function of a continuous random variable is

$$f_X(t) = \begin{cases} \frac{1}{2} & 0 \leq t \leq 1 \\ \frac{1}{4} & 2 \leq t \leq 4 \\ 0 & \text{otherwise} \end{cases}$$

(a) Describe F_X, the c.d.f. of X.
Compute the following:
(b) $P\{.75 \leq X \leq 2\}$
(c) $P\{.5 \leq X \leq 3.2\}$.

2.17. Suppose that the waiting time in minutes for a bus in a certain station is found to be a random variable with the following c.d.f.:

$$F_X(t) = 0 \qquad t < 0$$
$$= 1 - e^{-.1t} \qquad 0 \leq t \leq \infty$$

(a) Find f_X, the p.d.f. of X.
(b) What is the probability that the time a passenger will have to wait for a bus will be (1) more than 10 min, (2) less than 5 min, and (3) between 2 and 8 min?

2.18. Given that random variable X has the p.d.f.

$$f_X(t) = \frac{5}{\theta} \qquad -.1\theta < t < .1\theta$$
$$= 0 \qquad \text{otherwise}$$

and $P\{|X| < 2\} = 2P\{|X| > 2\}$, find the value of θ.

2.19. Suppose that the p.d.f. of variate X is specified by

$$f_X(t) = kte^{-t^2/2} \qquad t \geq 0$$
$$= 0 \qquad \text{otherwise}$$

(a) Find the value of k such that f_X is a p.d.f.
(b) Describe F_X, the c.d.f. of X.
(c) Compute $P\{X > \sqrt{2}\}$.

2.20. A newspaper boy buys his papers at \$.07 each, he sells them for \$.10, and he receives \$.03 for each unsold paper. Suppose that the probability of demand is specified by

$$P_X(k) = \frac{e^{-5}5^k}{k!} \qquad k = 0, 1, \ldots$$

(a) Describe the random variable that represents the daily profit if he buys 10 papers each day.

(b) Compute the following: $P\{X \geq 2\}$, $P\{X = 3\}$, and $P\{X < 1\}$.

2.21. The probability function of variate X is given by

$$P_X(k) = \begin{cases} \frac{1}{4} & k = 0 \\ \frac{1}{2} & k = 1 \\ \frac{1}{8} & k = 2 \\ \frac{1}{8} & k = 3 \end{cases}$$

(a) Describe F_X, the c.d.f. of X.
(b) Compute the following: $P\{X > 1\}$, $P\{X \leq 2\}$, and $P\{X < 3\}$.

2.22. Let X be a discrete random variable. Assume the range of X is $\{0, 1, 2, 3, 4\}$ and

$$P\{1 < X < 4\} = .55$$
$$P\{X \leq 1\} = .25$$
$$P\{X = 2\} = 2P\{X = 1\}$$
$$P\{X > 3\} = .20$$
$$P\{0 < X \leq 2\} = .45$$

(a) Find the probability function of X.
(b) Graph F_X, the c.d.f. of X.

2.23. The number of misprints per page of a certain magazine is a random variable with the following probability function:

$$P_X(k) = \frac{e^{-2}2^k}{k!} \qquad k = 0, 1, \ldots$$

Compute the following:
(a) $P\{0 < X < 3\}$
(b) $P\{X \geq 1\}$
(c) $P\{0 \leq X < 2\}$
(d) $P\{X = 2\}$
(e) $P\{0 \leq X \leq 2\}$
(f) $P\{X > 3\}$.

2.24. The probability function of variate X is given by

$$P_X(k) = C(.2)^{k-1} \qquad k = 1, 2, 3, \ldots$$

(a) Determine the value of C such that P_X is a probability function.
(b) Describe F_X, the c.d.f. of X.
(c) Compute $P\{X > 3\}$.

2.25. Two fair dice (marked A and B) are cast. Let X and Y denote the numbers that appear on A and B, respectively. Assume that the following random variables also are of interest:

$$D(X, Y) = X - Y \qquad M(X, Y) = \text{minimum } (X, Y)$$

(a) List the outcomes of the sample space.
(b) Describe the probability functions of the random variables D and M.
(c) Find the c.d.f.'s of D and M.

2.26. The following are 100 measurements (in pounds) of the net weight of sugar bags:

5.10	5.00	5.10	5.08	5.09	5.00	5.05	4.99	5.00	5.00
5.14	4.99	5.00	4.82	5.10	5.08	4.86	5.00	4.96	5.07
4.90	5.06	4.96	4.96	4.98	4.97	4.95	5.05	5.07	4.97
5.05	5.00	4.83	5.08	5.07	4.89	4.92	5.11	4.91	4.94
5.00	4.90	5.07	4.89	4.97	4.98	4.95	5.22	5.12	5.07
4.95	5.09	4.91	4.93	4.91	4.93	5.17	5.03	5.07	5.12
5.08	4.87	4.96	4.86	5.03	5.18	5.05	4.94	4.96	5.00
4.95	5.01	5.07	4.96	5.19	5.06	5.11	4.95	5.00	4.98
5.10	4.99	5.01	5.15	5.04	5.11	5.04	5.06	5.02	5.02
5.11	5.01	5.03	5.10	5.01	5.02	5.02	5.03	5.04	5.04

(a) Construct a histogram (use class boundaries of 0.05).
(b) Find the cumulative "less than" distribution and plot its ogive.

2.27. The following data represent the number of defective components in 100 television sets:

0	3	1	2	2	1	2	1	0	1
3	4	2	1	2	4	2	0	1	1
2	0	3	3	1	2	3	3	3	1
0	1	0	1	0	5	1	2	2	3
1	5	2	4	1	3	4	1	3	2
1	3	3	3	2	0	0	1	1	0
4	1	1	1	0	4	1	4	5	4
2	3	4	2	6	3	1	2	4	2
2	2	1	1	3	2	2	1	0	1
6	2	2	2	4	1	2	2	2	3

(a) Plot the histogram of the above data.
(b) Find the cumulative "less than" distribution and graph its ogive.

2.28. Suppose that the following random numbers are read from Table A (in the Appendix): 306, 700, 477, 941, 255, ... Generate five random observations from each of the following probability distributions:

(a)
$$P_X(k) = \frac{e^{-1.2}(1.2)^k}{k!} \qquad k = 0, 1, \ldots$$

(b)
$$P_X(k) = \begin{cases} .155 & k = 1 \\ .375 & k = 2 \\ .280 & k = 3 \\ .190 & k = 4 \end{cases}$$

2.29. Use the random numbers in the preceding problem to generate five random observations from the following probability distribution:

$$\begin{aligned} f_X(t) &= 3t^2 \qquad 0 \leq t \leq 1 \\ &= 0 \qquad \text{otherwise} \end{aligned}$$

2.30. Time (in minutes) between arrivals of customers at a checkout station has the following p.d.f.:

$$\begin{aligned} f_X(t) &= \tfrac{1}{2}e^{-t/2} \qquad t \geq 0 \\ &= 0 \qquad \text{otherwise} \end{aligned}$$

The time required to service the customers by adding the bills, receiving payment, etc., has the following probability distribution:

Service Time (min)	*Probability*
0.5	.125
1.0	.205
1.5	.360
2.0	.190
2.5	.070
3.0	.050

Simulate 120 min of operation of this system and estimate
(a) The average waiting time of a customer.
(b) The average percent idle time of the server.

SUGGESTED REFERENCES

[1] Conway, R. W., "Some Tactical Problems in Digital Simulation," *Management Science*, **10**(1), 1963.

[2] Derman, C., and M. Klein, *Probability and Statistical Inference for Engineers*, Oxford University Press, Inc., New York, 1959.

[3] Ehrenfeld, S., and S. Ben-Tuvia, "The Efficiency of Statistical Simulation Procedures," *Technometrics*, **4**(2), 1962.

[4] Ehrenfeld, S., and S. B. Littauer, *Introduction to Statistical Method*, McGraw-Hill Book Company, New York, 1964.

[5] Feller, W., *An Introduction to Probability Theory and Its Applications*, 3rd ed., John Wiley & Sons, Inc., New York, 1968.

[6] Hoel, P. G., *Introduction to Mathematical Statistics*, 3rd ed., John Wiley & Sons, Inc., New York, 1965.

[7] Meyer, H. A. (ed.), *Symposium on Monte Carlo Methods*, John Wiley & Sons, Inc., New York, 1956.

[8] Parzen, E., *Modern Probability Theory and Its Applications*, John Wiley & Sons, Inc., New York, 1960.

[9] Schlaifer, R., *Probability and Statistics for Business Decisions*, McGraw-Hill Book Company, New York, 1959.

[10] Shreider, Y. A. (ed.), *Method of Statistical Testing* (*Monte Carlo Method*), Elsevier, Amstertam, 1964.

[11] "Symposium on Monte Carlo Methods," *Journal of the Royal Statistical Society, Series B*, **16**, 1954.

[12] Taussky, O., and J. Todd, "Generation of Pseudo Random Numbers," in Meyer, 1956 (Ref. [7]).

[13] Tocher, K. D., *The Art of Simulation*, D. Van Nostrand, Co., Inc., Princeton, N. J., 1963.

3

JOINT DISTRIBUTION FUNCTIONS

3.1 Basic Concepts

So far we have studied the distribution functions of a one-dimensional random variable described by a real-valued function defined on a sample space. In many engineering problems we might be interested to find the distribution of an *n-dimensional random variable described by a real-valued function defined on a sample space whose range is a collection of n-tuples of real numbers.* These distributions are called *multivariate distributions.* A special case of the multivariate distributions involves a two-dimensional random variable and is known as *bivariate distributions.* We will consider a concrete example to give the reader an insight into this concept:

Example 3.1 The following characteristics might be of interest if a survey is to be conducted in a certain community:

X_1: The number of children in the family.
X_2: The annual income of the family.
X_3: The family owns a car.
X_4: The family owns a TV set.

In such a case we might be interested not only in studying each characteristic separately but also in determining interrelationships that exist among these characteristics. For example, one could be interested in knowledge of the probability associated with joint events such as

$$E_1 = \{0 < X_1 \leq 3,\ 4{,}000 < X_2 \leq 6{,}000\}$$
$$E_2 = \{5{,}000 < X_2 \leq 8{,}000,\ X_3 = 1\}$$

where X_3 is defined as follows:

$$X_3 = \begin{cases} 0 & \text{if the family does not own a car} \\ 1 & \text{if the family owns a car} \end{cases}$$

The probability associated with event E_1 means the probability that the family has one, two, or three children and its annual income is greater than \$4,000 but less than or equal to \$6,000.

We will develop these fundamental concepts as we proceed.

3.2 Joint Distribution Functions

Definition 3.1. The *joint distribution function* of two random variables X_1, X_2 is defined by

$$F_{X_1, X_2}(a, b) = P\{X_1 \leq a, X_2 \leq b\}$$

where $-\infty < a < \infty$ and $-\infty < b < \infty$. It follows from the above definition that the following relationships hold true:

$$(3.1)\quad \begin{cases} (1) & F_{X_1, X_2}(a, \infty) = P\{X_1 \leq a, X_2 \leq \infty\} \\ & \qquad\qquad\quad\ = P\{X_1 \leq a\} = F_{X_1}(a) \\ (2) & F_{X_1, X_2}(\infty, b) = F_{X_2}(b) \\ (3) & F_{X_1, X_2}(-\infty, b) = F_{X_1, X_2}(a, -\infty) = P\{\varnothing\} = 0 \end{cases}$$

The function $F_{X_1}(\cdot)$ is called the *marginal distribution function* of the random variable X_1 corresponding to the joint distribution function $F_{X_1, X_2}(\cdot, \cdot)$. Similarly, $F_{X_2}(\cdot)$ is the *marginal distribution* function of the random variable X_2 corresponding to the joint distribution function $F_{X_1, X_2}(\cdot)$.

Theorem 3.1. *Let F_{X_1, X_2}, the joint distribution function of random variables X_1 and X_2 be known, and let event E be expressed as*

$$E = \{a_1 < X_1 \leq a_2,\ b_1 < X_2 \leq b_2\}$$

Then the probability of event E is given by

$$(3.2)\quad \begin{aligned} P\{E\} = F_{X_1, X_2}(a_2, b_2) - F_{X_1, X_2}(a_2, b_1) - F_{X_1, X_2}(a_1, b_2) \\ + F_{X_1, X_2}(a_1, b_1) \end{aligned}$$

Proof. To facilitate the proof we will locate the points having positive probabilities on the coordinate system shown in Fig. 3.1. The event $\{X_1 \leq a_2, X_2 \leq b_2\}$ is the union of four disjoint events labeled 1, 2, 3, and 4. Hence, by the addition Theorem 1.1 we have

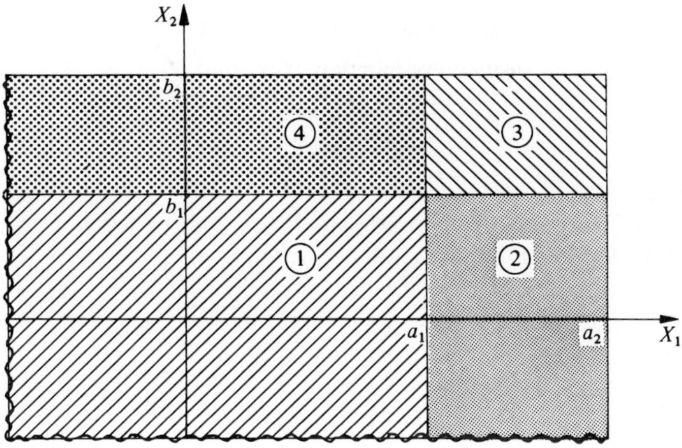

Fig. 3.1 View of the probability of event E.

$$\begin{aligned} P_{X_1,X_2}\{X_1 \le a_2, X_2 \le b_2\} = {} & P_{X_1,X_2}\{X_1 \le a_1, X_2 \le b_1\} \\ & + P_{X_1,X_2}\{a_1 < X_1 \le a_2, X_2 \le b_1\} \\ & + P_{X_1,X_2}\{a_1 < X_1 \le a_2, b_1 < X_2 \le b_2\} \\ & + P_{X_1,X_2}\{X_1 \le a_1, b_1 < X_2 \le b_2\} \end{aligned}$$

By Definition 3.1 the above expression can be written as

$$\begin{aligned} F_{X_1,X_2}(a_2, b_2) = {} & F_{X_1,X_2}(a_1, b_1) + P_{X_1,X_2}\{a_1 < X_1 \le a_2, X_2 \le b_1\} \\ & + P\{E\} + P_{X_1,X_2}\{X_1 \le a_1, b_1 < X_2 \le b_2\} \end{aligned} \tag{3.3}$$

The probability of the event labeled 2 is

$$P_{X_1,X_2}\{a_1 < X_1 \le a_2, X_2 \le b_1\} = F_{X_1,X_2}(a_2, b_1) - F_{X_1,X_2}(a_1, b_1) \tag{3.4}$$

Similarly, the probability of the event labeled 4 is

$$P_{X_1,X_2}\{X_1 \le a_1, b_1 < X_2 \le b_2\} = F_{X_1,X_2}(a_1, b_2) - F_{X_1,X_2}(a_1, b_1) \tag{3.5}$$

Substituting Eqs. (3.4) and (3.5) in Eq. (3.3) we obtain

$$\begin{aligned} F_{X_1,X_2}(a_2, b_2) = {} & F_{X_1,X_2}(a_1, b_1) + F_{X_1,X_2}(a_2, b_1) - F_{X_1,X_2}(a_1, b_1) \\ & + P\{E\} + F_{X_1,X_2}(a_1, b_2) - F_{X_1,X_2}(a_1, b_1) \end{aligned}$$

Therefore

$$P\{E\} = F_{X_1,X_2}(a_2, b_2) - F_{X_1,X_2}(a_2, b_1) - F_{X_1,X_2}(a_1, b_2) + F_{X_1,X_2}(a_1, b_1)$$

and the proof is complete.

Example 3.2 X_1 and X_2 are jointly distributed random variables that take on values $(-2, -2)$, $(0, 0)$, and $(2, 2)$. Furthermore,

$$P_{X_1}(-2) = \tfrac{1}{4} \qquad P_{X_2}(-2) = \tfrac{1}{4}$$
$$P_{X_1}(0) = \tfrac{1}{2} \qquad P_{X_2}(0) = \tfrac{1}{2}$$

$$P_{X_1}(2) = \tfrac{1}{4} \qquad P_{X_2}(2) = \tfrac{1}{4}$$

Describe the joint distribution function of X_1 and X_2.

Solution. The joint probability function of X_1 and X_2 is

$$P_{X_1,X_2}(-2, -2) = \tfrac{1}{4}$$
$$P_{X_1,X_2}(0, 0) = \tfrac{1}{2}$$
$$P_{X_1,X_2}(2, 2) = \tfrac{1}{4}$$

We will locate the points having positive probabilities on the coordinate system shown in Fig. 3.2. The joint distribution function is

Section Shown in Fig. **3.2**

$$F_{X_1,X_2}(a, b) = \begin{cases} 0 & a < -2 \quad b < -2 & (1) \\ 0 & -2 \le a \quad b < -2 & (2) \\ 0 & a < -2 \quad -2 \le b & (3) \\ \frac{1}{4} & a \ge -2 \quad -2 \le b < 0 & (4) \\ \frac{1}{4} & -2 \le a < 0 \quad b \ge 0 & (5) \\ \frac{3}{4} & 0 \le a < 2 \quad b \ge 0 & (6) \\ \frac{3}{4} & a \ge 2 \quad 0 \le b < 2 & (7) \\ 1 & 2 \le a \quad 2 \le b & (8) \end{cases}$$

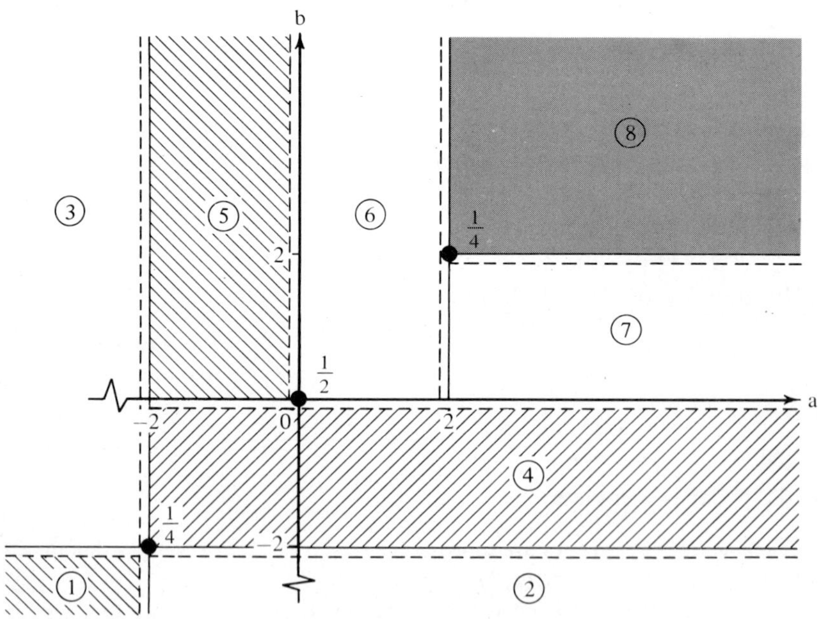

Fig. 3.2 View of the joint probability function of Example 3.2.

3.3 Joint Probability Functions

If X_1 and X_2 are discrete random variables we may define a new function P called the *joint probability function* of X_1 and X_2 as given by

$$P_{X_1, X_2}(a, b) = P\{X_1 = a, X_2 = b\} \tag{3.6}$$

for every pair of real numbers a and b. The distribution function is given by

$$F_{X_1, X_2}(u, v) = \sum_{a \leq u} \sum_{b \leq v} P_{X_1, X_2}(a, b) \tag{3.7}$$

where $-\infty < u < \infty$ and $-\infty < v < \infty$.

Example 3.3 A box contains 12 radio tubes of which 2 are defective. A random sample of 2 is drawn (a) with replacement and (b) without replacement. Let the random variables X_1 and X_2 be defined as follows:

$$X_1 = \begin{cases} 0 & \text{if the first tube is nondefective} \\ 1 & \text{if the first tube is defective} \end{cases}$$

$$X_2 = \begin{cases} 0 & \text{if the second tube is nondefective} \\ 1 & \text{if the second tube is defective} \end{cases}$$

(a) Describe the joint probability function of X_1 and X_2.
(b) Describe the marginal probability function of X_1 and X_2.

Solution. For convenience, the probabilities of all possible pairs of values of X_1 and X_2 will be presented in the following tables

b \ a	$P_{X_1, X_2}(a, b)$: 0	1	$P\{X_2 = b\}$
0	$\frac{10}{12} \cdot \frac{10}{12}$	$\frac{2}{12} \cdot \frac{10}{12}$	$\frac{5}{6}$
1	$\frac{10}{12} \cdot \frac{2}{12}$	$\frac{2}{12} \cdot \frac{2}{12}$	$\frac{1}{6}$
$P\{X_1 = a\}$	$\frac{5}{6}$	$\frac{1}{6}$	1

(a) with replacement

b \ a	$P_{X_1, X_2}(a, b)$: 0	1	$P\{X_2 = b\}$
0	$\frac{10}{12} \cdot \frac{9}{11}$	$\frac{2}{12} \cdot \frac{10}{11}$	$\frac{5}{6}$
1	$\frac{10}{12} \cdot \frac{2}{11}$	$\frac{2}{12} \cdot \frac{1}{11}$	$\frac{1}{6}$
$P\{X_1 = a\}$	$\frac{5}{6}$	$\frac{1}{6}$	1

(b) without replacement

We will extend further the analysis of this example with a view to derive the following *properties* of both the joint and marginal probability functions.

1. We note that both random variables X_1 and X_2 have possible values 0 and 1 with positive probabilities. In symbols,

$$P_{X_1}(a_i) = P\{X_1 = a_i\} > 0 \qquad i = 1, 2, \ldots, N$$
$$P_{X_2}(b_j) = P\{X_2 = b_j\} > 0 \qquad j = 1, 2, \ldots, M$$

2. Hence, from property 1, it follows that

$$\sum_{i=1}^{N} P_{X_1}(a_i) = 1 \qquad i = 1, 2, \ldots, N$$
$$\sum_{j=1}^{M} P_{X_2}(b_j) = 1 \qquad j = 1, 2, \ldots, M$$

This can be verified from Example 3.3, case (a) or case (b):

$$P_{X_1}(0) + P_{X_2}(1) = \tfrac{5}{6} + \tfrac{1}{6} = 1$$

If properties 1 and 2 are satisfied we say the random variables X_1 and X_2 are discrete or jointly discrete.

3. Since $P_{X_1}(a_i)$ and $P_{X_2}(b_j)$ are both positive,

$$P_{X_1, X_2}(a_i, b_j) = P\{X_1 = a_i, X_2 = b_j\}$$

is greater than zero if the joint event occurs and is equal to zero if

$$\{X_1 = a_i \cap X_2 = b_j\} = \varnothing$$

We indicate this by writing

$$P_{X_1, X_2}(a_i, b_j) \geq 0 \qquad i = 1, 2, \ldots, N \qquad j = 1, 2, \ldots, M$$

4. From Example 3.3, case (a), we have

$$\sum_{j=1}^{2} P_{X_1, X_2}(0, b_j) = P_{X_1, X_2}(0, 0) + P_{X_1, X_2}(0, 1)$$
$$= \tfrac{100}{144} + \tfrac{20}{144} = \tfrac{5}{6} = P_{X_1}(0)$$

Similarly,

$$\sum_{j=1}^{2} P_{X_1, X_2}(1, b_j) = \tfrac{20}{144} + \tfrac{4}{144} = \tfrac{1}{6} = P_{X_1}(1)$$

In general we can write

$$\sum_{j=1}^{M} P_{X_1, X_2}(a_i, b_j) = P_{X_1}(a_i) \qquad i = 1, 2, \ldots, N$$
$$\sum_{i=1}^{N} P_{X_1, X_2}(a_i, b_j) = P_{X_2}(b_j) \qquad j = 1, 2, \ldots, M$$

This means that it is always feasible to derive the marginal probabilities whenever the joint probability function is given. Is the *converse* true? We will answer this question as we proceed.

5. From properties 2 and 4 we have

$$\sum_{i=1}^{N} P_{X_1}(a_i) = \sum_{i=1}^{N} \sum_{j=1}^{M} P_{X_1, X_2}(a_i, b_j) = 1$$

$$\sum_{j=1}^{M} P_{X_2}(b_j) = \sum_{j=1}^{M} \sum_{i=1}^{N} P_{X_1, X_2}(a_i, b_j) = 1$$

Now, are the random variables X_1 and X_2 in Example 3.3 *independent*? To answer this question we need to derive the *conditional distributions* of these variates. The conditional probability of the event $\{X_1 = a_i\}$ given the occurrence of the event $\{X_2 = b_j\}$ is by definition

$$P\{X_1 = a_i \mid X_2 = b_j\} = \frac{P\{X_1 = a_i \cap X_2 = b_j\}}{P\{X_2 = b_j\}}$$

$$= \frac{P_{X_1, X_2}(a_i, b_j)}{P\{X_2 = b_j\}} \tag{3.8}$$

Applying the above formula to Example 3.3. we obtain the following results: (a) With replacement,

$$P\{X_1 = 0 \mid X_2 = 0\} = \frac{100/144}{5/6} = \frac{5}{6} = P\{X_1 = 0\}$$

$$P\{X_1 = 0 \mid X_2 = 1\} = \frac{20/144}{1/6} = \frac{5}{6} = P\{X_1 = 0\}$$

$$P\{X_1 = 1 \mid X_2 = 0\} = \frac{20/144}{5/6} = \frac{1}{6} = P\{X_1 = 1\}$$

$$P\{X_1 = 1 \mid X_2 = 1\} = \frac{4/144}{1/6} = \frac{1}{6} = P\{X_1 = 1\}$$

(b) Without replacement,

$$P\{X_1 = 0 \mid X_2 = 0\} = \frac{90/132}{5/6} = \frac{9}{11}$$

$$P\{X_1 = 0 \mid X_2 = 1\} = \frac{20/132}{1/6} = \frac{10}{11}$$

$$P\{X_1 = 1 \mid X_2 = 0\} = \frac{20/132}{5/6} = \frac{2}{11}$$

$$P\{X_1 = 1 \mid X_2 = 1\} = \frac{4/132}{1/6} = \frac{1}{11}$$

We note that in case (a), X_1 and X_2 are *independent* since the *conditional probability function* of X_1 and X_2 has the same values as the probability function of X_1. In other words, the outcome of X_2 in no way influences the outcome of X_1 and vice versa.

In case (b) it can readily be seen that X_1 and X_2 are not independent. The outcome of X_1 is affected by the outcome of X_2 (in a probability sense) and vice versa. Actually X_1 and X_2 are functionally independent since X_2 is not uniquely determined when X_1 is known; however, they are dependent in the probability sense. This type of *independence* is referred to as *statistical independence*. Accordingly, in sampling with replacement, the outcomes are *statistically independent*, whereas in sampling without replacement the outcomes are *statistically dependent*. Suppose now we wish to find the probability distribution provided that the second tube drawn (without replacement) is defective. This is given by

$$P\{X_1 = 0 \mid X_2 = 1\} = \tfrac{10}{11}$$
$$P\{X_1 = 1 \mid X_2 = 1\} = \tfrac{1}{11}$$

and is known as the *conditional distribution* of X_1 given $X_2 = 1$. It is noteworthy that the conditional distribution summed for all possible values of X_1 given a fixed value of X_2 is equal to one.

The results of this problem give us further insight into the definition of independence of random variables in terms of their joint probability function. From the definition of conditional probability we write

$$P\{X_1 = a_i \mid X_2 = b_j\} = \frac{P_{X_1,X_2}(a_i, b_j)}{P\{X_2 = b_j\}}$$

Now, if we know that X_1 and X_2 are independent we have

$$P\{X_1 = a_i \mid X_2 = b_j\} = P\{X_1 = a_i\}$$

Hence

$$P\{X_1 = a_i\}P\{X_2 = b_j\} = P_{X_1,X_2}(a_i, b_j) \tag{3.9}$$

The graph of the joint probability function is given in Fig. 3.3.

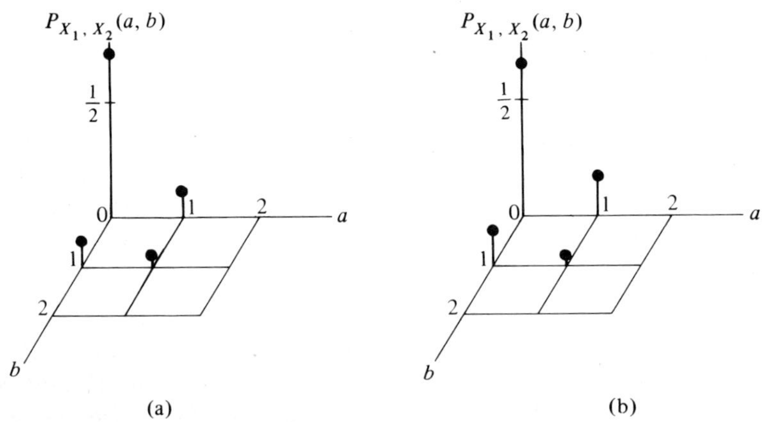

Fig. 3.3 Graph of the joint probability function of Example 3.3: (a) with replacement; (b) without replacement.

Theorem 3.2. *The random variables X_1 and X_2 defined on a sample space S are said to be independent if and only if their joint probability function can be written as the product of their marginal probabilities for all real values of X_1 and X_2.*

This result can be extended to n-dimensional random variables. In words, the random variables $X_1, X_2, \ldots, X_n$, where $n > 1$, are said to be independent if their joint probability function $P_{X_1, X_2, \ldots, X_n}(\cdot, \cdot, \ldots, \cdot)$ may be written as the product of their marginal probabilities $P_{X_1}(\cdot)\,P_{X_2}(\cdot), \ldots, P_{X_n}(\cdot)$ for all real values of $X_1, X_2, \ldots, X_n$. That is,

$$P_{X_1, X_2, \ldots, X_n}(a_1, a_2, \ldots, a_n) = P_{X_1}(a_1)P_{X_2}(a_2) \cdots P_{X_n}(a_n) \tag{3.10}$$

Since $F_{X_1, X_2}(u, v) = \sum_{a_1 \leqslant u} \sum_{a_2 \leqslant v} P_{X_1, X_2}(a_1, a_2)$ and X_1 and X_2 are independent, we can write

$$\begin{aligned} F_{X_1, X_2}(u, v) &= \sum_{a_1 \leqslant u} \sum_{a_2 \leqslant v} P_{X_1}(a_1)P_{X_2}(a_2) \\ &= \sum_{a_1 \leqslant u} P_{X_1}(a_1) \sum_{a_2 \leqslant v} P_{X_2}(a_2) \\ &= F_{X_1}(u)F_{X_2}(v) \end{aligned} \tag{3.11}$$

Equivalently, X_1 and X_2 are independent if their joint distribution function can be written as the product of their respective original distributions for all real values of X_1 and X_2. This result can be extended to n-dimensional random variables. That is,

$$F_{X_1, X_2, \ldots, X_n}(a_1, a_2, \ldots, a_n) = F_{X_1}(a_1)F_{X_2}(a_2) \cdots F_{X_n}(a_n) \tag{3.12}$$

The above result can be stated in the form of a theorem.

Theorem 3.3. *The random variables* $X_1, X_2, \ldots, X_n$, where $n > 1$, are *said to be independent* if *their joint distribution function may be written as the product of their individual distribution functions for all real values* of $X_1, X_2, \ldots, X_n$.

Now we can answer the question we raised previously (see property 4). The joint probability function of random variables *can* be derived from their marginal probabilities if and only if these random variables are independent. When the random variables are *dependent* we *cannot* synthesize their joint probability function from their marginal probabilities. In such cases we must either have *a priori* knowledge of the joint or conditional distribution or sufficient relevant data.

Example 3.4 X_1, X_2, and X_3 are independent, identically distributed random variables, each with probabilities of assuming the values $+1$ and -1 equal to $\frac{1}{2}$. The random variables U and V are further defined as follows:

$$U(X_1, X_2) = X_1 + X_2 \qquad V(X_2, X_3) = X_2 + X_3$$

(a) Describe P_U, P_V, and $P_{U,V}$.
(b) Are U and V statistically independent?

Solution

(a)

$$P_U(r) = \begin{cases} \frac{1}{4} & r = -2 \\ \frac{1}{2} & r = 0 \\ \frac{1}{4} & r = 2 \end{cases} \qquad P_V(s) = \begin{cases} \frac{1}{4} & s = -2 \\ \frac{1}{2} & s = 0 \\ \frac{1}{4} & s = 2 \end{cases}$$

The joint probability function of U and V can be derived by determining the compound probabilities by analytic deduction. To illustrate, when U takes on the value 2 variate X_2 must be equal to 1 and consequently V can be either 0 or 2. Therefore $P_{U,V}(2, -2)$ is equal to zero. We can proceed similarly and calculate the compound probabilities for each possible outcome. The joint probability function of U and V, then, is

$$P_{U,V}(a, b) = \begin{cases} 0 & a = -2 \quad b = 2 \\ 0 & a = 2 \quad b = -2 \\ \frac{1}{8} & a = -2 \quad b = -2 \\ \frac{1}{8} & a = -2 \quad b = 0 \\ \frac{1}{8} & a = 0 \quad b = -2 \\ \frac{1}{4} & a = 0 \quad b = 0 \\ \frac{1}{8} & a = 0 \quad b = 2 \\ \frac{1}{8} & a = 2 \quad b = 0 \\ \frac{1}{8} & a = 2 \quad b = 2 \end{cases}$$

(b) It is easily verified that U and V are not independent since the product of their probability functions is not equal to their joint probability function.

Example 3.5 X_1 and X_2 are two independent random variables. Let Y_1 and Y_2 be two random variables defined as follows:

$$Y_1 = c_1 X_1 \qquad Y_2 = X_2 + c_2$$

where c_1 and c_2 are constants. Prove that Y_1 and Y_2 are independent.

Solution. According to Theorem 3.2, if Y_1 and Y_2 are independent, their joint probability function must be equal to the product of their marginal probabilities. In symbols,

$$P_{Y_1,Y_2}(a, b) = P\{Y_1 = a\}\ P\{Y_2 = b\}$$

Now,

$$\begin{aligned} P_{Y_1,Y_2}(a, b) &= P\{Y_1 = a,\ Y_2 = b\} \\ &= P\{c_1 X_1 = a,\ X_2 + c_2 = b\} \\ &= P\left\{X_1 = \frac{a}{c_1},\ X_2 = b - c_2\right\} \end{aligned}$$

Since X_1 and X_2 are independent we have

$$\begin{aligned} P_{Y_1,Y_2}(a, b) &= P\left\{X_1 = \frac{a}{c_1}\right\}\ P\{X_2 = b - c_2\} \\ &= P\{X_1 c_1 = a\}\ P\{X_2 + c_2 = b\} \\ &= P\{Y_1 = a\}\ P\{Y_2 = b\} \end{aligned}$$

and the proof is complete. This is an important property of independent variables that we state in the form of a theorem.

Theorem 3.4. *Let $X_1, X_2, \ldots, X_n$ be independent random variables. Let $Y_1, Y_2, \ldots, Y_n$ be random variables such that $Y_1 = f_1(X_1)$, $Y_2 = f_2(X_2)$,*

$\ldots, Y_n = f_n(X_n)$. Then $Y_1, Y_2, \ldots, Y_n$ *also are independent random variables.* It should be noted that the *converse* of this theorem is not true.

3.4 Joint Density Functions

Definition 3.2. We say that random variables X_1 and X_2 are *jointly continuous* if a nonnegative function $f_{X_1,X_2}(.,.)$ exists such that the joint distribution function $F_{X_1,X_2}(.,.)$ can be computed by double integration of $f_{X_1,X_2}(.,.)$ for all real values of X_1 and X_2. $f_{X_1,X_2}(.,.)$ is called the *joint density function*. In symbols,

$$\begin{aligned} P\{X_1 \le a_1, X_2 \le a_2\} &= F_{X_1,X_2}(a_1, a_2) \\ &= \int_{-\infty}^{a_2} \int_{-\infty}^{a_1} f_{X_1,X_2}(u_1, u_2)\, du_1\, du_2 \end{aligned} \tag{3.13}$$

for all real numbers a_1 and a_2. The foregoing definition extends at once to the case of n random variables. Symbolically,

$$\begin{aligned} P\{X_1 \le a_1, \ldots, X_n \le a_n\} &= F_{X_1,\ldots,X_n}(a_1, \ldots, a_n) \\ &= \int_{-\infty}^{a_n} \cdots \int_{-\infty}^{a_1} f_{X_1,\ldots,X_n}(u_1, \ldots, u_n)\, du_1, \ldots, du_n \end{aligned}$$

Suppose now event E is given by

$$E = \{a_1 < X_1 \le a_1', a_2 < X_2 \le a_2'\}$$

then the probability of event E is

$$\begin{aligned} P\{E\} &= P\{a_1 < X_1 \le a_1', a_2 < X_2 \le a_2'\} \\ &= \int_{a_2}^{a_2'} \int_{a_1}^{a_1'} f_{X_1,X_2}(u_1, u_2)\, du_1\, du_2 \end{aligned}$$

In general, we write

$$\begin{aligned} &P\{a_1 < X_1 \le a_1', \ldots, a_n < X_n \le a_n'\} \\ &= \int_{a_n}^{a_n'} \cdots \int_{a_1}^{a_1'} f_{X_1,\ldots,X_n}(u_1, \ldots, u_n)\, du_1, \ldots, du_n \end{aligned}$$

If the joint distribution function is continuous and differentiable, the joint density function may be obtained by partial differentiation and is given by

$$f_{X_1,X_2,\ldots,X_n}(u_1, u_2, \ldots, u_n) = \left.\frac{\partial^n F_{X_1,X_2,\ldots,X_n}(a_1, a_2, \ldots, a_n)}{\partial a_1, \partial a_2, \ldots, \partial a_n}\right|_{u_1, u_2, \ldots, u_n} \tag{3.14}$$

Note:

$$\int_{-\infty}^{\infty} \cdots \int_{-\infty}^{\infty} f_{X_1,\ldots,X_n}(u_1, \ldots, u_n)\, du_1, \ldots, du_n = 1$$

It is always feasible to obtain the individual (marginal) density functions whenever the joint density function is known. To illustrate, let us derive the marginal density functions of random variables X_1 and X_2 provided that $f_{X_1,X_2}(u, v)$ is known. We write

$$F_{X_1,X_2}(a, \infty) = \int_{-\infty}^{a} \int_{-\infty}^{\infty} f_{X_1,X_2}(u, v)\, dv\, du$$
$$= F_{X_1}(a) = \int_{-\infty}^{a} f_{X_1}(u)\, du$$

for every a, so that

$$\int_{-\infty}^{\infty} f_{X_1,X_2}(u, v)\, dv = f_{X_1}(u)$$

Similarly,

$$\int_{-\infty}^{\infty} f_{X_1,X_2}(u, v)\, du = f_{X_2}(v)$$

Conversely, from a knowledge of marginal density functions one cannot derive the density function unless the variates are independent.

By Theorem 3.3, the random variables $X_1, X_2, \ldots, X_n$ are independent if their joint distribution function may be written as the product of their individual (marginal) distribution functions for all real numbers $a_1, a_2, \ldots, a_n$. In symbols,

$$(3.15) \qquad F_{X_1,X_2,\ldots,X_n}(a_1, a_2, \ldots, a_n) = F_{X_1}(a_1)F_{X_2}(a_2), \ldots, F_{X_n}(a_n)$$

By taking the appropriate nth partial derivatives of both sides of the above expression we obtain

$$(3.16) \qquad f_{X_1,X_2,\ldots,X_n}(a_1, a_2, \ldots, a_n) = f_{X_1}(a_1)f_{X_2}(a_2), \ldots, f_{X_n}(a_n)$$

Equivalently, the random variables are independent if their joint density function factors identically into the product of their individual (marginal) density functions. Let us now consider some illustrative examples.

Example 3.6 The joint density function of two random variables X and Y is

$$f_{X,Y}(u, v) = \tfrac{1}{2} \qquad 0 \le u \le 2 \qquad 0 \le v \le 1$$
$$= 0 \qquad \text{otherwise}$$

(a) Describe $F_{X,Y}$.
(b) Describe the marginal densities f_X and f_Y.
(c) Are X and Y independent?

Solution

(a) Here we have a two-dimensional, continuous random variable. The domain of definition is $0 \le X \le 2$, $0 \le Y \le 1$; that is, X can assume all values between 0 and 2 (inclusive) whereas Y can assume all values between 0 and 1 (inclusive). Since $f_{X,Y} = \frac{1}{2} =$ constant, the specified two-dimensional continuous random variable is uniformly distributed over a region R in the XY plane. Region R, the shaded area in Fig. 3.4, is sometimes called the admissible region. The joint distribution function has the form

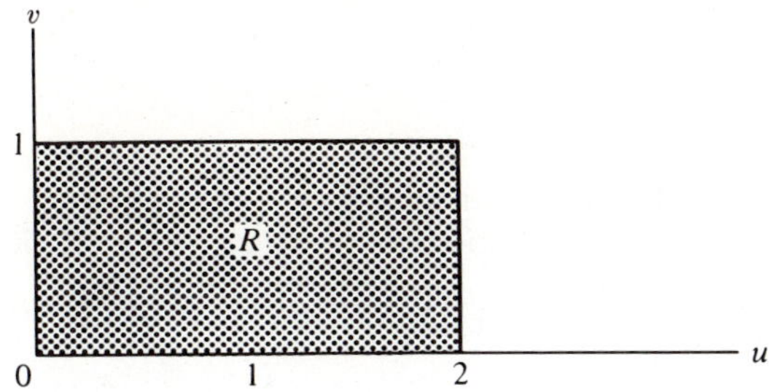

Fig. 3.4 Admissible region of Example 3.6.

$$F_{X,Y}(a, b) = \begin{cases} \int_0^b \int_0^a \frac{1}{2}\, du\, dv \\ 0 & 0 < a \quad b < 0 \\ \frac{1}{2}ab & 0 \le a \le 2 \quad 0 \le b \le 1 \\ b & a \ge 2 \quad 0 \le b \le 1 \\ \frac{1}{2}a & 0 \le a \le 2 \quad b \ge 1 \\ 1 & a \ge 2 \quad b \ge 1 \end{cases}$$

(b) By definition, the marginal densities of X and Y are

$$f_X(u) = \int_0^1 \tfrac{1}{2}\, dv = \tfrac{1}{2} \qquad 0 \le u \le 2$$
$$= 0 \qquad \text{otherwise}$$

$$f_Y(v) = \int_0^2 \tfrac{1}{2}\, du = 1 \qquad 0 \le v \le 1$$
$$= 0 \qquad \text{otherwise}$$

(c) We conclude that X and Y are independent, since

$$f_{X,Y}(u, v) = f_X(u) f_Y(v) = \tfrac{1}{2} \times 1 = \tfrac{1}{2}$$

Example 3.7 Let Y_1 denote the length of time in weeks until machine 1 fails. Let Y_2 denote the same for machine 2. Y_1 and Y_2 are assumed to be independent and identically distributed with density function given by

$$ye^{-1/2y^2} \qquad y \ge 0$$

Compute $P\{Y_1 \le 1,\ Y_2 \le 1\}$.

Solution. Since Y_1 and Y_2 are independent, the joint density function of Y_1 and Y_2 is given by

$$f_{Y_1,Y_2}(a, b) = (ae^{-1/2a^2})(be^{-1/2b^2}) \qquad a \ge 0 \qquad b \ge 0$$
$$= 0 \qquad \text{otherwise}$$

Hence, we have

$$P\{Y_1 \le 1,\ Y_2 \le 1\} = \int_{-\infty}^{1} \int_{-\infty}^{1} ab \exp[-\tfrac{1}{2}(a^2 + b^2)]\, da\, db$$

$$= \int_0^1 \int_0^1 ab \exp[-\tfrac{1}{2}(a^2 + b^2)]\, da\, db$$
$$= .1548$$

Example 3.8 An electronic device consists of two components. The length of time in hours until failure of each component is a random variable denoted by Y_1 and Y_2, respectively. The joint distribution of Y_1 and Y_2 is given by

$$F_{Y_1,Y_2}(a, b) = 1 - e^{-.01a} - e^{-.01b} + e^{-.01(a+b)} \qquad a \geq 0 \qquad b \geq 0$$

(a) Are Y_1 and Y_2 independent?
(b) What is the probability that each component has a lifetime greater than or equal to 120 hr?

Solution

(a) The joint density function of Y_1 and Y_2 is

$$f_{Y_1,Y_2}(a, b) = \frac{\partial^2}{\partial a \partial b}[1 - e^{-.01} - e^{-.01b} + e^{-.01(a+b)}]$$
$$= .0001e^{-.01(a+b)} \qquad a \geq 0 \qquad b \geq 0$$
$$= 0 \qquad \text{otherwise}$$

The density functions of Y_1 and Y_2 are

$$f_{Y_1}(a) = \int_0^\infty .0001e^{-.01(a+b)}\, db = .01e^{-.01a} \qquad a \geq 0$$
$$= 0 \qquad \text{otherwise}$$

$$f_{Y_2}(b) = \int_0^\infty .0001e^{-.01(a+b)}\, da = .01e^{-.01b} \qquad b \geq 0$$
$$= 0 \qquad \text{otherwise}$$

Since $f_{Y_1}(a)f_{Y_2}(b) = f_{Y_1,Y_2}(a, b)$ we conclude that Y_1 and Y_2 are independent.

(b) $P\{Y_1 \geq 120, Y_2 \geq 120\} = P\{Y_1 \geq 120\}P\{Y_2 \geq 120\}$

$$= \left(\int_{120}^\infty .01e^{-.01a}\, da\right)\left(\int_{120}^\infty .01e^{-.01b}\, db\right)$$
$$= e^{-2.4} = .091$$

Example 3.9 Suppose that the two-dimensional random variable (X, Y) has joint p.d.f.

$$f_{X,Y}(a, b) = 4.8b(2 - a) \qquad 0 \leq a \leq 1 \qquad 0 \leq b \leq a$$
$$= 0 \qquad \text{otherwise}$$

(a) Find the marginal densities of X and Y.
(b) Are X and Y independent?

Solution

(a) From the domain of definition we can graph the admissible region R as shown in Fig. 3.5(a). The path of integration for the marginal density of X and marginal density of Y is shown in Figs. 3.5(b) and 3.5(c), respectively. According to Fig. 3.5(a) the p.d.f. of X is

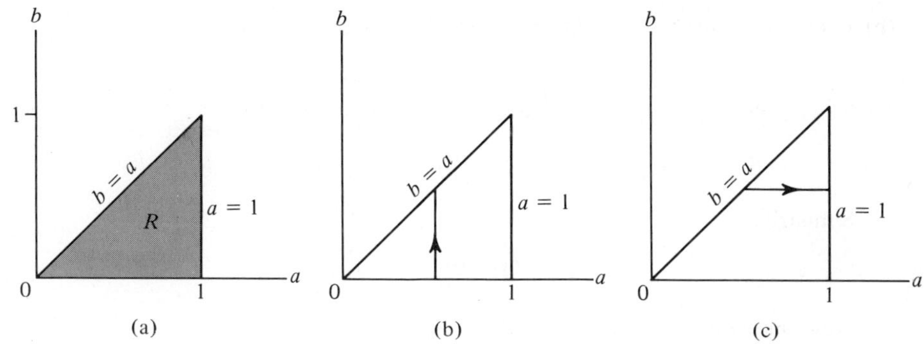

Fig. 3.5 (a) Admissible region; (b) path of integration for p.d.f. of X; (c) path of integration for p.d.f. of Y.

$$f_X(a) = \begin{cases} \int_0^a 4.8b(2-a)\,db \\ 2.4(2-a)a^2 & 0 \leq a \leq 1 \\ 0 & \text{otherwise} \end{cases}$$

Similarly,

$$f_Y(b) = \begin{cases} \int_b^1 4.8b(2-a)\,da \\ 2.4b(3-4b+b^2) & 0 \leq b \leq 1 \\ 0 & \text{otherwise} \end{cases}$$

(b) It is evident that $f_X(a)f_Y(b)$ is not equal to $f_{X,Y}(a, b)$. Hence, X and Y are not independent. In the present example, one can immediately determine that X and Y are not independent (without proving that the joint density function does not factor into pure functions of each variate), since the domain of definition is such that for given X, Y can assume only positive values less than or equal to X. *Thus, whenever the domain of one variate depends upon another variate, then the variates are dependent.*

Example 3.10 The joint density function of two random variables, X_1 and X_2, is given by

$$\begin{aligned} f_{X_1,X_2}(a, b) &= 6ab(2-a-b) \qquad 0 \leq a \leq 1 \qquad 0 \leq b \leq 1 \\ &= 0 \qquad \text{otherwise} \end{aligned}$$

Describe the following:
(a) f_{X_1}, the p.d.f. of X_1
(b) f_{X_2}, the p.d.f. of X_2
(c) Find the conditional distributions of X_1 and X_2, respectively.
(d) Are X_1 and X_2 independent?

Solution

(a) $$\begin{aligned} f_{X_1}(a) = \int_0^1 6ab(2-a-b)\,db &= 4a - 3a^2 \qquad 0 \leq a \leq 1 \\ &= 0 \qquad \text{otherwise} \end{aligned}$$

(b) $f_{X_2}(b) = \int_0^1 6ab(2 - a - b)\,da = 4b - 3b^2 \qquad 0 \leq b \leq 1$

$\qquad\qquad = 0 \qquad \text{otherwise}$

(c) The conditional p.d.f. of X_1 for given $X_2 = b$ is defined by

$$f_{X_1|X_2}(a\,|\,b) = \frac{f_{X_1, X_2}(a, b)}{f_{X_2}(b)} \qquad f_{X_2}(b) > 0$$

Similarly,

$$f_{X_2|X_1}(b\,|\,a) = \frac{f_{X_1, X_2}(a, b)}{f_{X_1}(a)} \qquad f_{X_1}(a) > 0$$

Now, applying the above definitions we obtain

$$f_{X_1|X_2}(a\,|\,b) = \frac{6ab(2 - a - b)}{4b - 3b^2} = \frac{6a(2 - a - b)}{4 - 3b} \qquad 0 \leq a \leq 1$$
$$= 0 \qquad \text{otherwise}$$

$$f_{X_2|X_1}(b\,|\,a) = \frac{6ab(2 - a - b)}{4a - 3a^2} = \frac{6b(2 - a - b)}{4 - 3a} \qquad 0 \leq b \leq 1$$
$$= 0 \qquad \text{otherwise}$$

The conditional distribution of X_1, say, for given $X_2 = b$ is the function $f_{X_1|X_2}(b\,|\,a)$, whose domain is the set of possible values of X_1 (from 0 to 1 inclusive) and whose value is $6a[(2 - a - b)/(4 - 3b)]$. Similarly, we can define the conditional distribution of X_2. It should be noted that the conditional distribution is a density function. That is, the function has nonnegative values and unit area. To prove that $f_{X_1|X_2}(a\,|\,b)$ is a p.d.f. we have

$$\int_0^1 \frac{6a(2 - a - b)}{4 - 3b}\,da = \frac{6}{4 - 3b}\left(\frac{2}{3} - \frac{b}{2}\right) = 1 \qquad 0 \leq b \leq 1$$

A similar computation can be made for $f_{X_1|X_2}(b\,|\,a)$.

(d) Since $f_{X_1}(a)f_{X_2}(b) = ab(4 - 3a)(4 - 3b)$ is not equal to $f_{X_1, X_2}(a, b)$, X and Y are not independent. We can also prove this fact by considering the conditional distribution of the variates. For instance, the conditional p.d.f. of X_1 that is,

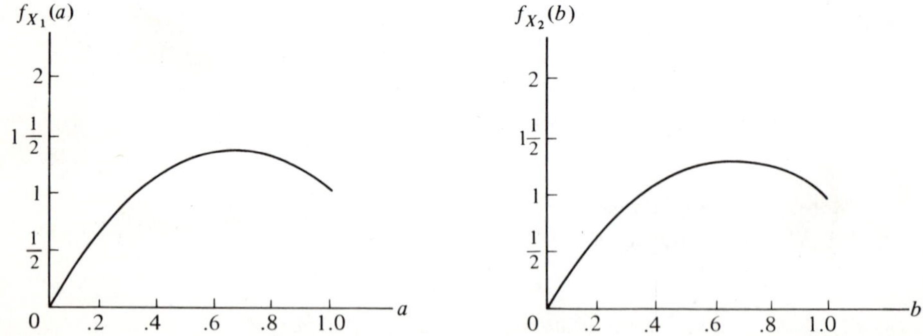

Fig. 3.6 Marginal density functions of Example 3.10.

$[6a(2 - a - b)]/(4 - 3b)$ depends on X_2. Similarly, the conditional p.d.f. of X_2 depends on X_1. Accordingly, whenever the conditional distribution of one variate depends upon another variate, then the variates are not independent. The graphs of the marginal and conditional densities are shown in Figs. 3.6 and 3.7, respectively.

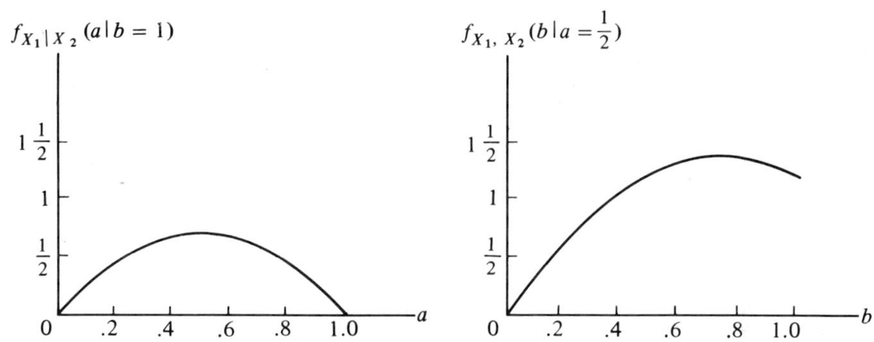

Fig. 3.7 Conditional density functions of Example 3.10.

PROBLEMS

3.1. Let X_1 and X_2 be two independent random variables, each with the same p.d.f. given by

$$.1e^{-.1t} \qquad t \geq 0$$

(a) Determine the p.d.f. of the two-dimensional random variable (X_1, X_2).
(b) Compute $P\{X_1 \leq 10,\ X_2 \leq 20\}$.

3.2. X_1 and X_2 are two independent random variables with p.d.f.'s

$$f_{X_1}(t_1) = 1 \quad 0 < t_1 < 1 \qquad f_{X_2}(t_2) = 1 \quad 0 < t_2 < 1$$
$$= 0 \quad \text{otherwise} \qquad\qquad = 0 \quad \text{otherwise}$$

(a) Determine the p.d.f. of the two-dimensional random variable (X_1, X_2).
(b) Compute $P\{X_1 \leq \frac{1}{2},\ X_2 \leq \frac{1}{2}\}$.

3.3. The variates X_1 and X_2 are jointly distributed as follows:

$$f_{X_1, X_2}(t_1, t_2) = t_1 + t_2 \qquad 0 \leq t_1 \leq 1 \qquad 0 \leq t_2 \leq 1$$
$$= 0 \qquad \text{otherwise}$$

(a) Describe F_{X_1, X_2}.
(b) Compute $P\{\frac{1}{4} \leq X_1 \leq \frac{1}{2},\ \frac{1}{4} \leq X_2 \leq \frac{1}{2}\}$.

3.4. Let X_1 and X_2 be two independent random variables, each with the same p.d.f. given by

$$te^{-(1/2)t^2} \qquad t \geq 0$$

(a) Determine the p.d.f. of the two-dimensional random variable (X_1, X_2).
(b) Compute $P\{X_1 \leq 1,\ X_2 \leq 2\}$.

3.5. The variates X_1 and X_2 are jointly distributed as follows:

$$f_{X_1, X_2}(t_1, t_2) = k \qquad 0 \leq t_1 \leq 2 \qquad 0 \leq t_2 \leq 2$$
$$= 0 \qquad \text{otherwise}$$

(a) Find k.
(b) Find f_{X_1}, the marginal density of X_1.
(c) Find f_{X_2}, the marginal density of X_2.
(d) Are X_1 and X_2 statistically independent?
(e) Compute $P\{X_1 > 1\}$.
(f) Compute $P\{X_1 = X_2\}$.

3.6. The variates X_1 and X_2 are jointly distributed as follows:

$$f_{X_1, X_2}(t_1, t_2) = t_1 t_2 e^{-(t_1+t_2)} \qquad t_1 \geq 0 \qquad t_2 \geq 0$$
$$= 0 \qquad \text{otherwise}$$

(a) Are X_1 and X_2 statistically independent?
(b) Are X_1^2 and X_2^2 independent random variables?
(c) Find the p.d.f. of X_1^2.

3.7. The discrete variable X has the following probability function:

$$P_X(k) = \begin{cases} .2 & k = 1 \\ .4 & k = 2 \\ .4 & k = 3 \end{cases}$$

Two random observations X_1 and X_2 are taken. Determine the distribution of Y when
(a) $Y = X_1 + X_2$.
(b) $Y = |X_1 - X_2|$.

3.8. X_1 and X_2 are identically distributed random variables which take on values -1 and 1 with probability

$$P_{X_1}(-1) = P_{X_2}(-1) = \tfrac{1}{4}$$
$$P_{X_1}(1) = P_{X_2}(1) = \tfrac{3}{4}$$

Describe the joint distribution function of X_1 and X_2.

3.9. Suppose that the joint probability function of X_1 and X_2 is

$$P_{X_1, X_2}(2, 1) = \tfrac{1}{4}$$
$$P_{X_1, X_2}(2, 2) = \tfrac{1}{2}$$
$$P_{X_1, X_2}(3, 2) = \tfrac{1}{8}$$
$$P_{X_1, X_2}(3, 3) = \tfrac{1}{8}$$

Describe the joint distribution function of X_1 and X_2.

3.10. The variates X_1 and X_2 are jointly distributed as follows:

$$f_{X_1, X_2}(t_1, t_2) = k \qquad 1 \leq t_1 \leq 3 \qquad 1 \leq t_2 \leq 2$$
$$= 0 \qquad \text{otherwise}$$

(a) Find k.
(b) Describe F_{X_1, X_2}.
(c) Compute $P\{X_1 > X_2\}$.

3.11. Suppose that the joint probability distribution of the discrete random variable (X_1, X_2) is as follows:

	$P_{X_1, X_2}(a_1, a_2)$		
a_1 / a_2	-1	0	1
-1	$\frac{1}{8}$	$\frac{1}{8}$	0
0	$\frac{1}{8}$	0	$\frac{1}{4}$
1	$\frac{1}{8}$	$\frac{1}{8}$	$\frac{1}{8}$

(a) Describe the marginal distribution of X_1 and of X_2.
(b) Are X_1 and X_2 statistically independent?
(c) Compute the following: $P\{X_1 \leq 0,\ X_2 \leq 0\}$, $P\{X_1 = X_2\}$, and $P\{X_1 < 2X_2\}$.

3.12. A box contains ten fuses of which two are defective. A sample of two fuses is drawn without replacement. Let the random variables X_1 and X_2 be defined as follows:

$$X_1 = \begin{cases} 0 \text{ if the first fuse is nondefective} \\ 1 \text{ if the first fuse is defective} \end{cases}$$

$$X_2 = \begin{cases} 0 \text{ if the second fuse is nondefective} \\ 1 \text{ if the second fuse is defective} \end{cases}$$

(a) Describe the joint probability function of X_1 and X_2.
(b) Describe the marginal probability function of X_1 and of X_2.

3.13. A factory has two departments A and B. Let X_A and X_B denote the number of industrial accidents per month in departments A and B, respectively. Assume that X_A and X_B are independent random variables with the same probability function given by

Number of accidents (x)	1	2	3	4	5	6	7	8	9	10
$P(x)$	.08	.12	.16	.13	.09	.12	.10	.09	.06	.05

(a) Describe the joint probability function of X_A and X_B.
(b) Find the probability that $X_A = 6$ conditional on $X_A \geq 3$.
(c) Find the probability that $X_A > 6$ conditional on $X_A \geq 3$.
(d) Compute $P\{X_A + X_B > 18\}$.

3.14. An urn contains one red ball and two white balls. A sample of size three is drawn with replacement. Let variate $X_i = 1$ or 0 if the ball drawn on the ith draw is red or white, respectively ($i = 1, 2, 3$).
(a) Describe the joint probability function of X_1, X_2, X_3.
(b) Find the probability function of Y where $Y = X_1 + X_2 + X_3$.

3.15. Suppose in the preceding problem the sample was drawn without replacement.
(a) Describe the joint probability function of X_1, X_2, X_3.
(b) Find the marginal probability function of X_1.

3.16. Independent variates X_1 and X_2 are distributed as follows:

$$P_{X_1}(a_1) = \frac{e^{-3}3^{a_1}}{a_1!} \qquad a_1 = 0, 1, \ldots \qquad P_{X_2}(a_2) = \frac{e^{-2}2^{a_2}}{a_2!} \qquad a_2 = 0, 1, \ldots$$

(a) Find the distribution of $Y = X_1 + X_2$.
(b) Find the distribution of X_1 conditional on Y.

3.17. The joint probability function of X_1 and X_2 is given by

$$P_{X_1,X_2}(a_1, a_2) = \tfrac{1}{19}(a_1 + a_2^2) \qquad a_1 = 1, 2 \qquad a_2 = 0, 1, 2$$

(a) Find the marginal probability function of X_1.
(b) Find the marginal probability function of X_2.
(c) Find the conditional probability function of X_1 given X_2.

3.18. Let X_1 be the smaller of the two numbers and X_2 be the absolute value of the difference between the two numbers showing when two fair dice are rolled.
(a) Describe the joint probability function of X_1 and X_2.
(b) Are X_1 and X_2 statistically independent?

3.19. A fair coin is tossed three independent times. Let X_1 be the number of heads obtained in the first two tosses and X_2 be defined as follows:

$$X_2 = \begin{cases} 0 \text{ if the last toss is a tail} \\ 1 \text{ if the last toss is a head} \end{cases}$$

(a) Describe the joint probability function of X_1 and X_2.
(b) Are X_1 and X_2 statistically independent?

3.20. The joint distribution of X_1 and X_2 is

$$\begin{aligned} f_{X_1,X_2}(a_1, a_2) &= \frac{1}{1 - a_1} \qquad 0 < a_1 < 1 \qquad a_1 < a_2 < 1 \\ &= 0 \qquad \text{otherwise} \end{aligned}$$

(a) Find the marginal distribution of X_1 and of X_2.
(b) Evaluate $f_{X_2|X_1}(a_2 \mid a_1)$.
(c) Are X_1 and X_2 statistically independent?

3.21. The joint distribution of X_1 and X_2 is

$$\begin{aligned} f_{X_1,X_2}(a_1, a_2) &= 3a_1 \qquad 0 < a_1 < 1 \qquad 0 < a_2 < a_1 \\ &= 0 \qquad \text{otherwise} \end{aligned}$$

(a) Find the marginal distribution of X_1 and of X_2.
(b) Evaluate $f_{X_2|X_1}(a_2 \mid a_1)$.
(c) Are X_1 and X_2 statistically independent?

3.22. Let X_1 and X_2 denote the diameter of a shaft and cam, respectively. Suppose the joint density of X_1 and X_2 is

$$\begin{aligned} f_{X_1,X_2}(a_1, a_2) &= 2{,}500 \qquad 1.00 < a_1 < 1.02 \qquad 1.01 < a_2 < 1.03 \\ &= 0 \qquad \text{otherwise} \end{aligned}$$

Suppose that a camshaft can be assembled if the diameter of the cam exceeds that of the shaft by at least .005 but not more than .01 in.
(a) Find the marginal distributions of X_1 and of X_2.
(b) What is the probability that a cam chosen at random will fit a shaft whose diameter is 1.01 in.?

3.23. The joint distribution of X_1 and X_2 is

$$\begin{aligned} f_{X_1,X_2}(a_1, a_2) &= 4a_1(1 - a_2) \qquad 0 < a_1 < 1 \qquad 0 < a_2 < 1 \\ &= 0 \qquad \text{otherwise} \end{aligned}$$

(a) Find the marginal distribution of X_1 and of X_2.
(b) Find $f_{X_2}(a_2 \mid a_2 < \tfrac{3}{4})$.

3.24. If

$$f_{X_1|X_2}(a_1 \mid a_2) = \frac{2(1 - a_1)}{(1 - a_2)^2} \qquad a_2 \leq a_1 \leq 1$$

and

$$f_{X_2}(a_2) = 12a_2(1 - a_2)^2 \qquad 0 \leq a_2 \leq 1$$

find $P\{X_1 \geq \frac{1}{2}\}$.

3.25. If

$$f_{X_1|X_2}(a_1 \mid a_2) = \frac{2(1 - a_1 - a_2)}{(1 - a_2)^2} \qquad 0 \leq a_1 \leq 1 - a_2$$

and

$$f_{X_2}(a_2) = 12a_2(1 - a_2)^2 \qquad 0 \leq a_2 \leq 1$$

find $P\{X_1 \geq \frac{3}{4}\}$.

3.26. Suppose that the joint p.d.f. of (X_1, X_2) is given by

$$\begin{aligned} f_{X_1, X_2}(a_1, a_2) &= 2e^{-(a_1+a_2)} \qquad a_1 > 0 \qquad 0 < a_2 < a_1 \\ &= 0 \qquad \text{otherwise} \end{aligned}$$

(a) Find the marginal distribution of X_1 and of X_2.
(b) Are X_1 and X_2 statistically independent?
(c) Evaluate $P\{X_1 > 1 \mid X_2 < 1\}$.

3.27. Find the conditional p.d.f. of X_1 given X_2 for the distribution of Prob. 3.26. What is the $P\{0 < X_1 < 2 \mid X_2 = 1\}$?

3.28. Suppose that the joint p.d.f. of (X_1, X_2, X_3) is given by

$$f_{X_1, X_2, X_3}(a_1, a_2, a_3) = ka_1a_2a_3 \qquad 0 < a_1 < 1 \qquad 0 < a_2 < \tfrac{1}{2} \qquad 0 < a_3 < 1$$

(a) Evaluate the constant k.
(b) Evaluate $P\{X_1 < X_2 < X_3\}$.

3.29. Given

$$f_{X_1}(a_1) = \frac{\ln a_1}{a_1^2} \qquad a_1 \geq 1$$

and

$$f_{X_2|X_1}(a_2 \mid a_1) = \frac{1}{2a_2 \ln a_1} \qquad \frac{1}{a_1} \leq a_2 \leq a_1$$

(a) Find the joint p.d.f. of (X_1, X_2).
(b) Find the marginal distribution of X_2.

3.30. Suppose that the joint p.d.f. of the two-dimensional random variable (X_1, X_2) is given by

$$\begin{aligned} f_{X_1, X_2}(a_1, a_2) &= 8a_1a_2 \qquad 0 < a_2 < 1 \qquad a_1 < a_2 < 1 \\ &= 0 \qquad \text{otherwise} \end{aligned}$$

(a) Find the marginal distribution of X_1 and of X_2.
(b) Are X_1 and X_2 statistically independent?
(c) Describe $f_{X_1|X_2}$.
(d) Compute $P\{2X_1 < X_2\}$.
(e) Compute $P\{X_1 < \frac{1}{2} \mid X_2 > \frac{1}{2}\}$.

SUGGESTED REFERENCES

[1] Brunk, H. D., *An Introduction to Mathematical Statistics*, Ginn and Company, Boston, 1960.

[2] Hoel, P. G., *Introduction to Mathematical Statistics*, 3rd ed., John Wiley & Sons, Inc., New York, 1965.

[3] Meyer, P., *Introductory Probability and Statistical Applications*, 2nd ed., Addison-Wesley Publishing Co., Reading, Mass., 1970.

[4] Mood, A. M., and F. A. Graybill, *Introduction to the Theory of Statistics*, McGraw-Hill Book Company, New York, 1963.

[5] Parzen, E., *Modern Probability Theory and Its Applications*, John Wiley & Sons, Inc., New York, 1960.

[6] Wadworth, G. P., and P. G. Bryan, *Introduction to Probability and Random Variables*, McGraw-Hill Book Company, New York, 1960.

4

DISTRIBUTIONS OF FUNCTIONS OF RANDOM VARIABLES

4.1 Introduction

In many situations one may be interested in the probability distribution of a function of a random variable or of a function of a number of random variables whose joint distributions are known. A physicist, for example, who knows the probability distribution of the velocity of a molecule in a gas may well be interested in knowing the probability distribution of the kinetic energy (function of velocity). An industrial engineer who knows the probability distribution of demand for a certain commodity for a prescribed interval of time may need to know what the probability distribution of demand will be for varying length time intervals. An electrical engineer who has obtained the probability distribution of a certain waveform may well be interested to know what the probability distribution will be if this waveform is passed through a half-wave rectifier.

A few examples will help give the reader insight into the concept of derived distributions. We shall consider the case of single and two-dimensional random variables. The application of these principles to n-dimensional random variables is straightforward.

4.2 Function of a Single Variate

Example 4.1 The probability function of a random variable X is given by

$$P_X(a) = \begin{cases} \frac{1}{4} & a = 1 \\ \frac{1}{2} & a = 2 \\ \frac{1}{4} & a = 3 \end{cases}$$

If a new random variable is given by $Y = X^2$, describe F_Y, the c.d.f. of Y.

Solution. The possible outcomes of random variable X are 1, 2, and 3. When X assumes the value 1, Y takes on the value 1 with probability $\frac{1}{4}$. Continuing in this way we find the outcomes and associated probabilities of random variable Y as listed below:

Possible Value of Y	*Probability of Y*
1	$P_Y(1) = P_X(1) = \frac{1}{4}$
4	$P_Y(4) = P_X(2) = \frac{1}{2}$
9	$P_Y(9) = P_X(3) = \frac{1}{4}$

Then the c.d.f. of Y is

$$F_Y(k) = \begin{cases} 0 & k < 1 \\ \frac{1}{4} & 1 \leq k < 4 \\ \frac{3}{4} & 4 \leq k < 9 \\ 1 & 9 \leq k \end{cases}$$

We note that the events $\{X = 1\}$ and $\{Y = 1\}$ are equivalent. Similarly, the events $\{X = 2\}$ and $\{X = 3\}$ are equivalent to $\{Y = 4\}$ and $\{Y = 9\}$, respectively. We also note that equivalent events have equal probabilities. Now we can introduce the following definition:

Definition 4.1. Let X be a random variable defined on the sample space S. Let random variable $Y = G(X)$, where G is a real-valued function, be defined on the sample space S'. Then for any event $E \subset S$ there is an *equivalent event* defined by $E' \subset S'$, whose probability is given by

$$P\{E' \subset S'\} = P\{E \subset S\} \tag{4.1}$$

We can solve this example in a slightly different way. By definition,

$$\begin{aligned} F_Y(k) &= P\{Y \leq k\} = P\{X^2 \leq k\} = P\{-\sqrt{k} \leq X \leq \sqrt{k}\} \\ &= P\{X \leq \sqrt{k}\} = F_X(\sqrt{k}) \end{aligned}$$

since X is never negative. The c.d.f. of X is

$$F_X(k) = \begin{cases} 0 & k < 1 \\ \frac{1}{4} & 1 \leq k < 2 \\ \frac{3}{4} & 2 \leq k < 3 \\ 1 & 3 \leq k \end{cases}$$

Consequently,

$$F_Y(k) = F_X(\sqrt{k}) = \begin{cases} 0 & \sqrt{k} < 1 \\ \frac{1}{4} & 1 \le \sqrt{k} < 2 \\ \frac{3}{4} & 2 \le \sqrt{k} < 3 \\ 1 & 3 \le \sqrt{k} \end{cases} = \begin{cases} 0 & k < 1 \\ \frac{1}{4} & 1 \le k < 4 \\ \frac{3}{4} & 4 \le k < 9 \\ 1 & 9 \le k \end{cases}$$

Example 4.2 Suppose that X has the following density function:

$$\begin{aligned} f_X(t) &= \tfrac{1}{4} \qquad 0 < t < 4 \\ &= 0 \qquad \text{otherwise} \end{aligned}$$

If a new discrete random variable G is given by

$$G(X) = \begin{cases} 2 & 3 < X \le 5 \\ 1 & \text{otherwise} \end{cases}$$

describe P_G and F_G.

Solution. The cumulative distribution function of X is

$$F_X(t) = \begin{cases} 0 & t \le 0 \\ \frac{1}{4}t & 0 < t < 4 \\ 1 & t \ge 4 \end{cases}$$

We note that the events $\{G = 2\}$ and $\{3 < X \le 5\}$ are equivalent. Then, by Definition 4.1,

$$\begin{aligned} P_G(2) &= P\{G = 2\} \\ &= P\{3 < X \le 5\} = F_X(5) - F_X(3) \\ &= 1 - \tfrac{3}{4} = \tfrac{1}{4} \end{aligned}$$

Similarly,

$$\begin{aligned} P_G(1) &= P\{G = 1\} \\ &= P\{X \le 3 \cup X > 5\} \\ &= F_X(3) + 1 - F_X(5) \\ &= \tfrac{3}{4} + 1 - 1 = \tfrac{3}{4} \end{aligned}$$

Therefore

$$P_G(k) = \begin{cases} \frac{3}{4} & k = 1 \\ \frac{1}{4} & k = 2 \end{cases}$$

and

$$F_G(k) = \begin{cases} 0 & k < 1 \\ \frac{3}{4} & 1 \le k < 2 \\ 1 & 2 \le k \end{cases}$$

Example 4.3 Suppose X has the p.d.f.

$$\begin{aligned} f_X(t) &= e^{-t} \qquad t \ge 0 \\ &= 0 \qquad \text{otherwise} \end{aligned}$$

If a new random variable is given by $Y = \frac{1}{4}X^2$, describe f_Y, the p.d.f. of Y. Prove that the events $\{X > 2\}$ and $\{Y > 1\}$ have equal probabilities.

Solution. By definition,

$$F_Y(t) = P\{Y \le t\} = P\left\{\frac{X^2}{4} \le t\right\}$$
$$= P\{-2\sqrt{t} \le X \le 2\sqrt{t}\} = F_X(2\sqrt{t})$$

since X is never negative.

Then the p.d.f. of Y is given by

$$f_Y(t) = \begin{cases} \dfrac{d}{dt}F_X(2\sqrt{t}) = f_X(2\sqrt{t}) \cdot \dfrac{1}{\sqrt{t}} \\ \dfrac{1}{\sqrt{t}}e^{-2\sqrt{t}} & t \ge 0 \\ 0 & \text{otherwise} \end{cases}$$

The p.d.f.'s of X and Y are shown in Figs. 4.1 and 4.2, respectively.

$$P\{X > 2\} = 1 - P\{X \le 2\} = 1 - \int_0^2 e^{-t}\,dt = e^{-2}$$

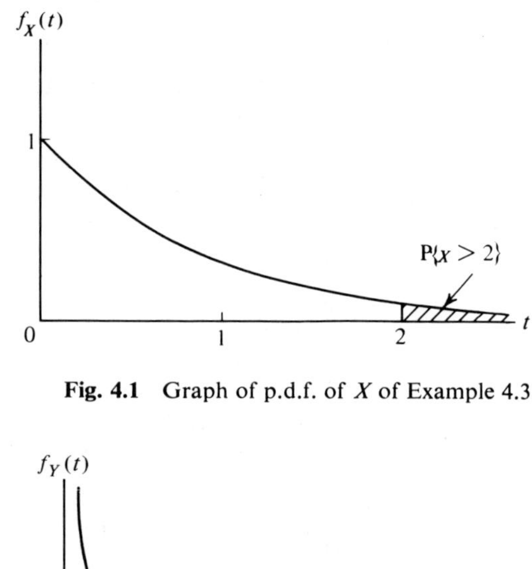

Fig. 4.1 Graph of p.d.f. of X of Example 4.3.

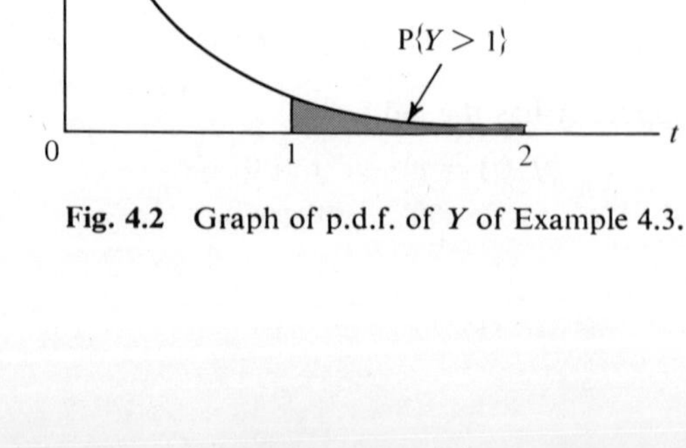

Fig. 4.2 Graph of p.d.f. of Y of Example 4.3.

$$P\{Y > 1\} = 1 - P\{Y \leq 1\} = 1 - \int_0^1 \frac{1}{\sqrt{t}} e^{-2\sqrt{t}}\,dt = e^{-2}$$

Hence, $P\{X > 2\}$ and $P\{Y > 1\}$ are equal. This is not surprising since the two events are equivalent.

Example 4.4 The density function of random variable X is given by

$$\begin{aligned} f_X(t) &= 2t \qquad 0 \leq t \leq 1 \\ &= 0 \qquad \text{otherwise} \end{aligned}$$

We wish to derive the density functions of Y_1 and Y_2 where

$$Y_1 = X^2 \quad \text{and} \quad Y_2 = \frac{1}{X}$$

Solution

(a) By definition,

$$F_{Y_1}(a_1) = P\{Y_1 \leq a_1\} = P\{X^2 \leq a_1\} = P\{-\sqrt{a_1} \leq X \leq \sqrt{a_1}\}$$

Since X is never negative, we can write

$$F_{Y_1}(a_1) = P\{X \leq \sqrt{a_1}\} = F_X(\sqrt{a_1})$$

Hence, the p.d.f. of Y_1 is

$$f_{Y_1}(a_1) = \begin{cases} f_X(\sqrt{a_1}) \cdot \dfrac{1}{2\sqrt{a_1}} \\ 1 \qquad 0 \leq a_1 \leq 1 \\ 0 \qquad \text{otherwise} \end{cases}$$

(b) $F_{Y_2}(a_2) = P\{Y_2 \leq a_2\} = P\left\{\dfrac{1}{X} \leq a_2\right\} = P\left\{X \geq \dfrac{1}{a_2}\right\} = 1 - F_X\left(\dfrac{1}{a_2}\right)$

Hence, the p.d.f. of Y_2 is

$$f_{Y_2}(a_2) = \begin{cases} -f_X\left(\dfrac{1}{a_2}\right) \cdot -(a_2)^{-2} \\ f_X\left(\dfrac{1}{a_2}\right) \cdot \dfrac{1}{a_2^2} \\ \dfrac{2}{a_2^3} \qquad 0 \leq a_2 \leq 1 \\ 0 \qquad \text{otherwise} \end{cases}$$

We will further extend the analysis of this example. We note that Y_1 is a real-valued function whose domain includes the range of random variable X. It is a *monotonically* increasing function of X. If we solve $Y_1 = X^2$ for X in terms of Y, we obtain $X = \sqrt{Y_1}$ or $X = \phi(Y_1)$, where ϕ is called the inverse function of Y_1. Since Y_1 is a *monotonically increasing* function of X, its inverse is also a monotonically increasing function. It follows that there is a *unique correspondence* between Y_1 and X. This means that when $Y_1 = a_1$, say, X is uniquely determined and equal to $\sqrt{a_1}$. Hence, we can assert that

$$P\{Y_1 \leq a_1\} = P\{X \leq \sqrt{a_1}\}$$

or

$$F_{Y_1}(a_1) = F_X(\sqrt{a_1}) = F_X[\phi(a_1)] \tag{4.2}$$

The graphs of p.d.f.'s of X and Y_1 are shown in Figs. 4.3 and 4.4, respectively. We note also that Y_2 is a continuous and monotonic decreasing function of X. Hence its inverse $\phi(Y_2) = 1/Y_2 = X$ is also a monotonic decreasing function. Then we have

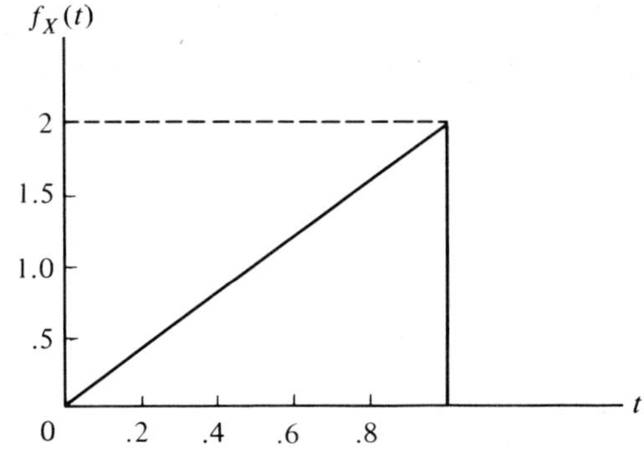

Fig. 4.3 Graph of p.d.f. of X of Example 4.4.

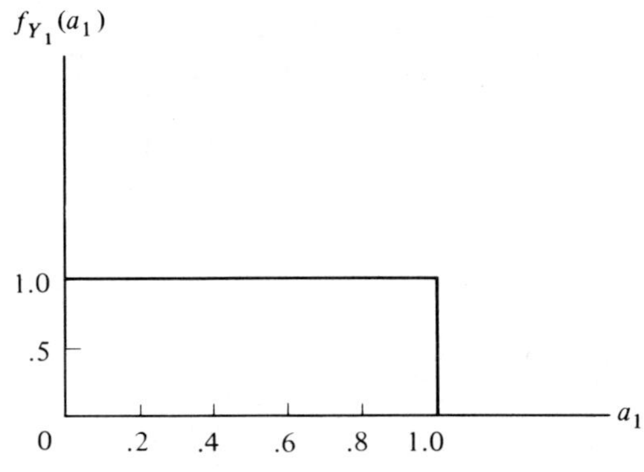

Fig. 4.4 Graph of p.d.f. of Y_1 of Example 4.4.

$$P\{Y_2 \leq a_2\} = P\left\{\frac{1}{X} \leq a_2\right\} = P\left\{X \geq \frac{1}{a_2}\right\}$$

or

$$F_{Y_2}(a_2) = 1 - F_X\left(\frac{1}{a_2}\right) = 1 - F_X[\phi(a_2)] \tag{4.3}$$

When Y_1 is differentiable, the first derivative of Eq. (4.2) is

(4.4) $$f_{Y_1}(a_1) = f_X[\phi(a_1)] \cdot \phi'(a_1)$$

Similarly, the first derivative of Eq. (4.3) is

(4.5) $$f_{Y_2}(a_2) = -f_X[\phi(a_2)] \cdot \phi'(a_2)$$

However, $\phi'(a_2)$ of Eq. (4.5) is always negative since Y_2 is a monotonically decreasing function of X. Therefore, Eq. (4.5) can be rewritten as

(4.6) $$f_{Y_2}(a_2) = f_X[\phi(a_2)] \cdot |\phi'(a_2)|$$

These results can be stated in the form of a theorem.

Theorem 4.1. *If random variable X is continuous with p.d.f. f_X, then the p.d.f. of a new random variable Y, which is either a monotonically increasing or monotonically decreasing function of X and differentiable for all values of X, is given by*

(4.7) $$f_Y(a) = f_X[\phi(a)] \cdot |\phi'(a)|$$

A word of caution is in order here. Equation (4.7) does *not* hold true if Y is *not* a monotonic function of X.

Example 4.5 Suppose that the p.d.f. of a random variable X is given by

$$\begin{aligned} f_X(t) &= \lambda e^{-\lambda t} \qquad \lambda > 0 \qquad t \geq 0 \\ &= 0 \qquad \text{otherwise} \end{aligned}$$

Derive the density function of Y where $Y = X + b$ (b is a positive constant).

Solution. Since Y is continuous and a monotonically increasing function of X, we can apply Theorem 4.1. Hence

$$f_Y(a) = \begin{cases} f_X[\phi(a)] \cdot |\phi'(a)| \\ f_X(a - b) \cdot 1 \\ \lambda e^{-\lambda(a-b)} \qquad \lambda > 0 \qquad a > b \geq 0 \\ 0 \qquad \text{otherwise} \end{cases}$$

Example 4.6 Solve Example 4.3 by applying Theorem 4.1.

Solution. Y is a continuous and an increasing function of X. Hence,

$$f_Y(a) = \begin{cases} f_X[\phi(a)] \cdot |\phi'(a)| \\ f_X(2\sqrt{a}) \cdot \phi'(2\sqrt{a}) \\ f_X(2\sqrt{a}) \cdot \dfrac{1}{\sqrt{a}} \qquad a \geq 0 \\ 0 \qquad \text{otherwise} \end{cases}$$

Example 4.7 The density function of variate X is given by

$$\begin{aligned} f_X(u) &= \tfrac{1}{4} \qquad -1 < u < 3 \\ &= 0 \qquad \text{otherwise} \end{aligned}$$

Suppose a new random variable Y is given by

$$Y = X^2 \qquad -1 \leq X \leq 3$$

Describe the following:
(a) F_Y, the c.d.f. of Y.
(b) f_Y, the p.d.f. of Y.

Solution

(a) We note that Y is not a monotonic function of X, since X can assume negative as well as positive values. Consequently, Theorem 4.1 cannot be applied. We derive

$$F_X(t) = \tfrac{1}{4}\int_{-1}^{t} du = \tfrac{1}{4}(t+1) \qquad -1 < t < 3$$

and

$$\begin{aligned} F_Y(a) &= P\{Y \leq a\} = P\{X^2 \leq a\} \\ &= P\{-\sqrt{a} \leq X \leq \sqrt{a}\} \\ &= F_X(\sqrt{a}) - F_X(-\sqrt{a}) \end{aligned}$$

The total range of a is from 0 to 9, but since $F_X(-\sqrt{a})$ has negative values from 0 to 1 only, we may write

$$\begin{aligned} F_Y(a) &= F_X(\sqrt{a}) - F_X(-\sqrt{a}) \qquad 0 \leq a \leq 1 \\ &= F_X(\sqrt{a}) \qquad 1 < a \leq 9 \end{aligned}$$

Or, replacing $F_X(\sqrt{a})$ by $\frac{1}{4}(\sqrt{a}+1)$ and $F_X(-\sqrt{a})$ by $\frac{1}{4}(-\sqrt{a}+1)$ we obtain

$$F_Y(a) = \begin{cases} 0 & a < 0 \\ \frac{1}{2}\sqrt{a} & 0 \leq a \leq 1 \\ \frac{1}{4}(\sqrt{a}+1) & 1 < a \leq 9 \\ 1 & a > 9 \end{cases}$$

(b) The probability density function of Y is

$$\begin{aligned} f_Y(a) &= \frac{1}{4\sqrt{a}} \qquad 0 \leq a \leq 1 \\ &= \frac{1}{8\sqrt{a}} \qquad 1 < a \leq 9 \end{aligned}$$

Example 4.8 Let $f_X(t) = \frac{1}{2} \qquad -1 < t < 1$ and $Y = |X|$. Derive F_Y and f_Y.

Solution. Since Y is not a monotonic function of X, we derive

$$F_X(t) = \tfrac{1}{2}(t+1) \qquad -1 < t < 1$$

and

$$\begin{aligned} F_Y(a) &= P\{Y \leq a\} = P\{|X| \leq a\} \\ &= P\{-a \leq X \leq a\} \\ &= F_X(a) - F_X(-a) \end{aligned}$$

Therefore

$$F_Y(a) = \begin{cases} 0 & a < 0 \\ a & 0 < a < 1 \\ 1 & a \geq 1 \end{cases}$$

and

$$f_Y(a) = \begin{cases} 1 & 0 < a < 1 \\ 0 & \text{otherwise} \end{cases}$$

4.3 The Distribution of a Function of Two or More Variates

Frequently one is interested in the knowledge of a new random variable Y that is a *function of n random variables $X_1, X_2, \ldots, X_n$*. In symbols we write

$$Y = G(X_1, X_2, \ldots, X_n)$$

If a joint density function of these variates exists, then the distribution of Y is given by

$$\begin{aligned} F_Y(a) &= P\{Y \leq a\} \\ &= \int_{G(u_1, \ldots, u_n) \leqslant a} \cdots \int f_{X_1, \ldots, X_n}(u_1, \ldots, u_n)\, du_1, \ldots, du_n \end{aligned} \tag{4.8}$$

For discrete variates the same equation holds with integrals replaced by sums.

We very often wish to find the distribution of new random variables as functions of variates whose distributions are known or directly measurable. We will consider a few typical examples to illustrate the method of deriving these distributions. To begin with, let us derive the distribution of the sum of two random variables X_1 and X_2 with known joint density function $f_{X_1, X_2}(\cdot\,, \cdot)$. In other words, given $f_{X_1, X_2}(\cdot\,, \cdot)$ we wish to derive the distribution of a new random variable Y where $Y = X_1 + X_2$. Then

$$\begin{aligned} F_Y(a) &= P\{Y \leq a\} \\ &= \int_{t_1 + t_2 \leqslant a} \int f_{X_1, X_2}(t_1, t_2)\, dt_1\, dt_2 \\ &= \int_{-\infty}^{\infty} dt_1 \int_{-\infty}^{a - t_1} f_{X_1, X_2}(t_1, t_2)\, dt_2 \end{aligned} \tag{4.9}$$

Differentiating both sides of Eq. (4.9) with respect to a we obtain

$$f_Y(a) = \int_{-\infty}^{\infty} f_{X_1, X_2}(t_1, a - t_1)\, dt_1 = \int_{-\infty}^{\infty} f_{X_1, X_2}(a - t_2, t_2)\, dt_2 \tag{4.10}$$

If X_1 and X_2 are independent random variables, then for any real number a we have

$$f_Y(a) = \int_{-\infty}^{\infty} f_{X_1}(t_1) f_{X_2}(a - t_1)\, dt_1 = \int_{-\infty}^{\infty} f_{X_1}(a - t_2) f_{X_2}(t_2)\, dt_2 \tag{4.11}$$

$f_Y(a)$, which is the density function of the sum of two independent continuous random variables and is known as the *convolution* of the density functions $f_{X_1}(.)$ and $f_{X_2}(.)$ of the variates.

Suppose now that X_1 and X_2 are two independent continuous random variables with probability density functions $f_{X_1}(.)$ and $f_{X_2}(.)$, respectively. We wish to derive the distribution of a new random variable Y, where $Y = X_1 X_2$. Then

$$\begin{aligned} F_Y(a) &= P\{Y \le a\} \\ &= \int_{-\infty}^{\infty} \int_{-\infty}^{a/t_2} f_{X_1}(t_1) f_{X_2}(t_2)\, dt_1\, dt_2 \\ &= \int_{-\infty}^{\infty} f_{X_2}(t_2) F_{X_1}\left(\frac{a}{t_2}\right) dt_2 \end{aligned} \tag{4.12}$$

Differentiating both sides of Eq. (4.12) with respect to a we obtain

$$f_Y(a) = \int_{-\infty}^{\infty} \frac{1}{|t_2|} f_{X_1}\left(\frac{a}{t_2}\right) f_{X_2}(t_2)\, dt_2 = \int_{-\infty}^{\infty} \frac{1}{|t_1|} f_{X_2}\left(\frac{a}{t_1}\right) f_{X_1}(t_1)\, dt_1 \tag{4.13}$$

Example 4.9 Two independent random variables are exponentially distributed with unequal parameters:

$$\begin{aligned} f_{X_1}(t_1) &= \lambda e^{-\lambda t_1} \qquad \lambda > 0 \qquad t_1 \ge 0 \\ f_{X_2}(t_2) &= \mu e^{-\mu t_2} \qquad \mu > \lambda > 0 \qquad t_2 \ge 0 \end{aligned}$$

Find the distribution of the sum $Y = X_1 + X_2$.

Solution. The density function of Y is

$$f_Y(a) = \begin{cases} \int_{-\infty}^{\infty} f_{X_1}(t_1) f_{X_2}(a - t_1)\, dt_1 \\ \int_0^a \lambda e^{-\lambda t_1} \mu e^{-\mu(a - t_1)}\, dt_1 \\ \lambda\mu e^{-\mu a} \int_0^a e^{t_1(\mu - \lambda)}\, dt_1 \\ \dfrac{\lambda\mu e^{-\mu a}}{\mu - \lambda}[e^{a(\mu - \lambda)} - 1] \\ \dfrac{\lambda\mu}{\mu - \lambda}(e^{-\lambda a} - e^{-\mu a}) \qquad \mu > \lambda > 0 \qquad a \ge 0 \end{cases}$$

Example 4.10 The joint density function of two continuous variates X_1 and X_2 is given by

$$\begin{aligned} f_{X_1, X_2}(t_1, t_2) &= 24 t_1^2 (1 - t_1) t_2 \qquad 0 \le t_1 \le 1 \\ & \qquad\qquad\qquad\qquad\quad 0 \le t_2 \le 1 \\ &= 0 \qquad \text{otherwise} \end{aligned}$$

(a) Are X_1 and X_2 independent?
(b) Find the distribution of $Y = X_1X_2$.

Solution. We derive

(a)
$$\begin{aligned} f_{X_1}(t_1) &= \int_0^1 24t_1^2(1-t_1)t_2\,dt_2 \\ &= 12t_1^2(1-t_1) \qquad 0 \le t_1 \le 1 \\ &= 0 \qquad \text{otherwise} \end{aligned}$$

and

$$\begin{aligned} f_{X_2}(t_2) &= \int_0^1 24t_1^2(1-t_1)t_2\,dt_1 \\ &= 2t_2 \qquad 0 \le t_2 \le 1 \\ &= 0 \qquad \text{otherwise} \end{aligned}$$

Since $f_{X_1}(t)\cdot f_{X_2}(t_2)$ is equal to $f_{X_1,X_2}(t_1, t_2)$, we conclude that variates X_1 and X_2 are independent.

(b)
$$f_Y(a) = \begin{cases} \displaystyle\int_{-\infty}^{\infty} \frac{1}{|t_1|} f_{X_1}(t_1) f_{X_2}\left(\frac{a}{t_1}\right) dt_1 \qquad 0 \le t_1 \le 1 \qquad 0 \le \frac{a}{t_1} \le 1 \\ \displaystyle\int_a^1 24a(1-t_1)\,dt_1 \\ \displaystyle 24a\int_a^1 (1-t_1)\,dt_1 \\ \displaystyle 24a\left(\frac{1}{2} - a + \frac{a^2}{2}\right) \\ 12a(1-a)^2 \qquad 0 \le a \le 1 \\ 0 \qquad \text{otherwise} \end{cases}$$

Example 4.11 Independent variates X_1 and X_2 are distributed as follows:

$$f_{X_1}(t_1) = \begin{cases} 1 & 0 < t_1 < 1 \\ 0 & \text{otherwise} \end{cases}$$

$$f_{X_2}(t_2) = \begin{cases} 1 & 0 < t_2 < 1 \\ 0 & \text{otherwise} \end{cases}$$

A new random variable is given by $Y = X_1X_2$.
(a) Describe F_Y, the c.d.f. of Y.
(b) Describe f_Y, the p.d.f. of Y.

Solution

(a) The admissible region (shaded) is shown in Fig. 4.5. By definition we have

$$\begin{aligned} F_Y(a) &= P\{Y \le a\} \\ &= P\{t_1t_2 \le a\} \\ &= \int_{t_1t_2 \leqslant a}\int 1\,dt_1\,dt_2 = \int_0^a\int_0^1 1\,dt_2\,dt_1 + \int_a^1\int_0^{a/t_1} 1\,dt_2\,dt_1 \\ &= a(1-\log a) \qquad 0 < a < 1 \end{aligned}$$

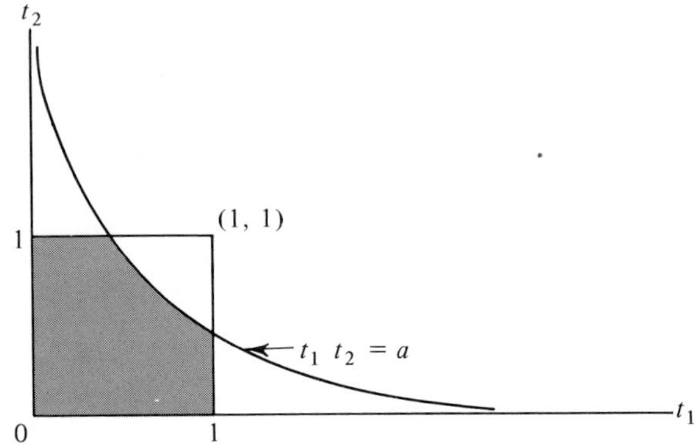

Fig. 4.5 Admissible region of Example 4.11.

Hence

$$F_Y(a) = \begin{cases} 0 & a < 0 \\ a(1 - \log a) & 0 < a < 1 \\ 1 & a \geq 1 \end{cases}$$

(b)
$$\begin{aligned} f_Y(a) &= -\log a \qquad 0 < a < 1 \\ &= 0 \qquad \text{otherwise} \end{aligned}$$

Example 4.12 The demand per week for a certain commodity is a random variable with the probability density function

$$\begin{aligned} f_Y(t) &= te^{-t} \qquad t \geq 0 \\ &= 0 \qquad \text{otherwise} \end{aligned}$$

Find the probability distribution of demand for
(a) Two weeks.
(b) Three weeks.
Assume weekly demands are independent.

Solution. In this problem we wish to derive the probability distribution of demand convoluted two and three times.
(a) Let Z be the variate that describes the demand for two weeks. Then

$$\begin{aligned} f_Z(a) &= \int_0^a (a - t)e^{-(a-t)} \cdot te^{-t}\, dt \\ &= e^{-a}\int_0^a (at - t^2)\, dt = \frac{a^3 e^{-a}}{6} \end{aligned}$$

Hence

$$\begin{aligned} f_Z(a) &= \frac{a^3 e^{-a}}{6} \qquad a \geq 0 \\ &= 0 \qquad \text{otherwise} \end{aligned}$$

(b) Let X be the variate that describes the demand for 3 weeks. Then the density function of X is the convolution of the functions $f_Y(.)$ and $f_Z(.)$. In symbols,

$$f_X(.) = f_Y(.) f_Z(.)$$

Hence

$$f_X(a) = \begin{cases} \displaystyle\int_0^a \frac{(a-t)^3 e^{-(a-t)} t e^{-t}}{6}\, dt = \frac{e^{-a}}{6} \int_0^a t(a-t)^3\, dt \\ \dfrac{a^5 e^{-a}}{5!} \qquad a \geq 0 \\ 0 \qquad \text{otherwise} \end{cases}$$

Example 4.13 The demand distribution for a certain commodity for 1 week is given by

Demand	*Probability*
0	.2
1	.5
2	.2
3	.1

What is the demand distribution for 2 weeks? Assume the demand level in the following periods is independent of the demand level in the preceding periods.

Solution. We note that the demand is a discrete random variable. Let Y be the random variable that describes the demand for a 2-week period. Then

$$P_Y(k) = \sum_{\text{over all } X_1} P_{X_1}(x_1) P_{X_2}(k - x_1)$$

and the probability function is

$$P_Y(k) = \begin{cases} .04 & k = 0 \\ .20 & k = 1 \\ .33 & k = 2 \\ .24 & k = 3 \\ .14 & k = 4 \\ .04 & k = 5 \\ .01 & k = 6 \end{cases}$$

4.4 Joint Distributions of Functions of Random Variables

In Sec. 4.3 we expounded the methods for deriving distributions of a single variate that is a function of two or more initial variates. We will consider next the problem of deriving the *joint distribution* of two or more functions. To illustrate, suppose X_1 and X_2 are continuous random variables and let $V = X_1 + X_2$ and $U = X_1 X_2$. What will be the joint distribution of U and V? We will proceed to develop the method for finding the joint distribution, but for simplicity we will confine our analysis to two functions.

Let X_1 and X_2 be two continuous random variables and their joint density functions $f_{X_1,X_2}(\cdot\,,\cdot)$ is defined over some region R in the X_1X_2 plane. Two new random variables U and V that are functions of X_1 and X_2 are defined as follows:

$$U = g_1(X_1, X_2) \qquad V = g_2(X_1, X_2)$$

Further assume that U and V are single-valued functions of X_1 and X_2. In other words, x_1 and x_2 are completely recovered if u and v are given (see

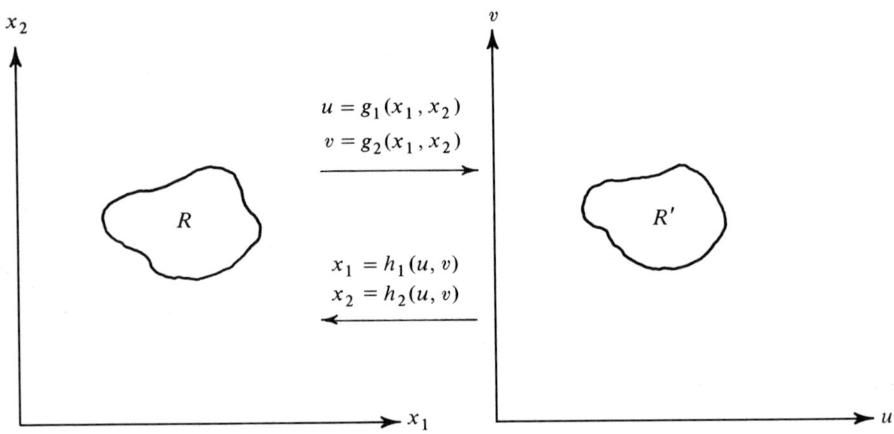

Fig. 4.6 Mapping region R in the x_1x_2-plane onto a region R' in the uv-plane.

Fig. 4.6). This means that we have now two functions, say h_1 and h_2, which connect x_1, x_2 with u, v as follows:

$$x_1 = h_1(u, v) \qquad x_2 = h_2(u, v)$$

Accordingly, region R in the X_1X_2 plane can be mapped onto a region R' in the UV plane. In advanced calculus the formula for transforming double integrals is given by

(4.14)

$$\int_R\int f_{X_1,X_2}(x_1, x_2)\, dx_1\, dx_2 = \int_{R'}\int g[h_1(u, v), h_2(u, v)]\left|\frac{\partial(h_1, h_2)}{\partial(u, v)}\right| du\, dv = 1$$

where $[\partial(h_1, h_2)]/[\partial(u, v)]$ is called the *Jacobian* of the mapping and is equal to the determinant

$$\text{(4.15)} \qquad \frac{\partial(h_1, h_2)}{\partial(u, v)} = \begin{vmatrix} \dfrac{\partial h_1}{\partial u} & \dfrac{\partial h_1}{\partial v} \\ \dfrac{\partial h_2}{\partial u} & \dfrac{\partial h_2}{\partial v} \end{vmatrix}$$

The formula for transforming the double integral is valid under the following conditions:

1. Continuity of X_1, X_2, U, and V.
2. Mapping is one-to-one.
3. The Jacobian is never zero.
4. The functions h_1 and h_2 have continuous partial derivatives.

The joint density function $f_{U,V}(u, v)$ is given by

$$g[h_1(u, v), h_2(u, v)]\left|\frac{\partial(h_1, h_2)}{\partial(u, v)}\right| \tag{4.16}$$

Example 4.14 Independent random variables X_1 and X_2 are distributed as follows:

$$f_{X_1}(t_1) = e^{-t_1} \quad t_1 \geq 0 \qquad f_{X_2}(t_2) = e^{-t_2} \quad t_2 \geq 0$$
$$= 0 \quad \text{otherwise} \qquad = 0 \quad \text{otherwise}$$

Further define random variable U and V as follows:

$$U = X_1 + X_2 \qquad V = \frac{X_1}{X_2}$$

Describe $f_{U,V}$.

Solution. Given

$$v = \frac{t_1}{t_2}$$

then

$$1 + v = \frac{t_1 + t_2}{t_2} = \frac{u}{t_2}$$

Hence

$$t_2 = \frac{u}{1 + v} = h_2(u, v)$$

and

$$t_1 = \frac{uv}{1 + v} = h_1(u, v)$$

Now we can obtain

$$\frac{\partial(h_1, h_2)}{\partial(u, v)} = \begin{vmatrix} \dfrac{\partial h_1}{\partial u} & \dfrac{\partial h_1}{\partial v} \\ \dfrac{\partial h_2}{\partial u} & \dfrac{\partial h_2}{\partial v} \end{vmatrix} = \begin{vmatrix} \dfrac{v}{1 + v} & \dfrac{u}{(1 + v)^2} \\ \dfrac{1}{1 + v} & \dfrac{-u}{(1 + v)^2} \end{vmatrix} = J$$

Therefore

$$J = \frac{-u}{(1 + v)^2} \qquad |J| = \frac{u}{(1 + v)^2}$$

The joint distribution, then, is

$$f_{U,V}(u, v) = e^{-(t_1+t_2)}\frac{u}{(1+v)^2} = \frac{ue^{-u}}{(1+v)^2} \qquad u \geq 0 \qquad v \geq 0$$
$$= 0 \qquad \text{otherwise}$$

Example 4.15 X_1 and X_2 are independent, each with the same density function as in Example 4.14. Let $U = X_1 + X_2$ and $V = X_1/(X_1 + X_2)$ be two new random variables.
(a) Compute $f_{U,V}$.
(b) Are U and V independent?

Solution
(a) We have

$$v = \frac{t_1}{t_1 + t_2} = \frac{t_1}{u}$$

whence

$$t_1 = uv = h_1(u, v)$$

Therefore

$$t_2 = u - uv = u(1 - v) = h_2(u, v)$$

and

$$J = \begin{vmatrix} v & u \\ 1 - v & -u \end{vmatrix} = -u$$

or

$$|J| = u$$

Since $u = t_1 + t_2$ and both t_1 and t_2 are greater than or equal to zero, the admissible region for u is from 0 to ∞. The admissible region for v is obtained from the original boundaries $t_1 = 0$ and $t_2 = 0$. Therefore, the new boundaries are

$$t_1 = 0 = uv \quad \text{or} \quad v = 0$$
$$t_2 = 0 = u(1 - v) \quad \text{or} \quad v = 1$$

The admissible region (shaded) is shown in Fig. 4.7. The joint distribution of U and V, then, is

$$f_{U,V}(u, v) = ue^{-u} \qquad u \geq 0 \qquad 0 < v < 1$$
$$= 0 \qquad \text{otherwise}$$

(b) The marginal distributions of U and V are derived as follows:

$$f_U(u) = \begin{cases} \int_0^1 ue^{-u}\,dv \\ ue^{-u} & u \geq 0 \\ 0 & \text{otherwise} \end{cases}$$

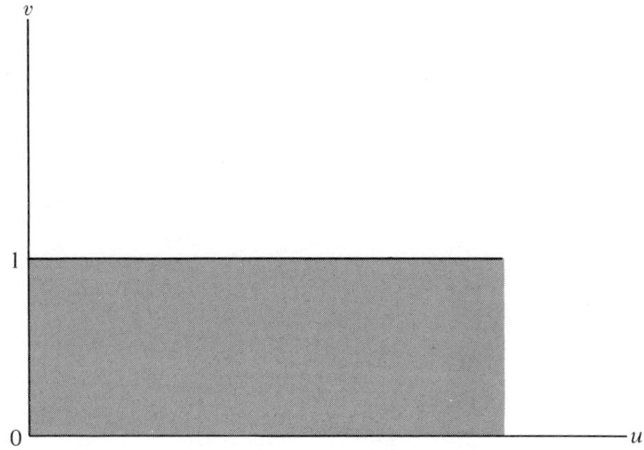

Fig. 4.7 Admissible region of Example 4.15.

$$f_V(v) = \begin{cases} \int_0^\infty ue^{-u}du \\ 1 & 0 < v < 1 \\ 0 & \text{otherwise} \end{cases}$$

Since $f_U(u) \cdot f_V(v) = f_{U,V}(u, v)$, we conclude that U and V are independent. The marginal p.d.f. of U and V are shown in Figs. 4.8 and 4.9, respectively.

We note that the derived distribution of a function of two or more variates is a special case of the joint distribution of functions of random variables described

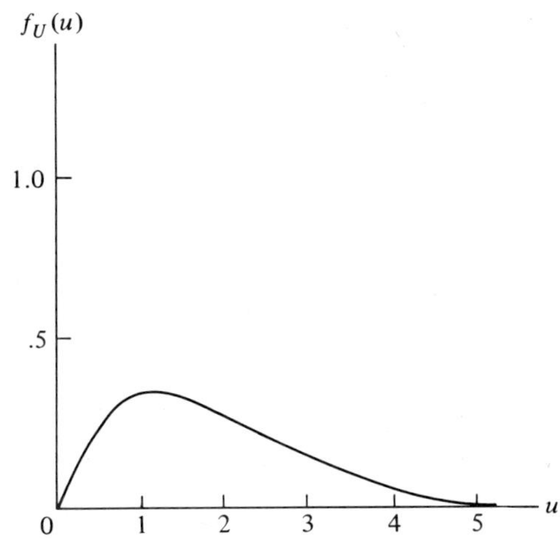

Fig. 4.8 Graph of p.d.f. f_U of Example 4.15.

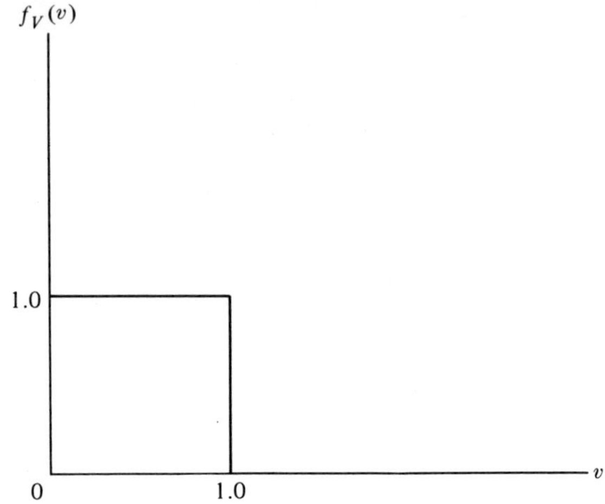

Fig. 4.9 Graph of p.d.f. f_V of Example 4.15.

in this section. To prove this assertion let us solve Example 4.9 by treating it as two functions of random variables.

Example 4.16 Solve Example 4.9 by using variate transformation. Here we have two independent random variables with the following p.d.f.'s:

$$\begin{aligned} f_{X_1}(t_1) &= \lambda e^{-\lambda t_1} \qquad \lambda > 0 \qquad t_1 \geq 0 \\ &= 0 \qquad \text{otherwise} \\ f_{X_2}(t_2) &= \mu e^{-\mu t_2} \qquad \mu > \lambda > 0 \qquad t_2 \geq 0 \\ &= 0 \qquad \text{otherwise} \end{aligned}$$

and we wish to derive the distribution of U where $U = X_1 + X_2$. Let us assume that V is a function of X_1, say $V = X_1$. Accordingly, $t_1 = v$ and $t_2 = u - t_1 = u - v$. The Jacobian is given by

$$J = \begin{vmatrix} 0 & 1 \\ 1 & -1 \end{vmatrix} = -1 \qquad |J| = 1$$

Since $u = t_1 + t_2$ and both t_1 and t_2 are greater than or equal to zero, the admissible region for u is from 0 to ∞. The admissible region for v is obtained from the original boundaries $t_1 = 0$ and $t_2 = 0$. Accordingly, the new boundaries are

$$\begin{aligned} t_1 &= v = 0 \\ t_2 &= u - v = 0 \quad \text{or} \quad v = u \end{aligned}$$

The joint distribution of U and V, then, is

$$\begin{aligned} f_{U,V}(u, v) &= \lambda\mu e^{-\lambda v} e^{-\mu(u-v)} \qquad u \geq 0 \qquad 0 < v < u \\ &= 0 \qquad \text{otherwise} \end{aligned}$$

Hence, the marginal density of U is given by

$$f_U(u) = \begin{cases} \int_0^u \lambda\mu e^{-\lambda v} e^{-\mu(u-v)}\, dv \\ \lambda\mu e^{-\mu u}\left(\dfrac{e^{\mu u - \lambda u} - 1}{\mu - \lambda}\right) \\ \dfrac{\lambda\mu}{\mu - \lambda}(e^{-\lambda u} - e^{-\mu u}) \qquad \mu > \lambda > 0 \qquad u > 0 \\ 0 \qquad \text{otherwise} \end{cases}$$

which proves the assertion.

PROBLEMS

4.1. The variate X is distributed as follows:

$$\begin{aligned} f_X(t) &= \tfrac{1}{2}e^{-t/2} \qquad t \geq 0 \\ &= 0 \qquad \text{otherwise} \end{aligned}$$

Derive the distribution of $Y = \sqrt{e^{-X}}$.

4.2. Suppose that random variable X has the p.d.f. given by

$$\begin{aligned} f_X(t) &= \tfrac{1}{2} \qquad -1 < t < 1 \\ &= 0 \qquad \text{otherwise} \end{aligned}$$

Derive the distribution of $Y = X^2$.

4.3. The variate X is distributed as follows:

$$\begin{aligned} f_X(t) &= 1 \qquad 0 < t < 1 \\ &= 0 \qquad \text{otherwise} \end{aligned}$$

Derive the distribution of $Y = e^X$.

4.4. Suppose that random variable X has the p.d.f. given in Prob. 4.3.

(a) Derive the distribution of $Y = \dfrac{1}{X}\cdot$

(b) Verify that the result obtained in (a) is a p.d.f.

4.5. The variate X has the following p.d.f.:

$$\begin{aligned} f_X(t) &= e^{-t} \qquad t \geq 0 \\ &= 0 \qquad \text{otherwise} \end{aligned}$$

Derive the distribution of $Y = 3X^2$.

4.6. Suppose that random variable X has the p.d.f. given by

$$\begin{aligned} f_X(t) &= 2t \qquad 0 \leq t \leq 1 \\ &= 0 \qquad \text{otherwise} \end{aligned}$$

Derive the distribution of $Y = 3X + 1$.

4.7. The density function of a random variable X is given by

$$\begin{aligned} f_X(t) &= 2t \qquad 0 \leq t \leq 1 \\ &= 0 \qquad \text{otherwise} \end{aligned}$$

If a new random variable G is given by

$$\begin{aligned} G(X) &= 5 & X &\leq -2 \\ &= 3X & X &> -2 \end{aligned}$$

describe F_G, the c.d.f. of G.

4.8. The density function of a random variable X is given by

$$\begin{aligned} f_X(t) &= \tfrac{1}{3} & 1 < t < 4 \\ &= 0 & \text{otherwise} \end{aligned}$$

If a new discrete random variable G is given by

$$\begin{aligned} G(X) &= 2 & X &\leq 2 \\ &= 1 & X &> 2 \end{aligned}$$

describe the probability function of G.

4.9. The probability function of a random variable X is given by

$$P_X(k) = \begin{cases} .20 & k = 0 \\ .30 & k = 1 \\ .50 & k = 2 \end{cases}$$

Derive the distribution of $Y = 2X$.

4.10. Suppose that variate X is distributed as follows:

$$P_X(k) = \begin{cases} .40 & k = -1 \\ .60 & k = 1 \end{cases}$$

Derive the distribution of $Y = X + 1$.

4.11. Suppose that X is a discrete variate with c.d.f.

$$F_X(k) = \begin{cases} 0 & k < 0 \\ .1 & 0 \leq k < 1 \\ .7 & 1 \leq k < 2 \\ 1 & 2 \leq k \end{cases}$$

Find the probability function of $Y = 3X$.

4.12. Suppose that X is a continuous variate with p.d.f.

$$f_X(t) = \frac{1}{\sqrt{2\pi}} e^{-(1/2)(t-10)^2} \qquad -\infty < t < \infty$$

Derive the distribution of $Y = 5X + 10$.

4.13. Assume X has the following c.d.f.:

$$F_X(t) = \begin{cases} 0 & t \leq 0 \\ t & 0 < t < 1 \\ 1 & 1 \leq t \end{cases}$$

Determine the c.d.f. of random variable Y if
(a) $Y = \ln |X|$
(b) $Y = 5X + 2$.

4.14. The p.d.f. of a random variable X is given by

$$\begin{aligned} f_X(t) &= \frac{1}{\sqrt{2\pi}} t^{1/2} e^{-t/2} & t \geq 0 \\ &= 0 & \text{otherwise} \end{aligned}$$

Derive the distribution of $Y = X^2$.

4.15. Assume X has the p.d.f. given by

$$f_X(t) = \begin{cases} \frac{2}{3} & 0 \le t \le 1 \\ \frac{1}{3} & 1 < t \le 2 \\ 0 & \text{otherwise} \end{cases}$$

Derive the distribution of $Y = \sqrt{X}$.

4.16. Assume X has the p.d.f. given by

$$f_X(t) = \frac{1}{\sqrt{2\pi}} e^{-t^2/2} \qquad -\infty < t < \infty$$

Derive the distribution of $Y = 1/X$.

4.17. Let X and Y be two independent random variables, each with the same probability density given by

$$\begin{aligned} f_X(t) = f_Y(t) &= 1 & 0 < t < 1 \\ &= 0 & \text{otherwise} \end{aligned}$$

Compute $P\{X - Y \le \frac{1}{2}\}$.

4.18. Independent variates X_1 and X_2 are distributed as follows:

$$\begin{aligned} f_{X_1}(t_1) &= 2t_1 \quad 0 \le t_1 \le 1 & f_{X_2}(t_2) &= \frac{t_2}{2} \quad 0 < t_2 \le 2 \\ &= 0 \quad \text{otherwise} & &= 0 \quad \text{otherwise} \end{aligned}$$

Derive the c.d.f. of $Y = X_1 + X_2$.

4.19. The joint distribution of variates X_1 and X_2 is as follows:

$$\begin{aligned} f_{X_1, X_2}(t_1, t_2) &= e^{-(t_1+t_2)} \qquad t_1 \ge 0 \qquad t_2 \ge 0 \\ &= 0 \qquad \text{otherwise} \end{aligned}$$

Compute the c.d.f. of $Y = X_1 - X_2$.

4.20. Independent variates X_1 and X_2 are distributed as follows:

$$\begin{aligned} f_{X_1}(t_1) &= e^{-t_1} \quad t_1 \ge 0 & f_{X_2}(t_2) &= \tfrac{3}{2} e^{-(3/2)t_2} \quad t_2 \ge 0 \\ &= 0 \quad \text{otherwise} & &= 0 \quad \text{otherwise} \end{aligned}$$

Find the p.d.f. of $Y = X_1/X_2$.

4.21. The weekly demand distribution for air conditioners in a certain department store is as follows:

Demand	*Probability*
0	.1
1	.2
2	.2
3	.3
4	.2

Derive the probability distribution of demand for a 2-week period (assume independence).

4.22. Suppose that X_1 and X_2 are two independent random variables with the following p.d.f.'s:

$$\begin{aligned} f_{X_1}(t_1) &= \frac{t_1^2}{9} \quad 0 \le t_1 \le 3 & f_{X_2}(t_2) &= \frac{t_2}{2} \quad 0 \le t_2 \le 2 \\ &= 0 \quad \text{otherwise} & &= 0 \quad \text{otherwise} \end{aligned}$$

Derive the c.d.f. of $Y = X_1 X_2$.

4.23. Independent variates X_1 and X_2 are distributed as follows:

$$f_{X_1}(t_1) = e^{-t_1} \quad t_1 \geq 0 \qquad f_{X_2}(t_2) = t_2 e^{-t_2} \quad t_2 \geq 0$$
$$= 0 \quad \text{otherwise} \qquad = 0 \quad \text{otherwise}$$

Find the distribution of their sum $Y = X_1 + X_2$.

4.24. Derive the distribution of $Y = X_1 X_2$, given that

$$f_{X_1, X_2}(t_1, t_2) = \tfrac{3}{4} t_1 t_2^2 \qquad 0 \leq t_1 \leq 1 \qquad 0 \leq t_2 \leq 2$$
$$= 0 \qquad \text{otherwise}$$

4.25. Derive the c.d.f. of $Y = X_1 + X_2$, given that

$$f_{X_1, X_2}(t_1, t_2) = \tfrac{1}{3} \qquad 0 < t_1 < 1 \qquad 0 < t_2 < 3$$
$$= 0 \qquad \text{otherwise}$$

4.26. The joint distribution of variates X_1 and X_2 is as follows:

$$f_{X_1, X_2}(t_1, t_2) = 4 t_1 t_2 \qquad 0 < t_1 < 1 \qquad 0 < t_2 < 1$$
$$= 0 \qquad \text{otherwise}$$

Let $U = X_1^2$ and $V = X_2^2$ be two new random variables. Derive $f_{U,V}$, the p.d.f. of (U, V).

4.27. Independent variates X_1 and X_2 are distributed as follows:

$$f_{X_1}(t_1) = \frac{1}{\sqrt{2\pi}} e^{-t_1^2/2} \qquad -\infty < t_1 < \infty \qquad f_{X_2}(t_2) = 2t_2 e^{-t_2^2} \qquad t_2 \geq 0$$
$$= 0 \qquad \text{otherwise}$$

Let $U = X_1/X_2$ and $V = X_2$ be two new random variables.
(a) Compute $f_{U,V}$, the p.d.f. of (U, V).
(b) Derive the marginal p.d.f. of U.

4.28. Let X_1 and X_2 be two independent determinations of variate X with the p.d.f. given by

$$f_X(t) = \tfrac{1}{2} e^{-(1/2)t} \qquad t \geq 0$$
$$= 0 \qquad \text{otherwise}$$

Let $U = X_1 + X_2$ and $V = X_1 - X_2$ be two new random variables. Find the p.d.f. of (U, V).

4.29. Let X_1 and X_2 be two independent random variables, each with the same p.d.f. given by

$$f_{X_1}(t) = f_{X_2}(t) = \begin{cases} 1 & 0 < t < 1 \\ 0 & \text{otherwise} \end{cases}$$

Let $U = X_1 + X_2$ and $V = X_1 - X_2$ be two new random variables.
(a) Find $f_{U,V}$, the joint p.d.f. of (U, V).
(b) Find f_U, the marginal p.d.f. of U.
(c) Find f_V, the marginal p.d.f. of V.
(d) Determine whether or not U and V are independent.

4.30. Let X_1 and X_2 be two independent random variables, each with the same p.d.f. given by

$$f_{X_1}(t) = f_{X_2}(t) = \begin{cases} \dfrac{2}{t^2} & t \geq 2 \\ 0 & \text{otherwise} \end{cases}$$

Let $U = X_1/X_2$ and $V = X_1 + X_2$ be two new random variables.
(a) Find $f_{U,V}$, the joint p.d.f. of (U, V).
(b) Find f_U, the marginal p.d.f. of U.

SUGGESTED REFERENCES

[1] Derman, C., and M. Klein, *Probability and Statistical Inference for Engineers*, Oxford University Press, Inc., New York, 1959.

[2] Ehrenfeld, S., and S. B. Littauer, *Introduction to Statistical Method*, McGraw-Hill Book Company, New York, 1964.

[3] Fry, T. C., *Probability and Its Engineering Uses*, 2nd ed., D. Van Nostrand Co., Inc., Princeton, N.J., 1964.

[4] Gnedenko, B. V., *The Theory of Probability*, Chelsea Publishing Co., Bronx, N.Y., 1962.

[5] Lindgren, B. W., and G. W. McElrath, *Introduction to Probability and Statistics*, The Macmillan Company, New York, 1959.

[6] McCord, J. R., and R. M. Moroney, Jr., *Introduction to Probability Theory*, The Macmillan Company, New York, 1964.

[7] Meyer, P., *Introductory Probability and Statistical Applications*, 2nd ed., Addison-Wesley Publishing Co., Reading, Mass., 1970.

[8] Mood, A. M., and F. A. Graybill, *Introduction to the Theory of Statistics*, 2nd ed., McGraw-Hill Book Company, New York, 1963.

[9] Parzen, E., *Modern Probability Theory and Its Applications*, John Wiley & Sons, Inc., New York, 1960.

[10] Wadsworth, G. P., and J. G. Bryan, *Introduction to Probability and Random Variables*, McGraw-Hill Book Company, New York, 1960.

5

EXPECTATION

In many scientific problems it is desirable to describe the probability distribution of a random variable in a *concise* form. Some of the properties of the distribution function may be used to describe some features of the random variable under consideration. In many situations knowledge of these functions or some of their properties is all that is needed in order to obtain the required probability calculations. In this chapter we shall introduce the concepts of expectation and variance of a random variable and analyze the properties of a random variable in terms of these concepts.

5.1 Measures of Location

Some important *measures of location* are the *median*, *mode*, and *mean*. The latter is the most important and is extensively used to specify the *central tendency* of the distribution.

Definition 5.1. The *median* of a random variable X is the value t such that $F_X(t) = \frac{1}{2}$. The median always *exists* if the random variable is continuous, whereas it may *not* always exist if the random variable is discrete.

Example 5.1 Let X be a random variable that describes the sum of numbers of two fair dice. Compute the median of X.

Solution. The sample space S is given by

$$S = \{2, 3, 4, 5, 6, 7, 8, 9, 10, 11, 12\}$$

It is easily confirmed that the median is equal to seven, since one-half of the remaining observations is above seven and another half is below seven.

Example 5.2 The density function of a random variable X is

$$\begin{aligned} f_X(t) &= 2t \qquad 0 \leq t \leq 1 \\ &= 0 \qquad \text{otherwise} \end{aligned}$$

Compute the median of X.

Solution. The cumulative distribution function of X is

$$F_X(t) = \begin{cases} 0 & t < 0 \\ t^2 & 0 \leq t \leq 1 \\ 1 & t > 1 \end{cases}$$

The median of X is equal to $1/\sqrt{2}$, since $t^2 = \frac{1}{2}$ determines the median.

Definition 5.2. The *mode* of a random variable X is the value that occurs *most frequently*, sometimes called the *most probable value*. We say that a is the mode of random variable X

$$\text{(5.1)} \qquad \begin{aligned} &\text{if} \quad f_X(a) = \max_t f_X(t) \quad \text{for continuous variate} \\ &\text{or} \quad P_X(a) = \max_t P_X(t) \quad \text{for discrete variate} \end{aligned}$$

It is noteworthy that a mode is *not* necessarily unique.

Example 5.3 The density function of a random variable X is

$$f_X(t) = \begin{cases} \frac{2}{3}t & 0 \leq t \leq 1 \\ \frac{1}{3} & 1 < t \leq 3 \end{cases}$$

Compute the mode of X and the median of X.

Solution. It is easily confirmed that $f_X(t)$ is a maximum when $t = 1$; therefore, the mode is equal to one. The cumulative distribution function of X is

$$F_X(t) = \begin{cases} 0 & t < 0 \\ \dfrac{t^2}{3} & 0 \leq t \leq 1 \\ \dfrac{1}{3}t & 1 < t \leq 3 \\ 1 & t > 3 \end{cases}$$

Therefore, the median is equal to $\frac{3}{2}$ since $t/3 = \frac{1}{2}$ determines the median.

Definition 5.3. The *mean* or the *expected value* of random variable X [denoted by $E(X)$] is defined as follows:

(5.2) $$E(X) = \int_{-\infty}^{\infty} tf_X(t)\,dt \quad \text{for a continuous variate}$$

or

$$E(X) = \sum_{\text{over all } k} kP_X(k) \quad \text{for a discrete variate}$$

The role of the mean in probability theory is the same as the role of the center of gravity in mechanics.

Example 5.4 Calculate the expected value of random variable X given in Example 5.3.

Solution. By definition,

$$E(X) = \int_0^1 \frac{2}{3}t^2\,dt + \int_1^3 \frac{t}{3}\,dt$$
$$= 1\frac{5}{9}$$

Example 5.5 Random variable X has the following probability function:

$$P_X(k) = \begin{cases} \frac{1}{8} & k = 2 \\ \frac{1}{8} & k = 3 \\ \frac{1}{4} & k = 4 \\ \frac{1}{2} & k = 5 \end{cases}$$

Compute $E(X)$.

Solution. By definition,

$$E(X) = (\tfrac{1}{8})(2) + (\tfrac{1}{8})(3) + (\tfrac{1}{4})(4) + (\tfrac{1}{2})(5) = 4\tfrac{1}{8}$$

If a new random variable Y is a function of variate X expressed as $Y = G(X)$, then its expectation is given by

(5.3) $$E(Y) = \begin{cases} \int_{-\infty}^{\infty} tf_Y(t)\,dt & \text{if } G \text{ is continuous} \\ \sum_{\text{over all } k} kP_Y(k) & \text{if } G \text{ is discrete} \end{cases}$$

However, the expectation $E(Y)$ can be easily computed in terms of the probability density function $f_X(.)$ without deriving $f_Y(.)$. Hence, we write

(5.4) $$E(Y) = \begin{cases} \int_{-\infty}^{\infty} G(t)f_X(t)\,dt & \text{if } G \text{ is continuous} \\ \sum_{\text{over all } k} G(k)P_X(k) & \text{if } G \text{ is discrete} \end{cases}$$

Example 5.6 The density function of random variable X is given by

$$f_X(t) = e^{-t} \qquad t \geq 0$$
$$= 0 \qquad \text{otherwise}$$

Of interest is a new random variable Y where $Y = 2X$. Compute $E(Y)$.

Solution. We shall use two methods of evaluation of $E(Y)$, namely, in terms of $f_Y(\cdot)$ and in terms of $f_X(\cdot)$, and will show that the two methods will lead to the same result:

First Alternative. By definition,

$$F_Y(a) = P\{Y \leq a\} = P\{2X \leq a\} = P\left\{X \leq \frac{a}{2}\right\} = F_X\left(\frac{a}{2}\right)$$

However,

$$F_X(a) = 1 - e^{-a} \qquad a \geq 0$$

Then

$$F_Y(a) = 1 - e^{-a/2} \qquad a \geq 0$$

and

$$\begin{aligned} f_Y(a) &= \tfrac{1}{2}e^{-a/2} \qquad a \geq 0 \\ &= 0 \qquad \text{otherwise} \end{aligned}$$

Hence

$$E(Y) = \int_0^\infty \tfrac{1}{2}ae^{-a/2}\,da = 2$$

Second Alternative.

$$\begin{aligned} E(Y) &= \int_{-\infty}^\infty G(a)f_X(a)\,da \\ &= \int_0^\infty 2ae^{-a}\,da = 2 \end{aligned}$$

It should be noted that the two integrals $af_Y(a)\,da$ and $G(a)f_X(a)\,da$ are equal. The proof of this equality is beyond the scope of this text. Both evaluations of the mean were given, indicating the truth of the equality.

The expectation of one-dimensional variates can be extended to n-dimensional variates. For example, for two-dimensional variates X and Y we write

$$(5.5) \qquad E(XY) = \begin{cases} \displaystyle\int_{-\infty}^\infty \int_{-\infty}^\infty uvf_{X,Y}(u, v)\,du\,dv & \text{for continuous variates} \\ \displaystyle\sum_{\text{over all } u,v} uvP_{X,Y}(u, v) & \text{for discrete variates} \end{cases}$$

Example 5.7 The joint density function of X_1 and X_2 is

$$\begin{aligned} f_{X_1,X_2}(u_1, u_2) &= u_1 + u_2 \qquad 0 \leq u_1 \leq 1 \qquad 0 \leq u_2 \leq 1 \\ &= 0 \qquad \text{otherwise} \end{aligned}$$

Compute $E(X_1X_2)$.

Solution. By definition,

$$\begin{aligned} E(X_1X_2) &= \int_0^1 \int_0^1 u_1u_2(u_1 + u_2)\,du_1\,du_2 \\ &= \int_0^1 \left(\frac{u_2}{3} + \frac{u_2^2}{2}\right) du_2 = \frac{1}{3} \end{aligned}$$

5.2 Properties of the Mean

Theorem 5.1. *If C is an arbitrary constant and X is a random variable, then*

$$E(CX)^k = C^k E(X^k) \tag{5.6}$$

Proof. Suppose X is a continuous random variable; then

$$\begin{aligned} E(CX)^k &= \int_{-\infty}^{\infty} C^k t^k f_X(t)\,dt \\ &= C^k \int_{-\infty}^{\infty} t^k f_X(t)\,dt \\ &= C^k E(X^k) \end{aligned}$$

In a similar way we can prove the theorem for a discrete variate.

Theorem 5.2. *If C_1 and C_2 are two arbitrary constants and X_1* and *X_2 are two random variables, then*

$$E(C_1X_1 + C_2X_2) = C_1E(X_1) + C_2E(X_2) \tag{5.7}$$

Proof (for continuous variates). By definition,

$$\begin{aligned} &E(C_1X_1 + C_2X_2) \\ &\quad= \int_{-\infty}^{\infty}\int_{-\infty}^{\infty} (C_1t_1 + C_2t_2) f_{X_1,X_2}(t_1, t_2)\,dt_1\,dt_2 \\ &\quad= \int_{-\infty}^{\infty}\int_{-\infty}^{\infty} C_1t_1 f_{X_1,X_2}(t_1, t_2)\,dt_1\,dt_2 + \int_{-\infty}^{\infty}\int_{-\infty}^{\infty} C_2t_2 f_{X_1,X_2}(t_1, t_2)\,dt_1\,dt_2 \\ &\quad= C_1\int_{-\infty}^{\infty} t_1\,dt_1 \int_{-\infty}^{\infty} f_{X_1,X_2}(t_1, t_2)\,dt_2 + C_2\int_{-\infty}^{\infty} t_2\,dt_2 \int_{-\infty}^{\infty} f_{X_1,X_2}(t_1, t_2)\,dt_1 \\ &\quad= C_1\int_{-\infty}^{\infty} t_1 f_{X_1}(t_1)\,dt_1 + C_2\int_{-\infty}^{\infty} t_2 f_{X_2}(t_2)\,dt_2 \\ &\quad= C_1E(X_1) + C_2E(X_2) \end{aligned}$$

and the proof is complete. In general, *if $C_1, \ldots, C_n$ are n arbitrary constants and $X_1, \ldots, X_n$ constitute an arbitrary set of n variates, then*

$$E(C_1X_1 + C_2X_2 + \cdots + C_nX_n) = C_1E(X_1) + C_2E(X_2) + \cdots + C_nE(X_n)$$

A *special case* of Theorem 5.2 is when $C_1 = C_2 = \cdots = C_n = 1$; then

$$E(X_1 + X_2 + \cdots + X_n) = E(X_1) + E(X_2) + \cdots + E(X_n) \tag{5.8}$$

Example 5.8 X_1 and X_2 are two exponentially distributed random variables; their density functions are

$$f_{X_1}(t_1) = 2e^{-2t_1} \qquad t_1 \geq 0 \qquad f_{X_2}(t_2) = 4e^{-4t_2} \qquad t_2 \geq 0$$

Compute $E(X_1 + X_2)$.

Solution. By Theorem 5.2,

$$\begin{aligned} E(X_1 + X_2) &= E(X_1) + E(X_2) \\ &= \int_0^\infty 2t_1 e^{-2t_1}\,dt_1 + \int_0^\infty 4t_2 e^{-4t_2}\,dt_2 \\ &= \tfrac{1}{2} + \tfrac{1}{4} = \tfrac{3}{4} \end{aligned}$$

Theorem 5.3. *If two random variables X_1 and X_2 are independent, then*

$$E(X_1 X_2) = E(X_1)E(X_2) \tag{5.9}$$

Proof. By definition,

$$E(X_1 X_2) = \int_{-\infty}^{\infty}\int_{-\infty}^{\infty} t_1 t_2 f_{X_1, X_2}(t_1, t_2)\,dt_1\,dt_2$$

Since X_1 and X_2 are independent, we have

$$f_{X_1, X_2}(t_1, t_2) = f_{X_1}(t_1) f_{X_2}(t_2)$$

Hence

$$\begin{aligned} E(X_1 X_2) &= \int_{-\infty}^{\infty}\int_{-\infty}^{\infty} t_1 t_2 f_{X_1}(t_1) f_{X_2}(t_2)\,dt_1\,dt_2 \\ &= \int_{-\infty}^{\infty} t_1 f_{X_1}(t_1)\,dt_1 \int_{-\infty}^{\infty} t_2 f_{X_2}(t_2)\,dt_2 \\ &= E(X_1)E(X_2) \end{aligned}$$

The converse of this theorem is *not* true. In words, if $E(X_1 X_2) = E(X_1)E(X_2)$, we *cannot* conclude that X_1 and X_2 are independent.

Example 5.9 X_1 and X_2 are two independent random variables; their density functions are

$$\begin{aligned} f_{X_1}(t_1) &= 2t_1 \quad && 0 \le t_1 \le 1 \qquad & f_{X_2}(t_2) &= e^{-(t_2-5)} \quad && t_2 > 5 \\ &= 0 && \text{otherwise} & &= 0 && \text{otherwise} \end{aligned}$$

Compute $E(X_1 X_2)$.

Solution. By Theorem 5.3,

$$\begin{aligned} E(X_1 X_2) &= E(X_1)E(X_2) \\ &= \int_0^1 2t_1^2\,dt_1 \cdot \int_5^\infty t_2 e^{-(t_2-5)}\,dt_2 \\ &= (\tfrac{2}{3}) \cdot (6) = 4 \end{aligned}$$

5.3 Conditional Expectation

Definition 5.4. The *conditional expectation* of a random variable X, given Y, is defined as follows:

$$(5.10)\qquad E(X \mid Y = y) = \begin{cases} \displaystyle\int_{-\infty}^{\infty} x f_{X|Y}(x \mid y)\, dx & \text{for continuous variates} \\ \displaystyle\sum_{\text{over all } x} x P_{X|Y}(x \mid y) & \text{for discrete variates} \end{cases}$$

If the conditional expectation of *X, given Y, is known we can derive the expectation of X.*

Theorem 5.4. *If X and Y are two random variables, then*

$$(5.11)\qquad E(X) = \begin{cases} \displaystyle\int_{-\infty}^{\infty} E(X \mid Y = y) f_Y(y)\, dy & \text{for continuous variates} \\ \displaystyle\sum_{\text{over all } y} E(X \mid Y = y) P_Y(y) & \text{for discrete variates} \end{cases}$$

Proof (*for the discrete case*)

$$\begin{aligned} \sum_y E(X \mid Y = y) P_Y(y) &= \sum_y \Big[\sum_x x P_{X|Y}(x \mid y)\Big] P_Y(y) \\ &= \sum_y \sum_x x P_{X|Y}(x \mid y) P_Y(y) \\ &= \sum_y \sum_x x P_{X,Y}(x, y) \\ &= \sum_x x\Big[\sum_y P_{X,Y}(x, y)\Big] \\ &= \sum_x x P_X(x) = E(X) \end{aligned}$$

Example 5.10 A process under the surveillance of an $\bar{X}$ chart is subject to the occurrence of an assignable cause of variation. The density function of the occurrence of the assignable cause is given by

$$\begin{aligned} f_X(t) &= \lambda e^{-\lambda t} \qquad \lambda > 0 \qquad t \geq 0 \\ &= 0 \qquad \text{otherwise} \end{aligned}$$

If samples are taken at intervals of h hr., find the mean time of occurrence of the assignable cause within an interval between samples.

Solution. Suppose the assignable cause has occurred in the interval between the ith and $i + 1$st sample, or between ih and $(i + 1)h$ hr. Let A denote this event and X denote the mean time of occurrence within an intersample interval; then

$$\begin{aligned} E(X \mid A) &= \frac{\displaystyle\int_{ih}^{(i+1)h} \lambda e^{-\lambda t}(t - ih)\, dt}{P(A)} \\ &= \frac{\displaystyle\int_{ih}^{(i+1)h} \lambda e^{-\lambda t}(t - ih)\, dt}{\displaystyle\int_{ih}^{(i+1)h} \lambda e^{-\lambda t} dt} \\ &= \frac{ihe^{-\lambda ih} - (i + 1)he^{-\lambda(i+1)h} + \displaystyle\int_{ih}^{(i+1)h} e^{-\lambda t}\, dt - ih \int_{ih}^{(i+1)h} \lambda e^{-\lambda t}\, dt}{\displaystyle\int_{ih}^{(i+1)h} \lambda e^{-\lambda t} dt} \\ &= \frac{1 - e^{-\lambda h}(1 + \lambda h)}{\lambda(1 - e^{-\lambda h})} \end{aligned}$$

5.4 Measures of Variation

Once the measure of central tendency is chosen, information about the *variability* of the variate under study seems essential. Though the median, mode, or the mean indicates the general location of the distribution, it gives us no information about the *spread* of the distribution around the chosen measure of location. Intuitively we may feel that the expected deviation of each possible value of the variate considered from its expected value is an appropriate measure of its variability. Now, is $E[X - E(X)]$ an appropriate measure of variability of variate X? Unfortunately not, simply because $E[X - E(X)]$ is equal to zero since some of these deviations are positive and some are negative. Actually we care about the *magnitudes* of these deviations irrespective of their algebraic signs; in other words, how far x_i is from $E(X)$. Hence, if we square $[x_i - E(X)]$ and compute the weighted average of these squared deviations we obtain an appropriate measure of variability called the *variance* (or dispersion), which is defined as follows:

Definition 5.5. Variance: The *variance* (denoted by Var) of a random variable X with mean $E(X)$ is given by

$$\text{Var}(X) = E[X - E(X)]^2 \tag{5.12}$$

The right-hand side is the second moment around the mean. Hence, the variance in probability theory plays the same role as the moment of inertia in mechanics. The variance of any random variable is always positive, whereas the mean may be positive or negative. When the variance is equal to zero, it means that the probability distribution is concentrated at one point, the mean. The variance of variate X (continuous or discrete) is expressed as follows:

$$\text{Var}(X) = \begin{cases} E[X - E(X)]^2 = \displaystyle\int_{-\infty}^{\infty} [t - E(X)]^2 f_X(t)\,dt \\ \qquad\qquad \text{for continuous variate} \\ \qquad = \displaystyle\sum_{\text{over all } t} [t - E(X)]^2 P_X(t) \\ \qquad\qquad \text{for discrete variate} \end{cases} \tag{5.13}$$

If $Y = G(X)$, then $\text{Var}(Y)$ is given by

$$\text{Var}(Y) = \begin{cases} \displaystyle\int_{-\infty}^{\infty} [t - E(Y)]^2 f_Y(t)\,dt = \int_{-\infty}^{\infty} [Y(t) - E(Y)]^2 f_X(t)\,dt \\ \qquad\qquad \text{if both } Y \text{ and } X \text{ are continuous} \\ \displaystyle\sum_{\text{over all } t} [t - E(Y)]^2 P_Y(t) = \sum_{\text{over all } t} [Y(t) - E(Y)]^2 P_X(t) \\ \qquad\qquad \text{if both } Y \text{ and } X \text{ are discrete} \end{cases} \tag{5.14}$$

Definition 5.6. The quantity $\sqrt{\text{Var}(X)}$ is called the *standard deviation* of variate X and is denoted by σ_X. Hence, by definition,

$$\sigma_X = \sqrt{\text{Var}(X)} = \sqrt{E[X - E(X)]^2} \tag{5.15}$$

σ_X is a nonnegative number and plays the same role as the radius of gyration in mechanics. It has the same physical units as the mean of the distribution.

Example 5.11 Random variable X has a Poisson distribution, i.e.,

$$P_X(k) = \frac{e^{-\lambda}\lambda^k}{k!} \qquad \lambda > 0 \qquad k = 0, 1, \ldots$$

Compute $\text{Var}(X)$.

Solution. By definition,

$$\text{Var}(X) = \sum_{k=0}^{\infty} [k - E(X)]^2 \frac{e^{-\lambda}\lambda^k}{k!}$$

To evaluate $\text{Var}(X)$ we need to compute $E(X)$ first. Hence

$$E(X) = \sum_{k=0}^{\infty} \frac{ke^{-\lambda}\lambda^k}{k!} = e^{-\lambda} \sum_{k=0}^{\infty} \frac{\lambda^k}{(k-1)!} = \lambda e^{-\lambda} \sum_{k=0}^{\infty} \frac{\lambda^{k-1}}{(k-1)!}$$

Setting $k - 1 = r$, we obtain

$$E(X) = \lambda e^{-\lambda} \sum_{r=0}^{\infty} \frac{\lambda^r}{r!} = \lambda e^{-\lambda} e^{\lambda} = \lambda$$

By Definition (5.13),

$$\begin{aligned}
\text{Var}(X) &= \sum_{k=0}^{\infty} (k - \lambda)^2 \frac{e^{-\lambda}\lambda^k}{k!} \\
&= \sum_{k=0}^{\infty} \frac{k^2 e^{-\lambda}\lambda^k}{k!} + \lambda^2 - 2\lambda \sum_{k=0}^{\infty} \frac{ke^{-\lambda}\lambda^k}{k!} \\
&= \sum_{k=1}^{\infty} \frac{k^2 e^{-\lambda}\lambda^k}{k!} - \lambda^2 \\
&= \sum_{k=1}^{\infty} \frac{ke^{-\lambda}\lambda^k}{(k-1)!} - \lambda^2 \\
&= \sum_{k=2}^{\infty} \frac{(k-1)e^{-\lambda}\lambda^k}{(k-1)!} + \sum_{k=1}^{\infty} \frac{e^{-\lambda}\lambda^k}{(k-1)!} - \lambda^2 \\
&= e^{-\lambda}\lambda^2 \sum_{k=2}^{\infty} \frac{\lambda^{k-2}}{(k-2)!} + \sum_{k=1}^{\infty} \frac{ke^{-\lambda}\lambda^k}{k!} - \lambda^2 \\
&= \lambda^2 + \lambda - \lambda^2 = \lambda
\end{aligned}$$

Example 5.12 The density function of a random variable X is given by

$$\begin{aligned}
f_X(t) &= \tfrac{1}{2} \qquad 0 < t < 2 \\
&= 0 \qquad \text{otherwise}
\end{aligned}$$

Compute the following:

(a) $E(X)$

(b) $\text{Var}(X)$, σ_X

(c) $\text{Var}(Y)$, where $Y = 2X - 1$.

Solution. By definition, we have

(a) $E(X) = \int_0^2 \frac{1}{2}t\,dt = 1$

(b) $\text{Var}(X) = \int_0^2 \frac{1}{2}(t-1)^2\,dt = \frac{1}{2}\int_0^2 (t^2 - 2t + 1)\,dt = \frac{1}{3}$

Hence

$$\sigma_X = \frac{1}{\sqrt{3}}$$

(c) $F_Y(a) = P\{Y \leq a\} = P\{2X - 1 \leq a\} = P\left\{X \leq \frac{1+a}{2}\right\} = F_X\left(\frac{1+a}{2}\right)$

The cumulative distribution of X is

$$F_X(t) = \begin{cases} 0 & t \leq 0 \\ \frac{1}{2}t & 0 < t < 2 \\ 1 & t \geq 2 \end{cases}$$

Hence

$$F_Y(a) = \begin{cases} 0 & a \leq -1 \\ \dfrac{1+a}{4} & -1 < a < 3 \\ 1 & a \geq 3 \end{cases}$$

and

$$\begin{aligned} f_Y(a) &= \tfrac{1}{4} && -1 < a < 3 \\ &= 0 && \text{otherwise} \end{aligned}$$

The expected value and variance of Y, then, are

$$E(Y) = \int_{-1}^{3} \tfrac{1}{4}a\,da = 1$$

$$\text{Var}(Y) = \int_{-1}^{3} (a-1)^2 \tfrac{1}{4}\,da = \tfrac{1}{4}\int_{-1}^{3} (a^2 - 2a + 1)\,da = 1\tfrac{1}{3}$$

The same result can be obtained directly by applying the definition of the variance of a function of a random variable; we write

$$\begin{aligned} \text{Var}(Y) &= \int_{-\infty}^{\infty} [Y(a) - E(Y)]^2 f_X(a)\,da \\ &= \int_0^2 [(2a-1) - 1]^2 \tfrac{1}{2}\,da \\ &= 2\int_0^2 (a^2 - 2a + 1)\,da = 1\tfrac{1}{3} \end{aligned}$$

which again agrees with our previous finding.

5.5 Properties of the Variance

Theorem 5.5. *If C is an arbitrary constant, then*

(5.16) $$\text{Var}(CX) = C^2\,\text{Var}(X)$$

Proof. By definition,

$$\begin{aligned}\operatorname{Var}(CX) &= E[CX - CE(X)]^2 \\ &= E\{C^2[X - E(X)]^2\} \\ &= C^2 E[X - E(X)]^2 \\ &= C^2 \operatorname{Var}(X)\end{aligned}$$

Theorem 5.6. *If X is a random variable, then*

$$\operatorname{Var}(X) = E(X^2) - [E(X)]^2$$

Proof. By definition,

$$\begin{aligned}\operatorname{Var}(X) &= E[X - E(X)]^2 \\ &= E\{X^2 + [E(X)]^2 - 2XE(X)\} \\ &= E(X^2) + [E(X)]^2 - 2[E(X)]^2 \\ &= E(X^2) - [E(X)]^2\end{aligned}$$

Theorem 5.7. *If X and Y are independent random variables, then*

$$\operatorname{Var}(X + Y) = \operatorname{Var}(X) + \operatorname{Var}(Y) \tag{5.17}$$

Proof

$$\begin{aligned}\operatorname{Var}(X + Y) &= E[X + Y - E(X) - E(Y)]^2 \\ &= E\{[X - E(X)]^2 + [Y - E(Y)]^2 \\ &\quad + 2[X - E(X)][Y - E(Y)]\} \\ &= E[X - E(X)]^2 + E[Y - E(Y)]^2 \\ &\quad + 2E\{[X - E(X)][Y - E(Y)]\} \\ &= \operatorname{Var}(X) + \operatorname{Var}(Y) \\ &\quad + 2E[X - E(X)]E[Y - E(Y)]\end{aligned}$$

The last term is equal to zero; hence

$$\operatorname{Var}(X + Y) = \operatorname{Var}(X) + \operatorname{Var}(Y)$$

Example 5.13 X and Y are two independent random variables, where

$$E(X) = E(Y) = 0 \quad \text{and} \quad \operatorname{Var}(X) = \operatorname{Var}(Y) = 1$$

Compute $E(X + Y)^2$.

Solution

$$\begin{aligned}E(X + Y)^2 &= E(X^2 + Y^2 + 2XY) \\ &= E(X^2) + E(Y^2) + 2E(X)E(Y) \\ &= E(X^2) + E(Y^2) \\ &= \operatorname{Var}(X) + [E(X)]^2 + \operatorname{Var}(Y) + [E(Y)]^2 \\ &= 1 + 0 + 1 + 0 \\ &= 2\end{aligned}$$

Example 5.14 A measurable characteristic is a random variable with mean μ and standard deviation σ. A sample of size n_1 is drawn with replacement. Let $\bar{X}_1$ denote the mean of the sample. Compute $\operatorname{Var}(\bar{X}_1)$. Suppose a second sample of size n_2 is drawn with replacement, and its sample mean is $\bar{X}_2$. Find $\operatorname{Var}(\bar{X}_1 - \bar{X}_2)$.

Solution. By definition,

$$\bar{X}_1 = \frac{1}{n_1}\sum_{i=1}^{n_1} X_i = \frac{1}{n_1}(X_1 + X_2 + \cdots + X_{n_1})$$

Then

$$\begin{aligned}\operatorname{Var}(\bar{X}_1) &= \frac{1}{n_1^2}[\operatorname{Var}(X_1) + \operatorname{Var}(X_2) + \cdots + \operatorname{Var}(X_{n_1})] \\ &= \frac{1}{n_1^2}[n_1 \operatorname{Var}(X)] \\ &= \frac{\sigma_X^2}{n_1}\end{aligned}$$

Similarly,

$$\operatorname{Var}(\bar{X}_2) = \frac{\sigma_X^2}{n_2}$$

Hence, by Theorem 5.7,

$$\begin{aligned}\operatorname{Var}(\bar{X}_1 - \bar{X}_2) &= \operatorname{Var}(\bar{X}_1) + \operatorname{Var}(\bar{X}_2) \\ &= \sigma_X^2\left(\frac{1}{n_1} + \frac{1}{n_2}\right)\end{aligned}$$

5.6 Covariance and Correlation

In Theorem 5.7 we proved that

$$\operatorname{Var}(X + Y) = \operatorname{Var}(X) + \operatorname{Var}(Y) + 2E\{[X - E(X)][Y - E(Y)]\}$$

It should be noted that if X and Y are independent the term $E\{[X - E(X)]\cdot[Y - E(Y)]\}$ *is equal to zero*. This term is called the *covariance* of the two random variables X and Y. It is a useful tool in studying the interrelationships among random variables.

Definition 5.7. If X and Y are two random variables, then the *covariance* of X and Y, denoted by $\operatorname{Cov}(X, Y)$, is given by

$$(5.18) \qquad \operatorname{Cov}(X, Y) = \begin{cases} E\{[X - E(X)][Y - E(Y)]\} \\ \displaystyle\int_{-\infty}^{\infty}\int_{-\infty}^{\infty} [u - E(X)][v - E(Y)]f_{X,Y}(u, v)\,du\,dv & \text{for continuous variates} \\ \displaystyle\sum_{\text{over all } u,v} [u - E(X)][v - E(Y)]P_{X,Y}(u, v) & \text{for discrete variates}\end{cases}$$

The covariance of two random variables can be expressed in a more convenient form. This is given in the form of a theorem:

Theorem 5.8. *If X and Y are two random variables, then*

$$\text{Cov}(X, Y) = E(XY) - E(X)E(Y) \tag{5.19}$$

Proof. By definition,

$$\begin{aligned}\text{Cov}(X, Y) &= E\{[X - E(X)][Y - E(Y)]\} \\ &= E[XY - XE(Y) - E(X)Y + E(X)E(Y)] \\ &= E(XY) - E(X)E(Y) - E(X)E(Y) + E(X)E(Y) \\ &= E(XY) - E(X)E(Y)\end{aligned}$$

If X and Y are independent, the right-hand side of Eq. (5.19) is zero. Hence, we conclude that *the covariance of two independent random variables is zero.* The converse is *not* true; namely, two random variables with *zero covariance* need *not* be independent. Having introduced the term covariance, we can write

$$\text{Var}(X + Y) = \text{Var}(X) + \text{Var}(Y) + 2\,\text{Cov}(X, Y) \tag{5.20}$$

The above expression can be extended to n arbitrary random variables and leads to the following theorem:

Theorem 5.9. *If $X_1, X_2, \ldots, X_n$ constitute an arbitrary set of n random variables, then*

$$\text{Var}(X_1 + X_2 + \cdots + X_n) = \sum_{i=1}^{n} \text{Var}(X_i) + 2 \sum_{k=i+1}^{n} \sum_{i=1}^{n-1} \text{Cov}(X_i, X_k) \tag{5.21}$$

Proof. By definition,

$$\begin{aligned}\text{Var}\left(\sum_{i=1}^{n} X_i\right) &= E\left[\sum_{i=1}^{n} (X_i) - E\left(\sum_{i=1}^{n} X_i\right)\right]^2 \\ &= E\left[\sum_{i=1}^{n} (X_i) - \sum_{i=1}^{n} E(X_i)\right]^2 \\ &= E\left\{\sum_{i=1}^{n} [(X_i) - E(X_i)]\right\}^2 \\ &= E\left\{\sum_{i=1}^{n} [(X_i) - E(X_i)]^2 + 2 \sum_{k=i+1}^{n} \sum_{i=1}^{n-1} [X_i - E(X_i)][X_k - E(X_k)]\right\} \\ &= \sum_{i=1}^{n} E[(X_i) - E(X_i)]^2 + 2 \sum_{k=i+1}^{n} \sum_{i=1}^{n-1} E\{[X_i - E(X_i)][X_k - E(X_k)]\} \\ &= \sum_{i=1}^{n} \text{Var}(X_i) + 2 \sum_{k=i+1}^{n} \sum_{i=1}^{n-1} \text{Cov}(X_i, X_k)\end{aligned}$$

Theorem 5.10. *If a and b are arbitrary constants and X and Y are random variables, then*

$$\text{Cov}(aX, bY) = ab\,\text{Cov}(X, Y) \tag{5.22}$$

Proof

$$\begin{aligned}\mathrm{Cov}(aX, bY) &= E(aXbY) - E(aX)E(bY)\\ &= ab\, E(XY) - abE(X)E(Y)\\ &= ab[E(XY) - E(X)E(Y)]\\ &= ab\,\mathrm{Cov}(X, Y)\end{aligned}$$

Definition 5.8. The *correlation coefficient* between two random variables X_1 and X_2 is defined by

$$\rho_{X_1,X_2} = \frac{\mathrm{Cov}(X_1, X_2)}{\sigma_{X_1}\sigma_{X_2}} \tag{5.23}$$

where

$$-1 \leq \rho_{X_1,X_2} \leq 1$$

If the relationship between X_1 and X_2 is *linear* and given by

$$X_1 = \theta_1 + \theta_2 X_2$$

then the correlation coefficient $\rho_{X_1,X_2} = 1$, whereas if the linear relationship between X_1 and X_2 is given by

$$X_1 = \theta_1 - \theta_2 X_2$$

the correlation coefficient $\rho_{X_1,X_2} = -1$. The converse of the above is true. That is,

$$X_1 = \theta_1 + \theta_2 X_2 \quad \text{if} \quad \rho_{X_1,X_2} = 1$$

and

$$X_2 = \theta_1 - \theta_2 X_2 \quad \text{if} \quad \rho_{X_1,X_2} = -1$$

Proof. Let X_1^* and X_2^* denote the standardized random variables corresponding to X_1 and X_2; i.e.,

$$X_1^* = \frac{X_1 - E(X_1)}{\sigma_{X_1}} \qquad X_2^* = \frac{X_1 - E(X_2)}{\sigma_{X_2}}$$

Hence

$$\mathrm{Var}(X_1^*) = \mathrm{Var}(X_2^*) = 1$$

and

$$\mathrm{Cov}(X_1^*, X_2^*) = \mathrm{Cov}\left[\frac{X_1 - E(X_1)}{\sigma_{X_1}}, \frac{X_2 - E(X_2)}{\sigma_{X_2}}\right]$$

Now, by Theorem 5.10,

$$\begin{aligned}\mathrm{Cov}(X_1^*, X_2^*) &= \frac{1}{\sigma_{X_1}\sigma_{X_2}}\mathrm{Cov}[X_1 - E(X_1), X_2 - E(X_2)]\\ &= \frac{1}{\sigma_{X_1}\sigma_{X_2}}\mathrm{Cov}(X_1, X_2)\\ &= \rho_{X_1,X_2}\end{aligned}$$

By Theorem 5.9, we have

$$\begin{aligned}\operatorname{Var}(X_1^* \pm X_2^*) &= \operatorname{Var}(X_1^*) + \operatorname{Var}(X_2^*) \pm 2\operatorname{Cov}(X_1^*, X_2^*)\\ &= 1 + 1 \pm 2\rho_{X_1,X_2} = 2(1 \pm \rho_{X_1,X_2})\end{aligned}$$

Since the left-hand side cannot be negative, we conclude that $-1 \le \rho_{X_1,X_2} \le 1$. Suppose now $\rho_{X_1,X_2} = 1$; then it is necessary that $\operatorname{Var}(X_1^* - X_2^*) = 0$, and it follows that $X_1^* - X_2^*$ has one value that occurs with unit probability. Thus we conclude that $X_1^* = X_2^*$ or, equivalently,

$$\frac{X_1 - E(X_1)}{\sigma_{X_1}} = \frac{X_2 - E(X_2)}{\sigma_{X_2}}$$

Hence

$$\begin{aligned}X_1 &= X_2\frac{\sigma_{X_1}}{\sigma_{X_2}} + E(X_1) - E(X_2)\frac{\sigma_{X_1}}{\sigma_{X_2}}\\ &= \theta_1 + \theta_2 X_2 \qquad \theta_2 > 0\end{aligned}$$

where

$$\frac{\sigma_{X_1}}{\sigma_{X_2}} = \theta_2 \quad \text{and} \quad E(X_1) - \theta_2 E(X_2) = \theta_1$$

Likewise, we can prove that

$$X_1 = \theta_1 - \theta_2 X_2 \quad \text{if} \quad \rho_{X_1,X_2} = -1$$

It is important to realize that *when two random variables are independent their correlation coefficient must be equal to zero*. The converse of this statement is *not* true. In fact, the correlation coefficient *can* be zero and the two random variables need *not* be independent.

Example 5.15 The probability functions of random variables X_1 and X_2 are as follows:

$$P_{X_1}(k_1) = \begin{cases}\frac{1}{4} & k_1 = 1\\ \frac{1}{8} & k_1 = 2\\ \frac{1}{2} & k_1 = 3\\ \frac{1}{8} & k_1 = 4\end{cases} \qquad P_{X_2}(2k_1) = \begin{cases}\frac{1}{4} & k_2 = 2\\ \frac{1}{8} & k_2 = 4\\ \frac{1}{2} & k_2 = 6\\ \frac{1}{8} & k_2 = 8\end{cases}$$

(a) Compute $\operatorname{Cov}(X_1, X_2)$.
(b) Compute ρ_{X_1,X_2}.

Solution. By definition, we obtain

$$\begin{aligned}E(X_1) &= (\tfrac{1}{4})(1) + (\tfrac{1}{8})(2) + (\tfrac{1}{2})(3) + (\tfrac{1}{8})(4) = 2\tfrac{1}{2}\\ E(X_2) &= (\tfrac{1}{4})(2) + (\tfrac{1}{8})(4) + (\tfrac{1}{2})(6) + (\tfrac{1}{8})(8) = 5\\ \operatorname{Var}(X_1) &= (1.5)^2(\tfrac{1}{4}) + (\tfrac{1}{2})^2(\tfrac{1}{8}) + (\tfrac{1}{2})^2(\tfrac{1}{2}) + (1.5)^2(\tfrac{1}{8}) = 1\\ \operatorname{Var}(X_2) &= (3)^2(\tfrac{1}{4}) + (1)^2(\tfrac{1}{8}) + (1)^2(\tfrac{1}{2}) + (3)^2(\tfrac{1}{8}) = 4\\ \operatorname{Cov}(X_1, X_2) &= E\{[X_1 - E(X_1)][X_2 - E(X_2)]\}\end{aligned}$$

$$= (1 - 2.5)(2 - 5)\tfrac{1}{4} + (2 - 2.5)(4 - 5)(\tfrac{1}{8}) + (3 - 2.5)(6 - 5)\tfrac{1}{2} + (4 - 2.5)(8 - 5)(\tfrac{1}{8}) = 2$$

$$\rho_{X_1, X_2} = \frac{\mathrm{Cov}(X_1, X_2)}{\sigma_{X_1}\sigma_{X_2}} = \frac{2}{1 \times 2} = 1$$

Hence, we conclude that the relationship between X_1 and X_2 is linear. This is easily verified from their respective probability functions. In fact, $X_2 = 2X_1$.

Example 5.16 The joint density function of random variables X_1 and X_2 is

$$f_{X_1, X_2}(u_1, u_2) = \tfrac{1}{8}(u_1 + u_2) \qquad 0 \le u_1 \le 2 \qquad 0 \le u_2 \le 2$$
$$= 0 \qquad \text{otherwise}$$

Compute $\mathrm{Cov}(X_1, X_2)$, ρ_{X_1, X_2}.

Solution. By definition,

$$E(X_1, X_2) = \int_{-\infty}^{\infty}\int_{-\infty}^{\infty} \frac{u_1u_2}{8}(u_1 + u_2)\, du_1\, du_2 = \int_0^2\int_0^2 \tfrac{1}{8}(u_1^2u_2 + u_1u_2^2)\, du_1\, du_2 = \int_0^2 (\tfrac{1}{3}u_2 + \tfrac{1}{4}u_2^2)\, du_2 = 1\tfrac{1}{3}$$

$$E(X_1) = \int_0^2\int_0^2 \frac{u_1}{8}(u_1 + u_2)\, du_1\, du_2 = \tfrac{7}{6}$$

$$E(X_2) = \int_0^2\int_0^2 \frac{u_2}{8}(u_1 + u_2)\, du_1\, du_2 = \tfrac{7}{6}$$

Consequently,

$$\mathrm{Cov}(X_1, X_2) = E(X_1X_2) - E(X_1)E(X_2) = \tfrac{4}{3} - \tfrac{7}{6}\cdot\tfrac{7}{6} = -\tfrac{1}{48}$$

However,

$$E(X_1^2) = \int_0^2\int_0^2 \frac{u_1^2}{8}(u_1 + u_2)\, du_1\, du_2 = 1\tfrac{2}{3}$$

and

$$E(X_2^2) = \int_0^2\int_0^2 \frac{u_2^2}{8}(u_1 + u_2)\, du_1\, du_2 = 1\tfrac{2}{3}$$

Hence

$$\sigma_{X_1}^2 = E(X_1^2) - [E(X_1)]^2 = 1\tfrac{2}{3} - \tfrac{49}{36} = \tfrac{11}{36}$$

Similarly,

$$\sigma_{X_2}^2 = \tfrac{11}{36}$$

Hence

$$\rho_{X_1, X_2} = \frac{\mathrm{Cov}(X_1, X_2)}{\sigma_{X_1}\sigma_{X_2}} = -\frac{1/48}{11/36} = -\tfrac{3}{44}$$

5.7 Chebyshev's Inequality

For many purposes one desires to make a probability statement regarding the absolute deviation of a random variable from its expected value. For example, if we know the expected demand for a certain commodity we may be interested in finding the probability that the random demand is less or more than the expected demand by a certain quantity. In other words, we wish to determine the probability

$$P\{|X - E(X)| > C\}$$

where C is a given positive number. Then

$$\begin{aligned} P\{|X - E(X)| > C\} &= 1 - P\{|X - E(X)| \leq C\} \\ &= 1 - P\{-C \leq X - E(X) \leq C\} \\ &= 1 - P\{E(X) - C \leq X \leq C + E(X)\} \\ &= 1 - F_X[C + E(X)] + F_X[E(X) - C] \end{aligned}$$

The right-hand side can be evaluated if and only if the distribution of variate X is known. Suppose we have no information whatsoever on the distribution of the variate under study but we know its variance; then, by *Chebyshev's inequality*, we can determine an upper bound for the left-hand side.

Theorem 5.11. *Chebyshev's inequality asserts that*

$$P\{|X - E(X)| > C\} \leq \frac{\text{Var}(X)}{C^2} \qquad (C > 0) \tag{5.24}$$

Proof (continuous case)

$$\begin{aligned} \text{Var}(X) &= \int_{-\infty}^{\infty} [u - E(X)]^2 f_X(u)\, du \\ &\geq \int_{-\infty}^{E(X)-C} [u - E(X)]^2 f_X(u)\, du + \int_{E(X)+C}^{\infty} [u - E(X)]^2 f_X(u)\, du \\ &\geq C^2 \int_{-\infty}^{E(X)-C} f_X(u)\, du + C^2 \int_{E(X)+C}^{\infty} f_X(u)\, du \end{aligned}$$

Hence

$$\begin{aligned} \frac{\text{Var}(X)}{C^2} &\geq F_X[E(X) - C] + 1 - F_X[E(X) + C] \\ &\geq P\{|X - E(X)| > C\} \end{aligned}$$

This completes the proof for the continuous case. In a similar way the discrete case can be proved. An alternate form of Chebyshev's inequality is obtained by replacing C by $k\sigma$. Thus we obtain

$$P\{|X - E(X)| > k\sigma\} \leq \frac{1}{k^2} \qquad (k > 0) \tag{5.25}$$

or, equivalently,

$$P\{|X - E(X)| \leq k\sigma\} \geq 1 - \frac{1}{k^2} \qquad (k > 0) \tag{5.26}$$

Needless to say, the upper bound, given by (5.25), is approximate. However, we can improve this approximation when more information is known about the variate under study. For example, if we know that the distribution is symmetrical the above inequality becomes

$$P\{|X - E(X)| > k\sigma\} \leq \frac{1}{2k^2} \qquad (k > 0) \tag{5.27}$$

Example 5.17 The distribution of demand for a certain commodity is unknown; it is known, however, that the mean demand is 100 and that the standard deviation is 5 units. Find the probability that the demand is greater than or equal to 85 but less than or equal to 115 units.

Solution. Let X denote the random demand. Then

$$\begin{aligned} P\{85 \leq X \leq 115\} &= P\{-15 \leq X - E(X) \leq 15\} \\ &= P\{|X - E(X)| \leq 15\} \\ &= P\{|X - E(X)| \leq 3\sigma\} \geq 1 - \tfrac{1}{9} = \tfrac{8}{9} \end{aligned}$$

Suppose we know that the demand distribution is symmetrical around the mean. This probability becomes

$$P\{|X - E(X)| \leq 3\sigma\} \geq 1 - \tfrac{1}{18} = \tfrac{17}{18}$$

which is greater than $\frac{8}{9}$.

5.8 The Law of Large Numbers

Theorem 5.12. *If $X_1, X_2, \ldots, X_n$ are n independent random variables, each having the same expectation and the same variance, in symbols,*

$$E(X_1) = E(X_2) = \cdots = E(X_n) = \mu$$
$$\operatorname{Var}(X_1) = \operatorname{Var}(X_2) = \cdots = \operatorname{Var}(X_n) = \sigma^2$$

and if we define a new random variable $\bar{X}$ (called the arithmetic mean) by the equation

$$\bar{X} = \frac{1}{n} \sum_{i=1}^{n} X_i$$

then, for any constant $\epsilon > 0$, we have

$$\lim_{n \to \infty} P\{|\bar{X} - \mu| > \epsilon\} = 0$$

In other words, the probability that the arithmetic mean ($\bar{X}$) is very close to the population mean (μ) can be made close to one when n is sufficiently large.

Proof. By definition,

$$\bar{X} = \frac{1}{n}(X_1 + X_2 + \cdots + X_n)$$

Then

$$E(\bar{X}) = \frac{1}{n}[E(X_1) + E(X_2) + \cdots + E(X_n)] = \mu$$

We also have

$$\text{Var}(\bar{X}) = \frac{1}{n^2}[\text{Var}(X_1) + \text{Var}(X_2) + \cdots + \text{Var}(X_n)] = \frac{\sigma^2}{n}$$

By Chebyshev's inequality,

$$P\{|\bar{X} - E(\bar{X})| > \epsilon\} \le \frac{\text{Var}(\bar{X})}{\epsilon^2}$$

or, equivalently,

$$P\{|\bar{X} - \mu| > \epsilon\} \le \frac{\sigma^2}{n\epsilon^2}$$

On taking the limit we obtain

$$\lim_{n\to\infty} P\{|\bar{X} - \mu| > \epsilon\} = 0$$

Example 5.18 In sampling without replacement, what sample size n will ensure that the sample mean differs at most from the population mean by 5%, with probability greater than or equal to 90%. Assume the *coefficient of variation* is equal to $\frac{1}{10}$.

Solution. The coefficient of variation is the ratio of the standard deviation to the population mean (σ/μ). Then

$$P\{|\bar{X} - \mu| \le .05\mu\} \ge 1 - \frac{\text{Var}(\bar{X})}{(.05\mu)^2} = 1 - \frac{400\sigma^2}{n\mu^2}$$

Hence, $n \ge 40$.

5.9 Moment-Generating Function

We shall discuss in this section a method of determining the moments (that do exist) of a probability law. Before we proceed let us postulate the definition of moments:

Definition 5.9. If $G(X) = (X - C)^k$, $k = 1, 2, \ldots,$ then $E[G(X)] = E(X - C)^k$ is called the kth *moment* of the random variable X around the *constant* C. If $C = 0$, then $E(X^k)$ is called the kth moment around the *origin* or simply the kth *raw* moment. If $C = E(X)$, then $E[X - E(X)]^k$ is called the

kth moment around the mean or simply the kth *central* moment. Note that the second central moment is the variance of the variate.

By the method of the moment-generating function, we shall evaluate all raw moments, when they exist, of the random variable under study.

Definition 5.10. The *moment-generating function* of a random variable X [denoted by $\Psi_X(\theta)$] may be defined as the expected value of the function $e^{\theta x}$, where θ is an arbitrary real number. In symbols,

$$\Psi_X(\theta) = \begin{cases} E(e^{\theta x}) \\ \int_{-\infty}^{\infty} e^{\theta x} f(x)\,dx & \text{for continuous variate} \\ \sum_{\text{over all } x} e^{\theta x} p(x) & \text{for discrete variate} \end{cases} \tag{5.28}$$

$\Psi_X(\theta)$ may not always exist for all values of θ. In other words, $\Psi_X(\theta)$ can be either finite or infinite (does not exist). It follows that a probability law may not always possess a moment-generating function. In such cases one may define a function, called the *characteristic function*, that always exists for all values of θ. The characteristic function is defined by $E(e^{i\theta x})$, where $i = \sqrt{-1}$. However, we shall restrict our analysis to the moment-generating function. The interested reader may consult advanced works on probability for further explication of this topic. To simplify the analysis, we shall always assume that the moment-generating function exists. [Note that if the moment-generating function exists for all values of θ, in the interval $(-L, L)$ where $L > 0$, then all moments exist and the distribution of the variate is uniquely determined by it.] To demonstrate how to evaluate the raw moments of random variable X we expand $e^{\theta x}$ into a power series:

$$e^{\theta x} = 1 + \theta x + \frac{1}{2!}\theta^2 x^2 + \frac{1}{3!}\theta^3 x^3 + \cdots$$

Hence

$$\Psi_X(\theta) = E(e^{\theta x}) = 1 + \theta E(x) + \frac{\theta^2}{2!}E(x^2) + \frac{\theta^3}{3!}E(x^3) + \cdots$$

Differentiating both sides with respect to θ we obtain

$$\Psi'_X(\theta) = E(x) + \theta E(x^2) + \frac{3\theta^2}{3!}E(x^3) + \cdots$$

Now, letting $\theta = 0$, we obtain

$$\Psi'_X(0) = E(x) = \text{the first raw moment} = \text{the mean}$$

Similarly,

$$\Psi''_X(\theta) = E(x^2) + \theta E(x^3) + \cdots$$

which gives

$$\Psi''_X(0) = E(x^2) = \text{the second raw moment}$$

Similarly,

$$\Psi_X^{(3)}(0) = E(x^3) = \text{the third raw moment}$$

$$\vdots$$

$$\Psi_X^{(n)}(0) = E(x^n) = \text{the } n\text{th raw moment}$$

Note that one also can derive the central moments directly from the moment-generating function. In a straightforward way, the central moments are obtained by differentiating $\Psi_{X-E(X)}(\theta) = E\{e^{[x-E(x)]\theta}\}$ with respect to θ. Thus the kth central moment is

$$\frac{d^k\Psi_{X-E(X)}(0)}{d\theta^k} = E[X - E(X)]^k \tag{5.29}$$

The moment-generating function is a useful tool in studying derived distributions. Before demonstrating this application we need to mention some important properties of the moment-generating function.

SOME PROPERTIES OF THE MOMENT-GENERATING FUNCTION

Theorem 5.13. *Suppose that variate X has moment-generating function Ψ_X. If a new variate is given by $Y = aX + b$, where a and b are constants, then*

$$\Psi_Y(\theta) = e^{b\theta}\Psi_X(a\theta) \tag{5.30}$$

Proof. By definition,

$$\begin{aligned}\Psi_Y(\theta) = E(e^{y\theta}) &= E[e^{(ax+b)\theta}] \\ &= e^{b\theta}E(e^{ax\theta}) = e^{b\theta}\Psi_X(a\theta)\end{aligned}$$

Note that when $b = 0$, $\Psi_Y(\theta) = \Psi_X(a\theta)$

Theorem 5.14. *If $X_1, X_2, \ldots, X_n$ are n independent random variables and $a_1, a_2, \ldots, a_n$ are any constants, then*

$$\Psi_{a_1X_1+\cdots+a_nX_n}(\theta) = \Psi_{X_1}(a_1\theta)\Psi_{X_2}(a_2\theta)\cdots\Psi_{X_n}(a_n\theta) \tag{5.31}$$

Proof

$$\begin{aligned}\Psi_{a_1X_1+\cdots+a_nX_n}(\theta) &= E[e^{(a_1x_1+\cdots+a_nx_n)\theta}] \\ &= Ee^{(a_1x_1)}Ee^{(a_2x_2)}\cdots Ee^{(a_nx_n)} \\ &= \Psi_{X_1}(a_1\theta)\Psi_{X_2}(a_2\theta)\cdots\Psi_{X_n}(a_n\theta)\end{aligned}$$

The following two theorems, stated without proof, are very useful in studying *derived* distributions:

Theorem 5.15. *Let X_1 and X_2 be two random variables with moment-generating functions $\Psi_{X_1}(\theta)$ and $\Psi_{X_2}(\theta)$, respectively. If $\Psi_{X_1}(\theta) = \Psi_{X_2}(\theta)$ for all values of θ, then the distributions of X_1 and X_2 are identical.*

Theorem 5.15. is known as the *uniqueness theorem.*

Theorem 5.16. *Let $X_1, X_2, \ldots, X_n$ be a sequence of random variables with c.d.f.'s $F_{X_1}, F_{X_2}, \ldots, F_{X_n}$ and moment-generating functions $\Psi_{X_1}, \Psi_{X_2}, \ldots, \Psi_{X_n}$. Let F_X be a c.d.f. with moment-generating function $\Psi_X(\theta)$. Then under certain conditions, if $\Psi_{X_n}(\theta) \to \Psi_X(\theta)$, $(-L < \theta < L$, where $L > 0)$ $F_{X_n} \to F_X$.* In other words, *if a sequence of c.d.f.'s has a limit that is a c.d.f., then this c.d.f. is uniquely determined by the limit of the corresponding sequence of moment-generating functions.* The following examples will illustrate the usefulness of these theorems.

Example 5.19 Given that $\Psi_X(\theta) = 1/[1 - (\theta/\lambda)]$, find the distribution of X.

Solution. The series expansion of $\Psi_X(\theta)$ exists for every value of θ in the interval $(-L, L)$, where $L < \lambda$; hence, the moment-generating function uniquely determines the distribution. Accordingly, X must have an exponential distribution with parameter λ (see Table 6.3, Chapter 6). Hence

$$
\begin{aligned}
f_X(t) &= \lambda e^{-\lambda t} \qquad \lambda > 0 \qquad t \geq 0 \\
&= 0 \qquad \text{otherwise}
\end{aligned}
$$

Example 5.20 Suppose that $\Psi_X(\theta) = (q + pe^\theta)^n$, where random variable X takes on integral values $0, 1, \ldots, n$. Find the probability function of X.

Solution. The series expansion of $\Psi_X(\theta)$ exists for every value of θ in the interval $(-\infty, \infty)$; hence, the moment-generating function uniquely determines the distribution. Expanding $\Psi_X(\theta)$ in a series, we have

$$
\begin{aligned}
\Psi_X(\theta) &= q^n + nq^{n-1}pe^\theta + \cdots + (pe^\theta)^n \\
&= \sum_{k=0}^{n} \binom{n}{k} p^k e^{\theta k} q^{n-k}
\end{aligned}
$$

However, by definition,

$$\Psi_X(\theta) = \sum_{k=0}^{n} e^{\theta k} P_X(k)$$

Therefore, equating general terms we have

$$\sum_{k=0}^{n} P_X(k) = \sum_{k=0}^{n} \binom{n}{k} p^k q^{n-k}$$

or, equivalently,

$$P_X(k) = \binom{n}{k} p^k q^{n-k} \qquad k = 0, 1, \ldots, n$$

which is the probability function of the binomial distribution with parameters n and p.

Example 5.21 Random variable X is exponentially distributed, i.e.,

$$
\begin{aligned}
f_X(t) &= \lambda e^{-\lambda t} \qquad \lambda > 0 \qquad t \geq 0 \\
&= 0 \qquad \text{otherwise}
\end{aligned}
$$

Find the mean and the variance of X by the method of moment-generating functions.

Solution. The moment-generating function of this exponential distribution is

$$\Psi_X(\theta) = E(e^{\theta x}) = \int_0^\infty e^{\theta t}\lambda e^{-\lambda t}\,dt$$
$$= \frac{\lambda}{\lambda - \theta} = \left(1 - \frac{\theta}{\lambda}\right)^{-1}$$

This means, for the exponential distribution with parameter λ, the moment-generating function exists for $\theta < \lambda$. From the first derivative we obtain

$$\Psi'_X(\theta) = \frac{\lambda}{(\lambda - \theta)^2}$$

and

$$\Psi'_X(0) = \frac{\lambda}{\lambda^2} = \frac{1}{\lambda} = E(X)$$

From the second derivative we obtain

$$\Psi''_X(\theta) = \frac{2\lambda}{(\lambda - \theta)^3}$$

and

$$\Psi''_X(0) = \frac{2}{\lambda^2} = E(X^2)$$

Hence

$$\text{Var}(X) = E(X^2) - [E(X)]^2 = \frac{2}{\lambda^2} - \frac{1}{\lambda^2} = \frac{1}{\lambda^2}$$

Thus, for the exponential distribution the mean and the standard deviation are equal, i.e.,

$$E(X) = \sigma_X = \frac{1}{\lambda}$$

Example 5.22 Given the moment-generating function of a normally distributed random variable X (with parameters μ, σ),

$$\Psi_X(\theta) = e^{\mu\theta + \sigma^2\theta^2/2}$$

find the distribution of random variable $Y = 10X + 5$.

Solution. By Theorem 5.13, the moment-generating function of Y is

$$\Psi_Y(\theta) = e^{5\theta}\Psi_X(10\theta) = e^{5\theta}(e^{10\mu\theta + 50\sigma^2\theta^2})$$
$$= e^{(5+10\mu)\theta}e^{50\sigma^2\theta^2}$$

which is the moment-generating function of a normally distributed random variable with parameters $5 + 10\mu$ and $100\,\sigma^2$. Thus, according to Theorem 10.15, random variable Y is normally distributed.

Example 5.23 Two independent variates X and Y are distributed as follows:

$$f_X(t_1) = t_1 e^{-t_1} \quad t_1 \geq 0 \qquad f_Y(t_2) = e^{-t_2} \quad t_2 \geq 0$$
$$= 0 \quad \text{otherwise} \qquad\qquad = 0 \quad \text{otherwise}$$

Derive the distribution of their sum $Z = X + Y$ by the method of moment-generating functions.

Solution. Let us derive the moment-generating functions of X and Y.

$$\Psi_X(\theta) = \int_0^\infty t_1 e^{\theta t_1} e^{-t_1}\, dt_1 = (1 - \theta)^{-2}$$

$$\Psi_Y(\theta) = \int_0^\infty e^{\theta t_2} e^{-t_2}\, dt_2 = (1 - \theta)^{-1}$$

By Theorem 5.14, we have

$$\Psi_{X+Y}(\theta) = \Psi_X(\theta)\Psi_Y(\theta) = (1 - \theta)^{-3}$$

However, $(1 - \theta)^{-3}$ is the moment-generating function of the gamma distribution (see Table 6.3, Chapter 6) with $\lambda = 1$ and $k = 3$. Thus the distribution of Z is

$$\begin{aligned} f_Z(t) &= \tfrac{1}{2}t^2 e^{-t} \qquad t \geq 0 \\ &= 0 \qquad \text{otherwise} \end{aligned}$$

Obviously the moment-generating function of Z is $\int_0^\infty \frac{1}{2}t^2 e^{-t(1-\theta)}\, dt$. This integral can be evaluated by letting $t(1 - \theta) = y$; thus

$$\begin{aligned} \Psi_Z(\theta) &= \int_0^\infty \frac{1}{2(1-\theta)}\left(\frac{y}{1-\theta}\right)^2 e^{-y}\, dy \\ &= \frac{1}{2(1-\theta)^3}\int_0^\infty y^2 e^{-y}\, dy = \frac{\Gamma(3)}{2(1-\theta)^3} = (1-\theta)^{-3} \end{aligned}$$

which agrees with our previous finding.

5.10 The Central Limit Theorem

Theorem 5.17. *If $X_1, X_2, \ldots, X_n$ are independent, identically distributed random variables with common mean and variance*, i.e.,

$$E(X_1) = E(X_2) = \cdots = E(X_n) = \mu$$

and

$$\text{Var}(X_1) = \text{Var}(X_2) = \cdots = \text{Var}(X_n) = \sigma^2$$

then

(5.32)

$$\begin{aligned} \lim_{n\to\infty} P\left\{a \leq \frac{S_n - n\mu}{\sigma\sqrt{n}} \leq b\right\} &= \lim_{n\to\infty} P\{a\sigma\sqrt{n} + n\mu \leq S_n \leq b\sigma\sqrt{n} + n\mu\} \\ &= \frac{1}{\sqrt{2\pi}}\int_a^b e^{-u^2/2} du \end{aligned}$$

(note that $(1/\sqrt{2\pi})e^{-u^2/2}$ is the density function of the normal distribution with $\mu = 0$ and $\sigma^2 = 1$, known as the *standard normal density*) where $S_n = X_1 +$

$X_2 + \cdots + X_n$. The expected value and variance of S_n are

$$E(S_n) = n\mu \qquad \text{Var}(S_n) = n\sigma^2$$

In words, *the central limit theorem indicates that the distribution of* $(S_n - n\mu)/\sigma\sqrt{n}$ *approaches a standard normal distribution as* $n \to \infty$.

Proof. Let Z and Z_i denote the random variables

$$\frac{S_n - n\mu}{\sigma\sqrt{n}} \quad \text{and} \quad \frac{X_i - \mu}{\sigma\sqrt{n}}$$

respectively. Hence

$$\begin{aligned} Z &= \frac{S_n - n\mu}{\sigma\sqrt{n}} = \frac{X_1 - \mu}{\sigma\sqrt{n}} + \frac{X_2 - \mu}{\sigma\sqrt{n}} + \cdots + \frac{X_n - \mu}{\sigma\sqrt{n}} \\ &= Z_1 + Z_2 + \cdots + Z_n \end{aligned}$$

The moment-generating function of Z_i is

$$\begin{aligned} \Psi_{Z_i}(\theta) &= E\left(1 + Z_i\theta + \frac{Z_i^2\theta^2}{2!} + \cdots\right) \\ &= 1 + \frac{\theta^2}{2n} + \frac{\theta^3}{3!}\frac{\mu_3}{n^{3/2}\sigma^3} + \frac{\theta^4}{4!}\frac{\mu_4}{n^2\sigma^4} + \cdots \\ &= 1 + \frac{\theta^2}{2n} + \sum_{k=3}^{\infty} \frac{\theta^k \mu_k}{k!n^{k/2}} \end{aligned}$$

where μ_k is the kth central moment.

Since the Z_i's are independent, by Theorem 5.14,

$$\begin{aligned} \Psi_Z(\theta) &= \Psi_{Z_1}(\theta)\Psi_{Z_2}(\theta) \cdots \Psi_{Z_n}(\theta) \\ &= \left(1 + \frac{\theta^2}{2n} + \sum_{k=3}^{\infty} \frac{\theta^k \mu_k}{k!n^{k/2}}\right)^n \end{aligned}$$

Hence

$$\ln \Psi_Z(\theta) = n \ln\left(1 + \frac{\theta^2}{2n} + \sum_{k=3}^{\infty} \frac{\theta^k \mu_k}{k!n^{k/2}}\right)$$

We recall that the Maclaurin expansion for $\ln(1 + x)$ is

$$\ln(1 + x) = x - \frac{x^2}{2} + \frac{x^3}{3} + \cdots$$

where $|x| < 1$. If we set

$$x = \frac{\theta^2}{2n} + \sum_{k=3}^{\infty} \frac{\theta^k \mu_k}{k!n^{k/2}}$$

provided that for large n the absolute value of this expression is less than one, then

$$\ln \Psi_Z(\theta) = n\left[\left(\frac{\theta^2}{2n} + \sum_{k=3}^{\infty} \frac{\theta^k \mu_k}{k!n^{k/2}}\right) - \frac{1}{2}\left(\frac{\theta^2}{2n} + \sum_{k=3}^{\infty} \frac{\theta^k \mu_k}{k!n^{k/2}}\right)^2 + \cdots\right]$$

$$= \left(\frac{\theta^2}{2} + \sum_{k=3}^{\infty} \frac{\theta^k \mu_k}{k! n^{(k/2)-1}}\right) - \frac{1}{2n}\left(\frac{\theta^2}{2} + \sum_{k=3}^{\infty} \frac{\theta^k \mu_k}{k! n^{(k/2)-1}}\right) + \cdots$$

As $n \longrightarrow \infty$, we have

$$\lim_{n\to\infty} \ln \Psi_Z(\theta) = \frac{\theta^2}{2}$$

Hence

$$\lim_{n\to\infty} \Psi_Z(\theta) = e^{\theta^2/2}$$

This is recognized as the moment-generating function of the standard normal distribution (see Table 6.3 in Chapter 6). Thus, by Theorem 5.16 we conclude that the limiting distribution of the random variable $(S_n - n\mu)/\sigma\sqrt{n}$ as $n \longrightarrow \infty$ is the standard normal distribution ($\mu = 0$, $\sigma^2 = 1$).

In fact, Theorem 5.17 is a special case of the central limit theorem. A more general version of the central limit theorem is applicable, *under certain conditions*, when the random variables are independent but not identically distributed.

Theorem 5.18. *Suppose $X_1, X_2, \ldots, X_n$ are independently distributed random variables with means $\mu_1, \mu_2, \ldots, \mu_n$ and variances $\sigma_1^2, \sigma_2^2, \ldots, \sigma_n^2$. Then, under certain conditions,*

$$\lim_{n\to\infty} P\left\{a \le \frac{S_n - E(S_n)}{\sqrt{\operatorname{Var}(S_n)}} \le b\right\} = \frac{1}{\sqrt{2\pi}} \int_a^b e^{-u^2/2}\, du \tag{5.33}$$

where $S_n = X_1 + X_2 + \cdots + X_n$

$$E(S_n) = \sum_{i=1}^{n} \mu_i$$

$$\operatorname{Var}(S_n) = \sum_{i=1}^{n} \sigma_i^2$$

It should be noted that Eq. (5.33) is valid under certain conditions. We shall not elaborate on these conditions in this text; however, the interested reader may consult advanced texts on probability theory such Gnedenko and Kolmogorov [5] and Loève [7] for further explication of this topic.

The central limit theorem plays an important role in theoretical and applied statistics. It shows that the limiting distribution of the sum of many independent random variables is asymptotically normal as the number of the variates tends to infinity. Consequently, it would be difficult to overestimate the importance of the normal distribution in modern statistical theory. The central limit theorem provides theoretical justification for the normal approximation to many arbitrary distributions encountered in engineering and scientific problems. This approximation has been found satisfactory for n as low as eight (for distributions that are approximately symmetrical). n tends to be larger than eight for skewed distributions.

Example 5.24 A fair coin is tossed 100 times. Compute (approximately) the probability that the number of heads is greater than 60.

Solution. In this coin-tossing experiment, let

$$\begin{aligned} X_i &= 1 \quad \text{if heads occurs on the } i\text{th trial} \\ &= 0 \quad \text{if tails occurs on the } i\text{th trial} \end{aligned}$$

and

$$S = \sum_{i=1}^{100} X_i$$

(Note that random variable S has a binomial distribution.) Now,

$$\begin{aligned} E(X_i) &= 0P_{X_i}(0) + 1P_{X_i}(1) \\ &= P_{X_i}(1) = P \end{aligned}$$

Then

$$E(S) = E \sum_{i=1}^{100} X_i = \sum_{i=1}^{100} E(X_i) = \sum_{i=1}^{100} p = 100\, p = 50$$

However,

$$\begin{aligned} \text{Var}(X_i) &= (0 - p)^2 p_{X_i}(0) + (1 - p)^2 p_{X_i}(1) \\ &= p^2(1 - p) + (1 - p)^2 p = p(1 - p) \end{aligned}$$

Hence

$$\text{Var}(S) = \sum_{i=1}^{100} p(1 - p) = 100p(1 - p) = 25$$

Applying the central limit theorem we have

$$\begin{aligned} P\{S \le 60\} &= P\left\{\frac{S - E(S)}{\sqrt{\text{Var}(S)}} \le \frac{60 - 50}{5}\right\} \\ &\sim \int_{-\infty}^{60-50/5} \frac{e^{-(1/2)u^2}}{\sqrt{2\pi}}\, du = .977 \end{aligned}$$

which is obtained from Table D in the Appendix. Therefore

$$P\{S > 60\} = 1 - .977 = .023$$

Example 5.25 An assembly consists of ten sections. The length of each section is a random variable having the same distribution with mean = 2 in. and standard deviation = .05 in. The specification of the total length is 20 ± .10 in. What is the probability that the assembly meets the specification requirement?

Solution. By the central limit theorem we have

$$\begin{aligned} P\left\{19.9 \le \sum_{i=1}^{10} X_i \le 20.1\right\} &= P\left\{\frac{19.9 - 20}{\sqrt{.025}} \le \frac{\sum_{i=1}^{10} X_i - 20}{\sqrt{.025}} \le \frac{20.1 - 20}{\sqrt{.025}}\right\} \\ &\cong \int_{19.9-20/\sqrt{.025}}^{20.1-20/\sqrt{.025}} \frac{e^{-u^2/2}}{\sqrt{2\pi}}\, du = .47 \end{aligned}$$

PROBLEMS

5.1. The c.d.f. of random variable X is

$$F_X(a) = 0 \qquad a < 0$$
$$= \frac{a}{1+a} \qquad a \geq 0$$

Compute the following:
(a) The mode of X.
(b) The median of X.
(c) $E(X)$.

5.2. A random variable X has the following density function:

$$f_X(t) = t \qquad 0 \leq t \leq 1$$
$$= \tfrac{1}{2} \qquad 1 \leq t \leq 2$$

Compute the following:
(a) The mode of X.
(b) The median of X.
(c) $E(X)$.

5.3. The probability function of variate X is

$$P_X(k) = \frac{1}{n+1} \qquad k = a, a+1, \ldots, a+n$$

Compute the following:
(a) $E(X)$
(b) $\text{Var}(X)$.

5.4. The p.d.f. of variate X is

$$f_X(t) = |t| \qquad |t| \leq 1$$
$$= 0 \qquad \text{otherwise}$$

Compute the mean and the variance of X.

5.5. Compute the mean and the variance if the p.d.f. of variate X is given by

$$f_X(t) = .01e^{-.01t} \qquad t \geq 0$$
$$= 0 \qquad \text{otherwise}$$

5.6. The probability function of a random variable X is

$$P_X(k) = \begin{cases} \frac{1}{4} & k = 0 \\ \frac{3}{4} & k = 1 \\ 0 & \text{otherwise} \end{cases}$$

Find the mean and the variance of X.

5.7. A random variable X has p.d.f. given by

$$f_X(t) = \frac{\theta - 1}{2|t|^{\theta}} \qquad |t| > 1 \qquad \theta > 1$$
$$= 0 \qquad \text{otherwise}$$

(a) Compute $P\{X < 7\}$ and $P\{X > 5\}$.
(b) For what values of θ does X have a finite mean and variance?

5.8. Assume X has the p.d.f. given by

$$f_X(t) = te^{-t^2/2} \qquad t \geq 0$$
$$= 0 \qquad \text{otherwise}$$

Compute the mean and variance of X.

5.9. A fair die is rolled. Let X denote the number of points on the upturned face. Compute $E(X)$ and $\text{Var}(X)$.

5.10. A product is sold in bags of nominal weight of 10 lb at a profit of \$5 per bag. The producer is penalized \$1 per pound for every bag in which the weight falls short of 10 lb but only on bags that weigh less than 9.98 lb. Assume that the actual weight of a bag has the following p.d.f.:

$$f_W(a) = 5 \qquad m - .1 < a < m + .1$$
$$= 0 \qquad \text{otherwise}$$

where m is established by adjusting the process that stays fixed once it is set.
(a) Compute the expected profit per bag if $m = 9.9$.
(b) Find m such that the expected profit is maximum.

5.11. A random variable X has the following p.d.f.:

$$f_X(a) = \begin{cases} \frac{3}{2}a^2 & 0 \leq a \leq 1 \\ \frac{1}{2} & 2 \leq a \leq 3 \\ 0 & \text{otherwise} \end{cases}$$

Compute the following:
(a) The mode of X.
(b) $E(X)$.

5.12. Suppose that the diameter X of a disk is a random variable with $E(X) = 5$, and $\text{Var}(X) = 3$. The thickness of the disk is equal to 1 in. and remains constant. What is the expected volume of the disk?

5.13. Assume X has the p.d.f. given by

$$f_X(t) = 1 \qquad 0 < t < 1$$
$$= 0 \qquad \text{otherwise}$$

Determine the expected value of the random value Y if
(a) $Y = 5X + 2$
(b) $Y = X^2$
(c) $Y = e^X$.

5.14. Assume X has a standard normal distribution. That is,

$$f_X(t) = \frac{1}{\sqrt{2\pi}} e^{-t^2/2} \qquad -\infty < t < \infty$$

Determine the expected value and variance of the variate Y when
(a) $Y = X^2$
(b) $Y = 3 - 2X$.

5.15. The p.d.f. of variate X is

$$f_X(t) = 2e^{-2t} \qquad t \geq 0$$
$$= 0 \qquad \text{otherwise}$$

Find $E(Y)$ where $Y = e^X$.

5.16. Let X and Y be two independent variates, each having the standard normal distribution. Compute the following:

(a) $E|X|$
(b) $E(X+Y)$
(c) $\text{Var}(X-Y)$
(d) $E(\sqrt{X^2+Y^2})$

5.17. Let X and Y be two independent random variables, each with the same p.d.f. given by

$$f_X(t) = f_Y(t) = \begin{cases} e^{-t} & t \geq 0 \\ 0 & \text{otherwise} \end{cases}$$

Compute the following:
(a) $E(X)$
(b) $\text{Var}(X+Y)$
(c) $E(XY)$.

5.18. The joint density function of random variables X_1 and X_2 is

$$\begin{aligned} f_{X_1X_2}(t_1, t_2) &= 2e^{-(t_1+2t_2)} \qquad t_1 \geq 0 \qquad t_2 \geq 0 \\ &= 0 \qquad \text{otherwise} \end{aligned}$$

(a) Compute $E(X_1, X_2)$.
(b) Determine whether X_1 and X_2 are independent.

5.19. The radius of a circle has been found to be uniformly distributed. That is,

$$\begin{aligned} f_R(a) &= 10 \qquad 1.95 < a < 2.05 \\ &= 0 \qquad \text{otherwise} \end{aligned}$$

Let A deno'· the area of the circle. Compute the following:
(a) $E(A)$
(b) $\text{Var}(A)$.

5.20. Two random variables X_1 and X_2 have the following probability functions:

$$P_{X_1}(k_1) = \begin{cases} \frac{1}{3} & k_1 = 1 \\ \frac{1}{3} & k_1 = 2 \\ \frac{1}{3} & k_1 = 3 \end{cases} \qquad P_{X_2}(k_1+1) = \begin{cases} \frac{1}{3} & k_2 = 2 \\ \frac{1}{3} & k_2 = 3 \\ \frac{1}{3} & k_2 = 4 \end{cases}$$

Compute the following:
(a) $\text{Cov}(X_1, X_2)$
(b) ρ_{X_1, X_2}
(c) $\text{Var}(X_1 + X_2)$.

5.21. The joint density function of variates X and Y is

$$\begin{aligned} f_{X,Y}(t_1, t_2) &= 6t_1t_2(2 - t_1 - t_2) \qquad 0 \leq t_1 \leq 1 \qquad 0 \leq t_2 \leq 1 \\ &= 0 \qquad \text{otherwise} \end{aligned}$$

Compute the following:
(a) $\text{Cov}(X, Y)$
(b) $\rho_{X,Y}$.

5.22. Suppose that X and Y are two random variables, with the following joint probability function:

X	Y	*Probability*
0	1	$\frac{1}{8}$
0	2	$\frac{1}{4}$
1	1	$\frac{1}{8}$
1	2	$\frac{1}{4}$
2	1	$\frac{1}{8}$
2	2	$\frac{1}{8}$

Compute the following:
(a) $E(X \mid Y = 1)$
(b) $E(X \mid Y = 2)$.

5.23. X and Y are two independent random variables with $\text{Var}(X) = \frac{1}{4}$ and $\text{Var}(Y) = \frac{1}{2}$. Use Chebyshev's inequality to compute an upper bound for

$$P\{|2(X - Y) - E[2(X - Y)]| > 2.0\}$$

5.24. A fair coin is tossed n times; the number of tails is denoted by X. With the use of Chebyshev's inequality, determine the smallest n such that

$$P\{.45 < \frac{X}{n} < .55\} > .95$$

5.25. Show that

$$P\{|X - E(X)| > k^r\sqrt{\lambda_r}\} \leq \frac{1}{k^r} \qquad k > 0$$

where λ_r is the absolute moment of order r around the mean. (Note that the above inequality reduces to Chebyshev's inequality when $r = 2$.)

5.26. The number of defective components in a TV set is known to have a Poisson distribution with $\lambda = 10$. Use the Chebyshev's inequality to compute a lower bound for the probability that a set chosen at random will have between 6 and 14 defective components.

5.27. The distribution of demand for a certain commodity is unknown; however, the approximate mean and variance of demand per week are 100 and 10, respectively. Use the Chebyshev's inequality to compute a lower bound for the probability that in a given week the demand will fall in the interval [90, 110].

5.28. The production rate of a special purpose machine is 10,000 units/hr. Suppose that the probability of producing a defective item is .01. With the use of Chebyshev's inequality, decide for what values of C does the probability exceed .95 that the total number of defective items in a given hour will fall in the interval $[100 - C, 100 + C]$.

5.29. The density function of variate X is as follows:

$$\begin{aligned} f_X(t) &= 3t^2 \qquad 0 < t < 1 \\ &= 0 \qquad \text{otherwise} \end{aligned}$$

(a) Determine the moment-generating function of X.
(b) Find the mean and variance of X from its moment-generating function.

5.30. The p.d.f. of variate X is given by

$$\begin{aligned} f_X(t) &= \lambda^2 t e^{-\lambda t} \qquad \lambda > 0 \qquad t \geq 0 \\ &= 0 \qquad \text{otherwise} \end{aligned}$$

(a) Determine the moment-generating function of X.
(b) Find the mean and variance of X from its moment-generating function.

5.31. Suppose that X has a p.d.f. given by

$$\begin{aligned} f_X(t) &= .01e^{-.01(t-10)} \qquad t \geq 10 \\ &= 0 \qquad \text{otherwise} \end{aligned}$$

(a) Determine the moment-generating function of X.
(b) Find the mean and variance of X from its moment generating function.

5.32. The p.d.f. of variate X is given by

$$f_X(t) = \tfrac{1}{4}te^{-t/2} \qquad t \geq 0$$
$$= 0 \qquad \text{otherwise}$$

By the method of moment-generating functions, find $E(X)$ and $\text{Var}(X)$.

5.33. The probability function of random variable X is given by

$$P_X(k) = \begin{cases} \frac{1}{8} & k = 0 \\ \frac{1}{4} & k = 1 \\ \frac{1}{2} & k = 2 \\ \frac{1}{8} & k = 3 \end{cases}$$

Determine the mean and variance of X from its moment-generating function.

5.34. Find the moment-generating function of variate X, given the p.d.f.

$$f_X(t) = .10 \qquad 1 < t < 11$$
$$= 0 \qquad \text{otherwise}$$

5.35. The repair time in hours of a certain type of pneumatic press in an assembly plant has been found to be uniformly distributed on the interval $[\frac{1}{2}, 1]$. That is,

$$f_X(t) = 2 \qquad \tfrac{1}{2} < t < 1$$
$$= 0 \qquad \text{otherwise}$$

Let T be the total repair time of n independent failures. By the method of moment-generating functions, find $E(T)$ and $\text{Var}(T)$.

5.36. The χ^2 distribution with parameter n_1 is specified by the p.d.f.

$$f(\chi^2) = \frac{1}{2^{n_1/2}(n_1/2 - 1)!}(\chi^2)^{n_1/2-1}e^{-\chi^2/2} \qquad \chi^2 \geq 0$$
$$= 0 \qquad \text{otherwise}$$

(a) Determine the moment-generating function of the χ^2 distribution.
(b) Prove that the distribution of the sum of two independent χ^2 distributions with parameters n_1 and n_2 is a χ^2 with parameter $n_1 + n_2$.

5.37. Random variable X is normally distributed with parameters μ and σ^2. That is,

$$f_X(t) = \frac{1}{\sigma\sqrt{2\pi}}\exp\left[-\frac{1}{2}\left(\frac{t-u}{\sigma}\right)^2\right] \qquad -\infty < t < \infty$$

Let $\bar{X}$ be the arithmetic mean of random sample of n observations. Prove by the method of moment-generating functions that $\bar{X}$ is normally distributed with parameters μ and σ^2/n.

5.38. The length of time to failure in hours of a certain type of light bulb is distributed as follows:

$$f_T(a) = .001e^{-.001a} \qquad a \geq 0$$
$$= 0 \qquad \text{otherwise}$$

Suppose that a bulb is instantaneously replaced when it fails. By the method of moment-generating functions, find
(a) The distribution of the length of time until the second replacement.
(b) The mean and variance of the distribution obtained in (a).

5.39. Suppose that X_i, $i = 1, 2, \ldots, 10$ are independent, indentically distributed random variables with common mean $= 10$ and common variance $= 4$. Using the central limit theorem, evaluate

$$P\{\sum_{i=1}^{10} X_i > 110\}$$

5.40. Suppose that in a certain circuit 20 resistors are connected in series. The mean and variance of each resistor are 5 and .20 ohms, respectively. What is the probability that the total resistance of the circuit will exceed 98 ohms (assume independence)?

SUGGESTED REFERENCES

[1] Brunk, H., *An Introduction to Mathematical Statistics*, Ginn and Company, Boston, 1960.

[2] Cramér, H., *Mathematical Methods of Statistics*, Princeton University Press, Princeton, N.J., 1946.

[3] Derman, C., and M. Klein, *Probability and Statistical Inference for Engineers*, Oxford University Press, New York, 1959.

[4] Ehrenfeld, S., and S. B. Littauer, *Introduction to Statistical Method*, McGraw-Hill Book Company, New York, 1964.

[5] Gnedenko, B. V., and A. N. Kolmogorov, *Limit Distributions for Sums of Independent Random Variables*, Addison-Wesley Publishing Co., Reading, Mass., 1954.

[6] Kendall, M. G., and A. Stuart, *The Advanced Theory of Statistics*, 3rd ed., Vol. 1, *Distribution Theory*, New York, Hafner Publishing Co., Inc., 1969.

[7] Loève, M., *Probability Theory: Foundations, Random Sequences*, D. Van Nostrand Co., Inc., Princeton, N.J., 1955.

[8] Meyer, P. L., *Introductory Probability and Statistical Applications*, Addison-Wesley Publishing Co., Reading, Mass., 1965.

[9] Mood, A. M., and F. A. Graybill, *Introduction to the Theory of Statistics*, McGraw-Hill Book Company, New York, 1963.

[10] Parzen, E., *Modern Probability Theory and Its Applications*, John Wiley & Sons, Inc., New York, 1960.

[11] Wadsworth, G. P., and J. G. Bryan, *Introduction to Probability and Random Variables*, McGraw-Hill Book Company, New York, 1960.

[12] Wilks, S. S., *Elementary Statistical Analysis*, Princeton University Press, Princeton N.J. 1948.

6

SOME IMPORTANT DISTRIBUTIONS

6.1 Introduction

The purpose of this chapter is to introduce a few of the *theoretical frequency distributions* that occur frequently in practice. In many situations the mathematical behavior of a random variable may be described by a theoretical distribution function. The distributions are applicable to many engineering, scientific, and industrial problems, of which we shall cite a few examples. It has been found that the number of accidents occurring in time, number of machine failures, number of defects per unit inspected, etc., tend to follow the Poisson distribution. Measurable characteristics tend to follow the normal distribution. Time to failure of electronic components, radioactive decay, periods between successive arrivals at the service facility, and time for repair and fault-finding, all tend to follow the exponential distribution. We shall now investigate more closely a few of the commonly used probability distributions. We begin with discrete distributions.

DISCRETE DISTRIBUTIONS

6.2 Binomial Distribution

The *binomial distribution* is one of the most frequently used discrete distributions that arise in many engineering problems. Suppose we are sampling with replacement from a lot that is $p\%$ defective. If an item is drawn at random, then, it can be classified as defective with probability p or non-defective with probability $q = 1 - p$. The outcomes of a single carrying out of the experiment then are

0 if the item drawn is nondefective

1 if the item drawn is defective

Now if n independent trials are performed the total number of defective items will be equal to k, where $k = 0, 1, \ldots, n$. Suppose that $n = 10$. What is the probability distribution of the number of defective items in n independent trials? Since the probability of a defective item is p, the probability that the sample will contain no defective items is, by the rule for the joint probability of independent events,

$$P\{0\} = q^{10}$$

Similarly, the probability that the sample will contain exactly one defective item is

$$P\{1\} = 10\,pq^9$$

The probability distribution of the number of defective items is given in Table 6.1 for two selected values of p (10 and 50%). The results shown in Table 6.1 can be obtained by expansion of the binomial formula $(p + q)^n$, which yields $(n + 1)$ terms. In the cited example, the binomial formula becomes

$$(6.1) \qquad (p + q)^n = (p + q)^{10} = q^{10} + 10pq^9 + 45p^2q^8 + \cdots + p^{10}$$

The eleven terms generated by the binomial expansion represent the probabilities of 0, 1, . . . , 10 defective items, in that order.

TABLE 6.1 Sampling Distribution for the Number of Defective Items in Samples of Size n = 10

Number of Defective Items (k)	*Probability* ($p = 10\%$)	*Probability* ($p = 50\%$)
0	.3487	.0010
1	.3874	.0098
2	.1937	.0439
3	.0574	.1172
4	.0112	.2051
5	.0015	.2461
6	.0001	.2051
7	.0000	.1172
8	.0000	.0439
9	.0000	.0098
10	.0000	.0010

We note that each trial (called *Bernoulli trial*) has only two outcomes: either defective or nondefective, good or bad, male or female, etc. In general, we classify the two outcomes by either success or failure where the probability of success is the same on each trial and is denoted by p. It is assumed that repeated trials are independent. The probability function of exactly k successes in n independent Bernoulli trials with probability p for success on each trial is called the *binomial probability function.* It follows that the number of successes in n independent Bernoulli trials is a binomial random variable.

Definition 6.1. Random variable X is said to be *binomially distributed* with parameters n and p if its probability function is given by

$$P_X(k) = \binom{n}{k} p^k q^{n-k} \quad (k = 0, 1, \ldots, n \qquad 0 \le p \le 1 \quad p + q = 1) \tag{6.2}$$

Since $P_X(k)$ is a probability function it follows that

$$\sum_{k=0}^{n} P_X(k) = \sum_{k=0}^{n} \binom{n}{k} p^k q^{n-k} = (p + q)^n = 1$$

The distribution of the number of defective items in a sample of ten items, which is presented in Table 6.1, is graphed for two selected values of p in Figs. 6.1 and 6.2 we note that the binomial distribution shown in Fig. 6.2

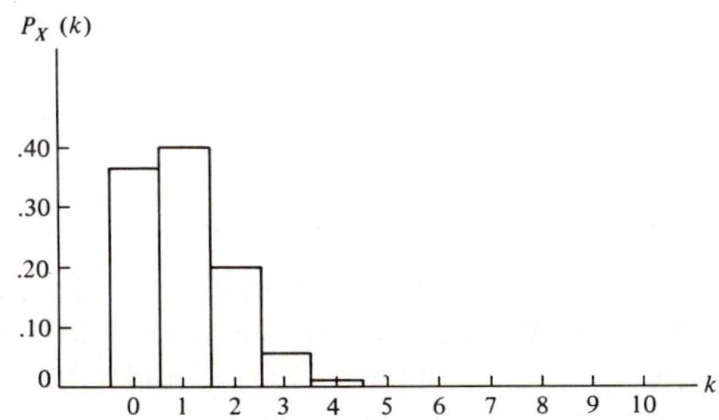

Fig. 6.1 Binomial probability function with $n = 10$ and $p = .1$.

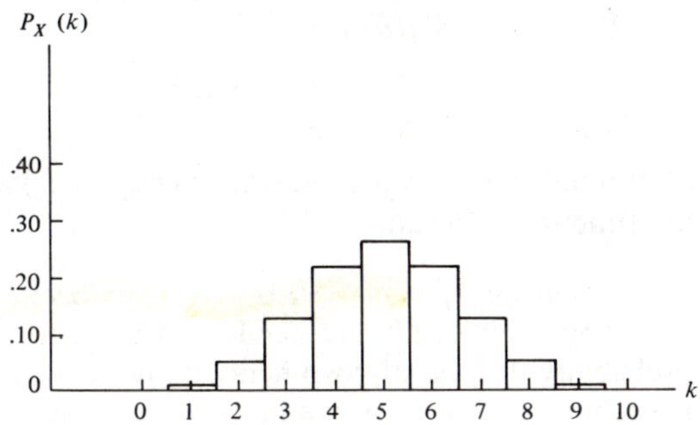

Fig. 6.2 Binomial probability function with $n = 10$ and $p = \frac{1}{2}$.

is symmetrical, whereas the distribution shown in Fig. 6.1 is skewed. Invariably we can state that the binomial distribution is skewed to the left when $p > \frac{1}{2}$, skewed to the right when $p < \frac{1}{2}$, and symmetrical when $p = q = \frac{1}{2}$. However, as n increases the binomial distribution tends to become more symmetrical even though p and q are not equal. For large n the binomial distribution can be approximated by the *normal distribution*.

If a variate X is binomially distributed with parameters n and p, then

$$E(X) = np \tag{6.3}$$

and

$$\text{Var}(X) = npq \tag{6.4}$$

Proof. The moment-generating function is

$$\Psi_X(\theta) = \sum_{k=0}^{n} e^{\theta k} P_X(k) = \sum_{k=0}^{n} \binom{n}{k} p^k q^{n-k} e^{\theta k}$$

$$= \sum_{k=0}^{n} \binom{n}{k} (pe^{\theta})^k q^{n-k} = (pe^{\theta} + q)^n$$

with derivatives

$$\Psi'_X(\theta) = npe^{\theta}(pe^{\theta} + q)^{n-1}$$

$$\Psi''_X(\theta) = npe^{\theta}(pe^{\theta} + q)^{n-1} + n(n-1)p^2 e^{2\theta}(pe^{\theta} + q)^{n-2}$$

Hence

$$E(X) = \Psi'_X(0) = np$$

and

$$E(X^2) = \Psi''_X(0) = np + n(n-1)p^2 = npq + n^2p^2$$

Therefore

$$\text{Var}(X) = E(X^2) - [E(X)]^2 = npq + n^2p^2 - n^2p^2 = npq$$

An important property of the binomial distribution is that its convolution is binomial. If X is a binomial variate with parameters n_1 and p and Y is a binomial variate, independent of X, with parameters n_2 and p, then Z is a binomial variate with parameters $n_1 + n_2$ and p where $Z = X + Y$.

Proof. The moment-generating function of variate Z is

$$\Psi_{X+Y}(\theta) = \Psi_X(\theta)\Psi_Y(\theta)$$

$$= (pe^{\theta} + q)^{n_1}(pe^{\theta} + q)^{n_2}$$

$$= (pe^{\theta} + q)^{n_1+n_2}$$

Therefore, Z is a binomial variate with parameters $(n_1 + n_2)$ and p since the moment-generating function is unique.

Example 6.1 A lot of material is 10% defective; four items are drawn with replacement (each drawing is physically independent of the other).

(a) Compute the probability that exactly two items are defective.

(b) Compute the probability that at least two items are defective.

Solution

(a) The binomial probability function is

$$P_X(k) = \binom{n}{k} p^k q^{n-k} \qquad (k = 0, 1, \ldots, 4 \qquad p + q = 1)$$

On substituting we obtain

$$P_X(2) = \binom{4}{2} .1^2 \, .9^2 = .0486$$

(b)

$$P\{X \geq 2\} = 1 - P\{X \leq 1\} = 1 - \sum_{k=0}^{1} \binom{4}{k} .1^k \, .9^{4-k}$$
$$= 1 - .9477 = .0523$$

To facilitate computations, the binomial distribution has been tabulated in tables of the *Cumulative Binomial Probability Distribution*, *Annals of the Computation Laboratory of Harvard University* (Vol. XXXV, Harvard University Press, Cambridge, 1955). Table J (in the Appendix) gives the cumulative binomial distribution function for values of n from $n = 1$ through $n = 20$ for selected values of p. The highest value of p given in the table is .50; however, for values of p higher than .50 we can use the columns headed $(1 - p)$. To illustrate, suppose that a lot of material is 60% defective and we wish to find the probability that there will be exactly one defective in five independent drawings. This is equivalent to the probability that there will be four nondefective items in five independent drawings when the probability of a nondefective is .40. From Table J this probability is found to be .0768.

6.3 Poisson Distribution

The *Poisson distribution* has been named after Seméon D. Poisson, a well-known pioneer in the theory of probability. The Poisson distribution is extensively used in queueing theory, reliability, acceptance sampling, and statistical quality control.

Definition 6.2. A random variable X is said to have a *Poisson distribution* if its probability function is given by

$$P_X(k) = \frac{e^{-\lambda}\lambda^k}{k!} \qquad \lambda > 0 \qquad k = 0, 1, \ldots \tag{6.5}$$

where e is the base of natural logarithms, a constant, $e = 2.71828$.

Since $P_X(k)$ is a probability function it follows that

$$\sum_{k=0}^{\infty} P_X(k) = \sum_{k=0}^{\infty} \frac{e^{-\lambda}\lambda^k}{k!} = e^{-\lambda} \sum_{k=0}^{\infty} \frac{\lambda^k}{k!} = e^{-\lambda} e^{\lambda} = 1$$

In fact *the Poisson distribution is the limiting distribution of the binomial.*

Proof. The probability function of the binomial distribution is

$$P_X(k) = \binom{n}{k} p^k q^{n-k} \tag{6.6}$$

Letting $p = \lambda/n$, Eq. (6.6) becomes

$$\begin{aligned} P_X(k) &= \binom{n}{k}\left(\frac{\lambda}{n}\right)^k\left(1-\frac{\lambda}{n}\right)^{n-k} \qquad k = 0, 1, 2, \ldots, \qquad \lambda > 0 \\ &= \frac{\lambda^k}{k!}\frac{n(n-1)\cdots(n-k+1)}{n^k}\left(1-\frac{\lambda}{n}\right)^{-k}\left(1-\frac{\lambda}{n}\right)^n \\ &= \frac{\lambda^k}{k!}\left(1-\frac{\lambda}{n}\right)^n \frac{1[1-(1/n)]\cdots[1-(k-1)/n]}{[1-(\lambda/n)]^k} \end{aligned} \tag{6.7}$$

However

$$\lim_{n\to\infty}\left(1-\frac{\lambda}{n}\right)^n = e^{-\lambda} \quad \text{and} \quad \lim_{n\to\infty}\frac{1[1-(1/n)]\cdots[1-(k-1)/n]}{[1-(\lambda/n)]^k} = 1$$

Hence

$$\lim_{n\to\infty} P_X(k) = e^{-\lambda}\frac{\lambda^k}{k!}$$

This means that when the probability of success in a Bernoulli trial p approaches zero and the number of trials n becomes infinite, so that $\lambda = np$ remains unchanged, the binomial distribution will approach the Poisson distribution with parameter λ.

If X is a Poisson variate with parameter λ, then

$$E(X) = \text{Var}(X) = \lambda \tag{6.8}$$

Proof. The moment-generating function is

$$\Psi_X(\theta) = \sum_{k=0}^{\infty} e^{\theta k} P_X(k) = \sum_{k=0}^{\infty} e^{\theta k}\frac{e^{-\lambda}\lambda^k}{k!} = e^{-\lambda}\sum_{k=0}^{\infty}\frac{(\lambda e^{\theta})^k}{k!} = e^{-\lambda}e^{\lambda e^{\theta}} = e^{\lambda(e^{\theta}-1)}$$

with derivatives,

$$\Psi'_X(\theta) = e^{\lambda(e^{\theta}-1)}\lambda e^{\theta}$$

$$\Psi''(\theta) = e^{\lambda(e^{\theta}-1)}\lambda^2 e^{2\theta} + \lambda e^{\theta} e^{\lambda(e^{\theta}-1)}$$

Hence

$$E(X) = \Psi'_X(0) = \lambda \quad \text{and} \quad E(X^2) = \Psi''_X(0) = \lambda^2 + \lambda$$

Therefore

$$\text{Var}(X) = E(X^2) - [E(X)]^2 = \lambda^2 + \lambda - \lambda^2 = \lambda$$

An important property of the Poisson distribution is *additivity;* i.e., *if X_1 and X_2 are two independent Poisson variates with parameters λ_1 and λ_2, respectively, their sum $X_1 + X_2$ has the Poisson distribution with parameter $\lambda_1 + \lambda_2$.*

Proof. The moment-generating function is

$$\begin{aligned}\Psi_{X_1+X_2}(\theta) &= \Psi_{X_1}(\theta)\Psi_{X_2}(\theta)\\ &= e^{\lambda_1(e^\theta-1)}e^{\lambda_2(e^\theta-1)}\\ &= e^{-(\lambda_1+\lambda_2)(1-e^\theta)}\end{aligned}$$

which is a Poisson with parameter $(\lambda_1 + \lambda_2)$. *In general we state that if $X_1, X_2, \ldots, X_n$ are independent Poisson variates with parameters $\lambda_1, \lambda_2, \ldots, \lambda_n$, then $\sum_{i=1}^{n} X_i$ is a Poisson variate with parameter $\sum_{i=1}^{n} \lambda_i$.*

Example 6.2 The number of machine failures per day in a certain plant has a Poisson distribution with parameter $\lambda = 2$. Present facilities at the plant can repair two machines per day. Failures in excess of two are repaired by an outside contractor.

(a) What is the probability, on a given day, of having machines repaired by the contractor?

(b) What is the expected number of machines repaired by the contractor per day?

Solution

(a) The number of machine failures per day is a random variable (X); its probability function is

$$P_X(k) = \frac{e^{-\lambda}\lambda^k}{k!} = \frac{e^{-2}2^k}{k!} \qquad k = 0, 1, \ldots$$

hence

$$\begin{aligned}P\{X > 2\} &= 1 - P\{X \leq 2\}\\ &= 1 - \sum_{k=0}^{2} \frac{e^{-2}2^k}{k!} = .323\end{aligned}$$

(b) The expected number of machines repaired by the contractor per day is

$$\sum_{k=3}^{\infty} (k-2)P_X(k) = \sum_{k=3}^{\infty} (k-2)\frac{e^{-2}2^k}{k!} = .5409$$

The Poisson distribution is tabulated by E. C. Molina in *Poisson's Exponential Limit* (D. Van Nostrand Co., Inc., Princeton, N.J., 1942). The cumulative distribution function of the Poisson distribution is given in Table I in the Appendix. The following example will illustrate the use of this table.

Example 6.3 The number of weekly accidents in a certain industrial plant follows the Poisson distribution with $\lambda = .75$. Graph the probability distribution of the number of weekly accidents.

Solution. From Eq. (6.5) we have

$$P_X(k) = \frac{e^{-.75}.75^k}{k!} \qquad k = 0, 1, \ldots$$

However,

$$P_X(k) = F_X(k) - F_X(k-1) \qquad k = 0, 1, \ldots$$

Hence

$$P_X(0) = F_X(0) = .472$$

and

$$P_X(1) = F_X(1) - F_X(0) = .827 - .472 = .355$$

Similarly, we obtain

$$P_X(2) = .132 \qquad P_X(3) = .034 \qquad P_X(4) = .006 \qquad P_X(5) = .001$$

The probability distribution is shown in Fig. 6.3. Note that this distribution is positively skewed. In fact, the Poisson distribution is positively skewed for any value of $\lambda > 0$; however, for values of $\lambda > 10$ the Poisson distribution tends to look like the normal distribution, which is symmetrical.

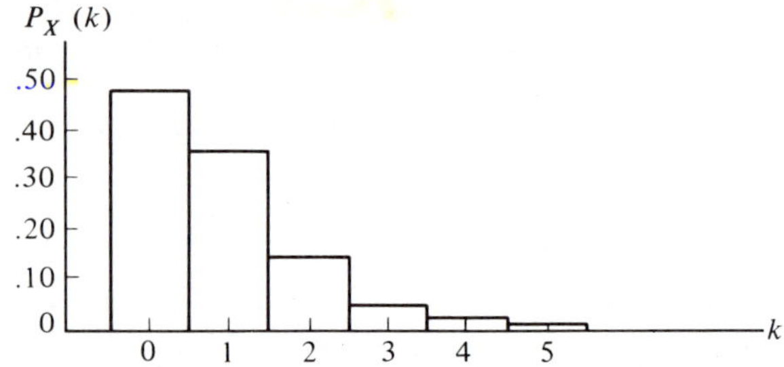

Fig. 6.3 Poisson probability function with $\lambda = .75$.

6.4 Multinomial Distribution

The *multinomial distribution* is a multivariate distribution involving $k - 1$ independent discrete random variables. In other words, the joint distribution of any $k - 1$ of k discrete random variables is known as a multinomial distribution. To simplify the discussion we shall consider an illustrative example: Suppose a statistically controlled process is used for the production of shear pins. Furthermore, suppose that the probability that a shear pin will conform to specifications is .70 and the probability that it will be short is .20 while the probability that it will be long is .10. If three shear pins are produced, what is the probability that one will conform to specifications, one will be short, and one will be long? In this case we make the following assumptions:

1. The outcomes, conforming to specifications, short, or long (denoted by c, s, or l), are mutually exclusive for each shear pin coming off the production line.
2. For each of the outcomes the chance of its occurrence is the same for each item produced.
3. Classification of each item produced as c, s, or l is independent of previous classifications.

We shall now find the desired probability under these assumptions. The event "one conforming to specifications, one short, and one long" is the subset

$$\{(s, l, c), (s, c, l), (c, l, s), (c, s, l), (l, s, c), (l, c, s)\}$$

Hence, the desired probability is

$$6(.7)(.2)(.1) = .084$$

The above probability can be obtained from the expansion of the multinomial expression $(.7 + .2 + .1)^3$. To expand this expression we write

$$\begin{aligned}(.7 + .2 + .1)^3 = (.7)^3 + (.2)^3 + (.1)^3 + 3(.7)^2(.2) \\ + 3(.7)(.2)^2 + 3(.7)^2(.1) + 3(.7)(.1)^2 \\ + 3(.2)^2(.1) + 3(.2)(.1)^2 + 6(.7)(.2)(.1) = 1\end{aligned}$$

The terms in the expansion give the probabilities of all possible combinations of classifications. For example, the first term gives the probability that all pins conform to specifications, the fifth term gives the probability that one pin conforms to specifications and two pins are short, etc. The coefficients of the terms represent the number of ways in which a sample of three items can be formed into three different groups of items, respectively. For example, the coefficient of the last term is

$$\frac{3!}{1!\,1!\,1!} = 6$$

In general, *suppose that there are k possible outcomes and let the probabilities of these outcomes be denoted by* $p_1, p_2, \ldots, p_k$ (with $\sum_{i=1}^{k} p_i = 1$). Furthermore, *suppose that the experiment is repeated n times and that* x_1 *is the number of times the outcome associated with* p_1 *occurs and that* x_2 *is the number of times the outcome associated with* p_2 *occurs and so forth* (with $\sum_{i=1}^{k} x_i = n$). *The distribution of possible outcome, then, is*

$$P\{x_1, x_2, \ldots, x_k\} = \frac{n!}{x_1!\,x_2!\,\ldots\,x_k!} p_1^{x_1} p_2^{x_2} \ldots p_k^{x_k} \tag{6.9}$$

Definition 6.3. The random variable $(X_1, X_2, \ldots, X_k)$ with the probability function given by Eq. (6.9) is said to have a *multinomial distribution*. Note that $k - 1$ of the x_i's are functionally independent since x_k is exactly determined when $x_1, \ldots, x_{k-1}$ are specified.

If the distribution of possible outcomes is multinomial, then

$$E(X_i) = np_i \tag{6.10}$$

and

$$\text{Var}(X_i) = np_i(1 - p_i) \tag{6.11}$$

The above results are derived from the fact that each X_i is binomially distributed with parameters p_i and n.

Example 6.4 Consider the previous example. Suppose ten shear pins were produced. What is the probability that among ten such pins six conform to specifications, two are short, and two are long?

Solution. By Eq. (6.9) we have

$$\begin{aligned} P\{x_1 = 6, x_2 = 2, x_3 = 2\} &= P\{6, 2, 2\} \\ &= \frac{10!}{6!\,2!\,2!}(.7)^6(.2)^2(.1)^2 \\ &= .059 \end{aligned}$$

6.5 Hypergeometric Distribution

Definition 6.4. A random variable X is said to have the *hypergeometric distribution* if the probability function is given by

$$P_X(k) = \frac{\binom{c}{k}\binom{N-c}{n-k}}{\binom{N}{n}} \qquad \begin{array}{l} n, c, N = 1, 2, \ldots \\ k = 0, 1, \ldots, n \\ N \geq c, n \end{array} \tag{6.12}$$

The following example will help to clarify the meaning of these symbols.

Example 6.5 Suppose a lot of size N is 5% defective. A sample of size n is drawn *without replacement*. We are now interested to find the probability of k defective items in the sample. Naturally k will not exceed either c or n. Now what is c? c is the number of defective items in the lot of size N. Hence, $c = .05N$. Suppose for simplicity that $N = 100$ and the lot is 5% defective. A sample of size five is drawn without replacement. What is the probability that the sample contains exactly one defective item?

Solution. Given $c = .05N = 5$, $n = 5$, $N = 100$, and $k = 1$, the required probability is

$$P_X(1) = \frac{\binom{5}{1}\binom{95}{4}}{\binom{100}{5}} = \frac{5!\,95!\,5!\,95!}{4!\,1!\,4!\,91!\,100!} = .2114$$

Since $P_X(k)$ is a probability function it follows that

$$\sum_{k=0}^{n} P_X(k) = 1$$

Proof. Consider the identity

$$(1 + \alpha)^N = (1 + \alpha)^c(1 + \alpha)^{N-c}$$

Applying the binomial theorem we have

$$\sum_{n=0}^{N}\binom{N}{n}\alpha^n = \left[\sum_{u=0}^{c}\binom{c}{u}\alpha^u\right]\left[\sum_{v=0}^{N-c}\binom{N-c}{v}\alpha^v\right]$$

$$= 1 + \alpha\left[\binom{c}{1} + \binom{N-c}{1}\right] + \alpha^2\left[\binom{c}{0}\binom{N-c}{2} + \binom{c}{1}\binom{N-c}{1}\right.$$

$$\left. + \binom{c}{2}\binom{N-c}{0}\right] + \cdots$$

$$= 1 + \alpha\sum_{k=0}^{1}\binom{c}{k}\binom{N-c}{1-k} + \alpha^2\sum_{k=0}^{2}\binom{c}{k}\binom{N-c}{2-k} + \cdots$$

Now, equating coefficients of α^n on both sides, we obtain

$$\binom{N}{n} = \sum_{k=0}^{n}\binom{c}{k}\binom{N-c}{n-k} = \sum_{k=0}^{n} P_X(k)\binom{N}{n}$$

Hence

$$\sum_{k=0}^{n} P_X(k) = 1$$

The probability function of the hypergeometric distribution can be written in the alternative form:

$$P_X(k) = \frac{\binom{Np}{k}\binom{Nq}{n-k}}{\binom{N}{n}} \qquad \begin{aligned} k &= 0, 1, \ldots, n \\ N &= 1, 2, \ldots \\ n &= 1, 2, \ldots, N \\ p &= 0, 1/N, 2/N, \ldots, 1 \\ p + q &= 1 \end{aligned} \tag{6.13}$$

Note that in this alternative form Np supersedes c.

The mean of the hypergeometric distribution is equal to np, whereas the variance is equal to $npq[(N-n)/(N-1)]$.

Proof. Let p = percentage defective in the lot, N = population size, n = sample size, and Y = number of defective items in the sample. Define

$$X_i = \begin{cases} 1 & \text{if the } i\text{th item drawn is defective} \\ 0 & \text{if the } i\text{th item drawn is nondefective} \end{cases}$$

where $i \leq n$. Hence

$$E(X_i) = p(1) + (1-p)(0) = p$$

and

$$\text{Var}(X_i) = (1-p)^2 p + (0-p)^2(1-p) = p(1-p) = pq$$

Consequently,

$$E(Y) = E\left[\sum_{i=1}^{n}(X_i)\right] = \sum_{i=1}^{n} E(X_i) = np \tag{6.14}$$

which is the mean of the hypergeometric distribution. If the ith and jth elements of the sample are defective, we define

$$X_i = 1 \text{ and } X_j = 1$$

Then

$$X_i X_j = \begin{cases} 1 & \text{if both items are defective} \\ 0 & \text{otherwise} \end{cases}$$

Hence

$$P\{X_i X_j = 1\} = \frac{Np(Np-1)}{N(N-1)} = \frac{p(Np-1)}{(N-1)}$$

and

$$E\{X_i X_j\} = \frac{p(Np-1)}{N-1}$$

Consequently,

$$\begin{aligned} \operatorname{Cov}(X_i, X_j) &= E(X_i X_j) - E(X_i)E(X_j) \\ &= \frac{p(Np-1)}{N-1} - p^2 = \frac{-pq}{N-1} \end{aligned}$$

Thus

$$\begin{aligned} \operatorname{Var}(Y) &= \sum_{i=1}^{n} \operatorname{Var}(X_i) + 2 \sum_{j=i+1}^{n} \sum_{i=1}^{n-1} \operatorname{Cov}(X_i, X_j) \\ &= npq - 2 \sum_{j=i+1}^{n} \sum_{i=1}^{n-1} \frac{pq}{N-1} \\ &= npq - 2\binom{n}{2}\left(\frac{pq}{N-1}\right) \\ &= npq - n(n-1)\frac{pq}{N-1} \end{aligned}$$

(6.15) $$= npq\left(\frac{N-n}{N-1}\right)$$ and the proof is complete.

6.6 Geometric Distribution

The probability function of the *geometric distribution* is given by

(6.16) $$P_X(k) = pq^{k-1} \quad (k = 1, 2, \ldots \quad 0 \le p \le 1 \quad p + q = 1)$$

The sample space is infinite. Since $P_X(k)$ is a probability function, it follows that

$$\sum_{k=1}^{\infty} P_X(k) = \sum_{k=1}^{\infty} pq^{k-1} = p(1 + q + q^2 + \cdots) = \frac{p}{1-q} = 1$$

and

$$F_X(k) = \sum_{x=1}^{k} pq^{x-1} = p(1 + q + q^2 + \cdots + q^{k-1}) = 1 - q^k$$

If a variate X has a geometric distribution then

$$E(X) = \frac{1}{p} \tag{6.17}$$

$$\mathrm{Var}(X) = \frac{q}{p^2} \tag{6.18}$$

Proof. The moment-generating function is

$$\begin{aligned}\Psi_X(\theta) &= \sum_{k=1}^{\infty} e^{\theta k} P_X(k) = \sum_{k=1}^{\infty} e^{\theta k} pq^{k-1} \\ &= pe^{\theta} \sum_{k=1}^{\infty} (qe^{\theta})^{k-1} = pe^{\theta} \frac{1}{1 - qe^{\theta}}\end{aligned}$$

The first derivative is

$$\Psi'_X(\theta) = pe^{\theta} \frac{1}{(1 - qe^{\theta})} + \frac{pqe^{2\theta}}{(1 - qe^{\theta})^2}$$

Thus

$$\Psi'_X(0) = 1 + \frac{pq}{(1-q)^2} = \frac{1}{p} = E(X)$$

The second derivative is

$$\Psi''_X(\theta) = \frac{pe^{\theta}}{(1 - qe^{\theta})} + \frac{pqe^{2\theta}}{(1 - qe^{\theta})^2} + \frac{2pqe^{2\theta}}{(1 - qe^{\theta})^2} + \frac{pq^2e^{3\theta}}{(1 - qe^{\theta})^3}$$

Hence

$$\begin{aligned}\Psi''_X(0) &= 1 + \frac{pq}{(1-q)^2} + \frac{2pq}{(1-q)^2} + \frac{pq^2}{(1-q)^3} \\ &= 1 + \frac{3q}{p} + \frac{q^2}{p^2} = \frac{1+q}{p^2} = E(X^2)\end{aligned}$$

Thus

$$\mathrm{Var}(X) = \frac{1+q}{p^2} - \frac{1}{p^2} = \frac{q}{p^2}$$

Example 6.6 The probability of failure of an equipment when it is tested is equal to .2. What is the probability that failure occurs in less than four tests? Assume independence of tests.

Solution. The required probability is

$$\begin{aligned}P\{X \leq 3\} &= 1 - (1 - p)^3 \\ &= 1 - (.8)^3 = .488\end{aligned}$$

6.7 Negative Binomial

The probability function of the *negative binomial* distribution is specified by

$$(6.19)\qquad P_X(k) = \binom{-r}{k} p^r(-q)^k \qquad \begin{aligned} &k = 0, 1, \ldots \\ &r > 0 \\ &0 \leq p \leq 1 \\ &p + q = 1 \end{aligned}$$

Variate X is said to obey the negative binomial distribution with parameters r and p if the probability of exactly k failures precede the rth success in a sequence of repeated independent Bernoulli trials (with probability p of success at each trial) is given by Eq. (6.19). Note that the $(r + k)$th trial results in success.

When parameter r is equal to unity the negative binomial distribution reduces to the geometric distribution.

Since $P_X(k)$ is a probability function it follows that

$$\sum_{k=0}^{\infty} P_X(k) = \sum_{k=0}^{\infty} \binom{-r}{k} p^r(-q)^k = p^r \sum_{k=0}^{\infty} \binom{-r}{k}(-q)^k$$
$$= p^r(1 - q)^{-r} = 1$$

The mean and the variance of the negative binomial are

$$(6.20)\qquad \begin{cases} E(X) = \dfrac{rq}{p} \\ \operatorname{Var}(X) = \dfrac{rq}{p^2} \end{cases}$$

Proof. The moment-generating function is

$$\Psi_X(\theta) = \sum_{k=0}^{\infty} \binom{-r}{k} p^r(-q)^k e^{\theta k}$$
$$= p^r \sum_{k=0}^{\infty} \binom{-r}{k}(-qe^{\theta})^k = p^r(1 - qe^{\theta})^{-r}$$

The first derivative is given by

$$\Psi'_X(\theta) = p^r(1 - qe^{\theta})^{-r-1} rqe^{\theta}$$

Hence

$$\Psi'_X(0) = p^r(1 - q)^{-r-1} rq = \frac{rq}{p} = E(X)$$

The second derivative is

$$\Psi''_X(\theta) = rqp^r(1 - qe^{\theta})^{-r-1}e^{\theta} + rq^2p^re^{2\theta}(r + 1)(1 - qe^{\theta})^{-r-2}$$

Thus

$$\Psi''_X(0) = \frac{rq}{p} + \frac{r(r + 1)q^2}{p^2} = E(X^2)$$

Consequently,

$$\operatorname{Var}(X) = E(X^2) - [E(X)]^2 = \frac{rq}{p} + \frac{r(r + 1)q^2}{p^2} - \left(\frac{rq}{p}\right)^2 = \frac{rq}{p^2}$$

TABLE 6.2 Some Discrete Probability Functions and Their Properties

Distribution	Probability Function	Parameters and Conditions	Mean $E(X)$	Variance $Var(X)$	Moment-Generating Function $\Psi_X(\theta) = E(e^{\theta x})$
Binomial	$P_X(k) = \binom{n}{k} p^k q^{n-k}$ $(k = 0, 1, \ldots, n)$	$n = 1, 2, \ldots$ $p + q = 1 \quad 0 \leq p \leq 1$	np	npq	$(pe^{\theta} + q)^n$
Poisson	$P_X(k) = \frac{e^{-\lambda}\lambda^k}{k!}$ $(k = 0, 1, \ldots)$	$\lambda > 0$	λ	λ	$\exp[\lambda(e^{\theta} - 1)]$
Hypergeometric	$P_X(k) = \frac{\binom{Np}{k}\binom{Nq}{n-k}}{\binom{N}{n}}$ $(k = 0, 1, \ldots, n)$	$n \leq N \quad N = 1, 2, \ldots$ $p = 0, \frac{1}{N}, \frac{2}{N}, \ldots, 1$ $p + q = 1 \quad 0 \leq p \leq 1$	np	$npq\left(\frac{N-n}{N-1}\right)$	†
Geometric	$P_X(k) = pq^{k-1}$ $(k = 1, 2, \ldots)$	$p + q = 1 \quad 0 \leq p \leq 1$	$\frac{1}{p}$	$\frac{q}{p^2}$	$\frac{pe^{\theta}}{1 - qe^{\theta}}$
Negative binomial	$P_X(k) = \binom{-r}{k} p^r(-q)^k$ $(k = 0, 1, \ldots)$	$r > 0$ $p + q = 1 \quad 0 \leq p \leq 1$	$\frac{rq}{p}$	$\frac{rq}{p^2}$	$p^r(1 - qe^{\theta})^{-r}$

†See M. G. Kendall, *Advanced Theory of Statistics*, Charles Griffin, London, 1948, p. 127.

Example 6.7 Suppose that attempts to launch a special satellite are physically independent. The probability of a successful launching is equal to .8 on any given attempt. Find the probability that three successes will occur before two failures.

Solution. We have $p = .8$, $q = .2$, $k = 0$ or 1, and $r = 3$. Hence, the required probability is

$$\begin{aligned}\sum_{k=0}^{1} P_X(k) &= \sum_{k=0}^{1} \binom{-r}{k} p^r(-q)^k \\ &= \binom{-3}{0}(.8)^3 + \binom{-3}{1}(.8)^3(-.2) \\ &= \binom{2}{0}(.8)^3 + \binom{3}{1}(.8)^3(.2) \\ &= .8192\end{aligned}$$

Note $\binom{-r}{k} = (-1)^k \binom{r+k-1}{k}$.

It is sometimes more convenient to define a new random variable $Y = X + r$, where Y is the number of trials until r successes occur. Thus, from Eq. (6.19),

$$\begin{aligned}P_Y(j) &= P\{X = j - r\} \\ &= \binom{-r}{j-r} p^r(-q)^{j-r} \\ &= \binom{j-1}{j-r} p^r q^{j-r} \\ &= \binom{j-1}{r-1} p^r q^{j-r} \qquad j = r, r+1, \ldots \qquad (6.21)\end{aligned}$$

A random variable with the probability distribution given by Eq. (6.21) is said to have a *Pascal distribution*. Note that random variable Y has mean r/p and variance rq/p^2.

The probability laws and the properties of some important discrete distributions are presented in Table 6.2.

CONTINUOUS DISTRIBUTIONS

6.8 Normal Distribution

In 1733 DeMoivre discovered what is known nowadays as the *normal distribution*. Sometimes it is referred to as the *Gaussian distribution*.

Definition 6.5. A random variable X is said to be *normally distributed* if its density function is specified by

(6.22) $$f_X(t) = \frac{1}{\sigma\sqrt{2\pi}} \exp\left[-\frac{1}{2}\left(\frac{t-\mu}{\sigma}\right)^2\right] \qquad -\infty < t < \infty$$
$$\sigma > 0 \qquad -\infty < \mu < \infty$$

σ and μ are two parameters (constants) that denote the mean and the standard deviation, respectively. The density function and distribution function are shown in Figs. 6.4 and 6.5, respectively.

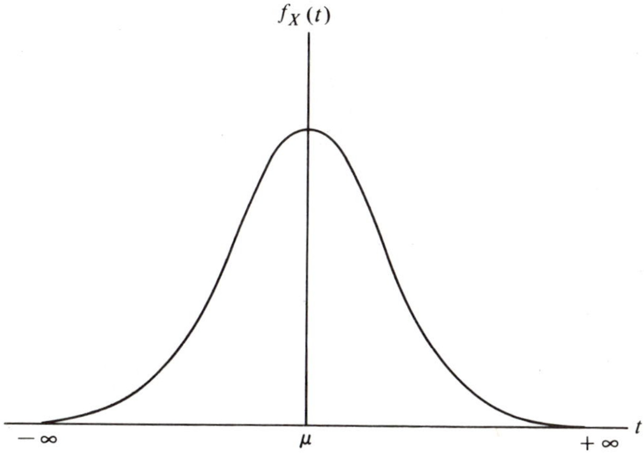

Fig. 6.4 Density function of the normal distribution.

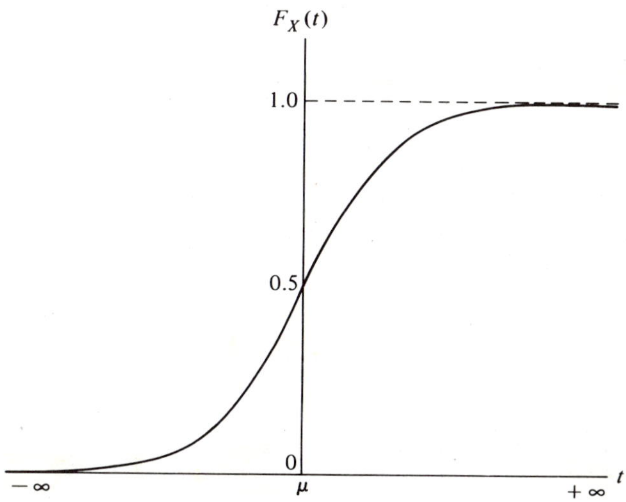

Fig. 6.5 Cumulative distribution function of the normal distribution.

If random variable X is normally distributed, then its mean $E(X) = \mu$ and its variance $\text{Var}(X) = \sigma^2$.

Proof

$$E(X) = \int_{-\infty}^{\infty} t f_X(t)\,dt = \int_{-\infty}^{\infty} \frac{t}{\sigma\sqrt{2\pi}} \exp\left[-\frac{1}{2}\left(\frac{t-\mu}{\sigma}\right)^2\right] dt$$

Let $y = t - \mu$; then

$$\begin{aligned} E(X) &= \int_{-\infty}^{\infty} \frac{(\mu + y)}{\sigma\sqrt{2\pi}} \exp\left[-\frac{1}{2}\left(\frac{y}{\sigma}\right)^2\right] dy \\ &= \int_{-\infty}^{\infty} \frac{y}{\sigma\sqrt{2\pi}} \exp\left[-\frac{1}{2}\left(\frac{y}{\sigma}\right)^2\right] dy + \mu \int_{-\infty}^{\infty} \frac{1}{\sigma\sqrt{2\pi}} \exp\left[-\frac{1}{2}\left(\frac{t-\mu}{\sigma}\right)^2\right] dt \\ &= 0 + \mu = \mu \end{aligned}$$

The variance is

$$\begin{aligned} \text{Var}(X) &= \int_{-\infty}^{\infty} (t-\mu)^2 f_X(t)\,dt \\ &= \int_{-\infty}^{\infty} \frac{y^2}{\sigma\sqrt{2\pi}} \exp\left[-\frac{1}{2}\left(\frac{y}{\sigma}\right)^2\right] dy \\ &= \frac{2}{\sigma\sqrt{2\pi}} \int_{0}^{\infty} y^2 \exp\left[-\frac{1}{2}\left(\frac{y}{\sigma}\right)^2\right] dy \\ &= \frac{2\sigma^2}{\sqrt{\pi}} \Gamma\left(\frac{3}{2}\right) = \sigma^2 \end{aligned}$$

where Γ represents the gamma function.

Suppose now we wish to find $P\{X \leq a\}$ where X is a normal variate with a mean $= \mu$ and a variance $= \sigma^2$. Then

$$P\{X \leq a\} = \int_{-\infty}^{a} f_X(t)\,dt = \int_{-\infty}^{a} \frac{1}{\sigma\sqrt{2\pi}} \exp\left[-\frac{1}{2}\left(\frac{t-\mu}{\sigma}\right)^2\right] dt$$

However, the density function of the normal distribution cannot be integrated explicitly. If the above function were tabulated in the above form a separate table would be required for each specified μ and σ. Thus an infinite number of tables would be required. This seems an infeasible task; however, by a simple transformation the computation of the above integral can be accomplished with a single table. Let

$$z = \frac{t-\mu}{\sigma}$$

whence

$$P\{X \leq a\} = \int_{-\infty}^{(a-\mu)/\sigma} \frac{e^{-(1/2)z^2}}{\sqrt{2\pi}}\,dz = \int_{-\infty}^{(a-\mu)/\sigma} f(z)\,dz = \Phi\left(\frac{a-\mu}{\sigma}\right)$$

$f(z)$ has a normal distribution with mean $= 0$ and variance $= 1$. $f(z)$ is known as the unit or standard normal distribution. Hence, the tabulated values of the standard normal distribution can be used to evaluate this integral. This tabulation is given in Table D (in the Appendix).

Example 6.8 The dimension of a certain item is normally distributed with $\mu = 2$ in. and $\sigma = .05$ in. The item is considered defective if its dimension exceeds 2.10 in. What is the probability of a defective item?

Solution

$$\begin{aligned} P\{X > 2.1\} &= \int_{2.1}^{\infty} f_X(t)\,dt \\ &= \int_{(2.1-2)/.05}^{\infty} f(z)\,dz = 1 - \Phi\left(\frac{2.1-2}{.05}\right) \\ &= 1 - \Phi(2) = .0228 \end{aligned}$$

An important property of the normal distribution is the sum of n independent random variates with means $\mu_1, \mu_2, \ldots, \mu_n$ and variances $\sigma_1^2, \sigma_2^2, \ldots, \sigma_n^2$ is a normal variate with mean equal to $\sum_{i=1}^{n} \mu_i$ and variance equal to $\sum_{i=1}^{n} \sigma_i^2$.

Proof. Let X_1 and X_2 be two independent normal variates with respective means μ_1 and μ_2 and respective variances σ_1^2 and σ_2^2. Then the moment-generating function is

$$\Psi_{X_1}(\theta) = \frac{1}{\sigma_1\sqrt{2\pi}} \int_{-\infty}^{\infty} e^{\theta t_1} \exp\left[-\frac{1}{2}\left(\frac{t_1-\mu_1}{\sigma_1}\right)^2\right] dt_1$$

Let

$$\frac{t_1 - \mu_1}{\sigma_1} = y$$

then

$$\begin{aligned} \Psi_{X_1}(\theta) &= e^{\mu_1\theta}\frac{1}{\sqrt{2\pi}} \int_{-\infty}^{\infty} e^{\theta y \sigma_1} e^{-(1/2)y^2} dy \\ &= \exp\left[\mu_1\theta + \frac{1}{2}\sigma_1^2\theta^2\right] \int_{-\infty}^{\infty} \frac{\exp[-(1/2)(y - \sigma_1\theta)^2]}{\sqrt{2\pi}}\,dy \\ &= \exp\left[\mu_1\theta + \frac{1}{2}\sigma_1^2\theta^2\right] \end{aligned}$$

since the integral is equal to unity. Hence

$$\begin{aligned} \Psi_{X_1+X_2}(\theta) &= \left[\exp\left(\mu_1\theta + \frac{1}{2}\sigma_1^2\theta^2\right)\right]\left[\exp\left(\mu_2\theta + \frac{1}{2}\sigma_2^2\theta^2\right)\right] \\ &= \exp\left[\theta(\mu_1 + \mu_2) + \frac{\theta^2}{2}(\sigma_1^2 + \sigma_2^2)\right] \end{aligned}$$

Since the moment-generating function is unique we conclude that the sum $(X_1 + X_2)$ is normally distributed with mean $(\mu_1 + \mu_2)$ and variance $(\sigma_1^2 + \sigma_2^2)$. Likewise, the proof can be extended to n variates.

Example 6.9 A delivery truck carries bags of cement. If the weight of each bag is normally distributed with mean 100 lb and standard deviation 5 lb, how many bags can the truck carry so that the probability of the total load exceeding 4,000 lb will be .05?

Solution. Let $n =$ the number of cement bags. Then the total load is normally distributed with mean $100n$ and variance $25n$. Thus

$$.05 = \int_{(4000-100n)/5\sqrt{n}}^{\infty} f(z)\,dz = 1 - \Phi\left(\frac{4000 - 100n}{5\sqrt{n}}\right)$$

Hence

$$\frac{4{,}000 - 100n}{5\sqrt{n}} = 1.645$$

whence $n \cong 39$.

6.9 Bivariate Normal Distribution

The *bivariate normal distribution* is a two-dimensional normal distribution. It is a generalization of the normal distribution for single random variable.

Definition 6.6. The random variable (X, Y) has a *two-dimensional normal distribution* if its joint density function is given by, for $-1 < \rho < 1$,

$$f_{X,Y}(t_1, t_2) = \frac{1}{2\pi\sigma_X\sigma_Y\sqrt{1-\rho^2}} \exp\left\{-\frac{1}{2(1-\rho^2)}\left[\left(\frac{t_1 - \mu_X}{\sigma_X}\right)^2 - \frac{2\rho(t_1 - \mu_X)(t_2 - \mu_Y)}{\sigma_X\sigma_Y} + \left(\frac{t_2 - \mu_Y}{\sigma_Y}\right)^2\right]\right\} \tag{6.23}$$

where μ_X, μ_Y, σ_X, σ_Y, and ρ are constants. The parameter ρ is known as the *correlation coefficient* of X and Y.

Since Eq. (6.23) is a density function it can be shown that

$$\int_{-\infty}^{\infty}\int_{-\infty}^{\infty} f_{X,Y}(t_1, t_2)\,dt_1\,dt_2 = 1 \tag{6.24}$$

The density function given by Eq. (6.23) is shown in Fig. 6.6. Any plane perpendicular to the XY plane intersects the normal surface in a curve of the normal form. Curves for which $f_{X,Y}(t_1, t_2)$ equal to a constant are ellipses.

The moment-generating function of the bivariate normal distribution has the form

$$\Psi_{X,Y}(\theta_1, \theta_2) = \exp[\theta_1\mu_X + \theta_2\mu_Y + \tfrac{1}{2}(\theta_1^2\sigma_X^2 + 2\rho\sigma_X\sigma_Y\theta_1\theta_2 + \theta_2^2\sigma_Y^2)] \tag{6.25}$$

The covariance is given by

$$\operatorname{Cov}(X, Y) = \rho\sigma_X\sigma_Y \tag{6.26}$$

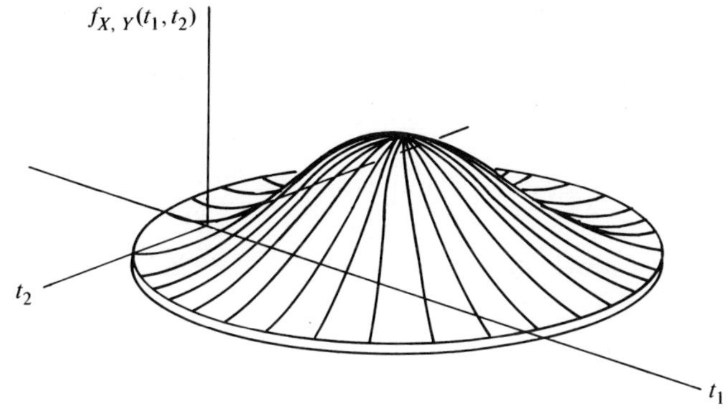

Fig. 6.6 The bivariate normal distribution.

It follows that the bivariate normal distribution depends on five parameters, namely, σ_X, σ_Y, μ_X, μ_Y, and ρ. Accordingly, if two random variables are jointly normally distributed, their joint density function is completely specified from a knowledge of their first and second moments.

If $\rho = \pm 1$, *there exists a linear correlation between random variables X and Y*. If $\rho = 0$, Eq. (6.23) takes the form

$$f_{X,Y}(t_1, t_2) = f_X(t_1) f_Y(t_2) \tag{6.27}$$

where $f_X(t_1)$ and $f_Y(t_2)$ are the density functions of X and Y, respectively. Thus, if $\rho = 0$, we say that X and Y are *uncorrelated* or *statistically independent*.

The marginal density of variate X is by definition

$$f_X(t_1) = \int_{-\infty}^{\infty} f_{X,Y}(t_1, t_2)\, dt_2 \tag{6.28}$$

It can be shown that Eq. (6.28) is equivalent to

$$f_X(t_1) = \frac{1}{\sqrt{2\pi}\sigma_X} \exp\left[-\frac{1}{2}\left(\frac{t_1 - \mu_X}{\sigma_X}\right)^2\right] \tag{6.29}$$

Similarly, the marginal density of variate Y is given by

$$f_Y(t_2) = \frac{1}{\sqrt{2\pi}\sigma_Y} \exp\left[-\frac{1}{2}\left(\frac{t_2 - \mu_Y}{\sigma_Y}\right)^2\right] \tag{6.30}$$

Thus the conditional density of Y provided that $X = t_1$ has the form

$$\begin{aligned} f_{Y|X}(t_2 \mid t_1) &= \frac{f_{X,Y}(t_1, t_2)}{f_X(t_1)} \\ &= \frac{1}{\sqrt{2\pi}\sigma_Y\sqrt{1-\rho^2}} \exp\left\{-\frac{1}{2\sigma_Y^2(1-\rho^2)} \right. \\ &\quad \left. \times \left[t_2 - \mu_Y - \frac{\rho\sigma_Y}{\sigma_X}(t_1 - \mu_X)\right]^2\right\} \end{aligned} \tag{6.31}$$

which is a univariate normal density with expected value $\mu_Y + (\rho\sigma_Y/\sigma_X)(t_1 - \mu_X)$ and with standard deviation $\sigma_Y\sqrt{1-\rho^2}$. The value of Eq. (6.31) is called the *regression function of* Y *on* X, which is a linear function of X. Thus the mean of Y for various values of X can be obtained from the regression equation

$$E(Y \mid X = t_1) = \mu_Y + \frac{\rho\sigma_Y}{\sigma_X}(t_1 - \mu_X) \tag{6.32}$$

The *conditional distribution* of X can be derived in a similar way. The regression line of X on Y is given by

$$E(X \mid Y = t_2) = \mu_X + \frac{\rho\sigma_X}{\sigma_Y}(t_2 - \mu_Y) \tag{6.33}$$

The *cumulative bivariate normal distribution*

$$F_{X,Y}(t_1, t_2) = \int_{-\infty}^{t_2}\int_{-\infty}^{t_1} f_{X,Y}(t_1, t_2)\, dt_1\, dt_2$$

is tabulated for selected values of $\rho = 0, .05, \ldots, .95$ in Karl Pearson's *Tables for Statisticians and Biometricians*, Part I (Cambridge University Press, London, 1914).

6.10 Log-Normal Distribution

The *log-normal distribution* is encountered in economic investigations involving the distributions of income and other economic indices.

Definition 6.7. A random variable X is said to have a *log-normal distribution* if its logarithm ($\ln X$) is normally distributed with parameters μ and σ.

Suppose that variate Y is normally distributed with parameters μ and σ. Now, if a new random variable is defined by $X = e^Y$, then the distribution function of X is

$$\begin{aligned} F_X(t) &= P\{X \le t\} = P\{e^Y \le t\} = P\{Y \le \ln t\} \\ &= \frac{1}{\sqrt{2\pi}}\int_{-\infty}^{(\ln t - \mu)/\sigma} e^{-z^2/2}\, dz \end{aligned}$$

Hence, the density function of X is

$$\begin{aligned} f_X(t) &= \frac{1}{\sqrt{2\pi}\sigma}t^{-1}\exp\left[-\frac{1}{2\sigma^2}(\ln t - \mu)^2\right] \qquad t > 0 \qquad \sigma > 0 \\ &= 0 \qquad \text{elsewhere} \end{aligned} \tag{6.34}$$

which is the density function of the log-normal distribution.

The kth moment of X is

$$E(X^k) = \frac{1}{\sqrt{2\pi}\sigma}\int_{-\infty}^{\infty} t^{k-1}\exp\left[-\frac{1}{2\sigma^2}(\ln t - \mu)^2\right] dt \tag{6.35}$$

and upon letting $\ln t = z$, Eq. (6.35) becomes

$$E(x^k) = \frac{1}{\sqrt{2\pi}\sigma}\int_{-\infty}^{\infty} e^{zk}\exp\left[-\frac{1}{2\sigma^2}(z-\mu)^2\right]dz$$

$$= \frac{\exp[k\mu + (k^2\sigma^2/2)]}{\sqrt{2\pi}\sigma}\int_{-\infty}^{\infty}\exp\left[-\frac{1}{2}\left(\frac{z-\mu}{\sigma} - k\sigma\right)^2\right]dz$$

$$= \exp\left(k\mu + \frac{k^2\sigma^2}{2}\right) \tag{6.36}$$

The mean and the variance of the log-normal distribution can be obtained from Eq. (6.36). Hence

$$E(X) = e^{\mu + (\sigma^2/2)}$$
$$\mathrm{Var}(X) = e^{2\mu + \sigma^2}(e^{\sigma^2} - 1) \tag{6.37}$$

Example 6.10 The amount of regular octane gasoline to be stored in tanks of an oil refinery at the start of each day can be treated as a random variable having the log-normal distribution with $\mu = 2.5$ (million) and $\sigma = .1$ (million). Each tank holds 10^5 gal. How many tanks should the refinery have so that the probability of exceeding its capacity at the start of any day is less than .025?

Solution. Let X denote the amount of gasoline to be stored in gallons (million). Then

$$P\{X > c\} = \int_c^{\infty} \frac{1}{\sqrt{2\pi}\sigma} a^{-1} e^{-(\ln a - \mu)^2/2\sigma^2} da$$

Changing variables by letting $\ln a = y$, we obtain

$$P\{X > c\} = \int_{\ln c}^{\infty} \frac{1}{\sqrt{2\pi}\sigma} e^{-(y-\mu)^2/2\sigma^2} dy$$
$$= 1 - \Phi\left(\frac{\ln c - 2.5}{.1}\right) < .025$$

whence $c > 14.8136$. That is, the refinery should have 149 tanks so that the probability of exceeding its capacity at the start of any day is less than .025. The number of required tanks seems to be large due to the fact that the log-normal distribution is positively skewed.

6.11 Exponential Distribution

Definition 6.8. A variate X is said to be *exponentially distributed* if its density function is given by

$$f_X(t) = \lambda e^{-\lambda t} \qquad \lambda > 0 \qquad t \geq 0$$
$$= 0 \qquad \text{otherwise} \tag{6.38}$$

The density function and distribution function of the exponential distribution are shown in Figs. 6.7 and 6.8, respectively.

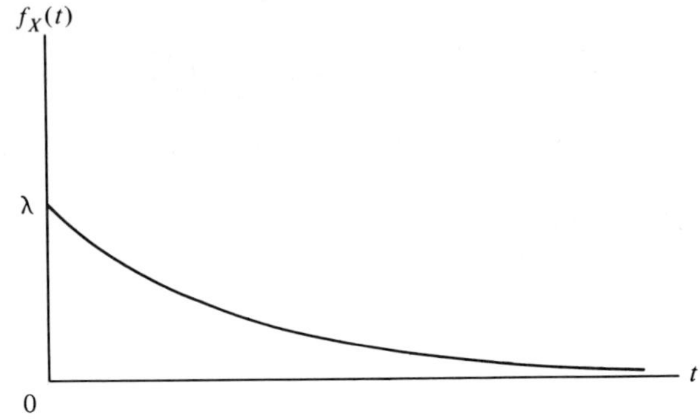

Fig. 6.7 Density function of the exponential distribution.

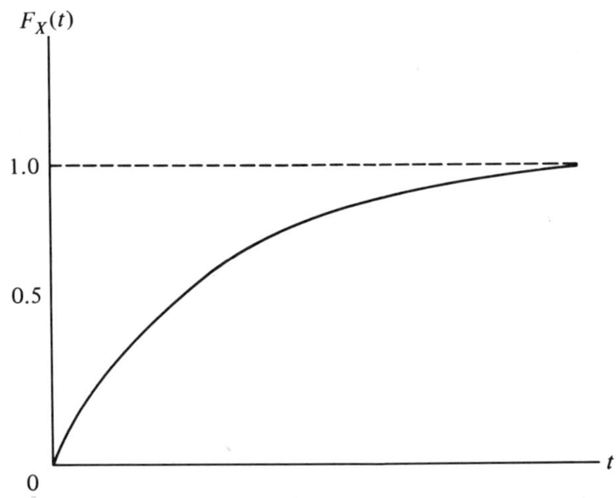

Fig. 6.8 Cumulative distribution function of the exponential distribution.

The exponential distribution is characterized by the following property (known as complete *lack of memory*):

$$P\{X > r + s \mid X > r\} = P\{X > s\}$$

where r and s are any positive numbers.

Proof

$$\begin{aligned} P\{X > r + s \mid X > r\} &= \frac{P\{X > r + s \cap X > r\}}{P\{X > r\}} \\ &= \frac{P\{X > r + s\}}{P\{X > r\}} \end{aligned}$$

$$= \frac{\int_{r+s}^{\infty} \lambda e^{-\lambda t}\, dt}{\int_{r}^{\infty} \lambda e^{-\lambda t}\, dt}$$

$$= \frac{e^{-\lambda(r+s)}}{e^{-\lambda r}} = e^{-\lambda s} = P\{X > s\}$$

This means that $P\{X > s\}$ is independent of r. In other words, *if a piece of equipment has not failed during r time units its conditional probability of serving r + s or more time units is independent of r and is equal to the probability of serving s or more time units.* Stated differently, if time to failure of a piece of equipment follows the exponential distribution, then aging of the equipment is immaterial.

The mean and the variance of the exponential distribution have already been found in Example 5.21 and are $E(X) = 1/\lambda$ and $\text{Var}(X) = 1/\lambda^2$. We note that the mean and standard deviation of the exponential distribution are equal.

Example 6.11 Suppose the life of automobile batteries is exponentially distributed with parameter $\lambda = .001$ days.

(a) What is the probability that a battery will last more than 1,200 days?

(b) What is the probability that a battery will last more than 1,200 days given that it has already served 1,000 days?

Solution.

(a) $$P\{X > 1200\} = \int_{1200}^{\infty} \lambda e^{-\lambda t}\, dt = \int_{1200}^{\infty} .001^{-.001t}\, dt$$
$$= e^{-1.2} = .301$$

(b) $$P\{X > 1200 \mid X > 1000\} = P\{X > 200\}$$
$$= e^{-.2} = .819$$

6.12 Uniform Distribution

The density function of the *uniform distribution* is given by

(6.39) $$f_X(t) = \frac{1}{b-a} \qquad a < t < b \qquad -\infty < a < b < \infty$$
$$= 0 \qquad \text{otherwise}$$

The mean of the uniform distribution over interval a to b is $(b + a)/2$ and the variance $(b - a)^2/12$.

Proof

(6.40) $$E(X) = \int_{b}^{a} \frac{t}{b-a}\, dt = \frac{b^2 - a^2}{2(b-a)} = \frac{b+a}{2}$$

and

$$E(X^2) = \int_a^b \frac{t^2}{b-a}\,dt = \frac{b^3 - a^3}{3(b-a)}$$

Hence

$$\text{Var}(X) = E(X^2) - [E(X)]^2 = \frac{b^3 - a^3}{3(b-a)} - \frac{(b+a)^2}{4} = \frac{(b-a)^2}{12} \tag{6.41}$$

Example 6.12 A variate X obeys the uniform distribution with the density function given by

$$\begin{aligned} f_X(t) &= \frac{1}{100} \qquad 0 < t < 100 \\ &= 0 \qquad \text{otherwise} \end{aligned}$$

(a) Compute $P\{X > 60\}$.
(b) Compute $P\{20 \leq X \leq 40\}$.

Solution

(a) $P\{X > 60\} = \int_{60}^{100} \frac{1}{100}\,dt = .4$

(b) $P\{20 \leq X \leq 40\} = \int_{20}^{40} \frac{1}{100}\,dt = .2$

6.13 Gamma Distribution

Suppose $X_1, X_2, \ldots, X_k$ are independent and identically distributed random variables. Furthermore, suppose that these variates obey the exponential distribution with parameter λ. What is the distribution of their sum X where $X = X_1 + X_2 + \cdots + X_k$? The mean and variance of X can be easily derived since these variates are independent. Thus we have

$$E(X) = E\sum_{i=1}^{k} X_i = \sum_{i=1}^{k} E(X_i) = \frac{k}{\lambda} \tag{6.42}$$

$$\text{Var}(X) = \text{Var}\sum_{i=1}^{k} X_i = \sum_{i=1}^{k} \text{Var}(X_i) = \frac{k}{\lambda^2} \tag{6.43}$$

The density function of X can be derived in several ways. We will derive the density function of X by the convolution method. The density function of $(X_1 + X_2)$ is given by

$$\begin{aligned} f_{X_1+X_2}(t) &= \int_0^t \lambda e^{-\lambda(t-t_1)} \lambda e^{-\lambda t_1} dt_1 \\ &= \lambda^2 e^{-\lambda t} \int_0^t dt_1 \\ &= \lambda^2 e^{-\lambda t} t \qquad t \geq 0 \end{aligned}$$

Similarly,

$$\begin{aligned} f_{X_1+X_2+X_3}(t) &= \int_0^t \lambda^2 e^{-\lambda(t-t_1)}(t-t_1)\lambda e^{-\lambda t_1}\,dt_1 \\ &= \lambda^3 e^{-\lambda t}\int_0^t (t-t_1)\,dt_1 \\ &= \frac{\lambda^3 e^{-\lambda t} t^2}{2!} \qquad t \geq 0 \end{aligned}$$

Thus, by induction we conclude that the density function of X is

$$\begin{aligned} f_X(t) &= \frac{\lambda^k e^{-\lambda t} t^{k-1}}{\Gamma(k)} \qquad t \geq 0 \qquad \lambda, k > 0 \\ &= 0 \qquad \text{otherwise} \end{aligned} \tag{6.44}$$

where $\Gamma(k)$ is a value of the gamma function, defined by

$$\Gamma(k) = \int_0^\infty t^{k-1} e^{-t}\,dt$$

Integration by parts shows that

$$\Gamma(k) = (k-1)\Gamma(k-1)$$

for any $k > 0$ and hence that $\Gamma(k) = (k-1)!$ when k is a positive integer. Thus when k is a positive integer Eq. (6.44) becomes

$$\begin{aligned} f_X(t) &= \frac{\lambda^k e^{-\lambda t} t^{k-1}}{(k-1)!} \qquad t \geq 0 \qquad \lambda > 0 \qquad k = 1, 2, \ldots \\ &= 0 \qquad \text{otherwise} \end{aligned}$$

and is known as the *Erlang distribution*.

Definition 6.9. A random variable X is said to have a *gamma distribution* if its density function is given by Eq. (6.44). The mean and the variance of the gamma distribution are given by Eqs. (6.42) and (6.43), respectively. Figure 6.9 shows several gamma distributions. As we see, these distributions

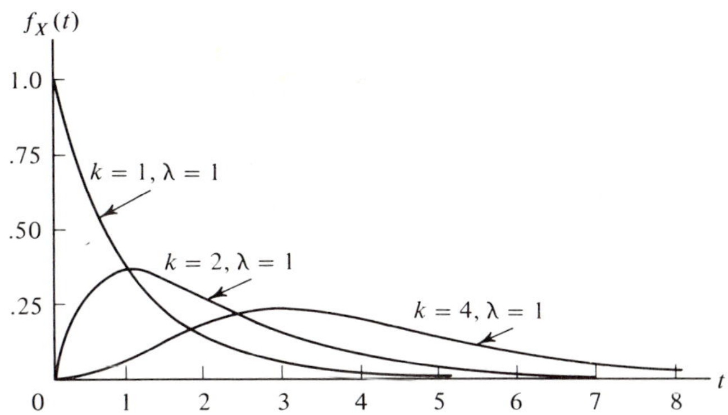

Fig. 6.9 Graphs of the gamma probability density functions.

are positively skewed. In fact skewness decreases as k increases for any fixed value of λ.

Note that if $k = 1$ the gamma distribution is identical to the exponential distribution.

Example 6.13 An electric network has N relays. If a relay is tripped the repair time is a random variable that obeys the exponential distribution with parameter $\lambda = 19$ hr. If one repair man is available and ten relays have tripped simultaneously, what is the probability that power failure will last more than $\frac{3}{4}$ hr?

Solution. Total repair time obeys the gamma distribution with parameter $k = 10$ and $\lambda = 19$. Thus

$$\begin{aligned} P\{X > \tfrac{3}{4}\} &= 1 - P\{X \leq \tfrac{3}{4}\} \\ &= 1 - \int_0^{3/4} \frac{19^9 e^{-19t} t^9}{9!}\, dt \\ &= .10 \end{aligned}$$

We state without proof an important property of the gamma distribution. *If $X_1, X_2, \ldots, X_r$ are independent gamma variates with common λ and parameters $k_1, k_2, \ldots, k_r$, then their sum $X = X_1 + X_2 + \cdots X_r$ is a gamma variate with parameters λ and $k = k_1 + k_2 + \cdots + k_r$.*

6.14 Chi-Square Distribution

Definition 6.10. Chi-square (χ^2) distribution is defined as *the sum of the squares of n independent standardized normal variates.* If X_i is a normal variate with mean μ_i and standard deviation σ_i, then

$$X_i^* = \frac{X_i - \mu_i}{\sigma_i}$$

is a standardized normal variate and, by definition,

$$\chi^2 = \sum_{i=1}^{n} X_i^{*2} = \sum_{i=1}^{n} \left(\frac{X_i - \mu_i}{\sigma_i}\right)^2$$

We seek now the density function of the random variable χ^2. Let us first derive the distribution of $[(X_i - \mu_i)/\sigma_i]^2$. By definition we have

$$\begin{aligned} F_{X_i^{*2}}(t_i) &= P\{X_i^{*2} \leq t_i\} = P\left\{\left(\frac{X_i - \mu_i}{\sigma_i}\right)^2 \leq t_i\right\} \\ &= P\{\mu_i - \sigma_i\sqrt{t_i} \leq X_i \leq \mu_i + \sigma_i\sqrt{t_i}\} \\ &= F_{X_i}(\mu_i + \sigma_i\sqrt{t_i}) - F_{X_i}(\mu_i - \sigma_i\sqrt{t_i}) \\ &= \int_{-\infty}^{\mu_i+\sigma_i\sqrt{t_i}} \frac{\exp\{-\frac{1}{2}[(t_i - \mu_i)/\sigma_i]^2\}\, dt_i}{\sigma_i\sqrt{2\pi}} \\ &\quad - \int_{-\infty}^{\mu_i-\sigma_i\sqrt{t_i}} \frac{\exp\{-\frac{1}{2}[(t_i - \mu_i)/\sigma_i]^2\}\, dt_i}{\sigma_i\sqrt{2\pi}} \end{aligned}$$

On taking the derivatives of both sides with respect to t_i we obtain

$$f_{X_i^{*2}}(t_i) = \frac{1}{\sqrt{2\pi}}(t_i)^{-1/2}e^{-(1/2)t_i} \qquad t_i \geq 0$$
$$= 0 \qquad \text{otherwise}$$

The distribution of X_i^{*2} is in fact equivalent to a gamma distribution with $\lambda = \frac{1}{2}$ and $k = \frac{1}{2}$. In symbols, the gamma density function is given by

$$f_X(t) = \frac{\lambda e^{-\lambda t}(\lambda t)^{k-1}}{\Gamma(k)} \qquad \lambda, k > 0 \qquad t \geq 0$$
$$= 0 \qquad \text{otherwise}$$

Hence, the gamma density function with $\lambda = \frac{1}{2}$ and $k = \frac{1}{2}$ is given by

$$f_X(t) = \frac{\frac{1}{2}e^{-t/2}(t/2)^{-1/2}}{\Gamma(\frac{1}{2})} = \frac{1}{\sqrt{2\pi}}e^{-t/2}t^{-1/2} \qquad \lambda, k > 0 \qquad t \geq 0$$
$$= 0 \qquad \text{otherwise}$$

which is equivalent to $f_{X_i^{*2}}(t_i)$. Since $X_1^{*2}, X_2^{*2}, \ldots, X_n^{*2}$ are independent gamma variates, their sum $\sum_{i=1}^{n} X_i^{*2}$ has a gamma distribution with $\lambda = \frac{1}{2}$ and $k = n/2$

Hence, the probability density function of χ^2 is

$$f(\chi^2) = \frac{(\chi^2)^{(n-2)/2}}{2^{n/2}\Gamma(n/2)}e^{-\chi^2/2} \qquad \chi^2 \geq 0 \tag{6.45}$$
$$= 0 \qquad \text{otherwise}$$

The parameter n is called the *degrees of freedom* of chi-square. The mean and the variance of χ^2 with n degrees of freedom are

$$\begin{cases} E(\chi^2) = n \\ \text{Var}(\chi^2) = 2n \end{cases} \tag{6.46}$$

Proof. Since χ^2 is the sum of n independent gamma variates with $\lambda = \frac{1}{2}$ and $k = \frac{1}{2}$,

$$E(\chi^2) = n\left(\frac{k}{\lambda}\right) = n$$

and

$$\text{Var}(\chi^2) = n\left(\frac{k}{\lambda^2}\right) = 2n$$

An important property of the χ^2 distribution is additivity. If $\chi_1^2, \chi_2^2, \ldots, \chi_k^2$ are independent chi-square variates with $n_1, n_2, \ldots, n_k$ degrees of freedom, respectively, the sum of these variates ($\sum_{i=1}^{k} \chi_i^2$) has χ^2 distribution with $n_1 + n_2 + \cdots + n_k$ degrees of freedom.

Proof. Since χ_i^2 is the sum of squares of n_i independent unit normal variates, $\sum_{i=1}^{k} \chi_i^2$ is the sum of squares of $\sum_{i=1}^{k} n_i$ independent unit normal variates.

Hence, by definition, the sum of squares of $\sum_{i=1}^{k} n_i$ independent unit normal variates must have a χ^2 distribution with $n_1 + n_2 + \cdots + n_k$ degrees of freedom.

Graphs of several χ^2 distributions are shown in Fig. 6.10. In the figure are presented χ^2 curves with number of degrees of freedom n equal to 1, 4, 10, and 20. Figure 6.10 exhibits the fact that these distributions are positively skewed. As the degrees of freedom increase the curve approaches symmetry. Percentage points of the χ^2 distribution are presented in Table G in the Appendix.

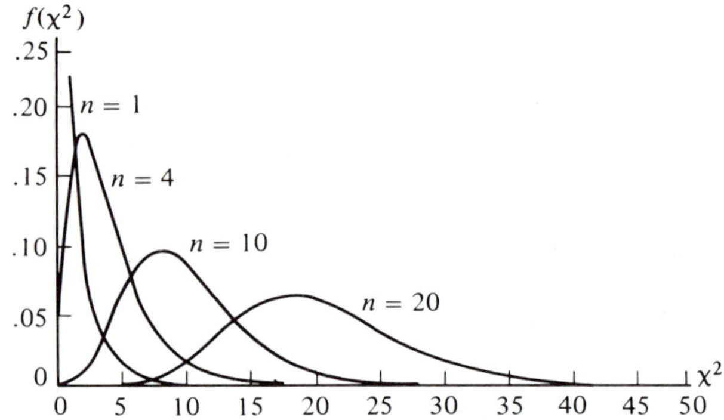

Fig. 6.10 Graphs of probability density functions of the χ^2 distribution.

Example 6.14 If χ_1^2, χ_2^2, and χ_3^2 have a chi-square distribution with 5, 4, and 3 degrees of freedom, respectively, find the probability that

$$\sum_{i=1}^{3} \chi_i^2 > 14.85$$

Solution. $\sum_{i=1}^{3} \chi_i^2$ has χ^2 with 12 degrees of freedom. Hence, from Table G (in the Appendix), we find

$$P\{\chi_{12}^2 > 14.85\} = .25$$

6.15 Student's *t* Distribution

Definition 6.11. If X^* is a unit normal variate and χ^2 is a variate that is independent of X^* and has a χ^2 distribution with n degrees of freedom, then the variate

$$t = \frac{X^* \sqrt{n}}{\sqrt{\chi^2}}$$

has a t distribution with n degrees of freedom. The density function of the t distribution with n degrees of freedom is given by

$$(6.47) \qquad f(t) = \frac{\Gamma[(n+1)/2]}{\sqrt{n\pi}\Gamma(n/2)}\left(1 + \frac{t^2}{n}\right)^{-(n+1)/2} \qquad -\infty < t < \infty$$

The t distribution is symmetric around its mean, which is equal to zero. We shall prove that the mean and the variance of the t variate are

$$(6.48) \qquad \begin{cases} E(t) = 0 & n > 1 \\ \operatorname{Var}(t) = \dfrac{n}{n-2} & n > 2 \end{cases}$$

Proof

$$E(t) = \frac{\Gamma[(n+1)/2]}{\sqrt{n\pi}\Gamma(n/2)} \int_{-\infty}^{\infty} \frac{t\,dt}{[1 + (t^2/n)]^{(n+1)/2}}$$

$$= \frac{n}{1-n}\frac{\Gamma[(n+1)/2]}{\sqrt{n\pi}\Gamma(n/2)}\left[\frac{1}{[1 + (t^2/n)]^{(n-1)/2}}\Bigg|_{-\infty}^{\infty}\right]$$

$$= 0 \qquad n > 1$$

$$\operatorname{Var}(t) = E(t^2) = \frac{\Gamma[(n+1)/2]}{\sqrt{n\pi}\Gamma(n/2)} \int_{-\infty}^{\infty} \frac{t^2\,dt}{[1 + (t^2/n)]^{(n+1)/2}}\,dt$$

Integrating by parts we obtain

$$\operatorname{Var}(t) = \frac{\Gamma[(n+1)/2]}{\sqrt{n\pi}\Gamma(n/2)}\left[\frac{nt}{(1-n)[1 + (t^2/n)]^{(n-1)/2}}\Bigg|_{-\infty}^{\infty} + \int_{-\infty}^{\infty} \frac{n}{(n-1)}\frac{dt}{[1 + (t^2/n)]^{(n-1)/2}}\right]$$

$$= \frac{\Gamma[(n+1)/2]}{\sqrt{n\pi}\Gamma(n/2)} \times \frac{\left\{\displaystyle\int_{-\infty}^{\infty} \frac{dt}{[1 + (t^2/n)]^{(n+1)/2}} - \left[\frac{t}{[1 + (t^2/n)]^{(n-1)/2}}\Bigg|_{-\infty}^{\infty}\right]\right\}\dfrac{n}{n-1}}{(n-2)/(n-1)}$$

$$= \frac{n}{n-2} \qquad n > 2$$

Example 6.15 If t has a t distribution with 10 degrees of freedom, find
(a) $P\{t \geq 1.81\}$
(b) $P\{1.37 \leq t \leq 2.22\}$.

Solution. From Table F (in the Appendix) we find that
(a) $P\{t_{10} \geq 1.81\} = .05$
(b) $P\{1.37 \leq t_{10} \leq 2.22\} = .975 - .9 = .075$

6.16 *F* Distribution

Definition 6.12. If χ_1^2 and χ_2^2 are two independent variates that have chi-square distributions with n_1 and n_2 degrees of freedom, respectively, then the variate

$$F = \frac{\chi_1^2/n_1}{\chi_2^2/n_2}$$

has an F distribution with n_1 and n_2 degrees of freedom. The density function of the F variate is given by

$$(6.49)\qquad \begin{aligned} f(F) &= \frac{\Gamma[(n_1 + n_2)/2]n_1^{n_1/2}n_2^{n_2/2}(F)^{[(n_1/2)-1]}}{\Gamma(n_1/2)\Gamma(n_2/2)(n_2 + n_1F)^{[(n_1+n_2)/2]}} \qquad F > 0 \\ &= 0 \qquad \text{otherwise} \end{aligned}$$

The mean and variance of an F variate are

$$(6.50)\qquad \begin{cases} E(F) = \dfrac{n_2}{n_2 - 2} & n_2 > 2 \\ \mathrm{Var}(F) = \dfrac{n_2^2(2n_2 + 2n_1 - 4)}{n_1(n_2 - 2)^2(n_2 - 4)} & n_2 > 4 \end{cases}$$

Proof. Let $(n_1/n_2)F = z$. Hence

$$g(z) = \frac{n_1}{n_2}\frac{\Gamma[(n_1 + n_2)/2]z^{[(n_1/2)-1]}}{\Gamma(n_1/2)\Gamma(n_2/2)(1 + z)^{[(n_1+n_2)/2]}}$$

However,

$$\int_0^\infty g(z)\,dz = 1$$

Therefore

$$\int_0^\infty \frac{z^{(n_1/2)-1}\,dz}{(1 + z)^{[(n_1+n_2)/2]}} = \frac{n_2}{n_1}\frac{\Gamma(n_1/2)\Gamma(n_2/2)}{\Gamma[(n_1 + n_2)/2]}$$

The expected value of F is given by

$$\begin{aligned} E(F) &= \frac{n_2}{n_1}E(z) = \frac{n_2}{n_1}\int_0^\infty zg(z)\,dz \\ &= \frac{\Gamma[(n_1 + n_2)/2]}{\Gamma(n_1/2)\Gamma(n_2/2)}\int_0^\infty \frac{z^{n_1/2}}{(1 + z)^{(n_1+n_2)/2}}\,dz \end{aligned}$$

Now let $n_1 = \nu_1 - 2$ and $n_2 = \nu_2 + 2$; then we have

$$\begin{aligned} E(F) &= \frac{\Gamma[(n_1 + n_2)/2]}{\Gamma(n_1/2)\Gamma(n_2/2)}\int_0^\infty \frac{z^{(\nu_1/2)-1}}{(1 + z)^{(\nu_1+\nu_2)/2}}\,dz \\ &= \frac{\Gamma[(n_1 + n_2)/2]}{\Gamma(n_1/2)\Gamma(n_2/2)}\frac{n_2\Gamma(\nu_1/2)\Gamma(\nu_2/2)}{n_1\Gamma[(\nu_1 + \nu_2)/2]} \\ &= \frac{n_2}{n_1}\frac{\Gamma[(n_1 + 2)/2]\Gamma[(n_2 - 2)/2]}{\Gamma(n_1/2)\Gamma(n_2/2)} \\ &= \frac{n_2}{n_2 - 2} \end{aligned}$$

and

$$E(F^2) = \int_\infty^\infty F^2f(F)\,dF$$

$$= \left(\frac{n_2}{n_1}\right)^2 \frac{\Gamma[(n_1 + n_2)/2]}{\Gamma(n_1/2)\Gamma(n_2/2)} \frac{\Gamma[(n_1 + 4)/2]\Gamma[(n_2 - 4)/2]}{\Gamma[(n_1 + n_2)/2]}$$

$$= \frac{n_2^2(n_1 + 2)}{n_1(n_2 - 2)(n_2 - 4)}$$

Hence

$$\begin{aligned}\text{Var}(F) &= E(F^2) - [E(F)]^2 \\ &= \frac{n_2^2(n_1 + 2)}{n_1(n_2 - 2)(n_2 - 4)} - \frac{n_2^2}{(n_2 - 2)^2} \\ &= \frac{2n_2^2(n_1 + n_2 - 2)}{(n_2 - 2)^2(n_2 - 4)n_1} \qquad n_2 > 4\end{aligned}$$

A useful property of the F distribution is the *reciprocal property of its percentiles.* In symbols,

(6.51)
$$F_{1-\alpha;\, n_1, n_2} = \frac{1}{F_{\alpha;\, n_2, n_1}}$$

Proof

$$P\left\{\frac{\chi_1^2/n_1}{\chi_2^2/n_2} \geq F_{1-\alpha;\, n_1, n_2}\right\} = P\left\{\frac{\chi_2^2/n_2}{\chi_1^2/n_1} \leq \frac{1}{F_{1-\alpha;\, n_1, n_2}}\right\} = \alpha$$

or

$$P\left\{\frac{\chi_2^2/n_2}{\chi_1^2/n_1} \geq \frac{1}{F_{1-\alpha;\, n_1, n_2}}\right\} = 1 - \alpha$$

Since by definition $(\chi_2^2/n_2)/(\chi_1^2/n_1)$ has an F distribution with n_2 and n_1 degrees of freedom, we write

$$P\left\{\frac{\chi_2^2/n_2}{\chi_1^2/n_1} \geq F_{\alpha;\, n_2, n_1}\right\} = 1 - \alpha$$

Hence

$$F_{\alpha;\, n_2, n_1} = \frac{1}{F_{1-\alpha;\, n_1, n_2}}$$

or

$$F_{1-\alpha;\, n_1, n_2} = \frac{1}{F_{\alpha;\, n_2, n_1}}$$

Example 6.16 Variate F has an F distribution with 15 and 12 degrees of freedom. Find

(a) $F_{.05}$
(b) $F_{.95}$
(c) $P\{F \geq 2.10\}$.

Solution. From Table H (in the Appendix) we find

(a) $F_{.05;\, 15, 12} = \dfrac{1}{F_{.95;\, 12, 15}} = \dfrac{1}{2.48} = .403$
(b) $F_{.95;\, 15, 12} = 2.62$
(c) $P\{F_{15, 12} \geq 2.10\} = .10$

A sketch of the density function of the F distribution is given in Fig. 6.11, and it can be seen from the figure that the distribution is positively skewed. Percentage points of the F distribution are given in Table H (in the Appendix).

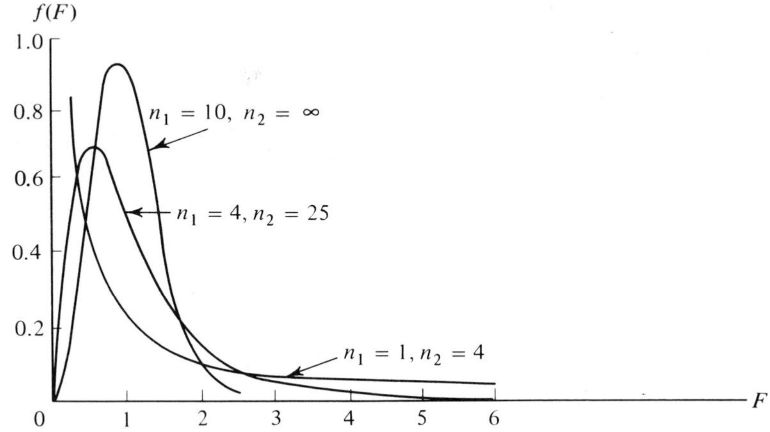

Fig. 6.11 Graphs of probability density functions of the F distribution.

6.17 Weibull Distribution

In 1951 W. Weibull [12] derived a three-parameter probability density function known as the *Weibull distribution.* The Weibull distribution was first applied to the fatigue of materials; however, empirical data collected in life testing of electronic components of various types substantiated the usefulness of this distribution. It is well suited to represent many empirical distributions. By adjustment of the three parameters, the Weibull distribution can be made to closely fit the random variables being studied.

Definition 6.13. A random variable X is said to have the *Weibull distribution* if its probability density function is given by

$$f_X(t) = \frac{\beta}{\eta}\left(\frac{t-\gamma}{\eta}\right)^{\beta-1} \exp\left[-\left(\frac{t-\gamma}{\eta}\right)^{\beta}\right] \qquad t \geq \gamma \qquad \eta, \beta > 0 \tag{6.52}$$
$$= 0 \qquad \text{otherwise}$$

The three parameters in Eq. (6.52) are

1. η called the *true scale parameter.*
2. β called the *shape parameter.*
3. γ called the *location or threshold parameter.*

The location parameter is usually set equal to zero. This procedure seems reasonable since failures can occur as soon as the experiment is started. Accordingly, Eq. (6.52) reduces to

$$f_X(t) = \frac{\beta}{\eta}\left(\frac{t}{\eta}\right)^{\beta-1} \exp\left[-\left(\frac{t}{\eta}\right)^{\beta}\right] \qquad t > 0 \qquad \eta > 0 \qquad \beta > 0 \tag{6.53}$$
$$= 0 \qquad \text{otherwise}$$

which is the density function of a two-parameter Weibull distribution. Figure 6.12 shows the probability density function of the Weibull distribution given by Eq. (6.53) for $\eta = 1$ and $\beta = \frac{1}{2}$, 1, 2, and 4. As illustrated in Fig. 6.12,

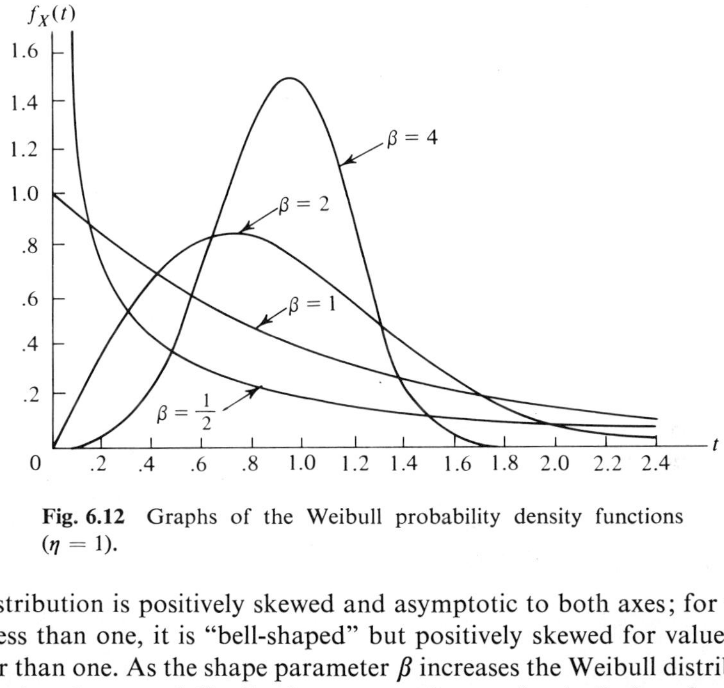

Fig. 6.12 Graphs of the Weibull probability density functions ($\eta = 1$).

the distribution is positively skewed and asymptotic to both axes; for values of β less than one, it is "bell-shaped" but positively skewed for values of β greater than one. As the shape parameter β increases the Weibull distribution approaches the normal distribution more and more closely. In fact, for $\beta = 4$ the Weibull and the normal are almost indistinguishable. For $\beta = 1$ the Weibull distribution coincides with the exponential distribution, whereas for $\beta = 2$ and $\eta = \sqrt{2}\,\alpha$ it coincides with the Rayleigh distribution (see Prob. 6.59).

Let us find the mean and variance of the two-parameter Weibull distribution given by Eq. (6.53). The kth moment around the origin is

$$E(X^k) = \int_0^\infty \frac{t^k \beta}{\eta}\left(\frac{t}{\eta}\right)^{\beta-1} e^{-(t/\eta)^\beta}\, dt \tag{6.54}$$

Letting $(t/\eta)^{\beta} = u$, Eq. (6.54) becomes

$$E(X^k) = \int_0^{\infty} \eta^k u^{k/\beta} e^{-u}\, du \tag{6.55}$$
$$= \eta^k \Gamma\left(\frac{k}{\beta} + 1\right)$$

Hence, the mean and the variance are

$$\begin{cases} E(X) = \eta\Gamma\left(\frac{1}{\beta} + 1\right) \\ \mathrm{Var}(X) = \eta^2\Gamma\left(\frac{2}{\beta} + 1\right) - \eta^2\left[\Gamma\left(\frac{1}{\beta} + 1\right)\right]^2 \\ \qquad = \eta^2\left\{\Gamma\left(\frac{2}{\beta} + 1\right) - \left[\Gamma\left(\frac{1}{\beta} + 1\right)\right]^2\right\} \end{cases} \tag{6.56}$$

Example 6.17 For a certain type of tube the values of the scale and shape parameters are 19.7 and 1.7 kwh, respectively. What is the probability that this type of tube will perform satisfactorily for at least 985 hr?

Solution. The desired probability is

$$P\{X > 985\} = 1 - \int_0^{.985} \frac{\beta}{\eta}\left(\frac{t}{\eta}\right)^{\beta - 1} \exp\left[-\left(\frac{t}{\eta}\right)^{\beta}\right] dt$$
$$= e^{-(.985/\eta)^{\beta}} = e^{-(.985/19.7)^{1.7}} = .994$$

6.18 The Beta Distribution

Definition 6.14. The *beta distribution* is defined by the equation

$$f_X(t) = \frac{\Gamma(\alpha + \beta + 2)}{\Gamma(\alpha + 1)\Gamma(\beta + 1)} t^{\alpha}(1 + t)^{\beta} \qquad 0 < t < 1 \qquad \alpha > -1 \tag{6.57}$$
$$= 0 \qquad \text{otherwise} \qquad\qquad \beta > -1$$

The beta distribution is a two-parameter family of distributions. The distribution coincides with the uniform distribution when $\alpha = \beta = 0$.

Since Eq. (6.57) is a density function, it can be shown that

$$\int_0^1 f_X(t)\, dt = 1 \tag{6.58}$$

It follows from Eq. (6.58) that

$$\int_0^1 t^{\alpha}(1 - t)^{\beta}\, dt = \frac{\Gamma(\alpha + 1)\Gamma(\beta + 1)}{\Gamma(\alpha + \beta + 2)} \tag{6.59}$$

The right-hand side of Eq. (6.59) is called the *beta function* and is usually denoted by $\mathrm{B}(\alpha + 1, \beta + 1)$.

The cumulative distribution, often called the *incomplete beta function*, is

$$(6.60) \qquad F_X(a) = \begin{cases} 0 & a \leq 0 \\ \displaystyle\int_0^a \frac{\Gamma(\alpha+\beta+2)}{\Gamma(\alpha+1)\Gamma(\beta+1)} t^\alpha (1-t)^\beta \, dt & 0 < a < 1 \\ 1 & a \geq 1 \end{cases}$$

For computing the cumulative distribution given by Eq. (6.60) we can use Karl Pearson's tables of the incomplete beta function published by Cambridge Press (London, 1932).

We can find the mean and variance of the beta distribution by first calculating the kth moment around the origin.

$$(6.61) \qquad \begin{aligned} E(X^k) &= \frac{\Gamma(\alpha+\beta+2)}{\Gamma(\alpha+1)\Gamma(\beta+1)} \int_0^1 t^{\alpha+k}(1-t)^\beta \, dt \\ &= \frac{\Gamma(\alpha+\beta+2)\Gamma(\alpha+k+1)}{\Gamma(\alpha+1)\Gamma(\alpha+\beta+k+2)} \end{aligned}$$

since the integral is equal to $[\Gamma(\alpha+k+1)\Gamma(\beta+1)/\Gamma(\alpha+\beta+k+2)]$. Thus the mean and variance are

$$(6.62) \qquad \begin{cases} E(X) = \dfrac{\alpha+1}{\alpha+\beta+2} \\ \operatorname{Var}(X) = \dfrac{(\alpha+1)(\beta+1)}{(\alpha+\beta+2)^2(\alpha+\beta+3)} \end{cases}$$

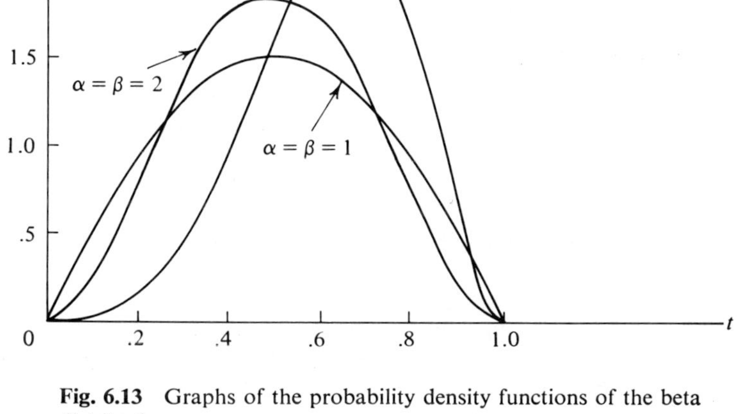

Fig. 6.13 Graphs of the probability density functions of the beta distribution.

Figure 6.13 represents the density of the beta distribution with $\alpha = \beta = 1$, $\alpha = \beta = 2$, and $\alpha = 4$, $\beta = 3$.

Example 6.18 The random variable X has the beta distribution with $\alpha = \beta = 1$. Compute the following:
(a) $E(X)$
(b) $P\{X > .6\}$.

Solution
(a) From Eq. (6.22) we find

$$E(X) = \tfrac{2}{4} = \tfrac{1}{2}$$

(b)

$$P\{X > .6\} = \int_{.6}^{1} \frac{\Gamma(4)}{\Gamma(2)\Gamma(2)} t(1-t)\,dt$$

$$= 6\left[\frac{t^2}{2} - \frac{t^3}{3}\right]_{.6}^{1} = .352$$

Table 6.3 presents the probability laws (and their properties) of some frequently encountered continuous distributions.

6.19 Approximations

6.19.1 Normal Approximation to the Binomial

It has been shown that the distribution of the sum of n independent binomially distributed variates each with parameters $n = 1$ and p is again a binomial with parameters n and p. Thus, applying the central limit theorem, we have

$$P\left\{a \le \frac{\sum_{i=1}^{n} X_i - np}{\sqrt{npq}} \le b\right\} \sim \frac{1}{\sqrt{2\pi}} \int_a^b \exp\left(-\frac{u^2}{2}\right) du \qquad \text{as } n \to \infty$$

where X_i is defined as

$$X_i = \begin{cases} 0 & \text{if success occurs on the } i\text{th trial} \\ 1 & \text{if failure occurs on the } i\text{th trial} \end{cases}$$

and

$$E(X_i) = p \qquad \text{Var}(X_i) = pq$$

The approximation tends to be good when p is in the neighborhood of $\frac{1}{2}$ and $npq > 9$.

Example 6.19 In sampling without replacement from a lot that is 10% defective the sample size is 18. Compute exactly the probability that the number of defective items in the sample is less than or equal to five but greater than or equal to three. Use the normal approximation and compare your results.

TABLE 6.3 Some Important Continuous Probability Laws and Their Properties

Distribution	*Density Function*	*Parameters and Conditions*	*Mean*	*Variance*	*Moment-Generating Function* $\Psi_X^{(\theta)}$
Normal	$f_X(t) = \frac{1}{\sigma\sqrt{2\pi}} \exp\left[-\frac{1}{2}\left(\frac{t-\mu}{\sigma}\right)^2\right]$ $-\infty < t < \infty$	$-\infty < \mu < \infty$ $\sigma > 0$	μ	σ^2	$e^{[\theta\mu + (\theta^2\sigma^2/2)]}$
Exponential	$f_X(t) = \lambda e^{-\lambda t} \quad t \geq 0$ $= 0 \quad t < 0$	$\lambda > 0$	$\frac{1}{\lambda}$	$\frac{1}{\lambda^2}$	$\frac{\lambda}{\lambda - \theta}$
Uniform over Interval a to b	$f_X(t) = \frac{1}{b-a} \quad a < t < b$ $= 0 \quad$ otherwise	$-\infty < a < b < \infty$	$\frac{a+b}{2}$	$\frac{(b-a)^2}{12}$	$\frac{e^{\theta b} - e^{\theta a}}{\theta(b-a)}$
Gamma	$f_X(t) = \frac{\lambda(\lambda t)^{k-1}e^{-\lambda t}}{\Gamma(k)} \quad t \geq 0$ $= 0 \quad$ otherwise	$\lambda > 0$ $k > 0$	$\frac{k}{\lambda}$	$\frac{k}{\lambda^2}$	$\left(1 - \frac{\theta}{\lambda}\right)^{-k}$
Chi-square	$f(\chi^2) = \frac{(\chi^2)^{[(n-2)/2]}e^{-\chi^2/2}}{2^{n/2}\Gamma(n/2)} \quad \chi^2 \geq 0$ $= 0 \quad$ otherwise	$n = 1, 2, \ldots$ (called degrees of freedom)	n	$2n$	$(1 - 2\theta)^{-(n/2)}$

Solution. The number of defective items in the sample is a binomially distributed random variable. Let d denote the number of defective items in the sample; then

$$P\{3 \leq d \leq 5\} = \sum_{k=3}^{5} \binom{18}{k}(.1)^k(.9)^{18-k}$$
$$= .168 + .07 + .0218 = .2598$$

Using the normal approximation we have

$$P\{3 \leq d \leq 5\} = P\left\{\frac{3-1.8}{\sqrt{(18)(.1)(.9)}} \leq \frac{d-1.8}{\sqrt{(18)(.1)(.9)}} \leq \frac{5-1.8}{\sqrt{(18)(.1)(.9)}}\right\}$$
$$\sim \int_{(3-1.8)/\sqrt{(18)(.1)(.9)}}^{(5-1.8)/\sqrt{(18)(.1)(.9)}} \frac{e^{-(u^2/2)}}{\sqrt{2\pi}}\,du = \Phi(2.521) - \Phi(.945) = .1665$$

which is not a good approximation since npq is relatively small.

6.19.2 Normal Approximation to the Poisson

The Poisson distribution approaches the normal for large values of λ. Consider a Poisson variate Y with large λ as the summation of a large number of Poisson variates such that the summation of their individual parameters is equal to λ. Hence, according to the central limit theorem, we may write

$$\lim_{\lambda\to\infty} P\left\{a \leq \frac{Y-\lambda}{\sqrt{\lambda}} \leq b\right\} = \frac{1}{\sqrt{2\pi}}\int_a^b e^{-(u^2/2)}\,du$$

Good approximation is obtained wherever $\lambda > 10$.

Example 6.20 The number of customers (N) arriving at a service counter per hour has been found to have a Poisson distribution with $\lambda = 10$. Compute both by the exact method and using the normal approximation: $P\{9 \leq N \leq 12\}$.

Solution

$$P\{9 \leq N \leq 12\} = \sum_{N=9}^{12} \frac{e^{-10}10^N}{N!} = .459$$

Using normal approximation we have

$$\int_{(9-10)/\sqrt{10}}^{(12-10)/\sqrt{10}} \frac{e^{-(u^2/2)}}{\sqrt{2\pi}}\,du = \Phi\left(\frac{2}{\sqrt{10}}\right) - \Phi\left(\frac{-1}{\sqrt{10}}\right) = .360$$

Note the approximation can be improved by correcting for continuity (see Sec. 6.19.5).

6.19.3 Poisson Approximation to the Binomial

In Sec. 6.2 we proved that the limiting distribution of the binomial is the Poisson distribution. In symbols,

$$\lim_{n\to\infty}\binom{n}{k}p^k(1-p)^{n-k} = \lim_{n\to\infty}\binom{n}{k}\left(\frac{\lambda}{n}\right)^k\left(1-\frac{\lambda}{n}\right)^{n-k} = \frac{e^{-\lambda}\lambda^k}{k!}$$

This approximation has been found satisfactory for $p < .1$ and n large.

Example 6.21 What is the probability of accepting a lot whose incoming quality is 5% defective using a sample of size 20 (with replacement) and accepting the lot if the number of defective items d in the sample is ≤ 1? Compare your answer using the Poisson approximation.

Solution. From the binomial distribution we have

$$P\{d \leq 1\} = \sum_{d=0}^{1} \binom{n}{d} p^d (1-p)^{n-d} = \sum_{d=0}^{1} \binom{20}{d} (.05)^d (.95)^{20-d} = .7358$$

Using Poisson approximation we have

$$P\{d \leq 1\} = \sum_{d=0}^{1} \frac{e^{-np}(np)^d}{d!} = \sum_{d=0}^{1} \frac{e^{-1}}{d!} = .736$$

which is almost identical to the exact value.

6.19.4 Binomial Approximation to the Hypergeometric

We shall now show that as the population size N increases without bound, the *hypergeometric distribution* with parameters n, p, and N approaches the *binomial distribution* with parameters n and p.

Proof. The probability function of the hypergeometric distribution is given by

$$\begin{aligned} P_X(k) &= \frac{\binom{Np}{k}\binom{N(1-p)}{(n-k)}}{\binom{N}{n}} \\ &= \frac{(Np)!}{(Np-k)!\,k!} \frac{[N(1-p)]!}{[N(1-p)-n+k]!\,(n-k)!} \frac{(N-n)!\,n!}{N!} \\ &= \frac{Np(Np-1)\cdots(Np-k+1)}{k!} \\ &\quad \times \frac{[N(1-p)][N(1-p)-1]\cdots[N(1-p)-n+k+1]}{(n-k)!} \\ &\quad \times \frac{n!}{N(N-1)\cdots(N-n+1)} \\ &= \binom{n}{k} \frac{p[p-(1/N)]\cdots[p-(k-1)/N]}{1[1-(1/N)]\cdots[1-(n-1)/N]} \\ &\quad \times \frac{(1-p)[1-p-(1/N)]\cdots[1-p-(n-k-1)/N]}{} \end{aligned} \tag{6.63}$$

If we let N increase without bound, p remains constant; we find that

$$\lim_{N\to\infty} P_X(k) = \binom{n}{k} p^k (1-p)^{n-k} \tag{6.64}$$

which is the probability function of the binomial distribution. The importance of Eq. (6.64) lies in the fact that when the sampling fraction n/N is small enough, binomial probabilities can be used as an approximation to the hypergeometric probabilities. In other words, *when the sampling fraction n/N is small, sampling with replacement gives approximately the same probabilities as sampling without replacement*. This approximation has been found satisfactory if the sampling fraction n/N is less than .1.

The following example will illustrate the closeness of the binomial approximation to the hypergeometric:

Example 6.22 A sample of three items is drawn without replacement from a lot of 50, which contains 6 defective items.
(a) What is the probability that the sample will contain at least two nondefectives?
(b) Use the binomial approximation to find the probability in (a).

Solution
(a) From the hypergeometric distribution we find

$$P\{X \leq 1\} = \sum_{k=0}^{1} \frac{\binom{Np}{k}\binom{Nq}{n-k}}{\binom{N}{n}} = \sum_{k=0}^{1} \frac{\binom{6}{k}\binom{44}{3-k}}{\binom{50}{3}} = .9653$$

(b) Using the binomial approximation we find

$$P\{X \leq 1\} = \sum_{k=0}^{1} \binom{n}{k} p^k q^{n-k} = \sum_{k=0}^{1} \binom{3}{k} (.12)^k (.88)^{3-k} = .9602$$

which is very close indeed to the true value despite the fact that the sampling fraction is comparatively large.

6.19.5 Correction for Continuity

In Secs. 6.19.1 and 6.19.2 we discussed the use of the normal distribution to approximate the probabilities associated with binomial and Poisson random variables. In other words, in using the normal approximation to the binomial and Poisson distributions we are approximating the distribution of discrete variates with the distribution of a continuous variate. Hence, some care must be exercised in determining the end points of the intervals involved. To illustrate, consider the probability function of the Poisson distribution with $\lambda = 11$, at values 9, 10, 11, and 12, as shown in Fig. 6.14. It can readily be seen from this figure that a correcting factor of $+\frac{1}{2}$ may be added to the upper bound and of $-\frac{1}{2}$ to the lower bound. These corrections for continuity have been found to yield a better approximation. To illustrate the extent of improvement, let us find the probability that X is either 10, 11, or 12, i.e., $P\{10 \leq X \leq 12\}$. From Fig. 6.14, we have

$$P\{10 \leq X \leq 12\} = .119 + .119 + .110 = .348$$

Using the normal approximation we obtain

$$\int_{(10-11)/\sqrt{11}}^{(12-11)/\sqrt{11}} \frac{e^{-(u^2/2)}\,du}{\sqrt{2\pi}} = \Phi\left(\frac{1}{\sqrt{11}}\right) - \Phi\left(\frac{-1}{\sqrt{11}}\right) = .237$$

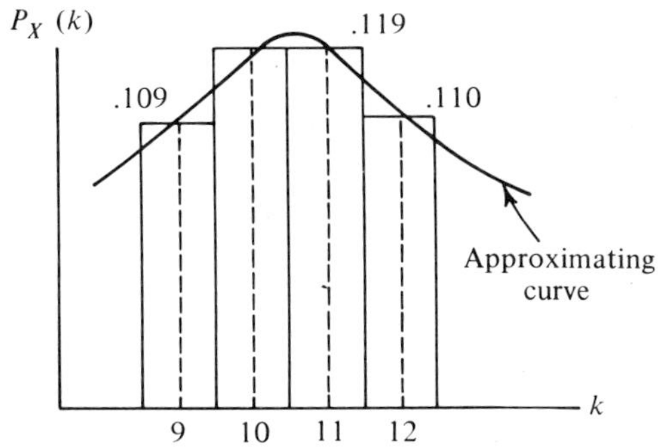

Fig. 6.14 Portion of the histogram of the Poisson distribution with $\lambda = 11$.

Now if we correct for continuity, this approximation becomes

$$\int_{(9.5-11)/\sqrt{11}}^{(12.5-11)/\sqrt{11}} \frac{e^{-(u^2/2)}du}{\sqrt{2\pi}} = \Phi\left(\frac{1.5}{\sqrt{11}}\right) - \Phi\left(\frac{-1.5}{\sqrt{11}}\right) = .338$$

which is much closer to the exact value. Similarly,

$$P\{X \geq 10\} \cong \int_{(9.5-11)/\sqrt{11}}^{\infty} \frac{e^{-(u^2/2)}du}{\sqrt{2\pi}} = 1 - \Phi\left(-\frac{1.5}{\sqrt{11}}\right) = .674$$

while the true value is .659.

The same procedure can be used for the binomial case. If X is binomially distributed with parameters n and p, then

$$P\{a \leq X \leq b\} \cong \Phi\left(\frac{b + \frac{1}{2} - np}{\sqrt{npq}}\right) - \Phi\left(\frac{a - \frac{1}{2} - np}{\sqrt{npq}}\right)$$

and

$$P\{X = a\} \cong \Phi\left(\frac{a + \frac{1}{2} - np}{\sqrt{npq}}\right) - \Phi\left(\frac{a - \frac{1}{2} - np}{\sqrt{npq}}\right)$$

Thus, by adding $\frac{1}{2}$ to the upper bound and subtracting $\frac{1}{2}$ from the lower bound the approximation tends to be closer to the actual value.

PROBLEMS

6.1. Let X be a binomially distributed random variable with mean $= 1$ and variance $= .90$. Compute the following:
(a) $P\{X > 5\}$
(b) $P\{1 < X < 4\}$
(c) $P\{3 \leq X \leq 5\}$.

6.2. Let X be a binomially distributed random variable with parameters $n = 5$ and $p = .30$. Compute the following:

(a) $P\{1 < X \leq 3\}$
(b) $P\{X \geq 4\}$
(c) $P\{2 \leq X \leq 4\}$.

6.3. A die is tossed ten times. Considering a one or two as a failure, what is the probability of getting six failures?

6.4. Suppose that random variable X is binomially distributed with parameters n and p. Use the moment-generating function to evaluate $E(X^3)$.

6.5. What is the most probable number of failures in ten binomial trials with failure probability $p = .10$?

6.6. Suppose the probability of hitting a target is .50 and 30 shots were fired independently.
(a) What is the probability that the target will be hit at least ten times?
(b) What is the expected number of hits?

6.7. Assuming that the sexes are equally likely, what is the distribution of the number of boys (X) in a family of eight children? Plot the cumulative distribution function of X. Compute $P\{X > 4\}$.

6.8. If X has a Poisson distribution with $\lambda = 10$, compute the following probabilities:
(a) $P\{X > 6\}$
(b) $P\{2 \leq X \leq 4\}$
(c) $P\{X \leq 8\}$
(d) $P\{6 \leq X \leq 12\}$.

6.9. Suppose X is a Poisson variate. If $P_X(0) = .135$, compute $P\{X > 3\}$.

6.10. Suppose X is a Poisson variate. If $P_X(1) = .0337$ and $P_X(2) = .08425$, compute $P\{X > 7\}$.

6.11. A newspaper boy buys his papers at \$.07 each and he sells them for \$.10 and receives \$.01 for an unsold paper. If the number of customers is a Poisson variate with $\lambda = 25$, how many papers should he buy in order to maximize his profit?

6.12. The number of defective components per television set has been found to follow the Poisson distribution with $\lambda = 2$. If five sets were tested, what is the probability that the number of defective components exceeds ten?

6.13. Assume that the demand for a certain commodity follows the Poisson distribution with $\lambda = 25$. How large a stock should be carried such that the demand is satisfied in 95% of the cases?

6.14. The number of machine failures per day in a certain plant has a Poisson distribution with $\lambda = 3$. Present maintenance facilities can repair three machines per day. Failures in excess of three are repaired by a contractor.
(a) On a given day, what is the probability of having machines repaired by the contractor?
(b) If present maintenance facilities can repair four machines per day, compute the probability in (a).
(c) What is the most probable number of machine failures per day?
(d) What is the expected number of machines repaired at the plant daily?
(e) What is the expected number of machines repaired by the contractor daily?

6.15. A fair die is tossed five times. What is the probability of obtaining 2 fours, 2 fives, and 1 six?

6.16. A camshaft assembly is classified as tight fit, loose fit, or close running fit. If the respective probabilities are .60, .20, and .20, find the probability that among five such assemblies one is tight fit, one is loose fit, and two are close running fits.

6.17. Suppose that in a production line on the average 5% of the items have major defects while 10% have minor defects. Furthermore, suppose that it costs \$10 and \$30 to repair an item with minor and major defects, respectively. If 100 items were produced find the following:
(a) The expected numbers of items with minor and major defects.
(b) The covariance of the number of items with minor and major defects.
(c) The expected loss due to defects.
(d) The variance of the loss due to defects.

6.18. What is the probability of obtaining 5 defectives in a sample of size 100 taken without replacement from a lot of 200 units containing 10 defectives?

6.19. In the preceding problem, find the probability that the number of defectives in the sample is less than or equal to three.

6.20. A box contains ten fuses, two of which are blown out. What is the probability of obtaining two defectives in a sample of size eight taken without replacement?

6.21. What is the probability of throwing a nine with a pair of dice in less than five tosses?

6.22. A shaft is subjected to repeated stress cycles. Assume that the probability of failure at any cycle is constant and equal to .0002. What is the probability that the shaft will fail on or before the one-thousandth stress cycle has been applied?

6.23. The probability of success in a certain experiment is .10. What is the probability of obtaining a success in less than three independent trials?

6.24. In the preceding problem, suppose that the costs associated with success and failure are \$800 and \$1,200, respectively. What is the expected cost until the first success is achieved?

6.25. It has been asserted that the geometric distribution has "no memory," i.e.,

$$P\{X > s + r \mid X > s\} = P\{X > r\}$$

where X has a geometric distribution and s and r are any positive numbers. Prove the assertion.

6.26. Suppose that the probability of success at each trial is .50. What is the probability that the second success occurs on or before the fifth trial? Assume trials are independent.

6.27. The probability of a successful launching of a special satellite is .20. Once two launchings are successful, no further launchings will be made.
(a) What is the probability that exactly ten attempts will be made?
(b) What is the probability that at most ten attempts will be required?

6.28. In the preceding problem, suppose that the costs associated with successful and unsuccessful launchings are \$500,000 and \$750,000, respectively. Compute the expected cost of the experiment.

6.29. Suppose that X is a normal variate with $\mu = 50$ and $\sigma = 5$. Find the following probabilities:
(a) $P\{X \geq 60\}$
(b) $P\{40 \leq X \leq 60\}$
(c) $P\{55 < X < 65\}$
(d) $P\{X \leq 45\}$.

6.30. Suppose that the diameter of a rivet produced by a certain process is a normally distributed random variable with $\mu = .200$ in. and $\sigma = .0033$ in. Rivet specifications call for diameters of $.205 \pm .005$ in. What percentage of these rivets will meet specifications?

6.31. A study showed that the life of army shoes is normally distributed with $\mu = 15$

months and $\sigma = 1.5$ months. If 30,000 pairs are issued, how many pairs would be replaced after 18 months?

6.32. The measurement of the width of a keyway is normally distributed with $\mu = .365$ in. and $\sigma = .005$ in. The specifications are given as $.366 \pm .002$ in. What is the probability that the width will not conform to specifications?

6.33. Three resistors are assembled in series. The resistors are known to be normally distributed with $\mu_1 = 5$, $\mu_2 = 8$, $\mu_3 = 12$, and $\sigma_1 = \sigma_2 = \sigma_3 = .5$ ohm. What is the probability that resistance of the assembly will exceed 26 ohms?

6.34. If X is a normal variate with parameters μ and σ, prove that the statistic

$$\bar{X} = \frac{1}{n} \sum_{i=1}^{n} X_i$$

is normally distributed with parameters μ and $\sigma/\sqrt{n}$.

6.35. The random variable X is normally distributed with $\mu = 2$ and $\sigma = 4$. What is the distribution of random variable $Y = 3X + 2$? Find $P\{Y > 20\}$.

6.36. The life in hours of a certain electronic tube is normally distributed with $\mu = 1{,}000$ hr. What is the maximum allowable value for σ if the life of a tube should exceed 900 hr in 90% of the cases?

6.37. Suppose that X represents a measurable characteristic. If X has normal distribution truncated to the left at $X = 0$, with $\mu = 10$ and $\sigma = 4$, find the probability density function of X.

6.38. Suppose that random variable X has the log-normal distribution. If $E(X) = 6.08$ and $\text{Var}(X) = .369$, what is the probability that X is greater than 7.39?

6.39. In a certain community the annual income of wage earners can be treated as a random variable having the log-normal distribution with $\mu = \$7.98$ and $\sigma = \$.2$.
(a) What is the expected annual income of a wage earner?
(b) What percentage of this community earns more than $4,000?

6.40. Suppose that X is exponentially distributed with $\lambda = .001$. What is the probability that X exceeds 1,200?

6.41. Suppose that ten components are put on life test. Furthermore, suppose that the time to failure of a component is exponentially distributed with $\lambda = .0046$ hr. What is the probability that at least eight of these components will still be operative after 150 hr?

6.42. The lifetime X in hours of a TV tube of a certain type obeys an exponential distribution with $\lambda = .001$ hr. Find the following probabilities:
(a) $P\{X > 1{,}000\}$
(b) $P\{700 \leq X \leq 1{,}000\}$
(c) $P\{X > 1{,}200 \mid X > 800\}$.

6.43. Suppose in the preceding problem that four such tubes are used in a TV set. What is the probability that exactly one tube will have a life greater than or equal to 1,200 hr, that exactly two tubes will have a life less than 1,200 but greater than 700 hr, and that exactly one tube will have a life less than or equal to 700 hr?

6.44. Let X have a uniform distribution over the interval (1, 3). Compute the following:
(a) $E(X)$
(b) $\text{Var}(X)$
(c) $P\{X \geq 2\}$
(d) $P\{1 < X \leq 2.6\}$.

6.45. The density function of a random variable X is

$$f_X(t) = \frac{1}{100} \qquad 100 < t < 200$$
$$= 0 \qquad \text{otherwise}$$

Compute the following:
(a) $E(X)$
(b) $\text{Var}(X)$
(c) $P\{X \geq 150\}$
(d) $P\{100 \leq X \leq 180\}$.

6.46. Suppose that X is a gamma variate with $\lambda = .1$ and $k = 2$. Compute the following:
(a) $E(X)$
(b) $\text{Var}(X)$
(c) $P\{X \geq 24\}$.

6.47. Prove that $\Gamma(\frac{1}{2}) = \sqrt{\pi}$.

6.48. The time interval between successive breakdowns of an electronic system has been found to be exponentially distributed with $\lambda = .001$. What is the probability that the third breakdown will occur after 4,250 hr?

6.49. In a certain locality the daily consumption of electrical power (in millions of kilowatt hours) has been found to have the gamma distribution with $\lambda = \frac{1}{4}$ and $k = 3$.
(a) What is the expected daily consumption?
(b) If the power station has a daily capacity of 15,000,000 kwh, what is the probability that the standby equipment will not be utilized?

6.50. If χ^2 has a chi-square distribution with 10 degrees of freedom, compute the following:
(a) $P\{\chi^2 \leq 4.86\}$
(b) $P\{\chi^2 \geq 12.5\}$
(c) $P\{7.26 \leq \chi^2 \leq 20.48\}$.

6.51. Suppose that the daily demand for a certain commodity follows the chi-square distribution. Furthermore, suppose that the expected daily demand is 6 units. What is the probability that the demand for 2 days exceeds 21 units? (Assume independence.)

6.52. If t has a student's t-distribution with 10 degrees of freedom, compute the following:
(a) $P\{t \geq 2.76\}$
(b) $P\{t \leq .26\}$
(c) $P\{.879 \leq t \leq 4.14\}$.

6.53. Show that $t^2_{1-\alpha/2;\, n} = F_{1-\alpha;\, 1,\, n}$.

6.54. If F has an F-distribution with 5 and 8 degrees of freedom, find the following:
(a) $P\{F \leq .529\}$
(b) $P\{F \leq 2.73\}$.

6.55. A random variable X has a Weibull distribution with $\beta = 1.6$ and $\eta = 20$. Compute $P\{X \geq 1\}$.

6.56. Suppose that variate X has a Weibull distribution with $\beta = 2.0$ and $\eta = 16$.
(a) Compute $E(X)$.
(b) Compute $\text{Var}(X)$.

6.57. The random variable X has the beta distribution with $\alpha = 1$ and $\beta = 2$.
(a) Compute $E(X)$
(b) Compute $\text{Var}(X)$
(c) Compute $P\{X \leq .1\}$.

6.58. A random variable X has a Cauchy distribution if its density function is given by

$$f_X(t) = \frac{1}{\pi\beta\{1 + [(t-\alpha)/\beta]^2\}} \qquad -\infty < \alpha < \infty \qquad \beta > 0 \qquad -\infty < t < \infty$$
$$= 0 \qquad \text{otherwise}$$

Find the cumulative distribution of X for the special case where $\beta = 1$ and $\alpha = 0$.

6.59. A random variable X has a Rayleigh distribution if its density function is given by

$$f_X(t) = \frac{1}{\alpha^2} t \exp\left[-\frac{1}{2}\left(\frac{t}{\alpha}\right)^2\right] \qquad \alpha, t > 0$$

Find $E(X)$ and $\mathrm{Var}(X)$ for the special case where $\alpha = 1$.

6.60. Let X have a binomial distribution with parameters $n = 1{,}600$ and $p = \frac{1}{2}$. Use the normal approximation to compute the following probabilities:
(a) $P\{X > 850\}$
(b) $P\{780 \leq X \leq 820\}$.

6.61. Suppose that X has the Poisson distribution with $\lambda = 16$. Use the normal approximation to compute the following probabilities:
(a) $P\{X > 21\}$
(b) $P\{12 \leq X \leq 20\}$.

6.62. Use the normal approximation (with correction for continuity) to find the probability of obtaining exactly 8 heads in 20 independent tosses of a fair coin.

6.63. The number of accidents per week in a large industrial plant has a Poisson distribution with $\lambda = 10$. Compute, both exactly and using the normal approximation (with correction for continuity), the probability that exactly eight accidents per week will occur.

6.64. Let X have a binomial distribution with parameters $n = 400$ and $p = .02$. Using the Poisson approximation, compute the following:
(a) $P\{X \leq 10\}$
(b) $P\{4 \leq X \leq 12\}$.

SUGGESTED REFERENCES

[1] Brunk, H., *An Introduction to Mathematical Statistics*, Ginn and Company, Boston, 1960.

[2] Ehrenfeld, S., and S. B. Littauer, *Introduction to Statistical Method*, McGraw-Hill Book Company, New York, 1964.

[3] Feller, W., *An Introduction to Probability Theory and Its Applications*, 3rd ed., John Wiley & Sons, Inc., New York, 1968.

[4] Fraser, D. A. S., *Statistics: An Introduction*, John Wiley & Sons, Inc., New York, 1958.

[5] Hodges, J. L., and E. L. Lehmann, *Elements of Finite Probability*, Holden-Day, Inc., San Francisco, 1965.

[6] Hoel, P. G., *Introduction to Mathematical Statistics*, 3rd ed., John Wiley & Sons, Inc., New York, 1965.

[7] Meyer, P. L., *Introductory Probability and Statistical Applications*, 2nd ed., Addison-Wesley Publishing Co., Reading, Mass., 1970.

[8] Mood, A. H., and F. A. Graybill, *Introduction to the Theory of Statistics*, McGraw-Hill Book Company, New York, 1963.

[9] Parzen, E., *Modern Probability Theory and Its Applications*, John Wiley & Sons, Inc., New York, 1960.

[10] Plait, A., "The Weibull Distribution," *Industrial Quality Control*, **19**, 1962.

[11] Wadsworth, G. P., and J. G. Bryan, *Introduction to Probability and Random Variables*, McGraw-Hill Book Company, New York, 1960.

[12] Weibull, W., "A Statistical Distribution of Wide Applicability," *Journal of Applied Mechanics*, **18**, 1951.

Part II

STATISTICAL INFERENCE AND APPLICATIONS

7

ORDER STATISTICS

7.1 Introduction

In this chapter we investigate the probability distributions of *order statistics*. The distributions of these statistics play an important role in probability theory and mathematical statistics. Glivenko, Kolmogorov, Loève, Smirnov, Wald, Wilks, and others have made important contributions to the theory of order statistics. Their work has a wide range of applications in engineering and the physical sciences.

We now define the notion of an order statistic. Let $x_1, x_2, \ldots, x_n$ be a sample of n elements drawn from a population having a probability density function $f(x)$. If we arrange these n observations in *ascending order of magnitude*, we obtain the sequence

$$x_{(1)}, x_{(2)}, \ldots, x_{(n)}$$

where $x_{(1)}$ is the smallest of $(x_1, x_2, \ldots, x_n)$ and $x_{(n)}$ is the largest of $(x_1, x_2, \ldots, x_n)$. In other words, the ordered sequence satisfies the inequalities

$$x_{(1)} < x_{(2)} < \ldots < x_{(n)}$$

(Note that the possibility of any of the x's being equal can be neglected because they are continuous random variables.) The *quantities* $x_{(1)}, x_{(2)}, \ldots, x_{(n)}$ *are random variables* and are called *the order statistics of the sample*. In the following sections we shall derive the probability distributions of these random variables and the distributions based on them such as the *range*. We will restrict ourselves to the case when the population has a distribution of the *continuous type*.

7.2 Distribution of the Largest Element in a Sample

Consider a sample of size n drawn from a population having a probability density function $f(x)$. Let $x_{(1)}, x_{(2)}, \ldots, x_{(n)}$ be the order statistics of this sample. The cumulative distribution function of the population is given by

$$F_X(a) = \int_{-\infty}^{a} f_X(t)\,dt \tag{7.1}$$

The largest element in the sample $x_{(n)}$ is a random variable defined by

$$x_{(n)} = G(x_1, x_2, \ldots, x_n) = \max(x_1, x_2, \ldots, x_n) \tag{7.2}$$

Then

$$\begin{aligned} P\{x_{(n)} \leq a\} &= P\{G \leq a\} \\ &= P\{\max(x_1, x_2, \ldots, x_n) \leq a\} \\ &= P\{x_1 \leq a, x_2 \leq a, \ldots, x_n \leq a\} \\ &= P\{x_1 \leq a\}\, P\{x_2 \leq a\} \cdots P\{x_n \leq a\} \\ &= \int_{-\infty}^{a} f_X(t)\,dt \int_{-\infty}^{a} f_X(t)\,dt \cdots \int_{-\infty}^{a} f_X(t)\,dt \\ &= [F_X(a)]^n \end{aligned}$$

That is, the *cumulative distribution function of the largest element in the sample* is given by

$$F_G(a) = [F_X(a)]^n \tag{7.3}$$

Differentiating Eq. (7.3) with respect to a we obtain the density function of the largest element in the sample:

$$f_G(a) = n[F_X(a)]^{n-1} f_X(a) \tag{7.4}$$

Example 7.1 Suppose that an electronic device consists of two components. The length of life of each component is exponentially distributed with the density function

$$\begin{aligned} f_X(t) &= e^{-t} & t &\geq 0 \\ &= 0 & t &< 0 \end{aligned}$$

Furthermore, suppose that the device fails to function when both components fail. What is the probability that the device will function for at least k units of time?

Solution. In this case the device will fail when the last component fails to function. Let t_1 and t_2 denote the times to failure of the first and second component, respectively. Then the probability that the device will function for at least k units of time is given by

$$P\{\max(t_1, t_2) \geq k\}$$

Applying Eq. (7.3) we obtain

$$\begin{aligned} F_G(a) = [F_X(a)]^n &= \left[\int_0^a e^{-t}\,dt\right]^n \\ &= (1 - e^{-a})^n = (1 - e^{-a})^2 \end{aligned}$$

Hence

$$\begin{aligned} P\{\max(t_1, t_2) \geq k\} &= 1 - F_G(k) \\ &= 1 - (1 - e^{-k})^2 \\ &= e^{-k}(2 - e^{-k}) \end{aligned}$$

7.3 Distribution of the Smallest Element in a Sample

We now wish to derive the distribution of $x_{(1)}$, which is the smallest of $(x_1, x_2, \ldots, x_n)$. $x_{(1)}$ is a random variable defined by

(7.5) $$x_{(1)} = W(x_1, x_2, \ldots, x_n) = \min(x_1, x_2, \ldots, x_n)$$

Then

$$\begin{aligned} P\{x_{(1)} \leq a\} = P\{W \leq a\} &= P\{\min(x_1, x_2 \ldots, x_n) \leq a\} \\ &= 1 - P\{\min(x_1, x_2, \ldots, x_n) > a\} \\ &= 1 - P\{x_1 > a, x_2 > a, \ldots, x_n > a\} \\ &= 1 - P\{x_1 > a\}\, P\{x_2 > a\} \cdots P\{x_n > a\} \\ &= 1 - [1 - F_X(a)]^n \end{aligned}$$

That is, the *cumulative distribution of the smallest element in the sample* is given by

(7.6) $$F_W(a) = 1 - [1 - F_X(a)]^n$$

Differentiating Eq. (7.6) with respect to a we obtain the density function of the smallest element in the sample:

(7.7) $$f_W(a) = n[1 - F_X(a)]^{n-1} f_X(a)$$

Example 7.2 Suppose in Example 7.1 that the electronic device fails to function when either component fails.

(a) What is the probability that the device will function for at least k units of time?
(b) What is the expected life of the device?

Solution

(a) Here the probability that the device will function for at least k units of time is $P\{\min(t_1, t_2) \geq k\}$. Applying Eq. (7.6) we obtain

$$\begin{aligned} F_W(t) = 1 - [1 - F_X(t)]^n &= 1 - (1 - 1 + e^{-t})^n \\ &= 1 - e^{-2t} \end{aligned}$$

Hence

$$
\begin{aligned}
P\{\min(t_1, t_2) \geq k\} &= 1 - F_W(k) \\
&= 1 - (1 - e^{-2k}) \\
&= e^{-2k}
\end{aligned}
$$

(b) The expected life of the electronic device is given by

$$
\begin{aligned}
E(T) &= \int_0^\infty t\, f_W(t)\, dt \\
&= \int_0^\infty tn[1 - F_X(t)]^{n-1} f_X(t)\, dt \\
&= \int_0^\infty 2t\, e^{-2t}\, dt \\
&= \tfrac{1}{2}
\end{aligned}
$$

7.4 Distribution of the Median of a Sample

Suppose that a sample $(x_1, x_2, \ldots, x_{2n+1})$ of size $2n + 1$ is drawn from a population having probability density function $f(x)$. The order statistics of this sample is given by

$$x_{(1)}, x_{(2)}, \ldots, x_{(n+1)}, \ldots, x_{(2n+1)}$$

Figure 7.1 shows the representation of the order statistics. The quantity $x_{(n+1)}$ is called the *median*. We now wish to determine the distribution of the median.

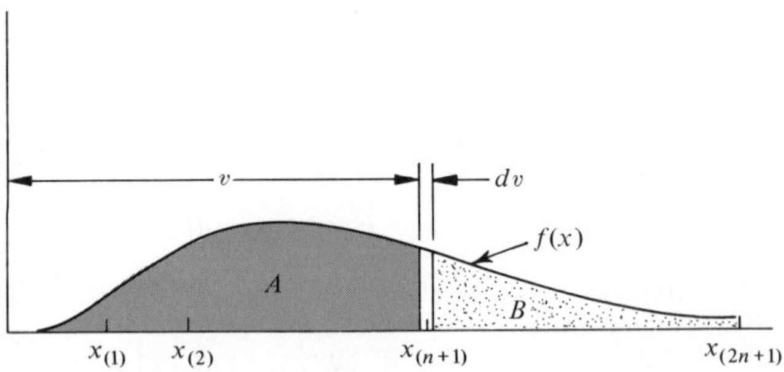

Fig. 7.1 The order statistics of a sample of size $2n + 1$ on a continuous random variable having p.d.f. $f(x)$.

The element $x_{(n+1)}$ will lie in the interval $(v, v + dv)$ if and only if n sample elements fall in A and n sample elements fall in B, as shown in Fig. 7.1. The probability that $x_{(n+1)}$ falls in the interval $(v, v + dv)$ given that n sample elements fall in A and n sample elements fall in B is obtained by applying the multinomial probability distribution. Thus we obtain

$$(7.8)\qquad P\{v < x_{(n+1)} < v + dv\} = \frac{(2n+1)!}{1!\,n!\,n!} P\{E_1\}^1 P\{E_2\}^n P\{E_3\}^n$$

where E_1 = the event that $x_{(n+1)}$ falls in the interval $(v, v + dv)$
E_2 = the event that n sample elements fall in A
E_3 = the event that n sample elements fall in B

The probabilities of events E_1, E_2, and E_3 are given by

$$(7.9)\qquad \begin{cases} P\{E_1\} = F_X(v + dv) - F_X(v) \\ \qquad\quad = F_X(v) + f_X(v)\,dv - F_X(v) \\ \qquad\quad = f_X(v)\,dv \\ P\{E_2\} = F_X(v) \\ P\{E_3\} = 1 - F_X(v + dv) \\ \qquad\quad = 1 - F_X(v) - f_X(v)\,dv \end{cases}$$

Substituting the values of $P\{E_1\}$, $P\{E_2\}$, $P\{E_3\}$ from Eq. (7.9) into Eq. (7.8) we obtain

$$\begin{aligned} P\{v < x_{(n+1)} < v + dv\} &\\ = \frac{(2n+1)!}{(n!)^2} f_X(v)\,dv[F_X(v)]^n\,[1 - F_X(v) - f_X(v)\,dv]^n &\\ = \frac{(2n+1)!}{(n!)^2} f_X(v)\,dv[F_X(v)]^n\,[1 - F_X(v)]^n &\\ (7.10)\qquad + \text{ multiples of } (dv)^2 \text{ and higher powers of } dv & \end{aligned}$$

Dividing both sides of Eq. (7.10) by dv we obtain

$$\begin{aligned} f_X(v) = \frac{(2n+1)!}{(n!)^2} f_X(v)[F_X(v)]^n\,[1 - F_X(v)]^n &\\ (7.11)\qquad + \text{ multiples of } dv \text{ and higher powers of } dv & \end{aligned}$$

On taking limiting values of both sides of Eq. (7.11) as dv tends to zero we obtain the *density function of the median* denoted by $g(v)$ where

$$(7.12)\qquad g(v) = \frac{(2n+1)!}{(n!)^2} [F_X(v)]^n\,[1 - F_X(v)]^n f_X(v)$$

The probability element of the median is $g(v)\,dv$, where

$$(7.13)\qquad g(v)\,dv = \frac{(2n+1)!}{(n!)^2} [F_X(v)]^n\,[1 - F_X(v)]^n f_X(v)\,dv$$

Example 7.3 Five observations are drawn at random from the density function

$$\begin{aligned} f_X(t) &= \lambda e^{-\lambda t} \qquad t \geq 0 \\ &= 0 \qquad t < 0 \end{aligned}$$

What is the probability that the median of these observations falls in the interval $(v, v + dv)$?

Solution. First we derive the cumulative distribution function $F_X(v)$ as follows:

$$\begin{aligned} F_X(v) &= 0 \qquad v < 0 \\ &= 1 - e^{-\lambda v} \qquad v \geq 0 \end{aligned}$$

The probability that the median falls in the interval $(v, v + dv)$ therefore is given by

$$\begin{aligned} g(v)\,dv &= \frac{5!}{(2!)^2}(1 - e^{-\lambda v})^2(e^{-\lambda v})^2\lambda e^{-\lambda v}\,dv \\ &= 30\lambda e^{-3\lambda v}(1 - e^{-\lambda v})^2\,dv \end{aligned}$$

7.5 Distribution of the *i*th-Order Statistic of a Sample

Let $x_{(i)}$ be the *ith-order statistic* of a sample of n elements, say $(x_1, x_2, \ldots, x_n)$, drawn from a population having the density function $f(x)$. The ith *smallest value* will lie in the interval $(v, v + dv)$ if and only if

1. $i - 1$ sample elements fall in $(-\infty, v)$
2. 1 sample element falls in $(v, v + dv)$
3. $n - i$ sample elements fall in $(v + dv, +\infty)$

The probability of this event is given by

$$\begin{aligned} &P\{v < x_{(i)} < v + dv\} \\ &\quad = \frac{n!}{(i-1)!\,(n-i)!}[F_X(v)]^{i-1}[1 - F_X(v)]^{n-i}f_X(v)\,dv \\ &\quad + \text{multiples of } (dv)^2 \text{ and higher powers of } dv \end{aligned} \tag{7.14}$$

Dividing both sides of Eq. (7.14) by dv and taking limiting values as dv tends to zero we obtain the *density function of the* ith *smallest of* $(x_1, x_2, \ldots, x_n)$ denoted by $g(v)$, where

$$g(v) = \frac{n!}{(i-1)!\,(n-i)!}[F_X(v)]^{i-1}[1 - F_X(v)]^{n-i}f_X(v) \tag{7.15}$$

Hence, the probability element of the ith order statistic is given by

$$g(v)\,dv = \frac{n!}{(i-1)!\,(n-i)!}[F_X(v)]^{i-1}[1 - F_X(v)]^{n-i}f_X(v)\,dv \tag{7.16}$$

When $i = 1$, Eq. (7.16) reduces to

$$g(v)\,dv = n[1 - F_X(v)]^{n-1}f_X(v)\,dv$$

or

$$g(v) = n[1 - F_X(v)]^{n-1}f_X(v) \tag{7.17}$$

The right-hand side of Eq. (7.17) is the *density function of the smallest element in the sample* and is *identical* to the right-hand side of *Eq. (7.7)*. Similarly, when $i = n$, Eq. (7.16) reduces to that on the right of Eq. (7.4), which is the density function of the largest element in the sample.

Example 7.4 Five observations are drawn at random from the density function

$$\begin{aligned} f_X(a) &= 2a \qquad 0 < a < 1 \\ &= 0 \qquad \text{otherwise} \end{aligned}$$

What is the probability that $x_{(3)}$ lies in the interval $(v, v + dv)$?

Solution. The cumulative distribution function of X is given by

$$F_X(v) = \begin{cases} 0 & v \leq 0 \\ v^2 & 0 < v < 1 \\ 1 & v \geq 1 \end{cases}$$

The probability that $x_{(3)}$ lies in the interval $(v, v + dv)$ therefore is given by

$$\begin{aligned} P\{v < x_{(3)} < v + dv\} &= \frac{n!}{(i-1)!\,(n-i)!}[F_X(v)]^{i-1}[1 - F_X(v)]^{n-i} f_X(v)\,dv \\ &= \frac{5!}{2!\,2!} v^2(1 - v^2)^2\, 2v\,dv \\ &= 60v^3(1 - v^2)^2\,dv \end{aligned}$$

Note that the same result can be obtained by applying Eq. (7.13) since $x_{(3)}$ is the median of the sample.

7.6 Distribution of the Range of a Sample

Let $x_{(1)}, x_{(2)}, \ldots, x_{(n)}$ be the order statistics of a sample of size n drawn from a population having a probability density function $f(x)$. The random variable $x_{(n)} - x_{(1)} = R$ is called the *sample range*. We now wish to derive the distribution of R. The distribution of R can be obtained by deriving the joint probability distribution of $x_{(n)}$ and $x_{(1)}$. Let v and u denote $x_{(n)}$ and $x_{(1)}$, respectively. The joint distribution function of u and v is found as follows: We must first determine the probability that u will fall between v and $v + dv$. This is equivalent to finding the probability that in a sample of size n one will obtain no observations between 0 and u, one observation between u and $u + du$, $(n - 2)$ observations between $u + du$ and v, one observation between v and $v + dv$, and none between $v + dv$ and $+\infty$. Substituting the individual probabilities in the multinomial distribution, we have the joint probability distribution of u and v as follows:

$$\frac{n!}{0!\,1!\,(n-2)!\,1!\,0!}\, p_1^0\, p_2^1\, p_3^{n-2}\, p_4^1\, p_5^0$$

which reduces to

$$n(n-1)p_2 p_3^{n-2} p_4 \tag{7.18}$$

where $p_2 = F_X(u + du) - F_X(u) = f_X(u)\,du$

$p_3 = F_X(v) - F_X(u + du) = F_X(v) - F_X(u) - f_X(u)\,du$

$p_4 = F_X(v + dv) - F_X(v) = f_X(v)\,dv$

If these values of p_2, p_3, and p_4 are inserted in Eq. (7.18) it becomes

$$n(n-1)f_X(u)\,du\,[F_X(v) - F_X(u) - f_X(u)\,du]^{n-2} f_X(v)\,dv \tag{7.19}$$

In order to find the joint probability density function $f(u, v)$ it is necessary to divide Eq. (7.19) by $du\,dv$ and to take the limit as du tends to zero. It follows from Eq. (7.19) that

$$f_{U,V}(u, v) = n(n-1)f_X(u)f_X(v)[F_X(v) - F_X(u)]^{n-2} \qquad -\infty < u < v < +\infty \tag{7.20}$$

Now from the joint probability density function of the random variables U and V we can easily determine the probability density function of a new random variable $R = V - U$ (the range of the sample). Hence, the probability density function of the range is given by

$$\begin{aligned} f_R(r) &= \int_{-\infty}^{\infty} f(u, u + r)\,du \\ &= n(n-1)\int_{-\infty}^{\infty} [F_X(u + r) - F_X(u)]^{n-2} f_X(u) f_X(u + r)\,du \end{aligned} \tag{7.21}$$

The probability element of R then is given by

$$\begin{aligned} &f_R(r)\,dr \\ &\quad = \{n(n-1)\int_{-\infty}^{\infty} [F_X(u + r) - F_X(u)]^{n-2} f_X(u) f_X(u + r)\,du\}\,dr \end{aligned} \tag{7.22}$$

If $F(x)$ is a simple function of x it may be possible to evaluate the integral explicitly; otherwise numerical integration is likely to be applied. For example, in the case of the normal distribution there is no simple explicit form for the probability distribution of the range. However, the probability distribution of the range of normal populations can be obtained from tables such as Harter [5] and Pearson and Hartley [9].

Example 7.5 The variate X is rectangularly distributed over the interval (0, 1). (a) Find the density function of the range R in repeated samples of n observations. (b) Determine the probability that the range of a sample of six will exceed .6.

Solution

(a) We derive first the cumulative distribution function of X. We have

$$F_X(a) = \begin{cases} 0 & a \le 0 \\ a & 0 < a < 1 \\ 1 & a \ge 1 \end{cases}$$

Then, by Eq. (7.21),

$$\begin{aligned} f_R(r) &= n(n-1)\int_{-\infty}^{\infty} [F_X(u+r) - F_X(u)]^{n-2} f_X(u) f_X(u+r)\, du \\ &= n(n-1)\int_0^{1-r} (u + r - u)^{n-2}\, du \\ &= n(n-1)\int_0^{1-r} r^{n-2}\, du \\ &= n(n-1)r^{n-2}(1-r) \qquad 0 < r < 1 \end{aligned}$$

(b)

$$\begin{aligned} P\{R > .6\} &= 1 - F_R(.6) \\ &= 1 - \int_0^{.6} n(n-1)r^{n-2}(1-r)\, dr \\ &= 1 - 30\int_0^{.6} (r^4 - r^5)\, dr = .7667 \end{aligned}$$

Example 7.6 Variate X has the density function

$$\begin{aligned} f_X(t) &= 2e^{-2t} \qquad t \geq 0 \\ &= 0 \qquad \text{otherwise} \end{aligned}$$

Find the density function of the range R in repeated random samples of five observations.

Solution. The cumulative distribution function of X is given by

$$\begin{aligned} F_X(t) &= 0 \qquad t < 0 \\ &= 1 - e^{-2t} \qquad t \geq 0 \end{aligned}$$

Then, by Eq. (7.21), we have

$$\begin{aligned} f_R(r) &= n(n-1)\int_{-\infty}^{\infty} [F_X(u+r) - F_X(u)]^{n-2} f_X(u) f_X(u+r)\, du \\ &= 20\int_0^{\infty} (1 - e^{-2(u+r)} - 1 + e^{-2u})^3 2e^{-2u} 2e^{-2(u+r)}\, du \\ &= 80\int_0^{\infty} (1 - e^{-2r})^3 e^{-10u-2r}\, du \\ &= 8(1 - e^{-2r})^3 e^{-2r} \qquad r \geq 0 \end{aligned}$$

7.7 Distribution-Free Tolerance Limits

A problem of frequent interest is to determine *how large a sample must be taken from a large lot to assure with specified odds that the range of the sample used will include at least a large specified proportion of the population.* The distribution of the population is unknown; however, it is assumed to be of the continuous type. In other words, we wish to determine the sample size n such that the range of the sample $x_{(n)} - x_{(1)}$ will include at least $100\beta\%$ of the values of x with a probability equal to $(1 - \alpha)$. $x_{(n)}$ and $x_{(1)}$ are sometimes known as the *upper and lower statistical tolerance limits*, respectively.

The interval $(x_{(1)}, x_{(n)})$ is known as the *two-sided 100β% statistical tolerance interval with confidence coefficient 100(1 − α)%*. The intervals $(-\infty, x_{(n)})$ and $(x_{(1)}, +\infty)$ are called *one-sided upper and lower 100β% statistical tolerance intervals*, respectively.

Suppose that sampling is done on a continuous random variable X having a density function $f(x)$ of any shape. Then for an x chosen at random, say x_i, we have

$$F(x_i) = \int_{-\infty}^{x_i} f(x)\,dx \tag{7.23}$$

The probability density function of $F(x_i)$ is

$$\Phi[F(x_i)] = f(x_i) \Big/ \frac{dF(x_i)}{dx_i} = 1 \qquad 0 < F(x_i) < 1 \tag{7.24}$$

It turns out that $F(x_i)$ has a uniform distribution in the interval (0, 1) independent of $f(x)$. Now if a sample of size n is drawn from the population, the fraction of the population that will lie within the interval $(x_{(1)}, x_{(n)})$ is

$$\int_{x_{(1)}}^{x_{(n)}} f(x)\,dx = F(x_{(n)}) - F(x_{(1)}) = W \tag{7.25}$$

$F(x_{(n)})$ and $F(x_{(1)})$ are random variables distributed as an *ordered sample* of size n from a population having a *uniform distribution* in the interval (0, 1). The joint distribution of random variables $F(x_{(1)})$ and $F(x_{(n)})$ in terms of u and v, that is at (u, v), is obtained from Eq. (7.20) by substituting random variable $F(X)$ for X. Thereby we have

$$\begin{aligned} f(u, v) &= n(n-1)(v-u)^{n-2}\Phi(u)\Phi(v) \\ &= n(n-1)(v-u)^{n-2} \qquad 0 < u < v < 1 \end{aligned} \tag{7.26}$$

The cumulative distribution of W, then, is

$$\begin{aligned} F(w) = P\{W \le w\} &= 1 - n(n-1)\int_w^1 \int_0^{v-w} (v-u)^{n-2}\,du\,dv \\ &= nw^{n-1} - (n-1)w^n \qquad 0 < w < 1 \end{aligned} \tag{7.27}$$

The density function of W is obtained by differentiating Eq. (7.27) with respect to w, yielding

$$f(w) = n(n-1)w^{n-2}(1-w) \qquad 0 < w < 1 \tag{7.28}$$

To find the probability that W is greater than β, where β is any number between 0 and 1, we therefore must evaluate the cumulative distribution of W at β. This gives

$$P\{W > \beta\} = 1 - P\{W \le \beta\} = 1 - \alpha$$

or

$$P\{W \le \beta\} = n\beta^{n-1} - (n-1)\beta^n = \alpha \tag{7.29}$$

We choose n as the smallest integer satisfying Eq. (7.29) for given α and β.

Equation (7.29) is a transcendental equation that can be solved by trial and error. It is possible, however, to reduce Eq. (7.29) to a binomial form that can be readily solved from the tables of cumulative binomial distribution. Equation (7.29) can be written as

$$\beta^n + n(1-\beta)\beta^{n-1} = \alpha \tag{7.30}$$

The left-hand side of Eq. (7.30) is the binomial summation

$$\sum_{k=0}^{1}\binom{n}{k}p^k(1-p)^{n-k} = (1-p)^n + np(1-p)^{n-1}$$

where $1 - p = \beta$. A nomograph for the direct solution of Eq. (7.29) is given by Nelson [8], where n can be obtained by entering the specified α and β. Thus n can be obtained from tables of the binomial distribution. However, a rather good approximation is obtained from the formula

$$n \cong \frac{1}{4}\chi^2_{1-\alpha;\,4}\left(\frac{1+\beta}{1-\beta}\right) + \frac{1}{2} \tag{7.31}$$

Example 7.7 A manufacturer wishes to determine tolerance limits within which at least 95% of the values of the measurable characteristic X should lie in 95% of the lots inspected. Find the sample size.

Solution. We wish to determine the size n of the sample for which $(x_{(1)}, x_{(n)})$ is a 95% statistical tolerance interval having 95% confidence. Hence, to find n we must solve the equation

$$n\beta^{n-1} - (n-1)\beta^n = \alpha$$

where $\alpha = .05$ and $\beta = .95$. Using the Poisson distribution as an approximation to the binomial we enter the cumulative Poisson distribution table (Table I in the Appendix) with $k = 1$ and $\alpha = .05$. We read $\lambda \cong 4.8$. The required sample size $n \cong 4.8/(1 - \beta) = 96$. Consulting the Nelson nomograph we find that the sample size is 92, which is close to the value given by the Poisson approximation. The chi-square approximation yields

$$n \cong \frac{1}{4}\chi^2_{.95;\,4}\left(\frac{1.95}{.05}\right) + \frac{1}{2} = \frac{1}{4}(9.488)(39) + \frac{1}{2} = 93$$

This result is interpreted as follows: If a sample of 92 items is drawn from a statistically controlled process, then in 95% of the times this experiment is performed at least 95% of the observations made on all future outcomes of the process will lie within $x_{(1)}$ and $x_{(92)}$.

Example 7.8 Suppose that a sample of 100 items drawn at random from a large lot showed a range of 1.87 to 2.00. Determine with 90% confidence the minimum proportion of the lot distribution that can be expected to lie within this range.

Solution. Given $x_{(1)} = 1.87$, $x_{(100)} = 2.00$, and $\alpha = .10$ we wish to determine β. Using the χ^2 approximation, the sample size n is given by

$$n \cong \frac{1}{4}\chi^2_{1-\alpha;\,4}\left(\frac{1+\beta}{1-\beta}\right) + \frac{1}{2}$$

or

$$\frac{1+\beta}{1-\beta} \cong \frac{4(99.5)}{\chi^2_{.90;\,4}} = 51.16$$

whence $\beta = .96$. That is, at least 96% of the lot distribution can be expected to lie within 1.87 to 2.00 in 90% of the times this experiment is performed.

Example 7.9 In Example 7.7 find the sample size for a one-sided statistical tolerance interval $(-\infty, x_{(n)})$. The fraction of the population that will lie in the interval $(-\infty, x_{(n)})$ is

$$\int_{-\infty}^{x_{(n)}} f(x)\,dx = F(x_{(n)}) \tag{7.32}$$

$F(x_{(n)})$ is a random variable distributed as an ordered sample of size n from a population having a uniform distribution in the interval (0, 1). The frequency distribution of $F(x_{(n)})$ in terms of v is obtained from Eq. (7.4) by substituting random variable $F(X)$ for X. Hence we have

$$f(v) = nv^{n-1} \qquad 0 < v < 1 \tag{7.33}$$

Thus we can assert that

$$P\{v > \beta\} = 1 - \int_0^\beta nv^{n-1}\,dv = 1 - \alpha$$

or, equivalently,

$$1 - \beta^n = 1 - \alpha$$

whence

$$\beta^n = \alpha \tag{7.34}$$

Now we can proceed to solve Example 7.9.

Solution. Given $\alpha = .05$, $\beta = .95$ we wish to determine the sample size n by solving the equation $.95^n = .05$, which yields $n \cong 58$.

This means that if a sample of 58 items is taken we can expect 95% of the lot distribution to lie within the interval $(-\infty, x_{(58)})$ in 95% of the times this experiment is performed.

Example 7.10 In Example 7.7 find the sample size for lower statistical tolerance interval $(x_{(1)}, +\infty)$. The fraction of the population that will lie in the interval $(x_{(1)}, +\infty)$ is

$$\int_{x_{(1)}}^{\infty} f(x)\,dx = 1 - F(x_{(1)}) \tag{7.35}$$

$F(x_{(1)})$ is a random variable distributed as an ordered sample of size n from a population having a uniform distribution in the interval (0, 1). The frequency distribution of $F(x_{(1)})$ in terms of u is obtained from Eq. (7.7) by substituting random variable $F(X)$ for X. Thereby we have

$$f(u) = n(1-u)^{n-1} \qquad 0 < u < 1 \tag{7.36}$$

Hence

$$P\{(1-u) > \beta\} = P\{u < 1 - \beta\} = \int_0^{1-\beta} n(1-u)^{n-1}\,du = 1 - \alpha$$

or, equivalently,

$$1 - \beta^n = 1 - \alpha$$

Whence

$$\beta^n = \alpha \tag{7.37}$$

which is identical to Eq. (7.34).

Now we can proceed to solve Example 7.10.

Solution. Given $\alpha = .05$ and $\beta = .95$, we wish to determine the sample size n by solving the equation

$$.95^n = .05$$

which yields $n \simeq 58$; hence, in a sample of 58 items the smallest element drawn, namely $x_{(1)}$, is a 95% lower statistical tolerance limit having confidence coefficient .95.

PROBLEMS

7.1. An electronic device is composed of three components. The length of life of each component is exponentially distributed with parameter $\lambda = \frac{2}{3}$ month. Suppose that the electronic device fails to function when all components fail.
(a) What is the probability that the device will function for at least 2.5 months?
(b) What is the expected life of the device?

7.2. Suppose that in Prob. 7.1 the length of life of each component has the following density function:

$$\begin{aligned} f_X(t) &= 2t \qquad 0 < t < 1 \\ &= 0 \qquad \text{otherwise} \end{aligned}$$

What is the expected life of the device?

7.3. If T denotes the time to failure in Prob. 7.1, find the following:
(a) The density function of T.
(b) The variance of T.

7.4. A continuous random variable X has the following density function:

$$\begin{aligned} f_X(a) &= 2ae^{-a^2} \qquad a \geq 0 \\ &= 0 \qquad \text{otherwise} \end{aligned}$$

If $x_{(1)}, x_{(2)}, \ldots, x_{(n)}$ are the order statistics of n independent observations, all having this density function, derive the density function for:
(a) The largest of these n observations.
(b) The smallest of these n observations.
(c) The ith smallest of these n observations.

7.5. A filling machine is filling tea bags. It is known from past experience that the density function of the weight in grams of filled bags is given by

$$\begin{aligned} f_W(a) &= 5 \qquad 1.8 < a < 2 \\ &= 0 \qquad \text{otherwise} \end{aligned}$$

Assume that ten bags are drawn at random.

(a) What is the probability that the weights of all ten bags exceed 1.9 g?
(b) What is the probability that the lightest of these ten bags will have a weight less than or equal to 1.9 g?
(c) What is the probability that the heaviest of these ten bags will have a weight greater than 1.95 g?

7.6. Suppose in Prob. 7.1 that the electronic device fails to function when any one of the three components fails.
(a) What is the probability that the device will function for at least one month?
(b) What is the expected life of the device?

7.7. Suppose that in Prob. 7.1 the electronic device fails to function when any two components fail.
(a) What is the probability that the device will function for at least one month?
(b) What is the expected life of the device?

7.8. A sample of n observations is taken at random from a population with density function

$$f_X(a) = c^2 a e^{-ca} \qquad c > 0 \qquad 0 \leq a \leq \infty$$

(a) Find the probability density function of the smallest observation.
(b) Find the probability density function of the largest observation.

7.9. Suppose that seven observations are taken at random from a population with probability density function

$$\begin{aligned} f_X(t) &= e^{-t} && t \geq 0 \\ &= 0 && \text{otherwise} \end{aligned}$$

(a) Find the distribution of the median of the observations.
(b) What is the probability that the median will exceed one?

7.10. Suppose we are given the density function

$$\begin{aligned} f_X(t) &= 2t && 0 < t < 1 \\ &= 0 && \text{otherwise} \end{aligned}$$

and that five observations are drawn at random from this distribution.
(a) Find the joint distribution of the second and fourth smallest of these five observations.
(b) Find the distribution of the median.

7.11. Suppose that three observations are taken at random from a population with probability density function

$$f_X(a) = \begin{cases} \frac{1}{2}a & 0 \leq a \leq 1 \\ \frac{1}{4} & 1 < a \leq 4 \\ 0 & \text{otherwise} \end{cases}$$

What is the distribution of the median?

7.12. If $x_{(1)}, \ldots, x_{(n)}$ are the order statistics of a sample of size n from a population

$$\begin{aligned} f_X(a) &= \tfrac{1}{2} + a && 0 \leq a \leq 1 \\ &= 0 && \text{otherwise} \end{aligned}$$

(a) Derive $F_{X_{(n)}}$.
(b) Find the mean and the variance of $F_{X_{(n)}}$.
(c) Discuss the results obtained under (b).

7.13. Suppose we are given the density function

$$f_X(a) = 2a \qquad 0 < a < 1$$
$$= 0 \qquad \text{otherwise}$$

and that four observations are drawn at random from this distribution. What is the distribution of the range?

7.14. If X is a random variable with probability density function given in Prob. 7.12, find the distribution of the range of a sample of size three.

7.15. Variate X has the probability density function given in Prob. 7.9. Find the minimum sample size required to ensure that in 90% of the times the range of the sample is greater than or equal to one.

7.16. Find the probability that in a sample of size five from the uniform distribution $f_X(a) = 1,\ 0 < a < 1$ the range will exceed .7.

7.17. Determine the sample size in Prob. 7.16 in order that the probability is greater than or equal to .90 that the range will exceed .85.

7.18. A manufacturer wishes to establish limits with which at least 90% of the values of measurable characteristic X should lie in 90% of the lots inspected. Find the sample size.

7.19. Suppose that a sample of 150 items drawn at random from a large lot showed a range of 2.63 to 2.78. Determine with 95% confidence the minimum proportion of the lot distribution that can be expected to lie within this range.

7.20. In Prob. 7.19, determine with 95% confidence the minimum proportion of the lot distribution that can be expected to lie within $(-\infty, 2.78)$.

SUGGESTED REFERENCES

[1] David and Johnson, "Some Tests of Significance with Ordered Variables," *Journal of the Royal Statistical Society, B*, **18**, 1956.

[2] Dixon, W. J., "Ratios Involving Extreme Values," *Annals of Mathematical Statistics*, **22**, 1951.

[3] Gumbel, E. J., "The Distribution of the Range," *Annals of Mathematical Statistics*, **18**, 1947.

[4] Gumbel, E. J., *Statistics of Extremes*, Columbia University Press, New York, 1958.

[5] Harter, H. L., "Tables of Range and Studentized Range," *Annals of Mathematical Statistics*, **31**, 1960.

[6] Mises, R. V., "La Distribution de la Plus Grande de *n* Valeurs," *Revue Mathematique de l'Union Interbalkanique*, **1**, 1, 1939.

[7] Mood, A. M., *Introduction to Mathematical Statistics*, McGraw-Hill Book Company, New York, 1950.

[8] Nelson, L. S., "Nomograph for Two-Sided Distribution-Free Tolernance Intervals," *Industrial Quality Control*, **19**, June 1963.

[9] Pearson and Hartley, *Biometrika Tables for Statisticians*, Vol. 1, Cambridge University Press, New York, 1958.

[10] Pearson, K. (ed.), *Tables for Incomplete Beta Function*, University College London, 1934.

[11] Sarhan, A. E., and B. G. Greenberg (eds.), *Contributions to Order Statistics*, John Wiley & Sons, Inc., New York, 1962.

[12] Tukey, J. W., "The Future of Data Analysis," *Annals of Mathematical Statistics*, **33**, 1962.

[13] Wald, A., "An Extension of Wilks' Method for Setting Tolerance Limits," *Annals of Mathematical Statistics*, **14**, 1943.

[14] Wilks, S. S., "Determination of Sample Size for Setting Tolerance Limits," *Annals of Mathematical Statistics*, **12**, 1941.

[15] Wilks, S. S., "Order Statistics," *Bulletin of the American Mathematical Society*, **54**, 1948.

8

SIGNIFICANCE TESTS

8.1 Introduction

One of the major applications of statistics is in *decision* problems. Statistics is concerned with decision making under *risk* or *uncertainty*. Decision making under risk depends upon *random variables whose probability distributions are known*, whereas decision making under uncertainty deals with *variates whose probability distributions are unknown*.

What decisions do we wish to make? On what *rules* are these decisions based? What *experiment* should be performed? What conditions should be specified prior to performing the experiment? Is the decision made optimal? These are some of the questions that may be posed by the decision maker. Instead of answering these questions explicitly, we shall analyze a typical example in detail in order to pinpoint the highlights of practical decision problems.

8.2 Introductory Example

Suppose the military is concerned with the acceptance of lots of brass cups supplied by a certain manufacturer. These brass cups are supplied in lots of 5 tons (500,000 cups approximately). The intended use of these brass cups is the manufacture of small arms cartridge cases. One critical dimension is of paramount significance, namely, the variation of wall thickness around the periphery denoted by X. Suppose management has set the following decision rule: Accept the lot if the expected variation in wall thickness around the periphery [denoted by $E(X)$] is $\leq .03$ mm; reject it otherwise. Obviously the decision management wishes to make is either to *accept* or *reject* the lot. The rule set by management for acceptance of a lot is

$$\text{Accept if} \quad E(X) \leq .03 \text{ mm}$$
$$\text{Reject if} \quad E(X) > .03 \text{ mm}$$

A naive approach would be to measure the variation in wall thickness of the whole lot, record these measurements and calculate their arithmetic mean $\bar{X}$, and then apply the decision rule set by management. Is this a practical approach? Obviously not: If we inspect each cup our inspection cost will be tremendous. Consequently, the cost of a cartridge case will be excessive. In the light of this analysis it is evident that the decision to accept or reject the lot will be based on the *result of an experiment* in which, say, a sample of size n is selected at random from the lot and then each cup is inspected and $\bar{X}$ is calculated. We expect that $\bar{X}$ will be too close to $E(X)$, and hence we are inclined to conclude that $E(X) \leq .03$ if and only if $\bar{X} \leq$ a prescribed constant, which should be $> .03$. How to determine the value of this constant is discussed later.

Actually the decision to accept or reject the lot will be based on the result of a specified experiment. In other words, our decision will be based on *statistical inference. Hence statistical inference can be defined as making inference about the population on the basis of samples.*

Now is this specified experiment the best? Or, in other words, is the choice of a sample of size n at random and observing the sample mean $\bar{X}$ as a criterion for decision the procedure that leads to the optimal decision? If the answer is yes, what is the value of n? If the answer is no, what other alternative procedures might be used in order to reach an optimal decision? An alternative procedure might be to inspect a sample of size n at random and to observe the largest measurement, $X_{\max}$. If $X_{\max} \leq$ a prescribed constant, conclude that $E(X)$ is $\leq .03$ and accordingly accept the lot or reject otherwise. Suppose this alternative procedure is better than the previous one. What is the value of n? In general, what is the basis for the selection of an *optimal procedure*? We shall answer these questions as we proceed.

8.3 Statistical Decision Theory

Considering again our introductory example, we have two *statistical hypotheses*. The first hypothesis is $E(X) \leq .03$, and the second hypothesis is $E(X) > .03$. The procedure by which a choice is made between these two statistical hypotheses is called *statistical hypothesis testing.*

The hypothesis that is tested $[E(X) \leq .03]$ is called the *null hypothesis* and is denoted by H_0; the other $[E(X) > .03]$ is called the *alternative hypothesis* and is denoted by H_1. Testing of statistical hypotheses involves *rejection* or *acceptance* of the null hypothesis. In other words, we wish to determine whether the null hypothesis is *true* or *false*. In symbols we write

$$H_0\colon\ E(X) \leq .03$$
$$H_1\colon\ E(X) > .03$$

The decision to accept or to reject the null hypothesis will be based on the outcome of our experiment. Suppose a sample of size n is drawn at random and the sample mean $\bar{X}$ is calculated. Furthermore, suppose that the following decision rule is specified: Accept the null hypothesis if and only if $\bar{X} \leq .035$ and reject otherwise. Accordingly, we shall reject the null hypothesis if and only if the observed outcome is greater than .035.

According to this decision rule we shall reject the null hypothesis if the value of the sample mean $\bar{X}$ falls in the *critical* region. *The critical region (sometimes known as the rejection region) is specified by the set of values of* $\bar{X}$ *that is greater than* .035. To simplify the analysis let us suppose that this critical dimension is normally distributed with unknown mean $E(X)$ and known standard deviation $\sigma = .006$. Accordingly, $\bar{X}$ is a random variable that is normally distributed with mean $E(X)$ and standard deviation $\sigma/\sqrt{n}$. Let us analyze further the outcomes based on this decision rule. If the mean of the lot under consideration is actually equal to .03, as shown in Fig. 8.1, then

1. The null hypothesis is *accepted* whenever the value of $\bar{X}$ does not fall in the critical region.
2. The null hypothesis is *rejected* whenever the value of $\bar{X}$ falls in the critical region.

If the mean of the lot under consideration is actually equal to .04, as shown in Fig. 8.2, then

3. The null hypothesis is *accepted* whenever the value of $\bar{X}$ does not fall in the critical region.
4. The null hypothesis is *rejected* whenever the value of $\bar{X}$ falls in the critical region.

Table 8.1 summarizes the outcomes based on this decision rule. Now in 1 and 4 we have made correct decisions whereas in 2 and 3 the decisions made are incorrect.

Obviously in case 2 we rejected the lot despite the fact that $E(X) = .03$. Thus we have committed an *error*. This error is known as an *error of the first*

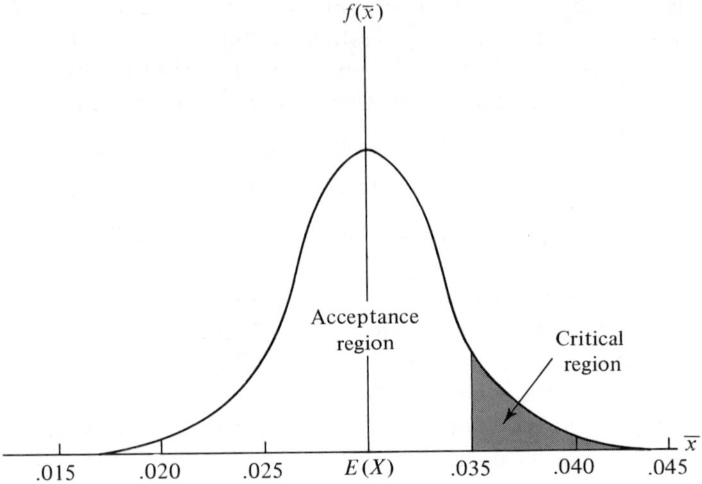

Fig. 8.1 Graphical description of critical and acceptance regions for the given decision rule [$E(X) = .03$].

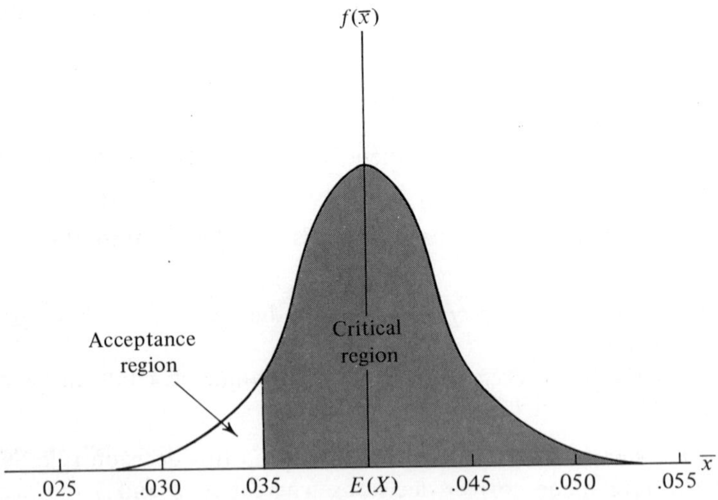

Fig. 8.2 Graphical description of critical and acceptance regions for the given decision rule [$E(X) = .04$].

TABLE 8.1

	Accept Null Hypothesis	Reject Null Hypothesis
Null hypothesis true	1	2
Null hypothesis false	3	4

kind or a *type I error*. Now let us evaluate the probability of occurrence of an error of the first kind (denoted by α). Then we write

$$\begin{aligned}\alpha &= P\{\bar{X} > .035 \mid E(X) \leq .03\} \\ &\leq P\left\{\frac{\bar{X} - E(X)}{\sigma/\sqrt{n}} > \frac{.035 - .03}{\sigma/\sqrt{n}}\right\} \\ &\leq P\left\{Z > \frac{.005}{.006/\sqrt{n}}\right\}\end{aligned}$$

Let us suppose that $\bar{X}$ is evaluated on the basis of a sample of size $n = 4$. Then the error of the first kind becomes

$$\begin{aligned}\alpha &\leq P\left\{Z > \frac{.010}{.006}\right\} \\ &\leq .0485\end{aligned}$$

This means (given the decision rule) the maximum error of the first kind is .0485. α is sometimes known as the *level of significance*.

In case 3 we accepted the lot despite the fact that $E(X) = .040$. Thus we have committed an *error of the second kind* or a *type II error*. Let us evaluate as well the probability of occurrence of an error of the second kind (denoted by β). Then we write

$$\beta\{E(X)\} = P\{\bar{X} \leq .35 \mid E(X) > .03\}$$

Obviously the error of the second type is not a constant then but depends on the value taken by $E(X)$. If $E(X) = .04$, we have

$$\begin{aligned}\beta\{E(X) = .04\} &= P\left\{\frac{\bar{X} - E(X)}{\sigma/\sqrt{n}} \leq \frac{.035 - .04}{\sigma/\sqrt{n}}\right\} \\ &= P\left\{Z \leq \frac{-.010}{.006}\right\} \\ &= .0485\end{aligned}$$

Usually it is more convenient to study the characteristics of the decision rule by defining a new function π called the *power of the test*. *The power of the test is the probability of rejecting the null hypothesis when the alternative is true*. Then we write

$$\begin{aligned}\pi\{E(X)\} &= 1 - \beta\{E(X)\} \\ &= P\{\bar{X} > .035 \mid E(X) > .03\}\end{aligned}$$

Table 8.2 shows the values of the power function for each possible value of the parameter $E(X)$ for the given decision rule. The graph of this power function is shown in Fig. 8.3. It can readily be seen from the graph that the chance of accepting a lot having an average variation in wall thickness greater than .03 mm decreases as this average value increases for the given decision rule.

TABLE 8.2

$E(X)$	.030	.033	.036	.039	.042	.045	.048
$\pi\{E(X)\}$	.0485	.2527	.6305	.9087	.9902	.9996	.9999

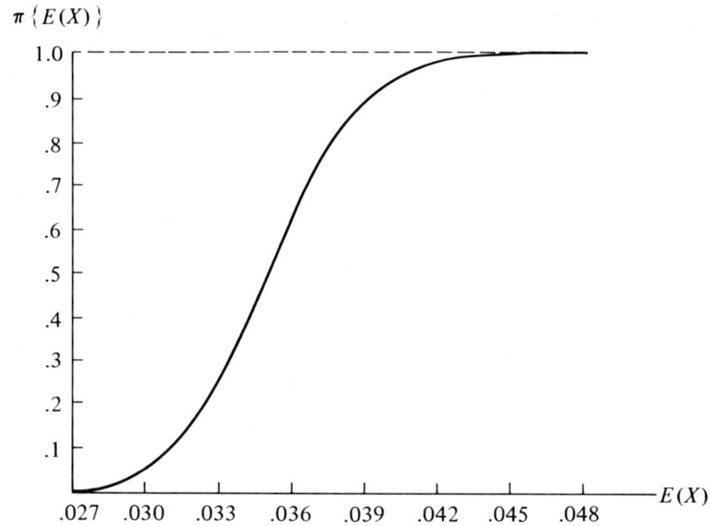

Fig. 8.3 Power curve for the given rule.

From our previous discussion the reader can see that we arbitrarily decided that $\bar{X}$ should be less than or equal to .035 and $n = 4$ in order to test the hypothesis given by management. Usually this will not be the appropriate procedure of testing statistical hypotheses. The inappropriateness of this procedure stems from the fact that the rule set by management was too *rigid*. Should management have decided that the *maximum probability of rejecting* a lot having $E(X) \leq .03$ is, say, .02, then we could find the appropriate test procedure in this case. In fact, then we would test the following hypothesis:

$$H_0\colon E(X) \leq .03 \quad \text{against} \quad H_1\colon E(X) > .03 \quad \text{given } \alpha_{\max} = .02$$

Then we write

$$\alpha_{\max} = P\{\bar{X} > k \mid E(X) = .03\}$$

where k is a constant to be determined. The constant k must satisfy the following equation:

$$.02 = P\left\{\frac{\bar{X} - E(X)}{\sigma/\sqrt{n}} > \frac{k - .03}{\sigma/\sqrt{n}}\right\}$$

$$= P\left\{Z > \frac{k - .03}{\sigma/\sqrt{n}}\right\}$$

$$= 1 - \Phi\left(\frac{k - .03}{\sigma/\sqrt{n}}\right)$$

Thus

$$k = .03 + \frac{2.055\sigma}{\sqrt{n}} = .03 + \frac{.01233}{\sqrt{n}}$$

Consequently, reject the null hypothesis if $\bar{X} > .03 + .01233/\sqrt{n}$. Now what is the sample size? The optimal procedure for determining the sample size will be treated in the next chapter; however, we will analyze the effect of the *sample size* on controlling risk of an error of the second kind. Let us assume that n can be either 4, 9, or 16 and evaluate the power function based on this assumption. If $n = 4$, then

$$\beta\{E(X)\} = P\left\{\bar{X} \leq .03 + \frac{.01233}{2} \,|\, E(X) > .03\right\}$$

Hence

$$\pi\{E(X)\} = 1 - P\{\bar{X} \leq .03616 \,|\, E(X) > .03\}$$

Similarly, we can evaluate the value of the power function for $n = 9$ and $n = 16$. Values of the power function for these three sample sizes are given in Table 8.3. The graphs for the three power functions are shown in Fig. 8.4.

TABLE 8.3

$E(X)$		.030	.033	.036	.039	.042	.045	.048
	$n = 4$	.02	.15	.48	.82	.97	.99	~ 1
$\pi\{E(X)\}$	$n = 9$	.02	.29	.82	.99	~ 1		
	$n = 16$	.02	.48	.97	~ 1			

The risk of making an error of the second kind decreases as the sample size increases. To illustrate, if the incoming lots have $E(X) = .039$ there will be a chance of accepting such lots 18 times out of 100 as having $E(X) \leq .030$ when $n = 4$, whereas there will be a chance of accepting such lots once out of 100 when $n = 9$. Referring to Table 8.2, we find that incoming lots with mean $= .039$ have a chance of being accepted 10 times out of 100 when $n = 4$ and

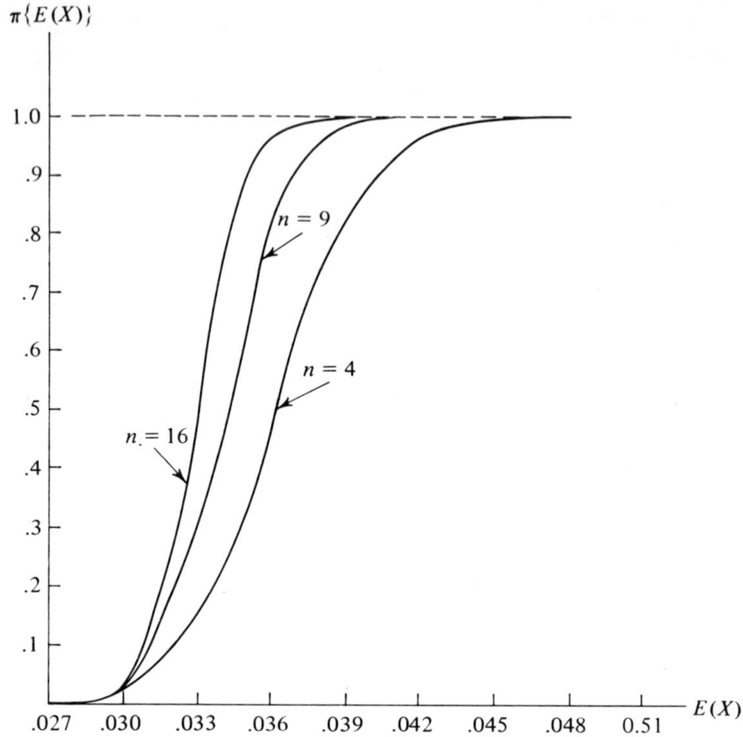

Fig. 8.4 Power functions for selected sample sizes ($\alpha = .02$).

$\alpha_{max} = .0485$. This means that *as the error of the first kind increases, the error of the second kind decreases and vice versa.* It follows that by *varying the sample size we can exercise control on the error of the second kind.*

Now if we plot $\beta\{E(X)\}$ against the true average variation in wall thickness for a fixed $\alpha = .02$ and sample sizes 4, 9, and 16, the resulting plot is called an *operating characteristic* (denoted by OC) *curve*. The result is represented by the curves in Fig. 8.5.

The level of significance and the sample size uniquely determine the OC curve for the given decision rule. It is evident that by increasing the sample size for a given level of significance the error of the second kind decreases. In practice a balance must be struck between the cost of additional observations and the advantage of decreasing the error of the second kind. In many situations it is not feasible to assess explicitly the cost parameters associated with alternative testing procedures. In the absence of knowledge of these cost parameters the criterion by which we can assess and compare tests of statistical hypothesis is found in the OC curves or power functions.

We have discussed so far a decision procedure that associated with it the outcome of random variable $\bar{X}$, sample size n, and acceptance region ($\bar{X} \leq$ a

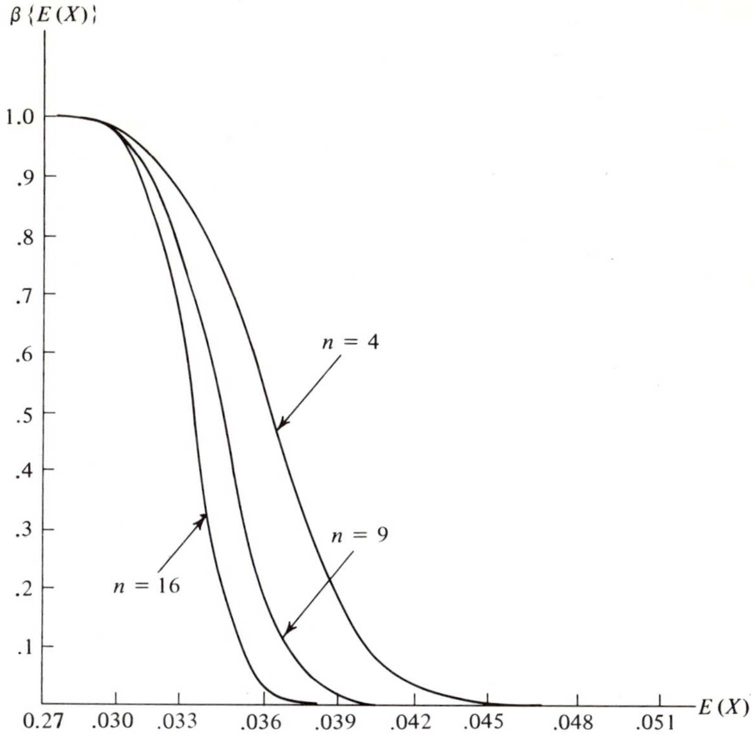

Fig. 8.5 *OC* curves for selected sample sizes ($\alpha = .02$).

prescribed constant). We shall refer to this decision procedure as the $\bar{X}$ procedure. Now we may raise the following question: Is this procedure superior to the $X_{\max}$ procedure? Note that the $X_{\max}$ procedure is specified by: draw at random a sample of size n; observe the largest measurement, $X_{\max}$; if $X_{\max} \leq$ a prescribed constant conclude that $E(X)$ is less than or equal to .03; otherwise conclude that $E(X)$ is greater than .03.

To compare these two procedures ($\bar{X}$ and $X_{\max}$) we will fix the level of significance at .02 and the sample size at 4 for both procedures and use the *OC* curves to provide a criterion of comparison. In other words, we will be comparing both procedures in probabilistic terms.

The alternative decision rule is to accept the null hypothesis if $X_{\max} \leq k$, hence

$$P\{X_{\max} \leq k\} = 1 - \alpha = .98$$

Since these measurements are independent, identically distributed, normal variates, we write

$$P\{X_1 \leq k\} P\{X_2 \leq k\} P\{X_3 \leq k\} P\{X_4 \leq k\} = .98$$

Hence

$$\int_{-\infty}^{\frac{k-.030}{.006}} \frac{e^{-(1/2)z^2}}{\sqrt{2\pi}}\, dz = (.98)^{1/4} = .995$$

whence $k = .0454$. This means the following: Reject the null hypothesis whenever $X_{\max}$ is greater than .0454. The probability of occurrence of the error of the second kind is given by

$$\beta\{E(X)\} = P\{X_{\max} \leq .0454 \mid E(X) > .03\}$$

Suppose now that $E(X) = .036$; then

$$\begin{aligned} \beta\{.036\} &= P\{X_{\max} \leq .0454 \mid E(X) = .036\} \\ &= \left[\int_{-\infty}^{\frac{.0454-.036}{.006}} \frac{1}{\sqrt{2\pi}} e^{-(1/2)z^2}\, dz\right]^4 \\ &= .78 \end{aligned}$$

Similarly, we can calculate the probability of occurrence of the error of the second kind for possible values of the true average $\{E(X)\}$. The result is tabulated in Table 8.4. The *OC* curves for the $\bar{X}$ and $X_{\max}$ procedures are as indicated in Fig. 8.6.

TABLE 8.4

$E(X)$	.030	.033	.036	.039	.042	.045	.048
$\beta\{E(X)\}$	.98	.92	.78	.53	.26	.06	.01

Obviously, the procedure based upon $\bar{X}$ gives better protection than the $X_{\max}$ procedure for a whole range of alternatives. Thus the $\bar{X}$ procedure is preferred to the $X_{\max}$ procedure. Now we proved that the $\bar{X}$ procedure is superior to the $X_{\max}$ procedure. Does this imply that the $\bar{X}$ procedure is the *optimal* one? On what basis is a procedure said to be optimal? *The procedure is said to be optimal if the rejection region for a fixed sample size and level of significance minimizes the probability of occurrence of the error of the second kind for a whole range of alternatives.* The optimal procedure is sometimes called a *uniformly most powerful test over a range of alternatives.* There may be, in some situations, more than one optimal procedure. The procedure based upon $\bar{X}$ is in fact the optimal procedure. This fact is proved by the Neyman–Pearson lemma, where justification for the use of the *likelihood ratio* has been established.

We have developed this example in order to introduce some fundamental concepts of statistical inference in decision making. In Chapter 9 we will take up the Neyman–Pearson lemma for testing statistical hypotheses about a single parameter.

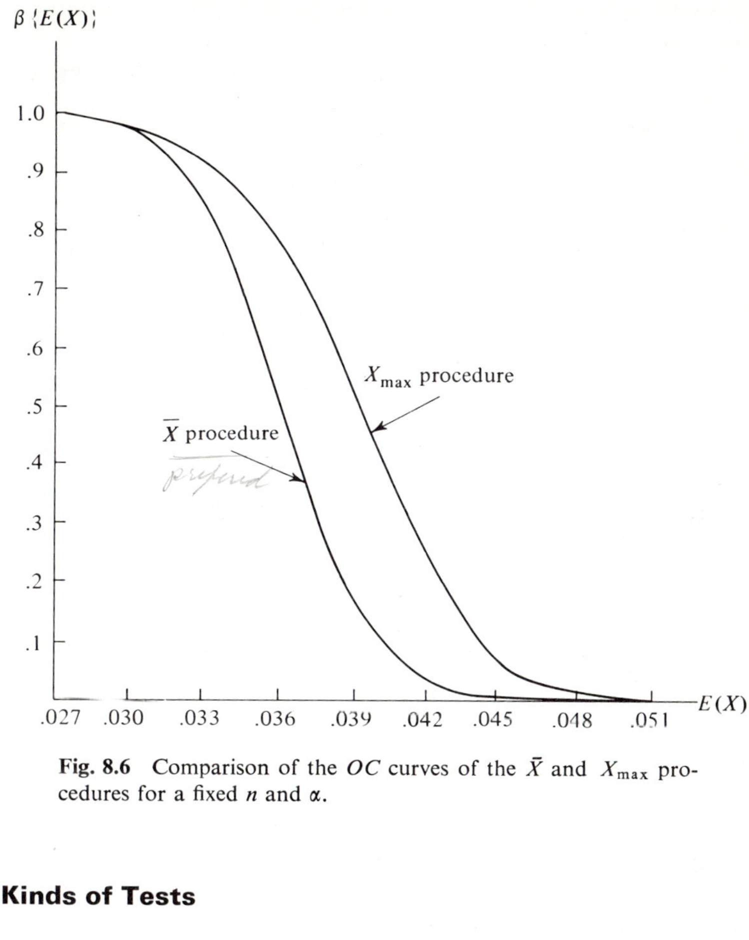

Fig. 8.6 Comparison of the *OC* curves of the $\bar{X}$ and $X_{\max}$ procedures for a fixed n and α.

8.4 Kinds of Tests

If a statistical hypothesis specifies the values of all the parameters of the distribution of the random variable under study it is called *simple hypothesis;* otherwise it is called a *composite hypothesis.* Suppose, for example, the random variable under study is normally distributed with mean μ and standard deviation σ. Hence the normal distribution is completely specified by two parameters, μ and σ. Testing if the mean is equal to, say, 100, given that σ is known and equal to, say, 5, the null hypothesis is $H_0 : \mu = 100$. It is called a *simple hypothesis.*

Another illustration is testing if the mean is greater than 100 given that σ is equal to 5. The null hypothesis in this case is given by $H_0 : \mu > 100$ is called a *composite hypothesis.*

Suppose the random variable under study has a Poisson distribution. The Poisson distribution is completely specified by a single parameter λ.

Now if we are interested in testing if $\lambda = 10$, then the null hypothesis H_0: $\lambda = 10$ is a simple hypothesis, whereas $H_0: \lambda < 10$ is a composite hypothesis. The study of testing hypotheses is usually classified in terms of the null hypothesis H_0 and alternate hypothesis H_1. Thus, if the variate studied has a Poisson distrubition, then the hypotheses

$$H_0: \lambda = 10 \qquad H_1: \lambda = 12$$

is a simple against simple, whereas

$$H_0: \lambda = 10 \qquad H_1: \lambda > 10$$

is a simple against composite. Finally

$$H_0: \lambda > 10 \qquad H_1: \lambda < 10$$

is a composite against composite. Simple hypotheses can be resolved, whereas some of the composite hypotheses defy analytical solution.

SUGGESTED REFERENCES

See the references given at the end of Chapter 10.

9

STATISTICAL HYPOTHESES TESTING (Single Parameter)

9.1 The Likelihood Function

The *likelihood function* developed by R. A. Fisher plays a significant role in testing statistical hypotheses. Suppose a manufacturer is interested in knowing whether the mean resistance of a lot of resistors is 1 ohm or greater than 1 ohm. Suppose further that the distribution of resistance is normal $N(\mu, \sigma_0^2)$, where σ_0^2 is known and μ is unknown. In order to test this conjecture, one takes a random sample of size n and measures the resistances of the resistors in the sample. The sample is known to have come from some normal distribution. Is it reasonable that the sample could have come from one having mean $\mu_0 = 1$ ohm? It could have come from some normal population having mean $\mu_1 > 1$ ohm. In other words, we wish to test the statistical hypothesis

$$H_\sigma\colon\ \mu_0 = 1 \text{ ohm} \quad \text{against} \quad H_1\colon\ \mu_1 > 1 \text{ ohm}$$

If the mean of the population is 1 ohm, what is the likelihood of the individual measurements of the sample $(r_1, r_2, \ldots, r_n)$?

Since the individual measurements are assumed to be independent, the likelihood of these measurements given H_0 is

$$\begin{aligned} L_0 &= f(r_1 \mid \mu_0 = 1) f(r_2 \mid \mu_0 = 1) \cdots f(r_n \mid \mu_0 = 1) \\ &= \left(\frac{1}{\sqrt{2\pi}\,\sigma_0}\right)^n \exp\left[-\frac{\sum(r_i - 1)^2}{2\sigma_0^2}\right] \end{aligned} \tag{9.1}$$

The likelihood of these measurements if H_1 is true is

$$L_1 = \left(\frac{1}{\sqrt{2\pi}\,\sigma_0}\right)^n \exp\left[\frac{-\sum(r_i - \mu_1)^2}{2\sigma_0^2}\right] \tag{9.2}$$

Now the question arises whether L_0 or L_1 can be considered as a satisfactory basis for choosing between H_0 and some alternative H_1. This dilemma has been resolved by the Neyman–Pearson lemma, which determines the appropriate rejection region of the hypothesis H_0. Once this appropriate rejection region is described, the acceptance or rejection of the null hypothesis H_0 is straightforward.

9.2 Neyman–Pearson Lemma

The assumptions underlying the Neyman–Pearson lemma for simple hypotheses are as follows:

1. The probability of individual measurements of the sample $(r_1, r_2, \ldots, r_n)$ falling in the critical region (R) when the null hypothesis is true should be less than or equal to the desired level of significance α. In symbols, the

$$P\{r_1, r_2, \ldots, r_n \text{ in } R \mid H_0\} \leq \alpha \tag{9.3}$$

2. The probability of individual measurements of the sample $(r_1, r_2, \ldots, r_n)$ falling in the critical region (R) when the alternative hypothesis is true should be maximum. In other words, the error of the second kind should be minimum. In symbols, the

$$P\{r_1, r_2, \ldots, r_n \text{ in } R \mid H_1\} \tag{9.4}$$

is a maximum among all critical regions satisfying Eq. (9.3). The Neyman–Pearson lemma states that assumptions 1 and 2 are satisfied if the critical region (R) is of the form

$$\frac{L_1(r_1, r_2, \ldots, r_n)}{L_0(r_1, r_2, \ldots, r_n)} > k \tag{9.5}$$

where k has to be adjusted so that

$$P\{r_1, r_2, \ldots, r_n \text{ in } R \mid H_0\} = \alpha \tag{9.6}$$

The Neyman–Pearson lemma asserts that the critical region given by (9.5) minimizes the error of the second kind for a whole range of alternatives (uniformly most powerful test).

Proof. Let R be the critical region derived by the Neyman–Pearson criterion. Suppose now that this critical region is not the best choice that yields minimum error of the second kind. Let R' be the critical region that yields minimum error of the second kind, as shown in Fig. 9.1. Since both critical regions are of size α, we have

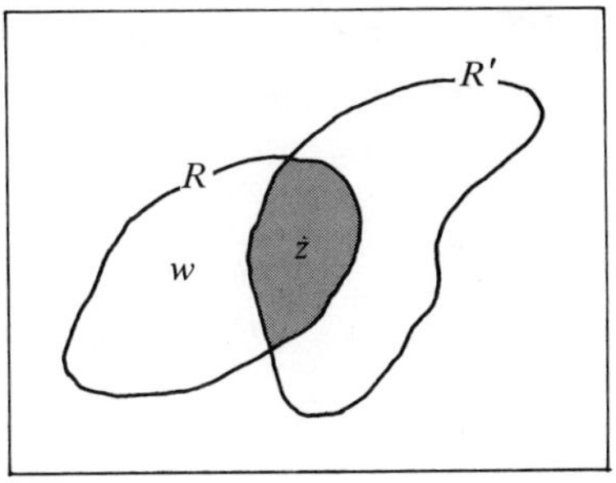

Fig. 9.1 Diagrammatic representation of two critical regions.

$$\int \underset{R}{\cdots} \int L_0(r_1, r_2, \ldots, r_n)\, dr_1\, dr_2 \cdots dr_n = \int \underset{R'}{\cdots} \int L_0(r_1, r_2, \ldots, r_n)\, dr_1\, dr_2 \cdots dr_n \tag{9.7}$$

Equation (9.7) reduces to

$$\int \underset{w}{\cdots} \int L_0(r_1, r_2, \ldots, r_n)\, dr_1\, dr_2 \cdots dr_n = \int \underset{v}{\cdots} \int L_0(r_1, r_2, \ldots, r_n)\, dr_1\, dr_2 \cdots dr_n \tag{9.8}$$

since the integral of z that is the common region of R and R' will cancel from both sides of Eq. (9.7). Multiplying both sides of Eq. (9.8) by k, we obtain

$$k \int \underset{w}{\cdots} \int L_0(r_1, r_2, \ldots, r_n)\, dr_1\, dr_2 \cdots dr_n = k \int \underset{v}{\cdots} \int L_0(r_1, r_2, \ldots, r_2)\, dr_1\, dr_2 \cdots dr_n \tag{9.9}$$

The power of the test given critical region R is

$$(1 - \beta) \mid H_1 = \int \underset{R}{\cdots} \int L_1(r_1, r_2, \ldots, r_n)\, dr_1\, dr_2 \cdots dr_n$$

The power of the test given critical region R' is

$$(1 - \beta') \mid H_1 = \int \underset{R'}{\cdots} \int L_1(r_1, r_2, \ldots, r_n)\, dr_1\, dr_2 \cdots dr_n$$

Since we assumed that critical region R' is a better choice, the increase in power (denoted by $\delta\pi$) is given by

$$\begin{aligned} \delta\pi &= (1 - \beta') \mid H_1 - (1 - \beta) \mid H_1 \\ &= \int \underset{v}{\cdots} \int L_1(r_1, r_2, \ldots, r_n)\, dr_1\, dr_2 \cdots dr_n \\ &\quad - \int \underset{w}{\cdots} \int L_1(r_1, r_2, \ldots, r_2)\, dr_1\, dr_2 \cdots dr_n \end{aligned} \tag{9.10}$$

Since region w lies in the original critical region R, it follows from the Neyman–Pearson lemma that

$$\frac{L_1(r_1, r_2, \ldots, r_n \text{ in } w)}{L_0(r_1, r_2, \ldots, r_n \text{ in } w)} > k$$

Hence

$$\int \underset{w}{\cdots} \int L_1(r_1, r_2, \ldots, r_n)\, dr_1\, dr_2 \cdots dr_n$$

$$> k \int \underset{w}{\cdots} \int L_0(r_1, r_2, \ldots, r_n)\, dr_1\, dr_2 \cdots dr_n \tag{9.11}$$

Since region v lies in the original acceptance region, it follows from the Neyman–Pearson lemma that

$$\frac{L_1(r_1, r_2, \ldots, r_n \text{ in } v)}{L_0(r_1, r_2, \ldots, r_n \text{ in } v)} < k$$

Hence

$$\int \underset{v}{\cdots} \int L_1(r_1, r_2, \ldots, r_n)\, dr_1\, dr_2 \cdots dr_n$$

$$< k \int \underset{v}{\cdots} \int L_0(r_1, r_2, \ldots, r_n)\, dr_1\, dr_2 \cdots dr_n \tag{9.12}$$

Now from (9.9) and (9.12) we obtain

$$\int \underset{v}{\cdots} \int L_1(r_1, r_2, \ldots, r_n)\, dr_1\, dr_2 \cdots dr_n$$

$$< k \int \underset{w}{\cdots} \int L_0(r_1, r_2, \ldots, r_n)\, dr_1\, dr_2 \cdots dr_n \tag{9.13}$$

From (9.11) and (9.13) we get

$$\int \underset{v}{\cdots} \int L_1(r_1, r_2, \ldots, r_n)\, dr_1\, dr_2 \cdots dr_n$$

$$\int \underset{w}{\cdots} \int L_1(r_1, r_2, \ldots, r_n)\, dr_1\, dr_2 \cdots dr_n < 0$$

The left-hand side of this inequality is equal to $\delta\pi$. Thus we have proven that $\delta\pi$ is negative. This means that the original test must have been optimum and the proof is complete.

9.3 Likelihood Ratio Tests of Composite Hypotheses

If the null or alternative hypotheses being tested are composite, we cannot apply the likelihood ratio method in the form described for simple hypotheses. A rather heuristic method is used to test composite hypotheses. Though this method has proven to be an excellent tool in deriving good tests, it is no longer guaranteed to provide most powerful tests. In this case each composite hypothesis can be regarded as a set of simple hypotheses. To illustrate, suppose that the random variable (X) under study is normally distributed with unknown mean μ and unknown standard deviation σ and it is required to test the following composite hypotheses, as shown in Fig. 9.2:

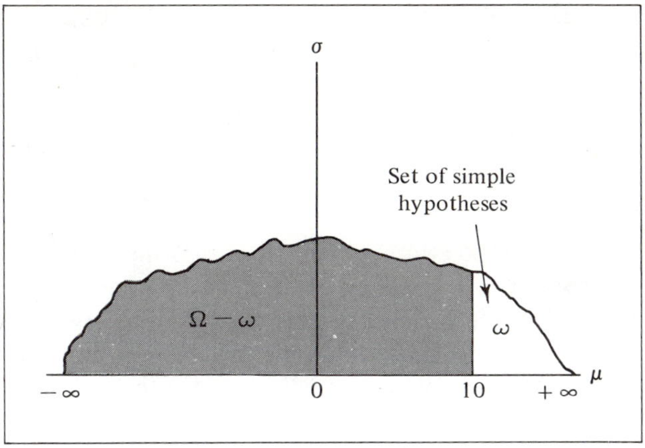

Fig. 9.2 Parameter space Ω.

Let ω denote the set of simple hypotheses defined by H_0 and Ω denote the set of all possible simple hypotheses including H_0 and all alternatives. A reasonable criterion of rejecting the null hypothesis is when *the ratio of the maximum likelihood of all simple hypotheses in* ω *to the maximum likelihood of all simple hypotheses including* H_0 *and all alternatives is sufficiently small.* In symbols, reject H_0 when

$$\lambda = \frac{\max\,[f(x_1)f(x_2)\cdots f(x_n)\,|\,\omega]}{\max\,[f(x_1)f(x_2)\cdots f(x_n)\,|\,\Omega]} < k \qquad k < 1 \tag{9.14}$$

where k is a number chosen such that the level of significance α is equal to some desired value.

The above criterion can be regarded as being equivalent to a special case when ω reduces to a simple hypothesis H_0 and Ω reduces to two simple hypotheses H_0 and H_1. If the above criterion does not yield a maximum the supremum is used.

9.4 Tests Concerning the Mean of a Normal Distribution (μ) with Known Variance (σ_0^2)

Case 1

$$H_0\colon \mu = \mu_0 \quad \text{against} \quad H_1\colon \mu = \mu_1 < \mu_0$$

The nature of the optimum rejection region given by the Neyman–Pearson lemma is

$$\frac{L_1(x_1, x_2, \ldots, x_n)}{L_0(x_1, x_2, \ldots, x_n)} > k$$

The ratio of the likelihood functions for the normal distribution is

$$\frac{L_1(x_1, x_2, \ldots, x_n)}{L_0(x_1, x_2, \ldots, x_n)} = \frac{(1/\sigma_0\sqrt{2\pi})^n \exp[-\sum_{i=1}^{n}(x_i - \mu_1)^2/2\sigma_0^2]}{(1/\sigma_0\sqrt{2\pi})^n \exp[-\sum_{i=1}^{n}(x_i - \mu_0)^2/2\sigma_0^2]}$$

$$= \exp\left[(\mu_1 - \mu_0)\frac{2n\bar{x} - n(\mu_1 + \mu_0)}{2\sigma_0^2}\right]$$

$$= \exp\left(\bar{x}\,\frac{\mu_1 - \mu_0}{\sigma_0^2/n}\right)\exp\left(-\frac{\mu_1^2 - \mu_0^2}{2\sigma_0^2/n}\right)$$

which is the product of two exponential terms; the first is a function of $\bar{x}$ whereas the second is a constant. Hence, the optimum rejection region is a monotonically decreasing function of $\bar{x}$ as long as $\mu_1 < \mu_0$. This implies that the set of values of the likelihood ratio greater than k is equivalent to the set of values of $\bar{x}$ less than or equal to some other constant k'. It is noteworthy that the present rejection region does not depend on the specific value of the alternative μ_1 but merely on the fact that $\mu_1 < \mu_0$. Thus the present rejection region constitutes a uniformly most powerful test for all one-sided alternatives $\mu_1 < \mu_0$. The rejection region of the test is thus given by

$$\bar{x} < k'$$

The test therefore is reject H_0 when $\bar{x} < k'$. The probability of a type I error is

$$P\{\bar{x} < k' \mid \mu = \mu_0\} = \alpha$$

which is equivalent to

$$P\left\{\frac{\bar{x} - \mu_0}{\sigma_0/\sqrt{n}} < \frac{k' - \mu_0}{\sigma_0/\sqrt{n}}\right\} = \alpha$$

Hence

$$k' = \mu_0 + \frac{Z_\alpha \sigma_0}{\sqrt{n}}$$

Thus the rejection region is the set of values of $\bar{x}$ for which

$$\bar{x} < \mu_0 + \frac{Z_\alpha \sigma_0}{\sqrt{n}} \tag{9.15}$$

The probability of accepting H_0 when H_1 is true is

$$\beta(\mu_1) = P\{\bar{x} \geq k' \mid \mu = \mu_1\}$$

$$= P\left\{\bar{x} \geq \mu_0 + \frac{Z_\alpha \sigma_0}{\sqrt{n}} \,\middle|\, \mu = \mu_1\right\}$$

$$= P\left\{\frac{\bar{x} - \mu_1}{\sigma_0/\sqrt{n}} \geq \frac{\mu_0 - \mu_1}{\sigma_0/\sqrt{n}} + Z_\alpha\right\}$$

$$= P\left\{Z \geq \frac{\mu_0 - \mu_1}{\sigma_0/\sqrt{n}} + Z_\alpha\right\}$$

$$= 1 - \Phi\left(\frac{\mu_0 - \mu_1}{\sigma_0/\sqrt{n}} + Z_\alpha\right)$$

$$= \Phi\left(\frac{\mu_1 - \mu_0}{\sigma_0/\sqrt{n}} - Z_\alpha\right) \tag{9.16}$$

The probability of the occurrence of the error of the second kind can be found from charts regardless of the values of μ_0 and σ_0 where the abscissa scale is changed to

$$d = \frac{\mu_0 - \mu_1}{\sigma_0}$$

Thus

$$\beta(\mu_1) = \Phi(-d\sqrt{n} - Z_\alpha)$$

The *OC* curves shown in Figs. 9.3 and 9.4 are plotted against d for a level of significance $\alpha = .01$ and $\alpha = .05$, respectively.

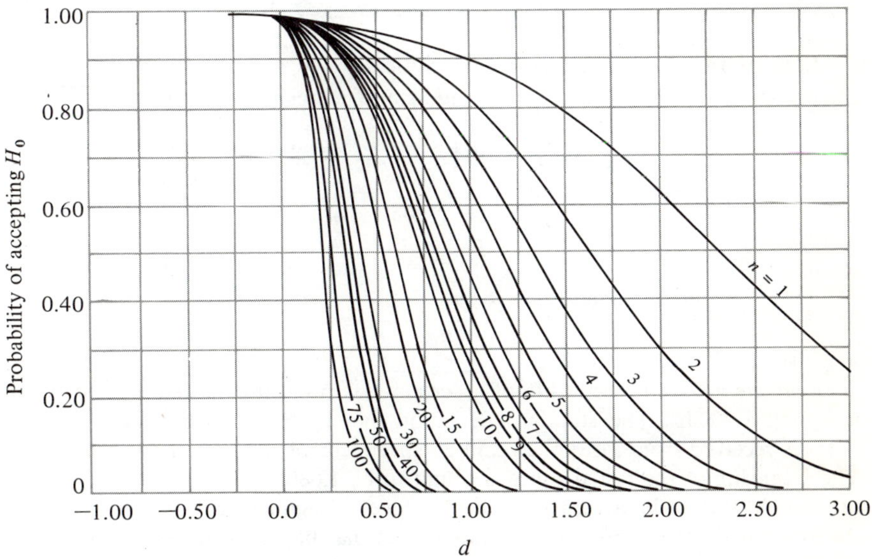

Fig. 9.3 *OC* curves for different values of n for the one-sided normal test for a level of significance = .01.

[Reproduced with permission from "Operating Characteristics for the Common Statistical Tests of Significance" by Charles L. Ferris, Frank E. Grubbs, and Chalmers L. Weaver, *Annals of Mathematical Statistics*, June 1946].

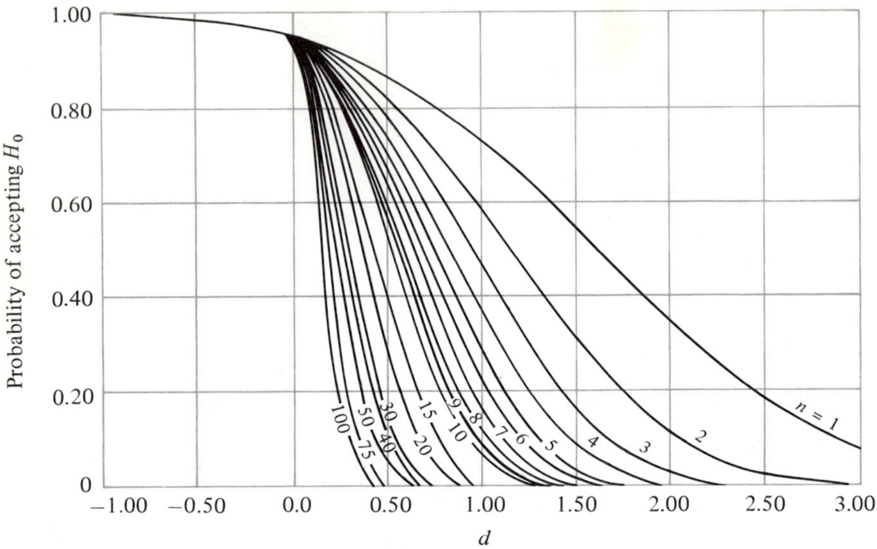

Fig. 9.4 *OC* curves for different values of n for the one-sided normal test for a level of significance $\alpha = .05$.

It is of interest to determine analytically the sample size that controls the errors of the first and second kind, α and β, respectively. Since

$$\beta(\mu_1) = \Phi\left(\frac{\mu_1 - \mu_0}{\sigma_0/\sqrt{n}} - Z_\alpha\right)$$

$$Z_\beta = \frac{\mu_1 - \mu_0}{\sigma_0/\sqrt{n}} - Z_\alpha$$

whence

(9.17) $$n = \frac{\sigma_0^2(Z_{1-\alpha} + Z_{1-\beta})^2}{(\mu_0 - \mu_1)^2}$$

Example 9.1 The mean life of a certain electronic tube is required to be not less than 1,200 hr. The standard deviation based on past experience is 150 hr. A lot is received from a supplier, and 25 specimens of the tubes have been tested. The mean life of the tested tubes is found to be 1,080 hr.

(a) Should the lot be accepted at the 5% level of significance?

(b) What is the probability of accepting a lot that has a mean life of 1,100 hr?

(c) If the mean life of a lot is 1,000 hours, the lot should be rejected with probability .99. If the mean life of the lot is 1,200, the lot should be accepted with probability .95. How many specimens should be tested?

Solution

(a) We wish to test

$$H_0: \mu_0 = 1200 \quad \text{against} \quad H_1: \mu_1 < 1200$$

(For purposes of illustration we assume a normal distribution.) The rejection rule is to reject H_0 when

$$\begin{aligned}\bar{x} &< \mu_0 + Z_\alpha \frac{\sigma_0}{\sqrt{n}} \\ &< 1200 + Z_{.05}\left(\frac{150}{5}\right) \\ &< 1150.65\end{aligned}$$

Since the value of $\bar{x}$ is 1,080, which falls in the critical region $(-\infty, 1150.65)$, we conclude that $\mu < 1{,}200$ hr.

(b)
$$\begin{aligned}\beta(\mu_1 = 1100) &= \Phi\left(\frac{\mu_1 - \mu_0}{\sigma_0/\sqrt{n}} - Z_\alpha\right) \\ &= \Phi\left(\frac{1100 - 1200}{150/5} + 1.645\right) \\ &= \Phi(-1.688) \\ &= .045\end{aligned}$$

Thus the probability of accepting a lot that has a mean life of 1,100 hr is .045.

(c)
$$\begin{aligned}n &= \frac{\sigma_0^2(Z_{1-\alpha} + Z_{1-\beta})^2}{(\mu_0 - \mu_1)^2} \\ &= \frac{22{,}500(1.645 + 2.33)^2}{40{,}000} \\ &= 9\end{aligned}$$

Hence, if we test nine specimens the conditions specified in (c) will be satisfied.

Case 2

$$H_0: \mu = \mu_0 \quad \text{against} \quad H_1: \mu = \mu_1 > \mu_0$$

It can be shown that the optimal rejection region is given by

$$\bar{x} > k'$$

The probability of type I error is

$$P\{\bar{x} > k' \mid \mu = \mu_0\} = \alpha$$

which is equivalent to

$$P\left\{\frac{\bar{x} - \mu_0}{\sigma_0/\sqrt{n}} > \frac{k' - \mu_0}{\sigma_0/\sqrt{n}}\right\} = \alpha$$

Hence

$$k' = \mu_0 + Z_{1-\alpha}\frac{\sigma_0}{\sqrt{n}}$$

Thus the rejection region is the set of values of $\bar{x}$ for which

(9.18)
$$\bar{x} > \mu_0 + Z_{1-\alpha}\frac{\sigma_0}{\sqrt{n}}$$

The probability of accepting H_0 when H_1 is true is

$$
\begin{aligned}
\beta(\mu_1) &= P\{\bar{x} \leq k' \mid \mu = \mu_1\} \\
&= P\left\{\bar{x} \leq \mu_0 + Z_{1-\alpha}\frac{\sigma_0}{\sqrt{n}} \mid \mu = \mu_1\right\} \\
&= P\left\{\frac{\bar{x} - \mu_1}{\sigma_0/\sqrt{n}} \leq \frac{\mu_0 - \mu_1}{\sigma_0/\sqrt{n}} + Z_{1-\alpha}\right\} \\
&= P\left\{Z \leq \frac{\mu_0 - \mu_1}{\sigma_0/\sqrt{n}} + Z_{1-\alpha}\right\} \\
&= \Phi\left(\frac{\mu_0 - \mu_1}{\sigma_0/\sqrt{n}} + Z_{1-\alpha}\right) \qquad (9.19)
\end{aligned}
$$

The *OC* curves shown in Figs. 9.3 and 9.4 can be used to find the probability of the occurrence of the error of the second kind in terms of d, where

$$d = \frac{\mu_1 - \mu_0}{\sigma_0}$$

Thus

$$\beta(\mu_1) = \Phi(-d\sqrt{n} + Z_{1-\alpha})$$

Similar to case 1, it can be shown that the sample size that controls the errors of the first and second kinds, α and β, respectively, is

$$n = \frac{\sigma_0^2(Z_{1-\alpha} + Z_{1-\beta})^2}{(\mu_1 - \mu_0)^2} \qquad (9.20)$$

(Note: Bear in mind that in this and the following examples in Secs. 9.4–9.6, the normal distribution is assumed unless explicitly given otherwise.)

Case 3

$$H_0\colon \mu = \mu_0 \quad \text{against} \quad H_1\colon \mu = \mu_1 \neq \mu_0$$

This test is sometimes referred to as the *double-sided test*, in contrast with the *single-sided tests* given in cases 1 and 2.

Since the null hypothesis is simple, whereas the alternative hypothesis is composite, the optimal rejection region is given by

$$\lambda = \frac{\max\,[f(x_1)f(x_2)\cdots f(x_n) \mid \omega]}{\max\,[f(x_1)f(x_2)\cdots f(x_n) \mid \Omega]} < k$$

Figure 9.5 shows that the numerator of the likelihood ratio is a fixed point (σ_0, μ_0) in space. In other words, no parameter is changing. Therefore,

$$
\begin{aligned}
&\max\,[f(x_1)f(x_2)\cdots f(x_n) \mid \omega] \\
&\quad = \left(\frac{1}{\sigma_0\sqrt{2\pi}}\right)^n \exp\left[\frac{-1}{2\sigma_0^2}\sum_{i=1}^{n}(x_i - \mu_0)^2\right]
\end{aligned}
$$

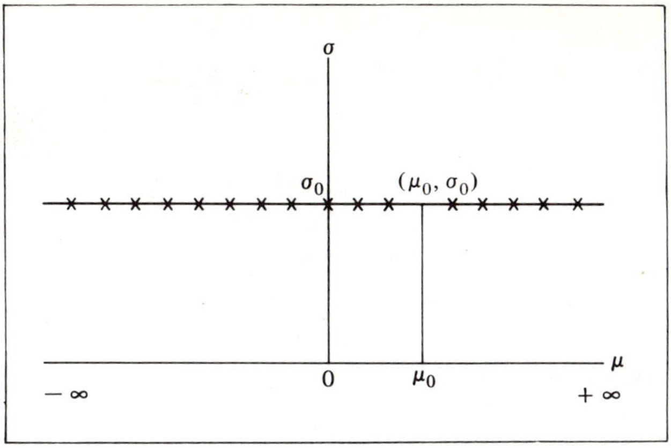

Fig. 9.5 Graph showing the parameters of the normal distribution under the null hypothesis.

The alternative hypothesis states that the parameter point is on the line ($\sigma = \sigma_0$), excluding (μ_0, σ_0). Thus μ is the variable parameter. Therefore,

$$\max [f(x_1)f(x_2) \cdots f(x_n) \mid \Omega] = \max \left(\frac{1}{\sigma_0\sqrt{2\pi}}\right)^n \exp\left[\frac{-1}{2\sigma_0^2}\sum_{i=1}^{n}(x_i - \mu)^2\right]$$

The value of μ that maximizes the right-hand side is equivalent to the value of μ that minimizes $\sum_{i=1}^{n}(x_i - \mu)^2$. On taking the partial derivative with respect to μ and equating to zero we obtain

$$\frac{\partial}{\partial \mu}\sum_{i=1}^{n}(x_i - \mu)^2 = -2\sum_{i=1}^{n}(x_i - \mu) = 0$$

whence

$$n\mu = n\bar{x} \quad \text{or} \quad \mu = \bar{x}$$

Hence, the optimal rejection region is given by

$$\lambda = \frac{\exp[-\sum_{i=1}^{n}(x_i - \mu_0)^2/2\sigma_0^2]}{\exp[-\sum_{i=1}^{n}(x_i - \bar{x})^2/2\sigma_0^2]} = \exp\left[\frac{-n}{2\sigma_0^2}(\bar{x} - \mu_0)^2\right] < k$$

It is evident that the rejection region is a monotonically decreasing function of $|\bar{x} - \mu_0|$ and equivalent to those values of $\bar{x}$ for which $|\bar{x} - \mu_0| > k'$.

The probability of a type I error is

$$P\{|\bar{x} - \mu_0| > k' \mid \mu = \mu_0\} = \alpha$$

That is,

$$P\left\{\frac{|\bar{x}-\mu_0|}{\sigma_0/\sqrt{n}} > \frac{k'}{\sigma_0/\sqrt{n}}\right\} = \alpha$$

hence

$$\alpha = 2\Phi\left(\frac{-k'}{\sigma_0/\sqrt{n}}\right)$$

Therefore

$$k' = Z_{1-\alpha/2}\frac{\sigma_0}{\sqrt{n}}$$

Whence the rejection region is the set of values of $\bar{x}$ for which

$$|\bar{x}-\mu_0| > Z_{1-\alpha/2}\frac{\sigma_0}{\sqrt{n}} \tag{9.21}$$

The probability of accepting H_0 when H_1 is true is

$$\begin{aligned}\beta(\mu_1) &= P\{|\bar{x}-\mu_0| \leq Z_{1-\alpha/2}\,\sigma_0/\sqrt{n} \mid \mu=\mu_1\}\\ &= P\left\{\frac{\mu_0-\mu_1}{\sigma_0/\sqrt{n}} - Z_{1-\alpha/2} \leq \frac{\bar{x}-\mu_1}{\sigma_0/\sqrt{n}} \leq \frac{\mu_0-\mu_1}{\sigma_0/\sqrt{n}} + Z_{1-\alpha/2}\right\}\\ &= \Phi\left(\frac{\mu_0-\mu_1}{\sigma_0/\sqrt{n}} + Z_{1-\alpha/2}\right) - \Phi\left(\frac{\mu_0-\mu_1}{\sigma_0/\sqrt{n}} - Z_{1-\alpha/2}\right) \qquad (9.22)\end{aligned}$$

The probability of the occurrence of the error of the second kind can be found from the *OC* curves in terms of d where

$$d = \left|\frac{\mu_0-\mu_1}{\sigma_0}\right|$$

The *OC* curves shown in Figs. 9.6 and 9.7 are plotted against d for a level of significance $\alpha = .01$ and $\alpha = .05$, respectively. We note that in Eq. (9.22) the second term on the right approaches zero when $\mu_1 > \mu_0$, whereas the first term on the right approaches one when $\mu_1 < \mu_0$. Thus Eq. (9.22) reduces to

$$\begin{aligned}\beta(\mu_1) &\cong \Phi\left(\frac{\mu_0-\mu_1}{\sigma_0/\sqrt{n}} + Z_{1-\alpha/2}\right) \qquad \mu_1 > \mu_0\\ &\cong 1 - \Phi\left(\frac{\mu_0-\mu_1}{\sigma_0/\sqrt{n}} - Z_{1-\alpha/2}\right) = \Phi\left(\frac{\mu_1-\mu_0}{\sigma_0/\sqrt{n}} + Z_{1-\alpha/2}\right) \qquad \mu_1 < \mu_0\end{aligned}$$

Hence

$$Z_\beta = -Z_{1-\beta} \cong Z_{1-\alpha/2} - \frac{|\mu_1-\mu_0|}{\sigma_0/\sqrt{n}} \tag{9.23}$$

for all values of μ_1. The sample size that controls the errors of the first and second kind for the double-sided case can be obtained by solving Eq. (9.23). Hence

$$n \cong \frac{\sigma_0^2(Z_{1-\alpha/2}+Z_{1-\beta})^2}{(\mu_1-\mu_0)^2} \tag{9.24}$$

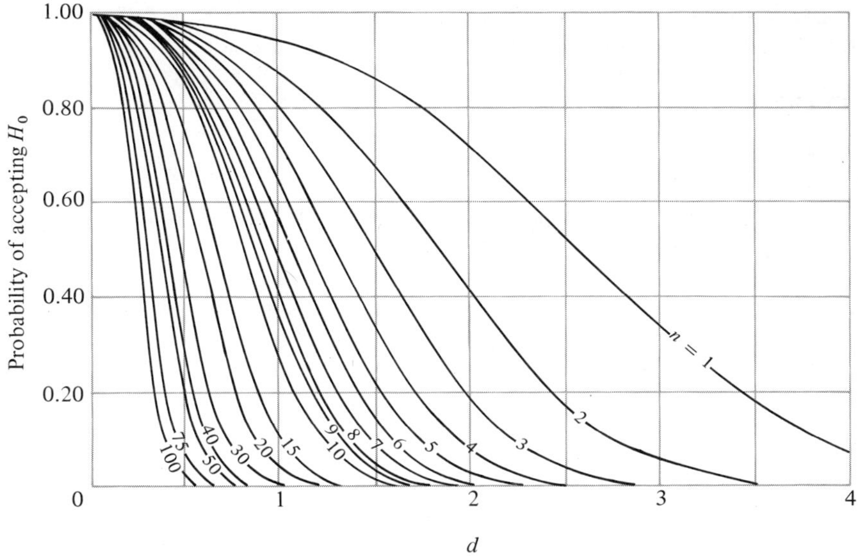

Fig. 9.6 *OC* curves for different values of n for the two-sided normal test for a level of significance $\alpha = .01$.

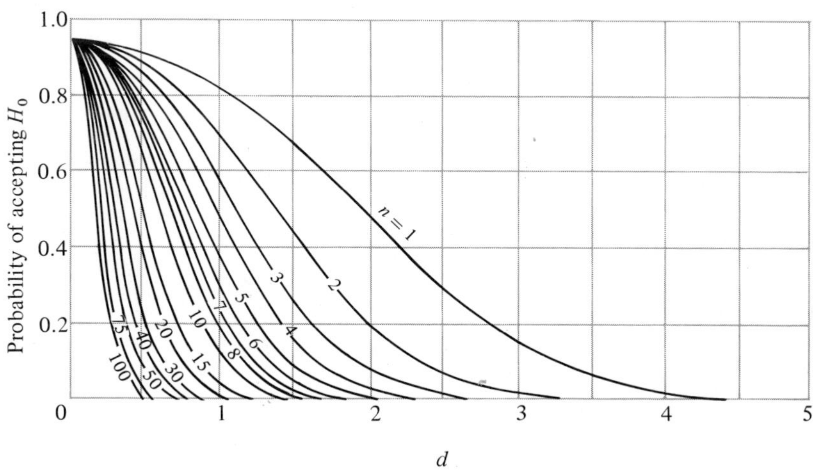

Fig. 9.7 *OC* curves for different values of n for the two-sided normal test for a level of significance $\alpha = .05$.

Example 9.2 A manufacturer of a certain food product sold in packages of 3 lb each is interested to know whether the mean weight of his product differs significantly from 3 lb. If the mean weight differs significantly at the 1% level of significance, the weighing machine has to be adjusted. Past experience indicates that the standard deviation is .03 lb. The mean weight of a random sample of nine packages was found to have a mean weight of 2.991 lb. Consider the following:

(a) Does the above result provide the basis for the decision that the true mean weight of the packages differs significantly from 3 lb?

(b) Graph the power function.

Solution

(a) We wish to test the hypothesis

$$H_0: \mu = \mu_0 = 3 \quad \text{against} \quad H_1: \mu = \mu_1 \neq 3$$

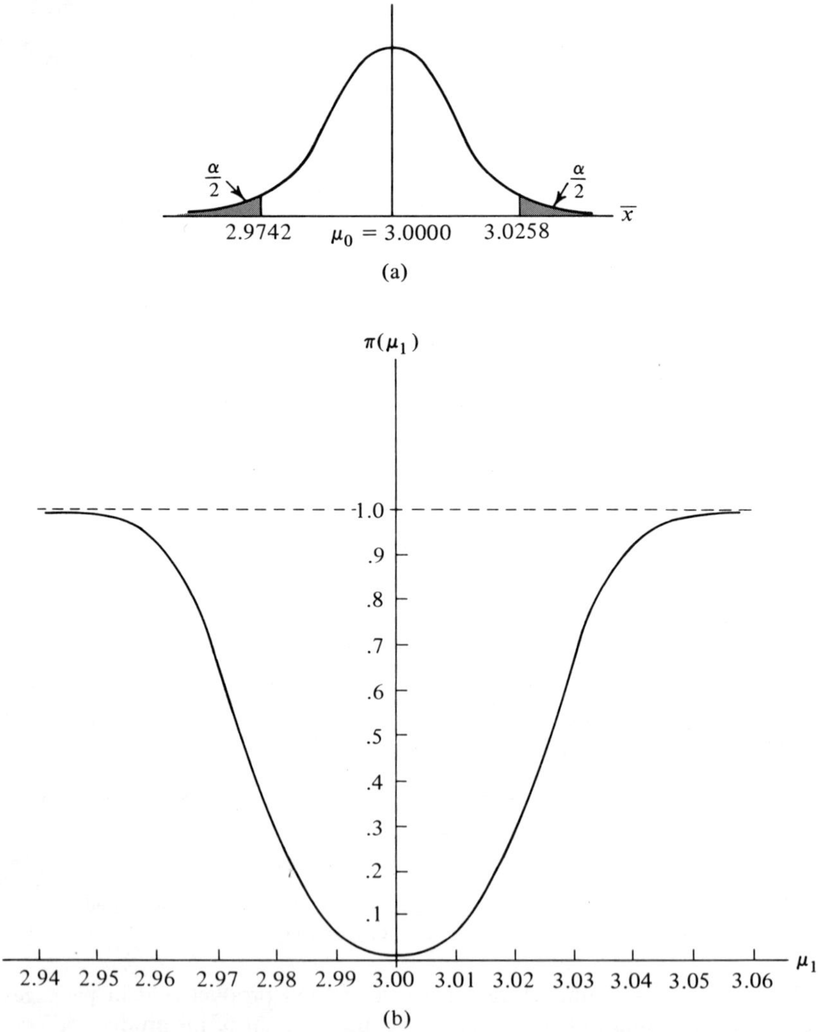

Fig. 9.8 (a) The rejection region (shaded) for the test of Example 9.2; (b) The power function for the test of Example 9.2.

The rejection region for $\bar{x}$ is the set of values for which

$$|\bar{x} - 3| > Z_{1-\alpha/2}\frac{\sigma_0}{\sqrt{n}}$$

That is,

$$|\bar{x} - 3| > .0258$$

The observed value of $\bar{x} = 2.991$ does not fall into this rejection region as shown in Fig. 9.8(a). Hence, we conclude that the mean does not differ significantly from the value $\mu_0 = 3$ at the 1% level of significance.

(b) $$\beta(\mu_1) = \Phi\left(\frac{\mu_0 - \mu_1}{\sigma_0/\sqrt{n}} + Z_{1-\alpha/2}\right) - \Phi\left(\frac{\mu_0 - \mu_1}{\sigma_0/\sqrt{n}} - Z_{1-\alpha/2}\right)$$

The power function is graphed for the following tabulated values of μ_1 and is shown in Fig. 9.8(b):

μ_1	3.000	3.005	3.010	3.015	3.020	3.025	3.030
$\beta(\mu_1)$	.99	.98	.94	.86	.72	.54	.33
$\pi(\mu_1)$	.01	.02	.06	.14	.28	.46	.67

μ_1	3.035	3.040	3.045	3.050	3.055	3.060
$\beta(\mu_1)$	.17	.07	.02	.007	.001	~ 0
$\pi(\mu_1)$	.83	.93	.98	.993	.999	~ 1

9.5 Tests Concerning the Mean of a Normal Distribution (μ) with Unknown Variance (σ^2)

Case 1

$$H_0: \mu = \mu_0 \quad \text{against} \quad H_1: \mu = \mu_1 \neq \mu_0$$

Since these hypotheses are composite, the optimal rejection region is given by

$$\lambda = \frac{\max\,[f(x_1)f(x_2)\cdots f(x_n)\,|\,\omega]}{\max\,[f(x_1)f(x_2)\cdots f(x_n)\,|\,\Omega]} < k$$

The likelihood of the numerator is

$$\left(\frac{1}{\sigma\sqrt{2\pi}}\right)^n \exp\left[-\frac{1}{2\sigma^2}\sum_{i=1}^{n}(x_i - \mu_0)^2\right]$$

We note that μ_0 is a fixed parameter; the only variable parameter is σ. We need, then, to find the value of σ that maximizes the numerator. This can be obtained by differentiating the numerator partially with respect to σ and equating the derivative to zero. Thus we have

$$\left(\frac{1}{\sigma\sqrt{2\pi}}\right)^n \exp\left[-\frac{1}{2\sigma^2}\sum_{i=1}^{n}(x_i-\mu_0)^2\right]\left[\frac{\sum_{i=1}^{n}(x_i-\mu_0)^2}{\sigma^3}-\frac{n}{\sigma}\right]=0$$

from which we obtain

$$\sigma^2=\frac{\sum_{i=1}^{n}(x_i-\mu_0)^2}{n}$$

The denominator has two variable parameters μ and σ. To find the values of these parameters that maximize the denominator, we differentiate the denominator partially with respect to σ and partially with respect to μ and equate each of the derivatives to zero. The maximum occurs for

$$\mu=\bar{x} \quad \text{and} \quad \sigma^2=\frac{\sum_{i=1}^{n}(x_i-\mu)^2}{n}$$

Thus the likelihood ratio is

$$\lambda=\left[\frac{\sum_{i=1}^{n}(x_i-\bar{x})^2}{\sum_{i=1}^{n}(x_i-\mu_0)^2}\right]^{n/2}$$

Hence

$$\begin{aligned}\lambda^{2/n}&=\frac{\sum_{i=1}^{n}(x_i-\bar{x})^2}{\sum_{i=1}^{n}(x_i-\mu_0)^2}=\frac{\sum_{i=1}^{n}(x_i-\bar{x})^2}{\sum_{i=1}^{n}(x_i-\bar{x})^2+n(\bar{x}-\mu_0)^2}\\&=\left[1+\frac{n(\bar{x}-\mu_0)^2}{\sum_{i=1}^{n}(x_i-\bar{x})^2}\right]^{-1}\\&=\left[1+\frac{(\bar{x}-\mu_0)^2}{(n-1)s^2/n}\right]^{-1}\end{aligned}$$

However,

$$\frac{\bar{x}-\mu_0}{s/\sqrt{n}}$$

is a random variable having a t distribution with $n-1$ degrees of freedom. On substituting t for $[(\bar{x}-\mu_0)/(s/\sqrt{n})]$ we obtain

$$\lambda^{2/n}=\left(1+\frac{t^2}{n-1}\right)^{-1}$$

Whence

$$\lambda = \left(1 + \frac{t^2}{n-1}\right)^{-n/2}$$

which is a monotonically decreasing function of t^2. Hence the optimal rejection region is equivalent to those values of $|t| > k'$.

The probability of type I error is

$$P\left\{\left|\frac{\bar{x} - \mu_0}{s/\sqrt{n}}\right| > k' \mid \mu = \mu_0\right\} = \alpha$$

Therefore

$$k' = t_{1-\alpha/2;\, n-1}$$

The optimal rejection region for a given α is therefore

$$|\bar{x} - \mu_0| > t_{1-\alpha/2;\, n-1} \frac{s}{\sqrt{n}} \tag{9.25}$$

The probability of accepting H_0 when H_1 is true is

$$\begin{aligned}
\beta(\mu_1, \sigma^2) &= P\left\{\left|\frac{\bar{x} - \mu_0}{s/\sqrt{n}}\right| \leq t_{1-\alpha/2;\, n-1} \mid \mu = \mu_1, \sigma^2\right\} \\
&= P\left\{\left|\frac{\bar{x} - \mu_1}{s/\sqrt{n}} + \frac{\mu_1 - \mu_0}{s/\sqrt{n}}\right| \leq t_{1-\alpha/2;\, n-1} \mid \sigma^2\right\} \\
&= P\left\{\left|\left[\frac{\sqrt{n}\,(\bar{x} - \mu_1)}{\sigma} + \frac{\sqrt{n}\,(\mu_1 - \mu_0)}{\sigma}\right]\Big/\frac{s}{\sigma}\right| \leq t_{1-\alpha/2;\, n-1}\right\} \\
&= P\left\{\left|\left[Z + \frac{\sqrt{n}\,(\mu_1 - \mu_0)}{\sigma}\right]\Big/\sqrt{\frac{\chi^2_{n-1}}{n-1}}\right| \leq t_{1-\alpha/2;\, n-1}\right\} \qquad (9.26)
\end{aligned}$$

The random variable $\{[Z + \sqrt{n}\,(\mu_1 - \mu_0)/\sigma]/\sqrt{\chi^2_{n-1}/(n-1)}\}$ has a *noncentral t* distribution with $n - 1$ degrees of freedom and depends on a *parameter of noncentrality* $[\sqrt{n}\,(\mu_1 - \mu_0)/\sigma]$ which is a measure of deviation from H_0. Extensive tabulation of the *noncentral t* distribution can be found in G. J. Resnikoff and G. J. Lieberman, *Tables of Non-Central t-Distribution* (Stanford University Press and Oxford University Press, Stanford and New York, 1957).

The probability of occurrence of the error of the second kind can be found from the *OC* curves in Figs. 9.9 and in 9.10 for $\alpha = .05$ and $\alpha = .01$, respectively, in terms of d where

$$d = \left|\frac{\mu_1 - \mu_0}{\sigma_0}\right|$$

Note that these *OC* curves are plotted as a function of $|(\mu_1 - \mu_0)/\sigma_0|$ for various values of n.

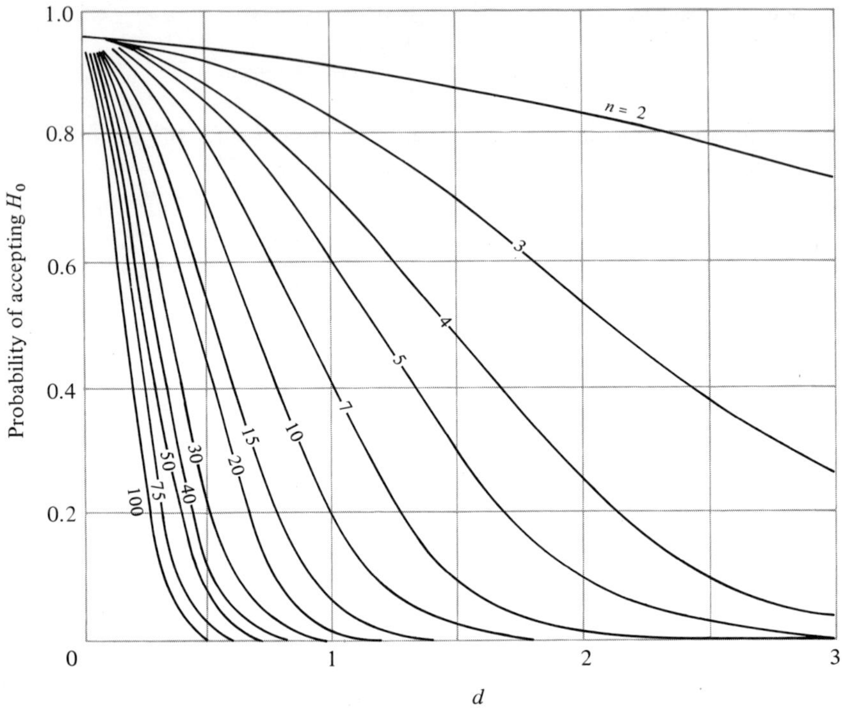

Fig. 9.9 *OC* curves for different values of n for the two-sided t test for a level of significance $\alpha = .05$.

Example 9.3 Suppose in Example 9.2 that the standard deviation were unknown and a sample of size ten were taken, with the results as follows:

3.036	3.031
3.085	3.064
2.991	2.942
2.935	3.051
2.999	3.011

Test the hypothesis

$$H_0: \mu = \mu_0 = 3 \quad \text{against} \quad H_1: \mu = \mu_1 \neq 3$$

at the 1% level of significance.

Solution

$$s^2 = \frac{\sum_{i=1}^{n} (x_i - \bar{x})^2}{n - 1}$$

$$= \frac{\sum_{i=1}^{10} (x_i - 3.0145)^2}{9} = .00243$$

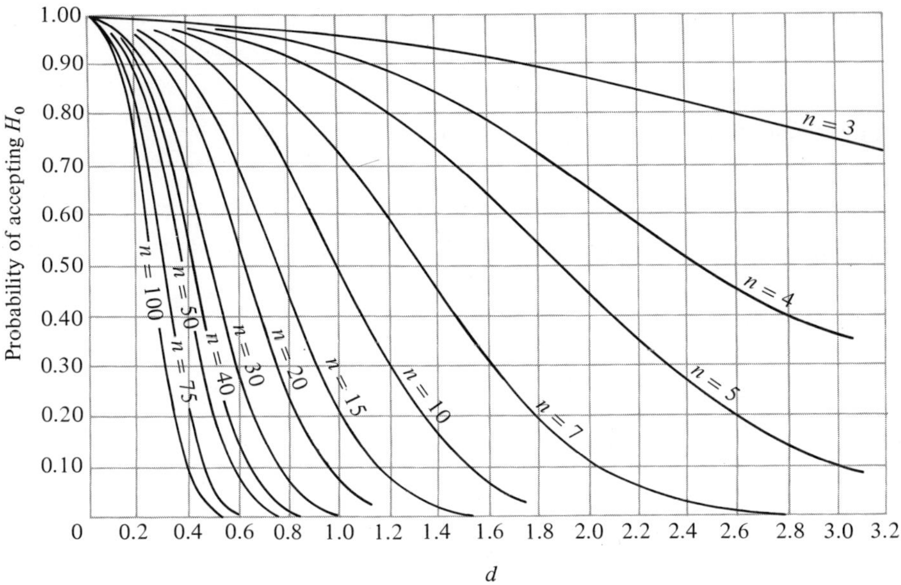

Fig. 9.10 *OC* curves for different values of n for the two-sided t test for a level of significance $\alpha = .01$.

whence $s = .0493$. The rejection region for $\bar{x}$ is the set of values for which

$$|\bar{x} - 3| > t_{1-\alpha/2;\, n-1} \frac{s}{\sqrt{n}}$$

That is,

$$|\bar{x} - 3| > .0506$$

The observed value of $\bar{x} = 3.0145$ does not fall into this critical region. Hence, we conclude that the mean does not differ significantly from the value $\mu_0 = 3$ at the 1% level of significance.

Case 2

Consider

$$H_0: \mu = \mu_0 \quad \text{against} \quad H_1: \mu = \mu_1 < \mu_0$$

It can be shown that the optimal rejection region is given by

$$\bar{x} < k'$$

The probability of type I error is

$$P\{\bar{x} < k' \mid \mu = \mu_0\} = \alpha$$

$$P\left\{\frac{\bar{x} - \mu_0}{s/\sqrt{n}} < \frac{k' - \mu_0}{s/\sqrt{n}}\right\} = \alpha$$

Whence

$$k' = \mu_0 + t_{\alpha;\, n-1} \frac{s}{\sqrt{n}}$$

Thus the optimal critical region for a given α is

$$\bar{x} < \mu_0 + t_{\alpha;\, n-1} \frac{s}{\sqrt{n}} \tag{9.27}$$

The OC curves shown in Figs. 9.11 and 9.12 for $\alpha = .05$ and $\alpha = .01$,

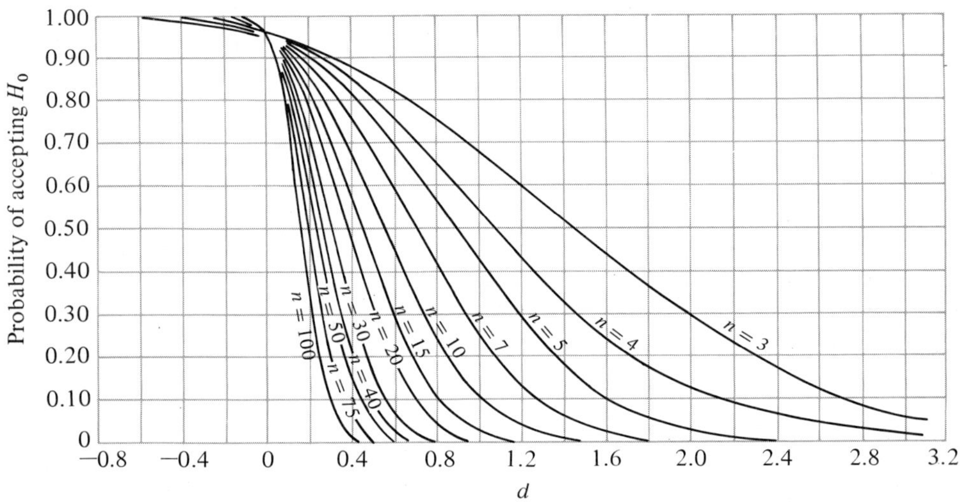

Fig. 9.11 *OC* curves for different values of n for the one-sided t test for a level of significance $\alpha = .05$.

respectively, can be used to find the probability of the occurrence of the error of the second kind in terms of d, where

$$d = \frac{\mu_0 - \mu_1}{\sigma_0}$$

Case 3

Consider

$$H_0\colon \mu = \mu_0 \quad \text{against} \quad H_1\colon \mu = \mu_1 > \mu_0$$

Application of the likelihood ratio test gives the following optimal rejection region:

$$\text{Reject } H_0 \text{ if } \bar{x} > k'$$

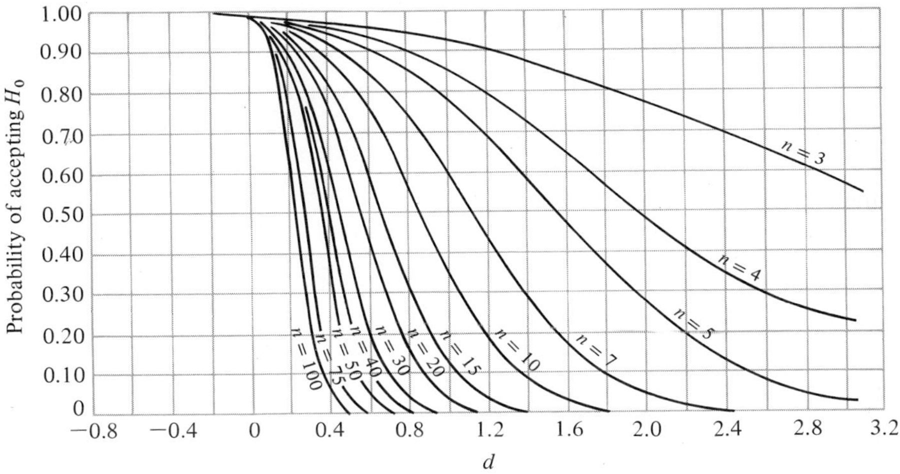

Fig. 9.12 *OC* curves for different values of *n* for the one-sided *t* test for a level of significance $\alpha = .01$.

The probability of a type I error is

$$P\{\bar{x} > k' \mid \mu = \mu_0\} = \alpha$$

$$P\left\{\frac{\bar{x} - \mu_0}{s/\sqrt{n}} > \frac{k' - \mu_0}{s/\sqrt{n}}\right\} = \alpha$$

Hence

$$k' = \mu_0 + t_{1-\alpha;\, n-1} \frac{s}{\sqrt{n}}$$

The optimal critical region for a given α is therefore

(9.28)
$$\bar{x} > \mu_0 + t_{1-\alpha;\, n-1} \frac{s}{\sqrt{n}}$$

The *OC* curves shown in Figs. 9.11 and 9.12 can be used to find the probability of the occurrence of the error of the second kind in terms of *d*, where

$$d = \frac{\mu_1 - \mu_0}{\sigma_0}$$

9.6 Tests Concerning the Variance of a Normal Distribution

Case 1

$$H_0 \colon \sigma^2 = \sigma_0^2 \quad \text{against} \quad H_1 \colon \sigma^2 = \sigma_1^2 \neq \sigma_0^2$$

H_0 is a composite hypothesis under the assumption that the mean of the distribution is unknown. H_1 is also a composite hypothesis since both the mean and the variance are unknown. Hence the optimal rejection region is given by

$$\lambda = \frac{\max [f(x_1)f(x_2) \cdots f(x_n) \mid \omega]}{\max [f(x_1)f(x_2) \cdots f(x_n) \mid \Omega]} < k$$

Under the null hypothesis the likelihood function is

$$\left(\frac{1}{\sigma_0\sqrt{2\pi}}\right)^n \exp\left[-\frac{1}{2\sigma_0^2}\sum_{i=1}^{n}(x_i - \mu)^2\right]$$

and its maximum value is

$$\left(\frac{1}{\sigma_0\sqrt{2\pi}}\right)^n \exp\left[-\frac{1}{2\sigma_0^2}\sum_{i=1}^{n}(x_i - \bar{x})^2\right] = \left(\frac{1}{\sigma_0\sqrt{2\pi}}\right)^n \exp\left[-\frac{1}{2\sigma_0^2}(n-1)s^2\right]$$

Under the alternative hypothesis the maximum value of the likelihood function is

$$\left[\frac{\sqrt{n}}{s\sqrt{2\pi(n-1)}}\right]^n \exp(-n/2)$$

Therefore

$$\lambda = \left[\frac{(n-1)s^2}{n\sigma_0^2}\right]^{n/2} \exp\left[\frac{n}{2} - \frac{(n-1)s^2}{2\sigma_0^2}\right]$$

We note that λ tends to zero as s^2/σ_0^2 tends to zero or as s^2/σ_0^2 tends to infinity. This means that the optimal acceptance region should lie between two prescribed constants k_1 and k_2, such that

$$P\left\{k_1 \leq \frac{s^2}{\sigma_0^2} \leq k_2 \mid \mu, \sigma_0^2\right\} = 1 - \alpha$$

or

$$P\left\{k_1(n-1) \leq \frac{(n-1)s^2}{\sigma_0^2} \leq k_2(n-1)\right\} = 1 - \alpha$$

It is customary to assume equal tail areas in the distribution of s^2. Accordingly the optimal acceptance region can be expressed as

$$\chi^2_{\alpha/2;\,n-1} \leq \chi^2_{n-1} \leq \chi^2_{1-\alpha/2;\,n-1}$$

In other words, the rejection region consists of the set of values of s^2 for which

$$\frac{(n-1)s^2}{\sigma_0^2} < \chi^2_{\alpha/2;\,n-1} \quad \text{or} \quad \frac{(n-1)s^2}{\sigma_0^2} > \chi^2_{1-\alpha/2;\,n-1} \tag{9.29}$$

The probability of accepting H_0 when H_1 is true is

$$\beta(\sigma_1^2) = P\left\{\chi^2_{\alpha/2;\,n-1} \leq \frac{(n-1)s^2}{\sigma_0^2} \leq \chi^2_{1-\alpha/2;\,n-1} \mid \sigma_1^2\right\}$$

$$= P\left\{\frac{\sigma_0^2}{\sigma_1^2}\chi^2_{\alpha/2;\,n-1} \leq \frac{(n-1)s^2}{\sigma_1^2} \leq \frac{\sigma_0^2}{\sigma_1^2}\chi^2_{1-\alpha/2;\,n-1}\right\}$$

$$= P\left\{\frac{\sigma_0^2}{\sigma_1^2}\chi^2_{\alpha/2;\,n-1} \leq \chi^2_{n-1} \leq \frac{\sigma_0^2}{\sigma_1^2}\chi^2_{1-\alpha/2;\,n-1}\right\} \tag{9.30}$$

Example 9.4 A sample of size 11 from a normal population gives $s^2 = 100$. (a) Test at the 5% level of significance the hypothesis

$$H_0: \sigma^2 = \sigma_0^2 = 80 \quad \text{against} \quad H_1: \sigma^2 = \sigma_1^2 \neq 80$$

(b) What is the probability of accepting H_0 when $\sigma_1^2 = 100$?

Solution

(a) Accept H_0 if

$$\frac{1}{n-1}\chi^2_{\alpha/2;\,n-1} \leq \frac{s^2}{\sigma_0^2} \leq \frac{1}{n-1}\chi^2_{1-\alpha/2;\,n-1}$$

Hence the acceptance region is given by

$$\frac{1}{10}\chi^2_{.025;\,10} \leq \frac{100}{80} \leq \frac{1}{10}\chi^2_{.975;\,10}$$

Since in substituting these values for χ^2 we obtain

$$.3247 \leq 1.25 \leq 2.0483$$

the null hypothesis that $\sigma^2 = 80$ is accepted.

(b)
$$\begin{aligned}\beta(\sigma_1^2 = 100) &= P\left\{\frac{\sigma_0^2}{\sigma_1^2}\chi^2_{\alpha/2;\,n-1} \leq \chi^2_{n-1} \leq \frac{\sigma_0^2}{\sigma_1^2}\chi^2_{1-\alpha/2;\,n-1}\right\} \\ &= P\{.8(3.247) \leq \chi^2_{10} \leq .8(20.483)\} \\ &= P\{2.597 \leq \chi^2_{10} \leq 16.386\} \\ &\cong .90\end{aligned}$$

We note that the probability of $(\sigma_0^2/\sigma_1^2)\,\chi^2_{\alpha/2;\,n-1}$ less than or equal to χ^2_{n-1} is almost insignificant and can be neglected. Therefore, from Eq. (9.30) we have

$$\chi^2_{\beta;\,n-1} \cong \frac{\sigma_0^2}{\sigma_1^2}\chi^2_{1-\alpha/2;\,n-1}$$

We can make use of the above expression to find the approximate sample size that controls the errors of the first and second kind, as shown in the following example:

Example 9.5 Suppose you wish to run a maximum risk of .20 of saying that σ_0^2 is different from 100 when it is actually equal to 100. On the other hand, you do not wish to run a risk of more than .25 of saying that $\sigma_0^2 = 100$ when it is actually equal to 176. Determine the sample size.

Solution

$$\frac{\sigma_1^2}{\sigma_0^2} \cong \frac{\chi^2_{1-\alpha/2;\,n-1}}{\chi^2_{\beta;\,n-1}}$$

$$1.76 \cong \frac{\chi^2_{.90;\,n-1}}{\chi^2_{.25;\,n-1}}$$

From the tables of the χ^2 distribution we find the degrees of freedom to satisfy the above equation. The required number of degrees of freedom is 23; therefore the sample size $n = 24$.

Case 2

$$H_0 : \sigma^2 = \sigma_0^2 \quad \text{against} \quad H_1 : \sigma^2 = \sigma_1^2 > \sigma_0^2$$

It can be shown that the optimal rejection region is given by

$$\frac{s^2}{\sigma_0^2} > k'$$

The probability of type I error is

$$P\left\{\frac{s^2}{\sigma_0^2} > k' \mid \sigma_0^2\right\} = \alpha$$

That is,

$$P\left\{\frac{(n-1)s^2}{\sigma_0^2} > k'(n-1)\right\} = \alpha$$

Hence

$$k' = \frac{1}{n-1}\chi^2_{1-\alpha;\,n-1}$$

Therefore, the optimal rejection region consists of the set of values of s^2 for which

$$\frac{(n-1)s^2}{\sigma_0^2} > \chi^2_{1-\alpha;\,n-1} \tag{9.31}$$

The probability of accepting H_0 when H_1 is true is

$$\begin{aligned} \beta(\sigma_1^2) &= P\left\{\frac{(n-1)s^2}{\sigma_0^2} \leq \chi^2_{1-\alpha;\,n-1} \mid \sigma_1^2\right\} \\ &= P\left\{\frac{(n-1)s^2}{\sigma_1^2} \leq \frac{\sigma_0^2}{\sigma_1^2}\chi^2_{1-\alpha;\,n-1}\right\} \end{aligned} \tag{9.32}$$

Case 3

$$H_0 : \sigma^2 = \sigma_0^2 \quad \text{against} \quad H_1 : \sigma^2 = \sigma_1^2 < \sigma_0^2$$

In this case the optimal critical region consists of the set of values of s^2 for which

$$\frac{(n-1)s^2}{\sigma_0^2} < \chi^2_{\alpha;\,n-1}$$

The probability of accepting H_0 when H_1 is true is

$$\begin{aligned} \beta(\sigma_1^2) &= P\left\{\frac{(n-1)s^2}{\sigma_0^2} \geq \chi^2_{\alpha;\,n-1} \mid \sigma_1^2\right\} \\ &= P\left\{\chi^2_{n-1} \geq \frac{\sigma_0^2}{\sigma_1^2}\chi^2_{\alpha;\,n-1}\right\} \end{aligned} \tag{9.33}$$

A summary of the procedures for testing hypotheses about the parameters of the normal distribution is given in Table 9.1.

TABLE 9.1 Summary of Tests of Hypotheses About the Parameters of the Normal Distribution

Hypotheses	*Unknown Parameters*	*Rejection Rule*	*Choice of Sample Size n for Given α and β*
$H_0: \mu = \mu_0$ $H_1: \mu = \mu_1 \neq \mu_0$	μ	$\lvert \bar{x} - \mu_0 \rvert > Z_{1-\alpha/2} \dfrac{\sigma_0}{\sqrt{n}}$	$n \cong \dfrac{\sigma_0^2(Z_{1-\alpha/2} + Z_{1-\beta})^2}{(\mu_1 - \mu_0)^2}$
$H_0: \mu = \mu_0$ $H_1: \mu = \mu_1 > \mu_0$	μ	$\bar{x} > \mu_0 + Z_{1-\alpha} \dfrac{\sigma_0}{\sqrt{n}}$	$n = \dfrac{\sigma_0^2(Z_{1-\alpha} + Z_{1-\beta})^2}{(\mu_1 - \mu_0)^2}$
$H_0: \mu = \mu_0$ $H_1: \mu = \mu_1 < \mu_0$	μ	$\bar{x} < \mu_0 + Z_{\alpha} \dfrac{\sigma_0}{\sqrt{n}}$	$n = \dfrac{\sigma_0^2(Z_{1-\alpha} + Z_{1-\beta})^2}{(\mu_0 - \mu_1)^2}$
$H_0: \mu = \mu_0$ $H_1: \mu = \mu_1 \neq \mu_0$	μ, σ	$\lvert \bar{x} - \mu_0 \rvert > t_{1-\alpha/2;\, n-1} \dfrac{s}{\sqrt{n}}$	$d = \left\lvert \dfrac{\mu_1 - \mu_0}{\sigma_0} \right\rvert$ Enter Fig. 9.9 or 9.10.
$H_0: \mu = \mu_0$ $H_1: \mu = \mu_1 > \mu_0$	μ, σ	$\bar{x} > \mu_0 + t_{1-\alpha;\, n-1} \dfrac{s}{\sqrt{n}}$	$d = \dfrac{\mu_1 - \mu_0}{\sigma_0}$ Enter Fig. 9.11 or 9.12.
$H_0: \mu = \mu_0$ $H_1: \mu = \mu_1 < \mu_0$	μ, σ	$\bar{x} < \mu_0 - t_{\alpha;\, n-1} \dfrac{s}{\sqrt{n}}$	$d = \dfrac{\mu_0 - \mu_1}{\sigma_0}$ Enter Fig. 9.11 or 9.12.
$H_0: \sigma^2 = \sigma_0^2$ $H_1: \sigma^2 = \sigma_1^2 \neq \sigma_0^2$	μ, σ	$(n-1)s^2/\sigma_0^2 < \chi^2_{\alpha/2;\, n-1}$ or $> \chi^2_{1-\alpha/2;\, n-1}$	$\dfrac{\chi^2_{\beta;\, n-1}}{\chi^2_{1-\alpha/2;\, n-1}} \cong \dfrac{\sigma_0^2}{\sigma_1^2}$
$H_0: \sigma^2 = \sigma_0^2$ $H_1: \sigma^2 = \sigma_1^2 > \sigma_0^2$	μ, σ	$\dfrac{(n-1)s^2}{\sigma_0^2} > \chi^2_{1-\alpha;\, n-1}$	$\dfrac{\chi^2_{\beta;\, n-1}}{\chi^2_{1-\alpha;\, n-1}} = \dfrac{\sigma_0^2}{\sigma_1^2}$
$H_0: \sigma^2 = \sigma_0^2$ $H_1: \sigma^2 = \sigma_1^2 < \sigma_0^2$	μ, σ	$\dfrac{(n-1)s^2}{\sigma_0^2} < \chi^2_{\alpha;\, n-1}$	$\dfrac{\chi^2_{\beta;\, n-1}}{\chi^2_{\alpha;\, n-1}} = \dfrac{\sigma_0^2}{\sigma_1^2}$

9.7 Tests Concerning the Parameters of Distributions Other Than the Normal

In Secs. 9.4–9.6 we discussed statistical tests concerning the parameters of the normal distribution. We are now in a position to extend our analysis to distributions other than the normal. We shall focus our analysis on three distributions, namely, the *exponential*, the *Poisson*, and the *binomial*.

9.7.1 Tests Concerning the Parameter (λ) of the Exponential Distribution

Suppose that X is an exponentially distributed random variable with parameter λ. That is,

$$f_X(t) = \lambda e^{-\lambda t} \qquad t \geq 0 \qquad \lambda > 0$$
$$= 0 \qquad \text{otherwise} \tag{9.34}$$

We wish to test

$$H_0: \lambda = \lambda_0 \quad \text{against} \quad H_1: \lambda = \lambda_1 < \lambda_0$$

The nature of the optimum rejection region given by the Neyman–Pearson lemma is

$$\frac{L_1(t_1, t_2, \ldots, t_n)}{L_0(t_1, t_2, \ldots, t_n)} = \frac{\lambda_1^n \exp[-\lambda_1(t_1 + t_2 + \cdots + t_n)]}{\lambda_0^n \exp[-\lambda_0(t_1 + t_2 + \cdots + t_n)]}$$
$$= \left(\frac{\lambda_1}{\lambda_0}\right)^n \exp\left[(\lambda_0 - \lambda_1)\sum_{i=1}^n t_i\right] > k \tag{9.35}$$

We note that the likelihood ratio given by Eq. (9.35) is a monotonically increasing function of $\sum_{i=1}^n t_i$ as long as $\lambda_1 < \lambda_0$. Accordingly, the set of values of the likelihood ratio greater than k is equivalent to the set of values of $\sum_{i=1}^n t_i$ greater than some other constant k'. Hence, *we reject H_0 when* $\sum_{i=1}^n t_i > k'$. *The rejection region* (specified by $\sum_{i=1}^n t_i > k'$) *does not depend on the specific value of the alternative λ_1 but merely on the fact that $\lambda_1 < \lambda_0$*. Consequently, this rejection region provides a uniformly most powerful test over the range of alternatives.

The probability of type I error is

$$P\left\{\sum_{i=1}^n t_i > k' \mid \lambda = \lambda_0\right\} = \alpha$$

which is equivalent to

$$P\left\{2\lambda \sum_{i=1}^n t_i > 2\lambda k' \mid \lambda = \lambda_0\right\} = \alpha \tag{9.36}$$

However, it can be shown that the statistic $2\lambda \sum_{i=1}^n t_i$ has a chi-square distribution with $2n$ degrees of freedom. Thus Eq. (9.36) becomes

$$P\{\chi_{2n}^2 > 2\lambda_0 k'\} = \alpha$$

Hence

$$k' = \frac{\chi^2_{1-\alpha;\,2n}}{2\lambda_0}$$

Therefore, the optimal rejection region is a set of values of $\sum_{i=1}^{n} t_1$ for which

$$\sum_{i=1}^{n} t_i > \frac{\chi^2_{1-\alpha;\,2n}}{2\lambda_0} \tag{9.37}$$

The probability of accepting H_0 when H_1 is true is

$$\beta(\lambda_1) = P\left\{\sum_{i=1}^{n} t_i \leq \frac{\chi^2_{1-\alpha;\,2n}}{2\lambda_0} \mid \lambda = \lambda_1\right\}$$

$$= P\left\{2\lambda_1 \sum_{i=1}^{n} t_i \leq \frac{\lambda_1}{\lambda_0} \chi^2_{1-\alpha;\,2n}\right\}$$

$$= P\left\{\chi^2_{2n} \leq \frac{\lambda_1}{\lambda_0} \chi^2_{1-\alpha;\,2n}\right\} \tag{9.38}$$

Example 9.6 Suppose that the time to failure of a certain electronic component obeys the exponential distribution with parameter λ. The mean time to failure of a random sample of ten components was found to be 106.8 hr.

(a) Does the result of this experiment provide the basis for the decision that the true mean time to failure is significantly greater than 100 hr? (Use $\alpha = .05$.)

(b) What is the power of the test if the true mean time to failure is 120 hr?

Solution

(a) The mean of the exponential distribution is $1/\lambda$. Accordingly, the hypothesis to be tested is

$$H_0: \frac{1}{\lambda_0} = 100 \quad \text{against} \quad H_1: \frac{1}{\lambda_1} > 100$$

which is equivalent to testing

$$H_0:\ \lambda_0 = .01 \quad \text{against} \quad H_1:\ \lambda_1 < .01$$

According to (9.37) we reject H_0 when

$$\sum_{i=1}^{n} t_i = n\bar{t} > \frac{\chi^2_{1-\alpha;\,2n}}{2\lambda_0}$$

That is,

$$\bar{t} > \frac{\chi^2_{1-\alpha;\,2n}}{2\lambda_0 n} = \frac{\chi^2_{.95;\,20}}{.2} = 157.05$$

The observed value of $\bar{t} = 106.8$ does not fall into the rejection region $[157.05, \infty]$. Hence, we conclude that the mean time to failure is not greater than 100 hr at the 5% level of significance.

(b) The power of test is

$$\pi = (1 - \beta) \mid \lambda_1 = 1 - P\left\{\chi^2_{2n} \leq \frac{\lambda_1}{\lambda_0} \chi^2_{1-\alpha;\,2n}\right\}$$

$$= 1 - P\left\{\chi^2_{20} \leq \frac{100}{120} \chi^2_{.95;\,20}\right\} = .17$$

That is, the probability that this test will reject H_0 when, indeed, $1/\lambda_1 = 120$ is .17. We note that the power of this test is comparatively low. The power of the test can be improved by increasing the sample size n.

Similarly, one can derive the rejection region and the power of the test for the alternative $H_1: \lambda = \lambda_1 > \lambda_0$ and $H_1: \lambda = \lambda_1 \neq \lambda_0$.

9.7.2 Tests Concerning the Parameter (λ) of the Poisson Distribution

Let X be a random variable having the Poisson distribution. That is,

$$P_X(r) = \frac{e^{-\lambda}\lambda^k}{r!} \qquad r = 0, 1, \ldots \qquad \lambda > 0 \tag{9.39}$$

As an illustration, let random variable X describe the number of truck arrivals per hour at a loading dock. Then the expected number of truck arrivals per hour is equal to λ. Now, what is the distribution of the number of truck arrivals per day (8-hr interval) at this loading dock? By the property of additivity, the distribution is a Poisson with parameter 8λ. In general, the distribution of the number of times the event occurs in a specified time interval t is a Poisson with parameter λt. Accordingly, Eq. (9.39) can be rewritten as

$$P_Y(r) = \frac{e^{-\lambda t}(\lambda t)^r}{r!} \qquad r = 0, 1, \ldots \qquad \lambda > 0 \tag{9.40}$$

(Note that in this formulation t is analogous to the sample size n.) Suppose the hypothesis to be tested is

$$H_0 : \lambda = \lambda_0 \quad \text{against} \quad H_1 : \lambda = \lambda_1 < \lambda_0$$

To test this hypothesis one observes the number of truck arrivals in a specified interval t. The likelihood function under the null hypothesis is

$$L_0(r) = \frac{e^{-\lambda_0 t}(\lambda_0 t)^r}{r!}$$

Similarly, the likelihood function under the alternative hypothesis is

$$L_1(r) = \frac{e^{-\lambda_1 t}(\lambda_1 t)^r}{r!}$$

Then the optimum rejection region has the form

$$\frac{L_1(r)}{L_0(r)} = \frac{e^{-\lambda_1 t}(\lambda_1 t)^r}{e^{-\lambda_0 t}(\lambda_0 t)^r} = \left(\frac{\lambda_1}{\lambda_0}\right)^r e^{(\lambda_0 - \lambda_1)t} > k \tag{9.41}$$

Since the likelihood ratio is a monotonically decreasing function of r, provided that $\lambda_1 < \lambda_0$, we can reject H_0 when

$$r < k'$$

The probability of type I error is

$$P\{r < k' \mid \lambda = \lambda_0\} = \sum_{v=0}^{k'-1} \frac{e^{-\lambda_0 t}(\lambda_0 t)^v}{v!} = \alpha \tag{9.42}$$

k' can be obtained from Table I in the Appendix. Since the Poisson variate is discrete, the values of k' that correspond exactly to the usual specified values of α will not always exist. In that event we choose the percentile that corresponds to the level closest to α. The probability of accepting H_0 when H_1 is true is

$$\beta(\lambda_1) = P\{r \geq k' \mid \lambda = \lambda_1\} = \sum_{v=k'}^{\infty} \frac{e^{-\lambda_1 t}(\lambda_1 t)^v}{v!} \tag{9.43}$$

When $\lambda_0 t > 10$ Eqs. (9.42) and (9.43) can be approximated by the normal distribution.

Example 9.7 Suppose that the number of unexcused absences per week in a certain plant follows the Poisson distribution with parameter $\lambda = 7$. Management improved the working conditions for a period of 3 weeks. At the end of this period (3 weeks), the observed number of absences was found to be equal to ten.

(a) Would you infer from this result that management's action has reduced the number of absences? Assume $\alpha = .025$.

(b) What is the power of the test if $\lambda_1 = 5$?

Solution

(a) Here we wish to test the hypothesis

$$H_0\colon \lambda = \lambda_0 \quad \text{against} \quad H_1\colon \lambda = \lambda_1 < \lambda_0$$

Hence, from Eq. (9.42) we have

$$\sum_{v=0}^{k'-1} \frac{e^{-21}(21)^v}{v!} = .025$$

Consulting Table I in the Appendix we find that $k' - 1 = 12$, i.e., $k' = 13$. Thus we reject H_0 when the number of observed absences is less than 13. Since the number of observed absences falls in the critical region (0, 12), we conclude that management's action has reduced the number of absences.

(b) The power of the test is

$$\pi(\lambda_1) = 1 - \sum_{v=13}^{\infty} \frac{e^{-15}(15)^v}{v!} = .268$$

The procedure of deriving the optimum rejection region and the power of the test for other alternatives $H_1 : \lambda = \lambda_1 > \lambda_0$ and $H_1 : \lambda = \lambda_1 \neq \lambda_0$ will be left as an exercise for the reader.

9.7.3 Tests Concerning the Parameter (p) of the Binomial Distribution

In this case the explicit hypothesis to be tested is

$$H_0 : p = p_0 \quad \text{against} \quad H_1 : p = p_1 < p_0$$

The probability function of the binomial distribution is

$$P_X(r) = \binom{n}{r} p^r (1-p)^{n-r} \qquad r = 0, 1, \ldots, n \qquad 0 \leq p \leq 1 \tag{9.44}$$

The nature of the optimum rejection region, then, is

$$\frac{L_1(r)}{L_0(r)} = \frac{p_1^r(1-p_1)^{n-r}}{p_0^r(1-p_0)^{n-r}} > k \tag{9.45}$$

We note that the likelihood ratio is a monotonically decreasing function for increasing r as long as $p_1 < p_0$. *Hence the optimum rejection region is equivalent to the set of values of r less than some other constant k'*. Accordingly, *we reject H_0 when*

$$r < k'$$

The probability of type I error, then, is

$$P\{r < k' \mid p = p_0\} = \sum_{v=0}^{k'-1} \binom{n}{v} p_0^v (1-p_0)^{n-v} = \alpha \tag{9.46}$$

The value of k' can be obtained from the tables of the cumulative binomial distribution (Table J in the Appendix). It should be noted that the exact value of k' for every α will not always exist because the binomial variate has a discrete distribution. In such cases we choose the percentile that corresponds to the level closest to α.

The probability of accepting H_0 when H_1 is true is

$$\beta(p_1) = P\{r \geq k' \mid p = p_1\} = \sum_{v=k'}^{\infty} \binom{n}{v} p_1^v (1-p_1)^{n-v} \tag{9.47}$$

When n is large, Eqs. (9.46) and (9.47) can be approximated by the normal distribution.

Example 9.8 A manufacturer claims that his product (submitted in large lots) is less than 25% defective. A random sample of size 20 is drawn from a large lot. The number of defective items observed in the sample was one.

(a) Would you substantiate or refute the manufacturer's claim? Use $\alpha = .025$.
(b) Find the probability of acceptance if the submitted lot is 10% defective.
(c) How large a sample is needed to make the answer in (b) equal .10? Use normal approximation.

Solution

(a) From Eq. (9.46) we have

$$\sum_{v=0}^{k'-1} \binom{20}{v} (.25)^v (.75)^{20-v} = .025$$

From Table J we find $k' = 2$. Actually, $k' = 2$ corresponds to the .0243 level of significance, which is close enough to the specified level. Hence, we reject H_0 when the number of defective items in the sample is less than two. Since the number of observed defectives falls into the critical region [0, 1], the null hypothesis can be rejected at the specified level of significance. This supports the manufacturer's claim.

(b) $$\beta(p_1 = .10) = \sum_{v=2}^{\infty} \binom{20}{v} (.10)^v (.90)^{20-v} = .6083$$

That is, the probability that this test will accept H_0 when actually $p_1 = .10$ is .6083.

(c) From Eqs. (9.46) and (9.47) we have

$$\sum_{v=0}^{k'-1} \binom{n}{v}(.25)^v(.75)^{n-v} = .025$$

$$\sum_{v=k'}^{\infty} \binom{n}{v}(.1)^v(.9)^{n-v} = .10$$

By trying successive values of n we can find from the tables of the cumulative binomial distribution the values of smallest n and k' that satisfy the above equations. However, since the largest sample size given in Table J is 20, we shall use the normal approximation to the binomial. Hence, we have

$$\sum_{v=0}^{k'-1} \binom{n}{v}(.25)^v(.75)^{n-v} \cong \int_{-\infty}^{k'-1} \frac{1}{\sigma\sqrt{2\pi}} \exp\left[-\frac{1}{2}\left(\frac{t-\mu}{\sigma}\right)^2\right] dt$$

where $\mu = np_0 = .25n$ and $\sigma = \sqrt{np_0(1-p_0)} = .433\sqrt{n}$. Therefore

$$\Phi\left[\frac{(k' - 1 + \frac{1}{2}) - .25n}{.433\sqrt{n}}\right] \cong .025$$

whence

$$k' \cong \tfrac{1}{2} + .25n - (1.96)(.433)\sqrt{n} \tag{9.48}$$

Similarly,

$$\Phi\left(\frac{.1n - k' + \frac{1}{2}}{.3\sqrt{n}}\right) \cong .10$$

whence

$$k' \cong \tfrac{1}{2} + .1n + (.3)(1.28)\sqrt{n} \tag{9.49}$$

Solving Eqs. (9.48) and (9.49), we obtain

$$n \cong \left[\frac{(1.96)(.433) + (.3)(1.28)}{.15}\right]^2 = 67$$

In general, the required sample size is given by

$$n \cong \left(\frac{\sqrt{p_0(1-p_0)}\,Z_{1-\alpha} + \sqrt{p_1(1-p_1)}\,Z_{1-\beta}}{p_1 - p_0}\right)^2 \tag{9.50}$$

Thus, if we take a sample of 67 items we can detect an alternative $p_1 = .10$ with probability 90%.

In a similar way the optimum rejection region and the power of the test can be found for the alternatives $H_1: p = p_1 > p_0$ and $H_1: p = p_1 \neq p_0$.

PROBLEMS

9.1. Compute the error of the second kind if you wish to test the following hypothesis:

$$H_0\colon \mu = \mu_0 = 10 \quad \text{against} \quad H_1\colon \mu = \mu_1 = 11$$

at the 5% level of significance. Assume that measurements are normally distributed with $\sigma = 2.5$ and a sample of size 16 is taken.

9.2. In Prob. 9.1 how many determinations should be made so that H_1 will be detected with probability .90?

9.3. A discrete random variable X has the following distributions under two possible hypotheses:

k	1	2	3	4	5	6	7	8	9	10
$P\{X = k \mid H_0\}$	.02	.18	.07	.03	.17	.24	.02	.08	.01	.18
$P\{X = k \mid H_1\}$	.10	.13	.04	.27	.01	.05	.14	.03	.11	.12

A proposed decision rule is the following: On the basis of one observation reject H_0 if $k = 1$ or 7 or 9; otherwise accept H_0.

(a) What is the error of the first kind and the power of the test using the above decision rule?

(b) Determine the optimal rejection region at the 5% level of significance.

9.4. $X_1, \ldots, X_9$ are nine independent normal random variables with common mean μ (unknown) and common variance equal to 4. It is of interest to test the hypothesis

$$H_0: \mu = \mu_0 = 20 \quad \text{against} \quad H_1: \mu = \mu_1 = 22$$

using the decision rule: Reject H_0 if $\bar{x} > 20.855$

if H_0 is rejected when true a loss of \$200 occurs
if H_0 is accepted when true no loss is sustained
if H_1 is rejected when true a loss of \$400 results
if H_1 is accepted when true no loss is sustained

(a) Compute the expected loss if $\mu = 20$.

(b) Compute the expected loss if $\mu = 22$.

(c) Compute the expected loss if μ takes on the values 20 or 22 with probability .4 and .6, respectively.

9.5. By means of the Neyman–Pearson lemma, determine the optimal rejection region based on a sample of size n for testing the hypothesis

$$H_0: \lambda = \lambda_0 \quad \text{against} \quad H_1: \lambda = \lambda_1 > \lambda_0$$

for random variable X with the density function

$$f_X(t) = (1 + \lambda)t^{\lambda} \qquad 0 \le t \le 1$$
$$= 0 \qquad \text{otherwise}$$

9.6. Use the Neyman–Pearson lemma to derive a likelihood ratio test for testing the hypothesis

$$H_0: \lambda = \lambda_0 \quad \text{against} \quad H_1: \lambda = \lambda_1 < \lambda_0$$

Suppose that the density function of the random variable under study is given by

$$f_X(t) = \lambda^2 t e^{-\lambda t} \qquad t \ge 0 \qquad \lambda > 0$$

9.7. The probability function of random variable X is given by

$$P_X(k) = p(1 - p)^r \qquad r = 0, 1, \ldots$$

Use the Neyman–Pearson lemma to determine the optimal rejection region based on a sample of size n for testing the hypothesis

$$H_0: p = p_0 = .30 \quad \text{against} \quad H_1: p = p_1 < p_0$$

9.8. Suppose it is desired to test the hypothesis that the probability of a successful launching (p) of a special satellite is .30. If p is less than .30 it is desired to change the design. It has been decided to perform an experiment until the third success occurs. If the third success is achieved at the twentieth trial, would you accept the hypothesis ($\alpha = .10$)? (*Hint*: use the result of Prob. 9.7.)

9.9. Construct a likelihood ratio test for testing the hypothesis

$$H_0\colon \sigma^2 = \sigma_0^2 \quad \text{against} \quad H_1\colon \sigma^2 = \sigma_1^2 < \sigma_0^2$$

for a normally distributed random variable with known mean μ_0.

9.10. It is known that the mean of a certain process is stable and equal to ten units. It is desired to test the hypothesis

$$H_0\colon \sigma^2 = \sigma_0^2 = 2 \quad \text{against} \quad H_1\colon \sigma^2 = \sigma_1^2 = 1.5$$

If $\sum_{i=1}^{n} (x_i - 10)^2 = 20$ and $n = 15$, would you accept the hypothesis at the 5% level of significance?

9.11. A manufacturer of an electronic device claims that the average life of his product will exceed 1,000 hr. A consumer is willing to buy a very large shipment if the manufacturer's claim is true. A random sample of 25 devices is tested, and the sample mean is found to be 1,050 hr. It is known that measurements are normally distributed with $\sigma = 100$ hr.

(a) Is it likely that the electronic devices will be purchased ($\alpha = .05$)?
(b) What is the power of the test when the average life is 1,050 hr?
(c) How large a sample size is needed to make the answer in (b) .90?

9.12. A company that sells ground coffee in cans claims that the average net weight of each can is 1 lb. A sample of 25 cans yields an average of 0.985 lb. Suppose it is known from past experience that measurements are normally distributed with $\sigma = .03$ lb. Using $\alpha = .05$, would you support the company's claim? What is the power of the test when the average net weight is .98 lb?

9.13. The lot average breaking strength of a certain yarn is required to be not less than 150 psi. The standard deviation based on past experience is 7 psi. Ten specimens were drawn at random from a large lot and tested with the following results:

145 142 151 153 138 131 146 153 131 150

(a) Using a 5% level of significance, should the lot be accepted?
(b) What is the probability of accepting a lot that has a lot average breaking strength of 145 psi?

9.14. In Prob. 9.13 suppose that the standard deviation is unknown. Would you accept the lot on the basis of the determinations given in the preceding problem?

9.15. The variance in tensile strength of a certain material is 36 units. A change in the manufacturing procedure is expected to reduce the variance. The following observations are recorded (coded data):

4	9	0	6	8
1	13	2	13	9
14	7	15	16	2

$$\Sigma\, x_i = 119 \qquad \Sigma\, x_i^2 = 1351$$

(a) Would you recommend the change in the manufacturing procedure ($\alpha = .05$)?
(b) What is the probability of accepting the null hypothesis when $\sigma = 4$ units?
(c) What sample size will reduce the probability obtained under (b) by 50%?

9.16. Suppose it is desired to test the hypothesis that the mean resistance of a certain resistor is 120 ohms. If the mean resistance differs by more than 2 ohms, it is desired to change the circuit design. Assume normality, $\alpha = .01$, $\beta = .05$, and $\sigma = 1.5$ ohms. How many determinations should be made? What will be the rejection rule? What is the power of the test given that the mean resistance is actually equal to 121.5 ohms?

9.17. XYZ Company is considering the replacement of a production machine that produces, on the average, 22.5 units per hour. It is felt that the procurement of a challenger is justified if it can average more than 22.5 units per hour. To make a decision regarding the replacement, 16 challengers are each used for 1 hr. Their average output is 22.6 units per hour with a standard deviation of 1.2 units. Assume that observations are normally distributed.
(a) Is it likely that the challenger will be adopted ($\alpha = .05$)?
(b) What is the probability of not adopting the challenger if its true average output is 23 units per hour? (Use the sample variance as the estimate of σ^2.)

9.18. Suppose in Prob. 9.17 the defender has a standard deviation of 1 unit per hour. Using $\alpha = .05$, what conclusions would be drawn concerning the standard deviation of the challenger?

9.19. Given a sample of size ten from a normal distribution with $s^2 = 100$, test the hypothesis

$$H_0: \sigma^2 = \sigma_0^2 = 70 \quad \text{against} \quad H_1: \sigma^2 = \sigma_1^2 \neq 70$$

with $\alpha = .01$. If $\sigma_2 = 112$, what is the probability of rejecting H_0?

9.20. A producer claims that the mean inside diameter of a certain aluminum tubing is 100 mm. A sample of size ten was measured with the results as follows:

100.36	100.31	99.99	100.11	100.64
100.85	99.42	99.91	99.35	100.51

(a) Would you accept the producer's claim at the 5% level of significance?
(b) What is the probability of accepting the producer's claim when in fact the mean diameter is 100.05 mm? Use s^2 as an estimate of σ^2.
(c) How many observations are required to make the answer in (b) equal to .10?

9.21. Work Prob. 9.20 assuming that the standard deviation $\sigma = .05$.

9.22. Ten specimens of nickel alloy music wire were tested for tensile strength (psi). The results are given below:

149,278	150,851	151,578	149,871	153,024
152,045	149,971	150,250	151,461	151,515

It is desired to test the hypothesis

$$H_0: \mu = \mu_0 = 150{,}000 \quad \text{against} \quad H_1: \mu = \mu_1 > 150{,}000$$

with $\alpha = .05$.
(a) Would you conclude that the tensile strength is greater than 150,000 psi?
(b) What is the probability of rejecting the hypothesis that $\mu = 150{,}000$ psi when the true tensile strength is 151,000 psi?

9.23. It is known that the melting point of zinc die-casting alloy (zamak) is a normal variable with mean 717°F. Nine determinations were made by a metallurgist with the following results:

708 716 711 720 723 712 726 703 707

Test at the 5% level of significance whether these results are consistent with the specified melting point.

9.24. In Prob. 9.23 how many determinations are required so that if the melting point differs by more than 5°F, there will be a probability of at least .99 of detecting it? Assume that 7°F is a good estimate of the standard deviation.

9.25. The density function of random variable X is given by

$$f_X(t) = \frac{1}{\lambda} \qquad 0 < t < \lambda$$
$$= 0 \qquad \text{otherwise}$$

The parameter λ is unknown, and it is desired to test the hypothesis

$$H_0\colon \lambda = \lambda_0 = 1 \quad \text{against} \quad H_1\colon \lambda = \lambda_1 < 1$$

using the following decision rule: Take three independent observations and reject H_0 if $\max(x_1, x_2, x_3) < .60$.
(a) Find the error of the first kind using the above decision rule.
(b) Find the power of the test when $\lambda = .8$.

9.26. A time-study engineer has developed a new sequence of operation elements that he hopes will reduce the mean cycle time of a certain production process. The results of a time study of 30 cycles are given below:

Cycle Time (*min*)					
11.97	11.94	12.09	12.04	11.90	12.25
12.15	11.89	12.15	12.11	11.86	12.16
12.08	12.16	12.14	12.25	12.48	12.19
12.31	12.25	12.47	12.15	12.23	12.31
12.24	12.04	11.98	12.34	12.07	12.30

(a) If the present mean cycle time is 12.50 min, should the time-study engineer adopt the new sequence? Assume normality, and use $\alpha = .05$.
(b) How many cycles should be timed so that if the mean cycle time is 12.40 min there will be a probability of at least .98 of detecting it? Use s^2 as an estimate of σ^2.

9.27. In Prob. 9.26 suppose that the standard deviation of the present sequence is .2 min. In view of the above determinations, would the time study engineer be justified in concluding that the new sequence will reduce the variance? Use $\alpha = .05$.

9.28. Suppose that variate X has the following p.d.f.:

$$f_X(t) = \frac{2t}{\lambda^2} \qquad 0 \leq t \leq \lambda$$
$$= 0 \qquad \text{otherwise}$$

The parameter λ is unknown, and it is desired to test

$$H_0\colon \lambda = \lambda_0 = 1 \quad \text{against} \quad H_1\colon \lambda = \lambda_1 < 1$$

using the following decision rule: Take two independent observations, say X_1 and X_2, and reject H_0 if $\max(X_1, X_2) < .7$.
(a) Compute the error of the first kind.
(b) What is the probability of concluding that $\lambda = 1$ when in fact it is equal to .84?

9.29. A manufacturer claims that the resistance of carbon film resistors he manufactures has a standard deviation of .5 ohm. The sample variance of 20 resistors has been found to be equal to .49 ohm.
(a) Would you reject the manufacturer's claim at the 5% level of significance?
(b) How many observations are required so that the probability of rejecting the manufacturer's claim when $\sigma_1 = 1.5\sigma_0$ will be greater than .80?

9.30. Ten determinations from a normal population yielded the following results:

$$\Sigma x_i = 30 \qquad \Sigma x_i^2 = 117$$

(a) Test at the 5% level of significance the hypothesis

$$H_0\colon \sigma = \sigma_0 = 1.5 \quad \text{against} \quad H_1\colon \sigma = \sigma_1 \neq \sigma_0$$

(b) How many observations are required so that the probability of rejecting H_0 when $\sigma_1 = 2\sigma_0$ will be greater than .95?

9.31. Suppose that random variable X has a Poisson distribution with parameter λt. It is desired to test the hypotheses

$$H_0\colon \lambda = \lambda_0 = 2 \quad \text{against} \quad H_1\colon \lambda = \lambda_1 < 2$$

by observing the number of events that occur in time interval $[0, t]$. Would you reject H_0 if the observed number of events during the interval $[0, 5]$ is 6? Use $\alpha = .01$.

9.32. During a 12-month period the numbers of unexcused absences per month in a small organization are 3, 0, 1, 2, 0, 1, 4, 1, 0, 1, 2, 1.

(a) Do these results contradict the hypothesis that the average number of unexcused absences is .75 per month? Use $\alpha = .05$ and assume Poisson distribution.

(b) Find the power of the test if the mean number of unexcused absences per month is one.

9.33. The demand per week for a certain commodity follows the Poisson distribution with parameter λ. It is desired to test the hypothesis

$$H_0\colon \lambda = \lambda_0 = 2 \quad \text{against} \quad H_1\colon \lambda_1 > \lambda_0$$

Would you support H_0 if during a 5-week period the demand is 0, 2, 4, 3, 1? Use $\alpha = .05$.

9.34. In Prob. 9.33 let n denote the number of weeks. How large does n have to be to achieve a power of .98 if $\lambda_1 = 2.5$? Use normal approximation.

9.35. The distribution of failure time of a certain electronic component is exponential with parameter λ. It is desired to test the hypothesis

$$H_0\colon \lambda = \lambda_0 = .02 \quad \text{against} \quad H_1\colon \lambda = \lambda_1 < \lambda_0$$

Five components are put on life test, and the following failure times are recorded: 60, 48, 50, 42, 55. Would you conclude that the mean life is greater than 50? Use $\alpha = .05$.

9.36. In Prob. 9.35 let n denote the number of components to be tested. How large does n have to be to achieve a power of .80 if $\lambda_1 = .010$?

9.37. A coin is tossed 100 times, and 58 tails are recorded. Would you conclude that the coin is unbiased? Use $\alpha = .035$.

9.38. XYZ Company is considering the replacement of an old production machine. Past experience indicates that the defender produces, on the average, 10% defective items. A challenger will be adopted if at least 90% of its production is satisfactory. A sample of 100 items produced by the challenger is examined, and 12 items are found to be defective. Would you recommend adopting the challenger? Use $\alpha = .04$.

9.39. Consider testing the hypothesis

$$H_0\colon p = p_0 = .5 \quad \text{against} \quad H_1\colon p = p_1 < p_0$$

for a binomial distribution at the 5% level of significance.

(a) Find the rejection region when the sample size is 50.

(b) Plot the power function.

9.40. A manufacturer suspects that the product of a special-purpose machine is more than 5% defective. To test this conjecture a sample of 400 items selected at random from a large lot was inspected. The number of defective items in the sample was found to be 32. (a) Is there an evidence that the percentage defective is greater than 5%? Use $\alpha = .05$. (b) What is the probability of H_0 being rejected if in fact the percentage defective is 10%? (*Hint*: use normal approximation.)

SUGGESTED REFERENCES

See the references given at the end of Chapter 10.

10

STATISTICAL HYPOTHESES TESTING (Two Parameters)

In the preceding chapter we were concerned with testing hypotheses about a *single parameter*. Situations in which the decision maker may wish to test hypotheses about *two parameters*, neither of which is known, frequently arise. For example, a quality control engineer may want to compare the products of two machines for the average value of a certain quality characteristic. A chemical engineer may be interested in comparing two processes for average yield. A manufacturer may be choosing between two new machines on the basis of the variance of a certain quality characteristic. In this chapter we shall consider statistical testing about the parameters of two independent *normal* and *binomial distributions.*

10.1 Tests Concerning the Difference of Means of Two Independent Normal Distributions with Known Variances

Case 1

Suppose we wish to test the hypothesis

$$H_0\colon \mu_X - \mu_Y = 0 \quad \text{against} \quad H_1\colon \mu_X - \mu_Y = \Delta \neq 0$$

In this case we have two independent normal random variables X and Y

with means μ_X and μ_Y (both unknown) and variances σ_X^2 and σ_Y^2 (both known), respectively. It is evident that both H_0 and H_1 are composite hypotheses since neither specifies the distributions of X and Y completely.

The likelihood function under the null hypothesis is given by

(10.1)

$$L_0 = \left(\frac{1}{\sigma_X\sqrt{2\pi}}\right)^n\left(\frac{1}{\sigma_Y\sqrt{2\pi}}\right)^m \exp\left[-\frac{\sum_{i=1}^{n}(x_i-\mu)^2}{2\sigma_X^2}\right]\exp\left[-\frac{\sum_{i=1}^{m}(y_i-\mu)^2}{2\sigma_Y^2}\right]$$

where $\mu = \mu_X = \mu_Y$, n = sample size taken from the x population, and m = sample size taken from the Y population. On taking the first derivative of Eq. (10.1) with respect to μ and equating to zero we obtain

$$L_0\left[\frac{1}{\sigma_X^2}\sum_{i=1}^{n}(x_i-\mu) + \frac{1}{\sigma_Y^2}\sum_{i=1}^{m}(y_i-\mu)\right]$$

(10.2)

$$= L_0\left[n\frac{(\bar{x}-\mu)}{\sigma_X^2} + m\frac{(\bar{y}-\mu)}{\sigma_Y^2}\right] = 0$$

Hence, the likelihood function under the null hypothesis (L_0) is maximized for

(10.3)

$$\mu = \frac{\bar{x}+c\bar{y}}{1+c}$$

where

$$c = \frac{m\sigma_X^2}{n\sigma_Y^2}$$

The likelihood function under the alternative hypothesis is

(10.4)

$$L_1 = \left(\frac{1}{\sigma_X\sqrt{2\pi}}\right)^n\left(\frac{1}{\sigma_Y\sqrt{2\pi}}\right)^m \exp\left[\frac{-\sum_{i=1}^{n}(x_i-\mu_X)^2}{2\sigma_X^2}\right]\exp\left[\frac{-\sum_{i=1}^{m}(y_i-\mu_Y)^2}{2\sigma_Y^2}\right]$$

which is maximized for

(10.5)

$$\mu_X = \bar{x}$$

and

(10.6)

$$\mu_Y = \bar{y}$$

The likelihood ratio, then, is

$$\lambda = \frac{\exp\left[\frac{-\sum_{i=1}^{n}\{x_i-[(\bar{x}+c\bar{y})/(1+c)]\}^2}{2\sigma_X^2}\right]\exp\left[\frac{-\sum_{i=1}^{m}\{y_i-[\bar{x}+c\bar{y})/(1+c)]\}^2}{2\sigma_Y^2}\right]}{\exp\left[\frac{-\sum_{i=1}^{n}(x_i-\bar{x})^2}{2\sigma_X^2}\right]\exp\left[\frac{-\sum_{i=1}^{m}(y_i-\bar{y})^2}{2\sigma_Y^2}\right]}$$

$$= \exp\left[\frac{-n\{\bar{x} - [(\bar{x} + c\bar{y})/(1 + c)]\}^2}{2\sigma_X^2}\right] \exp\left[\frac{-m\{\bar{y} - [(\bar{x} + c\bar{y})/(1 + c)]\}^2}{2\sigma_Y^2}\right]$$

(10.7)

Substituting $n/\sigma_X^2 = m/c\sigma_Y^2$ in Eq. (10.7) we obtain

$$\lambda = \exp\left\{\frac{-n}{2c\sigma_Y^2}\left[\bar{x} - \left(\frac{\bar{x} + c\bar{y}}{1 + c}\right)\right]^2\right\} \exp\left\{\frac{-m}{2c\sigma_Y^2}\left[\bar{y} - \left(\frac{\bar{x} + c\bar{y}}{1 + c}\right)\right]^2\right\}$$

$$= \exp\left[\frac{-m}{2c\sigma_Y^2}\left(\frac{c\bar{x} - c\bar{y}}{1 + c}\right)^2\right] \exp\left[\frac{-m}{2\sigma_Y^2}\left(\frac{\bar{y} - \bar{x}}{1 + c}\right)^2\right]$$

$$= \exp\left[-\frac{m(\bar{x} - \bar{y})^2}{2\sigma_Y^2(1 + c)}\right] \tag{10.8}$$

Since H_0 is rejected when λ is sufficiently small, this is equivalent to *rejecting H_0 when the set of values* of $(\bar{x} - \bar{y})^2$ *is too large*. The optimal rejection region, therefore, is given by

$$|\bar{x} - \bar{y}| > k'$$

The probability of a type I error is

$$P\{|\bar{x} - \bar{y}| > k' \mid \mu_X - \mu_Y = 0\} = \alpha$$

That is,

$$P\left\{\left|\frac{\bar{x} - \bar{y}}{\sqrt{\sigma_X^2/n + \sigma_Y^2/m}}\right| > \frac{k'}{\sqrt{\sigma_X^2/n + \sigma_Y^2/m}}\right\} = \alpha$$

Hence

$$k' = Z_{1-\alpha/2}\sqrt{\frac{\sigma_X^2}{n} + \frac{\sigma_Y^2}{m}}$$

The rejection region is then the set of values of $\bar{x} - \bar{y}$ for which

$$|\bar{x} - \bar{y}| > Z_{1-\alpha/2}\sqrt{\frac{\sigma_X^2}{n} + \frac{\sigma_Y^2}{m}} \tag{10.9}$$

The probability of accepting H_0 when $\mu_X - \mu_Y = \Delta$ is

$$\beta(\Delta) = P\left\{|\bar{x} - \bar{y}| \le Z_{1-\alpha/2}\sqrt{\frac{\sigma_X^2}{n} + \frac{\sigma_Y^2}{m}} \mid \mu_X - \mu_Y = \Delta\right\}$$

$$= P\left\{Z_{\alpha/2}\sqrt{\frac{\sigma_X^2}{n} + \frac{\sigma_Y^2}{m}} \le \bar{x} - \bar{y} \le Z_{1-\alpha/2}\sqrt{\frac{\sigma_X^2}{n} + \frac{\sigma_Y^2}{m}} \mid \mu_X - \mu_Y = \Delta\right\}$$

$$= P\left\{Z_{\alpha/2} - \frac{\Delta}{\sqrt{\sigma_X^2/n + \sigma_Y^2/m}} \le \frac{\bar{x} - \bar{y} - \Delta}{\sqrt{\sigma_X^2/n + \sigma_Y^2/m}} \le Z_{1-\alpha/2} - \frac{\Delta}{\sqrt{\sigma_X^2/n + \sigma_Y^2/m}}\right\}$$

$$= \Phi\left(Z_{1-\alpha/2} - \frac{\Delta}{\sqrt{\sigma_X^2/n + \sigma_Y^2/m}}\right) - \Phi\left(Z_{\alpha/2} - \frac{\Delta}{\sqrt{\sigma_X^2/n + \sigma_Y^2/m}}\right) \tag{10.10}$$

Example 10.1 It is suspected that two filling machines M_1 and M_2 in a plant are turning out a product of nonuniform weight. It is known from past experience that the standard deviations of weights of their respective fillings are .10 and .12 lb. An experiment is run by weighing the contents of 100 bags filled by each machine. Suppose the mean weights of the contents of 100 bags of machines M_1 and M_2 are 1.07 and 1.18 lb, respectively.

(a) Test whether the machines differ at the 5% level of significance.

(b) What is the probability of saying that the mean weights of both machines are the same when actually the mean weights differ by as much as .05 lb?

Solution

(a) We wish to test at the 5% level of significance the hypothesis

$$H_0\colon \mu_1 = \mu_2 \quad \text{against} \quad H_1\colon \mu_1 \neq \mu_2$$

where μ_1 and μ_2 are the mean weights of fillings of machines M_1 and M_2, respectively.

Reject H_0 if

$$|\bar{x} - \bar{y}| > Z_{1-\alpha/2}\sqrt{\frac{\sigma_X^2}{n} + \frac{\sigma_Y^2}{m}}$$

That is,

$$\begin{aligned} .11 &> 1.96\sqrt{\frac{.01}{100} + \frac{.0144}{100}} \\ &> .0306 \end{aligned}$$

Hence, we reject H_0 and conclude that the two machines differ significantly at the 5% level of significance.

(b) The error of the second kind can be obtained from Eq. (10.10). Thus we have

$$\begin{aligned} &\beta(\Delta = .05) \\ &= \Phi\left(Z_{.975} - \frac{.05}{\sqrt{\frac{.01}{100} + \frac{.0144}{100}}}\right) - \Phi\left(Z_{.025} - \frac{.05}{\sqrt{\frac{.01}{100} + \frac{.0144}{100}}}\right) \\ &= \Phi(1.96 - 3.20) - \Phi(-1.96 - 3.20) \cong .1075 \end{aligned}$$

In evaluating β we note that $\Phi[Z_{\alpha/2} - (\Delta/\sqrt{\sigma_X^2/n + \sigma_Y^2/m})]$ is very insignificant and tends to zero when Δ is large and positive. Accordingly, the error of the second kind can be approximated by

$$\beta(\Delta) \cong \Phi\left(Z_{1-\alpha/2} - \frac{\Delta}{\sqrt{\sigma_X^2/n + \sigma_Y^2/m}}\right)$$

hence

$$Z_\beta \cong Z_{1-\alpha/2} - \frac{\Delta}{\sqrt{\sigma_X^2/n + \sigma_Y^2/m}} \tag{10.11}$$

If $n = m$, the sample size that yields the prescribed power $(1 - \beta)$ can be obtained from Eq. (10.11). Thus, we write

$$n = m \cong (Z_{1-\alpha/2} + Z_{1-\beta})^2(\sigma_X^2 + \sigma_Y^2)/\Delta^2 \tag{10.12}$$

It should be noted that the same result holds true when Δ is assumed negative, since Eq. (10.12) is a function of Δ^2. The *OC* curves in Figs. 9.6 and 9.7 can be used for obtaining the sample size ($n = m$) that satisfies the prescribed errors of the first and second kind. In order to use these curves the abscissa scale d is defined as

$$d = \left| \frac{\mu_X - \mu_Y}{\sqrt{\sigma_X^2 + \sigma_Y^2}} \right|$$

If n is not equal to m an equivalent sample size denoted by n_e is computed from the expression

$$n_e = \frac{\sigma_X^2 + \sigma_Y^2}{\sigma_X^2/n + \sigma_Y^2/m} \tag{10.13}$$

To illustrate this procedure, suppose we wish to find β from the *OC* curves, given the following data: $\sigma_X^2 = 4$, $\sigma_Y^2 = 12$, $n = 10$, $m = 20$, $|\Delta| = 4$, and $\alpha = .01$. We compute the equivalent sample size from Eq. (10.13)

$$n_e = \frac{\sigma_X^2 + \sigma_Y^2}{\sigma_X^2/n + \sigma_Y^2/m} = 16$$

By definition,

$$d = \frac{|\Delta|}{\sqrt{\sigma_X^2 + \sigma_Y^2}} = 1$$

The *OC* curves in Fig. 9.6 is entered with $n_e = 16$ and $d = 1$, and β read from the curve is approximately equal to .08.

Case 2

$$H_0: \mu_X - \mu_Y = 0 \quad \text{against} \quad H_1: \mu_X - \mu_Y = \Delta > 0$$

The procedure for finding the optimal rejection region is quite similar to that for case 1. The likelihood ratio specifies the optimal rejection region by

$$\bar{x} - \bar{y} > k'$$

The probability of a type I error is

$$P\{\bar{x} - \bar{y} > k' \mid \mu_X - \mu_Y = 0\} = \alpha$$

That is,

$$P\left\{\frac{\bar{x} - \bar{y}}{\sqrt{\sigma_X^2/n + \sigma_Y^2/m}} > \frac{k'}{\sqrt{\sigma_X^2/n + \sigma_Y^2/m}}\right\} = \alpha$$

Therefore

$$k' = Z_{1-\alpha}\sqrt{\frac{\sigma_X^2}{n} + \frac{\sigma_Y^2}{m}}$$

The rejection region is then the set of values of $\bar{x} - \bar{y}$ for which

$$\bar{x} - \bar{y} > Z_{1-\alpha}\sqrt{\frac{\sigma_X^2}{n} + \frac{\sigma_Y^2}{m}} \tag{10.14}$$

The probability β of a type II error is the probability of accepting $\mu_X - \mu_Y = 0$ when $\mu_X - \mu_Y > 0$; i.e.,

$$\beta(\Delta > 0) = P\left\{\bar{x} - \bar{y} \leq Z_{1-\alpha}\sqrt{\frac{\sigma_X^2}{n} + \frac{\sigma_Y^2}{m}}\,\middle|\,\Delta > 0\right\}$$

or

$$\beta(\Delta > 0) = P\left\{\frac{\bar{x} - \bar{y} - \Delta}{\sqrt{\sigma_X^2/n + \sigma_Y^2/m}} \leq Z_{1-\alpha} - \frac{\Delta}{\sqrt{\sigma_X^2/n + \sigma_Y^2/m}}\right\}$$

$$= \Phi\left(Z_{1-\alpha} - \frac{\Delta}{\sqrt{\sigma_X^2/n + \sigma_Y^2/m}}\right) \tag{10.15}$$

If $n = m$, β can be found from the OC curves in Figs. 9.3 and 9.4 in terms of

$$d = \frac{\mu_X - \mu_Y}{\sqrt{\sigma_X^2 + \sigma_Y^2}}$$

If $n \neq m$, the OC curves can still be used by calculating the equivalent sample size n_e as described in case 1.

The expression for the sample size $n = m$ that yields the prescribed power $(1 - \beta)$ is given by

$$m = n = \frac{(Z_{1-\alpha} + Z_{1-\beta})^2(\sigma_X^2 + \sigma_Y^2)}{\Delta^2} \tag{10.16}$$

Case 3

$$H_0\colon \mu_X - \mu_Y = 0 \quad \text{against} \quad H_1\colon \mu_X - \mu_Y = \Delta < 0$$

Case 3 can be treated in a manner similar to case 2. It can be shown that the rejection region is the set of values of $\bar{x} - \bar{y}$ for which

$$\bar{x} - \bar{y} < Z_\alpha\sqrt{\frac{\sigma_X^2}{n} + \frac{\sigma_Y^2}{m}} \tag{10.17}$$

The probability β of a type II error is

$$\beta(\Delta < 0) = P\left\{\bar{x} - \bar{y} \geq Z_\alpha\sqrt{\frac{\sigma_X^2}{n} + \frac{\sigma_Y^2}{m}}\,\middle|\,\Delta < 0\right\}$$

$$= P\left\{\frac{\bar{x} - \bar{y} - \Delta}{\sqrt{\sigma_X^2/n + \sigma_Y^2/m}} \geq Z_\alpha - \frac{\Delta}{\sqrt{\sigma_X^2/n + \sigma_Y^2/m}}\right\}$$

$$= 1 - \Phi\left(Z_\alpha - \frac{\Delta}{\sqrt{\sigma_X^2/n + \sigma_Y^2/m}}\right) \tag{10.18}$$

If $n = m$, β can be found from the OC curves in Figs. 9.3 and 9.4 in terms of

$$d = \frac{\mu_Y - \mu_X}{\sqrt{\sigma_X^2 + \sigma_Y^2}}$$

If $n \neq m$, calculate the equivalent sample size n_e as described in case 1. The expression for the sample size n, which satisfies both errors α and β, is exactly the same as in case 2 and is given by Eq. (10.16).

10.2 Tests Concerning the Difference of Means of Two Independent Normal Distributions with Equal but Unknown Variances

Case 1

$$H_0: \mu_X - \mu_Y = 0 \quad \text{against} \quad H_1: \mu_X - \mu_Y = \Delta \neq 0$$

In this case the variances σ_X^2 and σ_Y^2 are both unknown but assumed to be equal to σ^2, i.e.,

$$\sigma_X^2 = \sigma_Y^2 = \sigma^2$$

It can be shown that the likelihood ratio in this case is given by

$$\lambda = \left(1 + \frac{t^2}{n + m - 2}\right)^{-(m+n)/2}$$

where

$$t = \frac{(\bar{x} - \bar{y}) - (\mu_X - \mu_Y)}{\sqrt{\dfrac{(n-1)s_X^2 + (m-1)s_Y^2}{(m+n-2)}}\sqrt{\dfrac{1}{n} + \dfrac{1}{m}}}$$

Since the likelihood ratio calls for rejecting H_0 when λ is sufficiently small, this is equivalent to rejecting H_0 when $|t|$ is too large. Thus the rejection region is given by

$$|t| > k'$$

The term

$$\frac{(n-1)s_X^2 + (m-1)s_Y^2}{n + m - 2}$$

is called the *pooled estimate* of the variance (σ^2), and it is an unbiased estimate of σ^2.

Proof

$$\frac{(n-1)s_X^2}{\sigma^2} + \frac{(m-1)s_Y^2}{\sigma^2} = \chi_{n-1}^2 + \chi_{m-1}^2$$

Since χ_{n-1}^2 and χ_{m-1}^2 are statistically independent we have

$$(n-1)s_X^2 + (m-1)s_Y^2 = \sigma^2 \chi_{n+m-2}^2$$

Dividing both sides by $n + m - 2$ we obtain

$$\frac{(n-1)s_X^2 + (m-1)s_Y^2}{n + m - 2} = \frac{\sigma^2 \chi_{n+m-2}^2}{n + m - 2}$$

On taking expectations of both sides we get

$$E\left[\frac{(n-1)s_X^2+(m-1)s_Y^2}{n+m-2}\right]=\frac{\sigma^2}{n+m-2}E(\chi^2_{n+m-2})=\sigma^2$$

The statistic

$$\frac{(\bar{x}-\bar{y})-(\mu_X-\mu_Y)}{\sqrt{\dfrac{(n-1)s_X^2+(m-1)s_Y^2}{n+m-2}}\sqrt{\dfrac{1}{n}+\dfrac{1}{m}}}$$

has a t distribution with $n+m-2$ degrees of freedom.

Proof. By definition, variate t has a t distribution with ν degrees of freedom if

$$t=\frac{X^*\sqrt{\nu}}{\sqrt{X_\nu^2}}$$

where X^* is unit normal variate independent of χ^2 with ν degrees of freedom. Since

$$\frac{(\bar{x}-\bar{y})-(\mu_X-\mu_Y)}{\sqrt{\sigma^2/n+\sqrt{\sigma^2/m}}}$$

is a unit normal variate and

$$\frac{(n-1)s_X^2+(m-1)s_Y^2}{\sigma^2}$$

has the χ^2_{n+m-2} distribution and they are statistically independent ($\bar{x}$ is independent of s_X^2 and $\bar{y}$ is independent of s_Y^2), by definition

$$\frac{\dfrac{(\bar{x}-\bar{y})-(\mu_X-\mu_Y)}{\sigma\sqrt{1/n+1/m}}\sqrt{n+m-2}}{\dfrac{\sqrt{(n-1)s_X^2+(m-1)s_Y^2}}{\sigma}}=\frac{(\bar{x}-\bar{y})-(\mu_X-\mu_Y)}{\sqrt{\dfrac{(n-1)s_X^2+(m-1)s_Y^2}{n+m-2}}\sqrt{\dfrac{1}{n}+\dfrac{1}{m}}}$$

has a t distribution with $n+m-2$ degrees of freedom.

When H_0 is true, the probability of a type I error is

$$P\{|t|>k'\mid\mu_X-\mu_Y=0\}=\alpha$$

That is,

$$P\left\{\frac{|\bar{x}-\bar{y}|}{s_p\sqrt{1/n+1/m}}>k'\right\}=\alpha$$

where

$$s_p^2=\frac{(n-1)s_X^2+(m-1)s_Y^2}{n+m-2}$$

Hence

$$k'=t_{1-\alpha/2;\,n+m-2}$$

Therefore, the optimal rejection region is specified by

$$|\bar{x} - \bar{y}| > t_{1-\alpha/2;\, n+m-2}\, s_p \sqrt{\frac{1}{n} + \frac{1}{m}} \tag{10.19}$$

The probability of accepting H_0 when $\mu_X - \mu_Y = \Delta$ is

$$\beta(\Delta) = P\left\{\frac{|\bar{x} - \bar{y}|}{s_p\sqrt{1/n + 1/m}} \leq t_{1-\alpha/2;\, n+m-2} \,|\, \Delta, \sigma^2\right\} \tag{10.20}$$

which can be evaluated by the aid of the noncentral t distribution tables for $n = m$ in terms of Δ/σ. However, when $n = m$ and s_p is used as an estimate of σ, the *OC* curves in Figs. 9.9 and 9.10 can be used for the purpose of an approximate evaluation of β in terms of d and n' where

$$d = \frac{|\mu_X - \mu_Y|}{2\sigma}$$

and

$$n' = 2n - 1 = 2m - 1$$

To illustrate this procedure the following example is solved.

Example 10.2 X and Y are two independent normal variates with unknown but equal variances. A good estimate of each variance is 25. The sample size taken from each population is equal to 7. What value of $|\mu_X - \mu_Y|$ will be detected with probability $= .90$? Assume $\alpha = .05$.

Solution. Given

$$1 - \beta = .90 \qquad n = m = 7$$

Whence $\beta = .10$ and $n' = 2(7) - 1 = 13$. If Fig. 9.9 is entered with $n' = 13$ and $\beta = .1$, then we read $d = 1.0$. That is,

$$1 = \frac{|\mu_X - \mu_Y|}{2\sigma} \cong \frac{|\mu_X - \mu_Y|}{10}$$

whence

$$|\mu_X - \mu_Y| \cong 10$$

Case 2

$$H_0\colon \mu_X - \mu_Y = 0 \quad \text{against} \quad H_1\colon \mu_X - \mu_Y = \Delta > 0$$

It can be shown that *the optimal rejection region is given by*

$$\bar{x} - \bar{y} > t_{1-\alpha;\, n+m-2}\, s_p \sqrt{\frac{1}{n} + \frac{1}{m}} \tag{19.21}$$

The *OC* curves in Figs. 9.11 and 9.12 can be used for an approximate evaluation of β in terms of

$$d = \frac{\mu_X - \mu_Y}{2\sigma}$$

and $n' = 2n - 1 = 2m - 1$ similar to case 1.

Example 10.3 A test is devised to determine whether a certain additive will increase the tensile strength of a certain material. Fifteen specimens without additive and 20 specimens with additive were used. The following results were obtained:

Type	$\bar{x}$	s^2	n
With additive	88.8	83.8	20
Without additive	75.8	126.6	15

The variances, although unknown, are assumed to be the same for both cases.
(a) Can we conclude that the tensile strength has increased ($\alpha = .05$)?
(b) What is the estimate of the variance?

Solution. The hypothesis to be tested is

(a) $$H_0\colon \mu_X - \mu_Y = 0 \quad \text{against} \quad H_1\colon \mu_X - \mu_Y = \Delta > 0$$

We reject H_0 if

$$\bar{x} - \bar{y} > t_{1-\alpha;\, n+m-2}\, s_p \sqrt{\frac{1}{n} + \frac{1}{m}}$$

On substituting, we obtain

$$13 > t_{.95,33} \sqrt{\frac{19(83.8) + 14(126.6)}{33}} \sqrt{\frac{1}{20} + \frac{1}{15}} = 5.8$$

Hence, we reject H_0 and conclude that the tensile strength has increased.
(b) The pooled estimate of the variance is

$$\sigma^2 = s_p^2 = \frac{19(83.8) + 14(126.6)}{33} = 101.96$$

Case 3

$$H_0\colon \mu_X - \mu_Y = 0 \quad \text{against} \quad H_1\colon \mu_X - \mu_Y = \Delta < 0$$

In this case the optimal rejection region is given by

(10.22) $$\bar{x} - \bar{y} < t_{\alpha;\, n+m-2}\, s_p \sqrt{\frac{1}{n} + \frac{1}{m}}$$

β can be obtained from the *OC* curves in Figs. 9.11 and 9.12 in terms of

$$d = \frac{\mu_X - \mu_Y}{2\sigma}$$

and $n' = 2n - 1 = 2m - 1$.

10.3 Tests Concerning the Difference of Means of Two Independent Normal Distributions with Unknown and Unequal Variances

Case 1

$$H_0\colon \mu_X - \mu_Y = 0 \quad \text{against} \quad H_0\colon \mu_X - \mu_Y = \Delta \neq 0$$

In this case the likelihood ratio procedure *fails* to give an optimal solution. Some *approximate solutions* had been developed. One of these solutions was developed by B. L. Welch (1947) and is known as the t' (t prime) test. The t' test specifies the rejection region by

$$|\bar{x} - \bar{y}| > t_{1-\alpha/2;\,\nu}\sqrt{\frac{s_X^2}{n} + \frac{s_Y^2}{m}} \tag{10.23}$$

where

$$\nu = \frac{(s_X^2/n + s_Y^2/m)^2}{[(s_X^2/n)^2/(n+1)] + [(s_Y^2/m)^2/(m+1)]} - 2 \tag{10.24}$$

The *OC* curves for the t' test have not been developed as yet.

Example 10.4 The following results are calculated for two samples, each of whose observations are normally distributed

$$n = 10 \qquad \sum_{i=1}^{n} x_i = 36 \qquad \sum_{i=1}^{n} x_i^2 = 136$$

$$m = 10 \qquad \sum_{i=1}^{m} y_i = 44 \qquad \sum_{i=1}^{m} y_i^2 = 195$$

Test whether $\mu_X \neq \mu_Y$ at the 5% level of significance.

Solution. We calculate

$$\bar{x} = 3.6 \qquad \bar{y} = 4.4$$

$$s_X^2 = \frac{\sum_{i=1}^{n} x_i^2 - n\bar{x}^2}{n-1} = \frac{136 - 129.6}{9} = .711$$

$$s_Y^2 = \frac{\sum_{i=1}^{m} y_i^2 - m\bar{y}^2}{m-1} = \frac{195 - 193.6}{9} = .155$$

Hence

$$t' = \frac{|\bar{x} - \bar{y}|}{\sqrt{s_X^2/n + s_Y^2/m}} = \frac{.8}{\sqrt{.0866}} \cong 2.72$$

$$\nu = \frac{(s_X^2/n + s_Y^2/m)^2}{[(s_X^2/n)^2/(n+1)] + [(s_Y^2/m)^2/(m+1)]} - 2 \cong 14$$

The rejection rule is of the form

$$|\bar{x} - \bar{y}| > t_{1-\alpha/2;\,\nu}\sqrt{\frac{s_X^2}{n} + \frac{s_Y^2}{m}}$$

or, equivalently,

$$t' > t_{.975;\,14}$$

That is,

$$2.72 > 2.145$$

Therefore, reject H_0 and conclude that the means of the treatments are not the same at the 5% level of significance.

Case 2

$$H_0\colon \mu_X - \mu_Y = 0 \quad \text{against} \quad H_1\colon \mu_X - \mu_Y = \Delta > 0$$

The rejection region is given by

$$\bar{x} - \bar{y} > t_{1-\alpha;\,\nu}\sqrt{\frac{s_X^2}{n} + \frac{s_Y^2}{m}} \tag{10.25}$$

Case 3

$$H_0\colon \mu_X - \mu_Y = 0 \quad \text{against} \quad H_1\colon \mu_X - \mu_Y = \Delta < 0$$

Reject H_0 if

$$\bar{x} - \bar{y} < t_{\alpha;\,\nu}\sqrt{\frac{s_X^2}{n} + \frac{s_Y^2}{m}} \tag{10.26}$$

10.4 Tests Concerning the Equality of Variances of Two Independent Normal Distributions

Case 1

$$H_0\colon \sigma_X = \sigma_Y \quad \text{against} \quad H_1\colon \sigma_X^2 \neq \sigma_Y^2$$

It can be proved that the optimal acceptance region is of the form

$$k_1 \leq \frac{s_X^2}{s_Y^2} \leq k_2$$

where

$$s_X^2 = \frac{\sum_{i=1}^{n}(x_i - \bar{x})^2}{n-1}$$

and

$$s_Y^2 = \frac{\sum_{i=1}^{m}(y_i - \bar{y})^2}{m-1}$$

By definition,

$$\frac{s_X^2/\sigma_X^2}{s_Y^2/\sigma_Y^2}$$

has an F distribution with $n-1$ and $m-1$ degrees of freedom. Hence, when H_0 is true, i.e., $\sigma_X = \sigma_Y$ the quantity s_X^2/s_Y^2 has an F distribution with $n-1$ and $m-1$ degrees of freedom.

The probability of a type I error is

$$P\left\{\frac{s_X^2}{s_Y^2} < k_1 \,|\, \sigma_X^2 = \sigma_Y^2\right\} + P\left\{\frac{s_X^2}{s_Y^2} > k_2 \,|\, \sigma_X^2 = \sigma_Y^2\right\} = \alpha$$

It is customary to assume equal tail areas in the distribution of the likelihood ratio; thus we can assert that

$$P\left\{\frac{s_X^2}{s_Y^2} < k_1 \,|\, \sigma_X^2 = \sigma_Y^2\right\} = P\left\{\frac{s_X^2}{s_Y^2} > k_2 \,|\, \sigma_X^2 = \sigma_Y^2\right\} = \frac{\alpha}{2}$$

Hence

$$k_1 = F_{\alpha/2;\, n-1,\, m-1}$$

and

$$k_2 = F_{1-\alpha/2;\, n-1,\, m-1}$$

Therefore, the rejection region consists of the set of values of s_X^2 and s_Y^2 for which

$$\text{(10.27)} \qquad \frac{s_X^2}{s_Y^2} < F_{\alpha/2;\, n-1,\, m-1} \quad \text{or} \quad \frac{s_X^2}{s_Y^2} > F_{1-\alpha/2;\, n-1,\, m-1}$$

The probability of accepting H_0 when H_1 is true is

$$\beta(\sigma_X^2, \sigma_Y^2) = \left\{F_{\alpha/2;\, n-1,\, m-1} \leq \frac{s_X^2}{s_Y^2} \leq F_{1-\alpha/2;\, n-1,\, m-1} \,|\, \sigma_X^2; \sigma_Y^2\right\}$$

$$= \left\{\frac{\sigma_Y^2}{\sigma_X^2} F_{\alpha/2;\, n-1,\, m-1} \leq \frac{s_X^2/\sigma_X^2}{s_Y^2/\sigma_Y^2} \leq \frac{\sigma_Y^2}{\sigma_X^2} F_{1-\alpha/2;\, n-1,\, m-1}\right\}$$

$$\text{(10.28)} \qquad = \left\{\frac{\sigma_Y^2}{\sigma_X^2} F_{\alpha/2;\, n-1,\, m-1} \leq F_{n-1,\, m-1} \leq \frac{\sigma_Y^2}{\sigma_X^2} F_{1-\alpha/2;\, n-1,\, m-1}\right\}$$

Case 2

$$H_0\colon \sigma_X^2 = \sigma_Y^2 \quad \text{against} \quad H_1\colon \sigma_X^2 > \sigma_Y^2$$

The rejection region for the test is the set of values of s_X^2 and s_Y^2 for which

$$\text{(10.29)} \qquad \frac{s_X^2}{s_Y^2} > F_{1-\alpha;\, n-1,\, m-1}$$

The probability of accepting H_0 when H_1 is true is

$$\beta(\sigma_X^2, \sigma_Y^2) = P\left\{\frac{s_X^2}{s_Y^2} \leq F_{1-\alpha;\, n-1,\, m-1} \,|\, \sigma_X^2, \sigma_Y^2\right\}$$

$$\text{(10.30)} \qquad = P\left\{F_{n-1,\, m-1} \leq \frac{\sigma_Y^2}{\sigma_X^2} F_{1-\alpha;\, n-1,\, m-1}\right\}$$

Case 3

$$H_0\colon \sigma_X^2 = \sigma_Y^2 \quad \text{against} \quad H_1\colon \sigma_X^2 < \sigma_Y^2$$

The rejection region for the test is the set of values of s_X^2 and s_Y^2 for which

$$\text{(10.31)} \qquad \frac{s_X^2}{s_Y^2} < F_{\alpha;\, n-1,\, m-1}$$

and

$$\beta(\sigma_X^2, \sigma_Y^2) = P\left\{F_{n-1,\, m-1} \geq \frac{\sigma_Y^2}{\sigma_X^2} F_{\alpha;\, n-1,\, m-1}\right\}$$

Example 10.5 It is argued that the variance of the resistance of wire A is greater than the variance of the resistance of wire B. Five test pieces of wire A, and seven test pieces of wire B gave the following results:

$$s_A^2 = .004 \qquad s_B^2 = .002$$

Settle the argument ($\alpha = .05$).

Solution. We wish to test the by hypothesis

$$H_0\colon \sigma_A^2 = \sigma_B^2 \quad \text{against} \quad H_1\colon \sigma_A^2 > \sigma_B^2$$

The rejection region is the set of values of s_A^2 and s_B^2 for which

$$\frac{s_A^2}{s_B^2} > F_{1-\alpha;\, n-1,\, m-1}$$

Substituting $s_A^2 = .004$, $s_B^2 = .002$, $n = 5$, and $m = 7$ in Eq. (10.29) we obtain

$$\frac{.004}{.002} < F_{.95;\, 4,6}$$

That is, $2.0 < 4.53$. Hence we accept H_0 and conclude that $\sigma_A^2 = \sigma_B^2$ at the 5% level of significance.

Example 10.6 Suppose you wish to run a maximum risk of .05 of saying that the variance of wire A is greater than B when they are actually the same. On the other hand you are not willing to run a risk of more than .05 of saying that they are the same when actually $\sigma_A^2 = 4\sigma_B^2$. Determine the sample size.

Solution. We have

$$\alpha = .05 \qquad \beta = .05 \qquad \sigma_A^2 = 4\sigma_B^2$$

$$\beta = P\left\{F_{n-1,\, n-1} \le \frac{\sigma_B^2}{\sigma_A^2} F_{1-\alpha;\, n-1,\, n-1}\right\}$$

On substituting we obtain

$$.05 = P\{F_{n-1,\, n-1} \le \tfrac{1}{4} F_{.95;\, n-1,\, n-1}\}$$

whence

$$F_{.05;\, n-1,\, n-1} = \tfrac{1}{4} F_{.95;\, n-1,\, n-1}$$

or

$$\frac{F_{.05;\, n-1,\, n-1}}{F_{.95;\, n-1,\, n-1}} = \frac{1}{4}$$

Table H (in the Appendix) of the F distribution can be used to find n by successive trials. The approximate sample size is $n = 24$.

10.5 Test Concerning the Equality of Variances of Several Independent Normal Distributions

The homogeneity of variance known as *homoscedasticity* for several normal populations is required for the application of many statistical techniques. For example, the *analysis of variance* technique assumes that the normal populations under study have common variance. Validation of the assumption of equal variances on the basis of statistical tests seems essential. Many tests have been devised to test the equality of variances. We will refer to one of these tests known as *Bartlett's test*. Actually we wish to test the hypothesis

$$H_0: \sigma_1^2 = \sigma_2^2 = \cdots = \sigma_k^2 \quad \text{against} \quad H_1: \sigma_1^2, \sigma_2^2, \ldots, \sigma_k^2 \text{ not all equal}$$
$$-\infty < \mu_i < \infty, i = 1, 2, \ldots, k)$$

where k = number of normal populations tested.

The likelihood function is given by

$$L = \left(\frac{1}{\sqrt{2\pi}}\right)^N \frac{1}{\sigma_1^{n_1}} \cdots \frac{1}{\sigma_k^{n_k}} \exp\left[-\frac{1}{2}\sum_{i=1}^{k}\sum_{j=1}^{n_i}\left(\frac{x_{ij}-\mu_i}{\sigma_i}\right)^2\right]$$

where n_i = sample size taken from population i and

$$N = \sum_{i=1}^{k} n_i$$

The likelihood function under the null hypothesis is maximized for

$$\mu_i = \bar{x}_i$$

and

$$\sigma_1^2 = \sigma_2^2 = \cdots = \sigma_k^2 = \sigma^2 = \frac{1}{N}\sum_{i=1}^{k}(n_i - 1)\, s_i^2$$

where

$$s_i^2 = \frac{1}{n_i - 1}\sum_{j=1}^{n_i}(x_{ij} - \bar{x}_i)^2$$

The likelihood function under the alternative hypothesis is maximized for

$$\mu_i = \bar{x}_i$$

and

$$\sigma_i^2 = \frac{(n_i - 1)\, s_i^2}{n_i}$$

The likelihood ratio, then, is

$$\lambda = \frac{\prod_{i=1}^{k}[(n_i - 1)\, s_i^2/n_i]^{n_i/2}}{\left[\sum_{i=1}^{k}(n_i - 1)\, s_i^2/N\right]^{N/2}}$$

The distribution of λ is too complicated; however, if we take logarithms of both sides we obtain

$$-2\ln\lambda = N\ln\left[\frac{\sum_{i=1}^{k}(n_i - 1)\, s_i^2}{N}\right] - \left[\sum_{i=1}^{k} n_i \ln\frac{(n_i - 1)\, s_i^2}{n_i}\right]$$

It has been proved that the distribution of $-2\ln\lambda$ converges to χ_{k-1}^2 for comparatively large n_i. The rejection region specified by the likelihood ratio can be written as

$$-2\ \ln\lambda > k'$$

or

$$\chi^2_{k-1} > k'$$

where k' should be adjusted to satisfy the desired level of significance α. Hence

$$k' = \chi^2_{1-\alpha;\ k-1}$$

and the rejection region, then, is

(10.32)
$$-2\ \ln\ \lambda > \chi^2_{1-\alpha;\ k-1}$$

A better approximation of the above result was found by Bartlett and is commonly referred to as Bartlett's test, which specifies the rejection region by

(10.33)

$$\frac{1}{\rho}\left\{\sum_{i=1}^{k}(n_i-1)\ln\left[\frac{\sum_{i=1}^{k}(n_i-1)\,s_i^2}{N-k}\right]-\sum_{i=1}^{k}(n_i-1)\ln s_i^2\right\} > \chi^2_{1-\alpha,\ k-1}$$

where

$$\rho = 1+\frac{1}{3(k-1)}\left[\left(\sum_{i=1}^{k}\frac{1}{n_i-1}\right)-\frac{1}{N-k}\right]$$

10.6 Paired Comparisons

In sampling from two populations it sometimes happens that *extraneous factors* cause a significant difference in means whereas there was no difference in effects we are trying to measure. Suppose we wish to compare the mean viscosity of two chemicals X and Y. If the average of n samples of chemical X is compared to the average of n samples of chemical Y, part of the difference may be due to temperature, humidity, etc., and to chemical effect. It might be that the chemicals cause a difference; this difference is obscured by other factors such as different temperatures or any other extraneous effects. *To obviate this difficulty observations are taken in pairs* (x_1, y_1), $(x_2, y_2), \ldots, (x_n, y_n)$, *each pair being observed under the same experimental conditions.* Let $d_1, d_2, \ldots, d_n$ be the differences between each pair, i.e.,

$$d_1 = x_1 - y_1, \qquad d_2 = x_2 - y_2, \qquad \ldots, \qquad d_n = x_n - y_n$$

It is assumed that the differences between each pair are independent normal variates with unknown mean μ_d and unknown variance σ_d^2. Then the mean difference

$$\bar{d} = \frac{1}{n}\sum_{i=1}^{n} d_i$$

is normally distributed with mean μ_d and unknown variance σ_d^2/n. Suppose we are interested to test the hypothesis

$$H_0\colon \mu_d = \mu_X - \mu_Y = 0 \quad \text{against} \quad H_1\colon \mu_d = \mu_X - \mu_Y \neq 0$$

The rejection region of the test consists of the pairs $(\bar{d}, s_d)$ for which

$$\left|\frac{\bar{d}\sqrt{n}}{s_d}\right| > t_{1-\alpha/2;\, n-1} \tag{10.34}$$

where

$$s_d = \frac{\sum_{i=1}^{n} (d_i - \bar{d})^2}{n - 1}$$

Similarly, for testing the hypothesis

$$H_0\colon \mu_d = \mu_X - \mu_Y = 0 \quad \text{against} \quad H_1\colon \mu_d = \mu_X - \mu_Y > 0$$

the rejection region of the test consists of the pairs $(\bar{d}, s_d)$ for which

$$\frac{\bar{d}\sqrt{n}}{s_d} > t_{1-\alpha;\, n-1} \tag{10.35}$$

Finally, for the hypothesis

$$H_0\colon \mu_d = \mu_X - \mu_Y = 0 \quad \text{against} \quad H_1\colon \mu_d = \mu_X - \mu_Y < 0$$

the rejection region is given by

$$\frac{\bar{d}\sqrt{n}}{s_d} < t_{\alpha;\, n-1} \tag{10.36}$$

Example 10.7 Sixteen paired observations were taken on treatments A and B with the following results (coded data):

Treatment A	*Treatment B*
12	10
14	16
11	15
9	10
8	9
13	15
12	10
11	15
14	13
15	15
13	16
13	11
12	14
11	9
12	14
13	13

Test at the 5% level of significance whether the two treatments differ.

Solution. We compute the following:

$$|\sum d_i| = 12$$
$$\sum d_i^2 = 76$$
$$|\bar{d}| = \tfrac{12}{16} = .75$$
$$s_d^2 = \frac{1}{n-1}(\sum d_i^2 - n\bar{d}^2)$$
$$= \frac{1}{15}\left(76 - \frac{16 \times 9}{16}\right)$$
$$= 4.46$$

Reject H_0 if

$$\left|\frac{\bar{d}\sqrt{n}}{s_d}\right| > t_{1-\alpha/2;\, n-1}$$

On substituting we obtain $1.43 < 2.131$. Therefore, accept H_0 and conclude that both treatments are the same at the 5% level of significance.

We shall demonstrate the merit of paired over unpaired comparisons by considering an illustrative example:

Example 10.8 An analyst wishes to compare two apparatus at the 1% level of significance. These apparatus measure the spectrochemical content of a certain metal alloy. Nine paired observations at different levels are presented below:

Apparatus, X:	.20	.30	.40	.50	.60	.70	.80	.90	1.00
Apparatus, Y:	.10	.21	.52	.32	.78	.59	.68	.77	.89
Difference, d_i:	.10	.09	−.12	.18	−.18	.11	.12	.13	.11

From the data we compute the following:

$$\sum d_i = .54 \qquad \bar{d} = \frac{1}{n}\sum d_i = \frac{.54}{9} = .06$$

$$\sum (d_i - \bar{d})^2 = .1204 \qquad s_d^2 = \frac{\sum (d_i - \bar{d})^2}{n-1} = \frac{.1204}{8} = .01505$$

$$\sum x_i = 5.4 \qquad \bar{x} = \frac{1}{n}\sum x_i = \frac{5.4}{9} = .60$$

$$\sum (x_i - \bar{x})^2 = .60 \qquad s_X^2 = \frac{\sum (x_i - \bar{x})^2}{n-1} = \frac{.60}{8} = .075$$

$$\sum y_i = 4.86 \qquad \bar{y} = \frac{1}{n}\sum y_i = \frac{4.86}{9} = .54$$

$$\sum (y_i - \bar{y})^2 = .6064 \qquad s_Y^2 = \frac{\sum (y_i - \bar{y})^2}{n-1} = \frac{.6064}{8} = .0758$$

Now if it can be assumed that d is a normal random variable, $N(\mu_d, \sigma_d^2)$, independent of the level of measurement, one can test the hypothesis

$$H_0\colon \mu_d = 0 \quad \text{against} \quad H_1\colon \mu_d \neq 0$$

by using the test statistic $|\bar{d}\sqrt{n}|/s_d$. Thus, H_0 is to be rejected if

$$\frac{|\bar{d}\sqrt{n}|}{s_d} = \frac{.06 \times 3}{\sqrt{.01505}} = 1.47 > t_{.995;\,8} = 3.355$$

Therefore, we conclude that the two apparatus do not differ at the 1% level of significance. The power $\pi(\Delta)$ of the test can be obtained from Fig. 9.10. It should be noted that this power is approximate since we are using s_d as an estimate of σ_d. The graph of the power function (paired comparison) is shown in Fig. 10.1.

Let us suppose that the experimenter has ignored the existence of the extraneous factors and, as a result, he used unpaired sampling procedure. Furthermore, he assumes normality, independence, and equal variances, i.e., $\sigma_X^2 = \sigma_Y^2 = \sigma_0^2$. Accordingly, the null hypothesis is to be rejected if

$$|\bar{x} - \bar{y}| > s_p\sqrt{\frac{1}{n} + \frac{1}{m}}\, t_{1-\alpha/2;\, n+m-2}$$

where

$$s_p^2 = \frac{(n-1)\,s_X^2 + (m-1)\,s_Y^2}{n+m-2} = \frac{.60 + .6064}{16} = .0754$$

On substituting one can verify that H_0 is accepted at the 1% level of significance. The power function can be obtained approximately from Fig. 9.10, with abcissa $\Delta/2s_p$ and $n' = 17$. This power function is approximate since the pooled estimate of the variance s_p^2 is used instead of σ_0^2. The graph of the power function (unpaired comparison) is shown in Fig. 10.1. The graphs in Fig. 10.1 show the advantage of paired over unpaired comparison for the given example. The paired procedure

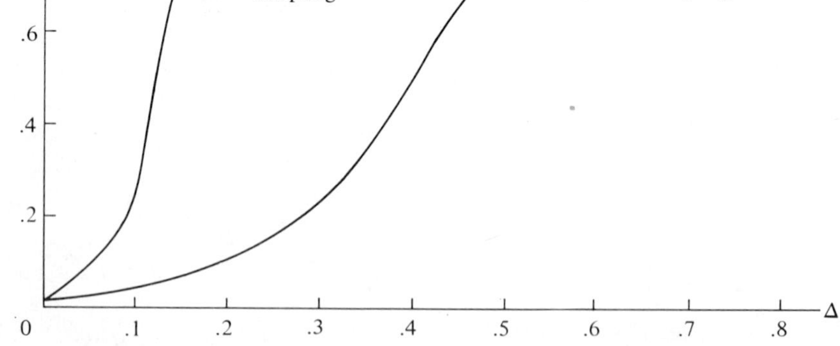

Fig. 10.1 Approximate comparison of paired and unpaired test for Example 10.8.

is more powerful since pairing has reduced the variance introduced by the extraneous factors. In the light of this example one can summarize the *merits and deficiencies* of paired comparisons.

Paired comparisons have these advantages:

1. It does not require the assumption that the two populations being compared have equal variances.
2. The two random samples need not be independent.
3. If extraneous factors exist, pairing will reduce the variance introduced by these factors and thereby increase the power of the test.
4. The efficiency of the pairing test increases if pairs are chosen so that units within a pair are similar and there are wide differences from pair to pair.

Paired comparisons have these deficiencies:

1. Sample sizes must be equal.
2. It is postulated that the variance of the difference (d) does not depend on the level of measurement.
3. Paired observations test has only $n - 1$ degrees of freedom, whereas an unpaired test has $2(n - 1)$ degrees of freedom. Thus, if pairing is not necessary there will be a big loss of information. As a result the accuracy of conclusions tends to be unreliable.

A summary of the procedures for testing hypotheses about the parameters of two independent normal distributions is given in Table 10.1.

10.7 Comparison of Two Binomial Populations

Suppose we have two independent *binomial* populations with unknown parameters p_1 and p_2. Let us assume that population 1 represents the production of machine 1, whereas population 2 represents the production of machine 2. In this situation, let p_1 and p_2 denote the fraction defective in populations 1 and 2, respectively. Furthermore, we assume the production processes are statistically controlled. As a result, p_1 and p_2 are fixed. Suppose the manufacturer wishes to test whether or not the fractions defective of both machines are the same. That is, he wishes to test the hypothesis

$$H_0: p_1 - p_2 = 0 \quad \text{against} \quad H_1: p_1 - p_2 \neq 0$$

To test this conjecture he takes a random sample of size n_1 and a random sample of size n_2 from populations 1 and 2, respectively (sampling is performed with replacement). Let x_1 denote the number of defectives in the first sample of size n_1, and let x_2 denote the number of defectives in the second sample of size n_2. The maximum likelihood estimators of p_1 and p_2

TABLE 10.1 Summary of Tests of Hypotheses About the Parameters of Two Independent Normal Distributions

Hypothesis	Unknown Parameters and Conditions	Rejection Rule	For Given α and β, Choice of Sample Size $n = m$
$H_0: \mu_X - \mu_Y = 0$ $H_1: \mu_X - \mu_Y = \Delta \neq 0$	μ_X, μ_Y	$\lvert \bar{x} - \bar{y} \rvert > Z_{1-\alpha/2} \sqrt{\frac{\sigma_X^2}{n} + \frac{\sigma_Y^2}{m}}$	$n = m \cong \frac{(Z_{1-\alpha/2} + Z_{1-\beta})^2(\sigma_X^2 + \sigma_Y^2)}{\Delta^2}$
$H_0: \mu_X - \mu_Y = 0$ $H_1: \mu_X - \mu_Y = \Delta > 0$	μ_X, μ_Y	$\bar{x} - \bar{y} > Z_{1-\alpha} \sqrt{\frac{\sigma_X^2}{n} + \frac{\sigma_Y^2}{m}}$	$n = m = \frac{(Z_{1-\alpha} + Z_{1-\beta})^2(\sigma_X^2 + \sigma_Y^2)}{\Delta^2}$
$H_0: \mu_X - \mu_Y = 0$ $H_1: \mu_X - \mu_Y = \Delta < 0$	μ_X, μ_Y	$\bar{x} - \bar{y} < Z_{\alpha} \sqrt{\frac{\sigma_X^2}{n} + \frac{\sigma_Y^2}{m}}$	$n = m = \frac{(Z_{1-\alpha} + Z_{1-\beta})^2(\sigma_X^2 + \sigma_Y^2)}{\Delta^2}$
$H_0: \mu_X - \mu_Y = 0$ $H_1: \mu_X - \mu_Y = \Delta \neq 0$	μ_X, μ_Y σ_X, σ_Y $\sigma_X = \sigma_Y$	$\lvert \bar{x} - \bar{y} \rvert > t_{1-\alpha/2;\, n+m-2}\, s_p† \sqrt{\frac{1}{n} + \frac{1}{m}}$	$d = \frac{\lvert \mu_X - \mu_Y \rvert}{2\sigma}$ Enter Fig. 9.9 or 9.10 to obtain n'. $n = m = \frac{n'+1}{2}$
$H_0: \mu_X - \mu_Y = 0$ $H_1: \mu_X - \mu_Y = \Delta > 0$	μ_X, μ_Y σ_X, σ_Y $\sigma_X = \sigma_Y$	$\bar{x} - \bar{y} > t_{1-\alpha;\, n+m-2}\, s_p \sqrt{\frac{1}{n} + \frac{1}{m}}$	$d = \frac{\mu_X - \mu_Y}{2\sigma}$ Enter Fig. 9.11 or 9.12 to obtain n'. $n = m = \frac{n'+1}{2}$
$H_0: \mu_X - \mu_Y = 0$ $H_1: \mu_X - \mu_Y = \Delta < 0$	μ_X, μ_Y σ_X, σ_Y $\sigma_X = \sigma_Y$	$\bar{x} - \bar{y} < t_{\alpha;\, n+m-2}\, s_p \sqrt{\frac{1}{n} + \frac{1}{m}}$	$d = \frac{\mu_X - \mu_Y}{2\sigma}$ Enter Fig. 9.11 or 9.12 to obtain n'. $n = m = \frac{n'+1}{2}$

† $s_p = \sqrt{\frac{(n-1)s_X^2 + (m-1)s_Y^2}{n+m-2}}$

TABLE 10.1—*Cont.*

Hypothesis	Unknown Parameters and Conditions	Rejection Rule	For Given α and β, Choice of Sample Size $n = m$
$H_0: \mu_X - \mu_Y = 0$ $H_1: \mu_X - \mu_Y = \Delta \neq 0$	μ_X, μ_Y σ_X, σ_Y $\sigma_X \neq \sigma_Y$	$\lvert \bar{x} - \bar{y} \rvert > t_{1-\alpha/2;\, \nu}\ddagger \sqrt{\frac{s_X^2}{n} + \frac{s_Y^2}{m}}$	
$H_0: \mu_X - \mu_Y = 0$ $H_1: \mu_X - \mu_Y = \Delta > 0$	μ_X, μ_Y σ_X, σ_Y $\sigma_X \neq \sigma_Y$	$\bar{x} - \bar{y} > t_{1-\alpha;\, \nu} \sqrt{\frac{s_X^2}{n} + \frac{s_Y^2}{m}}$	
$H_0: \mu_X - \mu_Y = 0$ $H_1: \mu_X - \mu_Y = \Delta < 0$	μ_X, μ_Y σ_X, σ_Y $\sigma_X \neq \sigma_Y$	$\bar{x} - \bar{y} < t_{\alpha;\, \nu} \sqrt{\frac{s_X^2}{n} + \frac{s_Y^2}{m}}$	
$H_0: \sigma_X^2 = \sigma_Y^2$ $H_1: \sigma_X^2 \neq \sigma_Y^2$	μ_X, μ_Y σ_X, σ_Y	$\frac{s_X^2}{s_Y^2} > F_{1-\alpha/2;\, n-1,\, m-1}$ $\quad \frac{s_X^2}{s_Y^2} < F_{\alpha/2;\, n-1,\, m-1}$	$\beta = \left\{ \frac{\sigma_Y^2}{\sigma_X^2} \frac{1}{F_{1-\alpha/2;\, n-1,\, m-1}} \leq F_{n-1,\, n-1} \leq \frac{\sigma_Y^2}{\sigma_X^2} F_{1-\alpha/2;\, n-1,\, m-1} \right\}$ $n = m$ is found by successive trials from F tables
$H_0: \sigma_X^2 = \sigma_Y^2$ $H_1: \sigma_X^2 > \sigma_Y^2$	μ_X, μ_Y σ_X, σ_Y	$\frac{s_X^2}{s_Y^2} > F_{1-\alpha;\, n-1,\, m-1}$	$\frac{F_{\beta;\, n-1,\, n-1}}{F_{1-\alpha;\, n-1,\, n-1}} = \frac{\sigma_Y^2}{\sigma_X^2}$ $n = m$ is found by successive trials from F tables
$H_0: \sigma_X^2 = \sigma_Y^2$ $H_1: \sigma_X^2 < \sigma_Y^2$	μ_X, μ_Y σ_X, σ_Y	$\frac{s_X^2}{s_Y^2} < F_{1-\alpha;\, n-1,\, m-1}$	$\frac{F_{1-\beta;\, n-1,\, n-1}}{F_{\alpha;\, n-1,\, n-1}} = \frac{\sigma_Y^2}{\sigma_X^2}$ $n = m$ is found by successive trials from F tables

‡ $\nu = \frac{(s_X^2/n + s_Y^2/m)^2}{[(s_X^2/n)^2/(n+1)] + [(s_Y^2/m)^2/(m+1)]} - 2$

are then x_1/n_1 and x_2/n_2, respectively. Now if $p_1 = p_2 = p_0$, we may pool these estimators to obtain the maximum likelihood estimator of p_0

$$\hat{p}_0 = \frac{x_1 + x_2}{n_1 + n_2} \tag{10.37}$$

If n_1 and n_2 are comparatively large, then the normal approximation can be used. That is, under H_0 the random variable $(x_1/n_2 - x_1/n_2)$ is approximately normally distributed, with mean 0 and variance

$$\begin{aligned}\frac{1}{n_1^2}\operatorname{Var}(x_1) + \frac{1}{n_2^2}\operatorname{Var}(x_2) &= \frac{1}{n_1^2} n_1 p_0(1-p_0) + \frac{1}{n_2^2} n_2 p_0(1-p_0)\\ &= \frac{n_1+n_2}{n_1 n_2}(1-p_0)p_0 \end{aligned} \tag{10.38}$$

Since p_0 is unknown, we can use its estimate $\hat{p}_0$. Accordingly, Eq. (10.38) becomes

$$\begin{aligned}\frac{1}{n_1^2}\operatorname{Var}(x_1) + \frac{1}{n_2^2}\operatorname{Var}(x_2) &\cong \frac{n_1+n_2}{n_1 n_2}\left(1 - \frac{x_1+x_2}{n_1+n_2}\right)\left(\frac{x_1+x_2}{n_1+n_2}\right)\\ &\cong \frac{x_1+x_2}{n_1 n_2}\left(1 - \frac{x_1+x_2}{n_1+n_2}\right)\end{aligned} \tag{10.39}$$

Finally, under H_0 the statistic

$$\frac{x_1/n_1 - x_2/n_2}{\sqrt{(x_1+x_2)/n_1 n_2\,[1-(x_1+x_2)/(n_1+n_2)]}}$$

is approximately normally distributed with mean 0 and variance 1. Thus H_0 is rejected at the α level of significance if

$$\frac{|x_1/n_1 - x_2/n_2|}{\sqrt{(x_1+x_2)/n_1 n_2\,[1-(x_1+x_2)/(n_1+n_2)]}} > Z_{1-\alpha/2} \tag{10.40}$$

Similarly, for testing the hypothesis

$$H_0: p_1 - p_2 = 0 \quad \text{against} \quad H_1: p_1 - p_2 > 0$$

the rejection rule, at the α level of significance, is

$$\frac{x_1/n_1 - x_2/n_2}{\sqrt{(x_1+x_2)/n_1 n_2\,[1-(x_1+x_2)/(n_1+n_2)]}} > Z_{1-\alpha} \tag{10.41}$$

and, finally, for testing the hypothesis

$$H_0: p_1 - p_2 = 0 \quad \text{against} \quad H_1: p_1 - p_2 < 0$$

the rejection rule, at the α level of significance, is

$$\frac{x_1/n_1 - x_2/n_2}{\sqrt{(x_1+x_2)/n_1 n_2\,[1-(x_1+x_2)/(n_1+n_2)]}} < Z_\alpha \tag{10.42}$$

Example 10.9 A manufacturer, who buys a certain electronic device from two suppliers A and B, has had poor relations with supplier A. He will switch to supplier B if the quality of his device is the same as that of A. 120 devices from supplier A and 80 devices from supplier B were put on life test. If 6 of A's devices and 5 of B's devices failed the test, should the manufacturer discontinue buying from supplier A? Use $\alpha = .01$.

Solution. We wish to test the hypothesis

$$H_0: p_A - p_B = 0 \quad \text{against} \quad H_1: p_A - p_B < 0$$

The calculations yield

$$\frac{x_A}{n_A} = \frac{1}{20} \qquad \frac{x_B}{n_B} = \frac{1}{16} \qquad \frac{x_A + x_B}{n_A + n_B} = \frac{11}{200} \qquad \frac{x_A + x_B}{n_A\, n_B} = \frac{11}{9600}$$

The null hypothesis is to be rejected if

$$\frac{x_A/n_A - x_B/n_B}{\sqrt{(x_A + x_B)/n_A\, n_B\,[1 - (x_A + x_B)/(n_A + n_B)]}} = \frac{1/20 - 1/16}{\sqrt{11/9600\,(1 - 11/200)}}$$
$$= -.379 < Z_{.01} = -2.33$$

Therefore, there are no grounds for rejecting the null hypothesis at the 1% level of significance. Consequently, the manufacturer may discontinue buying from supplier A.

PROBLEMS

10.1. Suppose that X and Y are two independent standard normal variates. How many observations are required to detect a difference of .50 in the expected values of X and Y with probability .95? Assume equal sample size and $\alpha = .10$.

10.2. In Prob. 10.1 suppose that the number of observations drawn from the X and Y populations are 36 and 25, respectively. What difference in the expected values of X and Y will be detected with probability .95? Use $\alpha = .10$.

10.3. X and Y are two independent normal populations with variances 4.5 and 8, respectively. Let n_X and n_Y denote the number of observations drawn from X and Y, respectively. A sample of size $N = n_X + n_Y = 70$ is to be chosen. Compute the optimal structure of the sample if it is desired to minimize the error of the second kind.

10.4. In Prob. 10.3 suppose that the sampling cost is $c_S = c_X n_X + c_Y n_Y$ where c_X and c_Y are unit inspection costs from the X and Y populations, respectively. Assume a sum of \$200 is available for taking a sample and $c_X = 4c_Y = \$8$. Compute the optimal structure of the sample if it is desired to minimize the error of the second kind.

10.5. Two bottle-filling machines are known to have a common variance of 2 cm³ of liquid. To test if the mean filling volumes of both machines are the same, the following samples were taken (volume in cubic centimeters)

Machine I:	991	993	993	998	997	997	998	989	989	996
Machine II:	997	998	996	997	992	993	993	995	995	999

Using a significance level of 1%, test whether the two machines differ in their mean filling volumes (assume normality).

10.6. Sulphur determinations were made on each of five standard grades of heating oil at each of two laboratories with the results shown below:

	Sulphur (%)				
Laboratory X:	2.6	3.1	3.5	3.8	4.1
Laboratory Y:	2.8	3.2	3.4	3.8	4.2

Assume that measurements are normally distributed with common variance .04%. Test the hypothesis that the results of the two laboratories are not significantly different at the 1% level of significance.

10.7. Two brands of liquors were tested to ascertain if the average alcohol contents (by volume) were different. From a sample of 25 of brand X, $\bar{x}$ was found to be 16.8%, while a sample of 16 from brand Y yielded an average of 18%. Assume normality and $\sigma_X = 1.2\%$ and $\sigma_Y = 1.5\%$.

(a) Test the hypothesis that there is no difference between the average alcohol contents. Use $\alpha = .05$.

(b) What is the probability of detecting a difference of 1.5% between the average alcohol contents?

10.8. It is suspected that waxing machine A is depositing more wax on the inside surface of certain paper bags than machine B. The average amount of wax deposited by machine A was found to be 3.4 g per bag as determined from a sample of 25 bags, while a sample of 36 bags from machine B yielded an average of 3.1 g of wax. Assume normality and $\sigma_A = \sigma_B = .3$.

(a) Test the hypothesis that machine A is depositing more wax, on the average, than machine B. Use $\alpha = .05$.

(b) Using the test procedure above, what is the probability of accepting the null hypothesis when machine A, on the average, deposits .2 g per bag more than machine B?

10.9. Random samples of 25 washers manufactured by machine A and 50 washers manufactured by machine B yielded a mean thickness of .152 and .143 in., respectively. The standard deviations for both machine are .0025.

(a) Test the hypothesis that both machines are producing washers of equal thickness. Use $\alpha = .01$.

(b) Using the test procedures in (a), compute the probability of accepting the null hypothesis when the true means differ by .002 in.

10.10. The following results are calculated for two samples, each of whose observations are normally distributed:

$$X:\ n_X = 18 \qquad \sum x_i = 36$$
$$Y:\ n_Y = 22 \qquad \sum y_i = 48$$

Assume that $\sigma_X = \sigma_Y = .5$. Test the hypothesis

$$H_0\colon \mu_X = \mu_Y \quad \text{against} \quad H_1\colon \mu_X < \mu_Y$$

at the 5% level of significance. Compute the probability of rejecting the null hypothesis when $\mu_Y - \mu_X = .25$.

10.11. You wish to determine whether the average lifetime of automobile batteries supplied by manufacturer A is longer than those supplied by manufacturer B. Past experience indicates that $\sigma_A = \sigma_B = 10$ weeks. You are willing to run a maximum risk of .10 of saying that the average lifetime of A is longer than B when they are actually the same. On the other hand, you do not wish to run a risk of more than .10 of saying that they are the same when actually the average lifetime of A exceeds that of B by as much as 10 weeks. Determine the sample size and state the acceptance rule. Assume normality.

10.12. XYZ Company desires to choose between two brands of batteries of transistor radios on the basis of mean lifetime. Brand A is slightly more expensive than brand B. The company is inclined to purchase brand B unless the mean lifetime of brand A is significantly longer at the 5% level of significance. A sample of 20 batteries from each brand was tested with the following results:

Lifetime (hr)

Brand A:	45.1	47.2	43.6	41.9	50.4	46.7	45.6	44.9	42.8	48.6
Brand B:	45.2	37.6	45.1	43.6	36.9	38.3	39.1	45.7	44.3	40.5
Brand A:	49.1	49.1	43.7	38.5	40.9	42.7	44.7	46.1	36.9	47.5
Brand B:	43.7	35.7	40.8	38.5	41.1	39.6	42.3	41.8	38.6	44.6

(a) What conclusions should be drawn? Assume normality and equal variances.
(b) What is the estimate of the variance?

10.13. In the preceding problem, how many observations from each brand (equal sample size) would be required to offer protection of $\beta = .10$ when the mean lifetime of brand A exceeds that of B by 5.2 hr? Assume the sample variance is a good estimate of the true variance.

10.14. Fuel consumption in an industrial plant is known to be normally distributed. It is of interest to determine whether or not the mean consumption during the night shift is greater than that during the day shift. The consumption during 30 shifts was as follows:

Day Shift	*Night Shift*
$x =$ consumption per day shift	$y =$ consumption per night shift
$\sum x_i = 7{,}500$ gal	$\sum y_i = 7{,}860$ gal
$\sum x_i^2 = 1{,}881{,}320$	$\sum y_i^2 = 2{,}068{,}000$

Would you conclude that, on the average, more fuel has been consumed during the night shift? Use $\alpha = .05$, and assume equal variances.

10.15. The following results are calculated for two samples, each of whose observations are normally distributed:

$$X:\ n_X = 5 \qquad \sum x_i = 50 \qquad \sum x_i^2 = 522$$
$$Y:\ n_Y = 7 \qquad \sum y_i = 84 \qquad \sum y_i^2 = 1036$$

Assuming equal variances, test the hypothesis

$$H_0\colon \mu_X = \mu_Y \quad \text{against} \quad H_1\colon \mu_X < \mu_Y$$

at the 5% level of significance. Assuming equal sample size, how many observations would be required to offer a protection at $\beta = .05$ when $\mu_X - \mu_Y = 2$? Use the pooled sample variance as an estimate of the true variance.

10.16. A new compounding technique has been developed for vinylidene chloride. Tests are performed to determine if the new compounding technique will increase the dielectric strength. Tests under standard and new compounding techniques yielded the following results:

Dielectric Strength: Step-by-step (volts/mil)

Standard Compounding:	310	318	298	316	286	316	305	299	311
New Compounding:	302	322	315	338	314	327	335	308	321

(a) What conclusions should be drawn? Assume normality and equal variances. Use $\alpha = .05$.
(b) Assuming that a common value of σ is approximately 11 and using the procedure in (a), find the sample sizes necessary to detect an increase in dielectric strength of 12 volts per mil with probability .90.

10.17. A study of the flexural strength of phenol formaldehyde using two different fillers yielded the following results:

	Flexural Strength (psi)					
Filler A:	8750	9200	9560	8850	9030	9150
Filler B:	9180	9530	9360	9480	8950	9210

(a) Assuming equal variances, test the hypothesis for equality of means at the 5% level of significance.
(b) Find the error of the second kind when the means differ by 500 psi. Assume that the pooled sample variance is a good estimate of the true common variance.

10.18. In the preceding problem, assuming equal sample size, how many observations are required to detect a difference in the flexural strength means equal to 250 psi with probability .90? Assume that the calculated pooled variance is a good estimate of the true common variance.

10.19. In Prob. 10.17 verify the assumption of equality of variances. Use $\alpha = .05$.

10.20. In Prob. 10.12 verify the assumption of equality of variances. What is the probability of concluding that the variances are equal when one variance is 2.5 times the other? Use $\alpha = .05$.

10.21. In Prob. 10.14 verify the assumption of equality of variances. Determine the sample size that yields protection of $\beta = .05$ when one variance is 3.2 times the other. Use $\alpha = .05$.

10.22. In Prob. 10.16 verify the assumption of equality of variances. Use $\alpha = .05$.

10.23. An experiment is performed to test the effectiveness of an advertising campaign on the demand of heating oil in a certain locality. The records of weekly demand for 10 weeks before and after the campaign are as follows:

	Demand (1000 gal)									
Before the campaign X:	62.3	57.6	55.1	63.1	64.1	58.1	59.6	62.7	60.5	56.9
After the campaign Y:	61.5	67.3	62.7	61.7	60.4	66.8	63.7	59.7	67.9	68.3

(a) Test the hypothesis

$$H_0: \sigma_X^2 = \sigma_Y^2 \quad \text{against} \quad H_1: \sigma_X^2 \neq \sigma_Y^2$$

at the 5% level of significance. Assume that demand is normally distributed.
(b) Would you infer from the above data that the advertising campaign has been effective? Use $\alpha = .05$.

10.24. A sample of 30 resistors supplied by vendor X and a sample of 40 resistors supplied by vendor Y yielded the following results:

Vendor X	*Vendor Y*
x = resistance in ohms/unit	y = resistance in ohms/unit
$\sum x_i = 156$	$\sum y_i = 204$
$\sum x_i^2 = 811.345$	$\sum y_i^2 = 1040.4585$

(a) Test the hypothesis of no difference in variances of the resistors supplied by the two vendors, using a level of significance of 1%. Assume normality.
(b) Test the hypothesis of no difference in the two populations' means using $\alpha = .01$.

10.25. X and Y are two normally distributed independent random variables. Two samples are drawn, yielding the following results:

$$X: \quad n_X = 10 \qquad \sum x_i = 200 \qquad \sum x_i^2 = 4009$$
$$Y: \quad n_Y = 18 \qquad \sum y_i = 396 \qquad \sum y_i^2 = 8780$$

(a) Test the hypothesis

$$H_0: \sigma_X^2 = \sigma_Y^2 \quad \text{against} \quad H_1: \sigma_X^2 \neq \sigma_Y^2$$

at the 5% level of significance.

(b) Test the hypothesis

$$H_0: \mu_X - \mu_Y = 0 \quad \text{against} \quad H_1: \mu_X - \mu_Y < 0$$

at the 5% level of significance.

10.26. A dairy desires to compare the butterfat content of the milk supplied by two farms. A sample of 20 qt from farm A yielded an average of .012 lb of butterfat with a standard deviation of .0015 lb, while a sample of 16 qt from farm B yielded an average of .014 lb with a standard deviation of .0010 lb.

(a) Test the hypothesis of no difference in variances of the butterfat using a level of significance of 5%. Assume normality.

(b) Test the null hypothesis that there is no difference between the true average butterfat content of the two farms against the alternative that farm B is superior. Use $\alpha = .05$.

10.27. X and Y are two independent normal variates. From population X the sample variance is calculated as 26.5, and from population Y the sample variance is found to be 29.15. Test the hypothesis of equal variances of the two populations if the sample sizes were $n_X = n_Y = 10$. Use $\alpha = .10$.

10.28. Two Brinell hardness-testing machines X and Y are compared by measuring the deviations from the correct indenter hold diameter (in millimeters) on ten specimens of known hardness. From past data it is known that measurements are normally distributed with $E(X) = E(Y) = 0$. The following data were obtained:

Testing Machine	*Specimen Number*									
	1	2	3	4	5	6	7	8	9	10
X: deviation (mm)	−.10	0	−.15	−.20	.05	.10	.15	.20	−.05	.10
Y: deviation (mm)	.03	−.04	.05	−.07	.20	−.02	−.08	.06	.05	−.04

(a) Test the hypothesis

$$H_0: \sigma_X^2 = \sigma_Y^2 \quad \text{against} \quad H_1: \sigma_X^2 > \sigma_Y^2$$

at the 10% level of significance.

(b) Test the hypothesis

$$H_0: \sigma_X^2 = \sigma_Y^2 \quad \text{against} \quad H_1: \sigma_X^2 \neq \sigma_Y^2$$

at the 5% level of significance.

10.29. In Prob. 10.28, using the test procedure in (a), determine the sample size to use if you desire to run a maximum risk of .10 of saying $\sigma_X^2 = \sigma_Y^2$ when $\sigma_X^2 = 2.25\sigma_Y^2$.

10.30. In Prob. 10.28, using the test procedure in (b) determine the sample sizes to use if you desire to run a maximum risk of .05 of saying $\sigma_X^2 = \sigma_Y^2$ when one variance is .148 of the other.

10.31. The amount of tin deposited on a steel sheet surface measured in pounds per base box is known to be normally distributed. An experiment was run to determine whether two different lines have the same variability in amount of tin deposited on the surface of the steel. The following data were obtained:

Line X:	.30	.32	.28	.36	.26	.35	.31	.31	.34	.27
Line Y:	.27	.31	.22	.36	.29	.25	.33	.31	.27	.29

(a) Test the hypothesis of equal variances. Use $\alpha = .05$.

(b) Test the hypothesis that $\sigma_X > \sigma_Y$ using $\alpha = .10$.

10.32. Two different detergents are compared by measuring the amount of dirt removed from soiled loads of clothes under the same experimental conditions. The following data are obtained:

Dirt Removal in Grams

Sample number:	1	2	3	4	5	6	7	8	9
Detergent X:	41	47	28	35	22	39	53	48	33
Detergent Y:	36	41	27	32	18	37	48	45	29

Would you conclude from these data that detergent X is superior to Y? Assume normality, and use $\alpha = .05$.

10.33. In Prob. 10.32 how many observations would be required to have a probability of .90 of rejecting the null hypothesis if the average dirt removal of detergent X exceeds that of Y by 1.5 g? Assume that the standard deviation of differences (σ_d) is about 1.5 g.

10.34. The following data were obtained in an experiment designed to check whether there is a systematic difference in the determinations of temperature made by two different optical pyrometers. The following readings were obtained by sighting the pyrometers on different glowing-filaments:

Temperature (°F)

Filament:	1	2	3	4	5	6	7	8	9	10
Pyrometer X:	1850	1630	1518	1980	2010	1760	1300	1415	2510	2730
Pyrometer Y:	1890	1620	1550	1985	2030	1800	1293	1455	2530	2710

Test whether there is a difference in the true average readings obtained with the two pyrometers. Assume normality and use $\alpha = .01$.

10.35. In Prob. 10.34 how many observations would be required to detect a difference of 20°F between the true average readings with a probability of .90? Assume that the standard deviation of differences (σ_d) is approximately 20°F.

10.36. It is claimed that the viscosity of lubricant X is higher than that of lubricant Y. Ten paired observations yielded the following results:

Viscosity (sec Saybolt)

Sample:	1	2	3	4	5	6	7	8	9	10
Lubricant X:	120	130	128	170	190	210	230	250	270	290
Lubricant Y:	105	115	121	175	183	207	216	230	261	275

Assuming normality, what conclusions should be drawn? Use $\alpha = .05$.

10.37. In Prob. 10.36 what is the probability of accepting the null hypothesis when the average viscosity of lubricant X exceeds that of Y by 8 sec Saybolt? How large a sample is needed to reduce the probability of the error of the second kind by 50%? Assume σ_d is approximately 8 sec Saybolt.

10.38. An experiment was run to determine the effect of a new type drug on blood pressure. Fifteen persons have their blood pressure measured before and after the drug is given. The results of the experiment are as follows:

Blood Pressure

Person:	1	2	3	4	5	6	7	8	9	10	11	12	13	14	15
Before:	116	118	120	124	128	130	131	134	136	137	137	140	143	146	147
After:	119	124	126	128	121	135	137	138	139	135	143	146	150	149	142

(a) Test the hypothesis that the new drug does not raise blood pressure. Assume normality and use $\alpha = .05$.

(b) Test the hypothesis

$$H_0\colon \mu_d = 5 \quad \text{against} \quad H_1\colon \mu_d < 5$$

where

$$\mu_d = \text{mean of differences}$$

at the 5% level of significance.

10.39. In part (a) of the preceding problem perform a nonparametric test. Graph the approximate power functions of the nonparametric and paired normal tests. (Assume σ_d is approximately 4 for this calculation.)

10.40. A manufacturer of photographic equipment wishes to compare two brands of flashbulbs. Of 100 (brand A) flashbulbs 15 were found defective, and of 150 (brand B) flashbulbs 33 were found defective. Using a level of significance of .05, test whether there is a difference between the two brands.

10.41. A tobacco distributor claims that the proportion of smokers in locality A is greater than that of locality B. An experiment was run with the following results:

	Locality A	*Locality B*
Size of sample	500	700
Number of smokers	225	245

Would you justify the distributor's claim? Use $\alpha = .05$.

10.42. An experiment was run to compare the effectiveness of two types of flu vaccines in an industrial plant. 250 workers were inoculated by type A, and 250 workers were inoculated by type B. The results were as follows:

Vaccine Type	*Caught Flu*
A	16
B	8

Test the hypothesis that vaccine B is superior to vaccine A. Use $\alpha = .05$.

10.43. A metallurgist claims that he can reduce the proportion of defective castings by including a certain additive in the cupola charge. Tests are made on the proportion of defective castings produced with and without additive. The following results are obtained:

	Castings Tested	
	With Additive	*Without Additive*
Good	145	165
Defective	15	35

Would you justify the metallurgist's claim? Use $\alpha = .05$.

10.44. It is known that the number of deaths from motor vehicle accidents has the Poisson distribution. A vigorous safety campaign in a certain locality was conducted for a period of 1 yr. During the period the observed number of accidents was found to be 218. The number of accidents in the preceding year was 296.

(a) Would you conclude that the safety campaign has been effective? Use $\alpha = .05$.

(b) Find the power of the test if the expected numbers of accidents during and before the campaign are 200 and 300, respectively. (*Hint:* Use the square root transformation; i.e., if X is a Poisson variate with mean λt for period t, then $X = \sqrt{k}$ has an approximately normal distribution with mean $\sqrt{\lambda t}$ and variance $\frac{1}{4}$.)

10.45. Suppose that X_1, X_2, and X_3 are independent normal populations. The following results are calculated for a sample drawn from each population:

$$X_1:\ n_{X_1} = 10 \qquad \sum x_{1i} = 50 \qquad \sum x_{1i}^2 = 304$$
$$X_2:\ n_{X_2} = 10 \qquad \sum x_{2i} = 54 \qquad \sum x_{2i}^2 = 395.1$$
$$X_3:\ n_{X_3} = 15 \qquad \sum x_{3i} = 90 \qquad \sum x_{3i}^2 = 820$$

Test the hypothesis that the variances are equal at the 10% level of significance.

10.46. The variability of eight filling machines is to be tested. Ten observations are taken on each machine. The sample variances are found to be

9.2 11.6 7.6 4.6 12.1 8.9 5.6 10.4

Do all filling machines have the same variance? Use $\alpha = .05$ and assume normality.

SUGGESTED REFERENCES

[1] Alexander, H. W., *Elements of Mathematical Statistics*, New York, John Wiley & Sons, Inc., 1961.

[2] Benett, C. A., and N. L. Franklin, *Statistical Analysis in Chemistry and the Chemical Industry*, New York, John Wiley & Sons, Inc., 1954.

[3] Bowker, A. H., and G. J. Lieberman, *Engineering Statistics*, Englewood Cliffs, N.J., Prentice-Hall, Inc., 1959.

[4] Brownlee, K. A., *Statistical Theory and Methodology in Science and Engineering*, 2nd ed., New York, John Wiley & Sons, Inc., 1965.

[5] Brunk, H. D., *An Introduction to Mathematical Statistics*, Boston, Ginn and Company, 1960.

[6] Cramér, H., *Mathematical Methods of Statistics*, Princeton, N. J., Princeton University Press, 1951.

[7] Dixon, W. J., and F. J. Massey, *Introduction to Statistical Analysis*, 3rd ed., New York, McGraw-Hill Book Company, 1969.

[8] Ehrenfeld, S., and S. B. Littauer, *Introduction to Statistical Method*, New York, McGraw-Hill Book Company, 1964.

[9] Fraser, D. A. S., *Statistics*, New York, John Wiley & Sons, Inc., 1958.

[10] Guttman, I., S. S. Wilks, and J. S. Hunter, *Introductory Engineering Statistics*, 2nd ed., New York, John Wiley & Sons, Inc., 1971.

[11] Hald, A., *Statistical Theory with Engineering Applications*, New York, John Wiley & Sons, Inc., 1952.

[12] Hoel, P. G., *Introduction to Mathematical Statistics*, 3rd ed., New York, John Wiley & Sons, Inc., 1965.

[13] Kendall, M. G., and A. Stuart, *The Advanced Theory of Statistics*, Vol. 1, *Distribution Theory*, 3rd ed., 1969, and Vol. 2, *Statistical Inference and Statistical Relationship*, 2nd ed., 1967, New York, Hafner Publishing Co., Inc.

[14] Kenney, J. F., and E. S. Keeping, *Mathematics of Statistics*, Vols. 1 and 2, 3rd ed., Princeton, N.J., D. Van Nostrand Co., Inc., 1956.

[15] Mood, A. M., *Introduction to the Theory of Statistics*, New York, McGraw-Hill Book Company, 1950.

[16] Neyman, J., *First Course in Probability*, Holt, Rinehart & Winston, Inc., New York, 1950.

[17] Neyman, J., and E. S. Pearson, "On the Problem of the Most Efficient Type of Statistical Hypothesis," *Philosophical Transactions of the Royal Society*, *A*, **231**, 289–337, 1933.

[18] Pearson, E. S., and H. O. Hartley, *Biometrika Tables for Statisticians*, New York, Cambridge University Press, 1958.

[19] Resnikoff, G. J., and G. J. Lieberman, *Tables of the Non-Central t-Distribution*, Stanford, Stanford University Press, 1957.

[20] Steel, R. G. D., and J. H. Torrie, *Principles and Procedures of Statistics*, New York, McGraw-Hill Book Company, Inc., 1960.

[21] Wald, A., *Statistical Decision Functions*, New York, John Wiley & Sons, Inc., 1950.

[22] Wilks, S. S., "The Large Sample Distribution of the Likelihood Ratio for Testing Composite Hypotheses," *Annals of Mathematical Statistics*, **9**, 60–62, 1938.

[23] Wilks, S. S., *Mathematical Statistics*, New York, John Wiley & Sons, Inc., 1962.

[24] Yule, F. U., and M. G. Kendall, *An Introduction to the Theory of Statistics*, London, Charles Griffin, and New York, Hafner Publishing Co., Inc., 1948.

11

SKIP

NONPARAMETRIC TESTS

11.1 Introduction

In previous chapters we discussed various statistical tests on the assumption that original observations are drawn from a normal population. In many situations the distribution of the population may be unknown and, consequently, the assumption of normality may be inappropriate. In such cases the application of the previous statistical tests is invalid. Other procedures were developed that do not require a knowledge of the distribution of the random variables under study. These procedures are usually called *nonparametric tests* that are valid for observations drawn from a population whose distribution is of any form. These tests usually assume that the probability distributions of the populations under study are continuous. As a rule nonparametric tests have a lower efficiency than the corresponding parametric tests. The advantages and disadvantages of nonparametric tests can be summarized as follows:

ADVANTAGES OF NONPARAMETRIC TESTS

1. Probability statements can be made with a relatively high degree of accuracy regardless of the shape of the distribution.
2. Data can be obtained easily and at low cost.
3. Small samples statistics are usually easy to compute and apply.

4. Nonparametric tests can be applied to samples made up of observations from serveral different populations.
5. Nonparametric tests can easily be applied to data that are naturally in ranks.
6. Nonparametric tests are extensively used in assessing the adequacy of random number generators in simulation.

DISADVANTAGES OF NONPARAMETRIC TESTS

1. Nonparametric tests are generally less efficient than the corresponding parametric tests.
2. Nonparametric tests usually require larger samples than parametric tests for a specified power efficiency.
3. Inferences about the populations under study by the nonparametric methods are usually limited. Often we can test whether the populations are the same or different. However, if a difference exists we might not be able to specify the nature of the difference.
4. Nonparametric tests are not as well elaborated as are parametric tests. Very little has been done to compare the efficiencies of parametric and nonparametric tests on general types of populations.

Recently many nonparametric tests were developed: Exposition of all these tests is beyond the scope of this work. However, in this chapter we shall consider several of the most important and extensively used nonparametric tests.

11.2 The Sign Test

Suppose we have two populations denoted by X and Y, respectively. Furthermore, we assume that their distributions are unknown but continuous. Consider the hypothesis

$$H_0: f_1(x) \equiv f_2(y)$$

where $f_1(x)$ and $f_2(y)$ are the density functions of X and Y, respectively. If H_0 is true, then the probability that a random observation drawn from X will exceed a random observation drawn from Y is equal to $\frac{1}{2}$. (Note that theoretically the probability of a tie is zero.) In symbols,

$$P\{x_i > y_i\} = \int_{-\infty}^{\infty} \int_{y_i}^{\infty} f_1(x_i) f_2(y_i)\, dx_i\, dy_i = \tfrac{1}{2} \quad \text{for all } i \tag{11.1}$$

It follows that

$$P\{x_i > y_i\} = P\{x_i - y_i > 0\} = P\{x_i - y_i < 0\} = p = \tfrac{1}{2}$$

It is assumed that these observations are taken in pairs and are independent.

Let $(x_1, y_1), (x_2, y_2), \ldots, (x_n, y_n)$ denote n independent paired observations drawn from the two populations. Their differences $d_1, d_2, \ldots, d_n$ have continuous distribution since the observations on x_i and y_i are continuous. Thus a test of the null hypothesis (H_0) is equivalent to testing that each difference between the paired observations has a probability distribution with median equal to zero. The probability distribution of each difference need not be the same if each paired observation is taken under a different environmental condition. In this case the null hypothesis (H_0) is that

$$f_{11}(x) \equiv f_{12}(y) \qquad f_{21}(x) \equiv f_{22}(y) \qquad \cdots \qquad f_{n1}(x) \equiv f_{n2}(y)$$

If only the signs of the differences are considered, the number of positive and negative signs in a sample of size n is a random variable having a binomial distribution with parameters n and p. When H_0 is true, $p = \frac{1}{2}$. If it can be assumed that the underlying distributions of individual observations are symmetric and furthermore that the underlying distributions of each pair of observations may differ only in their means (μ), the *sign test* can be used to test the hypothesis $\mu_X = \mu_Y$ against the alternatives $\mu_X \neq \mu_Y$ or $\mu_X > \mu_Y$ or $\mu_X < \mu_Y$, which is equivalent to testing $p = \frac{1}{2}$ against the alternatives $p \neq \frac{1}{2}$ or $p > \frac{1}{2}$ or $p < \frac{1}{2}$, respectively.

Now consider the hypothesis formulated in terms of p.

Case 1

$$H_0 : p = \tfrac{1}{2} \quad \text{against} \quad H_1 : p > \tfrac{1}{2}$$

According to the Neyman–Pearson lemma the optimum rejection region is of the form

$$r^{(+)} \geq k'$$

where $r^{(+)}$ = the number of positive differences and k' = prescribed constant that satisfies the level of significance α.
The probability of type I error is

$$P\{r^{(+)} \geq k' \mid H_0\} = \sum_{j=k'}^{n} \frac{n!}{j!\,(n-j)!}\left(\frac{1}{2}\right)^n = \alpha \tag{11.2}$$

k' can be obtained from the tables of the cumulative binomial distribution (Table J in the Appendix).

Thus we reject H_0 when the number of positive signs is greater than or equal to k' at the required level of significance.

Case 2

$$H_0 : p = \tfrac{1}{2} \qquad \text{against} \qquad H_1 : p < \tfrac{1}{2}$$

Similarly to case 1 it can be shown that H_0 is rejected when the number of negative signs $r^{(-)}$ is greater than or equal to k' at the required level of significance.

Case 3

$$H_0 : p = \tfrac{1}{2} \qquad \text{against} \quad H_1 : p \neq \tfrac{1}{2}$$

This is a two-sided test. In this case the null hypothesis is rejected when the number of times the less frequent sign (r) occurs is less than or equal to k' at the required level of significance. Table 11.1 shows the critical values of r for the sign test for a two-sided procedure based upon the binomial distribution with parameter $p = \frac{1}{2}$. Table 11.1 also can be used for single-sided tests with the following modifications:

TABLE 11.1 Critical Values of r for the Sign Test (Two-tail Percentage Points for the Binomial with p = .5)

n	1%	2%	5%	10%	n	1%	2%	5%	10%
6			0	0	26	6	6	7	8
7		0	0	0	27	6	7	7	8
8	0	0	0	1	28	6	7	8	9
9	0	0	1	1	29	7	7	8	9
10	0	0	1	1	30	7	8	9	10
11	0	1	1	2	31	7	8	9	10
12	1	1	2	2	32	8	9	9	10
13	1	1	2	3	33	8	9	10	11
14	1	2	2	3	34	9	9	10	11
15	2	2	3	3	35	9	10	11	12
16	2	2	3	4	36	9	10	11	12
17	2	3	4	4	37	10	11	12	13
18	3	3	4	5	38	10	11	12	13
19	3	4	4	5	39	11	11	12	13
20	3	4	5	5	40	11	12	13	14
21	4	4	5	6	41	11	12	13	14
22	4	5	5	6	42	12	13	14	15
23	4	5	6	7	43	12	13	14	15
24	5	5	6	7	44	13	13	15	16
25	5	6	7	7	45	13	14	15	16

1. The null hypothesis is rejected at a level of significance equal to half the values (1%, 2%, . . . , 10%) given in the table heading.
2. The null hypothesis is rejected when r is less than or equal to the critical value in the table, where

$$r = n - r^{(+)} \quad \text{for case 1}$$
$$= n - r^{(-)} \quad \text{for case 2}$$

If n is larger than 30 the normal approximation to the binomial can be used. Since in practice measurements are recorded to a finite number of decimals ties may occur though theoretically the probability of a tie occurring is zero. In such cases ties are ignored.

It is noteworthy that the sign test is less powerful than the paired test procedure based on the normality assumption since the sign and not the magnitude of the difference is taken into consideration. However, it has broad range of application when the underlying distribution is unknown.

Example 11.1 Table 11.2 shows the measurements of wear effects under various conditions (lubrication and speed) for two types of piston rings.

TABLE 11.2

Piston Ring Type A	*Piston Ring Type B*	*Piston Ring Type A*	*Piston Ring Type B*	*Piston Ring Type A*	*Piston Ring Type B*
1.472	1.407	1.583	1.483	1.318	1.328
1.581	1.592	1.683	1.596	1.375	1.391
1.736	1.768	1.542	1.583	1.418	1.297
1.918	1.927	1.463	1.412	1.573	1.593
1.937	1.908	1.295	1.315	1.712	1.748
1.816	1.819	1.938	1.900	1.716	1.736
1.718	1.763	1.871	1.936	1.738	1.793
1.663	1.598	1.736	1.816	1.781	1.873
1.618	1.673	1.582	1.511	1.875	1.890
1.713	1.778	1.836	1.912	1.811	1.821
1.728	1.709	1.348	1.351	1.416	1.426
1.517	1.537	1.376	1.401	1.518	1.531

(a) Use the sign test to test the hypothesis that the type of piston ring has no effect on wear ($\alpha = .05$). Assume that the distributions are symmetric and differ in their means.

(b) Compare your results using normal approximation.

Solution

(a) The hypothesis to be tested is

$$H_0: \mu_A = \mu_B \quad \text{against} \quad H_1: \mu_A \neq \mu_B$$

The number of positive signs $(A > B) = 10$. The number of negative signs $(A < B) = 26$. Hence, r = least frequent sign $= 10$. For $n = 36$ and $\alpha = .05$, the critical value of r given by Table 11.1 is 11. Therefore, we reject H_0 when $r \leq 11$. Since $r = 10$, we reject the null hypothesis at the 5% level of significance and conclude that the type of piston ring does affect wear.

(b) Using the normal approximation we obtain

$$P\{r \leq k' \mid H_0\} \cong \Phi\left(\frac{k' + \frac{1}{2} - np}{\sqrt{np(1-p)}}\right)$$

$$\cong \Phi\left(\frac{k' + \frac{1}{2} - 18}{3}\right) = \frac{\alpha}{2} = .025$$

Hence

$$k' = 17.5 + 3Z_{.025} = 11.62$$

Since $r = 10 < 11.62$ we reject H_0 and conclude that the type of piston ring does affect wear at the 5% level of significance.

We note that the critical value derived by using the normal approximation is close indeed to that obtained from Table 11.1. In general, when n is larger than 30 the normal approximation is satisfactory.

11.3 The Wilcoxon Two-Sample Rank Test

The sign test discussed in Sec. 11.1 requires that samples from the two populations be taken in pairs: equal sample size from both populations. The Wilcoxon test does not require that samples be taken in pairs; i.e., samples are not necessarily of the same size. The Wilcoxon test *is a test of the null hypothesis that two populations are identical against the alternative that they differ by a linear translation either to the left or to the right or in both directions.* This means that the distributions of the two populations are assumed to differ only in their means.

Let $x_1, x_2, \ldots, x_n$ and $y_1, y_2, \ldots, y_m$ denote two random independent samples of sizes n and m $(n \leq m)$ from populations having continuous density functions $f_1(x)$ and $f_2(y)$, respectively. If we pool the two samples into a combined sample of size $n + m = N$ and arrange the pooled sample in order of increasing magnitude, a typical arrangement is given by the ordered set $x_1, y_3, y_5, x_2, \ldots,$ where the ranks of $x_1, y_3, \ldots$ are the integers $1, 2, \ldots, N$.

When the null hypothesis is true, i.e., $f_1(x)$ and $f_2(y)$ are identical, the above ordered set represents a random sample of size N from one (common) population. If $n = m$ we expect the difference between the sum of ranks for both samples to be insignificant.

When the alternative is true—i.e., one distribution is displaced relative to the other—the low ranks will tend to fall in one sample whereas the high ranks fall in the other sample. Accordingly the sum of ranks will be relatively high or low.

Under the null hypothesis the sampling distribution of the sum of ranks will depend on the sample sizes n and m of the two samples. To illustrate, suppose random observations from two populations yield the following results:

Population X	*Rank*	*Population Y*	*Rank*
17.06 (x_1)	1	18.73 (y_1)	2
18.86 (x_2)	3	18.92 (y_2)	4
		19.01 (y_3)	5

When the null hypothesis is true the ordered set x_1, y_1, x_2, y_2, y_3 is a random sample from a common population. Therefore, the occurrence of any particular order of these sample values is equiprobable. Thus the sum of ranks (R_1) of observations drawn from the X population can be as low as $n(n+1)/2 = 3$ and as high as $n(2m+n+1)/2 = 9$. The probability that the sum of ranks is equal to 3 is derived as follows:

There is only one way to obtain a rank sum equal to 3, namely, $1 + 2 = 3$. Thus the probability that the sum of ranks is equal to 3 is

$$P\{R_1 = 3\} = 1\Big/\binom{5}{2} = .10$$

Similarly, the probability that the sum of ranks, say, is equal to 7 can be obtained. The number of ways we can obtain a rank sum equal to 7 is

$$2 + 5 = 7 \qquad 3 + 4 = 7$$

i.e., two ways. Thus the probability that the sum of ranks is equal to 7 is

$$P\{R_1 = 7\} = 2\Big/\binom{5}{2} = .2$$

Proceeding in the same way the cumulative distribution of R_1 can be shown to be as below:

$$F_{R_1}(a) = \begin{cases} 0 & a < 3 \\ .1 & 3 \leq a < 4 \\ .2 & 4 \leq a < 5 \\ .4 & 5 \leq a < 6 \\ .6 & 6 \leq a < 7 \\ .8 & 7 \leq a < 8 \\ .9 & 8 \leq a < 9 \\ 1.0 & a \geq 9 \end{cases} \tag{11.3}$$

Obviously the distribution of R_1 depends on sample sizes n and m. Looking at the cumulative distribution of R_1 we find that the probability of R_1 being less than or equal to 3 is .10 and the probability of R_1 being greater than or equal to 9 is .10. Thus $R_1 \leq 3$ and $R_1 \geq 9$ form a 20% critical region for our hypothesis.

The distribution of R_1 has been calculated by combinatorial methods for various sample sizes. Tables 11.3 and 11.4 give the critical points of rank sums R_1 for the 5% and 1% level of significance, respectively. To test the hypothesis

$$H_0 : \mu_X = \mu_Y \quad \text{against} \quad H_1 : \mu_X \neq \mu_Y$$

calculate

$$R_1 = \text{sum of ranks of the smaller sample (size } n)$$

and

$$R_2 = \text{sum of ranks of the larger sample (size } m)$$

TABLE 11.3 5% Critical Points of Rank Sums (R_1^{++})†

n_2 \ n_1	2	3	4	5	6	7	8	9	10	11	12	13	14	15
4			10											
5		6	11	17										
6		7	12	18	26									
7		7	13	20	27	36								
8	3	8	14	21	29	38	49							
9	3	8	15	22	31	40	51	63						
10	3	9	15	23	32	42	53	65	78					
11	4	9	16	24	34	44	55	68	81	96				
12	4	10	17	26	35	46	58	71	85	99	115			
13	4	10	18	27	37	48	60	73	88	103	119	137		
14	4	11	19	28	38	50	63	76	91	106	123	141	160	
15	4	11	20	29	40	52	65	79	94	110	127	145	164	185
16	4	12	21	31	42	54	67	82	97	114	131	150	169	
17	5	12	21	32	43	56	70	84	100	117	135	154		
18	5	13	22	33	45	58	72	87	103	121	139			
19	5	13	23	34	46	60	74	90	107	124				
20	5	14	24	35	48	62	77	93	110					
21	6	14	25	37	50	64	79	95						
22	6	15	26	38	51	66	82							
23	6	15	27	39	53	68								
24	6	16	28	40	55									
25	6	16	28	42										
26	7	17	29											
27	7	17												
28	7													

†Reprinted by permission from Colin White, "The Use of Ranks in a Test of Significance for Comparing Two Treatments," *Biometrics*, **8**, 37, 1952.

Reject the null hypothesis if either R_1 or R_2 is less than or equal to the critical value shown in the table at the desired level of significance.

Tables 11.3 and 11.4 also can be used for one-sided tests. To illustrate, consider the hypothesis

$$H_0 : \mu_X = \mu_Y \quad \text{against} \quad H_1 : \mu_X < \mu_Y$$

Reject H_0 if R_1 is less than or equal to the tabled critical value at the desired level of significance and for the hypothesis

$$H_0 : \mu_X = \mu_Y \quad \text{against} \quad H_1 : \mu_X > \mu_Y$$

reject H_0 if R_2 is less than or equal to the tabled critical value at the desired level of significance. Note that for one-sided tests the level of significance equals half the values ($2\frac{1}{2}\%$ and $\frac{1}{2}\%$) given in the table heading.

For large sample sizes the distribution of R_1 (sum rank of the smaller

TABLE 11.4 1% Critical Points of Rank Sums (R_1^{++})

n_2 \ n_1	2	3	4	5	6	7	8	9	10	11	12	13	14	15
5				15										
6			10	16	23									
7			10	17	24	32								
8			11	17	25	34	43							
9		6	11	18	26	35	45	56						
10		6	12	19	27	37	47	58	71					
11		6	12	20	28	38	49	61	74	87				
12		7	13	21	30	40	51	63	76	90	106			
13		7	14	22	31	41	53	65	79	93	109	125		
14		7	14	22	32	43	54	67	81	96	112	129	147	
15		8	15	23	33	44	56	70	84	99	115	133	151	171
16		8	15	24	34	46	58	72	86	102	119	137	155	
17		8	16	25	36	47	60	74	89	105	122	140		
18		8	16	26	37	49	62	76	92	108	125			
19	3	9	17	27	38	50	64	78	94	111				
20	3	9	18	28	39	52	66	81	97					
21	3	9	18	29	40	53	68	83						
22	3	10	19	29	42	55	70							
23	3	10	19	30	43	57								
24	3	10	20	31	44									
25	3	11	20	32										
26	3	11	21											
27	4	11												
28	4													

sample of size n) can be approximated by the normal distribution. It can be proved that

$$E(R_1) = \frac{n(N+1)}{2} \tag{11.4}$$

$$\text{Var}(R_1) = \frac{nm(N+1)}{12} \tag{11.5}$$

Then the statistic

$$\frac{R_1 - n(N+1)/2}{\sqrt{nm(N+1)/12}} = \frac{2R_1 - n(N+1)}{\sqrt{nm(N+1)/3}} \tag{11.6}$$

is asymptotically a unit normal random variable. A good approximation is obtained when $n > 8$.

Example 11.2 Two manufacturers A and B supply compression springs used in a specific mechanism. It is essential that the distribution of the compressed length at a certain load be the same for the two manufacturers. The distributions of compressed length are unknown, although it is believed that these distributions differ in their means only. The following observations (compressed length in inches) were taken:

Manufacturer A	*Rank*	*Manufacturer B*	*Rank*
12.05	4	11.95	3
12.61	8	11.20	2
12.83	11	12.60	7
12.75	9	13.18	16
12.90	12	12.96	14
12.92	13	12.53	5
12.58	6	13.60	17
13.01	15	12.78	10
		11.03	1

Determine whether springs supplied by the two manufacturers are significantly different. Use $\alpha = .05$.

Solution. Given $n = 8$, $m = 9$, and $N = 17$ we have

$$R_1 = 4 + 8 + 11 + 9 + 12 + 13 + 6 + 15 = 78$$
$$R_2 = 3 + 2 + 7 + 16 + 14 + 5 + 17 + 10 + 1 = 75$$

Hence, the observed value of

$$\frac{2R_1 - n(N+1)}{\sqrt{nm(N+1)/3}} = \frac{156 - 144}{\sqrt{432}} = .577$$

Since .577 lies between $Z_{.025}(-1.96)$ and $Z_{.975}(1.96)$ we accept the null hypothesis that the two lots do not differ significantly in their means.

Consulting Table 11.3 we find the critical value is 51. Since neither R_1 nor R_2 is less than or equal to 51 we accept the null hypothesis. Let us derive the critical value using the normal approximation and compare it with the value given by Table 11.3. We calculate

$$E(R_1) = \frac{n(N+1)}{2} = \frac{8 \times 18}{2} = 72$$

$$\sigma_{R_1} = \sqrt{\frac{nm(N+1)}{12}} = \sqrt{\frac{8 \times 9 \times 18}{12}} = 10.39$$

Hence the critical values for a two-sided test are

$$72 \pm (10.39)(1.96) = [51.64, 92.36]$$

Thus we find that the lower critical value is almost the same as given by Table 11.3. (Note that Table 11.3 specifies the lower critical values only.)

11.3.1 Adjustments for Ties in the Wilcoxon Two-Sample Rank Test

If *ties* occur, average ranks are assigned. A simple example will illustrate the procedure:

Example 11.3 An experiment was run to test whether treatments A and B differ in their means. The following results are recorded:

Treatment A	*Rank*	*Treatment B*	*Rank*
86.5	14	84.1	8
84.2	10	84.1	8
83.8	3.5	83.8	3.5
85.9	12	84.6	11
86.1	13	83.1	2
87.3	15	84.0	6
84.1	8	83.9	5
		83.0	1

Test whether treatments differ at the 5% level of significance.

In this example we have two groups of ties: group one (83.8, 83.8) has two tied observations, and group two (84.1, 84.1, 84.1) has three tied observations. It can be proved that when ties occur $E(R_1)$ remains unchanged whereas $\text{Var}(R_1)$ will decrease by

$$\sum j\frac{t(t^2-1)}{12} \tag{11.7}$$

where j = number of *tied groups* with the same number of *tied ranks* and t = number of *ties* in groups with the same number of *tied ranks*. Hence,

$$\frac{R_1 - n(N+1)/2}{\sqrt{[nm(N+1)/12] - [\sum jt(t^2-1)/12]}} \tag{11.8}$$

is an asymptotically normally distributed variate with mean 0 and variance 1.

Now we proceed to solve Example 11.3.

Solution. Given $n = 7$, $m = 8$, and $N = 15$ we have

$$R_1 = 14 + 10 + 3.5 + 12 + 13 + 15 + 8 = 75.5$$

We have one group with two ties ($j = 1$, $t = 2$) and one group with three ties ($j = 1$, $t = 3$). Hence,

$$\frac{\sum jt(t^2-1)}{12} = \frac{1}{12}[2(4-1) + 3(9-1)] = 2.5$$

and the observed value of

$$\frac{R_1 - n(N+1)/2}{\sqrt{(1/12)[nm(N+1) - \sum jt(t^2-1)]}} = \frac{75.5 - 56}{\sqrt{(1/12)(896-30)}} = 2.29$$

Since 2.29 does not lie between $Z_{.025}(-1.96)$ and $Z_{.975}(1.96)$ we reject the null hypothesis. We conclude that treatments differ in their means at the 5% level of significance.

11.4 The *H* Test

The H test was developed by Kruskal and Wallis [13] in 1952. It is a generalization of the Wilcoxon two-sample rank test. In this case we have k populations ($k > 2$); from each a sample of size $n_i (i = 1, 2, \ldots, k)$ is taken.

The entire set of observations N is ranked, and the sum of ranks of each population (R_i) is calculated. Kruskal and Wallis proved that the statistic H has approximately a χ^2 distribution with $k - 1$ degrees of freedom. In symbols,

$$H = \frac{12}{N(N+1)} \sum_{i=1}^{k} \left(\frac{R_i^2}{n_i}\right) - 3(N+1) \sim \chi^2_{k-1} \tag{11.9}$$

where $N = \sum_{i=1}^{k} n_i$ and R_i = Sum of ranks of the ith population. The null hypothesis that k samples of sizes $n_1, n_2, \ldots, n_k$ are randomly drawn from identically distributed populations is rejected at the α level of significance when

$$H > \chi^2_{1-\alpha;\, k-1}$$

It has been proved that H is distributed asymptotically as χ^2_{k-1} for large k and n_i. However, when $k = 3$ and $n_i \leq 5$, the approximation is not very satisfactory. The authors provided tables for the exact distribution when $k = 3$ and $n_i \leq 5$.

If ties occur the value of statistic H is given by

$$\frac{[12/N(N+1)] \sum_{i=1}^{k} (R_i^2/n_i) - 3(N+1)}{1 - [\sum jt(t^2 - 1)/(N^3 - N)]} \sim \chi^2_{k-1} \tag{11.10}$$

where j = number of *tied groups* with same number of *tied ranks* and t = number of *ties* in groups with same number of *tied ranks*. When $k = 2$ it can be proved that statistic H is identical with the Wilcoxon two-sample rank test. The Wilcoxon test is less powerful than the t test based on the assumption of normality. For nonnormal populations the Wilcoxon test may be more powerful than the t test. The Wilcoxon test is superior to the paired-sign test and compares favorably with Student's t test. It can be shown that when the variates are normal the Wilcoxon test based on 100 observations compares favorably with the t test based on 95 observations.

Example 11.4 A factory has six machines for producing thread gauges. It is essential that the distribution of the pitch diameter be the same for the gauges produced on the six machines. Though the distributions of the pitch diameters are unknown it is believed that these distributions differ in their means only. An experiment was run, and the following results were obtained:

	Machine 1 Observations	Rank	*Machine 2* Observations	Rank
	1.006	$19\frac{1}{2}$	1.003	$7\frac{1}{2}$
	1.005	$15\frac{1}{2}$	1.003	$7\frac{1}{2}$
	1.006	$19\frac{1}{2}$	1.004	$11\frac{1}{2}$
	1.007	$23\frac{1}{2}$	1.006	$19\frac{1}{2}$
	1.004	$11\frac{1}{2}$	1.007	$23\frac{1}{2}$
	1.008	28		
n_i	6		5	
R_i		$117\frac{1}{2}$		$69\frac{1}{2}$

	Machine 3 Observations	Rank	*Machine 4* Observations	Rank
	1.001	1	1.008	28
	1.002	$3\frac{1}{2}$	1.008	28
	1.002	$3\frac{1}{2}$	1.009	$31\frac{1}{2}$
	1.002	$3\frac{1}{2}$	1.111	34
	1.002	$3\frac{1}{2}$	1.007	$23\frac{1}{2}$
	1.004	$11\frac{1}{2}$	1.006	$19\frac{1}{2}$
n_i	6		6	
R_i		$26\frac{1}{2}$		$164\frac{1}{2}$

	Machine 5 Observations	Rank	*Machine 6* Observations	Rank
	1.004	$11\frac{1}{2}$	1.110	33
	1.003	$7\frac{1}{2}$	1.009	$31\frac{1}{2}$
	1.003	$7\frac{1}{2}$	1.008	28
	1.005	$15\frac{1}{2}$	1.007	$23\frac{1}{2}$
	1.005	$15\frac{1}{2}$	1.008	28
	1.005	$15\frac{1}{2}$		
n_i	6		5	
R_i		73		144

Determine whether these measurements are drawn from identically distributed populations. Use $\alpha = .05$.

Solution

$$H = \frac{12}{N(N+1)} \sum_{i=1}^{6} \frac{R_i^2}{n_i} - 3(N+1)$$

On substituting we obtain

$$H = \frac{12}{34(34+1)} \left[\frac{(117\frac{1}{2})^2}{6} + \frac{(69\frac{1}{2})^2}{5} + \frac{(26\frac{1}{2})^2}{6} + \frac{(164\frac{1}{2})^2}{6} + \frac{(73)^2}{6} + \frac{(144)^2}{5} \right] - 3(34+1) = 25.38$$

However, we need correction for ties; we have six groups of four tied ranks, one group of five tied ranks, and one group of two tied ranks. Hence we have the correction factor

$$1 - \frac{\sum jt(t^2 - 1)}{N^3 - N} = 1 - \frac{6 \times 4(16 - 1) + 5(25 - 1) + 2(4 - 1)}{34^3 - 34}$$
$$= .9877$$

The corrected H is $25.38/.9877 = 25.69$ with $6 - 1 = 5$ degrees of freedom. $\chi^2_{.95;\,5} = 11.07$, so the null hypothesis is rejected at the 5% level of significance. We conclude that the distributions of the pitch diameters are significantly different.

11.5 Run Tests

In the previous sections on the sign test and the Wilcoxon two-sample rank test the time order of the observations was ignored on the assumption that these observations constitute a *random sample* from a fixed population. However, it seems logical to test whether this assumption of *randomness* in sequences of observations is a valid one. To test the validity of this assumption on the sign test it is reasonable to examine whether the sequences of signs exhibit a homogeneous scattering of the positive signs among the negative ones. Now the question may be raised as to how we can ascertain lack of homogeneity (nonrandomness) in a series of time-ordered observations. To answer this question let us consider a simple example.

Suppose that the sequence of signs in a series of observations is given by

$$+ + - - - + + - - - + + + - -$$

namely, two positive signs followed by three negative signs, etc. We say *the above sequence has a run of length two followed by a run of length three*, etc. The total number of runs is six. *Thus a run of length j is defined as a sequence of j identical signs that is preceded or followed by a different sign or no sign.* In the cited example the sequence of 15 differences have 6 runs. This sequence may have only one run of length 15 should all the signs be positive or negative and may have 15 runs of length 1 if the signs alternate. Too few or too many runs can serve as evidence for rejecting the hypothesis of random arrangement of plus and minus signs. In other words, the number of runs can be used to derive tests for randomness. It is also possible to use runs in testing whether two random samples come from identically distributed populations. These tests will be discussed in the following sections.

11.5.1 Runs Above and Below the Median

The theory of runs can be used to test whether observations drawn from a single population exhibit *randomness*. A simple example will illustrate this procedure.

Example 11.5 Suppose a manufacturing process is turning out gudgeon pins. We wish to test whether the diameters of these pins vary randomly. Suppose further that the diameters of 20 pins were measured in order of production (consecutive rows) and yielded the following results:

.7241(6)	.7262(17)	.7253(11)	.7259(16)	.7235(2)
.7246(8)	.7243(7)	.7255(13)	.7248(9)	.7258(15)
.7256(14)	.7265(18)	.7268(19)	.7239(5)	.7249(10)
.7254(12)	.7230(1)	.7237(3)	.7269(20)	.7238(4)

The parentheses refer to the ordering in magnitude of the samples.

The median of these measurements is .7251. Writing (+) if a sample diameter is greater than the median and (−) if it is less than the median, we obtain the sequence

$$- + + + - - - + - + + + + - - + - - + -$$

The following notation will be used: u = the total number of runs both above and below the median, n_1 = the number of positive signs, and n_2 = the number of minus signs. Since the number of samples greater than the median is equal to the number of samples less than the median, $n_1 = n_2 = n$. If the number of observations is odd, we ignore the observation which coincides with the median.

It can be shown that the probability function $P(u)$ of u, assuming that these $(2n)$ measurements constitute a random sample from a continuous distribution, is given by

(11.11)
$$P(u) = 2\frac{\binom{n-1}{u/2-1}^2}{\binom{2n}{n}} \qquad u = 2, 4, \ldots, 2n$$

$$P(u) = 2\frac{\binom{n-1}{u/2-1/2}\binom{n-1}{u/2-3/2}}{\binom{2n}{n}} \qquad u = 3, 5, \ldots, 2n-1$$

The mean and variance of u are given by

(11.12) $$E(u) = n + 1$$

(11.13) $$\text{Var}(u) = \sigma_u^2 = n(n-1)/(2n-1)$$

When the value of n is larger than ten, u has approximately a normal distribution with mean $= n + 1$ and standard deviation $= \sqrt{n(n-1)/(2n-1)}$. This point will be taken up in solving Example 11.5.

Now we proceed to solve Example 11.5. We compute

$$E(u) = n + 1 = 11$$

$$\sigma_u^2 = \frac{n(n-1)}{2n-1} = \frac{10 \times 9}{19} = 4.74$$

whence $\sigma_u = 2.17$. The hypothesis of randomness is rejected if u is either too small or too large, or equivalently, if

$$|u - E(u)| \geq \sigma_u Z_{1-\alpha/2}$$

On substituting we obtain

$$0 < (2.17)(1.96) = 4.25$$

so that we do not reject the hypothesis of randomness at the 5% level of significance. We conclude that we ought not act as though the diameters vary nonrandomly.

11.5.2 The Wald-Wolfowitz Run Test

Run theory can be used to *test whether two random samples come from continuous and identically distributed populations.* Suppose we have two continuous populations denoted by X and Y, respectively. A sample of size n_1 is drawn from population X, and a sample of size n_2 is drawn from population Y. Now on the basis of these random samples we wish to test the hypothesis

$$H_0 : F_1(X) \equiv F_2(Y) \quad \text{against} \quad H_1 : F_1(X) \not\equiv F_2(Y)$$

where $F(\cdot)$ denotes the cumulative distribution function.

The following example will illustrate this procedure.

Example 11.6 In an experiment on dirt removal two detergents X and Y are used. The experiment was run under identical environmental conditions (temperature, time, etc.), and the amount of dirt removed was measured in grams. The results of the experiment are recorded in the following tabulation:

Detergent X	*Detergent Y*
67.3	80.1
78.2	80.7
75.1	79.8
79.2	82.7
80.3	74.6
77.6	78.8
69.8	82.1
74.7	81.9
81.1	72.8
76.9	79.6
76.1	83.0
75.7	80.9
	76.4
	78.0

Suppose we wish to test by the theory of runs the hypothesis that the two samples come from populations having identical cumulative distribution functions ($\alpha = .05$). If we combine the two samples into one pooled sample and rearrange the individual values sampled in order of magnitude, we have

67.3*	69.8*	72.8	74.6	74.7*
75.1*	75.7*	76.1*	76.4	76.9*
77.6*	78.0	78.2*	78.8	79.2*
79.6	79.8	80.1	80.3*	80.7
80.9	81.1*	81.9	82.1	82.7
83.0				

where the asterisk denotes observations from population X.

Now if we denote an X observation by a positive sign (+) and a Y observation by a negative sign (−) we obtain the following sequence:

$$++--+++++-++-+-+---+---+----$$

Here we find that $n_1 = 12$, $n_2 = 14$, and $u = 14$.

It can be proved that the probability function of u is given by

(11.14)

$$P(u) = 2\frac{\binom{n_1 - 1}{u/2 - 1}\binom{n_2 - 1}{u/2 - 1}}{\binom{n_1 + n_2}{n_1}} \qquad u \text{ even}$$

$$P(u) = \frac{\binom{n_1 - 1}{u/2 - 1/2}\binom{n_2 - 1}{u/2 - 3/2} + \binom{n_1 - 1}{u/2 - 3/2}\binom{n_2 - 1}{u/2 - 1/2}}{\binom{n_1 + n_2}{n_1}} \qquad u \text{ odd}$$

The mean and variance of u are as follows:

(11.15) $$E(u) = \frac{2n_1n_2}{n_1 + n_2} + 1$$

(11.16) $$\text{Var}(u) = \sigma_u^2 = \frac{2n_1n_2(2n_1n_2 - n_1 - n_2)}{(n_1 + n_2)^2(n_1 + n_2 - 1)}$$

If $n_1 = n_2 = n$ the above formulas reduce to those given in Sec. 11.5.1. Furthermore, if n_1 and n_2 are both larger than ten, the cumulative sampling distribution of u can be approximated by the normal distribution. Probabilities associated with the number of runs (u) have been tabulated by Swed and Eisenhart. Table 11.5 (reproduced from Swed and Eisenhart) shows the significance limits for a two-sided test ($\alpha = .05$).

Solution. Consulting Table 11.5 (for $n_1 = 12$, $n_2 = 14$) we find the critical values for a two-sided test ($\alpha = .05$) to be 8 and 19. Since the observed $u = 14$ falls between 8 and 19 the result of the run test just used asserts that the evidence does not give a basis for denying the hypothesis that dirt removal by these two detergents is equivalent at the 5% level of significance. Using the normal approximation we find that

$$E(u) = \frac{2n_1n_2}{n_1 + n_2} + 1$$

$$= \frac{2 \times 12 \times 14}{26} + 1 = 13.92$$

$$\sigma_u^2 = \frac{2n_1n_2(2n_1n_2 - n_1 - n_2)}{(n_1 + n_2)^2(n_1 + n_2 - 1)}$$

$$= \frac{(2 \times 12 \times 14)(2 \times 12 \times 14 - 12 - 14)}{(26)^2(25)} = 6.16$$

so that

$$\sigma_u = 2.48$$

Here we reject the null hypothesis at the 5% level of significance if

$$|u - 13.92| \geq Z_{.975}(2.48)$$

But since

$$.08 < (1.96)(2.48) = 4.86$$

we conclude that the two samples could come from populations having identical cumulative distributions.

It is interesting to note that Table 11.5 can also be used to test whether observations drawn from a single population exhibit randomness. To illustrate, consider again Example 11.5, where $n_1 = n_2 = n = 10$. From Table 11.5, we find that for $n_1 = n_2 = 10$, the critical values at the 5% level are 6 and 15. Since the observed number $u = 11$ falls between 6 and 15 there is not sufficient evidence to reject the hypothesis of randomness at the 5% level of significance.

TABLE 11.5 Significance Limits for Tests Based on Runs†

	$u_{0.025}$																		
n_2 \ n_1	2	3	4	5	6	7	8	9	10	11	12	13	14	15	16	17	18	19	20
2																			
3																			
4																			
5			2	2															
6		2	2	3	3														
7		2	2	3	3	3													
8		2	3	3	3	4	4												
9		2	3	3	4	4	5	5											
10		2	3	3	4	5	5	5	6										
11		2	3	4	4	5	5	6	6	7									
12	2	2	3	4	4	5	6	6	7	7	7								
13	2	2	3	4	5	5	6	6	7	7	8	8							
14	2	2	3	4	5	5	6	7	7	8	8	9	9						
15	2	3	3	4	5	6	6	7	7	8	8	9	9	10					
16	2	3	4	4	5	6	6	7	8	8	9	9	10	10	11				
17	2	3	4	4	5	6	7	7	8	9	9	10	10	11	11	11			
18	2	3	4	5	5	6	7	8	8	9	9	10	10	11	11	12	12		
19	2	3	4	5	6	6	7	8	8	9	10	10	11	11	12	12	13	13	
20	2	3	4	5	6	6	7	8	9	9	10	10	11	12	12	13	13	13	14

***TABLE 11.5**—Cont.*

$n_2 \backslash n_1$	2	3	4	5	6	7	8	9	10	11	12	13	14	15	16	17	18	19	20
									$u_{0.975}$										
1																			
2	4																		
3	5	6																	
4	5	7	8																
5	5	7	8	9															
6	5	7	8	9	10														
7	5	7	9	10	11	12													
8	5	7	9	10	11	12	13												
9	5	7	9	11	12	13	13	14											
10	5	7	9	11	12	13	14	15	15										
11	5	7	9	11	12	13	14	15	16	16									
12	5	7	9	11	12	13	15	15	16	17	18								
13	5	7	9	11	13	14	15	16	17	18	18	19							
14	5	7	9	11	13	14	15	16	17	18	19	19	20						
15	5	7	9	11	13	14	15	17	17	18	19	20	21	21					
16	5	7	9	11	13	15	16	17	18	19	20	20	21	22	23				
17	5	7	9	11	13	15	16	17	18	19	20	21	22	22	23	24			
18	5	7	9	11	13	15	16	17	18	19	20	21	22	23	24	24	25		
19	5	7	9	11	13	15	16	17	19	20	21	22	23	23	24	25	25	26	
20	5	7	9	11	13	15	16	17	19	20	21	22	23	24	24	25	26	26	27

†Reproduced with permission from Swed and Eisenhart, "Tables for Testing Randomness of Grouping in a Sequence of Alternatives," *Annals of Mathematical Statistics*, **14**, 66, 1943.

The tests for randomness based on run theory have some inherent weaknesses. The run tests are not very sensitive; however, the effectiveness of run tests is discovered only when lack of nonrandomness produces too few or too many runs. When the total number of runs does not differ significantly from the expected number of runs of a random sequence the run test fails to detect some types of nonrandomness. We wish to emphasize that the theory of runs sometimes fails to detect nonrandomness when observations are drawn from a population having a cyclical nature. The reason for this failure may be attributed to the inadequate spacing of the observations. In this instance tests based on correlation prove to be more effective than run tests. In the next section we shall describe one of these techniques.

11.6 Bivariate Techniques

In this section we will discuss how *bivariate* data may be analyzed by *nonparametric methods*. The derivation of these techniques is rather complicated; consequently, only the results will be given here. Although there are many bivariate methods, such as rank correlation, coefficient of concordance, Kendall's t, . . . , etc., only the *Rank Correlation* technique will be presented here. The interested reader may consult advanced works in statistics for exposition of these techniques.

11.6.1 Rank Correlation

Suppose we have a sample of size n and that two measurements x and y for the ith individual are made. The distributions of x and y are unknown. We are interested in testing the hypothesis that x and y are independent. The following example will illustrate the meaning of two measurements:

Example 11.7 The relationship between the weight (x) of crankshaft bearings before lining with white metal and the weight (y) after lining was a subject of a recent experiment. A sample of ten bearings was chosen at random, and each bearing was weighed before and after lining. The following table represents the results of the experiment. We want to know whether x and y are independent at the 5% level of significance:

Weight before lining in grams (x):

120	125	128	115	131	137	127	133	110	118

Weight after lining in grams (y):

153	151	162	150	158	168	156	170	141	140

Now assigning rank 1 to the largest value of x, 2 to the second largest, etc., and, similarly, ranking each of ten values of y, we have

Rank $x = x'$:	7	6	4	9	3	1	5	2	10	8
Rank $y = y'$:	6	7	3	8	4	2	5	1	9	10
$d = x' - y'$:	1	-1	1	1	-1	-1	0	1	1	-2

The statistic

$$r_s = 1 - \frac{6 \sum d_i^2}{n(n^2 - 1)} \tag{11.17}$$

is called the *rank correlation coefficient*. It can be used to test the hypothesis that x and y are independent. The exact distribution of $\sum d_i^2$ has been studied by E. G. Olds [16], and [17], and critical values when there is complete independence have been found.

The results were published in *Annals of Mathematical Statistics* [**9** (1938) and **20** (1949)]. These critical values also can be found in terms of r_s as shown in Table 11.6. A value of $+1$ for r_s indicates complete agreement; a value of -1 indicates complete disagreement; and a value of 0 indicates complete independence between x and y.

For a double-sided test the hypothesis of independence is rejected if the absolute value of r_s as given by Eq. (11.17) is found to be greater than or equal to the critical value given by Table 11.6 at the desired level of significance.

TABLE 11.6 Critical Values of the Rank Correlation Coefficient† (Double-Sided Test)

Sample Size (n)	$\alpha = .05$	$\alpha = .01$
8	.738	.881
9	.683	.833
10	.648	.794
11	.623	.818
12	.591	.780
13	.566	.745
14	.545	.716
15	.525	.689
16	.507	.666
17	.490	.645
18	.476	.625
19	.462	.608
20	.450	.591

†Compiled from the distribution of $\sum d_i^2$. See E. G. Olds [16] and [17].

Continuing Example 11.6, we compute

$$r_s = 1 - \frac{6 \times 12}{10 \times 99} = .928 > .648$$

which is highly significant. Consequently, we reject the hypothesis of independence at the 5% level of significance.

PROBLEMS

11.1. Two treatments are to be compared at the 5% level of significance. The distributions of both treatments are unknown. The following observations are taken in pairs:

Treatment X	*Treatment Y*
19.05	21.62
18.01	18.26
23.94	24.76
28.24	29.26
31.51	32.83
27.61	27.01
27.83	27.63
26.11	27.45
41.82	38.96

Use the sign test to determine whether the distributions of both treatments differ.

11.2. It has been suspected that the average resistance of wire A is greater than the average resistance of wire B. The distributions are unknown; however, it can be assumed that they differ in their means. Tests made on samples of each wire yielded the following results:

Wire A (ohm)	*Wire B (ohm)*
.124	.129
.125	.122
.126	.123
.124	.124
.127	.122
.121	.126
.128	.125
.129	.124

Use the Wilcoxon two-sample rank test to determine whether the resistance of wire A is greater than the resistance of wire B at the 5% level of significance.

11.3. The data below show measurements of the corrosion effects in various soils for steel pipe with two kinds of coating:

Coating X:	72	63	53	47	38	56	55	75	68	57	61	52	67
Coating Y:	69	69	58	52	33	57	54	66	72	71	69	53	60

Use the sign test to test the hypothesis at the 5% level of significance that the kind of coating used has no effect on corrosion.

11.4. Nicotine determinations were made by two methods, say A and B, on a standard batch of tobacco. A batch of tobacco was sent to 30 laboratories, in each of which both methods (A and B) were used to obtain the percentage of nicotine. The data reported by the 30 laboratories are as follows:

Laboratory	*Percent Nicotine*		*Laboratory*	*Percent Nicotine*	
	A	*B*		*A*	*B*
1	.028	.026	16	.023	.023
2	.022	.024	17	.027	.025
3	.027	.023	18	.032	.030
4	.024	.026	19	.031	.029
5	.030	.030	20	.030	.021
6	.029	.027	21	.031	.021
7	.029	.028	22	.026	.028
8	.027	.026	23	.025	.026
9	.026	.029	24	.028	.027
10	.026	.024	25	.028	.022
11	.024	.029	26	.028	.030
12	.021	.020	27	.027	.031
13	.030	.031	28	.026	.029
14	.028	.030	29	.026	.027
15	.021	.027	30	.025	.024

Using the sign test and ignoring ties, test whether methods differ at the 5% level of significance.

11.5. Two analysts took repeated readings on the viscosity of a certain lubricant. Assume the distributions are unknown but differ in their means. Determine whether one analyst has a tendency to read higher than the other ($\alpha = .05$).

Viscosity Readings SUV

Analyst A	*Analyst B*
80	82
76	73
92	91
86	84
74	77
96	98
83	81
79	

11.6. In Prob. 11.4 determine the power of the test when either method yields higher results than the other in 75% of the cases ($\alpha = .006$).

11.7. When the sign test is used to compare two populations, say X and Y, how would the total number of sign changes be influenced by each of the following situations?
(a) The populations differ in their means.
(b) The populations differ in their variances.

11.8. The following two sets of observations were obtained from sampling two populations:

Population I:	12.6	12.8	15.7	13.1	14.2	15.9	15.2	13.3	13.8	14.2
Population II:	12.5	12.0	16.3	11.8	12.9	15.2	15.0	13.6		

Test the hypothesis that the two populations possess the same frequency function at the 1% level of significance.

11.9. Two types of propellants were tested on a certain type of small arms ammunition. The barrel pressure (psi) has been coded. The data are given below:

Propellant A	*Propellant B*
1250	1210
1060	980
890	930
725	725
1010	680
865	780
690	1100
930	945
1300	735
780	1030

Use the Wilcoxon two-sample rank test to test whether the propellants are different ($\alpha = .05$).

11.10. A sequence of 100 measurements on a process supposedly under statistical control was analyzed for runs above and below the median. The total number of runs above and below the median was found to be 41. Test whether this sequence exhibits significant nonrandomness at the 5% level of significance.

11.11. Given the sequence of numbers 1, 1, 2, 1, 2, 2, 2, 1, 2, 1, 1, 1, 2, 2, 1, 2, 2, 2, 1, 1, 2, 2, 2, 1, 1, 1, 1, 2, 1, 1, 2, 2, test for randomness at the 5% level of significance.

11.12. The following are means of 22 subgroups, the data pertain to the outside diameter of steel bushing; and the measurements are in inches:

Subgroup Number	$\bar{X}$	*Subgroup Number*	$\bar{X}$
1	1.2275	12	1.2100
2	1.2525	13	1.2375
3	1.2225	14	1.2750
4	1.2675	15	1.2250
5	1.2600	16	1.2700
6	1.2356	17	1.2300
7	1.2625	18	1.2325
8	1.2400	19	1.2425
9	1.2528	20	1.2350
10	1.2775	21	1.2850
11	1.2725	22	1.2410

All sample averages are found to fall within the control limits. By using runs above and below the median, test the hypothesis at the 5% level of significance that the variation in this sequence exhibits randomness.

11.13. The following data represent the number of defective transistors for samples of 200 each from daily production:

Day	*Number Defective*	*Day*	*Number Defective*
1	6	11	10
2	7	12	2
3	11	13	9
4	3	14	6
5	8	15	8
6	4	16	4
7	10	17	8
8	7	18	0
9	5	19	10
10	8	20	0

Test for randomness at the 5% level of significance.

11.14. The following two samples of measurements were obtained from sampling two populations X and Y:

Population X: 31 19 22 26 15 18 36 30 29 34 33 26 19 19 26 28 31 30
Population Y: 35 14 19 30 8 14 34 28 23 24 27 28 20 21 21 26

Use the Wald–Wolfowitz test to test the hypothesis at the 5% level of significance that the two populations have identical distribution functions.

11.15. The following measurements represent the thickness of steel washers produced by two presses:

Press I: .117 .135 .126 .139 .132 .128 .120 .123 .118 .126 .125 .129
Press II: .130 .127 .133 .115 .137 .127 .130 .131 .131 .127 .121 .122

Using the Wald–Wolfowitz test, test whether the two samples are from populations having identical density functions ($\alpha = .05$).

11.16. A factory has three machines for producing thread gauges. It is essential that the distribution of the pitch diameter be the same for the gauges produced on the three machines. An experiment was run and the following results were obtained:

Machine 1	*Machine 2*	*Machine 3*
.756	.759	.747
.759	.742	.749
.761	.751	.752
.763	.753	.756
.758	.754	.755
.757	.746	.763
.760	.748	.764
.762	.749	.749
.750		

Test at the 5% level of significance that these measurements are drawn from identically distributed populations.

11.17. The following data represent the length of life in hours of an electronic device supplied by three manufacturers.

Manufacturer A	*Manufacturer B*	*Manufacturer C*
1294	1293	1074
1154	1218	1135
1107	1092	1017
1290	1159	1203
1136	1063	1105
1197	1138	1023
1206	1291	1298
1230	1024	1149
1238	1292	1042
1194	1147	1129

Test whether the distribution of length of life is the same for all manufacturers ($\alpha = .05$).

11.18. Two instructors ranked the final examination of ten students as follows:

Instructor I:	9	7	6	3	10	1	2	4	5	8
Instructor II:	8	4	7	2	9	3	1	6	5	10

Test for independence in the rankings at the 5% level of significance.

11.19. State the difference between a conventional correlation technique and the rank correlation technique.

11.20. The following data represent the weight of diesel engine pistons in pounds before and after machining (denoted by X and Y, respectively).

X:	2.710	2.700	2.730	2.718	2.698	2.691	2.716
Y:	2.538	2.520	2.541	2.536	2.518	2.510	2.526
X:	2.690	2.721	2.725	2.699	2.712	2.723	2.695
Y:	2.501	2.513	2.529	2.517	2.533	2.540	2.511

Test whether X and Y are independent at the 5% level of significance.

SUGGESTED REFERENCES

[1] Box, G. E. P., and S. L. Anderson, "Permutation Theory in the Derivation of Robust Criteria and the Study of Departure from Assumptions," *Journal of the Royal Statistical Society, B (Methodological)*, **17**, 1–26, 1955.

[2] Bradley, T. V., *Distribution-Free Statistical Tests*, Wright-Patterson Air Force Base (WADC Technical Report 60-661), 1960.

[3] Cochran, W. G., "The Comparison of Percentages in Matched Samples," *Biometrika*, **37**, 256–266, 1950.

[4] Cramér, H., *Mathematical Methods of Statistics*, Princeton, Princeton University Press, 1951.

[5] Dean, G., "The Atomic Weight of Carbon and Silver," *Journal of the Chemical Society of London*, **125**, 2656–2672, 1924.

[6] Fisz, M., *Probability Theory and Mathematical Statistics*, New York, John Wiley & Sons, Inc., 1963.

[7] Fraser, D. A. S., *Nonparametric Methods in Statistics*, New York, John Wiley & Sons, Inc., 1957.

[8] Friedman, M., "The Use of Ranks to Avoid the Assumption of Normality in the Analysis of Variance," *Journal of the American Statistical Association*, **32**, 675–701, 1937.

[9] Kendall, M. G., *Rank Correlation Methods*, 2nd ed., London, Charles Griffin, 1955.

[10] Kendall, M. G., and A. Stuart, *The Advanced Theory of Statistics*, Vol. 2, Chap. 31, London, Charles Griffin, 1961.

[11] Kruskal, W. H., "Historical Notes on the Wilcoxon Unpaired Two-Sample Test," *Journal of the American Statistical Association*, **52**, 1957.

[12] Kruskal, W. H., "Ordinal Measures of Association," *Journal of the American Statistical Association*, **53**, 1958.

[13] Kruskal, W. H., and W. A. Wallis, "Use of Ranks in One-Criterion Analysis of Variance," *Journal of the American Statistical Association*, **47**, 583–621, 1952.

[14] Levene, H., "Robust Tests for Equality of Variances," *Contribution to Probability and Statistics: Essays in Honor of H. Hotelling*, ed. I. Olkin, Stanford, Stanford University Press, 1960.

[15] Mann, H. B., and D. R. Whitney, "On a Test of Whether One of Two Random Variables Is Stochastically Larger than the Other," *Annals of Mathematical Statistics*, **18**, 50–60, 1947.

[16] Olds, E. G., "Distribution of the Sums of Squares of Rank Differences for Small Numbers of Individuals," *Annals of Mathematical Statistics*, **9**, 1938.

[17] Olds, E. G., "The 5 Per Cent Significance Levels for Sums of Squares of Rank Differences and a Correction," *Annals of Mathematical Statistics*, **20**, 1949.

[18] Putter, J., "The Treatment of Ties in Some Nonparametric Tests," *Annals of Mathematical Statistics*, **26**, 368–386, 1955.

[19] Quenouille, M. H., *Rapid Statistical Calculations*, New York: Hafner Publishing Co., Inc., 1960.

[20] Siegel, S., *Nonparametric Statistics for the Behavioral Science*, New York, McGraw-Hill Book Company, 1956.

[21] Somerville, P. N., "Tables for Obtaining Nonparametric Tolerance Limits," *Annals of Mathematical Statistics*, **29**, 1956.

[22] Wallace, D. L., "Simplified Beta-Approximations to Kruskal–Wallis H Test," *Journal of the American Statistical Association*, **54**, 225–230, 1959.

[23] Walsh, J. E., *Handbook of Nonparametric Statistics*, Princeton, N.J., D. Van Nostrand Co., Inc., 1962.

[24] Wilcoxon, F., *Some Rapid Approximation Procedures*, Stamford, Conn., American Cynamid Company, 1949.

[25] Wilcoxon, F., "Individual Comparisons by Ranking Methods," *Biometrics Bulletin*, **1**, 80–83, 1947.

[26] Wilcoxon, F., "Probability Tables for Individual Comparisons by Ranking Methods," *Biometrics*, **3**, 119–122, 1947.

(1) $\sum x_i = 119$, $\sum x_i^2 = 1351$ $n = 15$ $H_0: \sigma^2 = 36$ $H_1: \sigma^2 < 36$ $\alpha = .05$

$s^2 = \frac{\sum x_i^2 - \frac{1}{n}[\sum(x_i)]^2}{n-1}$ or $\frac{\sum x_i^2 - \bar{x}\sum x_i}{n-1}$

$\frac{(n-1)s^2}{\sigma^2} < \chi^2_{\alpha; n-1}$ or $\frac{(n-1)s^2}{\sigma^2} > \chi^2_{1-\alpha; n-1}$

(2) sample $\bar{x} = 714$ $s^2 = 60.5$ $H_0: \mu = \mu_0$ $H_1: \mu \neq \mu_0$ $n = 9$
$\mu = 717$ $\alpha = .05$

$|\bar{x} - \mu_0| > t_{1-\alpha/2; n-1} \, s/\sqrt{n}$

(3) $H_0: \mu_x - \mu_y = 0$ $H_1: \mu_x - \mu_y \neq 0$ $\alpha = .01$

$|\bar{x} - \bar{y}| > z_{1-\alpha/2}\sqrt{\frac{\sigma_x^2}{n} + \frac{\sigma_y^2}{m}}$

x: 2.6 3.1 . . . common $\sigma_x = \sigma_y = .64$ $n = m = 5$
y: 2.8 3.2 . . .

(4) | | A | B |
|---|---|---|
| size sam | 500 | 700 |
| # smokers | 225 | 245 |

$H_0: p_A - p_B = 0$ $H_1: p_A - p_B > 0$ $\alpha = .0$

$\left(\frac{x_1}{n_1} - \frac{x_2}{n_2}\right) / \left[\frac{x_1+x_2}{n_1 n_2}\left(1 - \frac{x_1+x_2}{n_1+n_2}\right)\right]^{1/2} > z_{1-\alpha}$

$\left(\frac{225}{500} - \frac{245}{700}\right) / \ldots > z_{.95}$

(5) prob. density func. $f(x) = a^2 x e^{-ax}$ $x \geq 0$

max Likelihood $L = a^{2n} \prod x_i \, e^{-a\sum x_i}$

$\ln L = 2n \ln a + a\sum x_i$

$\frac{\partial \ln L}{\partial a} = \frac{2n}{a} - \sum x_i = 0$

$a = \frac{2}{\bar{x}}$

(6) graph $\frac{x}{n} = \frac{30}{50} = .6$ on lines 50 [.47, .72]
Find 90% confid. ; catch of 50, 20 defect
for nondefect

(7) 2 sample results $s_x^2 = 142.5$ $s_y^2 = 114$
$\bar{x} = 320.27$ $\bar{y} = 306.55$ $H_0: \sigma_x^2 = \sigma_y^2$ $H_1: \sigma_x^2 \neq \sigma_y^2$

$\frac{s_x^2}{s_y^2} > F_{1-\alpha/2; n-1, m-1}$ $\frac{s_x^2}{s_y^2} < F_{\alpha/2; n-1, m-1}$ 1.25 ≤ 3. accept

(8) $s^2 = \frac{\sum x_i^2 - \frac{1}{n}(\sum x_i)^2}{n-1}$ or $\frac{\sum x_i^2 - \bar{x}\sum x_i}{n-1} = \frac{\sum(x_i - \bar{x})^2}{n-1}$

12

ESTIMATION

12.1 Introduction

Inference about the values of parameters involved in statistical problems is known as *estimation*. For example, the chemical engineer might be interested in estimating the mean yield of a manufacturing process from the information contained in the measurements made in a random sample or the electrical engineer might be interested in estimating the mean life of an electronic component based on the failure times of a random sample placed on life test. Once an *estimator* has been considered it seems essential to investigate how close this estimator would be to the true value of the parameter being estimated. Most of the problems of estimation involve the *estimation of parameters of distribution functions* or *some quantity whose value depends on the parameters*.

Two categories of estimators are in common use: One is called a *point estimator of the parameter concerned*—a point estimator provides a *single* value; the other is called an *interval estimator*. It *brackets* the true value of the parameter with certain specified odds. In the following sections we will discuss some aspects of both point and interval estimators.

12.2 Point Estimation

Point estimators are obtained by forming a function whose domain is the set of observed values of the random variable under study and whose range is the set of possible values of the parameter being estimated. Suppose that a random sample of size n is drawn from a normal distribution. The normal distribution is characterized by two parameters; the mean (μ) and the variance (σ^2). We may wish, e.g., to estimate the mean regardless of whether the standard deviation is known or unknown. The density function of the normal distribution is denoted in terms of its parameters by

$$f\{x: \mu, \sigma\}$$

Using the conventional notation for parameters to be estimated we write

$$f\{x: \theta_1, \theta_2\}$$

Generalizing the notation we have θ_j, ($j = 1, 2, \ldots, k$), a parameter to be estimated, where

$$\hat{\theta}_j(X_1, X_2, \ldots, X_n)$$

is known as an *estimator* and the numerical value of $\hat{\theta}_j(x_1, x_2, \ldots, x_n)$ is known as an *estimate*.

Since $X_1, X_2, \ldots, X_n$ are random variables the distribution of the estimator will depend on the distribution of these random variables and hence on the parameters of the model. Thus $\hat{\theta}_j$ is a random variable and is sometimes called a *statistic*. Often there are many possible functions capable of providing estimators: e.g., the *arithmetic mean*, the *median*, and the *mode* may be used as estimators of the mean of a normal distribution. Which of these estimators is to be preferred? Intuitively the objective would be to find an estimator $\hat{\theta}$ whose distribution is centered on the value of θ. Although this is a very desirable characteristic for $\hat{\theta}$, it is by no means a sufficient criterion that provides optimal choice of an estimator. In this connection it should be pointed out that in practice our choice is sometimes influenced by economic considerations. Apart from such considerations there are certain desirable characteristics for *good* estimators. These characteristics will be considered in some detail.

12.2.1 Unbiasedness

Definition 12.1. An estimator $\hat{\theta}$ of the parameter θ is said to be *unbiased* if

$$E(\hat{\theta}) = \theta \tag{12.1}$$

If $E(\hat{\theta}) \neq \theta$, then $\hat{\theta}$ is said to be *biased*. $[E(\hat{\theta}) - \theta]$ is called the *bias of the estimate*.

Example 12.1 Prove that the statistic

$$S^2 = \frac{\sum_{i=1}^{n} (X_i - \bar{X})^2}{n - 1}$$

is an unbiased estimator of $\text{Var}(X) = \sigma_X^2$.

Proof.

$$
\begin{aligned}
E(S^2) &= E\left[\frac{\sum_{i=1}^{n}(X_i - \bar{X})^2}{n-1}\right] \\
&= \frac{1}{n-1}E\left[\sum_{i=1}^{n}(X_i - EX + EX - \bar{X})^2\right] \\
&= \frac{1}{n-1}E\left[\sum_{i=1}^{n}(X_i - EX)^2 - n(\bar{X} - EX)^2\right] \\
&= \frac{1}{n-1}\left[\sum_{i=1}^{n}E(X_i - EX)^2 - nE(\bar{X} - EX)^2\right] \\
&= \frac{n\sigma^2 - \sigma^2}{n-1} \\
&= \sigma^2
\end{aligned}
$$

Thus the statistic S^2 is an unbiased estimator of σ_X^2.

Example 12.2 Find an unbiased estimator for the percentage defective p in n binomial trials.

Solution. Let X be a random variable having the binomial distribution corresponding to n trials, where p denotes the probability that an item drawn at random is defective. It seems intuitively reasonable to use the random variable X/n as an estimator of p. The random variable X/n is unbiased since

$$
\begin{aligned}
E\left(\frac{X}{n}\right) &= \frac{1}{n}E(X) \\
&= \frac{1}{n}(np) = p
\end{aligned}
$$

Thus X/n is an unbiased estimator of the parameter p.

12.2.2 Consistency

This criterion involves the *closeness* of an estimator $(\hat{\theta})$ to the true value of the parameter (θ) being estimated, in a probability sense, as the number of observations increases indefinitely.

Definition 12.2. An estimator is said to be *consistent* for any small quantity $\delta > 0$ if the probability of the absolute deviation of $\hat{\theta}$ from θ is less than δ and approaches unity as the number of observations n tends to infinity.

(12.2) $$P\{|\hat{\theta} - \theta| < \delta\} \longrightarrow 1 \text{ as } n \longrightarrow \infty$$

Example 12.3 Prove that the statistic

$$\bar{X} = \frac{1}{n}\sum_{i=1}^{n} X_i$$

is a consistent estimator of $E(X)$.

Proof. Let

$$\hat{\theta} = \bar{X} = \frac{1}{n}\sum_{i=1}^{n} X_i$$

then

$$P\{|\hat{\theta} - \theta| < \delta\} = P\{|\bar{X} - E(X)| < \delta\}$$

By Chebyshev's inequality

$$P\{|\bar{X} - E(\bar{X})| < \delta\} \geq 1 - \frac{\text{Var}(\bar{X})}{\delta^2}$$

$$\geq 1 - \frac{\sigma_X^2}{n\delta^2}$$

Hence

$$\lim_{n\to\infty} P\{|\bar{X} - E(\bar{X})| < \delta\} = 1$$

Therefore $\bar{X}$ is a consistent estimator of $E(X)$.

12.2.3 Minimum Variance

Suppose we wish to estimate the parameter θ that is the mean (μ) of a measurable characteristic X on the basis of three independent observations. The following two estimators are possible:

$$\hat{\theta}_1(X_1, X_2, \ldots, X_n) = \tfrac{1}{4}x_1 + \tfrac{1}{2}x_2 + \tfrac{1}{4}x_3$$

and

$$\hat{\theta}_2(X_1, X_2, \ldots, X_n) = \tfrac{1}{3}(x_1 + x_2 + x_3)$$

we note that both $\hat{\theta}_1$ and $\hat{\theta}_2$ give unbiased estimates of μ. Now it is evident that on the basis of unbiasedness alone we cannot tell which estimator is better. One would naturally prefer $\hat{\theta}_1$ to $\hat{\theta}_2$ if $\hat{\theta}_1$ is *consistently closer* to μ in repeated samples of the same size n. Since an estimator $\hat{\theta}$ that is consistently close to θ will usually be tightly concentrated about θ, the variability of $\hat{\theta}$ about θ is used as a measure of closeness. Thus the second moments of estimators about θ can be compared to determine which estimate is better. In the example cited above, the parameter to be estimated is the mean of the distribution; thus by comparing the variances of $\hat{\theta}_1$ and $\hat{\theta}_2$ one can tell which estimator is better. Since Var $(\hat{\theta}_2) <$ Var $(\hat{\theta}_1)$, $\hat{\theta}_2$ is a better estimator than $\hat{\theta}_1$. Now we shall formalize the definition of *minimum variance unbiased estimators.*

Definition 12.3. A *minimum variance unbiased estimator* is one whose variance is least for all possible choices of unbiased estimators.

12.2.4 Mean Square Error

The variance can be viewed as a measure of the precision of the estimator. The combination of precision and bias is a useful criterion for comparing estimators including biased estimators. This criterion is known as the *mean square error.*

Definition 12.4. *Mean square error* (MSE) is defined as the expected value of the square of the deviation of the estimate from the parameter being estimated. It is equal to the variance of the estimate plus the square of the bias.

Proof

$$\begin{aligned}\text{MSE} &= E(\hat{\theta} - \theta)^2 \\ &= E\{[\hat{\theta} - E(\hat{\theta})] + [E(\hat{\theta} - \theta)]\}^2 \\ &= E[\hat{\theta} - E(\hat{\theta})]^2 + E[E(\hat{\theta}) - \theta]^2 + 2E[\hat{\theta} - E(\hat{\theta})][E(\hat{\theta}) - \theta] \\ &= \text{Var}(\hat{\theta}) + (\text{bias})^2 \qquad (12.3)\end{aligned}$$

since $E[\hat{\theta} - E(\hat{\theta})] = 0$.

Example 12.4 Suppose $\hat{\theta}_1$ and $\hat{\theta}_2$ are two possible estimators of θ. Furthermore, suppose that $E(\hat{\theta}_1) = \theta$, $E(\hat{\theta}_2) = .9\theta$, $\text{Var}(\hat{\theta}_1) = 3$, and $\text{Var}(\hat{\theta}_2) = 2$. Which estimator would you prefer?

Solution

$$\begin{aligned}\text{MSE of } \hat{\theta}_1 &= \text{Var}(\hat{\theta}_1) + (\text{bias})^2 \\ &= 3 + 0 = 3 \\ \text{MSE of } \hat{\theta}_2 &= \text{Var}(\hat{\theta}_2) + (\text{bias})^2 \\ &= 2 + .01\theta^2\end{aligned}$$

Hence, $\hat{\theta}_1$ is preferred if $|\theta| > 10$. But we note that $\hat{\theta}_2$ is to be preferred if $|\theta| < 10$ in spite of the fact that it is a biased estimator. Figure 12.1 shows that the biased estimator may be preferred if $|\theta| < 10$.

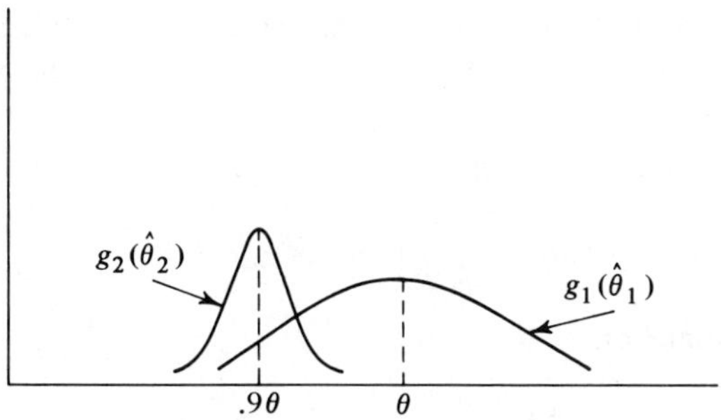

Fig. 12.1 Densities of two estimators of the parameter θ.

12.2.5 Sufficiency

A further criterion known as *sufficiency* was introduced by R. A. Fisher. Full exposition of this notion is beyond the scope of this work. A brief analysis, however, is in order.

An estimator of parameter θ is said to be *sufficient* (denoted by $\hat{\theta}_s$) if $\hat{\theta}_s$ contains all the information about θ and no other estimator computed from the same outcomes can provide additional information about θ.

Definition 12.5. An estimator $\hat{\theta}_s$ is said to be *sufficient* if the likelihood function has the form

$$L(x_1, x_2, \ldots, x_n: \theta) = g(\hat{\theta}_s \mid \theta) f(x_1, x_2, \ldots, x_n) \tag{12.4}$$

where $g(\hat{\theta}_s \mid \theta)$ is a probability density function of $(x_1, x_2, \ldots, x_n)$ only in terms of $\hat{\theta}_s$ and $f(x_1, x_2, \ldots, x_n)$ is a function of $(x_1, x_2, \ldots, x_n)$ independent of the parameter θ. The notion given by Eq. (12.4) also can be generalized to the case when several parameters are unknown.

It can be shown that the maximum likelihood of θ depends only on the value of $\hat{\theta}_s$ and does not depend on the set of outcomes $(x_1, x_2, \ldots, x_n)$ that yield the value of $\hat{\theta}_s$.

Also, it can be proved that when $\hat{\theta}_s$ exists the maximum likelihood estimator will be a function of $\hat{\theta}_s$ only.

The significance of the notion of sufficiency lies in the fact that all possible nonsufficient estimators may be ignored whenever a sufficient estimator exists.

Example 12.5 Suppose X is a random variable whose density function is given by

$$f_X(t) = \frac{1}{\theta} \qquad 0 < t < \theta$$
$$= 0 \qquad \text{otherwise}$$

where θ is a parameter to be estimated on the basis of n independent observations. Prove that the statistic $U = \max(X_1, X_2, \ldots, X_n)$ is a sufficient estimator of the parameter θ.

Solution. The cumulative distribution function of U is

$$F_U(a) = \left(\frac{a}{\theta}\right)^n \qquad 0 < a < \theta$$

Hence the density function of U is

$$f_U(a) = \frac{na^{n-1}}{\theta^n} \qquad 0 < a < \theta$$

The likelihood function, then, is

$$L(t_1, t_2, \ldots, t_n: \theta) = \frac{1}{\theta^n} = f_U(a)\left(\frac{1}{na^{n-1}}\right)$$

The right-hand side of the above expression is a product of two terms, of which the first is a probability density function in terms of U while the second does not depend on θ. Thus by definition the statistic U is a sufficient estimator of the parameter θ.

12.2.6 Efficiency

Efficiency is an important criterion in the theory of estimation. This notion hinges on the fact that the smaller the variance of the estimator the greater the probability that the estimate will be close to the true value of the parameter being estimated. Thus the smaller the variance the higher the *efficiency* of the estimator.

Many important theorems concerning the efficiency of unbiased estimators were developed by Fisher, Cramér, Blackwell, and others. We state two important theorems without proof:

Theorem 12.1. *If $\hat{\theta}$ is an unbiased estimator of the parameter θ, then the Rao–Cramér inequality states, assuming that some regularity conditions hold, that*

$$\text{Var}(\hat{\theta}) \geq \frac{1}{n\int_{-\infty}^{\infty} [\partial \ln f(x:\theta)/\partial\theta]^2 f(x:\theta)\, dx} \tag{12.5}$$

where X is a continuous random variable. If X is a discrete random variable we use the probability function instead of the density function and summation instead of integration. The above inequality states that the variance of the unbiased estimator is greater than or equal to a certain value that is constant for a fixed number (n) of independent observations and that it is determined by the distribution of the random variable X. The right-hand side of the inequality is equal to the minimal variance of $\hat{\theta}$. Thus if $\hat{\theta}$ is unbiased and its variance is given by the right of (12.5), then $\hat{\theta}$ is said to be the most efficient estimator among the class of all unbiased estimators. It follows from (12.5) that a most efficient estimator must satisfy certain conditions that we state without proof in the form of a theorem:

Theorem 12.2. *If $\hat{\theta}$ is an unbiased estimator of the parameter θ, then $\hat{\theta}$ is said to be the most efficient estimator if and only if the following two conditions are satisfied:*

(12.6)

(a) $\hat{\theta}$ is sufficient

(b) $$\frac{\partial \ln f(\hat{\theta}:\theta)}{\partial\theta} = k(\hat{\theta} - \theta) \qquad f(\hat{\theta}:\theta) > 0$$

where k does not depend on $\hat{\theta}$.

Example 12.6 Prove that $\bar{X}$ the arithmetic mean is the most efficient estimator of the mean μ of the normal distribution with known variance σ_0^2.

Proof. Since $E(\bar{X}) = \mu$, $\bar{X}$ is an unbiased estimator of μ. Now we have to prove that the conditions of Theorem 12.2 are satisfied. First we have to prove that $\bar{X}$ is a sufficient estimator. The likelihood function is given by

$$
\begin{aligned}
L &= f(x_1 : \mu) f(x_2 : \mu) \cdots f(x_n : \mu) \\
&= \left(\frac{1}{\sigma_0\sqrt{2\pi}}\right)^n \exp\left[-\frac{1}{2\sigma_0^2}\sum_{i=1}^{n}(x_i - \mu)^2\right] \\
&= \left(\frac{1}{\sigma_0\sqrt{2\pi}}\right)^n \exp\left[-\frac{1}{2\sigma_0^2}\sum_{i=1}^{n}(x_i - \bar{x} + \bar{x} - \mu)^2\right] \\
&= \left\{\left(\frac{1}{\sigma_0\sqrt{2\pi}}\right)^n \exp\left[-\frac{n}{2\sigma_0^2}(\bar{x} - \mu)^2\right]\right\}\left\{\exp\left[-\frac{1}{2\sigma_0^2}\sum_{i=1}^{n}(x_i - \bar{x})^2\right]\right\}
\end{aligned}
$$

Since the first term on the right-hand side is a probability density function of the observations in terms of $\bar{x}$ and the second term does not depend on the parameter μ, by definition of sufficiency $\bar{X}$ is a sufficient estimator of μ. It should be noted that sufficient estimators have been readily found for both parameters of the normal distribution without specifying a particular value for either one. To prove that condition (b) is satisfied we have

$$f(\hat{\theta} : \mu) = f(\bar{x} : \mu) = \frac{\sqrt{n}}{\sigma_0\sqrt{2\pi}} \exp\left[-\frac{n}{2\sigma_0^2}(\bar{x} - \mu)^2\right]$$

or, equivalently,

$$\ln f(\bar{x} : \mu) = \ln \frac{\sqrt{n}}{\sigma_0\sqrt{2n}} - \frac{n}{2\sigma_0^2}(\bar{x} - \mu)^2$$

whence

$$
\begin{aligned}
\frac{\partial \ln f(\bar{x} : \mu)}{\partial \mu} &= \frac{n}{\sigma_0^2}(\bar{x} - \mu) \\
&= k(\bar{x} - \mu)
\end{aligned}
$$

where $k = n/\sigma_0^2$, we note that k does not depend on $\bar{x}$. Thus $\bar{X}$ is the most efficient estimator of the mean μ of the normal distribution. An alternative proof would be as follows: Since $\bar{X}$ is an unbiased estimator of μ, we need to prove that $\mathrm{Var}(\bar{X}) = \sigma_0^2/n$ is equivalent to the lower bound of the Rao–Cramér inequality. The lower bound of Rao–Cramér is

$$
\begin{aligned}
&\frac{1}{n\int_{-\infty}^{\infty}[\partial \ln f(x : \mu)/\partial\mu]^2 f(x : \mu)\,dx} \\
&\qquad = \frac{1}{n\int_{-\infty}^{\infty}[(x - \mu)/\sigma_0^2]^2 f(x : \mu)\,dx} \\
&\qquad = \frac{\sigma_0^4}{n\,\mathrm{Var}(X)} = \frac{\sigma_0^2}{n}
\end{aligned}
$$

which is equivalent to the variance of $\bar{X}$. This means that $\bar{X}$ is an unbiased estimator of μ with minimum variance. Accordingly $\bar{X}$ is the most efficient estimator of the mean μ of the normal distribution.

Example 12.7 Consider the Poisson distribution

$$P_X(k) = \frac{e^{-\lambda}\lambda^k}{k!} \qquad k = 0, 1, \ldots \qquad \lambda > 0$$

with unknown parameter λ. Using Theorem 12.1 prove that the minimal variance of the estimator of λ based on n independent observations is λ/n.

Proof. By Theorem 12.1 the minimal variance is

$$\frac{1}{n \sum_{k=0}^{\infty} [d \ln P_X(k)/d\lambda]^2 P_X(k)}$$

However,

$$\ln P_X(k) = -\lambda + k \ln \lambda - \ln (k!)$$

Hence,

$$\frac{d \ln P_X(k)}{d\lambda} = \left(\frac{k}{\lambda} - 1\right)$$

Substituting, the lower bound of the Rao–Cramér inequality becomes

$$\frac{1}{n \sum_{k=0}^{\infty} [(k/\lambda) - 1]^2 P_X(k)}$$

$$= \frac{1}{(n/\lambda^2) \sum_{k=0}^{\infty} (k - \lambda)^2 P_X(k)} = \frac{1}{(n/\lambda^2)\operatorname{Var}(X)} = \frac{\lambda}{n}$$

12.2.7 Relative Efficiency

A measure of *relative efficiency* η of two possible estimators $\hat{\theta}_1$ and $\hat{\theta}_2$ for the parameter θ can be assessed from the expression

$$\eta = \frac{\text{MSE}(\hat{\theta}_1)}{\text{MSE}(\hat{\theta}_2)} \tag{12.7}$$

If η turns out to be > 1, $\hat{\theta}_2$ may be considered better than $\hat{\theta}_1$. If $\eta < 1$, $\hat{\theta}_1$ may be considered better than $\hat{\theta}_2$. If both $\hat{\theta}_1$ and $\hat{\theta}_2$ are *unbiased estimators* of θ, η reduces to

$$\eta = \frac{\operatorname{Var}(\hat{\theta}_1)}{\operatorname{Var}(\hat{\theta}_2)} \tag{12.8}$$

12.2.8 Asymptotically Most Efficient Estimators

If $\hat{\theta}_1$ is an *unbiased and most efficient estimator* of θ and $\hat{\theta}_2$ is an *unbiased estimator* of θ, then a measure of *relative efficiency* can be assessed from the expression

$$\eta' = \frac{\operatorname{Var}(\hat{\theta}_1)}{\operatorname{Var}(\hat{\theta}_2)} \tag{12.9}$$

$\hat{\theta}_2$ *is* called an *asymptotically most efficient estimator* if

$$\lim_{n\to\infty} \eta' = 1 \tag{12.10}$$

This means that we may have several asymptotically most efficient estimators for a given parameter. Asymptotically most efficient estimators are unbiased when, as n tends to infinity, they are also consistent estimators.

Example 12.8 Random variable X is normally distributed with unknown variance σ^2. Prove that the statistic

$$\hat{\theta}_2 = \frac{\sum_{i=1}^{n} (X_i - \bar{X})^2}{n-1}$$

is an asymptotically most efficient estimator of σ^2.

Proof. Let $\hat{\theta}_1$ be the unbiased most efficient estimator of σ^2. By Theorem 12.1 the minimal variance is given by

$$\operatorname{Var}(\hat{\theta}_1) = \frac{1}{n\int_{-\infty}^{\infty} [\partial \ln f(x:\sigma^2)/\partial\sigma^2]^2 f(x:\sigma^2)\,dx} = \frac{2\sigma^4}{n}$$

The variance of $\hat{\theta}_2$ is

$$\begin{aligned}
\operatorname{Var}(\hat{\theta}_2) &= \frac{1}{(n-1)^2}\operatorname{Var}\left[\sum_{i=1}^{n} (X_i - \bar{X})^2\right] \\
&= \frac{\sigma^4}{(n-1)^2}\operatorname{Var}(\chi^2_{n-1}) \\
&= \frac{2\sigma^4}{n-1}
\end{aligned}$$

By Eq. (12.9) the relative efficiency of $\hat{\theta}_2$ is

$$\eta' = \frac{\operatorname{Var}(\hat{\theta}_1)}{\operatorname{Var}(\hat{\theta}_2)} = \frac{n-1}{n}$$

Therefore

$$\lim_{n\to\infty} \eta' = \lim_{n\to\infty}\left(1 - \frac{1}{n}\right) = 1$$

Hence, $\hat{\theta}_2$ is an asymptotically most efficient estimator of σ^2.

12.2.9 Estimation by Maximum Likelihood

The most important technique of estimation so far known is the method of *maximum likelihood*. This technique was introduced and developed by R. A. Fisher. It hinges on the concept of the *likelihood function* (L) discussed in Chapter 9. The likelihood function (L) of a sample of n independent observations drawn from a stable population is given for a continuous random variable by

$$(12.11) \qquad L(x_1, x_2, \ldots, x_n : \theta) = f(x_1 : \theta) f(x_2 : \theta) \cdots f(x_n : \theta)$$

For a discrete random variable it is

$$(12.12) \qquad L(x_1, x_2, \ldots, x_n : \theta) = (P_{i_1} : \theta)(P_{i_2} : \theta) \cdots (P_{i_n} : \theta)$$

where θ is a parameter to be estimated. The objective of the maximum likelihood technique is to find the value of the estimator of θ that renders the likelihood function as large as possible for the given values of $(X_1, X_2, \ldots, X_n)$.

If the distribution of X is known in terms of the parameter or the parameters to be estimated, we can maximize the likelihood function by regarding the observed x's as constants and the parameters as variables. It is often more convenient to work with $\ln L$ than with L since $\ln L$ attains its maximum for the same value of θ as L. Accordingly, we write

$$\ln L = \ln f(x_1 : \theta) + \ln f(x_2 : \theta) + \cdots + \ln f(x_n : \theta)$$

Thus we have to solve the likelihood equation

$$(12.13) \qquad \frac{\partial (\ln L)}{\partial \theta} = \frac{1}{f(x_1 : \theta)} \frac{\partial f(x_1 : \theta)}{\partial \theta} + \cdots + \frac{1}{f(x_n : \theta)} \frac{\partial f(x_n : \theta)}{\partial \theta} = 0$$

which yields a maximum likelihood estimate of θ in terms of the x's. If more than one parameter is to be estimated, the partial derivatives with respect to each parameter are set equal to zero and the resulting equations are solved for the estimates. A word of caution is in order here: It must be ascertained whether the solution is actually a maximum and not a minimum or a point of inflection. If a maximum does not exist, the supremum may be used. It is noteworthy that maximum likelihood estimators are often *biased.* They possess, however, the properties of being *consistent*, *asymptotically normal*, and *asymptotically efficient* for large sample sizes under unrestrictive conditions.

Some of these properties can be summarized as follows (we omit the proof, for which the interested reader may consult advanced texts such as Cramér [2], Mood [10], and Yule and Kendall [15]):

1. *The maximum likelihood estimator $\hat{\theta}_L$ is an asymptotically unbiased estimator of θ*, i.e.,

$$\lim_{n \to \infty} E(\hat{\theta}_L : \theta) = \theta$$

2. *The transformed chance variable $\sqrt{n}(\hat{\theta}_L - \theta)$ is asymptotically normally distributed as n tends to infinity.*

3. $$\lim_{n \to \infty} n \operatorname{Var}(\hat{\theta}_L) = -\left\{E\left[\frac{\partial^2 \ln L(x_1, x_2, \ldots, x_n : \theta)}{\partial \theta^2}\right]\right\}^{-1}$$

4. *The maximum likelihood estimators possess the property of invariance. That is, if $h(\theta)$ is a function of θ, with a single inverse, then the maximum likelihood estimator of $h(\theta)$ is $h(\hat{\theta}_L)$.*
5. *When a sufficient estimator for θ exists, then $\hat{\theta}_L$ is a function of observations $(x_1, x_2, \ldots, x_n)$ only in terms of the sufficient estimator.*

Example 12.9 Suppose X is a random variable that is exponentially distributed. Its density function is given by

$$f_X(t) = \lambda e^{-\lambda t} \qquad t \geqq 0 \qquad \lambda > 0$$

On the basis of n independent observations determine the maximum likelihood estimator of λ.

Solution. Computing the likelihood function we have

$$L = \lambda^n e^{-\lambda \sum_{i=1}^{n} t_i} \qquad \text{or} \qquad \ln L = n \ln \lambda - \lambda \sum_{i=1}^{n} t_i$$

Differentiating with respect to λ and equating the derivative to zero we obtain

$$\frac{\partial}{\partial \lambda}(\ln L) = \frac{n}{\lambda} - \sum_{i=1}^{n} t_i = 0$$

Whence

$$\hat{\lambda}_L = \frac{n}{\sum_{i=1}^{n} t_i} = \frac{1}{\bar{t}}$$

is the maximum likelihood estimator of λ.

Example 12.10 Consider the binomial distribution with unknown parameter p given by

$$p_X(k) = \binom{n}{k} p^k (1-p)^{n-k} \qquad k = 1, 2, \ldots, n \qquad 0 \leq p \leq 1$$

What is the maximum likelihood estimate of p based on n independent observations?

Solution. Let the outcome of n observations be such that the result $x = 1$ is obtained k times and the result $x = 0$ is obtained $n - k$ times. This can be expressed as

$$P_{X_i}(1) = p \qquad i = 1, \ldots, n$$
$$P_{X_i}(0) = 1 - p$$

Hence

$$L = \binom{n}{k} p^k (1-p)^{n-k}$$

$$\ln L = \ln\binom{n}{k} + k \ln p + (n-k) \ln (1-p)$$

$$\frac{\partial(\ln L)}{\partial p} = \frac{k}{p} - \frac{(n-k)}{1-p} = 0$$

Whence $p = k/n$ and the maximum likelihood estimator of p is

$$\hat{p}_L = \frac{k}{n}$$

Example 12.11 Determine the maximum likelihood estimator of the parameters μ and σ^2 of a normal distribution from a sample $x_1, \ldots, x_n$.

Solution. The likelihood function is given by

$$L = \frac{1}{\sqrt{2\pi}\sigma} e^{-(x_1-\mu)^2/2\sigma^2} \cdots \frac{1}{\sqrt{2\pi}\sigma} e^{-(x_n-\mu)^2/2\sigma^2}$$

$$= \left(\frac{1}{2\pi\sigma^2}\right)^{n/2} e^{-\Sigma(x_i-\mu)^2/2\sigma^2}$$

Hence

$$\ln L = -\frac{1}{2\sigma^2} \sum (x_i - \mu)^2 - \frac{1}{2} n \ln \sigma^2 - \frac{1}{2} n \ln 2\pi$$

Setting the derivative with respect to μ equal to zero yields

$$\frac{\partial(\ln L)}{\partial \mu} = \frac{1}{\sigma^2} \sum (x_i - \mu) = 0$$

Whence

$$\hat{\mu}_L = \frac{\sum x_i}{n} = \bar{x}$$

To obtain the maximum likelihood estimator of σ^2 provided that μ is known, we differentiate the likelihood function with respect to σ^2 and equate the result to zero. Thus we have

$$\frac{\partial(\ln L)}{\partial \sigma^2} = \frac{1}{2\sigma^4} \sum (x_i - \mu)^2 - \frac{n}{2\sigma^2} = 0$$

Whence

$$\hat{\sigma}^2_L = \frac{\sum (x_i = \mu)^2}{n}$$

Example 12.12 Consider the Poisson distribution with unknown parameter λ given by

$$P_X(k) = \frac{e^{-\lambda} \lambda^k}{k!} \qquad k = 0, 1, \ldots, \qquad \lambda > 0$$

Suppose that $x_1, x_2, \ldots, x_n$ are n observations of the number of arrivals at a toll gate per hour, taken at hours supposed to be homogeneous with respect to arrivals. What is the maximum likelihood estimator of λ?

Solution

$$L = \frac{e^{-\lambda} \lambda^{k_1}}{k_1!} \frac{e^{-\lambda} \lambda^{k_2}}{k_2!} \cdots \frac{e^{-\lambda} \lambda^{k_n}}{k_n!}$$

$$= \frac{e^{-n\lambda} \lambda^{\Sigma k_i}}{k_1! \, k_2! \cdots k_n!}$$

$$\ln L = -n\lambda + \sum k_i \ln \lambda - \sum \ln k_i!$$

$$\frac{\partial(\ln L)}{\partial \lambda} = -n + \frac{\sum k_i}{\lambda} = 0$$

Whence

$$\hat{\lambda}_L = \frac{\sum k_i}{n} = \bar{x}$$

12.3 Interval Estimation

In the previous sections we have considered point estimators of parameters. The point estimator is a single value. This estimated value rarely coincides with the true value of the parameter being estimated. Actually point estimators are random variables distributed in some fashion around the true value of the estimator. For example, if the mean (μ) of the normal distribution is the parameter to be estimated, then $\bar{X}$ is the maximum likelihood estimator. $\bar{X}$ is a random variable that is normally distributed with mean equal to the parameter. Since $\bar{X}$ will rarely coincide with μ it is desirable to know the magnitude of the absolute deviation of $\bar{X}$ from μ, with some specified odds. Actually what is needed is an *interval* that is expected to bracket the true value of the parameter with some specified odds. This interval is called the *confidence interval*, and the specified odds are known as *confidence coefficient*. Thus a 90% confidence interval for a given parameter implies that in the long run the computed limits of the interval will include the true value of the parameter 90 times in 100. In the following sections we will derive the confidence limits for parameters of the normal and binomial distributions.

12.3.1 Confidence Limits for the Mean of a Normal Population

Case 1. Known Variance

Let $x_1, \ldots, x_n$ be a random sample drawn from a normal distribution with unknown mean μ and known variance σ^2. Then $\bar{x}$ is the most efficient estimator of the mean and is normally distributed with mean μ and variance σ^2/n. Hence we have

$$P\left\{-Z_{1-\alpha/2} \le \frac{\sqrt{n}\,(\bar{x}-\mu)}{\sigma} \le Z_{1-\alpha/2}\right\} = 1-\alpha$$

or

$$P\left\{\bar{x} - \frac{Z_{1-\alpha/2}\sigma}{\sqrt{n}} \le \mu \le \bar{x} + \frac{Z_{1-\alpha/2}\sigma}{\sqrt{n}}\right\} = 1-\alpha \tag{12.14}$$

where Z is the standard normal variate.

The above expression has the following interpretation: Given the assumptions made above and a random independent sample of observations, $x_1, x_2, \ldots, x_n$, the mean μ can be said to lie within the interval

$$\left[\bar{x} \pm Z_{1-\alpha/2}\frac{\sigma}{\sqrt{n}}\right] \tag{12.15}$$

with a probability of $1 - \alpha$. On repeating such a procedure we may be wrong $100\alpha\%$ of the time. This interval is called a $(1 - \alpha)$ *confidence interval* while

$$\bar{x} - Z_{1-\alpha/2}\frac{\sigma}{\sqrt{n}} \qquad \text{and} \qquad \bar{x} + Z_{1-\alpha/2}\frac{\sigma}{\sqrt{n}}$$

are referred to as *confidence limits.*

Example 12.13 A random sample of size nine is drawn from a normal distribution with unknown mean and variance = 4. $\bar{x}$ was calculated from nine observations and found to be equal to 100.8.

(a) Find a 90% confidence interval for μ.
(b) Find a 90% lower confidence interval for μ.
(c) What sample size will ensure that the estimate is within $\pm .5$ of the true mean in 90% of the cases?

Solution

(a) We note that $1 - \alpha = .90$ and $1 - \alpha/2 = .95$. Hence the 90% confidence interval for μ is $100.8 \pm \frac{2}{3}Z_{.95} = 100.8 \pm \frac{2}{3}(1.645) = [99.704, 101.896]$.
(b) A 90% lower confidence interval for μ is

$$100.8 - \tfrac{2}{3}Z_{.90} = 100.8 - \tfrac{2}{3}(1.28) = [99.947, +\infty]$$

(c) Here we have

$$P\{\mu - .5 \leq x \leq \mu + .5\} = 1 - \alpha = .90$$

Hence

$$Z_{.95}\sigma/\sqrt{n} = .5$$

whence

$$n = (1.645)^2(4)/.25 \simeq 44$$

Case 2. Unknown Variance

If σ^2 is unknown, s^2 the sample variance is used as an estimator of σ^2. The random variable $[(\bar{x} - \mu)/(s/\sqrt{n})]$ then has the Student's t distribution with $n - 1$ degrees of freedom. Hence we can assert that

$$P\left\{-t_{1-\alpha/2;\,n-1} \leq \frac{\sqrt{n}\,(\bar{x} - \mu)}{s} \leq t_{1-\alpha/2;\,n-1}\right\} = 1 - \alpha$$

or

$$P\left\{\bar{x} - t_{1-\alpha/2;\,n-1}\frac{s}{\sqrt{n}} \leq \mu \leq \bar{x} + t_{1-\alpha/2;\,n-1}\frac{s}{\sqrt{n}}\right\} = 1 - \alpha \tag{12.16}$$

Whence

$$\left[\bar{x} \pm t_{1-\alpha/2;\,n-1}\frac{s}{\sqrt{n}}\right] \tag{12.17}$$

is a $100\,(1 - \alpha)\%$ confidence interval for μ.

12.3.2 Confidence Limits for the Difference of Means of Two Normal Populations

Case 1. The Variances of Both Populations Are Known

Let X and Y be two independent normally distributed random variables with means μ_X and μ_Y and known variances σ_X^2 and σ_Y^2, respectively. Let $\bar{x}$ and $\bar{y}$ be the sample means of n and m observations taken from the X and Y populations, respectively. Since $\bar{x}$ and $\bar{y}$ are independent unbiased point estimators of μ_X and μ_Y, respectively, then their difference $\bar{x} - \bar{y}$ also will be an unbiased estimator of $\mu_X - \mu_Y$. Furthermore, $\bar{x} - \bar{y}$ has the normal distribution with mean $\mu_X - \mu_Y$ and variance $\sigma_X^2/n + \sigma_Y^2/m$. Then we may state that

$$P\left\{-Z_{1-\alpha/2} \leq \frac{\bar{x} - \bar{y} - (\mu_X - \mu_Y)}{\sqrt{\sigma_X^2/n + \sigma_Y^2/m}} \leq Z_{1-\alpha/2}\right\} = 1 - \alpha$$

or

$$P\left\{\bar{x} - \bar{y} - Z_{1-\alpha/2}\sqrt{\frac{\sigma_X^2}{n} + \frac{\sigma_Y^2}{m}} \leq \mu_X - \mu_Y \leq \bar{x} - \bar{y} + Z_{1-\alpha/2}\sqrt{\frac{\sigma_X^2}{n} + \frac{\sigma_Y^2}{m}}\right\} = 1 - \alpha \tag{12.18}$$

Whence a $100(1 - \alpha)\%$ confidence interval for $\mu_X - \mu_Y$ is given by

$$\left[\bar{x} - \bar{y} \pm Z_{1-\alpha/2}\sqrt{\frac{\sigma_X^2}{n} + \frac{\sigma_Y^2}{m}}\right] \tag{12.19}$$

Case 2. The Variances of Both Populations Are Unknown but Equal

Here both σ_X^2 and σ_Y^2 are *unknown but equal*. Stated differently, $\sigma_X^2 = \sigma_Y^2 = \sigma^2$, where σ^2 is unknown. It was proved in Chapter 9 that the random variable

$$\frac{\bar{x} - \bar{y} - (\mu_X - \mu_Y)}{\sqrt{[(n-1)s_X^2 + (m-1)s_Y^2]/(n+m-2)}\sqrt{1/n + 1/m}}$$

has the Student's t distribution with $n + m - 2$ degrees of freedom. Therefore, we can state that

$$P\left\{-t_{1-\alpha/2;\,n+m-2} \leq \frac{\bar{x} - \bar{y} - (\mu_X - \mu_Y)}{s_p\sqrt{1/n + 1/m}} \leq t_{1-\alpha/2;\,n+m-2}\right\} = 1 - \alpha \tag{12.20}$$

where

$$s_p = \sqrt{\frac{(n-1)s_X^2 + (m-1)s_Y^2}{n+m-2}}$$

or

$$P\left\{\bar{x} - \bar{y} - t_{1-\alpha/2;\,n+m-2}\,s_p\sqrt{\frac{1}{n} + \frac{1}{m}} \leq \mu_X - \mu_Y \leq \bar{x} - \bar{y} + t_{1-\alpha/2;\,n+m-2}\,s_p\sqrt{\frac{1}{n} + \frac{1}{m}}\right\} = 1 - \alpha \tag{12.21}$$

Whence a $100(1-\alpha)\%$ confidence interval for $\mu_X - \mu_Y$ is given by

$$\left[\bar{x} - \bar{y} \pm t_{1-\alpha/2;\, n+m-2}\, s_p \sqrt{\frac{1}{n} + \frac{1}{m}}\right] \tag{12.22}$$

Example 12.14 A manufacturer has two annealing furnaces used for annealing brass strips. A sample of $n = 5$ strips from furnace A gave a mean Rockwell B hardness number of $\bar{x} = 30$ with standard deviation $s_X = 1$. A sample of $m = 7$ strips from furnace B gave a mean Rockwell B hardness number of $\bar{y} = 28$ with standard deviation $s_Y = 1.2$. It is assumed that the hardness readings of each furnace are independent normally distributed random variables with common unknown standard deviation. Find a 99% confidence interval for $\mu_X - \mu_Y$.

Solution. We note that $1 - \alpha = .99$, $\alpha/2 = .005$, $n = 5$, $m = 7$, $n + m - 2 = 10$, and

$$s_p = \sqrt{\frac{(5-1)(1)^2 + (7-1)(1.2)^2}{10}} = 1.12$$

Hence the confidence interval for $\mu_X - \mu_Y$ is

$$\begin{aligned} 30 - 28 \pm (3.169)(1.12)\sqrt{\frac{1}{5} + \frac{1}{7}} &= 2 \pm 2.08 \\ &= [-.08, 4.08] \end{aligned}$$

Case 3. The Variances of Both Populations Are Unknown but Unequal

Here we cannot assume that $\sigma_X = \sigma_Y$. To obtain an approximate confidence interval we first compute

$$\nu = \frac{(s_X^2/n + s_Y^2/m)^2}{[(s_X^2/n)^2/(n+1)]\,[(s_Y^2/m)^2/(m+1)]} - 2$$

Then an approximate $100(1-\alpha)\%$ confidence interval for $\mu_X - \mu_Y$ is given by

$$\left[\bar{x} - \bar{y} \pm t_{1-\alpha/2;\, \nu} \sqrt{\frac{s_X^2}{n} + \frac{s_Y^2}{m}}\right] \tag{12.23}$$

12.3.3 Confidence Limits for the Variance of a Normal Population

Let X be a normally distributed random variable with unknown mean μ and unknown variance σ^2. The unbiased point estimator of σ^2 can be computed from a sample of size n and is given by

$$s^2 = \frac{\sum_{i=1}^{n} (x_i - \bar{x})^2}{n-1}$$

We recall that $(n-1)s^2/\sigma^2$ has the chi-square distribution with $n - 1$ degrees of freedom. Thus

$$P\left\{\chi^2_{\alpha/2;\, n-1} \le \frac{(n-1)s^2}{\sigma^2} \le \chi^2_{1-\alpha/2;\, n-1}\right\} = 1 - \alpha \tag{12.24}$$

or, alternatively, we have

$$P\left\{\frac{(n-1)s^2}{\chi^2_{1-\alpha/2;\, n-1}} \leq \sigma^2 \leq \frac{(n-1)s^2}{\chi^2_{\alpha/2;\, n-1}}\right\} = 1 - \alpha \tag{12.25}$$

Whence a $100(1-\alpha)\%$ confidence interval for σ is given by

$$\left[\frac{(n-1)s^2}{\chi^2_{1-\alpha/2;\, n-1}},\ \frac{(n-1)s^2}{\chi^2_{\alpha/2;\, n-1}}\right] \tag{12.26}$$

Example 12.15 Find the 90% lower confidence interval for σ^2 given that a random sample of ten observations from a normal population yields

$$\sum x_i = 50 \qquad \sum x_i^2 = 380$$

Solution. We compute the following:

$$\bar{x} = 5 \qquad \sum (x_i - \bar{x})^2 = \sum x_i^2 - \bar{x} \sum x_i$$
$$= 380 - 250 = 130$$

$$s^2 = \frac{\sum (x_i - \bar{x})^2}{n-1} = \frac{130}{9} = 14.44$$

Hence the lower 90% confidence interval for σ^2 is

$$\left[\frac{(n-1)s^2}{\chi^2_{1-\alpha;\, n-1}}, +\infty\right] = \left[\frac{(9)(14.44)}{\chi^2_{.90;\, 9}}, +\infty\right]$$
$$= [8.85, +\infty]$$

12.3.4 Confidence Limits for the Ratio of Variances of Two Normal Populations

Let X and Y denote two independent normally distributed random variables with unknown variances σ_X^2 and σ_Y^2, respectively. We wish to find the confidence interval for the ratio σ_X^2/σ_Y^2. Let s_X^2 be the sample variance based on n observations drawn from the X population, and let s_Y^2 be the sample variance based on m observations drawn from the Y population. We recall that $(n-1)s_X^2/\sigma_X^2$ and $(m-1)s_Y^2/\sigma_Y^2$ are two independent random variables having chi-square distributions with $n-1$ and $m-1$ degrees of freedom, respectively. Consequently,

$$\frac{[(n-1)s_X^2/\sigma_X^2]/(n-1)}{[(m-1)s_Y^2/\sigma_Y^2]/(m-1)} = \frac{s_X^2/\sigma_X^2}{s_Y^2/\sigma_Y^2}$$

has the F distribution with $n-1$ and $m-1$ degrees of freedom. Hence, we can state that

$$P\left\{F_{\alpha/2;\, n-1,\, m-1} \leq \frac{s_X^2/\sigma_X^2}{s_Y^2/\sigma_Y^2} \leq F_{1-\alpha/2;\, n-1,\, m-1}\right\} = 1 - \alpha$$

TABLE 12.1 Summary of Point and Interval Estimates of the Parameters of the Normal Distribution

Parameters	Conditions	Point Estimates	Confidence Interval
μ	known σ	$\bar{x}$	$\bar{x} - Z_{1-\alpha/2}\dfrac{\sigma}{\sqrt{n}} \leq \mu \leq \bar{x} + Z_{1-\alpha/2}\dfrac{\sigma}{\sqrt{n}}$
μ	unknown σ	$\bar{x}$	$\bar{x} - t_{1-\alpha/2;\, n-1}\dfrac{s}{\sqrt{n}} \leq \mu \leq \bar{x} + t_{1-\alpha/2;\, n-1}\dfrac{s}{\sqrt{n}}$
$\mu_X - \mu_Y$	known σ_X, σ_Y	$\bar{x} - \bar{y}$	$\bar{x} - \bar{y} - Z_{1-\alpha/2}\sqrt{\dfrac{\sigma_X^2}{n\dagger} + \dfrac{\sigma_Y^2}{m\ddagger}} \leq \mu_X - \mu_Y \leq \bar{x} - \bar{y} + Z_{1-\alpha/2}\sqrt{\dfrac{\sigma_X^2}{n} + \dfrac{\sigma_Y^2}{m}}$
$\mu_X - \mu_Y$	$\sigma_X = \sigma_Y = \sigma$ σ unknown	$\bar{x} - \bar{y}$	$\bar{x} - \bar{y} - t_{1-\alpha/2;\, n+m-2}\sqrt{\dfrac{(n-1)s_X^2 + (m-1)s_Y^2}{n+m-2}}\sqrt{\dfrac{1}{n} + \dfrac{1}{m}}$ $\leq \mu_X - \mu_Y \leq \bar{x} - \bar{y} + t_{1-\alpha/2;\, n+m-2}\sqrt{\dfrac{(n-1)s_X^2 + (m-1)s_Y^2}{n+m-2}}\sqrt{\dfrac{1}{n} + \dfrac{1}{m}}$
$\mu_X - \mu_Y$	$\sigma_X \neq \sigma_Y$ both unknown	$\bar{x} - \bar{y}$	$\bar{x} - \bar{y} - t_{1-\alpha/2;\, v\S}\sqrt{\dfrac{s_X^2}{n} + \dfrac{s_Y^2}{m}} \leq \mu_X - \mu_Y \leq \bar{x} - \bar{y} + t_{1-\alpha/2;\, v}\sqrt{\dfrac{s_X^2}{n} + \dfrac{s_Y^2}{m}}$
σ^2		s^2	$\dfrac{(n-1)s^2}{\chi^2_{1-\alpha/2;\, n-1}} \leq \sigma^2 \leq \dfrac{(n-1)s^2}{\chi^2_{\alpha/2;\, n-1}}$
$\dfrac{\sigma_X^2}{\sigma_Y^2}$		$\dfrac{s_X^2}{s_Y^2}$	$\dfrac{s_X^2}{s_Y^2}\dfrac{1}{F_{1-\alpha/2;\, n-1,\, m-1}} \leq \dfrac{\sigma_X^2}{\sigma_Y^2} \leq \dfrac{s_X^2}{s_Y^2}\dfrac{1}{F_{\alpha/2;\, n-1,\, m-1}}$

† n = Number of observations taken from X population.
‡ m = Number of observations taken from Y population.
§ $v = \dfrac{(s_X^2/n + s_Y^2/m)^2}{\dfrac{(s_X^2/n)^2}{n+1} + \dfrac{(s_Y^2/m)^2}{m+1}} - 2$

which can be rewritten as

$$P\left\{\frac{s_X^2}{s_Y^2}\,\frac{1}{F_{1-\alpha/2;\,n-1,\,m-1}} \leq \frac{\sigma_X^2}{\sigma_Y^2} \leq \frac{s_X^2}{s_Y^2}\,\frac{1}{F_{\alpha/2;\,n-1,\,m-1}}\right\} = 1-\alpha \tag{12.27}$$

Whence a $100(1-\alpha)\%$ confidence interval for σ_X^2/σ_Y^2 is given by

$$\left[\frac{s_X^2}{s_Y^2}\,\frac{1}{F_{1-\alpha/2;\,n-1,\,m-1}},\ \frac{s_X^2}{s_Y^2}\,\frac{1}{F_{\alpha/2;\,n-1,\,m-1}}\right] \tag{12.28}$$

A summary for point estimates and confidence intervals is given in Table 12.1.

12.4 Confidence Limits for the Parameter p of the Binomial Distribution

The binomial distribution is characterized by two parameters n and p. Its probability function is given by

$$P_X(k) = \binom{n}{k} p^k (1-p)^{n-k} \qquad k = 1, 2, \ldots, n \qquad 0 \leq p \leq 1 \tag{12.29}$$

where p = proportion of the population having a certain common characteristic and n = sample size.

To illustrate, suppose a sample of size n is drawn with replacement from a very large lot. Each item is inspected and classified as defective or non-defective. Now if the number of defective items found in n independent trials is denoted by X, X is a random variable having a binomial distribution with parameters n (the sample size) and p (percentage defective of the lot). p is usually unknown and can be estimated.

In Sec. 12.2.9 we proved that the maximum likelihood estimator of p is k/n, where k is the observed value of a binomial variate X. We wish to find a $100(1-\alpha)\%$ confidence interval for p. Let p_L and p_U denote the lower and upper confidence limits, respectively. Exact confidence limits for p can be obtained in terms of the percentiles of the F distribution. When the sample size is not comparatively large ($n < 100$), we can make use of the relationship developed by Hald,† namely,

$$\begin{aligned} P\{X \leq k\} &= \sum_{r=0}^{k} \binom{n}{r} p^r (1-p)^{n-r} \\ &= 1 - P\left\{F < \frac{(n-k)p}{(k+1)(1-p)}\right\} \end{aligned} \tag{12.30}$$

†See A. Hald, *Statistical Theory with Engineering Applications*, New York, John Wiley & Sons, 1952.

where F is a random variable having an F distribution with $2(k + 1)$ and $2(n - k)$ degrees of freedom. Using Hald's relationship we write

$$\sum_{r=0}^{k} \binom{n}{r} p_U^r (1 - p_U)^{n-r} = 1 - P\left\{F_{2(k+1),\,2(n-k)} < \frac{(n-k)p_U}{(k+1)(1-p_U)}\right\} = \frac{\alpha}{2}$$

or, equivalently,

$$1 - \frac{\alpha}{2} = P\left\{F_{2(k+1),\,2(n-k)} < \frac{(n-k)p_U}{(k+1)(1-p_U)}\right\}$$

Hence

$$\frac{(n-k)p_U}{(k+1)(1-p_U)} = F_{1-\alpha/2;\,2(k+1),\,2(n-k)}$$

whence

$$p_U = \frac{(k+1)F_{1-\alpha/2;\,2(k+1),\,2(n-k)}}{(n-k) + (k+1)F_{1-\alpha/2;\,2(k+1),\,2(n-k)}} \tag{12.31}$$

Likewise

$$\sum_{r=k}^{n} \binom{n}{r} p_L (1 - p_L)^{n-r} = 1 - \sum_{r=0}^{k-1} \binom{n}{r} p_L (1 - p_L)^{n-r}$$
$$= P\left\{F_{2k,\,2(n-k+1)} < \frac{(n-k+1)p_L}{k(1-p_L)}\right\} = \frac{\alpha}{2}$$

Hence

$$\frac{(n-k+1)p_L}{k(1-p_L)} = F_{\alpha/2;\,2k,\,2(n-k+1)}$$
$$= \frac{1}{F_{1-\alpha/2;\,2(n-k+1),\,2k}}$$

whence

$$p_L = \frac{k}{k + (n-k+1)F_{1-\alpha/2;\,2(n-k+1),\,2k}} \tag{12.32}$$

When n is large and p is neither close to 0 nor to 1, we can obtain an approximate confidence interval by using normal approximation to the binomial distribution. Thus we consider the binomial variate as approximately normally distributed with mean np and variance $np(1 - p)$. Hence we assert that

$$P\left\{Z_{\alpha/2} \le \frac{k - np}{\sqrt{np(1-p)}} \le Z_{1-\alpha/2}\right\} \cong 1 - \alpha \tag{12.33}$$

which is equivalent to

$$P\left\{\frac{(k-np)^2}{np(1-p)} \le Z_{1-\alpha/2}^2\right\} \cong 1 - \alpha$$

which, in turn, is equivalent to the event

$$p^2(n + Z_{1-\alpha/2}^2) - p(2k + Z_{1-\alpha/2}^2) + \frac{k^2}{n} \le 0 \tag{12.34}$$

Now it is apparent that we seek the values of p so that the above inequality is satisfied. If we set the left-hand side of (12.34) equal to zero, then we obtain a quadratic equation in p with the roots

$$\frac{2k + Z_{1-\alpha/2}^2 \pm \sqrt{(2k + Z_{1-\alpha/2}^2)^2 - 4(k^2/n)(n + Z_{1-\alpha/2}^2)}}{2(n + Z_{1-\alpha/2}^2)} \tag{12.35}$$

We note that the left-hand side of (12.34) is negative when p lies between the roots given by (12.35). Hence the $100(1 - \alpha)\%$ confidence limits for p are

$$p_L = \text{smaller root} \qquad p_U = \text{larger root}$$

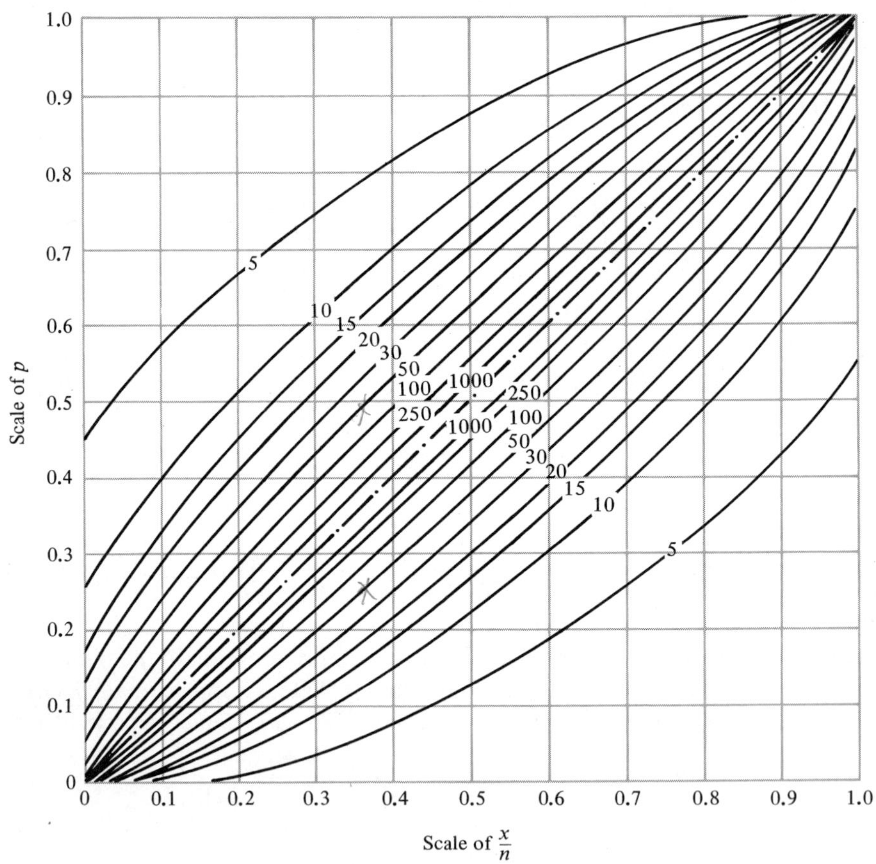

Fig. 12.2 Confidence belts for proportions (confidence coefficient .90).

Reproduced with permission of Professor E. S. Pearson from C. J. Clopper and E. S. Pearson, "The Use of Confidence or Fiducial Limits Illustrated in the Case of the Binomial," *Biometrika*, **26**, 404, 1934.

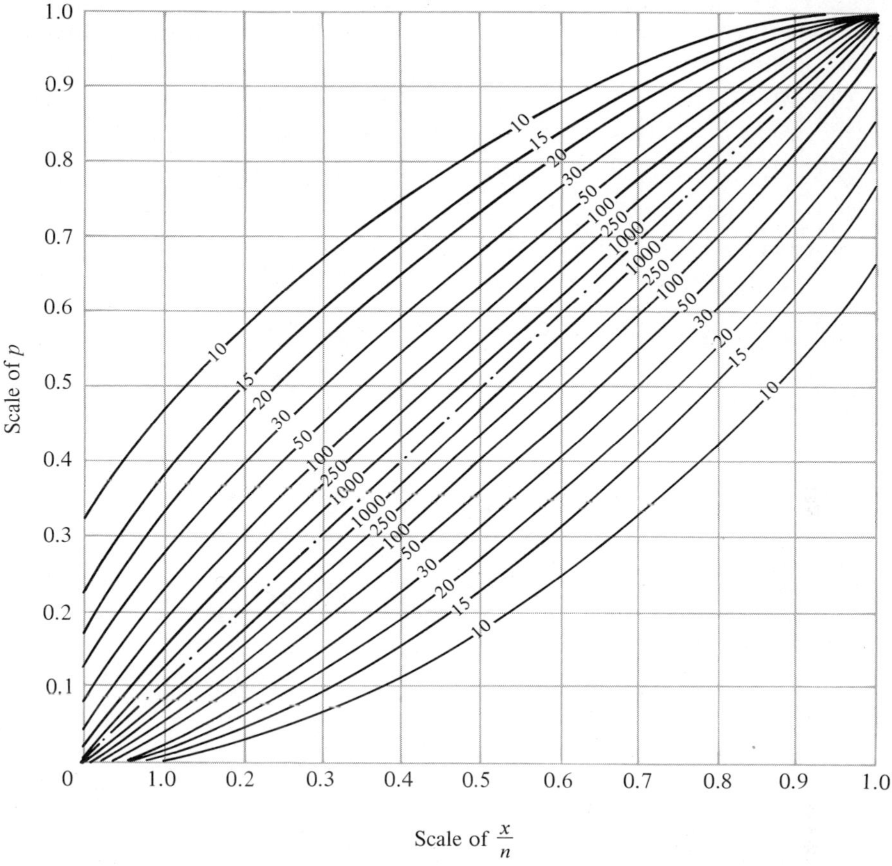

Fig. 12.3 Confidence belts for proportions (confidence coefficient .95).

Confidence limits for various sample sizes have been calculated in terms of x/n by Clopper and Pearson. The charts in Figs. 12.2, 12.3, and 12.4 give confidence belts for .10, .05, and .01 levels of significance, respectively.

Example 12.16 Find a 90% confidence interval for the percentage defective in a very large lot of material on the basis of a random sample of 50 items. The number of defective items found in the sample is 18.

Solution. The exact confidence limits are

$$p_U = \frac{19F_{.95;\,38,64}}{32 + 19F_{.95;38,64}} = .48$$

$$p_L = \frac{18}{18 + 33F_{.95;66,36}} = .25$$

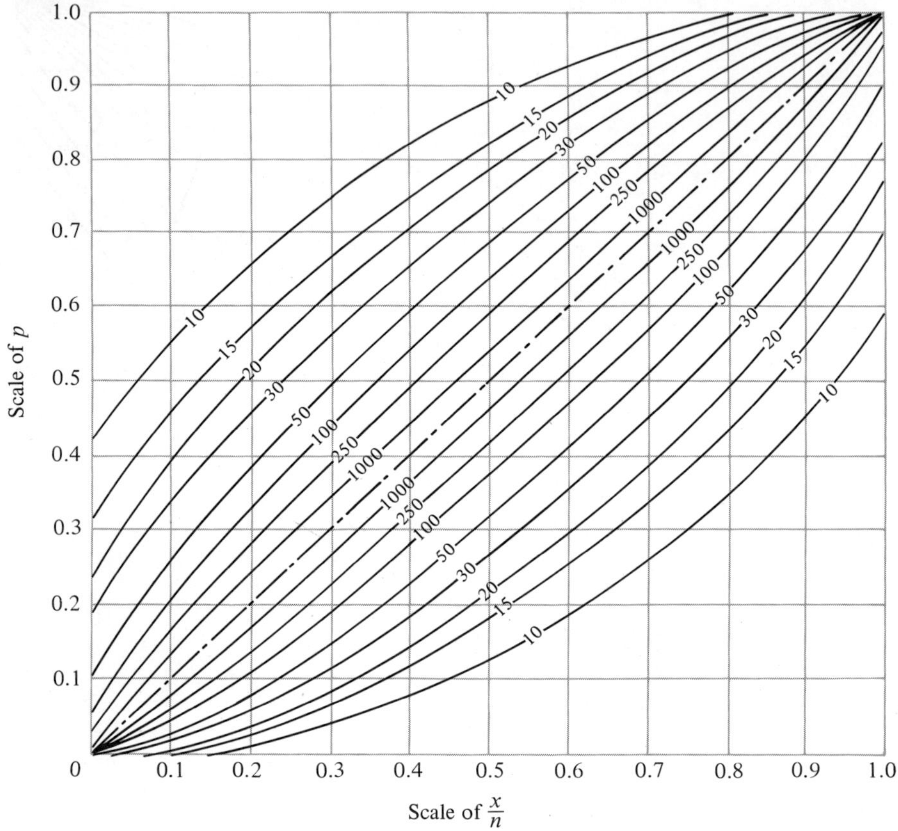

Fig. 12.4 Confidence belts for proportions (confidence coefficient .99).

The confidence limits for p can be obtained from Fig. 12.2 by entering $x/n = \frac{18}{50} = .36$ on the abscissa and reading the values of p associated with the sample size 50. We find that the confidence interval for p is $.26 < p < .47$, which is very close indeed to the confidence interval obtained by using the F distribution.

We proceed now to calculate the 90% confidence interval using the normal approximation. We have

$$k = 18 \qquad (Z_{.95})^2 = 2.706 \qquad n = 50$$

Hence by (12.35) the roots (p_L, p_U) are

$$\frac{38.706 \pm \sqrt{(38.706)^2 - (25.92)(52.706)}}{2(52.706)}$$

i.e., $p_L = .26$ and $p_U = .47$. Thus the approximate 90% confidence interval for p using the normal approximation is [.26, .47] which is identical to the interval given by Fig. 12.2.

PROBLEMS

12.1. If $X_1, X_2, \ldots, X_n$ are independent identically distributed random variables, prove that the statistic

$$\frac{1}{2(n-1)} \sum_{i=1}^{n-1} (X_{i+1} - X_i)^2$$

is an unbiased estimator of the variance.

12.2. Two possible estimators for the parameter θ are $\hat{\theta}_1$ and $\hat{\theta}_2$. $E(\hat{\theta}_1) = \theta$ and $E(\hat{\theta}_2) = \theta + .5$. The variances of $\hat{\theta}_1$ and $\hat{\theta}_2$ are 6 and 4, respectively. Which estimator is better? Justify your answer.

12.3. Suppose X_1, X_2, X_3, and X_4 are four independent random variables with common mean and variances 1, 1.5, 2, and 2.5, respectively. Given observations x_1, x_2, x_3, and x_4, find an unbiased linear estimator for the mean with minimum variance.

12.4. Prove that the statistic

$$\frac{\sum_{i=1}^{n} (X_i - \bar{X})^2}{n}$$

is not an unbiased estimator of the variance of a normally distributed random variable X.

12.5. Suppose X is a random variable with mean μ and variance σ^2. Given four independent observations find the following:
(a) An unbiased linear estimator for μ with minimum variance.
(b) A linear estimator for μ with minimum mean square error. Assume $\sigma/\mu = \frac{1}{2}$.

12.6. If X is a normally distributed random variable with unknown mean μ and variance $\sigma^2 = 1$, prove that the statistic

$$\bar{X} = \frac{1}{n} \sum_{i=1}^{n} X_i$$

is a sufficient estimator of μ.

12.7. The variance of a normally distributed random variable X with known mean μ is estimated by the statistic

$$S^2 = \frac{1}{n} \sum_{i=1}^{n} (X_i - \mu)^2$$

Show that S^2 is the most efficient estimator of the variance.

12.8. Suppose S^2 in the preceding problem is evaluated on the basis of ten observations. Furthermore, suppose that the estimator

$$S_1^2 = \frac{1}{n-1} \sum_{i=1}^{n} (X_i - \bar{X})^2$$

is suggested as an alternative. Find the efficiency of S_1^2 when $n = 10$.

12.9. Variate X is normally distributed with known mean μ and unknown standard deviation σ. Find the minimal variance of an unbiased estimator of σ based on a sample of size n.

12.10. Variate X has the following density function:

$$f_X(t) = \lambda^2 t e^{-\lambda t} \qquad \lambda > 0,\ t \geqq 0$$

The parameter λ is estimated from a sample of size n.

(a) Show that

$$\hat{\theta} = \frac{2n - 1}{\sum_{i=1}^{n} t_i}$$

is an unbiased and consistent estimator of λ.

(b) Is $\hat{\theta}$ the most efficient estimator of λ?

(c) Is $\hat{\theta}$ an asymptotically most efficient estimator of λ?

12.11. An electronic device consists of three components. The device fails to function when all components fail. Each component has an independent exponential lifetime distribution with parameter λ. Let t_i be the time of the ith failure ($t_1 \leq t_2 \leq t_3$).

(a) What is the expected failure time of the device?

(b) Show that $\frac{1}{3} \sum_{i=1}^{3} t_i$ is an unbiased estimator for $1/\lambda$.

(c) Find the variance of the failure time of the device.

(d) Show that $2\lambda \sum_{i=1}^{3} t_i$ has a chi-square distribution with six degrees of freedom.

(e) Find a 90% confidence interval for $1/\lambda$.

12.12. Suppose in the preceding problem the device consists of n components. Let t_i be the time of the ith failure ($t_1 \leq t_2 \leq \cdots \leq t_n$). Show that

$$\hat{\theta} = \frac{1}{n}(t_1 + t_2 + \cdots + t_n)$$

is the most efficient estimator for $1/\lambda$.

12.13. Find the maximum likelihood estimator of λ in Prob. 12.10.

12.14. Variate X has the density function given by

$$f_X(a) = \frac{1}{k} \qquad 0 < a < k$$

Find the maximum likelihood estimator of k given a sample of size n.

12.15. The distribution of random variable X is given by

$$f_X(t) = \tfrac{1}{2}\lambda^3 t^2 e^{-\lambda t} \qquad \lambda > 0 \qquad t \geq 0$$

What is the maximum likelihood estimator of σ_X^2?

12.16. Random variable X has the following density function:

$$f_X(t) = (1 + \lambda)t^{\lambda} \qquad 0 < t < 1$$

Prove that the maximum likelihood estimator of λ for a sample of size n is

$$-1 - \frac{n}{\sum_{i=1}^{n} \ln t_i}$$

12.17. The probability function of variate X is given by

$$P_X(k) = (1 - q)q^k \qquad k = 0, 1, \ldots$$

Find the maximum likelihood estimator of q.

12.18. Find a 90% confidenence interval for the mean length of a certain nail based on the data below (assume normality).

1.14	1.10	1.13	1.15
1.13	1.12	1.13	1.10
1.15	1.12	1.14	1.10
1.13	1.11	1.14	1.11

12.19. In the preceding problem assume that the standard deviation of the length of a nail is .01. Find a 90% confidence interval.

12.20. Suppose that a random sample of 15 observations from a normal population yields

$$\sum x_i = 39.60 \qquad \sum x_i^2 = 281.83$$

(a) Find a 90% confidence interval for the mean.
(b) Find a 90% confidence interval for the standard deviation.

12.21. Ten cigarettes were picked at random from a large lot. The individual weights in units of .01 oz were as follows:

3.79	3.66	3.82	4.21	3.97	4.15	4.04	4.25	3.86	4.10

Find a 90% confidence interval for the mean weight (assume normality).

12.22. Suppose in the preceding problem σ^2 is known and equal to .002 oz and it is required to obtain an estimate within .001 oz of the true average weight in 95% of the cases. What is the required sample size?

12.23. A factory has two sawing machines, both of which cut steel bars to a specified length of 2 in. Lengths of independent samples of each machine were measured to the nearest hundredth of an inch with the following results:

Machine I:	1.98	2.10	1.90	2.05	2.08	
Machine II:	1.99	2.00	1.93	2.11	2.10	2.13

Find a 99% confidence interval for the expected difference between the length of bars produced by the two machines (assume normality).

12.24. The following results are calculated for two samples from normal distributions:

$$X: \quad n_X = 5 \quad \sum x_i = 64 \quad \sum x_i^2 = 830$$
$$Y: \quad n_Y = 6 \quad \sum y_i = 75 \quad \sum y_i^2 = 950$$

Find a 95% confidence interval for $\mu_X - \mu_Y$.

12.25. In the preceding problem find a 90% confidence interval for σ_X/σ_Y.

12.26. Ten pipes were coated with coating *A*, and ten pipes with coating *B*. Pairs of specimens were placed in several types of very heterogeneous soil. Depth of maximum pits was measured in thousandths of an inch. The results were as follows:

Coating A	*Coating B*	*Coating A*	*Coating B*
33	51	53	59
38	40	37	43
51	38	41	36
48	60	56	46
37	42	56	49

Find a 90% confidence interval for the expected difference in corrosion rates between the two coatings (assume normality).

12.27. How large a sample is required in order to have a 99% confidence that the estimate of the standard deviation of measurable characteristic be at most 1.3 times the true standard deviation?

12.28. Variate X has a Poisson distribution with unknown mean. If ten observations yield the values 8, 9, 22, 10, 3, 6, 20, 5, 11, and 6, find an approximate 90% confidence interval for λ. Use normal approximation.

12.29. A lake contains N fish of a certain kind. A netting experiment yielded 60 fish, which were tagged and thrown back into the lake. A second experiment yielded 95 fish of which 4 were found to be tagged. The fish was thrown back into the lake. A third experiment yielded 80 fish of which 3 were found to be tagged.
(a) Estimate N.
(b) What is the approximate standard deviation of the estimate obtained in (a)?

12.30. Fifty items were drawn at random from a large lot. Five items were found to be defective. Find a 95% confidence interval for the true percent defective of the lot.

SUGGESTED REFERENCES

[1] Blackwell, D., "Conditional Expectation and Unbiased Sequential Estimation," *Annals of Mathematical Statistics*, **18**, 1947.

[2] Cramér, H., *Mathematical Methods in Statistics*, Princeton, N.J., Princeton University Press, 1946.

[3] David, F. N., and J. Neyman, "Extension of the Markoff Theorem on Least Squares," *Statistical Research Memoirs*, **2**, 1938.

[4] Dixon, W. J. and F. J. Massey, *Introduction to Statistical Analysis*, 3rd ed., New York, McGraw-Hill Book Company, 1969.

[5] Eisenhart C., M. W. Hastay, and W. A. Wallis (eds.), *Selected Techniques of Statistical Analysis*, New York, McGraw-Hill Book Company, 1947.

[6] Halmos, P., "The Theory of Unbiased Estimation," *Annals of Mathematical Statistics*, **17**, 1946.

[7] Halmos, P., and L. J. Savage, "Application of the Radon–Nikodym Theorem to the Theory of Sufficient Statistics," *Annals of Mathematical Statistics*, **20**, 1949.

[8] Le Cam, L., "On the Asymptotic Theory of Estimation and Testing Hypotheses," *Proceedings of the Third Berkeley Symposium*, **1**, 1956.

[9] Lehmann, E. L., and H. Scheffé, "Completeness, Similar Regions and Unbiased Estimation," Part I, *Sankhya*, **10**, 1950.

[10] Mood, A. M., *Introduction to the Theory of Statistics*, New York, McGraw-Hill Book Company, 1950.

[11] Rao, C. R., "Information and Accuracy Obtainable in an Estimation of a Statistical Parameter," *Bulletin of the Calcutta Mathematics Society*, **37**, 1945.

[12] Roy, S. N., *Some Aspects of Multivariate Analysis*, New York, John Wiley & Sons, Inc., 1957.

[13] Stein, C., "Unbiased Estimates with Minimum Variance," *Annals of Mathematical Statistics*, **21**, 1950.

[14] Wilks, S. S., *Mathematical Statistics*, 2nd ed., New York, John Wiley & Sons, Inc., 1962.

[15] Yule, G. U., and M. G. Kendall, *An Introduction to the Theory of Statistics*, New York, Hafner Publishing Co., Inc., 1958.

13

ANALYSIS OF VARIANCE

13.1 Introduction

In Chapter 10 we developed procedures for testing hypotheses about two population means. In particular, we referred to the case of testing the equality of two population means under the assumptions that the populations were independent, normally distributed, and with equal but unknown variances. In that case the t test was used to test the equality of means. Now consider the situation in which we have more than two, say k, populations with means $\mu_1, \mu_2, \ldots, \mu_k$ and we wish to test the hypothesis

$$H_0: \mu_1 = \mu_2 = \cdots = \mu_k \quad \text{against} \quad H_1: \text{at least two means are not equal}$$

at a prescribed level of significance. An appealing procedure would be to enumerate every possible pair and test whether each pair differs using the t test. Accordingly, one is required to run a series of t tests. More precisely, one would require three such tests for three means, six such tests for four means, and in general $k(k - 1)/2$ such tests for k means. Obviously the number of paired comparisons increases as the number of the means increases. Apart from the difficulty imposed by the large number of t tests, each individual t test has associated with it an α risk of committing an error of the first kind. As a result, the overall error of the first kind would be much greater than the prescribed level of significance. To illustrate, suppose we wish to test the equality of three means at the $\alpha\%$ level of significance. In symbols, the hypothesis to be tested is

$$H_0: \mu_1 = \mu_2 = \mu_3 \quad \text{against} \quad H_1: \text{at least two means are not equal}$$

If we were to perform a t test for each of the following hypotheses:

$$\begin{array}{llll} & H_0': \mu_1 = \mu_2 & H'': \mu_1 = \mu_3 & H_0''': \mu_2 = \mu_3 \\ \text{against} & H_1': \mu_1 \neq \mu_2 & H_1'': \mu_1 \neq \mu_3 & H_1''': \mu_2 \neq \mu_3 \end{array}$$

at the $\alpha\%$ level of significance, then the probability of drawing a wrong conclusion in at least one test would be much greater than the prescribed level of significance α. It can be shown that for testing three means with a prescribed level of significance $\alpha = 5\%$ the overall error becomes 13%, and for six means the overall error becomes 40%. What we are seeking is a method that permits us to perform one test with a *single* α risk. The method is called *analysis of variance*. It is a powerful tool for detecting differences among the means of several populations. The analysis of variance technique, however, can be applied to a considerable variety of experimental patterns other than this simple pattern. The analysis of variance is actually a collection of statistical methods used to analyze data from a designed experiment involving several *sources of variation*, to partition the total variation into components, and to estimate and to test the significance of the effects of these components. In applying the analysis of variance technique to experimental data, one must first formulate an appropriate *mathematical model* and decide on the information he wishes to obtain in terms of this model. The following sections will deal with formulation, analysis, and interpretation of various models.

13.2 One-Way Classification; Fixed Effects Model

Suppose a factory has three machines, each of which makes $\frac{1}{4}$-in. aluminum washers from flat strips. It is desired to know if the average thickness of washers produced by each machine is the same. The measure of performance is the thickness to the nearest thousandth of an inch. The measure of performance is usually called the *response variable*. Before we proceed to analyze this example we shall introduce the following nomenclature:

13.2.1 Nomenclature

Factor is a quality of property according to which the data are classified. In our example the machines are the *factor* under study.

Level describes the particular property defining each category in the classification. Thus the three machines are the three *levels* of the factor being studied. A particular level or a combination of levels is called a "treatment."

Replication is the repetition of a test.

Experimental unit is the smallest division of the experimental material. Thus a strip is an *experimental unit*.

Randomization is the assignment of treatments to experimental units in a random order such that each treatment has an equal chance of being tested under the most favorable conditions and possible sources of variations may not cause systematic error. In one-way classification experimental units are assigned *at random* to any one of the treatments under investigation. Thus, in our example, the strips are assigned at random to any one of the three machines. For this reason the one-way classification is sometimes called a *completely randomized design*.

The experimental pattern is called *one-way classification* because we are comparing a number of levels of a single factor. Sometimes this pattern is referred to as *single-factor experiments with no restrictions on randomization*.

13.2.2 Notation

The observations are recorded as shown in Table 13.1. The first column represents a random sample of size n_1 from the population that has received treatment 1; the second, a random sample of size n_2 from the population that has received treatment 2; the jth column, a random sample of size n_j from the population that has received treatment j. The symbols are defined as follows

TABLE 13.1 Sample Layout for One-Way Classification

	Treatments						
	1	2	...	j	...	k	
Observations	y_{11}	y_{12}	...	y_{1j}	...	y_{1k}	
	y_{21}	y_{22}	...	y_{2j}	...	y_{2k}	
	...	...	...	...	...	...	
	y_{i1}	y_{i2}	...	y_{ij}	...	y_{ik}	
	...	...	...	...	...	...	
	$y_{n_1 1}$	...	...	$y_{n_j j}$	...	...	
	...	$y_{n_2 2}$	...	...	...	$y_{n_k k}$	
Totals	$T_{.1}$	$T_{.2}$	...	$T_{.j}$	...	$T_{.k}$	$T_{..}$
Number	n_1	n_2	...	n_j	...	n_k	N
Means	$\bar{y}_{.1}$	$\bar{y}_{.2}$	...	$\bar{y}_{.j}$	...	$\bar{y}_{.k}$	$\bar{y}_{..}$

(note that the n_j's are not necessarily equal):

The jth observation receiving

$$\text{treatment } j = y_{ij},\ i = 1, 2, \ldots, n_j,\ j = 1, 2, \ldots, k$$

$$\text{Total of the } j\text{th column} = T_{.j} = \sum_{i=1}^{n_j} y_{ij}$$

$$\text{Mean of the } j\text{th column} = \bar{y}_{.j} = \frac{T_{.j}}{n_j} = \frac{\sum_{i=1}^{n_j} y_{ij}}{n_j}$$

$$\text{Total of all observations} = T_{..} = \sum_{j=1}^{k} T_{.j} = \sum_{j=1}^{k}\sum_{i=1}^{n_j} y_{ij}$$

$$\text{Mean of all observations} = \bar{y}_{..} = \frac{T_{..}}{N} = \frac{\sum_{j=1}^{k}\sum_{i=1}^{n_j} y_{ij}}{\sum_{j=1}^{k} n_j}$$

13.2.3 The Model

Consider the situation in which three machines are tested to ascertain whether the average thickness of washers produced by each machine is the same. We wish to formulate a mathematical model that will represent the real situation as truly as possible and will enable us to choose between the hypotheses to be tested. Thus, in this situation three specific machines are used in the investigation. Let n_1, n_2, and n_3 denote the number of washers punched by machines 1, 2, and 3, respectively. The flat strips, from which the washers are punched, are assumed to be homogeneous and hence can be discounted as a possible source of variability. The strips are, however, assigned to machines at random. This random choice will help to eliminate the coincidence of effects, which might distort our conclusions. Therefore, using our previous notation, $y_{11}, y_{21}, \ldots, y_{n_1 1}$ represent the thicknesses of n_1 washers punched by machine 1. These observations are random variables distributed in some fashion around their expected value (μ_1). It follows that the thickness of the second washer punched by machine 1 can be expressed as

$$y_{21} = \mu_1 + e_{21}$$

where e_{21}, called a *chance error*, represents the random variation of the y_{21} observation from μ_1. More generally, we can express the y_{ij} observation as

$$y_{ij} = \mu_j + e_{ij} \tag{13.1}$$

Now if the punching operation has no effect on the thickness of the strips fed into each machine, we shall expect that $\mu_1 = \mu_2 = \mu_3 = \mu$ (the overall mean). Consequently, we say in this situation that there is no *machine effect*. However, if the machine effect is significant, we shall expect that $\mu_1 \neq \mu_2 \neq \mu_3 \neq \mu$ (strictly speaking, at least two means are not equal), and it follows that

(13.2) $$\mu_j = \mu + \delta_j$$

where δ_j, called *fixed effect peculiar to the jth machine*, represents the deviation of μ_j from the overall mean μ. The overall mean, then, is

$$\mu = \frac{\sum_{j=1}^{k} n_j \mu_j}{\sum_{j=1}^{k} n_j}$$

Combining Eqs. (13.1) and (13.2) we find that the model then is

(13.3) $$y_{ij} = \mu + \delta_j + e_{ij} \qquad i = 1, 2, \ldots, n_j \qquad j = 1, 2, \ldots, k$$

so that

$$\sum_{j=1}^{k} n_j \delta_j = \sum_{j=1}^{k} n_j(\mu_j - \mu) = \sum_{j=1}^{k} n_j \mu_j - N\mu = N\mu - N\mu = 0$$

It should be noted that the δ_j's are *not* random variables but rather *systematic components*, each peculiar to a particular treatment. Accordingly, this model is usually called *systematic component* or *parametric* or *fixed effects model.* The conclusion derived from the experiment will concern the treatments (three machines in this example) considered.

The analysis of variance technique is based on certain plausible assumptions that can be summarized as follows:

1. The y_{ij}'s are normally distributed random variables.
2. The e_{ij}'s are mutually independent and normally distributed random variables with $E(e_{ij}) = 0$ and $\text{Var}(e_{ij}) = \sigma_e^2$ for all i and j.
3. The observations are drawn from a statistically controlled process.

Although the test of significance given in Sec. 13.2.6 is derived under these assumptions, empirical studies have shown that violations of some of these assumptions have little effect on the test of significance. Some remarks on the validity of these assumptions are noted below:

1. The χ^2 *goodness-of-fit test* can be used for testing the validity of assumption 1. However, empirical studies have shown that the test of significance is not seriously affected by departures from normality.
2. The validity of the assumption that the e_{ij}'s are mutually independent depends upon how careful the experimenter is to ensure the random assignment of treatments to experimental units. Randomization provides an excellent tool by which the correlations among results tend to cancel, leaving treatment effects independent of any systematic bias caused by any other source of variability.
3. The validity of the assumption that the variance of the e_{ij}'s is the same for all treatment populations can be tested by the Bartlett's test described in Chapter 10. Note that this test depends on normality. Empirical studies

have shown that moderate heterogeneity of variance has small effect on the test of significance. However, in cases where extreme heterogeneity of variance exists, an appropriate *transformation* of the data can be found, provided that the relationship between the population mean and variance is continuous and known. For example, if the populations have the Poisson distribution, then the *square root transformation* of the data will stabilize the variance. In other words, $\sqrt{y_{ij}}$ is found to be approximately normally distributed with variance $\cong \frac{1}{4}$. The interested reader may consult the references given at the end of this chapter for further explication of this topic.

13.2.4 The Analysis

Consider the identity

$$(y_{ij} - \bar{y}_{..}) = (y_{ij} - \bar{y}_{.j}) + (\bar{y}_{.j} - \bar{y}_{..})$$

Squaring and summing over i we get

$$\begin{aligned}\sum_i (y_{ij} - \bar{y}_{..})^2 &= \sum_i (y_{ij} - \bar{y}_{.j})^2 + \sum_i (\bar{y}_{.j} - \bar{y}_{..})^2 \\ &\quad + 2\sum_i (y_{ij} - \bar{y}_{.j})(\bar{y}_{.j} - \bar{y}_{..}) \\ &= \sum_i (y_{ij} - \bar{y}_{.j})^2 + \sum_i (\bar{y}_{.j} - \bar{y}_{..})^2 \qquad (13.4)\end{aligned}$$

since

$$\sum_i (y_{ij} - \bar{y}_{.j}) = \sum_i \left(y_{ij} - \frac{T_{.j}}{n_j}\right) = T_{.j} - T_{.j} = 0, \quad \text{for all } j$$

Summing Eq. (13.4) over j gives us

$$\begin{aligned}\sum_j \sum_i (y_{ij} - \bar{y}_{..})^2 &= \sum_j \sum_i (y_{ij} - \bar{y}_{.j})^2 + \sum_j \sum_i (\bar{y}_{.j} - \bar{y}_{..})^2 \\ &= \sum_j \sum_i (y_{ij} - \bar{y}_{.j})^2 + \sum_j n_j(\bar{y}_{.j} - \bar{y}_{..})^2 \qquad (13.5)\end{aligned}$$

To simplify the presentation, let

$$\text{SST} = \sum_j \sum_i (y_{ij} - \bar{y}_{..})^2$$

which will be referred to as the *total sum of squares*,

$$\text{SSW} = \sum_j \sum_i (y_{ij} - \bar{y}_{.j})^2$$

which will be called *within samples sum of squares*, and

$$\text{SSG} = \sum_j n_j(\bar{y}_{.j} - \bar{y}_{..})^2$$

which will be known as *among samples sum of squares*. Hence, Eq. (13.5) may be written

$$\text{SST} = \text{SSW} + \text{SSG}$$

i.e., the total sum of squares has been partitioned into two components: the sum of squares within samples and the sum of squares among samples.

Now we proceed to derive the expected value of SSW. By definition,

$$\text{SSW} = \sum_j \sum_i (y_{ij} - \bar{y}_{.j})^2 = \sum_j (n_j - 1)s_j^2$$

where s_j^2 is the sample variance of the jth population. Then the expected value of SSW is

$$E(\text{SSW}) = \sum_j (n_j - 1)E(s_j^2) = \sum_j (n_j - 1)\sigma_e^2$$
$$= (N - k)\sigma_e^2$$

since according to assumption 2 the k populations have common variance σ_e^2. Hence, it follows that

$$\sigma_e^2 = E\left(\frac{\text{SSW}}{N - k}\right) \tag{13.6}$$

Therefore, $\text{SSW}/(N - k)$, usually called *mean sum of squares within samples* (denoted by MSW), is an unbiased estimate of σ_e^2. To evaluate the expected value of the sum of squares among samples we obtain from the model of Eq. (13.3)

$$\bar{y}_{.j} = \frac{\sum_i y_{ij}}{n_j} = \mu + \delta_j + \frac{\sum_i e_{ij}}{n_j} = \mu + \delta_j + \bar{e}_{.j} \tag{13.7}$$

where $\bar{e}_{.j} = \sum_i e_{ij}/n_j$. Also,

$$\bar{y}_{..} = \mu + \frac{\sum_j \sum_i e_{ij}}{N} = \mu + \bar{e}_{..} \tag{13.8}$$

where

$$\bar{e}_{..} = \frac{\sum_j \sum_i e_{ij}}{N}$$

since $\sum_j n_j \delta_j = 0$. Subtracting Eq. (13.8) from Eq. (13.7) we get

$$\bar{y}_{.j} - \bar{y}_{..} = \delta_j + (\bar{e}_{.j} - \bar{e}_{..})$$

Squaring and multiplying by n_j gives us

$$n_j(\bar{y}_{.j} - \bar{y}_{..})^2 = n_j \delta_j^2 + n_j(\bar{e}_{.j} - \bar{e}_{..})^2 + 2\delta_j n_j(\bar{e}_{.j} - \bar{e}_{..})$$

Summing over j and taking expected value we obtain

$$E[\sum_j n_j(\bar{y}_{.j} - \bar{y}_{..})^2]$$
$$= E(\sum_j n_j \delta_j^2) + E[\sum_j n_j(\bar{e}_{.j} - \bar{e}_{..})^2] + 2E[\sum_j \delta_j n_j(\bar{e}_{.j} - \bar{e}_{..})]$$

The first term is

$$E(\sum_j n_j \delta_j^2) = \sum_j E(n_j \delta_j^2) = \sum_j n_j \delta_j^2$$

The second term is

$$E[\sum_j n_j(\bar{e}_{.j} - \bar{e}_{..})^2] = E(\sum_j n_j\bar{e}_{.j}^2 + \sum_j n_j\bar{e}_{..}^2 - 2\sum_j n_j\bar{e}_{.j}\bar{e}_{..})$$
$$= \sum_j n_j E(\bar{e}_{.j}^2) + \sum_j n_j E(\bar{e}_{..}^2) - 2E(N\bar{e}_{..}^2)$$
$$= k\sigma_e^2 + \sigma_e^2 - 2\sigma_e^2 = (k-1)\sigma_e^2$$

since

$$\text{Var}(e_{ij}) = E(e_{ij}^2) - [E(e_{ij})]^2 = E(e_{ij}^2) = \sigma_e^2$$

Finally, the third term is

$$2E[\sum_j \delta_j n_j(\bar{e}_{.j} - \bar{e}_{..})] = 0$$

Hence

$$E(\text{SSG}) = E[\sum_j n_j(\bar{y}_{.j} - \bar{y}_{..})^2] = \sum_j n_j\delta_j^2 + (k-1)\sigma_e^2$$

or

$$E\left(\frac{\text{SSG}}{k-1}\right) = \sigma_e^2 + \sum_j \frac{n_j\delta_j^2}{k-1} = \sigma_e^2 + \sum_j \frac{n_j(\mu_j - \mu)^2}{k-1} \tag{13.9}$$

SSG/$(k-1)$ is usually called *mean sum of squares among samples* (denoted by MSG).

It is customary to summarize the results as shown in Table 13.2, usually referred to as the ANOVA (*Analysis of Variance*) table.

TABLE 13.2 ANOVA for the One-Way Classification; Fixed Effects Model

Source of Variation	*Sum of Squares*	*Degrees of Freedom*	*Mean Square*	*Expected Mean Square*
Among treatments	SSG	$k-1$	$\frac{\text{SSG}}{k-1} = \text{MSG}$	$\sigma_e^2 + \sum_{j=1}^{k} \frac{n_j(\mu_j - \mu)^2}{k-1}$
Within treatments	SSW	$N-k$	$\frac{\text{SSW}}{N-k} = \text{MSW}$	σ_e^2
Total	SST	$N-1$		

13.2.5 Computing Formulas and Procedure

The actual computing of sums of squares indicated in Table 13.2 can be simplified by rewriting the formulas that define SSG, SSW, and SST in a form more adaptable to computation. First we consider

$$\text{SST} = \sum_j\sum_i (y_{ij} - \bar{y}_{..})^2 = \sum_j\sum_i y_{ij}^2 + \sum_j\sum_i \bar{y}_{..}^2 - 2\sum_j\sum_i y_{ij}y_{..}$$
$$= \sum_j\sum_i y_{ij}^2 + N\bar{y}_{..}^2 - 2N\bar{y}_{..}^2 = \sum_j\sum_i y_{ij}^2 - N\bar{y}_{..}^2$$

However,

$$\bar{y}_{..} = \frac{T_{..}}{N}$$

Thus

(13.10) $$\text{SST} = \sum_j \sum_i y_{ij}^2 - \frac{T_{..}^2}{N}$$

Next, consider

$$\begin{aligned}\text{SSG} &= \sum_j n_j(\bar{y}_{.j} - \bar{y}_{..})^2 = \sum_j n_j \bar{y}_{.j}^2 + \sum_j n_j \bar{y}_{..}^2 - 2\sum_j n_j \bar{y}_{.j}\bar{y}_{..} \\ &= \sum_j n_j \bar{y}_{.j}^2 + N\bar{y}_{..}^2 - 2N\bar{y}_{..}^2 = \sum_j n_j \bar{y}_{.j}^2 - N\bar{y}_{..}^2\end{aligned}$$

However,

$$\bar{y}_{.j} = \frac{T_{.j}}{n_j}$$

Thus

(13.11) $$\text{SSG} = \sum_j \frac{T_{.j}^2}{n_j} - \frac{T_{..}^2}{N}$$

Finally, by subtraction we obtain

(13.12) $$\text{SSW} = \text{SST} - \text{SSG}$$

The term $T_{..}^2/N$ is referred to as the *correction factor*.

The computational procedure can be performed in a systematic fashion by following the sequence of steps described below:

1. Compute totals for each treatment: $T_{.1}, T_{.2}, \ldots, T_{.k}$.
2. Compute the overall total: $T_{..} = T_{.1} + T_{.2} + \cdots + T_{.k}$.
3. Compute $\sum_j \sum_i y_{ij}^2 = y_{11}^2 + y_{12}^2 + \cdots + y_{n_k k}^2$.
4. Compute $\sum_j \frac{T_{.j}^2}{n_j} = \frac{T_{.1}^2}{n_1} + \frac{T_{.2}^2}{n_2} + \cdots + \frac{T_{.k}^2}{n_k}$.
5. Compute the correction factor $\frac{T_{..}^2}{N}$.
6. Compute $\text{SST} = \sum_j \sum_i y_{ij}^2 - \frac{T_{..}^2}{N}$.
7. Compute $\text{SSG} = \sum_j \frac{T_{.j}^2}{n_j} - \frac{T_{..}^2}{N}$.
8. Compute $\text{SSW} = \text{SST} - \text{SSG}$.

13.2.6 Testing the Hypothesis of Equal Means

We wish to test that k mutually independent populations have equal means. Each of these populations is assumed to be normally distributed with mean μ_j and variance σ_e^2 (both unknown). In symbols, we wish to test

H_0: $\mu_1 = \mu_2 = \cdots = \mu_k$ against H_1: at least two means are not equal

If H_0 is true, all k effects are equal to zero. Accordingly, the null and alternative hypotheses can be expressed as

$$H_0\colon \delta_1 = \delta_2 = \cdots = \delta_k = 0 \quad \text{against} \quad H_1\colon \text{not all } \delta_j\text{'s are zero}$$

In order to construct a formal test for H_0, we shall make use of the previously cited distribution assumptions. First consider the sample variance of population j:

$$s_j^2 = \frac{\sum_i (y_{ij} - \bar{y}_{.j})^2}{n - 1}$$

Since all k populations have the same variance, and are independent, the ratios

$$\sum_j \frac{(n_j - 1)s_j^2}{\sigma_e^2} = \sum_j \sum_i \frac{(y_{ij} - \bar{y}_{.j})^2}{\sigma_e^2} = \frac{\text{SSW}}{\sigma_e^2} \tag{13.13}$$

have the χ^2 distribution with $N - k$ degrees of freedom.

Next consider the among samples sum of squares

$$\text{SSG} = \sum_j n_j(\bar{y}_{.j} - \bar{y}_{..})^2$$

Dividing both sides by σ_e^2 we obtain

$$\frac{\text{SSG}}{\sigma_e^2} = \sum_j \left(\frac{\bar{y}_{.j} - \bar{y}_{..}}{\sigma_e/\sqrt{n_j}}\right)^2 \tag{13.14}$$

When H_0 is true, the $\bar{y}_{.j}$'s are normally distributed with mean μ and variance σ_e^2/n_j. Therefore, SSG/σ_e^2 has the χ^2 with $k - 1$ degrees of freedom. However, when H_1 is true SSG/σ_e^2 is distributed as *noncentral* χ^2 with $k - 1$ degrees of freedom and *noncentrality parameter* equal to $\frac{1}{2} \sum_j n_j\, \delta_j^2/\sigma_e^2$. It should be noted that SSW is a function only of s_j^2's whereas SSG is a function only of $\bar{y}_j$'s. It follows that SSW and SSG are statistically independent since for a normal distribution the estimates of the mean and variance are independent. By definition, the F distribution is the distribution of the ratio of two independently distributed chi-squares divided by their respective degrees of freedom. Thus, when H_0 is true we can assert that

$$\frac{(\text{SSG}/\sigma_e^2)/(k - 1)}{(\text{SSW}/\sigma_e^2)/(N - k)} = \frac{\text{MSG}}{\text{MSW}} \tag{13.15}$$

has the F distribution with $k - 1$ and $N - k$ degrees of freedom. Furthermore, when H_1 is true, the expected ratio of mean squares

$$E\left(\frac{\text{MSG}}{\text{MSW}}\right) = \frac{E(\text{MSG})}{E(\text{MSW})} = \frac{\sigma_e^2 + \sum_j [n_j(\mu_j - \mu)^2]/(k - 1)}{\sigma_e^2} \tag{13.16}$$

is greater than one and increases with $\sum_j \delta_j^2$ whereas when H_0 is true

$$\sum_j \frac{n_j(\mu_j - \mu)^2}{k - 1} = 0 \quad \text{and} \quad E\left(\frac{\text{MSG}}{\text{MSW}}\right) = 1$$

Therefore, *the magnitude of the ratio of mean squares can be expected to reflect the validity of* H_0. The F distribution indicates how large a ratio of mean squares we can expect when the population value actually is one. Thus, when H_0 is true the rejection region for a specified level of significance α is

$$\frac{\text{MSG}}{\text{MSW}} > F_{1-\alpha;\, k-1;\, N-k} \tag{13.17}$$

Example 13.1 Consider again the previously cited example. Suppose that a completely randomized arrangement is used to test if the average thickness of washers produced by each machine is the same. Using an α risk of .05, test for a significant difference among machines on the basis of the following data:

Machine	1	2	3	
	.236 .238 .248 .245 .243	.257 .253 .255 .254 .261	.258 .264 .259 .267 .262	
$T_{.j}$:	1.210	1.280	1.310	$T_{..} = 3.800$
nj:	5	5	5	$N = 15$
$\sum_i y_{ij}^2$:	.292918	.327720	.343274	$\sum_j \sum_i y_{ij}^2 = .963912$

Solution. We compute

$$\sum_j \frac{T_{.j}^2}{n_j} = \frac{1}{5}(1.4641 + 1.6384 + 1.7161) = 96{,}372 \times 10^{-5}$$

$$\frac{T_{..}^2}{N} = \frac{(3.8)^2}{15} = 962667 \times 10^{-6}$$

$$\text{SST} = \sum_j \sum_i y_{ij}^2 - \frac{T_{..}^2}{N} = .963912 - .962667 = 1245 \times 10^{-6}$$

$$\text{SSG} = \sum_j \frac{T_{.j}^2}{n_j} - \frac{T_{..}^2}{N} = .96372 - .962667 = 1053 \times 10^{-6}$$

$$\text{SSW} = \text{SST} - \text{SSG} = (1245 - 1053)10^{-6} = 192 \times 10^{-6}$$

The results are presented in ANOVA table as shown below:

ANOVA FOR EXAMPLE 13.1

Source of Variation	*Sum of Squares*	*Degrees of Freedom*	*Mean Square*	*Expected Mean Square*
Among machines	1053×10^{-6}	2	526.5×10^{-6}	$\sigma_e^2 + \frac{5}{2} \sum_j \delta_j^2$
Within machines	192×10^{-6}	12	16×10^{-6}	σ_e^2
Total	1245×10^{-6}	14		

The hypothesis of equal means is rejected if

$$\frac{\text{MSG}}{\text{MSW}} > F_{1-\alpha;\, k-1;\, N-k}$$

From the data we have

$$\frac{\text{MSG}}{\text{MSW}} = \frac{526.5}{16} = 32.9 > F_{.95;\, 2,12} = 3.89$$

Hence, we reject the hypothesis of equal means at the 5% level of significance and conclude that the machines are significantly different.

13.2.7 The Power of the F Test; Fixed Effects Model

The power of the F test plays a significant role in the design of experiments. Prior to the running of an experiment it is important to determine the odds with which differences among the means of k groups, as a given multiple of σ, can be detected. Thus the power of the test will indicate whether a satisfactory test of a hypothesis exists. In some cases the power of the test might be considered unsatisfactory, in which case the experimenter might wish to change the design of the experiment in order to increase the power of the test. Usually this would involve such changes as a choice of more groups of some factors or increasing the number of replications.

It can be shown that under the alternative hypothesis SSG/σ_e^2 is distributed as *noncentral* χ^2 with $k - 1$ degrees of freedom and *noncentrality parameter* equal to $\frac{1}{2}\sum_j n_j\delta_j^2/\sigma_e^2$. Hence, under the alternative hypothesis, the ratio

$$\frac{(\text{SSG}/\sigma_e^2)/(k-1)}{(\text{SSW}/\sigma_e^2)/(N-k)} = \frac{\text{MSG}}{\text{MSW}}$$

has the *noncentral F* distribution with $k - 1$ and $N - k$ degrees of freedom and *noncentrality parameter* equal to $\frac{1}{2}\sum n_j\delta_j^2/\sigma_e^2$. Tang [25] developed extensive tabulation for the noncentral F distribution. Pearson and Hartley [20] constructed charts from Tang's tables. Some of these charts are shown in Figs. 13.1–13.6. In order to use these charts, the following information is required:

1. ν_1 = the degrees of freedom associated with the sum of squares used for the test. Thus, when testing $H_0\colon \delta_1 = \delta_2 = \cdots = \delta_k = 0$, the degrees of freedom are $k_1 - 1 = \nu_1$. ν_1 is found in the upper left-hand corner of Figs. 13.1 through 13.6.
2. ν_2 = the degrees of freedom associated with the sum of squares within samples, SSW, usually referred to as the sum of squares due to chance error. Accordingly, $\nu_2 = N - k = k(n - 1)$ where $n = n_1 = n_2 = \cdots n_k$. (It is to be noted that these charts are derived for populations with equal sample sizes.) On all charts the curves are indexed with selected values of ν_2.

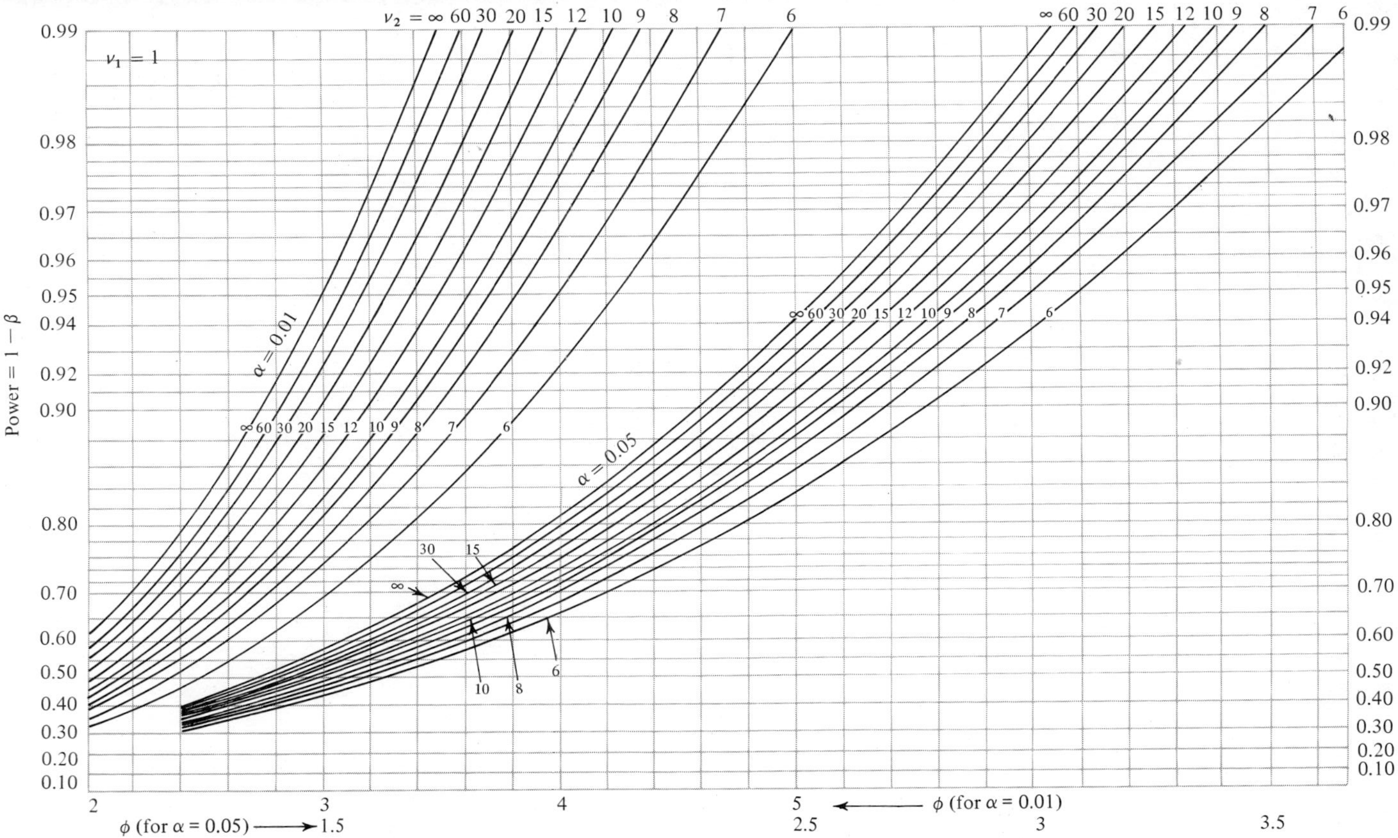

Fig. 13.1 Power function for analysis of variance test with $\nu_1 = 1$.

[Reproduced with permission from E. S. Pearson and H. O. Hartley, "Charts of the Power Function for Analysis of Variance Tests, Derived from the Non-Central *F*-Distribution," *Biometrika*, **38**, 112, 1951.]

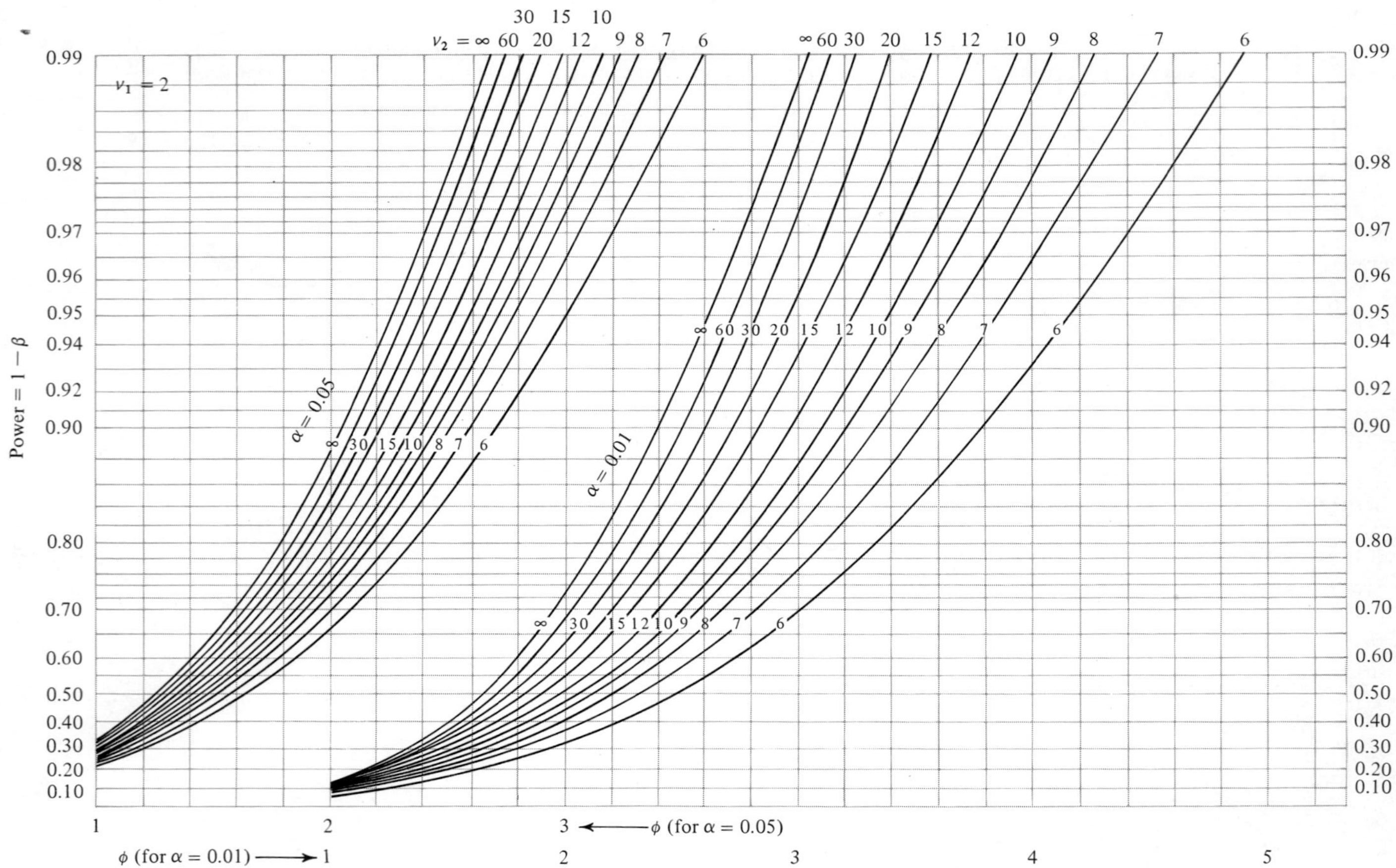

Fig. 13.2 Power function for analysis of variance test with $\nu_1 = 2$.

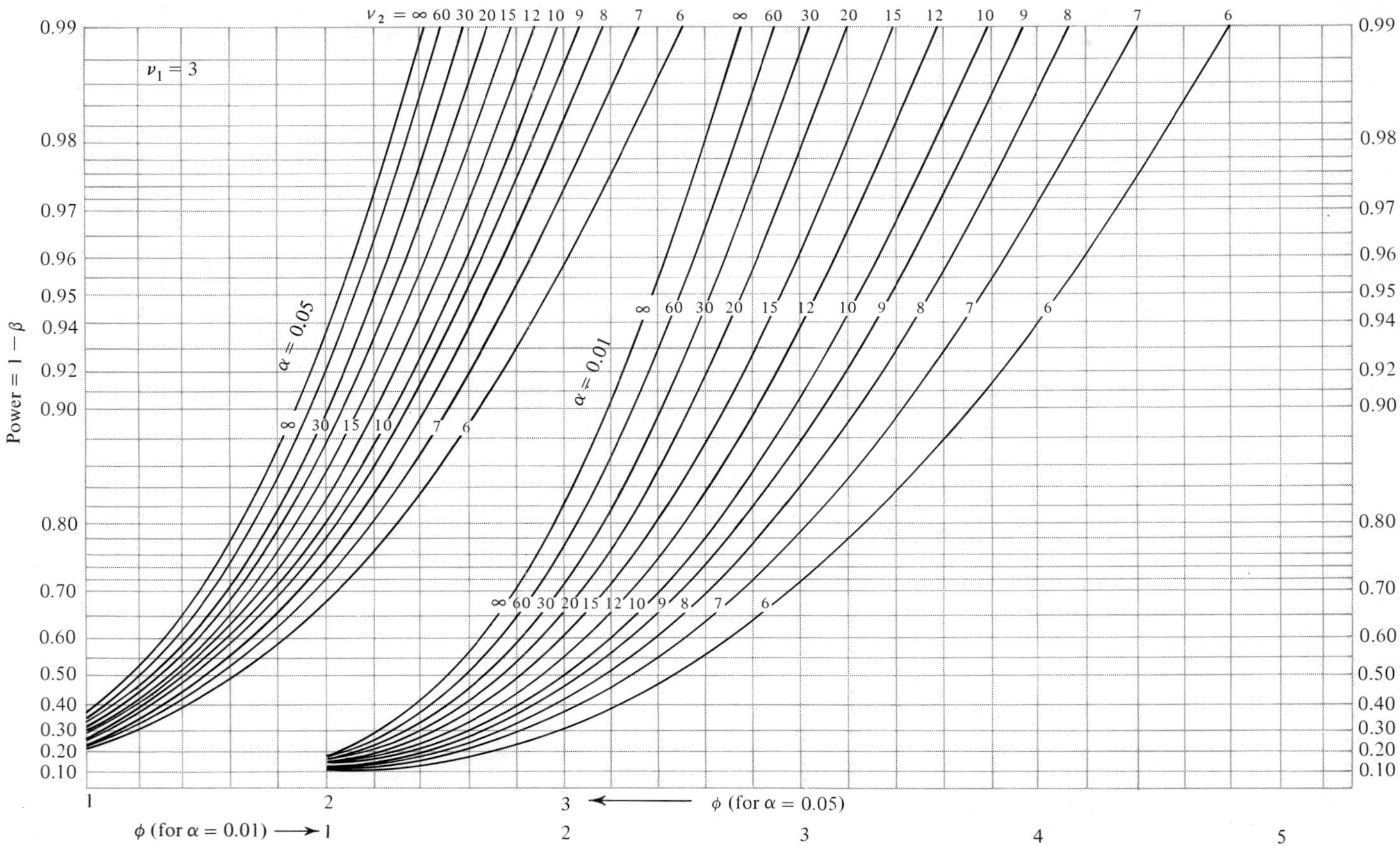

Fig. 13.3 Power function for analysis of variance test with $\nu_1 = 3$.

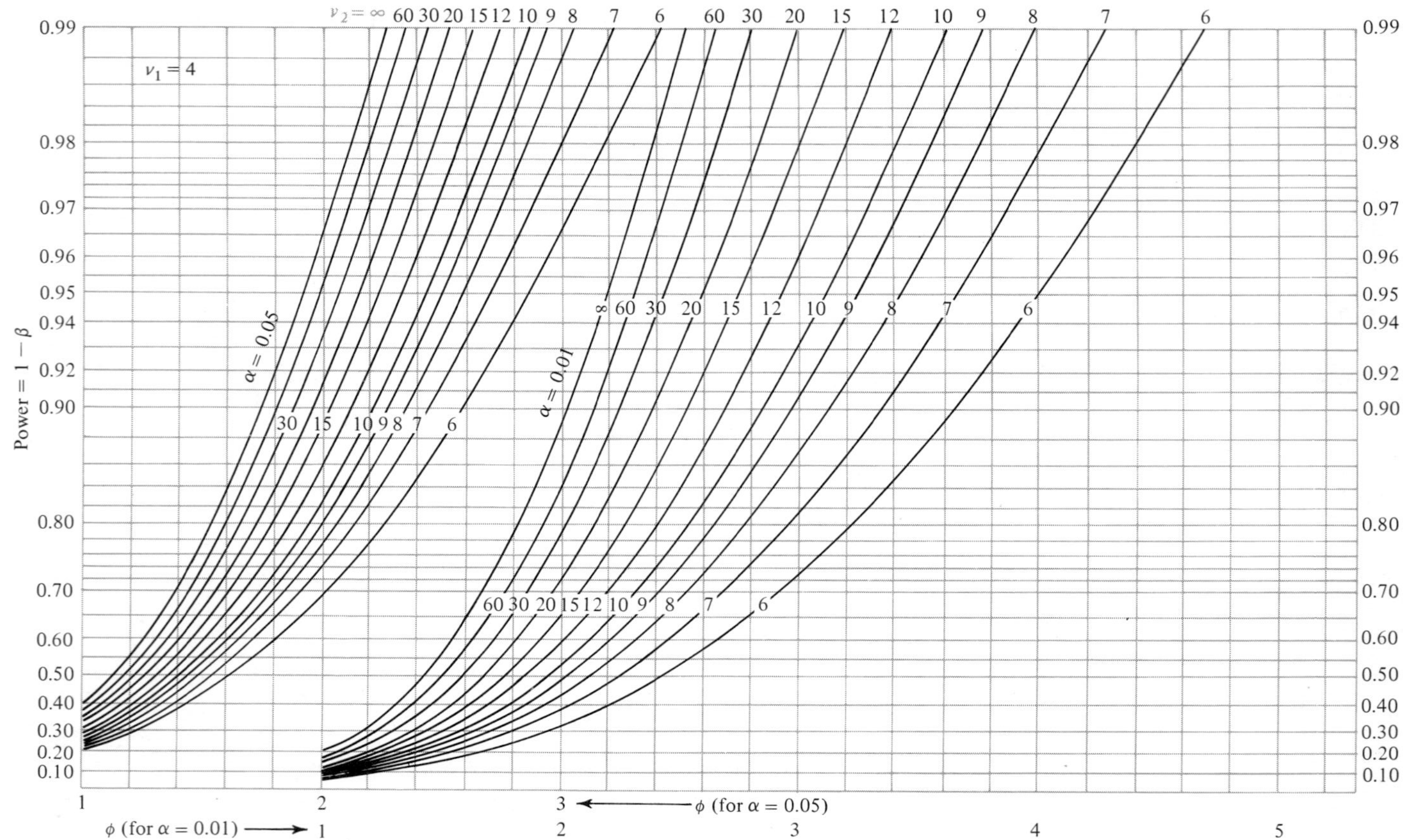

Fig. 13.4 Power function for analysis of variance test with $\nu_1 = 4$.

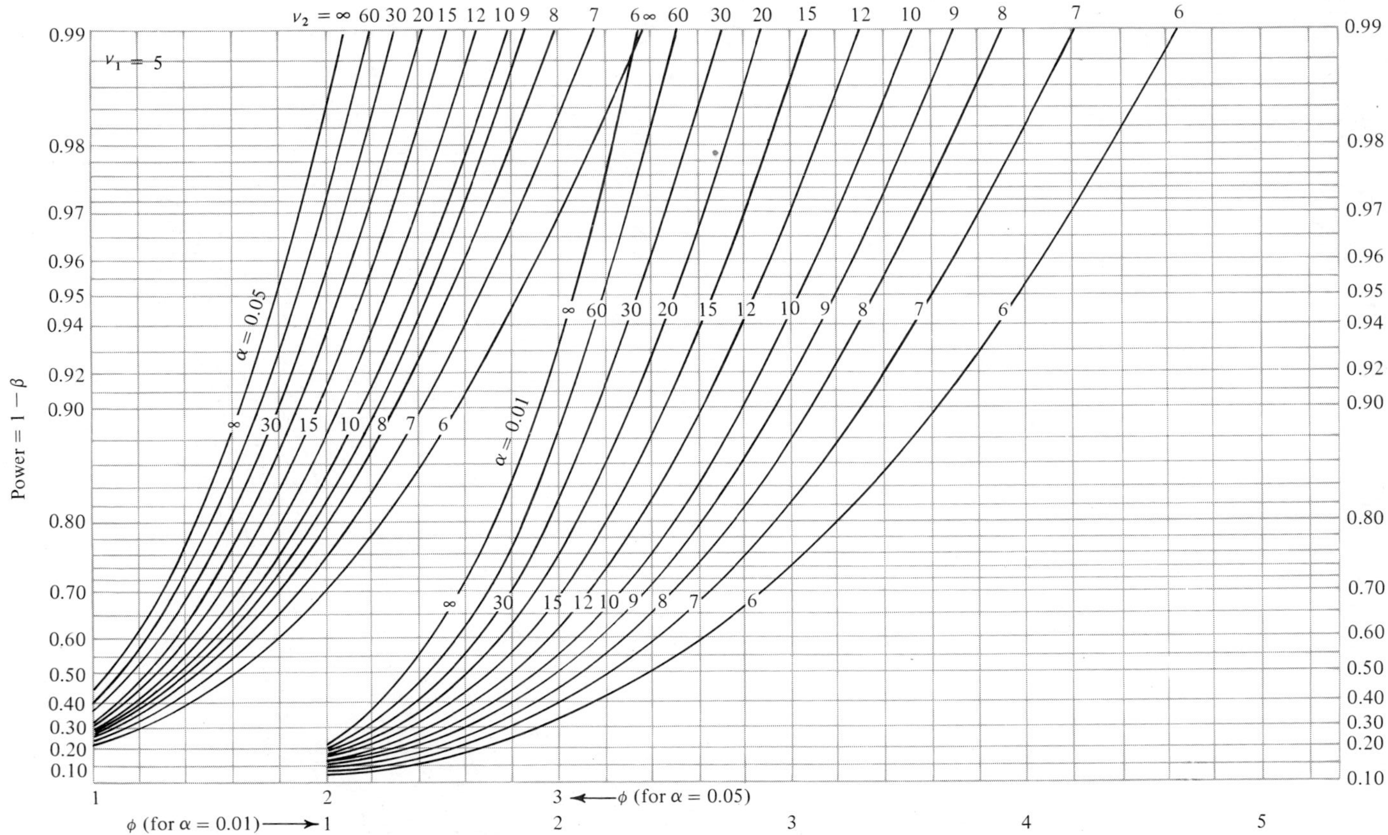

Fig. 13.5 Power function for analysis of variance test with $\nu_1 = 5$.

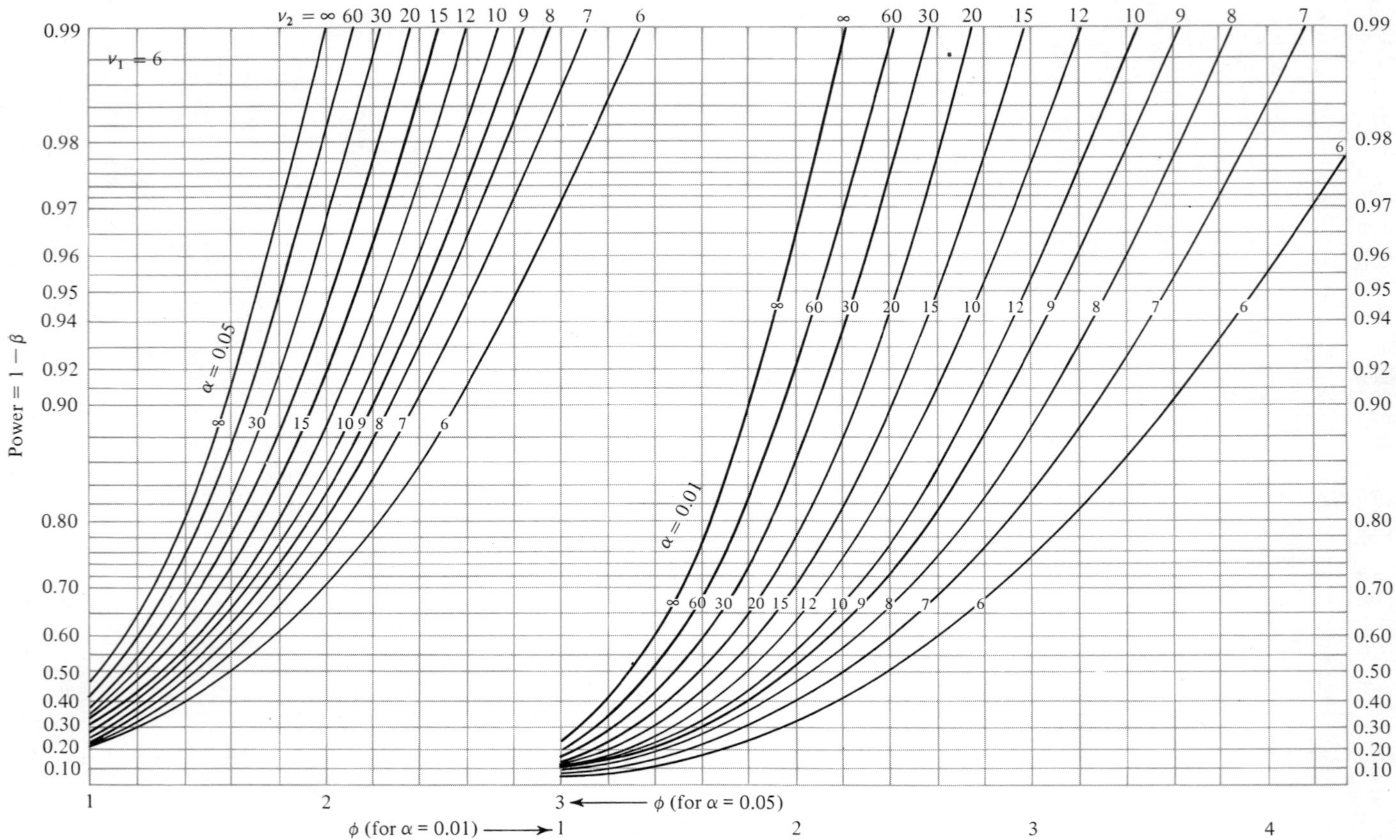

Fig. 13.6 Power function for analysis of variance test with $\nu_1 = 6$.

3. α = the level of significance of the test. The curves on the left side are for $\alpha = .05$; those on the right, for $\alpha = .01$.
4. ϕ = a parameter that depends on the parameter of noncentrality and the degrees of freedom. The value of this parameter for the one-way classification, fixed effects model is

$$\phi = \frac{\sqrt{\sum_j \delta_j^2/(\nu_1 + 1)}}{\sigma_e/\sqrt{n}} \qquad \nu_1 = k - 1 \quad \text{and} \quad \nu_2 = k(n - 1) \tag{13.18}$$

It should be noted that ϕ as given by Eq. (13.18) depends on σ_e. Since σ_e^2 is usually unknown, its estimate MSW may be used. However, power calculations can be made prior to running of an experiment by expressing δ_j's as multiples of σ_e. We shall now demonstrate how the power of a test can be determined by the aid of these charts.

Example 13.2 In Example 13.1 find the probability of accepting the null hypothesis if two means are equal and the third exceeds the others by $1.5\sigma_e$. Use $\alpha = .05$.

Solution. Let $\mu_1 = \mu_2 = c$. Hence, $\mu_3 = c + 1.5\sigma$ and the overall mean is

$$\mu = \frac{\mu_1 + \mu_2 + \mu_3}{3} = c + .5\sigma$$

Thus

$$\sum_j \delta_j^2 = (-.5\sigma_e)^2 + (-.5\sigma_e)^2 + \sigma_e^2 = 1.5\sigma_e^2$$

so that (since $n = 5$, $\nu_1 = 2$, and $\nu_2 = 12$)

$$\phi = \frac{\sqrt{\sum_j \delta_j^2/(\nu_1 + 1)}}{\sigma_e/\sqrt{n}} = \frac{\sqrt{1.5\sigma_e^2/3}}{\sigma_e/\sqrt{5}} = \sqrt{2.50} = 1.58$$

Consulting the chart in Fig. 13.2 we find that

$$1 - \beta \cong .575 \quad \text{whence} \quad \beta \cong .425$$

Consequently, in 42.5% of the cases, we shall conclude that the means are equal when in fact two means are equal and the third exceeds the others by $1.5\sigma_e$ at the 5% level of significance.

Example 13.3 In Example 13.1 determine the sample size n so that the power of the test is at least 95% when $\mu_1 = .250$, $\mu_2 = .255$, $\mu_3 = .260$, and $\alpha = .05$. Assume that σ_e is approximately 3.9×10^{-3}.

Solution. We now compute

$$\mu = \tfrac{1}{3}(.250 + .255 + .260) = .255$$

$$\sum_j \delta_j^2 = (-.005)^2 + 0 + (.005)^2 = 50 \times 10^{-6}$$

$$\phi = \frac{\sqrt{(50/3) \times 10^{-6} n}}{3.9 \times 10^{-3}} = \sqrt{1.095n}$$

Now we read from the chart in Fig. 13.2, for $\alpha = .05$, $\nu_1 = 2$, $\nu_2 = 3(n - 1)$, and

$1 - \beta = .95$, the ϕ values corresponding to several values of n, each time comparing the value read with $\sqrt{1.095n}$ until an agreement is obtained. For $n = 6$, $\nu_2 = 15$, and the corresponding value of ϕ from chart is $\simeq 2.56$. Since $\sqrt{1.095 \times 6} \simeq 2.56$ we conclude that six washers should be tested from each machine.

13.2.8 Contrasts

Occasionally a situation will arise where we are interested in testing hypotheses having certain empirical significance. In practice these hypotheses are formulated before conducting the experiment. The formulation is meaningful, particularly if the experimenter is almost certain that not all means are equal. To illustrate, suppose that we are concerned with four operators. We wish to run an experiment to determine whether a difference in operators due to age or sex has an effect on the assembly time of a certain manual operation. Let the four operators be designated by

O_1: young female operator
O_2: old female operator
O_3: young male operator
O_4: old male operator

We may now seek an answer to the following question: Is the assembly time affected by different operators? The answer to this question would be "No" if the hypothesis

$$H_0: \mu_{O_1} = \mu_{O_2} = \mu_{O_3} = \mu_{O_4}$$

is not rejected. However, if H_0 is rejected, we may be concerned with whether the operator's age or sex or both age and sex affect the assembly time. Accordingly, one may formulate the following hypotheses:

(a) $H_0': \dfrac{\mu_{O_1} + \mu_{O_3}}{2} = \dfrac{\mu_{O_2} + \mu_{O_4}}{2}$ against $H_1': \dfrac{\mu_{O_1} + \mu_{O_3}}{2} \neq \dfrac{\mu_{O_2} + \mu_{O_4}}{2}$

(b) $H_0'': \mu_{O_1} = \mu_{O_3}$ against $H_1'': \mu_{O_1} \neq \mu_{O_3}$

(c) $H_0''': \mu_{O_2} = \mu_{O_4}$ against $H_1''': \mu_{O_2} \neq \mu_{O_4}$

The null hypothesis in (a) may be written

$$H_0': \mu_{O_1} - \mu_{O_2} + \mu_{O_3} - \mu_{O_4} = 0$$

If H_0' is rejected, we infer that the age of the operator has an effect on the assembly time at the prescribed level of significance. Similarly, if

$$H_0'': \mu_{O_1} - \mu_{O_3} = 0$$

is rejected, we conclude that the sex of young operators has an effect on the assembly time. Finally, if

$$H_0''': \mu_{O_2} - \mu_{O_4} = 0$$

is rejected, we conclude that the effect of the sex of old operators is significant.

We note that each of H_0', H_0'', and H_0''' is a linear combination of the treatment means of the form

$$L = c_1\mu_1 + c_2\mu_2 + \cdots + c_k\mu_k \tag{13.19}$$

where the c_j's are constants and $\sum_{j=1}^{k} c_j = 0$. If these two conditions are satisfied Eq. (13.19) is called a *linear contrast*. If the c_j's are constant, $\sum_{j=1}^{k} c_j = 0$, and $\sum_{j=1}^{k} c_j^2 = 1$, Eq. (13.19) is known as a *normalized contrast*. Hence, by definition, each of H_0', H_0'', and H_0''', denoted by

$$\begin{cases} L_1 = \mu_{O_1} - \mu_{O_2} + \mu_{O_3} - \mu_{O_4} \\ L_2 = \mu_{O_1} - \mu_{O_3} \\ L_3 = \mu_{O_2} - \mu_{O_4} \end{cases} \tag{13.20}$$

respectively, is a linear contrast. The coefficients of the μ_{O_j}'s ($j = 1, 2, \ldots, 4$) in L_1, L_2, and L_3 are represented in Table 13.3.

TABLE 13.3 Coefficients of the Parameter Components (μ_{Oj})

	μ_{O_1}	μ_{O_2}	μ_{O_3}	μ_{O_4}
L_1	+1	−1	+1	−1
L_2	+1	0	−1	0
L_3	0	+1	0	−1

It can be seen from Table 13.3 that if the coefficients of any pair of contrasts are multiplied and then added, the result is zero. To verify this property we note that

$$\sum_j c_{j1}c_{j2} = (+1)(+1) + (-1)(0) + (+1)(-1) + (-1)(0) = 0$$

$$\sum_j c_{j1}c_{j3} = (+1)(0) + (-1)(+1) + (+1)(0) + (-1)(-1) = 0$$

$$\sum_j c_{j2}c_{j3} = (+1)(0) + (0)(+1) + (-1)(0) + (0)(-1) = 0$$

Any pair of contrasts having this property is said to be *independent* or *orthogonal*. Many orthogonal sets other than the one shown in Table 13.3 could be obtained. In general, for k treatment means many orthogonal sets of contrasts can be formed; however, each set can have *no* more than $k - 1$ mutually orthogonal contrasts. Again, we would like to emphasize that contrasts are tested because they have *empirical significance*, not because they are independent of each other. Now if we normalize the set of orthogonal contrasts given in Eq. (13.20) we obtain

$$\text{(13.21)} \qquad \begin{cases} \tilde{L}_1 = \frac{1}{2}(\mu_{O_1} - \mu_{O_2} + \mu_{O_3} - \mu_{O_4}) \\ \tilde{L}_2 = \dfrac{1}{\sqrt{2}}(\mu_{O_1} - \mu_{O_3}) \\ \tilde{L}_3 = \dfrac{1}{\sqrt{2}}(\mu_{O_2} - \mu_{O_4}) \end{cases}$$

where $\tilde{L}_i$ denotes a normalized contrast. Note that a linear contrast can always be normalized by dividing by $1/\sqrt{\sum_j c_j^2}$. The unbiased estimates of the orthogonal normalized contrasts given in Eq. (13.21) are

$$\text{(13.22)} \qquad \begin{cases} \tilde{l}_1 = \frac{1}{2}(\bar{y}_{.01} - \bar{y}_{.02} + \bar{y}_{.03} - \bar{y}_{.04}) \\ \tilde{l}_2 = \dfrac{1}{\sqrt{2}}(\bar{y}_{.01} - \bar{y}_{.03}) \\ \tilde{l}_3 = \dfrac{1}{\sqrt{2}}(\bar{y}_{.02} - \bar{y}_{.04}) \end{cases}$$

where the $\bar{y}_{.0j}$'s are sample averages, each of which is based on n observations. Now if we square $\tilde{l}_1$, $\tilde{l}_2$, and $\tilde{l}_3$ and add them up, we obtain

$$\begin{aligned} \tilde{l}_1^2 + \tilde{l}_2^2 + \tilde{l}_3^2 &= \tfrac{1}{4}(\bar{y}_{.01} - \bar{y}_{.02} + \bar{y}_{.03} - \bar{y}_{.04})^2 + \tfrac{1}{2}(\bar{y}_{.01} - \bar{y}_{.03})^2 \\ &\quad + \tfrac{1}{2}(\bar{y}_{.02} - \bar{y}_{.04})^2 \\ &= \bar{y}_{.01}^2 + \bar{y}_{.02}^2 + \bar{y}_{.03}^2 + \bar{y}_{.04}^2 - 4\bar{y}_{..}^2 \end{aligned}$$

$$\text{(13.23)} \qquad = \sum_{j=1}^{4} (\bar{y}_{.0j} - \bar{y}_{..})^2$$

where

$$\bar{y}_{..} = \tfrac{1}{4}\sum_{j=1}^{4} \bar{y}_{.0j}$$

We note that the right-hand side of Eq. (13.23) is equal to SSG/n where SSG is the sum of squares among samples. Hence Eq. (13.23) may be written as

$$\tilde{l}_1^2 + \tilde{l}_2^2 + \tilde{l}_3^2 = \frac{1}{n}(\text{SSG})$$

or

$$\text{(13.24)} \qquad n\tilde{l}_1^2 + n\tilde{l}_2^2 + n\tilde{l}_3^2 = \text{SSG}$$

Equation (13.24) exhibits an important property: The sum of squares among samples can be partitioned into three components, each of which represents a contrast of particular interest. Is this a coincidence? Of course not; it can be shown (the demonstration is beyond the scope of this text) that *any sum of squares among samples with $k - 1$ degrees of freedom can always be subdivided into $k - 1$ components.* In symbols, this important theorem is

$$\text{(13.25)} \qquad \text{SSG} = n\sum_{i=1}^{k-1} \tilde{l}_i^2$$

where n = sample size, k = number of treatment means, $\tilde{l}_i$ = normalized contrast among the means, and $\tilde{l}_1, \ldots, \tilde{l}_{k-1}$ are mutually orthogonal.

If the assumptions underlying the analysis of variance technique are valid, then $\tilde{l}_i$ is a normally distributed variate with mean

$$E(\tilde{l}_i) = \tilde{L}_i = \sum_{j=1}^{k} c_{ji}\mu_j \tag{13.26}$$

and variance

$$\mathrm{Var}(\tilde{l}_i) = \mathrm{Var}\left(\sum_{j=1}^{k} c_{ji}\bar{y}_{\cdot j}\right) = \sum_{j=1}^{k} c_{ji}^2 \frac{\sigma_e^2}{n} = \frac{\sigma_e^2}{n}$$

Therefore

$$\frac{\tilde{l}_i - \tilde{L}_i}{\sqrt{\sigma_e^2/n}} \sim N(0, 1) \tag{13.27}$$

and

$$\frac{n(\tilde{l}_i - \tilde{L}_i)^2}{\sigma_e^2} \sim \chi_1^2 \tag{13.28}$$

However, when the null hypothesis is true, i.e., $\mu_1 = \mu_2 = \cdots = \mu_k$, $\tilde{L}_i = 0$. Thus, under the null hypothesis Eq. (13.28) becomes

$$\frac{n\tilde{l}_i^2}{\sigma_e^2} \sim \chi_1^2 \tag{13.29}$$

Since the $\tilde{l}_i$'s are orthogonal, it follows that when the null hypothesis is true

$$\frac{n\tilde{l}_1^2}{\sigma_e^2} + \cdots + \frac{n\tilde{l}_{k-1}^2}{\sigma_e^2} \sim \chi_{k-1}^2 \tag{13.30}$$

It is noteworthy that this important relationship agrees with our finding in Eq. (13.25). Since SSW/σ_e^2 has the χ^2 distribution with $k(n-1)$ degrees of freedom, $n\tilde{l}_i^2/\sigma_e^2$ has the χ^2 distribution with one degree of freedom, and $\tilde{l}_i$ is independent of SSW, it follows from the definition of the F distribution that

$$\frac{(n\tilde{l}_i^2/\sigma_e^2)/1}{(\mathrm{SSW}/\sigma_e^2)/k(n-1)} = \frac{n\tilde{l}_i^2}{\mathrm{MSW}} \sim F_{1,\, k(n-1)} \tag{13.31}$$

Hence, the hypothesis that $L_i = 0$ is rejected at the α level of significance if

$$\frac{n\tilde{l}_i^2}{\mathrm{MSW}} > F_{1-\alpha;\, 1,\, k(n-1)} \tag{13.32}$$

Before we can present the results of this analysis in table form we need to evaluate the expected mean square due to contrast i when H_0 is not necessarily true. In symbols, we write

$$\begin{aligned} E\left(\frac{n\tilde{l}_i^2}{1}\right) &= nE(\tilde{l}_i^2) \\ &= n\{\mathrm{Var}(\tilde{l}_i) + [E(\tilde{l}_i)]^2\} = \sigma_e^2 + n\tilde{L}_i^2 \end{aligned} \tag{13.33}$$

Table 13.4 shows the results of this analysis.

TABLE 13.4 Analysis of Variance

Source of Variation	*Sum of Squares*	*Degrees of Freedom*	*Mean Square*	*Expected Mean Square*
Contrast 1	$n\hat{l}_1^2$	1	$n\hat{l}_1^2$	$\sigma_e^2 + nL_1^2$
Contrast 2	$n\hat{l}_2^2$	1	$n\hat{l}_2^2$	$\sigma_e^2 + nL_2^2$
⋮	⋮	⋮	⋮	⋮
Contrast $(k-1)$	$n\hat{l}_{k-1}^2$	1	$n\hat{l}_{k-1}^2$	$\sigma_e^2 + nL_{k-1}^2$
Among treatments	SSG	$k-1$	MSG	$\sigma_e^2 + \frac{n}{k-1}\sum_{j=1}^{k}(\mu_j - \mu)^2$
Within treatments	SSW	$k(n-1)$	MSW	σ_e^2
Total	SST	$nk-1$		

The previous analysis merits further discussion. In practice contrasts are formulated before running the experiment. Usually the chosen contrasts exhibit empirical significance in terms of the experimental variables; whether these contrasts are orthogonal makes little or no difference. In this connection we wish to point out that in a one-way classification meaningful contrasts are likely to arise frequently; however, they may not be mutually orthogonal.

When an overall F test is significant, one may be interested in identifying the sources of variation by testing meaningful contrasts. If a set of $(k-1)$ mutually orthogonal contrasts is formulated before running the experiment, each of these contrasts can be tested against $F_{1,\,k(n-1)}$ at a desirable level of significance α. Hence, some of the decisions that reject the null hypothesis may be due to type I error. For example, consider the situation when $k = 4$ and $\alpha = .10$. If the equality of means is rejected and three independent tests are each made at the 10% level of significance, then the probability of a type I error is $1 - (1 - .10)^3 \cong .27$. This means that the probability of rejecting one or more of these contrasts is less than 27% when all are true. Obviously, as the number of contrasts increases, the probability of at least one type I error increases. It should be noted that if the contrasts are not orthogonal, the resulting level of significance may be difficult to establish. In the light of this discussion *we need a criterion for significance such that the probability of error for a set of comparisons is constant*. Only in this way would it be possible to compare the results of the same comparison in different experiments. If such a criterion can be established one can make sequenced testing for contrasts formulated before conducting the experiment and for certain particular contrasts suggested by the data themselves. Many procedures have been developed to handle such situations. In the following section two such procedures will be considered.

Example 13.4 Using the data of Example 13.1, work out the following:

(a) Verify that $\mu_3 - \mu_1$ and $\mu_1 - 2\mu_2 + \mu_3$ are mutually orthogonal contrasts.
(b) Verify that Eq. (13.24) holds true.
(c) Test whether the contrast $\mu_3 - \mu_1$ is significant.
(d) Find the power of the test in (c) when $\mu_3 - \mu_1 = .01$. Use $\alpha = .01$ and MSW as an estimate of σ_e^2.

Solution

(a) To verify that the contrasts $L_1 = \mu_3 - \mu_1$ and $L_2 = \mu_1 - 2\mu_2 + \mu_3$ are orthogonal, we shall tabulate their coefficients:

	μ_1	μ_2	μ_3
L_1	-1	0	$+1$
L_2	$+1$	-2	$+1$

Since $(-1)(+1) + (0)(-2) + (+1)(+1) = 0$, we conclude that these contrasts are mutually orthogonal. The first contrast compares the average thickness of washers produced by machine 1 with the average of machine 3, whereas the second contrast compares the average thickness of washers produced by machines 1 and 3 with the average of machine 2.

(b) First we have to normalize these contrasts. Thus

$$\tilde{L}_1 = \frac{1}{\sqrt{2}}(\mu_3 - \mu_1) \qquad \tilde{L}_2 = \frac{1}{\sqrt{6}}(\mu_1 - 2\mu_2 + \mu_3)$$

The unbiased estimate of $\tilde{L}_1$ and $\tilde{L}_2$ are

$$\tilde{l}_1 = \frac{1}{\sqrt{2}}(\bar{y}_{.3} - \bar{y}_{.1}) \qquad \tilde{l}_2 = \frac{1}{\sqrt{6}}(\bar{y}_{.1} - 2\bar{y}_{.2} + \bar{y}_{.3})$$

$$\tilde{l}_1 = \frac{1}{\sqrt{2}}(.262 - .242) \qquad \tilde{l}_2 = \frac{1}{\sqrt{6}}[.242 - 2(.256) + .262]$$

$$= \frac{.02}{\sqrt{2}} \qquad\qquad = \frac{-.008}{\sqrt{6}}$$

Hence,

$$n\tilde{l}_1^2 + n\tilde{l}_2^2 = 5\left(\frac{.0004}{2}\right) + 5\left(\frac{.000064}{6}\right) = 1053 \times 10^{-6}$$

which is equal to SSG.

(c) We wish to test

$$H_0'\colon \mu_3 - \mu_1 = 0 \quad \text{against} \quad H_1'\colon \mu_3 - \mu_1 \neq 0$$

The null hypothesis is rejected, since

$$\frac{n\tilde{l}_1^2}{\text{MSW}} = \frac{.001}{16 \times 10^{-6}} = 62.5 > F_{.99;\,1,12} = 9.33$$

Hence we conclude that the contrast is highly significant.

(d) To find the power of the test we evaluate

$$\phi = \frac{\sqrt{\sum_j \delta_j^2/(\nu_1 + 1)}}{\sigma_e/\sqrt{n}} = \frac{\sqrt{\hat{L}_i^2/2}}{\sigma_e/\sqrt{n}} = \frac{\sqrt{(.01)^2/2}}{\sqrt{(16 \times 10^{-6})/5}} = 3.95$$

Consulting Fig. 13.1 we find that the power of the test is .987.

13.2.9 Multiple Comparisons

An overall F test may reject the null hypothesis that $\mu_1 = \mu_2 = \cdots = \mu_k$; however, it does not tell us which means are significantly different from which. An appealing procedure would be comparing all possible pairs by an ordinary two-sample t test. Accordingly, two specified means μ_i and μ_j are different at the α level of significance when

$$|\bar{y}_{i.} - \bar{y}_{.j}| > \sqrt{\frac{2\text{MSW}}{n}}\, t_{1-\alpha/2;\, k(n-1)} \tag{13.34}$$

The right-hand side of (13.34) is known as the *least significant difference*, usually denoted by LSD. Unfortunately this appealing procedure is valid only for one prechosen comparison before the data are examined; since for a sequence of comparisons the significance level experimentwise becomes excessively large. Several procedures have been developed whereby it is possible to assign an overall level of significance for a class of comparisons chosen after the data are available. We will discuss those developed by Tukey [26] and Scheffé [23].

Tukey's Procedure

Let $L = \sum_{j=1}^{k} c_j \mu_j$ be a linear contrast among k means and $l = \sum_{j=1}^{k} c_j \bar{y}_{.j}$ be its unbiased estimate. Furthermore, assume that each sample average is based on a sample of size n. Tukey's procedure is based on the distribution of the studentized range statistic, defined by

$$q_{n,\nu} = \frac{y_{\max} - y_{\min}}{s} \tag{13.35}$$

where s^2 = an unbiased estimate of σ_e^2 based upon ν degrees of freedom and $y_1, \ldots, y_n$ are independently normally distributed random variables. Furthermore, Tukey assumes that the $\bar{y}_{.j}$'s have equal variances. The sampling distribution of q depends upon k and ν. Tukey has shown that the probability is $1 - \alpha$ that

$$l - q_{1-\alpha;\, k,\, k(n-1)} \sqrt{\frac{\text{MSW}}{n}} \left(\frac{1}{2} \sum_j |c_j|\right) \le L \le l + q_{1-\alpha;\, k,\, k(n-1)} \sqrt{\frac{\text{MSW}}{n}} \left(\frac{1}{2} \sum_j |c_j|\right) \tag{13.36}$$

where $q_{1-\alpha;\, k,\, k(n-1)}$ is tabulated in Table K is the Appendix. [Note in Table K that $\nu = k(n - 1)$.]

If the null hypothesis ($L = 0$) is true, the confidence limits given by (13.36) must include zero. Accordingly, the null hypothesis is rejected at the α level when

$$|l| > q_{1-\alpha;\, k,\, k(n-1)} \sqrt{\frac{\text{MSW}}{n}} \left(\frac{1}{2} \sum_j |c_j|\right) \tag{13.37}$$

It should be noted that the use of the Tukey's procedure is not necessarily limited to tests on differences between pairs of means; it may also be used in testing contrasts involving three or more means. Tukey's procedure exhibits this important property: *All tests on differences between pairs have an overall level of significance that is at most equal to* α.

Scheffé's Procedure

Scheffé has proven that the probability is $1 - \alpha$ that the values of all contrasts simultaneously lie within the confidence intervals given by

$$l - S\sqrt{\text{MSW}}\sqrt{\frac{\sum_j c_j^2}{n_j}} \leq L \leq l + S\sqrt{\text{MSW}}\sqrt{\frac{\sum_j c_j^2}{n_j}} \tag{13.38}$$

where $S^2 = (k - 1)F_{1-\alpha;\, k-1,\, k(n-1)}$, k = the number of means that might be entering into the contrast rather than the number of means actually entering the contrast, and n_j = the sample size drawn from population j (note that Tukey's procedure requires that the n_j's be all equal to n). When the null hypothesis ($L = 0$) is true, the confidence limits given by (13.38) must include zero. It follows that the null hypothesis is rejected at the α level when

$$|l| > S\sqrt{\text{MSW}}\sqrt{\frac{\sum_j c_j^2}{n_j}} \tag{13.39}$$

Thus, again, the probability of at least one type I error within the set of possible contrasts is α. An interesting feature of the Scheffé's procedure is that *if an overall F test rejects the equality of means at the* α *level, then at least the maximum possible contrast will be significant at the* α *level.* We shall illustrate these procedures with a numerical example.

Example 13.5 Using the data of Example 13.1, work out the following:

(a) Determine which machines differ at the 5% level of significance.

(b) Compute a 95% confidence band for the contrast $\mu_1 - 2\mu_2 + \mu_3$.

Solution

(a) We shall answer this question by using both procedures. Here $k = 3$, $n = 5$, and MSW $= 16 \times 10^{-6}$. To determine which machines differ we have to test for significant differences between all possible pairs of machines. For $k = 3$, the number of possible pairs $= k(k - 1)/2 = 3$.

Accordingly, machines 1 and 2 are different if $L_1 = \mu_1 - \mu_2 \neq 0$, machines 1 and 3 are different if $L_2 = \mu_1 - \mu_3 \neq 0$, and machines 2 and 3 are different if $L_3 = \mu_2 - \mu_3 \neq 0$. Now for Tukey's procedure we have to compute

$$q_{1-\alpha;\, k,\, k(n-1)} \sqrt{\frac{\text{MSW}}{n}} \left(\frac{1}{2} \sum |c_j|\right)$$

From Table K we find that $q_{.95;\, 3,12} = 3.77$. Hence

$$q_{.95;\, 3,12} \sqrt{\frac{16 \times 10^{-6}}{5}} \left[\frac{1}{2}(|1| + |-1|)\right] = .00674$$

and the confidence limits for L_1 are

$$l_1 \pm .00674 = (\bar{y}_{.1} - \bar{y}_{.2}) \pm .00674 = [-.02074, -.00726]$$

which do not include zero. Thus we conclude that machines 1 and 2 are significantly different. We can obtain the same result by using (13.37), i.e.,

$$|l_1| = .014 > .00674$$

Hence, we conclude that machines 1 and 2 are different. Similarly,

$$|l_2| = .020 > .00674$$

Hence, machines 1 and 3 are different. And, finally,

$$|l_3| = .006 < .00674$$

Hence, machines 2 and 3 are not different.

The Scheffé limits involve calculating

$$S^2 = (k-1)F_{1-\alpha;\, (k-1),\, k(n-1)} = 2F_{.95;\, 2,12} = 2(3.89) = 7.78$$

and

$$S\sqrt{\text{MSW}} \sqrt{\frac{\sum_j c_j^2}{n_j}} = \sqrt{7.78(16 \times 10^{-6})} \sqrt{\frac{1+1}{5}} = .00705$$

Since

$$|l_1| = .014 > .00705$$
$$|l_2| = .020 > .00705$$
$$|l_3| = .006 < .00705$$

we conclude that machines 1 and 2 and 1 and 3 are different whereas machines 2 and 3 are not.

(b) The Tukey limits are

$$(\bar{y}_{.1} - 2\bar{y}_{.2} + \bar{y}_{.3}) \pm q_{1-\alpha;\, k,\, k(n-1)} \sqrt{\frac{\text{MSW}}{n}} \left(\frac{1}{2} \sum_j |c_j|\right)$$

$$= (.242 - .512 + .262) \pm q_{.95;\, 3,12} \sqrt{\frac{6 \times 10^{-6}}{5}} \left[\frac{1}{2}(1 + 2 + 1)\right]$$

$$= -.008 \pm .01348 = [-.02148, .00548]$$

Since the confidence interval covers zero we conclude that the contrast $(\mu_1 - 2\mu_2 + \mu_3)$ is not significantly different from zero.

The Scheffé limits are

$$-.008 \pm S\sqrt{\text{MSW}}\sqrt{\frac{\sum_j c_j^2}{n_j}} = -.008 \pm \sqrt{7.78(16 \times 10^{-6})}\sqrt{\frac{1+4+1}{5}}$$
$$= -.008 \pm .0122 = [-.0202, .0042]$$

Again we conclude that the complex contrast is not significantly different from zero.

If the contrast $(\mu_1 - 2\mu_2 + \mu_3)$ were chosen before we examined the data and were the only contrast for which we wanted confidence limits, then it would be appropriate to use the t distribution, which yields

$$-.008 \pm t_{1-\alpha/2;\, k(n-1)}\sqrt{\frac{\text{MSW}}{n}}\sqrt{\sum_j c_j^2}$$
$$= -.008 \pm t_{.975;12}\sqrt{\frac{16 \times 10^{-6}}{5}}\sqrt{6}$$
$$= [-.0175; .0015]$$

a much shorter interval than given either by Tukey's or Scheffé's procedure. This result is not surprising since the confidence limits devised by Tukey or Scheffé are supposed to bracket the expected values for all possible contrasts rather than a prechosen one.

We note that Scheffé's procedure is more powerful for more complex contrasts involving more than two means while Tukey's procedure is more powerful for paired comparisons (involving two means). An important feature of the Scheffé procedure is that the type I error is at most α for any of the possible comparisons. Furthermore, the Scheffé procedure is less sensitive to violations of normality and equal variances assumptions.

13.3 One-Way Classification; Random Effects Model

In many situations we wish to compare a large population of levels of a single factor. In such cases sampling is carried out in two *stages:* The first stage involves the choice of a random sample of k levels from the population of levels, and the second stage involves the choice of a random sample of size n_j from the jth level ($j = 1, 2, \ldots, k$). To illustrate, consider the machines example; suppose the factory has a large number of machines, each of which makes $\frac{1}{4}$-inch aluminum washers. However, only a sample of k machines, representative of the entire population of such machines, would be tested. The distinction between fixed effects and random effects models is important to our inferential processes. In the latter, having assumed that the factory has only three machines that were included in the experiment, it is possible to extend the conclusion derived from the experiment only to these three machines. In the former it is possible to draw inferences about all the machines in the factory from the results based on a sample of k machines chosen randomly from a large population of machines. That is, in the random effects model our inferential results extend *beyond* the actual levels sampled to the entire population from which they were randomly chosen. In subsequent sections we discuss the assumptions, analysis, and testing procedure for the random effects model.

13.3.1 The Model

The mathematical model is now

$$y_{ij} = \mu + \delta_j + e_{ij} \qquad i = 1, 2, \ldots, n_j \qquad j = 1, 2, \ldots, k \tag{13.40}$$

where $\mu =$ the overall mean and $\delta_j = \mu_j - \mu$. The δ_j's are assumed to be normally distributed with zero mean and variance σ_δ^2. The e_{ij}'s are independent chance errors with identical normal distributions with zero mean and variance σ_e^2. Finally, the δ_{ij}'s and e_{ij}'s are assumed to be statistically independent.

13.3.2 The Analysis

In this case we use the same formal analysis as for the fixed effects model. The total sum of squares is partitioned into two components SSW and SSG. As before, the expected value of the sum of squares within samples is

$$E(\text{SSW}) = \sum_j (n_j - 1)E(s_j^2) = \sum_j (n_j - 1)\sigma_e^2 = (N - k)\sigma_e^2$$

Therefore

$$E\left(\frac{\text{SSW}}{N - k}\right) = E(\text{MSW}) = \sigma_e^2 \tag{13.41}$$

Hence MSW is an unbiased estimate of σ_e^2.

Now we proceed to derive the estimate of σ_δ^2. From Eq. (13.40) we have

$$\bar{y}_{.j} = \mu + \delta_j + \frac{\sum_i e_{ij}}{n_j} = \mu + \delta_j + \bar{e}_{.j}$$

and

$$\bar{y}_{..} = \mu + \sum_j \frac{n_j\delta_j}{N} + \frac{\sum_j \sum_i e_{ij}}{N} = \mu + \sum_j \frac{n_j\delta_j}{N} + \bar{e}_{..}$$

whence

$$(\bar{y}_{.j} - \bar{y}_{..}) = \left(\delta_j - \sum_j \frac{n_j\delta_j}{N}\right) + (\bar{e}_{.j} - \bar{e}_{..})$$

Squaring and multiplying by n_j we obtain

$$n_j(\bar{y}_{.j} - \bar{y}_{..})^2 = n_j\left(\delta_j - \sum_j \frac{n_j\delta_j}{N}\right)^2 + n_j(\bar{e}_{.j} - \bar{e}_{..})^2 + 2n_j\left(\delta_j - \sum_j \frac{n_j\delta_j}{N}\right)(\bar{e}_{.j} - \bar{e}_{..})$$

Summing over j and taking the expected values gives us

$$E[\sum_j n_j(\bar{y}_{.j} - \bar{y}_{..})^2] = E\left[\sum_j n_j\left(\delta_j - \sum_j \frac{n_j\delta_j}{N}\right)^2\right] + E[\sum_j n_j(\bar{e}_{.j} - \bar{e}_{..})^2] + 2E\left[\sum_j n_j\left(\delta_j - \sum_j \frac{n_j\delta_j}{N}\right)(\bar{e}_{.j} - \bar{e}_{..})\right]$$

The first term

$$\begin{aligned} E\left[\sum_j n_j\left(\delta_j - \sum_j \frac{n_j\delta_j}{N}\right)^2\right] &= E\left[\sum_j n_j\delta_j^2 + \sum_j n_j\left(\sum_j \frac{n_j\delta_j}{N}\right)^2 - 2\sum_j n_j\left(\delta_j \sum_j \frac{n_j\delta_j}{N}\right)\right] \\ &= N\sigma_\delta^2 + \frac{1}{N}E(\sum_j n_j\delta_j)^2 - \frac{2}{N}E(\sum_j n_j\delta_j)^2 \\ &= N\sigma_\delta^2 - \frac{1}{N}E(n_1\delta_1 + n_2\delta_2 + \cdots + n_k\delta_k)^2 \end{aligned}$$

(13.42) $$= N\sigma_\delta^2 - \frac{1}{N}\sum_j n_j^2\sigma_\delta^2 = \sigma^2\left(\frac{N^2 - \sum_j n_j^2}{N}\right)$$

The second term is

$$E[\sum_j n_j(\bar{e}_{.j} - \bar{e}_{..})^2] = (k-1)\sigma_e^2$$

Finally, the third term is

$$2E\left[\sum_j n_j\left(\delta_j - \sum_j \frac{n_j\delta_j}{N}\right)(\bar{e}_{.j} - \bar{e}_{..})\right] = 0$$

Hence

$$E(\text{SSG}) = E[\sum_j n_j(\bar{y}_{.j} - \bar{y}_{..})^2] = \sigma_\delta^2\left(\frac{N^2 - \sum_j n_j^2}{N}\right) + (k-1)\sigma_e^2$$

or

(13.43) $$E\left(\frac{\text{SSG}}{k-1}\right) = E(\text{MSG}) = \sigma_e^2 + n_0\sigma_\delta^2 \quad \text{where} \quad n_0 = \frac{N^2 - \sum_j n_j^2}{N(k-1)}$$

(Note when the n_j's are all equal to n, $n_0 = n$.) These results are summarized in Table 13.5.

Subtracting Eq. (13.41) from Eq. (13.43) yields

$$E(\text{MSG}) - E(\text{MSW}) = E(\text{MSG} - \text{MSW}) = n_0\sigma_\delta^2$$

TABLE 13.5 ANOVA for the One-Way Classification; Random Effects Model

Source of Variation	*Sum of Squares*	*Degrees of Freedom*	*Mean Square*	*Expected Mean Square*
Among treatments	SSG	$k-1$	$\frac{\text{SSG}}{k-1} = \text{MSG}$	$\sigma_e^2 + n_0\sigma_\delta^2$
Within treatments	SSW	$N-k$	$\frac{\text{SSW}}{k-1} = \text{MSW}$	σ_e^2
Total	SST	$N-1$		

whence

$$\sigma_\delta^2 = \frac{1}{n_0} E(\text{MSG} - \text{MSW}) \tag{13.44}$$

Therefore, $(\text{MSG} - \text{MSW})/n_0$ is an unbiased estimate of σ_δ^2.

The computing formulas and procedure are exactly the same as those of the fixed effects model.

13.3.3 Testing the Hypothesis of Equal Means

If all k effects are equal, then the variance of δ_j's should be equal to zero. Consequently, the hypothesis of equal means may be expressed as

$$H_0: \sigma_\delta^2 = 0$$

$$\text{against} \quad H_1: \sigma_\delta^2 > 0 \quad (\text{since } \sigma_\delta^2 \text{ is necessarily positive})$$

In order to construct a formal test for H_0 it seems reasonable to consider the ratio MSG over MSW. To simplify the analysis we shall assume an equal number of observations in each group, i.e., $n_1 = n_2 = \cdots = n_k = n$. The among samples sum of squares, therefore, is

$$\text{SSG} = n\sum_j (\bar{y}_{.j} - \bar{y}_{..})^2$$

The $\bar{y}_{.j}$'s are mutually independent normal random variables with mean μ and variance $\sigma_\delta^2 + \sigma_e^2/n$. Therefore, the ratios

$$\frac{\sum_j (\bar{y}_{.j} - \bar{y}_{..})^2}{\sigma_\delta^2 + \sigma_e^2/n} = \frac{n\sum_j (\bar{y}_{.j} - \bar{y}_{..})^2}{n\sigma_\delta^2 + \sigma_e^2} = \frac{\text{SSG}}{n\sigma_\delta^2 + \sigma_e^2} \tag{13.45}$$

are distributed as χ^2 with $k-1$ degrees of freedom. Furthermore, SSW is statistically independent of SSG and distributed as $\chi^2_{k(n-1)}\sigma_e^2$. Hence, when H_1 is true the ratio

$$\frac{[\text{SSG}/(n\sigma_\delta^2 + \sigma_e^2)]/(k-1)}{(\text{SSW}/\sigma_e^2)/k(n-1)} = \frac{\text{MSG}}{\text{MSW}}\left(\frac{\sigma_e^2}{\sigma_e^2 + n\sigma_\delta^2}\right) \tag{13.46}$$

has the *central F* distribution with $k - 1$ and $k(n - 1)$ degrees of freedom. Accordingly, we can use the *central F* distribution to evaluate the power function. Under $H_0(\sigma_\delta^2 = 0)$, MSG/MSW also has the *central F* distribution with $k - 1$ and $k(n - 1)$ degrees of freedom. Therefore, when $H_0(\sigma_\delta^2 = 0)$ is true, the rejection region for a specified level of significance α is

$$\frac{\text{MSG}}{\text{MSW}} > F_{1-\alpha;\, k-1,\, k(n-1)} \tag{13.47}$$

which is the same as that of the random effects model. However, the distribution of the power functions is different. In the fixed effects case the power function has the *noncentral F* distribution, while in the random effects case the power functuon has the *central F* distribution.

13.3.4 The Power of F Test; Random Effects Model

The power of the test, with equal sample sizes, is given by

$$\begin{aligned}
\pi(\sigma_\delta^2) &= P\left\{\frac{\text{MSG}}{\text{MSW}} > F_{1-\alpha;\, k-1,\, k(n-1)} \,\middle|\, \sigma_\delta^2\right\} \\
&= P\left\{\frac{\text{MSG}}{\text{MSW}}\left(\frac{\sigma_e^2}{\sigma_e^2 + n\sigma_\delta^2}\right) > \frac{\sigma_e^2}{\sigma_e^2 + n\sigma_\delta^2} F_{1-\alpha;\, k-1,\, k(n-1)}\right\} \\
&= P\left\{F_{k-1;\, k(n-1)} > \frac{F_{1-\alpha;\, k-1,\, k(n-1)}}{1 + n(\sigma_\delta^2/\sigma_e^2)}\right\} \qquad (13.48)
\end{aligned}$$

which can be evaluated by using the standard tables of the F distribution. The exact power of the test can be evaluated if σ_δ is given as a multiple of σ_e; however, an approximate value can be obtained by using MSW as an estimate of σ_e^2.

Example 13.6 In Example 13.1 consider the three machines as a random sample chosen from a large population of machines.
(a) Find the power of the test when $\sigma_\delta = 2\frac{2}{3}\sigma_e$. Use $\alpha = .05$.
(b) Find a 95% confidence interval for $\sigma_\delta^2/\sigma_e^2$.

Solution
(a) Using Eq. (13.48) we obtain the power of the test

$$\begin{aligned}
\pi\left(\frac{\sigma_\delta^2}{\sigma_e^2} = 7.11\right) &= P\left\{F_{k-1,\, k(n-1)} > \frac{F_{1-\alpha;\, k-1,\, k(n-1)}}{1 + n(\sigma_\delta^2/\sigma_e^2)}\right\} \\
&= P\left\{F_{2,12} > \frac{F_{.95;2,12}}{1 + 5(\frac{8}{3})^2}\right\} = P\left\{F_{2,12} > \frac{3.89}{36.55}\right\} \\
&= P\{F_{2,12} > .106\}
\end{aligned}$$

whence $F_{1-\pi;\, 2,12} = 1/F_{\pi;\, 12,2} = .106$ or $F_{\pi;\, 12,2} \simeq 9.4$. Hence, from the F tables, we find $\pi \simeq .90$.

(b) We recall that the $(1 - \alpha)\%$ confidence interval for random variable F with $k - 1$ and $k(n - 1)$ degrees of freedom is

$$[F_{\alpha/2;\,k-1,\,k(n-1)},\, F_{1-\alpha/2;\,k-1,\,k(n-1)}]$$

i.e.,

$$P\{F_{\alpha/2;\,k-1,\,k(n-1)} \leq F_{k-1,\,k(n-1)} \leq F_{1-\alpha/2;\,k-1,\,k(n-1)}\} = 1 - \alpha$$

However,

$\dfrac{\text{MSG}}{\text{MSW}}\left[\dfrac{1}{1 + n(\sigma_\delta^2/\sigma_e^2)}\right]$ is a random variable having the F distribution

with $k - 1$ and $n(k - 1)$ degrees of freedom. Therefore

$$P\left\{F_{\alpha/2;\,k-1,\,k(n-1)} \leq \frac{\text{MSG}}{\text{MSW}}\frac{1}{1 + n(\sigma_\delta^2/\sigma_e^2)} \leq F_{1-\alpha/2;\,k-1,\,k(n-1)}\right\} = 1 - \alpha$$

so that

$$P\left\{\frac{1}{n}\left(\frac{\text{MSG/MSW}}{F_{1-\alpha/2;\,k-1,\,k(n-1)}} - 1\right) \leq \frac{\sigma_\delta^2}{\sigma_e^2} \leq \frac{1}{n}\left(\frac{\text{MSG/MSW}}{F_{\alpha/2;\,k-1,\,k(n-1)}} - 1\right)\right\} = 1 - \alpha$$

(13.49)

Therefore, by (13.49)

$$\text{The upper confidence limit} = \frac{1}{n}\left(\frac{\text{MSG/MSW}}{F_{\alpha/2;\,k-1,\,k(n-1)}} - 1\right)$$

$$= \frac{1}{5}\left(\frac{32.9}{F_{.025;\,2,12}} - 1\right) = 259.05$$

and

$$\text{The lower confidence limit} = \frac{1}{n}\left(\frac{\text{MSG/MSW}}{F_{1-\alpha/2;\,k-1,\,k(n-1)}} - 1\right)$$

$$= \frac{1}{5}\left(\frac{32.9}{F_{.975;\,2,12}} - 1\right) = 1.09$$

This means, in the long run, 95 times in 100 the computed limits [1.09, 259.05] will include $\sigma_\delta^2/\sigma_e^2$. If the lower confidence limit turns out to be negative it is customary to replace the negative end point by zero, since $\sigma_\delta^2/\sigma_e^2$ must be positive.

We now consider the situation where the investigator knows beforehand that differences among the means of the populations under study are almost certain to exist: consequently, he may not be concerned with testing the hypothesis of equal means. Instead he might wish to know whether the ratio $\sigma_\delta^2/\sigma_e^2$ is greater than, less than, or equal to a prescribed constant. Let us consider the case where it is desired to test the hypothesis

$$H_0\colon \frac{\sigma_\delta^2}{\sigma_e^2} = a \quad \text{against} \quad H_1\colon \frac{\sigma_\delta^2}{\sigma_e^2} > a$$

where $\boldsymbol{a}$ is a prescribed constant. To construct a formal test we shall make use of the result that the ratio

$$\frac{\text{MSG}}{\text{MSW}}\left(\frac{\sigma_e^2}{\sigma_e^2 + n\sigma_\delta^2}\right)$$

has the *central F* distribution with $k - 1$ and $k(n - 1)$ degrees of freedom. It is reasonable to reject H_0 when the ratio MSG/MSW is greater than a prescribed constant c. Hence, the probability of rejecting H_0 at the α level of significance is

$$P\left\{\frac{\text{MSG}}{\text{MSW}} > c \mid \frac{\sigma_\delta^2}{\sigma_e^2} = \boldsymbol{a}\right\} = \alpha$$

i.e.,

$$P\left\{\frac{\text{MSG}}{\text{MSW}}\left(\frac{1}{1 + n\boldsymbol{a}}\right) > \frac{c}{1 + n\boldsymbol{a}}\right\} = P\left\{F_{k-1,\, k(n-1)} > \frac{c}{1 + n\boldsymbol{a}}\right\} = \alpha$$

whence

$$c = (1 + n\boldsymbol{a})F_{1-\alpha;\, k-1,\, k(n-1)}$$

Thus H_0 is to be rejected at the α level of significance if

$$\frac{\text{MSG}}{\text{MSW}} > (1 + n\boldsymbol{a})F_{1-\alpha;\, k-1,\, k(n-1)} \tag{13.50}$$

Example 13.7 Using the data of Example 13.1 test the hypothesis that

$$H_0: \frac{\sigma_\delta^2}{\sigma_e^2} \leq 5 \quad \text{against} \quad H_1: \frac{\sigma_\delta^2}{\sigma_e^2} > 5$$

Assume that the three machines constitute a random sample chosen from a large population of machines. Use $\alpha = .10$.

Solution. The null hypothesis is to be rejected when MSG/MSW $> c$ where

$$c = (1 + n\boldsymbol{a})F_{1-\alpha;\, k-1,\, k(n-1)} = (1 + 5 \times 5)F_{.90;\, 2,\, 12} = 26 \times 2.81 = 73.06$$

Since MSG/MSW $= 32.9 < 73.06$, we accept the null hypothesis and conclude that $\sigma_\delta^2/\sigma_e^2$ is less than or equal to 5 at the 10% level of significance.

13.4 Two-Way Classification; Fixed Effects Model; One Observation per Cell

In this section we shall discuss the problem of designing an experiment to study the effects of two factors, say A and B. We shall assume that factor A has r levels, $A_1, A_2, \ldots, A_r$ and that factor B has c levels, $B_1, B_2, \ldots, B_c$. Furthermore, we shall assume that exactly one observation is made for each combination of levels. When observations represent all possible combinations of the levels of the factors under study, the experiment is called a *factorial*

TABLE 13.6 Sample Layout for Two-Way Classification One Observation per Cell

B *levels* / A *levels*	B_1	B_2	$\cdots$	B_j	$\cdots$	B_c	*Totals*	*Row Means*
A_1	y_{11}	y_{12}	$\cdots$	y_{1i}	$\cdots$	y_{1c}	$T_{1.}$	$\bar{y}_{1.}$
A_2	y_{21}	y_{22}	$\cdots$	y_{2j}	$\cdots$	y_{2c}	$T_{2.}$	$\bar{y}_{2.}$
$\cdots$	$\cdots$	$\cdots$	$\cdots$	$\cdots$	$\cdots$	$\cdots$	$\cdots$	$\cdots$
A_i	y_{i1}	y_{i2}	$\cdots$	y_{ij}	$\cdots$	y_{ic}	$T_{i.}$	$\bar{y}_{i.}$
$\cdots$	$\cdots$	$\cdots$	$\cdots$	$\cdots$	$\cdots$	$\cdots$	$\cdots$	$\cdots$
A_r	y_{r1}	y_{r2}	$\cdots$	y_{rj}	$\cdots$	y_{rc}	$T_{r.}$	$\bar{y}_{r.}$
Totals	$T_{.1}$	$T_{.2}$	$\cdots$	$T_{.j}$	$\cdots$	$T_{.c}$	$T_{..}$	
Column means	$\bar{y}_{.1}$	$\bar{y}_{.2}$	$\cdots$	$\bar{y}_{.j}$	$\cdots$	$\bar{y}_{.c}$		$\bar{y}_{..}$

experiment. The data for two-way classification are recorded as shown in Table 13.6. The intersection of the ith row with the jth column is known as the ijth cell. In the present case we are considering only one observation per cell.

The symbols are defined as follows:

The observation corresponding to the ijth treatment $= y_{ij};\ i = 1, \ldots, r;\ j = 1, \ldots, c$

Mean of all observations made

$$\text{with level } A_i = \frac{\sum_{j=1}^{c} y_{ij}}{c} = \frac{T_{i.}}{c} = \bar{y}_{i.}$$

Mean of all observations made

$$\text{with level } B_j = \frac{\sum_{i=1}^{r} y_{ij}}{r} = \frac{T_{.j}}{r} = \bar{y}_{.j}$$

$$\text{Mean of all observations} = \sum_{i=1}^{r} \sum_{j=1}^{c} \frac{y_{ij}}{rc} = \frac{T_{..}}{rc} = \bar{y}_{..}$$

Note that r and c are equivalent to the number of rows and columns, respectively.

13.4.1 The Model

The development of this model will be illustrated by an example. Suppose we wish to run an experiment to assess the effects of machine and operator

TABLE 13.7 Expected Values for Number of Defective Items

Machines / Operators	1	2	3	4	Row Means	Deviations
1	μ_{11}	μ_{12}	μ_{13}	μ_{14}	$\mu_{1.}$	$\mu_{1.} - \mu = \alpha_1$
2	μ_{21}	μ_{22}	μ_{23}	μ_{24}	$\mu_{2.}$	$\mu_{2.} - \mu = \alpha_2$
3	μ_{31}	μ_{32}	μ_{33}	μ_{34}	$\mu_{3.}$	$\mu_{3.} - \mu = \alpha_3$
4	μ_{41}	μ_{42}	μ_{43}	μ_{44}	$\mu_{4.}$	$\mu_{4.} - \mu = \alpha_4$
Column Means	$\mu_{.1}$	$\mu_{.2}$	$\mu_{.3}$	$\mu_{.4}$	μ	
Deviations	$\mu_{.1} - \mu = \beta_1$	$\mu_{.2} - \mu = \beta_2$	$\mu_{.3} - \mu = \beta_3$	$\mu_{.4} - \mu = \beta_4$		

on the number of defective items produced. Furthermore, suppose that we are concerned with four particular machines and only four particular operators. Thus the machines and the operators are considered as *fixed factors.* To make the analysis easier to understand we shall tabulate the expected values of the observations made for each combination of levels as displayed in Table 13.7. The symbols in this table may be defined as follows:

$$\text{Row mean} = \mu_{i.} = \frac{\sum_{j=1}^{4} \mu_{ij}}{4} \qquad i = 1, \ldots, 4$$

$$\text{Column mean} = \mu_{.j} = \frac{\sum_{i=1}^{4} \mu_{ij}}{4} \qquad j = 1, \ldots, 4$$

$$\text{Grand mean} = \mu = \sum_{i=1}^{4} \sum_{j=1}^{4} \mu_{ij} = \frac{\sum_{i=1}^{4} \mu_{i.}}{4} = \frac{\sum_{j=1}^{4} \mu_{.j}}{4}$$

$$\text{Row effect (operator effect)} = \mu_{i.} - \mu = \alpha_i$$

$$\text{Column effect (machine effect)} = \mu_{.j} - \mu = \beta_j$$

Obviously both row and column effects sum to zero:

$$\sum_{i=1}^{4} \alpha_i = \sum_{i=1}^{4} (\mu_{i.} - \mu) = 4\mu - 4\mu = 0$$

$$\sum_{j=1}^{4} \beta_j = \sum_{j=1}^{4} (\mu_{.j} - \mu) = 4\mu - 4\mu = 0$$

Thus the true mean for the *ij*th cell can be written

$$\begin{aligned}\mu_{ij} &= \mu + (\mu_{i.} - \mu) + (\mu_{.j} - \mu) + (\mu_{ij} - \mu_{i.} - \mu_{.j} + \mu) \\ &= \mu + \alpha_i + \beta_j + \gamma_{ij} \qquad (13.51)\end{aligned}$$

where $\gamma_{ij} = \mu_{ij} - \mu_{i.} - \mu_{.j} + \mu$ and is known as the *interaction effect*. In this situation γ_{ij} represents the interaction effect between the *i*th operator and the *j*th machine. We say *two effects do not interact if their joint effect is not different from the sum of their separate effects.* If all machines are identical the performance of all operators is the same and machine–operator interactions do not exist; then, discounting other sources of variability, we shall expect that the true mean for each cell is a constant equal to μ. However, if operator, machine, and interaction effects are likely to be significant, then the true mean for the *ij*th cell is given by Eq. (13.51).

In two-way classification, one observation per cell, we must assume that *no interaction occurs* since we have no estimate of error against which we can test the significance of possible interaction. Accordingly, we shall assume that $\gamma_{ij} = 0$ for all *i* and *j*. This means that the main effects are *additive*. In symbols, we write

$$\mu_{ij} = \mu + \alpha_i + \beta_j \qquad (13.52)$$

Furthermore, we shall assume that the y_{ij}'s are independently normally distributed with mean μ_{ij} and variance σ_e^2 for all *i* and *j*. Thus the mathematical model for this two-way classification, fixed effects, one observation per cell experiment is given by

$$y_{ij} = \mu + \alpha_i + \beta_j + e_{ij} \qquad i = 1, \ldots, r \qquad j = 1, \ldots, c \qquad (13.53)$$

with the restrictions

$$\sum_{i=1}^{r} \alpha_i = 0 \quad \text{and} \quad \sum_{j=1}^{c} \beta_j = 0$$

The e_{ij}'s are assumed to be independent chance components with identical normal distributions $N(0, \sigma_e^2)$.

13.4.2 The Analysis

Consider the identity

$$y_{ij} - \bar{y}_{..} = (\bar{y}_{i.} - \bar{y}_{..}) + (\bar{y}_{.j} - \bar{y}_{..}) + (y_{ij} - \bar{y}_{i.} - \bar{y}_{.j} + \bar{y}_{..})$$

Squaring and summing over *i* and *j* we obtain

$$\begin{aligned}\sum_{i=1}^{r}\sum_{j=1}^{c}(y_{ij} - \bar{y}_{..})^2 &= \sum_{i=1}^{r}\sum_{j=1}^{c}(\bar{y}_{i.} - \bar{y}_{..})^2 + \sum_{i=1}^{r}\sum_{j=1}^{c}(\bar{y}_{.j} - \bar{y}_{..})^2 \\ &\quad + \sum_{i=1}^{r}\sum_{j=1}^{c}(y_{ij} - \bar{y}_{i.} - \bar{y}_{.j} + \bar{y}_{..})^2 \qquad (13.54) \\ &\quad + 3 \text{ cross product terms}\end{aligned}$$

Let us evaluate these cross product terms. One of them is

$$2\sum_{i=1}^{r}\sum_{j=1}^{c}(\bar{y}_{i.}-\bar{y}_{..})(\bar{y}_{.j}-\bar{y}_{..}) = 2\sum_{i=1}^{r}(\bar{y}_{i.}-\bar{y}_{..})\sum_{j=1}^{c}(\bar{y}_{.j}-\bar{y}_{..})$$
$$= 2(r\bar{y}_{..}-r\bar{y}_{..})(c\bar{y}_{..}-c\bar{y}_{..}) = 0$$

Proceeding similarly we can show that the other two cross product terms are equal to zero. Thus Eq. (13.54) can be written

$$\sum_{i=1}^{r}\sum_{j=1}^{c}(y_{ij}-\bar{y}_{..})^2 = \sum_{i=1}^{r}\sum_{j=1}^{c}(\bar{y}_{i.}-\bar{y}_{..})^2 + \sum_{i=1}^{r}\sum_{j=1}^{c}(\bar{y}_{.j}-\bar{y}_{..})^2 + \sum_{i=1}^{r}\sum_{j=1}^{c}(y_{ij}-\bar{y}_{i.}-\bar{y}_{.j}+\bar{y}_{..})^2$$

(13.55)
$$= c\sum_{i=1}^{r}(\bar{y}_{i.}-\bar{y}_{..})^2 + r\sum_{j=1}^{c}(\bar{y}_{.j}-\bar{y}_{..})^2 + \sum_{i=1}^{r}\sum_{j=1}^{c}(y_{ij}-\bar{y}_{i.}-\bar{y}_{.j}+\bar{y}_{..})^2$$

Using our customary notation Eq. (13.55) can be expressed as

(13.56)
$$\text{SST} = \text{SSA} + \text{SSB} + \text{SSE}$$

where SST, SSA, SSB, and SSE represent the total sum of squares, sum of squares for the A factor, sum of squares for the B factor, and sum of squares for error, respectively. Thus we have partitioned the total variation into three components. Using the model of Eq. (13.53),

(13.57)
$$\bar{y}_{.j} = \frac{\sum_{i=1}^{r} y_{ij}}{r} = \frac{1}{r}\left(\mu r + \sum_{i=1}^{r}\alpha_i + r\beta_j + \sum_{i=1}^{r} e_{ij}\right)$$
$$= \mu + \beta_j + \bar{e}_{.j}$$

where $\bar{e}_{.j} = \frac{1}{r}\sum_{i=1}^{r} e_{ij}$

since $\sum_{i=1}^{r}\alpha_i = 0$ for a fixed effects model. Similarly,

(13.58)
$$\bar{y}_{i.} = \mu + \alpha_i + \bar{e}_{i.}$$

where $\bar{e}_{i.} = \frac{1}{c}\sum_{j=1}^{c} e_{ij}$ and,

(13.59)
$$\bar{y}_{..} = \mu + \bar{e}_{..}$$

where $\bar{e}_{..} = \frac{1}{rc}\sum_{i=1}^{r}\sum_{j=1}^{c} e_{ij}$,

so that SSA, SSB, and SSE may be written

(13.60)
$$\text{SSA} = c\sum_{i=1}^{r}(\bar{y}_{i.}-\bar{y}_{..})^2 = c\sum_{i=1}^{r}(\alpha_i+\bar{e}_{i.}-\bar{e}_{..})^2$$

(13.61)
$$\text{SSB} = r\sum_{j=1}^{c}(\bar{y}_{.j}-\bar{y}_{..})^2 = r\sum_{j=1}^{c}(\beta_j+\bar{e}_{.j}-\bar{e}_{..})^2$$

$$\begin{aligned} \text{SSE} &= \sum_{i=1}^{r}\sum_{j=1}^{c}(y_{ij} - \bar{y}_{i.} - \bar{y}_{.j} + \bar{y}_{..})^2 \\ &= \sum_{i=1}^{r}\sum_{j=1}^{c}(e_{ij} - \bar{e}_{i.} - \bar{e}_{.j} + e_{..})^2 \end{aligned} \tag{13.62}$$

The expected value of SSA, then, is

$$\begin{aligned} E(\text{SSA}) &= cE\left\{\sum_{i=1}^{r}[\alpha_i^2 + (\bar{e}_{i.} - \bar{e}_{..})^2 + 2\alpha_i(\bar{e}_{i.} - \bar{e}_{..})]\right\} \\ &= c\sum_{i=1}^{r}\alpha_i^2 + (r-1)\sigma_e^2 \end{aligned} \tag{13.63}$$

since $E(\bar{e}_{i.} - \bar{e}_{..}) = 0$. Proceeding similarly we find

$$E(\text{SSB}) = r\sum_{j=1}^{c}\beta_j^2 + (c-1)\sigma_e^2 \tag{13.64}$$

$$E(\text{SSE}) = (r-1)(c-1)\sigma_e^2 \tag{13.65}$$

The results are presented in the analysis of variance scheme in Table 13.8.

TABLE 13.8 ANOVA for the Two-Way Classification; Fixed Effects Model; One Observation per Cell

Source of Variation	*Sum of Squares*	*Degress of Freedom*	*Mean Square*	*Expected Mean Square*
Factor A	SSA	$r-1$	$\frac{\text{SSA}}{r-1} = \text{MSA}$	$\sigma_e^2 + \frac{c}{r-1}\sum_{i=1}^{r}\alpha_i^2$
Factor B	SSB	$c-1$	$\frac{\text{SSB}}{c-1} = \text{MSB}$	$\sigma_e^2 + \frac{r}{c-1}\sum_{j=1}^{c}\beta_j^2$
Error	SSE	$(r-1)(c-1)$	$\frac{\text{SSE}}{(r-1)(c-1)} = \text{MSE}$	σ_e^2
Total	SST	$rc-1$		

13.4.3 Testing the Hypothesis of Equal Treatment Effects

In the analysis of a two-way classification we may wish to test the significance of the effects of factors A and B. Henceforth it is necessary to test the hypothesis

$$H_0'\colon \alpha_1 = \alpha_2 = \cdots = \alpha_r = 0 \quad \text{against} \quad H_1'\colon \text{not all } \alpha_i\text{'s are zero}$$

$$H_0''\colon \beta_1 = \beta_2 = \cdots = \beta_c = 0 \quad \text{against} \quad H_1''\colon \text{not all } \beta_j\text{'s are zero}$$

If H_0' is true, then there is no difference between the means of the various levels of factor A. Similarly, if H_0'' is true there is no difference between the means of the various levels of factor B.

When H_0' is true the sum of squares for the A factor is

$$\text{SSA} = c \sum_{i=1}^{r} (\bar{e}_{i.} - \bar{e}_{..})^2$$

since $\alpha_1 = \alpha_2 = \cdots = \alpha_r = 0$.

Similarly, when H_0'' is true the sum of squares for B factor is

$$\text{SSB} = r \sum_{j=1}^{c} (\bar{e}_{.j} - \bar{e}_{..})^2$$

since $\beta_1 = \beta_2 = \cdots = \beta_c = 0$. If the conditions specified by Eq. (13.53) are satisfied and H_0' and H_0'' are true, then it can be proven, by the Partition Theorem, that the ratios

$$\frac{c \sum_{i=1}^{r} (\bar{e}_{i.} - \bar{e}_{..})^2}{\sigma_e^2} \qquad \frac{r \sum_{j=1}^{c} (\bar{e}_{.j} - \bar{e}_{..})^2}{\sigma_e^2} \qquad \frac{\sum_{i=1}^{r} \sum_{j=1}^{c} (e_{ij} - \bar{e}_{i.} - \bar{e}_{.j} + \bar{e}_{..})^2}{\sigma_e^2}$$

are independent random variables having the χ^2 distribution with $(r-1)$, $(c-1)$, and $(r-1)(c-1)$ degrees of freedom, respectively. Thus, when H_0' is true the ratio

$$\frac{c \sum_{i=1}^{r} (\bar{e}_{i.} - \bar{e}_{..})^2}{\sigma_e^2(r-1)} \div \frac{\sum_{i=1}^{r} \sum_{j=1}^{c} (e_{ij} - \bar{e}_{i.} - \bar{e}_{.j} + \bar{e}_{..})^2}{\sigma_e^2(r-1)(c-1)} = \frac{\text{MSA}}{\text{MSE}} \tag{13.66}$$

is distributed as an F random variable with $r-1$ and $(r-1)(c-1)$ degrees of freedom. Therefore, H_0' is rejected at the α level of significance if

$$\frac{\text{MSA}}{\text{MSE}} > F_{1-\alpha;\,(r-1),\,(r-1)(c-1)} \tag{13.67}$$

Similarly, H_0'' is rejected at the α level of significance if

$$\frac{\text{MSB}}{\text{MSE}} > F_{1-\alpha;\,(c-1),\,(r-1)(c-1)} \tag{13.68}$$

Also, it can be shown that MSE is an unbiased estimate of σ_e^2. If H_0' is rejected, then $(\bar{y}_{i.} - \bar{y}_{..})$ is an unbiased estimate of α_i. Similarly, if H_0'' is rejected, $(\bar{y}_{.j} - \bar{y}_{..})$ is an unbiased estimate of β_j.

Here we wish to emphasize that the significance tests are derived under the assumption that *main effects are additive* or, equivalently, *interaction does not exist*. However, if appreciable interaction is present, then conclusions drawn from these significance tests may be misleading.

If either A or B effects are found to be significant, Tukey's or Scheffé's procedure may be used to test the significance of meaningful contrasts. The change in notation used in these procedures becomes fairly obvious. For two-way analysis a contrast among the A means will be of the form

$$L = c_1\mu_{1.} + c_2\mu_{2.} + \cdots + c_r\mu_{r.} = \sum_{i=1}^{r} c_i\mu_{i.} \tag{13.69}$$

which would be estimated by

$$l = c_1\bar{y}_{1.} + c_2\bar{y}_{2.} + \cdots + c_r\bar{y}_{r.} = \sum_{i=1}^{r} c_i\bar{y}_{i.} \tag{13.70}$$

Now if Tukey's procedure is used, L is significant at the α level if

$$|l| > \sqrt{\frac{\text{MSE}}{c}}\, q_{1-\alpha;\, r,\, (r-1)(c-1)}\left(\frac{1}{2}\sum_{i=1}^{r}|c_i|\right) \tag{13.71}$$

Similarly, if Scheffé's procedure is used, L is significant at the α level if

$$|l| > S\sqrt{\frac{\text{MSE}}{c}}\sqrt{\sum_{i=1}^{r} c_i^2} \tag{13.72}$$

where $S^2 = (r-1)F_{1-\alpha;\,(r-1),\,(r-1)(c-1)}$. Likewise, a contrast among the B means will be of the form

$$L = c_1\mu_{.1} + c_2\mu_{.2} + \cdots + c_c\mu_{.c} = \sum_{j=1}^{r} c_j\mu_{.j}$$

Hence (13.71) and (13.72) are of the form

$$|l| > \sqrt{\frac{\text{MSE}}{r}}\, q_{1-\alpha;\, c,\, (r-1)(c-1)}\left(\frac{1}{2}\sum_{j=1}^{c}|c_j|\right) \tag{13.73}$$

and

$$|l| > S\sqrt{\frac{\text{MSE}}{r}}\sqrt{\sum_{j=1}^{c} c_j^2} \tag{13.74}$$

respectively, where $S^2 = (c-1)\,F_{1-\alpha;\,(c-1),\,(r-1)(c-1)}$.

The power function of the F test has the *noncentral* F distribution. Hence, for A effects we compute

$$\phi = \frac{\sqrt{\sum_{i=1}^{r}\alpha_i^2/(\nu_1+1)}}{\sigma_e/\sqrt{c}} \qquad \nu_1 = r-1 \quad \text{and} \quad \nu_2 = (r-1)(c-1) \tag{13.75}$$

and use the charts given in Figs. 13.1–13.6 to find the power of the test. Similarly, for B effects we compute

$$\phi = \frac{\sqrt{\sum_{j=1}^{c}\beta_j^2/(\nu_1+1)}}{\sigma_e/\sqrt{r}} \qquad \nu_1 = c-1 \quad \text{and} \quad \nu_2 = (r-1)(c-1) \tag{13.76}$$

A numerical illustration is given in Example 13.8.

13.5 Two-Way Classification; Random Effects Model; One Observation per Cell

13.5.1 The Model

If in the example cited in Sec. 13.4.1 the four operators and four machines were chosen at random from a large population of operators and machines, then the conclusions drawn from the experiment will concern the populations of workers and machines rather than be confined to the random samples used in the experiment.

The assumptions for this model are

(13.77) $$y_{ij} = \mu + \alpha_i + \beta_j + e_{ij} \qquad i = 1, \ldots, r \qquad j = 1 \ldots, c$$

e_{ij}'s are independently $N(0, \sigma_e^2)$

α_i's are independently $N(0, \sigma_\alpha^2)$

β_j's are independently $N(0, \sigma_\beta^2)$

The interaction effect, which is random, will be assumed to be zero.

13.5.2 The Analysis

As before, the total variation can be partitioned into three components: SST = SSA + SSB + SSE. Now consider

$$\text{SSA} = c \sum_{i=1}^{r} (\bar{y}_{i.} - \bar{y}_{..})^2$$

where $$\bar{y}_{i.} = \frac{\sum_{j=1}^{c} y_{ij}}{c} = \frac{1}{c}\left(\mu c + c\alpha_i + \sum_{j=1}^{c} \beta_j + \sum_{j=1}^{c} e_{ij}\right)$$

$$= \mu + \alpha_i + \frac{1}{c} \sum_{j=1}^{c} \beta_j + \bar{e}_{i.}$$

and $$\bar{y}_{..} = \mu + \frac{\sum_{i=1}^{r} \alpha_i}{r} + \frac{1}{c} \sum_{j=1}^{c} \beta_j + \bar{e}_{..}$$

so that

(13.78) $$\text{SSA} = c \sum_{i=1}^{r} \left[\left(\alpha_i - \frac{\sum_{i=1}^{r} \alpha_i}{r}\right) + (\bar{e}_{i.} - \bar{e}_{..})\right]^2$$

Expectation yields

$$E(\text{SSA}) = cE\left\{\sum_{i=1}^{r} \left[\left(\alpha_i - \sum_{i=1}^{r} \frac{\alpha_i}{r}\right)^2 + (\bar{e}_{i.} - \bar{e}_{..})^2 \right.\right.$$

$$\left.\left. + 2(\bar{e}_{i.} - \bar{e}_{..})\left(\alpha_i - \frac{\sum_{i=1}^{r} \alpha_i}{r}\right)\right]\right\}$$

(13.79) $$= c(r - 1)\sigma_\alpha^2 + (r - 1)\sigma_e^2$$

since the expected value of the cross product term vanishes. Proceeding similarly we find that

(13.80) $$E(\text{SSB}) = r(c - 1)\sigma_\beta^2 + (c - 1)\sigma_e^2$$

(13.81) $$E(\text{SSE}) = (r - 1)(c - 1)\sigma_e^2$$

The results are presented in the analysis of variance scheme in Table 13.9.

TABLE 13.9 ANOVA for the Two-Way Classification; Random Effects Model; One Observation per Cell

Source of Variation	Sum of Squares	Degrees of Freedom	Mean Square	Expected Mean Square
Factor A	SSA	$r-1$	$\frac{\text{SSA}}{r-1} = \text{MSA}$	$\sigma_e^2 + c\sigma_\alpha^2$
Factor B	SSB	$c-1$	$\frac{\text{SSB}}{c-1} = \text{MSB}$	$\sigma_e^2 + r\sigma_\beta^2$
Error	SSE	$(r-1)(c-1)$	$\frac{\text{SSE}}{(r-1)(c-1)} = \text{MSE}$	σ_e^2
Total	SST	$rc-1$		

13.5.3 Testing the Hypotheses of Equal Treatment Effects

In the random effects model, testing the significance of the effects of factor A is equivalent to testing the hypothesis

$$H_0': \sigma_\alpha^2 = 0 \quad \text{against} \quad H_1': \sigma_\alpha^2 > 0 \quad (\text{since } \sigma_\alpha^2 \text{ cannot be negative})$$

Similarly, for factor B

$$H_0'': \sigma_\beta^2 = 0 \quad \text{against} \quad H_1'': \sigma_\beta^2 > 0$$

As before, it can be shown that

$$\frac{\text{MSA}}{\text{MSE}} \sim F_{(r-1),\,(r-1)(c-1)} \qquad \frac{\text{MSB}}{\text{MSE}} \sim F_{(c-1),\,(r-1)(c-1)}$$

Thus H_0' is rejected at the α level of significance if

$$\frac{\text{MSA}}{\text{MSE}} > F_{1-\alpha;\,(r-1),\,(r-1)(c-1)} \tag{13.82}$$

Similarly, H_0'' is rejected at the α level of significance if

$$\frac{\text{MSB}}{\text{MSE}} > F_{1-\alpha;\,(c-1),\,(r-1)(c-1)} \tag{13.83}$$

If H_0' is rejected, then an unbiased estimate of σ_α^2 is given by

$$\hat{\sigma}_\alpha^2 = \frac{1}{c}(\text{MSA} - \text{MSE}) \tag{13.84}$$

Similarly, if H_0'' is rejected, an unbiased estimate of σ_β^2 is given by

$$\hat{\sigma}_\beta^2 = \frac{1}{r}(\text{MSB} - \text{MSE}) \tag{13.85}$$

An important feature of the random effects model is that when interaction is present the significance tests for main effects are unchanged. However, we have postulated the absence of interaction because interaction effects are confounded with the random effects and, accordingly, we cannot test for interaction.

The power of the F tests associated with (13.82) and (13.83) can easily be computed. First we consider the power of the test when σ_α^2 is significant. This is given by

$$\pi(\sigma_\alpha^2) = P\left\{\frac{\text{MSA}}{\text{MSE}} > F_{1-\alpha;\,(r-1),\,(r-1)(c-1)} \mid \sigma_\alpha^2\right\}$$

$$= P\left\{\frac{\text{MSA}/(\sigma_e^2 + c\sigma_\alpha^2)}{\text{MSE}/\sigma_e^2} > \frac{\sigma_e^2}{\sigma_e^2 + c\sigma_\alpha^2} F_{1-\alpha;\,(r-1),\,(r-1)(c-1)}\right\}$$

$$(13.86) \qquad = P\left\{F_{(r-1),\,(r-1)(c-1)} > \frac{\sigma_e^2}{\sigma_e^2 + c\sigma_\alpha^2} F_{1-\alpha;\,(r-1),\,(r-1)(c-1)}\right\}$$

which can be evaluated by using the standard tables of the F distribution. Likewise, the power of the test when σ_β^2 is significant is given by

$$(13.87) \qquad \pi(\sigma_\beta^2) = P\left\{F_{(c-1),\,(r-1)(c-1)} > \frac{\sigma_e^2}{\sigma_e^2 + r\sigma_\beta^2} F_{1-\alpha;\,(c-1),\,(r-1)(c-1)}\right\}$$

13.6 Two-Way Classification; Mixed Effects Model; One Observation per Cell

Here one effect is fixed and one effect is random. The decision as to whether a factor is at fixed or random level is based on practical considerations. To illustrate, in the example of machines and operators we may be interested in the population of workers but only in four specific machines. It follows that workers are considered as a random factor whereas the four machines are considered as a fixed factor. If we consider factor A as a random factor and factor B as fixed, then the mathematical model is given by

$$(13.88) \qquad y_{ij} = \mu + \alpha_i + \beta_j + e_{ij}$$

with the restriction

$$\sum_{j=1}^{c} \beta_j = 0$$

The e_{ij}'s are independently $N(0, \sigma_e^2)$, and α_i's are independently $N(0, \sigma_\alpha^2)$. The interaction effects, which are random, will be assumed to be zero. The analysis of variance for the mixed effects model is shown in Table 13.10. Testing procedures for factor A are identical with those in Sec. 13.5 and for factor B with those in Sec. 13.4. When appreciable interaction is present, the significance test for the effects of the fixed factor is unchanged whereas the test for the effects of the random factor may be misleading.

TABLE 13.10 ANOVA for Two-Way Classification; Mixed Effects Model (B Fixed); One Observation per Cell

Source of Variation	*Sum of Squares*	*Degrees of Freedom*	*Mean Square*	*Expected Mean Square*
Factor A	SSA	$r-1$	$\dfrac{\text{SSA}}{r-1} = \text{MSA}$	$\sigma_e^2 + c\sigma_\alpha^2$
Factor B	SSB	$c-1$	$\dfrac{\text{SSB}}{c-1} = \text{MSB}$	$\sigma_e^2 + \dfrac{r}{c-1}\sum_{j=1}^{c}\beta_j^2$
Error	SSE	$(r-1)(c-1)$	$\dfrac{\text{SSE}}{(r-1)(c-1)} = \text{MSE}$	σ_e^2
Total	SST	$rc-1$		

13.7 Computing Formulas and Procedure; One Observation per Cell

The computational procedure is the same for fixed, random, and mixed effects models. As before,

$$\text{SST} = \sum_{i=1}^{r}\sum_{j=1}^{c} y_{ij}^2 - \frac{T_{..}^2}{rc} \tag{13.89}$$

$$\text{SSA} = c\sum_{i=1}^{r}(\bar{y}_{i.} - \bar{y}_{..})^2 = c\sum_{i=1}^{r}\bar{y}_{i.}^2 - rc\bar{y}_{..}^2$$

$$= \sum_{i=1}^{r}\frac{T_{i.}^2}{c} - \frac{T_{..}^2}{rc} \tag{13.90}$$

and

$$\text{SSB} = r\sum_{j=1}^{c}(\bar{y}_{.j} - \bar{y}_{..})^2 = \sum_{j=1}^{c}\frac{T_{.j}^2}{r} - \frac{T_{..}^2}{rc} \tag{13.91}$$

Finally, by subtraction we obtain

$$\text{SSE} = \text{SST} - \text{SSA} - \text{SSB} \tag{13.92}$$

The term $T_{..}^2/rc$ is known as the correction factor. The computational procedure can be performed in a systematic fashion by following the sequence of steps described below.

1. Compute row totals: $T_{1.}, T_{2.}, \ldots, T_{r.}$.
2. Compute column totals: $T_{.1}, T_{.2}, \ldots, T_{.c}$.
3. Compute overall total: $T_{..} = \sum_i T_{i.} = \sum_j T_{.j}$.
4. Compute $\sum_i \sum_j y_{ij}^2 = y_{11}^2 + \cdots + y_{rc}^2$.

5. Compute the correction factor: $\dfrac{T_{..}^2}{rc}$.

6. Compute $\sum_i \dfrac{T_{i.}^2}{c}$.

7. Compute $\sum_j \dfrac{T_{.j}^2}{r}$.

8. Compute SST $= \sum_i \sum_j y_{ij}^2 - \dfrac{T_{..}^2}{rc}$.

9. Compute SSA $= \sum_i \dfrac{T_{i.}^2}{c} - \dfrac{T_{..}^2}{rc}$.

10. Compute SSB $= \sum_j \dfrac{T_{.j}^2}{r} - \dfrac{T_{..}^2}{rc}$.

11. Compute SSE = SST − SSA − SSB.

Example 13.8 In a textile company that manufactures cotton yarn it is noticed that there is considerable variability in the breaking strength of the yarn produced by three different looms. The raw material is known to be homogeneous and hence can be discounted as a possible source of variability. An experiment was run to determine the effects of the loom and operator on the breaking strength. Four operators were picked at random and each was paired with each loom. The data were as follows:

BREAKING STRENGTH (lb)

Loom \ *Operator*	O_1	O_2	O_3	O_4	$T_{i.}$
L_1	111	116	110	110	447
L_2	112	116	113	116	457
L_3	113	119	118	117	467
$T_{.j}$	336	351	341	343	$T_{..} = 1371$

Assume that no interaction between the loom and operator is likely to exist.

(a) Write the mathematical model for the experiment.

(b) Write out the ANOVA table.

(c) Does the operator have a significant effect on the breaking strength? Use $\alpha = .05$.

(d) Does the loom have a significant effect on the breaking strength? Use $\alpha = .05$.

(e) If there are differences in breaking strength due to the looms, determine which differ.

(f) What is the probability of accepting the hypothesis of no loom effects when the effect of one loom is higher than the other two by 3 lb? Use MSE as an estimate of σ_e^2.

Solution

(a) Since the four operators were chosen at random from a large population of operators and assigned to operate, in turn, three specific looms, a mixed effects model would fit this experiment. Hence the mathematical model is

$$y_{ij} = \mu + \alpha_i + \beta_j + e_{ij} \qquad i = 1, \ldots, 3 \qquad j = 1, \ldots, 4$$

where μ = grand mean, α_i = loom effects $\left(\sum_{i=j}^{3} \alpha_i = 0\right)$, β_j = operator effects $[N(0, \sigma_\beta^2)]$, and e_{ij} = chance components $[N(0, \sigma_e^2)]$. The experimenter must take measures to assure that operators are actually chosen at random from the large number of operators available. Another point that merits attention is avoiding systematic errors due to nontested factors; *randomizing* the order of assigning raw materials to looms is necessary, as well as eliminating possible sources of variation in the working conditions and in measuring the response variable.

(b) We compute

$$\sum_{i=1}^{3} \sum_{j=1}^{4} y_{ij}^2 = (111)^2 + (116)^2 + \cdots + (117)^2 = 156{,}745$$

$$\frac{T_{..}^2}{rc} = \frac{(1371)^2}{12} = 156{,}636.75$$

$$\frac{\sum_{i=1}^{3} T_{i.}^2}{c} = \frac{(447)^2 + \cdots + (467)^2}{4} = 156{,}686.75$$

$$\frac{\sum_{j=1}^{4} T_{.j}^2}{r} = \frac{(336)^2 + \cdots + (343)^2}{3} = 156{,}675.67$$

$$\text{SST} = 156{,}745 - 156{,}636.75 = 108.25$$

$$\text{SSA} = 156{,}686.75 - 156{,}636.75 = 50$$

$$\text{SSB} = 156{,}675.67 - 156{,}636.75 = 38.92$$

$$\text{SSE} = 108.25 - 50 - 38.92 = 19.33$$

The results are displayed in the following table:

ANOVA TABLE FOR EXAMPLE 13.8

Source of Variation	*Sum of Squares*	*Degrees of Freedom*	*Mean Square*
Looms	50.00	2	25.00
Operators	38.92	3	12.97
Error	19.33	6	3.22
Total	108.25	11	

(c) Since $12.97/3.22 = 4.03$ is less than $F_{.95;\,3,6} = 4.76$ we accept the hypothesis of no significant operator effects.

(d) Since $25/3.22 = 7.76$ exceeds $F_{.95;\,2,6} = 5.14$ we conclude that the looms differ significantly.

(e) We shall use Tukey's procedure. Hence, for a paired comparison

$$\sqrt{\frac{\text{MSE}}{c}}\, q_{1-\alpha;\, r,\, (r-1)(c-1)} \left(\frac{1}{2}\sum_{i=1}^{2} |c_i|\right) = \sqrt{\frac{3.22}{4}}\, q_{.95;\,3,6}\, \frac{1}{2}(|1| + |1|)$$
$$= (.897)(4.34) = 3.89$$

Now, by Eq. (13.71),

$$|\bar{y}_{1.} - \bar{y}_{2.}| = 2.5 < 3.89$$
$$|\bar{y}_{1.} \quad \bar{y}_{3.}| = 5 \quad > 3.89$$
$$|\bar{y}_{2.} \quad \bar{y}_{3.}| = 2.5 < 3.89$$

Hence, we conclude that looms one and three are different.

(f) Let

$$\mu_{3.} = 3 + \mu_{2.} = 3 + \mu_{1.}$$

Hence

$$\mu = \tfrac{1}{3}(\mu_{1.} + \mu_{1.} + \mu_{1.} + 3) = \mu_{1.} + 1 = \mu_{2.} + 1 = \mu_{3.} - 2$$

so that

$$\alpha_1^2 = 1 \qquad \alpha_2^2 = 1 \qquad \alpha_3^2 = 4$$

By Eq. (13.75),

$$\phi = \frac{\sqrt{\sum_{i=1}^{3} \alpha_i^2/(\nu_1 + 1)}}{\sigma_e/\sqrt{c}} = \frac{\sqrt{6/3}}{\sqrt{3.22/4}} = 1.57 \quad N_1 = 2 \quad \text{and} \quad N_2 = 6$$

and from Fig. 13.2 we find β to be about 0.55.

13.8 Two-Way Classification; Fixed Effects Model; *n* Observations per Cell

In a two-way analysis of variance, one observation per cell, we postulated that *no* interaction occurs since we have no estimate of error against which we can test the possible interaction.

In practice there are many situations in which the effects of the two (or more) factors may not be independent. In other words, the effects of one factor may vary with the levels of the other factor. In such cases interaction is said to occur. To illustrate, suppose the yield of a chemical process depends on two factors: the concentration of the chemical and the operating temperature. Now if the yield at different concentration levels varies with the level of the operating temperature, we would say that interaction was present. Accordingly, we seek an experimental design that will allow us to test for the presence of interaction. This can be accomplished if some previous estimate of random error is available or more than one observation per cell can be made. The latter means *repetition of test procedures for the ijth combination*, sometims referred to as *replication of the experiment for the ijth combination.*

TABLE 13.11 Sample Layout for Two-Way Classification, n Observations per Cell

B *Levels* / A *Levels*	B_1	B_2	$\cdots$	B_j	$\cdots$	B_c	*Totals*	*Row Means*
A_1	y_{111} $\cdots$ y_{11n}	y_{121} $\cdots$ y_{12n}	$\cdots$	y_{1j1} $\cdots$ y_{1jn}	$\cdots$	y_{1c1} $\cdots$ y_{1cn}	$T_{1..}$	$\bar{y}_{1..}$
A_2	y_{211} $\cdots$ y_{21n}	y_{221} $\cdots$ y_{22n}	$\cdots$	y_{2j1} $\cdots$ y_{2jn}	$\cdots$	y_{2c1} $\cdots$ y_{2cn}	$T_{2..}$	$\bar{y}_{2..}$
$\cdots$	$\cdots$	$\cdots$	$\cdots$	$\cdots$	$\cdots$	$\cdots$	$\cdots$	$\cdots$
A_i	y_{i11} $\cdots$ y_{i1n}	y_{i21} $\cdots$ y_{i2n}	$\cdots$	y_{ij1} $\cdots$ y_{ijn}	$\cdots$	y_{ic1} $\cdots$ y_{icn}	$T_{i..}$	$\bar{y}_{i..}$
$\cdots$	$\cdots$	$\cdots$	$\cdots$	$\cdots$	$\cdots$	$\cdots$	$\cdots$	$\cdots$
A_r	y_{r11} $\cdots$ y_{r1n}	y_{r21} $\cdots$ y_{r2n}	$\cdots$	y_{rj1} $\cdots$ y_{rjn}	$\cdots$	y_{rc1} $\cdots$ y_{rcn}	$T_{r..}$	$\bar{y}_{r..}$
Totals	$T_{.1.}$	$T_{.2.}$	$\cdots$	$T_{.j.}$	$\cdots$	$T_{.c.}$	$T_{...}$	
Column means	$\bar{y}_{.1.}$	$\bar{y}_{.2.}$	$\cdots$	$\bar{y}_{.j.}$	$\cdots$	$\bar{y}_{.c.}$		$\bar{y}_{...}$

This replication requires measuring the response variable corresponding to two or more independent samples prepared at the same experimental conditions. The differences in these measurements are attributed to chance factors and thereby are used to assess the random error. The observations for this experimental design are recorded as shown in Table 13.11. The symbols are defined as follows:

The kth observation corresponding to the ijth treatment $= y_{ijk}\ i = 1, \ldots, r;\ j = 1, \ldots, c$
$k = 1, \ldots, n$

Mean of all observations corresponding to the ijth cell $= \dfrac{1}{n}\sum_{k=1}^{n} y_{ijk} = \dfrac{T_{ij.}}{n} = \bar{y}_{ij.}$

$$\text{Mean of all observations made with level A}_i = \frac{1}{cn}\sum_{j=1}^{c}\sum_{k=1}^{n} y_{ijk} = \frac{T_{i..}}{cn} = \bar{y}_{i.}$$

$$\text{Mean of all observations made with level B}_j = \frac{1}{rn}\sum_{i=1}^{r}\sum_{k=1}^{n} y_{ijk} = \frac{T_{.j.}}{rn} = \bar{y}_{.j}$$

$$\text{Mean of all observations} = \frac{1}{rcn}\sum_{i=1}^{r}\sum_{j=1}^{c}\sum_{k=1}^{n} y_{ijk} = \frac{T_{...}}{rcn} = \bar{y}_{...}$$

13.8.1 The Model

As before, the true mean for the *ij*th cell is

$$\begin{aligned}\mu_{ij} &= \text{grand mean} + \text{row effects} + \text{column effects} \\ &\quad + \text{interaction effects} \\ &= \mu + (\mu_{i..} - \mu) + (\mu_{.j.} - \mu) + (\mu_{ij} - \mu_{i..} - \mu_{.j.} + \mu)\end{aligned}$$

(13.93) $$= \mu + \alpha_i + \beta_j + \gamma_{ij}$$

Thus, if the factors are fixed the underlying mathematical model for two-way classification, n observations per cell, is given by

(13.94) $$y_{ijk} = \mu + \alpha_i + \beta_j + \gamma_{ij} + e_{ijk}$$
$$i = 1, \ldots, r \quad j = 1, \ldots, c \quad k = 1, \ldots, n$$

with the following restrictions:

$$\sum_{i=1}^{r} \alpha_i = 0 \qquad \sum_{j=1}^{c} \beta_j = 0 \qquad \sum_{i=1}^{r}\sum_{j=1}^{c} \gamma_{ij} = 0$$

The e_{ijk}'s are assumed to be independent chance components with identical normal distributions $N(0, \sigma_e^2)$.

It is evident that for a fixed effects model both row and column effects sum to zero since

$$\sum_{i=1}^{r} \alpha_i = \sum_{i=1}^{r} (\mu_{i..} - \mu) = r(\mu - \mu) = 0$$

$$\sum_{j=1}^{c} \beta_j = \sum_{j=1}^{c} (\mu_{.j.} - \mu) = c(\mu - \mu) = 0$$

We shall prove that for a fixed effects model $\sum_{i=1}^{r}\sum_{j=1}^{c} \gamma_{ij}$ sums to zero.

Proof. From Eq. (13.93) we have

$$\gamma_{ij} = \mu_{ij} - \mu - \alpha_i - \beta_j$$

Summing over i gives us

$$\begin{aligned}\sum_{i=1}^{r} \gamma_{ij} &= \sum_{i=1}^{r} \mu_{ij} - r\mu - \sum_{i=1}^{r} \alpha_i - r\beta_j \\ &= \sum_{i=1}^{r} \mu_{ij} - r\mu - r\beta_j\end{aligned}$$

since for a fixed effects model $\sum_{i=1}^{r} \alpha_i = 0$. However, $\beta_j = \mu_{.j.} - \mu$. Therefore

$$\sum_{i=1}^{r} \gamma_{ij} = r\mu_{.j.} - r\mu - r(\mu_{.j.} - \mu) = 0$$

Furthermore,

$$\sum_{j=1}^{c} \sum_{i=1}^{r} \gamma_{ij} = 0$$

and the proof is complete.

13.8.2 The Analysis

Consider the identity

$$\begin{aligned} y_{ijk} - \bar{y}_{...} = (\bar{y}_{i..} - \bar{y}_{...}) + (\bar{y}_{.j.} - \bar{y}_{...}) \\ + (\bar{y}_{ij.} - \bar{y}_{i..} - \bar{y}_{.j.} + \bar{y}_{...}) + (y_{ijk} - \bar{y}_{ij.}) \end{aligned}$$

Squaring and summing over i, j, and k gives

$$\begin{aligned} \sum_{i=1}^{r} \sum_{j=1}^{c} \sum_{k=1}^{n} (y_{ijk} - \bar{y}_{...})^2 &= \sum_{i=1}^{r} \sum_{j=1}^{c} \sum_{k=1}^{n} (\bar{y}_{i..} - \bar{y}_{...})^2 + \sum_{i=1}^{r} \sum_{j=1}^{c} \sum_{k=1}^{n} (\bar{y}_{.j.} - \bar{y}_{...})^2 \\ &+ \sum_{i=1}^{r} \sum_{j=1}^{c} \sum_{k=1}^{n} (\bar{y}_{ij.} - \bar{y}_{i..} - \bar{y}_{.j.} + \bar{y}_{...})^2 \\ &+ \sum_{i=1}^{r} \sum_{j=1}^{c} \sum_{k=1}^{n} (y_{ijk} - y_{ij.})^2 \end{aligned}$$

(since all sums of cross product terms vanish)

$$\begin{aligned} &= cn \sum_{i=1}^{r} (\bar{y}_{i..} - \bar{y}_{...})^2 - rn \sum_{j=1}^{c} (\bar{y}_{.j.} - \bar{y}_{...})^2 \\ &+ n \sum_{i=1}^{r} \sum_{j=1}^{c} (\bar{y}_{ij.} - \bar{y}_{i..} - \bar{y}_{.j.} + \bar{y}_{...})^2 \qquad (13.95) \\ &+ \sum_{i=1}^{r} \sum_{j=1}^{c} \sum_{k=1}^{n} (y_{ijk} - \bar{y}_{ij.})^2 \end{aligned}$$

Using our customary notation, Eq. (13.95) can be written

$$\text{SST} = \text{SSA} + \text{SSB} + \text{SSI} + \text{SSW} \qquad (13.96)$$

where SST = total sum of squares, SSA = sum of squares for the A factor, SSB = sum of squares for the B factor, SSI = sum of squares for interaction, and SSW = sum of squares within samples. That is, we have partitioned the total variation into four components as given by Eq. (13.96). By direct manipulation it can be shown that

$$\text{SSA} + \text{SSB} + \text{SSI} = n \sum_{i=1}^{r} \sum_{j=1}^{c} (\bar{y}_{ij.} - \bar{y}_{...})^2 = \text{SSG}$$

where SSG = sum of squares among cell means. Now consider

$$\text{SSA} = cn \sum_{i=1}^{r} (\bar{y}_{i..} - \bar{y}_{...})^2$$

where $\bar{y}_{i..} = \frac{1}{cn}\sum_{j=1}^{c}\sum_{k=1}^{n} y_{ijk} = \frac{1}{cn}\sum_{j=1}^{c}\sum_{k=1}^{n}(\mu + \alpha_i + \beta_j + \gamma_{ij} + e_{ijk})$

$$= \mu + \alpha_i + \frac{1}{c}\sum_{j=1}^{c}\beta_j + \frac{1}{c}\sum_{j=1}^{c}\gamma_{ij} + \bar{e}_{i..}$$

$$= \mu + \alpha_i + \bar{e}_{i..}$$

since for a fixed effects model

$$\sum_{i=1}^{c}\beta_j = 0 \quad \text{and} \quad \sum_{j=1}^{c}\gamma_{ij} = 0$$

and

$$\bar{y}_{...} = \frac{1}{r}\sum_{i=1}^{r}(\mu + \alpha_i + \bar{e}_{i..}) = \mu + \frac{1}{r}\sum_{i=1}^{r}\alpha_i + \bar{e}_{...} = \mu + \bar{e}_{...}$$

since for a fixed effects model $\sum_{i=1}^{r}\alpha_i = 0$, so that

$$\text{SSA} = cn\sum_{i=1}^{r}(\alpha_i + \bar{e}_{i..} - \bar{e}_{...})^2$$

(13.97) $$= cn\sum_{i=1}^{r}[\alpha_i^2 + (\bar{e}_{i..} - \bar{e}_{...})^2 + 2\alpha_i(\bar{e}_{i..} - \bar{e}_{...})]$$

Expectation yields

$$E(\text{SSA}) = cn\sum_{i=1}^{r}\alpha_i^2 + cnE\left[\sum_{i=1}^{r}(\bar{e}_{i..} - \bar{e}_{...})^2\right]$$

(13.98) $$= cn\sum_{i=1}^{r}\alpha_i^2 + (r-1)\sigma_e^2$$

since

$$E(\bar{e}_{i..} - \bar{e}_{...}) = 0$$

Proceeding similarly we find that

(13.99) $$E(\text{SSB}) = rn\sum_{j=1}^{c}\beta_j^2 + (c-1)\sigma_e^2$$

(13.100) $$E(\text{SSI}) = n\sum_{i=1}^{r}\sum_{j=1}^{c}\gamma_{ij}^2 + (r-1)(c-1)\sigma_e^2$$

The results are presented in the analysis of variance scheme in Table 13.12.

13.8.3 Testing the Hypothesis of Equal Treatment Effects

There are three basic hypotheses to be tested under the present model:

H_0': $\alpha_1 = \alpha_2 = \cdots = \alpha_r = 0$ against H_1': not all α_i's are zero
H_0'': $\beta_1 = \beta_2 = \cdots = \beta_c = 0$ against H_1'': not all β_j's are zero
H_0''': $\gamma_{11} = \gamma_{12} = \cdots = \gamma_{rc} = 0$ against H_1''': not all γ_{ij}'s are zero

TABLE 13.12 ANOVA for the Two-Way Classification; Fixed Effects Model; n Observations per Cell

Source of Variation	*Sum of Squares*	*Degrees of Freedom*	*Mean Square*	*Expected Mean Square*
Factor A	SSA	$r-1$	$\frac{\text{SSA}}{r-1} = \text{MSA}$	$\sigma_e^2 + \frac{cn}{r-1}\sum_{i=1}^{r}\alpha_i^2$
Factor B	SSB	$c-1$	$\frac{\text{SSB}}{c-1} = \text{MSB}$	$\sigma_e^2 + \frac{rn}{c-1}\sum_{j=1}^{c}\beta_j^2$
Interaction	SSI	$(r-1)(c-1)$	$\frac{\text{MSI}}{(r-1)(c-1)} = \text{MSI}$	$\sigma_e^2 + \frac{n}{(r-1)(c-1)}\sum_{i=1}^{r}\sum_{j=1}^{c}\gamma_{ij}^2$
Within cells	SSW	$rc(n-1)$	$\frac{\text{SSW}}{rc(n-1)} = \text{MSW}$	σ_e^2
Total	SST	$rcn-1$		

It can be shown, by the Partition Theorem, that when H_0', H_0'', and H_0''' are true the ratios

$$\frac{\text{SSA}}{\sigma_e^2} \qquad \frac{\text{SSB}}{\sigma_e^2} \qquad \frac{\text{SSI}}{\sigma_e^2}$$

are independent random variables having the χ^2 distribution with $(r-1)$, $(c-1)$, and $(r-1)(c-1)$ degrees of freedom, respectively. Hence, when H_0' is true, the ratio

$$\frac{\text{SSA}/(r-1)\sigma_e^2}{\text{SSW}/rc(n-1)\sigma_e^2} = \frac{\text{MSA}}{\text{MSW}} \tag{13.101}$$

is distributed as an F random variable with $(r-1)$ and $rc(n-1)$ degrees of freedom. Therefore, H_0' is rejected at the α level of significance if

$$\frac{\text{MSA}}{\text{MSW}} > F_{1-\alpha;\,(r-1),\,rc(n-1)} \tag{13.102}$$

Similarly, H_0'' is rejected at the α level of significance if

$$\frac{\text{MSB}}{\text{MSW}} > F_{1-\alpha;\,(c-1),\,rc(n-1)} \tag{13.103}$$

Finally, H_0''' is rejected at the α level of significance if

$$\frac{\text{MSI}}{\text{MSW}} > F_{1-\alpha;\,(r-1)(c-1),\,rc(n-1)} \tag{13.104}$$

If H_0''' is rejected we conclude that the effects of A and B are *not* additive. In other words, factors A and B *interact*. If this happens, testing the significance of H_0' and H_0'' becomes *meaningless* under the present formulation of this model. Note in the present formulation accepting H_0' means that there are no differences in the various levels of A when averaged over the levels of B. However, when interaction is present this interpretation is irrelevant. The presence of interaction means that the effect of one factor is dependent on the level of the other. Thus, when quoting the effects of one factor it is necessary to specify the level of the other. Similarly, rejecting H_0' when interaction is present is meaningless. The same argument holds true for H_0''.

Acceptance of H_0''' means that factors A and B do *not* interact and, consequently, testing the significance of the main effects is appropriate. *In this case inferences concerning the effects of factor A are the same for all levels of B whereas inferences concerning the effects of factor B are the same for all levels of factor A.*

If interaction effects are assumed to be zero prior to performing the experiment, then it is appropriate to *pool* SSI and SSW. Accordingly, we replace SSI and SSW by their sum = SSI + SSW and MSI and MSW by

$$\frac{\text{MSI} + \text{MSW}}{(r-1)(c-1) + rc(n-1)}$$

It follows that

$$E\left[\frac{\text{MSI} + \text{MSW}}{(r-1)(c-1) + rc(n-1)}\right] = \sigma_e^2$$

The significance of meaningful contrasts can be tested by means of the Tukey's or Scheffé's procedure. To illustrate, if Scheffé's procedure is used, a contrast L among the A means is significant at the α level when the absolute value of its estimate

$$|l| > S\sqrt{\frac{\text{MSW}}{cn}}\sqrt{\sum_{i=1}^{r} c_i^2} \tag{13.105}$$

where $S^2 = (r-1)\,F_{1-\alpha;\,(r-1),\,rc(n-1)}$. If Tukey's procedure is used, then L is significant when the absolute value of its estimate

$$|l| > \sqrt{\frac{\text{MSW}}{cn}}\, q_{1-\alpha;\,r,\,rc(n-1)}\left(\frac{1}{2}\sum_{i=1}^{r}|c_i|\right) \tag{13.106}$$

Power of the test can be computed for each of the three F tests by specifying appropriately the values of ϕ, ν_1, and ν_2. Thus we have for A effects

$$\phi = \frac{\sqrt{\sum_{i=1}^{r} \alpha_i^2/(\nu_1 + 1)}}{\sigma_e/\sqrt{cn}} \qquad \nu_1 = r - 1 \quad \text{and} \quad \nu_2 = rc(n-1) \tag{13.107}$$

and for B effects

$$\phi = \frac{\sqrt{\sum_{j=1}^{c} \beta_j^2/(\nu_1 + 1)}}{\sigma_e/\sqrt{rn}} \qquad \nu_1 = c - 1 \quad \text{and} \quad \nu_2 = rc(n-1) \tag{13.108}$$

Finally, for interaction effects,

$$\phi = \frac{\sqrt{\sum_{i=1}^{r} \sum_{j=1}^{c} \gamma_{ij}^2/(\nu_1 + 1)}}{\sigma_e/\sqrt{n}} \tag{13.109}$$

$$\nu_1 = (r - 1)(c - 1) \quad \text{and} \quad \nu_2 = rc(n - 1)$$

13.9 Two-Way Classification; Random Effects Model; *n* Observations per Cell

13.9.1 The Model

Under the present model, both the levels of factors A and B are random samples chosen from a large number of levels. The assumptions underlying the random effects model are

$$y_{ijk} = \mu + \alpha_i + \beta_j + \gamma_{ij} + e_{ijk} \tag{13.110}$$

$$i = 1, \ldots, r \quad j = 1, \ldots, c \quad k = 1, \ldots, n$$

e_{ijk}'s are independently $N(0, \sigma_e^2)$
α_i's are independently $N(0, \sigma_\alpha^2)$
β_j's are independently $N(0, \sigma_\beta^2)$
γ_{ij}'s are independently $N(0, \sigma_\gamma^2)$

TABLE 13.13 *ANOVA for the Two-Way Classification; Random Effects Model; n Observations per Cell*

Source of Variation	Sum of Squares	Degrees of Freedom	Mean Square	Expected Mean Square
Factor A	SSA	$r - 1$	$\frac{\text{SSA}}{r-1} = \text{MSA}$	$\sigma_e^2 + n\sigma_\gamma^2 + cn\sigma_\alpha^2$
Factor B	SSB	$c - 1$	$\frac{\text{SSB}}{c-1} = \text{MSB}$	$\sigma_e^2 + n\sigma_\gamma^2 + rn\sigma_\beta^2$
Interaction	SSI	$(r - 1)(c - 1)$	$\frac{\text{SSI}}{(r-1)(c-1)} = \text{MSI}$	$\sigma_e^2 + n\sigma_\gamma^2$
Within cells	SSW	$rc(n - 1)$	$\frac{\text{SSW}}{rc(n-1)} = \text{MSW}$	σ_e^2
Total	SST	$rcn - 1$		

13.9.2 The Analysis

The analysis is identical to that of the fixed effects model, and the results are given in Table 13.13.

13.9.3 Testing the Hypothesis of Equal Treatment Effects

There are three basic hypotheses to be tested under the present model:

$$H_0': \sigma_\alpha^2 = 0 \quad \text{against} \quad H_1': \sigma_\alpha^2 > 0$$
$$H_0'': \sigma_\beta^2 = 0 \quad \text{against} \quad H_1'': \sigma_\beta^2 > 0$$
$$H_0''': \sigma_\gamma^2 = 0 \quad \text{against} \quad H_1''': \sigma_\gamma^2 > 0$$

The expected mean square column of Table 13.13 can be used to indicate how the tests of hypotheses should be run. Since MSA and MSI are independent, it follows that

$$E\left(\frac{\text{MSA}}{\text{MSI}}\right) = \frac{E(\text{MSA})}{E(\text{MSI})} = \frac{\sigma_e^2 + n\sigma_\gamma^2 + cn\sigma_\alpha^2}{\sigma_e^2 + n\sigma_\gamma^2} \tag{13.111}$$

If H'_0 is true, $\sigma_\alpha^2 = 0$; then

$$\frac{E(\text{MSA})}{E(\text{MSI})} = \frac{\sigma^2 + n\sigma_\gamma^2}{\sigma^2 + n\sigma_\gamma^2} = 1 \tag{13.112}$$

However, if H_0' is false, $\sigma_\alpha^2 > 0$ and $E(\text{MSA})/E(\text{MSI}) > 1$. Thus the magnitude of the ratio of mean squares can be used to reflect the validity of H_0'. It can be shown that when H_0' is true the ratio of mean squares is an F random variable with $(r - 1)$ and $(r - 1)(c - 1)$ degrees of freedom. Hence, if the observed ratio of mean squares is too large, one may conclude that $\sigma_\alpha^2 > 0$. Consequently, we reject H_0' at the α level of significance if

$$\frac{\text{MSA}}{\text{MSI}} > F_{1-\alpha,\ (r-1);\ (r-1)(c-1)} \tag{13.113}$$

Similarly, H_0'' is rejected if

$$\frac{\text{MSB}}{\text{MSI}} > F_{1-\alpha;\ (c-1),\ (r-1)(c-1)} \tag{13.114}$$

Finally, H_0''' is rejected if

$$\frac{\text{MSI}}{\text{MSW}} > F_{1-\alpha;\ (r-1)(c-1),\ rc(n-1)} \tag{13.115}$$

Note carefully that the test statistics for H_0' and H_0'' in the random effects model are different from those given in the fixed effects model. In the fixed effects model we compare MSA and MSB against MSW for testing the significance of H_0' and H_0'', respectively.

In the random effects model it is still appropriate to test the significance of H'_0 and H''_0 when the interaction effects are found to be significant.

13.10 Two-Way Classification; Mixed Effects Model; *n* Observations per Cell

If the levels of factor B are assumed as fixed and of factor A as random, then the mathematical model is given by

$$y_{ijk} = \mu + \alpha_i + \beta_j + \gamma_{ij} + e_{ijk} \qquad (13.116)$$

$$i = 1, \ldots, r \qquad j = 1, \ldots, c \qquad k = 1, \ldots, n$$

with the restriction

$$\sum_{j=1}^{c} \beta_j = 0$$

The e_{ij}'s are independently $N(0, \sigma_e^2)$, α_i's are independently $N(0, \sigma_\alpha^2)$, and γ_{ij}'s are independently $N(0, \sigma_\gamma^2)$. The analysis of variance for this model is shown in Table 13.14.

TABLE 13.14 ANOVA for Two-Way Classification, Mixed Effects Model (B Fixed); n Observations per Cell

Source of Variation	*Sum of Squares*	*Degrees of Freedom*	*Mean Square*	*Expected Mean Square*
Factor A	SSA	$r - 1$	$\frac{\text{SSA}}{r-1} = \text{MSA}$	$\sigma_e^2 + nc\sigma_\alpha^2$
Factor B	SSB	$c - 1$	$\frac{\text{SSB}}{c-1} = \text{MSB}$	$\sigma_e^2 + n\sigma_\gamma^2 + \frac{rn}{c-1}\sum_{j=1}^{c}\beta_j^2$
Interaction	SSI	$(r-1)(c-1)$	$\frac{\text{SSI}}{(r-1)(c-1)} = \text{MSI}$	$\sigma_e^2 + n\sigma_\gamma^2$
Within cells	SSW	$rc(n-1)$	$\frac{\text{SSW}}{rc(n-1)} = \text{MSW}$	σ_e^2
Total	SST	$rcn - 1$		

As before, the expected mean square column can be used to indicate the appropriate statistic for testing the hypotheses. Thus, reject

$$\begin{aligned} &H'_0 \quad \text{if} \quad \frac{\text{MSA}}{\text{MSW}} > F_{1-\alpha;\,(r-1),\,rc(n-1)} \\ &H''_0 \quad \text{if} \quad \frac{\text{MSB}}{\text{MSI}} > F_{1-\alpha;\,(c-1),\,(r-1)(c-1)} \qquad (13.117) \\ &H'''_0 \quad \text{if} \quad \frac{\text{MSI}}{\text{MSW}} > F_{1-\alpha;\,(r-1)(c-1),\,rc(n-1)} \end{aligned}$$

13.11 Computing Formula and Procedure; *n* Observations per Cell

The computational procedure is the same for fixed, random, and mixed models. As before,

$$\text{SST} = \sum_{i=1}^{r} \sum_{j=1}^{c} \sum_{k=1}^{n} y_{ijk}^2 - \frac{T_{...}^2}{rcn} \tag{13.118}$$

where $T_{...}^2/rcn$ is the correction factor.

$$\text{SSA} = \sum_{i=1}^{r} \frac{T_{i..}^2}{cn} - \frac{T_{...}^2}{rcn} \tag{13.119}$$

$$\text{SSB} = \sum_{j=1}^{c} \frac{T_{.j.}^2}{rn} - \frac{T_{...}^2}{rcn} \tag{13.120}$$

$$\text{SSG} = \sum_{i=1}^{r} \sum_{j=1}^{c} \frac{T_{ij.}^2}{n} - \frac{T_{...}^2}{rcn} \tag{13.121}$$

$$\text{SSI} = \text{SSG} - \text{SSA} - \text{SSB}$$

$$\text{SSW} = \text{SST} - \text{SSG} \tag{13.122}$$

The computational procedure can be performed in a systematic fashion by the following sequence of steps:

1. Compute cell totals: $T_{11.}, \ldots, T_{ij.}$.
2. Compute row totals: $T_{1..}, T_{2..}, \ldots, T_{i..}$.
3. Compute column totals: $T_{.1.}, T_{.2.}, \ldots, T_{.j.}$.
4. Compute overall total: $T_{...} = \sum_i T_{i..} = \sum_j T_{.j.}$.
5. Compute $\sum_i \sum_j \sum_k y_{ijk}^2 = y_{111}^2 + \cdots + y_{rcn}^2$.
6. Compute the correction factor $\dfrac{T_{...}^2}{rcn}$.
7. Compute $\sum_i \dfrac{T_{i..}^2}{cn}$.
8. Compute $\sum_j \dfrac{T_{.j.}^2}{rn}$.
9. Compute $\sum_i \sum_j \dfrac{T_{ij.}^2}{n}$.
10. Compute $\text{SST} = \sum_i \sum_j \sum_k y_{ijk}^2 - \dfrac{T_{...}^2}{rcn}$.
11. Compute $\text{SSG} = \sum_i \sum_j \dfrac{T_{ij.}^2}{n} - \dfrac{T_{...}^2}{rcn}$.
12. Compute $\text{SSA} = \sum_i \dfrac{T_{i..}^2}{cn} - \dfrac{T_{...}^2}{rcn}$.
13. Compute $\text{SSB} = \sum_j \dfrac{T_{.j.}^2}{rn} - \dfrac{T_{...}^2}{rcn}$.
14. Compute SSI = SSG − SSA − SSB.
15. Compute SSW = SST − SSG.

Example 13.9 Suppose in Example 13.8 that two independent observations were made for each combination of loom and operator with the following results:

BREAKING STRENGTH (lb)

Loom \ Operator	O_1	O_2	O_3	O_4	$T_{i..}$
L_1	111 226 115	116 234 118	110 221 111	110 222 112	903
L_2	112 224 112	116 235 119	113 228 115	116 234 118	921
L_3	113 228 115	119 240 121	118 232 114	117 237 120	937
$T_{.j.}$	678	709	681	693	$T_{...} = 2761$

(a) Write out the ANOVA table.
(b) Test for main effects and interaction. Use $\alpha = .05$.
(c) If the looms differ, determine which are different.

Solution
(a) To simplify the computations, 110 is subtracted from each observation. Using the coded data we compute

$$\sum_{i=1}^{3}\sum_{j=1}^{4}\sum_{k=1}^{2} y_{ijk}^2 = (1)^2 + \cdots + (10)^2 = 855$$

$$\frac{T_{...}^2}{rcn} = \frac{(121)^2}{24} = 610.04$$

$$\sum_{i=1}^{3}\frac{T_{i..}^2}{cn} = \frac{(23)^2 + \cdots + (57)^2}{8} = 682.375$$

$$\sum_{j=1}^{4}\frac{T_{.j.}^2}{rn} = \frac{(18)^2 + \cdots + (33)^2}{6} = 709.17$$

$$\sum_{i=1}^{3}\sum_{j=1}^{4}\frac{T_{ij.}^2}{n} = \frac{(6)^2 + \cdots + (17)^2}{2} = 817.50$$

$$\text{SST} = 855 - 610.04 = 244.96$$

$$\text{SSG} = 817.5 - 610.04 = 207.46$$

$$\text{SSA} = 682.375 - 610.04 = 72.335$$

$$\text{SSB} = 709.17 - 610.04 = 99.13$$

$$\text{SSI} = 207.46 - 72.335 - 99.13 = 35.995$$

$$\text{SSW} = 244.96 - 207.46 = 37.50$$

The results are tabulated and shown below:

ANOVA TABLE FOR EXAMPLE 13.9

Source of Variation	*Sum of Squares*	*Degrees of Freedom*	*Mean Square*
Looms	72.335	2	36.167
Operators	99.130	3	33.043
Interaction	35.995	6	5.999
Within	37.500	12	3.125
Total	244.960	23	

(b) To test the significance of loom effects we use

$$F_{2,6} = \frac{36.167}{5.999} = 6.02$$

and since $F_{.95;\,2,6} = 5.14$, we conclude that the looms are different. Furthermore, to test the hypothesis that effects due to operators are zero we use

$$F_{3,12} = \frac{33.043}{3.125} = 10.57$$

Since $10.57 > F_{.95;\,3,12} = 3.49$, we conclude that operators are significantly different.

Finally, to test for interaction we use

$$F_{6,12} = \frac{5.999}{3.125} = 1.92$$

Since $1.92 < F_{.95;\,6,12} = 3.00$ there is no evidence that interaction effects are significant. Since σ_β^2 is significant we may wish to find its estimate

$$\hat{\sigma}_\beta^2 = \frac{\text{MSA} - \text{MSW}}{rn} = \frac{36.167 - 3.125}{3 \times 2} = 5.507$$

(c) We shall use Tukey's procedure and compute for a paired comparison:

$$\sqrt{\frac{\text{MSW}}{cn}}\, q_{1-\alpha;\, r,\, rc(n-1)}\left(\frac{1}{2}\sum_{i=1}^{2} |c_i|\right) = \sqrt{\frac{3.125}{8}}\, q_{.95;\,3,\,12} = (.624)(3.77) = 2.35$$

Now, by (13.106),

$$|\bar{y}_{1..} - \bar{y}_{2..}| = 2.25 < 2.35$$
$$|\bar{y}_{1..} - \bar{y}_{3..}| = 4.25 > 2.35$$
$$|\bar{y}_{2..} - \bar{y}_{3..}| = 2.00 < 2.35$$

Thus we conclude that looms one and three are different. Several comments concerning this example seem to be worthwhile. Replication in this experiment enabled us to analyze the data more fully. As a result, the term *interaction* is used in the analysis in place of error (two-way classification; single observation per cell) since we have a further estimate of the variance: the within-cells mean square.

If interaction were present and nonreplicated experiment were used, then our conclusion concerning the effects of operator would be misleading.

Although interaction effects are not significant at the 5% level of significance, it seems advisable to run an additional experiment with more replication per cell to ascertain whether interaction effects are likely to occur.

In this experiment measurements are made in sequence on two independent samples drawn from the output of each loom–operator combination. Samples (per cell) are regarded as duplicates and are assumed to be as nearly alike as possible. The variability within samples in each cell is attributed to chance causes. This variability is likely to be *inflated* if measurements are made at *different* times and perhaps by *different* technicians. In general, *when the entire experiment is replicated n times it would be difficult to duplicate exactly the experimental conditions n times.* As a result the experimental error may become excessively large.

13.12 Concluding Remarks

The two-way classification permits the experimenter to study efficiently the effects of two independent factors within the same experiment. To demonstrate the advantages of this design, consider again the example of the yield of a chemical process that depends on two factors: concentration and temperature. If one-way classification were used to study the effects of one factor, say concentration, one would fix the temperature at a certain level and measure the yield at two or more concentration levels. If three concentration levels are chosen, then the layout of this one-way classification would be as shown below:

	Concentration Levels		
	$c_1\%$	$c_2\%$	$c_3\%$
Observations	y_{11}	y_{21}	y_{31}
	y_{12}	y_{22}	y_{32}
	$\cdots$	$\cdots$	$\cdots$
	y_{1n}	y_{2n}	y_{3n}

Another similar experiment would be performed to study the effects of temperature. However, these controlled experiments have many disadvantages. First, the effects of concentration are known only at an arbitrarily chosen temperature. In general, the effects of one factor are determined at one *arbi-*

trarily chosen level of the other factor. To find out whether the effects of concentration are independent of the temperature level would necessitate performing *additional* experiments at *various* temperature levels. Second, it would be difficult to perform these additional experiments under the *same* conditions. As a result the experimental error may become excessively large. Third, we are unaware of the joint effects of these two factors.

Accordingly, the two-way classification analysis of variance has the following three major advantages:

1. It is possible to test the effects of two factors within the same experiment, thereby saving the time and effort of the experimenter.
2. It permits investigation of the presence of possible interaction between the two factors, thereby avoiding misleading conclusions.
3. It enables us to make conclusions of greater validity. This is due to the fact that the effect of one factor is estimated at several levels of the other factor. Thus the conclusions hold over a wide range of conditions.

PROBLEMS

13.1. A random sample is selected from each of three makes of nylon threads, and their breaking strengths (in pounds) are measured with the following results:

I	II	III
25	31	30
36	38	28
30	39	24
38	42	28
31	35	25

Test whether there are any differences in the breaking strengths of the threads. Use $\alpha = .05$.

13.2. Four machines (A, B, C, and D) are used to produce certain plastic gears. Random samples of 1,000 gears are selected from the output of each machine, and the number of defectives are counted with the following results:

A	B	C	D
15	27	34	45
24	24	47	16
20	28	39	33
35	36	48	39
	35	52	22
	30		

Test whether the machines are appreciably different at the 10% level of significance.

13.3. A manufacturer wishes to test whether there are any differences in the mean life of three makes of batteries. A random sample of five batteries is tested from each make with the following results:

A_1	A_2	A_3
40	26	39
48	34	40
38	30	43
42	28	50
45	32	50

(a) Write down the analysis of variance table.
(b) What conclusions can you draw? Use $\alpha = .05$.

13.4. An experiment was run to determine the effects of four grades of lubricants on the wear of certain wire drawing dies. The dies were known to be homogeneous and hence could be discounted as a possible source of variability. The measure of wear was the loss of die weight (in $g \times 100$). The results of the experiment were as follows:

Lubricant Grade	*Loss of Die Weight* ($g \times 100$)				
A	1.78	1.56	1.62	1.68	1.71
B	1.63	1.59	1.51	1.70	1.64
C	1.58	1.47	1.42	1.38	1.36
D	1.42	1.30	1.19	1.21	1.23

(a) Set up the ANOVA table.
(b) Does the grade of lubricant have any effect on wear ($\alpha = .01$)?
(c) If effects are significant, estimate the effect of lubricant D.

13.5. The data below represent the amount of wax deposited on the inner surface of paper bags measured in grams. The experiment was conducted on four waxing machines with the objective of determining whether any machine was depositing an abnormal amount of wax on the inner surface of paper bags.

Machine	*Amount of Wax Deposited* (g)							
A	1.4	2.0	3.2	1.4	2.3	4.0	5.0	4.7
B	3.2	6.8	5.0	2.5	6.1	4.8	4.6	4.2
C	6.2	3.1	3.2	4.0	4.5	6.4	4.4	4.1
D	4.8	6.6	6.5	5.9	5.9	3.0	5.9	5.6

(a) Set up the ANOVA table.
(b) Test whether machines differ at the 5% level of significance.
(c) Would the analysis of variance have a 90% chance of detecting the circumstance where three of the machines had equal means but the other one had a mean two units higher? Use the "within treatments" mean square as an estimate of σ_e^2.

13.6. In Prob. 13.4 use Tukey's procedure to determine which lubricants differ ($\alpha = .01$). Set up a 90% confidence limits for the average wear when lubricant A is used.

13.7. An experiment was carried out on four different tin-plating lines with the objective of determining whether any line was depositing an abnormal amount of tin on the surface of steel sheets. The amount of tin deposited was measured in grams per base box on a random sample of 20 sheets per each line. It is desired to test the hypothesis at the 5% level of significance that the mean amount of tin deposited is the same for all four lines. Past experience with the test has indicated that $\sigma_e^2 = 100$.
(a) Would you accept the hypothesis if SSG = 1,000?
(b) Would the analysis of variance have a 90% chance of detecting a situation where two lines' means were 10 g higher than the other two?
(c) How many sheets per line are needed to satisfy the requirement in (a)?

13.8. In Prob. 13.5, if the four machines had been a random sample from a population of machines, how would you use the previous data to estimate the variability of the population of machines?

13.9. Management has decided to run an experiment to determine whether their large number of filling machines turn out a product of nonuniform weight. Management is particularly interested in detecting a variation among machines as large as 2.2 times the variation within machines at the 5% level of significance. It is decided that 28 observations may be taken according to one of two schemes: (a) seven machines are chosen at random and four determinations are made per each machine *or* (b) four machines are chosen at random and seven determinations are made per each machine. Which scheme is better? Justify your answer analytically.

13.10. The rate of output of an operator, equally efficient on each of four special purpose machines, was measured with the following results:

Machines			*Output*		
A	47	37	44	41	36
B	40	45	49	45	44
C	48	48	41	43	41
D	39	37	37	44	41

Is there evidence of a difference among the special purpose machines? Use $\alpha = .05$.

13.11. A small laboratory has many voltmeters that are used interchangeably to make voltage measurements. A sample of four voltmeters was chosen at random. Each voltmeter was used to measure a constant emf of 100 volts five times with the following results:

Voltmeter			*Voltage* (volts)		
A	0.9	1.1	0.8	0.9	0.4
B	0.2	0.9	1.0	0.6	0.3
C	0.8	0.7	0.7	0.4	0.0
D	0.4	0.1	0.3	0.2	0.0

(a) Are voltmeters homogeneous ($\alpha = .05$)?
(b) Estimate the variability of the population of voltmeters.

13.12. In the preceding problem assume that the laboratory has only four voltmeters. Use Scheffé's procedure to determine which voltmeters differ at the 5% level of significance.

13.13. The output of a certain type of step-down transformer was the subject of a recent experiment. Five different testers and three different transformers were chosen at random. The output of each transformer was tested by each tester three times. The following ANOVA table was compiled:

Source of Variation	SS	*df*	MS	EMS
Transformers	10.44	...	...	...
Testers	5.45	...	...	...
...	...	...	...	...
Within	...	...	1.44	...
Total	79.32	...		

(a) Complete the ANOVA table.
(b) Test for transformer effects. Use $\alpha = .05$.
(c) What is the probability of accepting the hypothesis of no transformer effects when the transformer variance is actually equal to $2\frac{1}{3}$? Use estimates of the variabilities from the experiment ($\alpha = .05$).

13.14. An experiment was run to study the effects of launchers and propellants on the resulting ranges (in nautical miles). Three different launchers and four types of propellants were used in the experiment, and two readings were taken for each combination. The following ANOVA table was compiled.

Source of Variation	SS	*df*	MS	EMS
Propellants	260	...	...	...
Launchers	...	...	...	$\sigma_e^2 + 2\sigma_I^2 + 8\sigma_L^2$
Interaction	768	...	...	$\sigma_e^2 + 2\sigma_I^2$
Within	...	...	...	
Total	2938	...	...	

where I = interaction, L = launchers, $\sum_i \sum_j \sum_k y_{ijk} = 1{,}320$, and $\sum_i \sum_j T_{ij}^2 = 149{,}960$.

(a) Identify the model you decided to use.
(b) Set up the ANOVA table.
(c) Test for launchers and propellants effects ($\alpha = .05$).
(d) What is the estimate of the mean range?

13.15. As part of new fiberglass boat performance tests, mileage per gallon of light diesel oil was tested on original and modified engine design. Six boats are chosen at random and run under the same environmental conditions. The following table represents the results in coded units (mileage in miles per gallon):

Boat	1	2	3	4	5	6
Original engine:	2	4	2	0	1	3
Modified engine:	6	4	8	0	8	2

(a) Write down the analysis of variance table.
(b) Test whether main effects are significant ($\alpha = .05$).
(c) Is it possible to test for interaction ($\alpha = .05$)?
(d) What value of the boat variability will be detected with probability .95?

13.16. Four types of aircraft screws are produced by an aircraft company. The pitch diameter, which is a critical dimension, was the subject of a recent experiment. Five inspectors were chosen at random from a large group of inspectors. Each inspector measured the pitch diameter of each screw. The following results are obtained (coded data):

	Inspector				
Screw	1	2	3	4	5
1	25	22	35	17	29
2	27	29	31	23	33
3	21	18	39	20	28
4	25	24	35	21	21

(a) Test whether the screws are homogeneous ($\alpha = .01$)?
(b) Estimate the variability of the inspector population.

13.17. The yield of a certain chemical process was measured three times at five levels of temperature. The results were as follows (coded data):

Temperature (°F)				
30	60	90	120	150
7.62	8.14	7.76	7.17	7.46
8.00	8.15	7.73	7.57	7.68
7.93	7.87	7.74	7.80	7.21

(a) Set up the analysis of variance table and test the null hypothesis that the level of temperature has no effect on the yield ($\alpha = .05$).
(b) The linear response is defined as

$$-2\mu_1 - \mu_2 + 0\mu_3 + \mu_4 + 2\mu_5$$

Test whether the above contrast is significant ($\alpha = .05$).
(c) Find the estimate of the variance of the linear response.

13.18. The following data represent the number of units of production per day turned out by four randomly chosen operators using three milling machines:

	Machines		
Operators	M_1	M_2	M_3
1	150	151	156
2	147	149	155
3	141	146	153
4	154	152	159

(a) Set up the ANOVA table.
(b) Test whether main effects are significant. Use $\alpha = .05$.
(c) Estimate the variability due to the operators.

13.19. Use the data of the preceding problem to perform the pairwise test for the milling machines (by Tukey's procedure). Assuming no interaction, what value of operator variability will be detected with probability .80? Use the error mean square as an estimate of σ_e^2 ($\alpha = .05$).

13.20. The XYZ firm has four time study engineers. The firm wishes to ascertain whether the time standard set for a certain assembly operation is not affected by the chosen operator and/or the time study engineer. To obtain this information five operators are selected at random and timed by each time study engineer. The results of the experiment are as follows (cycle time in seconds):

	Time-Study Engineer			
Operators	A	B	C	D
1	149	144	142	139
2	146	138	138	132
3	144	143	138	133
4	152	134	140	136
5	147	146	136	143

(a) Set up the ANOVA table.
(b) Test whether main effects are significant. Use $\alpha = .05$.
(c) If the effects of engineers are significant, determine which differ ($\alpha = .05$).
(d) What is the estimate of the operator variability?

13.21. What is the rationale behind the completely randomized, two-way classification, n observations per cell model?

13.22. Explain how you would use the analysis of variance technique to test that several means of exponentially distributed populations are equal.

13.23. The following table gives the yield of a certain chemical process at three concentration and four temperature levels (coded data):

Concentration (%)	*Temperature* (°F) 50	75	100	125
2	14 10	11 11	13 9	10 12
4	9 7	10 8	7 11	6 10
6	5 11	13 14	12 13	14 10

(a) Set up the ANOVA table.
(b) Test for main effects and interaction at the 5% level of significance.

13.24. Three methods (A, B, and C) are applied to the fabrication of a certain mechanism on two different special purpose machines. The following rejection rates (measured by the number of acceptable mechanisms per machine per hour) occurred:

Machine	*Method* A	B	C
I	18 14	12 8	22 21
II	12 10	8 6	23 20

(a) Set up the ANOVA table.
(b) The the hypothesis that there is no difference in rejection rate among the methods and machines ($\alpha = .05$).
(c) Test for interaction ($\alpha = .05$).
(d) If methods are different, determine which differ at the 5% level of significance.

13.25. Verify Eq. (13.64).

13.26. Verify Eq. (13.100).

13.27. Metal- and glass-enclosed electronic tubes of the same type are life tested at regular and severe conditions. Failure times in hours are given below:

Condition	*Electronic Tube* *Metal*	*Glass*
Regular	69 51 60	40 51 49
Severe	53 48 39	40 32 38

(a) Set up the analysis of variance table.
(b) Test for the significance of interaction, tube, and condition effects. Use $\alpha = .05$.

13.28. In the preceding problem work out the following:
(a) Estimate the contrasts that have empirical significance.
(b) Use Scheffé's procedure to compute a 90% confidence interval for the contrasts found in (a).
(c) Estimate and draw the power curve.

13.29. A paint manufacturer has developed three formulas (A, B, and C) for painting die cast parts. The painted parts are oven dried. An experiment was run to investigate the effects due to formulas and ovens on the hardness of paint as determined by a penetrometer. Three randomly selected ovens are used for the experiment. The data given below for five replications per combination represent penetrations in tenth of a millimeter:

Formula	*Ovens* I	II	III
A	16 18 14 19 23	17 16 19 14 20	13 17 19 21 14
B	23 20 18 21 21	24 21 20 29 12	24 17 21 24 25
C	27 26 18 17 14	20 21 20 27 21	18 13 17 16 28

(a) Write out the ANOVA table.
(b) Test for the significance for interaction, formula, and oven effects. Use $\alpha = .05$.
(c) If formulas have different effects, determine which differ. Use $\alpha = .05$.

13.30. In the preceding problem consider three ovens as the only ovens in the factory. Test for the significance of the formula and oven effects at the 5% level of significance.

13.31. Suppose that a manufacturer purchases his raw material in lots from two suppliers, A and B. He receives K_A and K_B lots from suppliers A and B, respectively, so that $K_A + K_B = K$. He desires to test the following hypotheses, based on samples of size n taken from each lot:
(a) The means of the K_A lots are homogeneous.
(b) The means of the K_B lots are homogeneous.
(c) The means of K_A and K_B lots are equal.
Using the customary notation, derive the ANOVA table.

13.32. The data below represent test measurements with dynamometers chosen at random from a large number of available dynamometers. Three diesel engines of specific model are chosen at random from a large population of engines and tested with the results shown. Duplicate measurements are obtained. The results are as follows:

Dynamometer Load (lb)	*Diesel Engine* I	II	III
A	90 96	81 82	88 89
B	92 91	95 86	91 89
C	94 92	87 89	93 90

(a) Write out the analysis of variance table.
(b) Test for interaction and main effects. Use $\alpha = .05$.
(c) How many observations are required per combination to detect a dynamometer variance of 100 with probability .90? Use estimates of the variabilities from the experiment.

13.33. In the manufacture of a certain electronics component the parts are acid etched. The effects of two different compositions of the etching compound and two times of etch were tested with the results shown below (coded data). Measurements are replicated five times. The response variable is the reverse voltage.

Time of Etch	Etching Compound E_0		Etching Compound E_1	
T_0	2	0	5	9
	5	2	7	4
	6		5	
T_1	3	10	11	7
	10	8	12	6
	7		13	

(a) Set up the ANOVA table.
(b) Test for the significance of effects ($\alpha = .05$).
(c) Compute 95% confidence intervals for all possible contrasts having empirical significance. Use Scheffé's procedure.

SUGGESTED REFERENCES

[1] Bennett, C. A., and N. L. Franklin, *Statistical Analysis in Chemistry and the Chemical Industry*, New York, John Wiley & Sons, Inc., 1954.

[2] Bowker, A. H., and G. J. Lieberman, *Engineering Statistics*, Englewood Cliffs, N. J., Prentice-Hall, Inc., 1959.

[3] Brownlee, K. A., *Statistical Theory and Methodology in Science and Engineering*, 2nd ed., New York, John Wiley & Sons, Inc., 1965.

[4] Chew, V., *Experimental Designs in Industry*, New York, John Wiley & Sons, Inc., 1958.

[5] Cochran, W. G., and G. M. Cox, *Experimental Designs*, New York, John Wiley & Sons, Inc., 1958.

[6] Davies, O. L., *Design and Analysis of Industrial Experiments*, New York, Hafner Publishing Co., Inc., 1963.

[7] Dixon, W. J., and T. J. Massey, *Introduction to Statistical Analysis*, 3rd ed., New York, McGraw-Hill Book Company, 1969.

[8] Duncan, A. J., *Quality Control and Industrial Statistics*, 3rd ed., Homewood, Ill., Richard D. Irwin, Inc., 1965.

[9] Duncan, D. B., "Multiple Range and Multiple *F* Tests," *Biometrics*, **11**, 1956.

[10] Ehrenfeld, S., and S. B. Littauer, *Introduction to Statistical Method*, New York, McGraw-Hill Book Company, Inc., 1964.

[11] Federrer, W. T., *Experimental Design, Theory and Application*, New York, The Macmillan Company, 1955.

[12] Finney, D. J., *An Introduction to the Theory of Experimental Design*, Chicago, University of Chicago Press, 1960.

[13] Fisher, R. A., and F. Yates, *Statistical Tables for Biological, Agricultural, and Medical Research*, New York, Hafner Publishing Co., Inc., 1957.

[14] Graybill, F. A., *An Introduction to Linear Statistical Models*, New York, McGraw-Hill Book Company, 1961.

[15] Guenther, W. C., *Analysis of Variance*, Englewood Cliffs, N. J., Prentice-Hall, Inc., 1964.

[16] Hicks, C. R., *Fundamental Concepts in the Design of Experiments*, New York, Holt, Rinehart and Winston, Inc., 1964.

[17] Johnson, N. L., and F. C. Leone, *Statistical and Experimental Design in Engineering and the Physical Sciences*, New York, John Wiley & Sons, Inc., 1964.

[18] Kempthorne, O., *The Design and Analysis of Experiments*, New York, John Wiley & Sons, Inc., 1952.

[19] Li, C. C., *Introduction to Experimental Statistics*, New York, McGraw-Hill Book Company, 1964.

[20] Pearson, E. S., and H. O. Hartley, "Charts of the Power Function for Analysis of Variance Tests Derived from the Non-Central *F*-Distribution," *Biometrika*, **38**, 1951.

[21] Rickmers, A. D., and H. N. Todd, *Statistics: An Introduction*, New York, McGraw-Hill Book Company, 1967.

[22] Satterthwaite, F. E., "An Approximate Distrubution of Estimates of Variance Components," *Biometrics*, **2**, 1946.

[23] Scheffé, H., *The Analysis of Variance*, New York, John Wiley & Sons, Inc., 1959.

[24] Schultz, E. F., Jr., "Rules of Thumb for Determining Expectations of Means Squares in Analysis of Variance," *Biometrics*, **11**, 1955.

[25] Tang, P. C., "The Power Function of the Analysis of Variance Tests with Tables and Illustrations of Their Use," *Statistical Research Memoirs*, **2**, 1938.

[26] Tukey, J. W., "Comparing Individual Means in the Analysis of Variance," *Biometrics*, **5**, 1949.

[27] Yates, F., *The Design and Analysis of Factorial Experiments*, London, Imperial Bureau of Soil Science, 1937.

14

LINEAR REGRESSION AND CORRELATION

14.1 Introduction to Regression Analysis

Regression and correlation techniques are extensively applied to a wide range of problems in engineering and the physical and social sciences. In previous chapters we were concerned with the statistical analysis of univariate data. In other words, the analysis has been focused on repeated measurements $(x_1, x_2, \ldots, x_n)$ pertaining to one random variable (X). In practice, situations arise where measurements are taken on a random variable of interest that depends on one or more *controlled variables.* To illustrate, suppose the yield (Y) of a chemical process depends upon temperature (T) and concentration (C) of the chemical being used. Thus, in this situation Y is a *dependent* random variable while T and C are controlled variables. Strictly speaking, Y depends on T, C, and a set of parameters we wish to estimate on the basis of triple measurements. To simplify the analysis we will restrict ourselves to illustrations concerning one controlled variable. For example, suppose the yield Y of a chemical process is recorded at predetermined levels of temperature T. The data are given in Table 14.1 and are shown plotted in Fig. 14.1.

The graphical representation called a *scatter diagram* shows a functional relationship between the controlled variable T and the random variable Y.

TABLE 14.1 Yield Readings for Various Temperature Levels

Yield (Y)	*Temperature* (t°F)
20.1	100
20.9	110
22.0	120
23.0	130
24.2	140
24.8	150
26.2	160
27.0	170
28.0	180
28.9	190

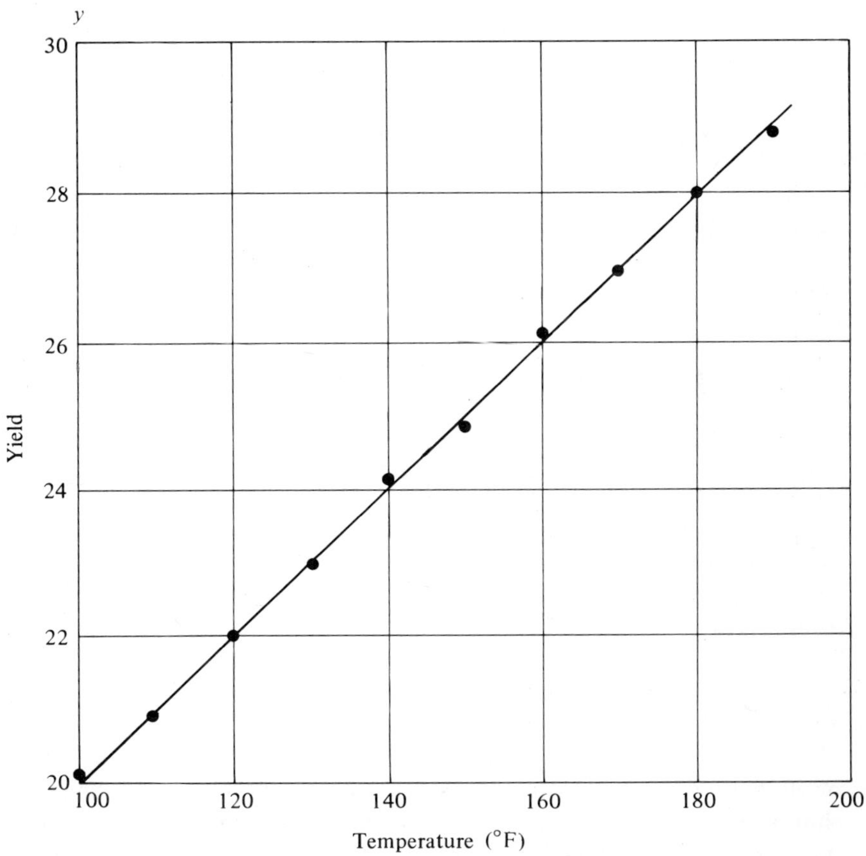

Fig. 14.1 Relationship between yield and temperature.

In fact, a straight line can be "fitted" by eye without resort to statistical analysis. The fitted straight line is called the *regression line of Y on T* and serves the purpose of predicting the value of y from a new observed t or vice versa. From the graph given in Fig. 14.1 it appears that Y and T are approximately linearly related within the range of temperature (100–190°F). However, in many situations the functional form of the relationship between the independent and dependent variates cannot be determined with sufficient accuracy from the plotted data. As a further illustration consider the relationship between the ultimate tensile strength of carbon steels and hardness. The data are given in Table 14.2 and are shown plotted in Fig. 14.2.

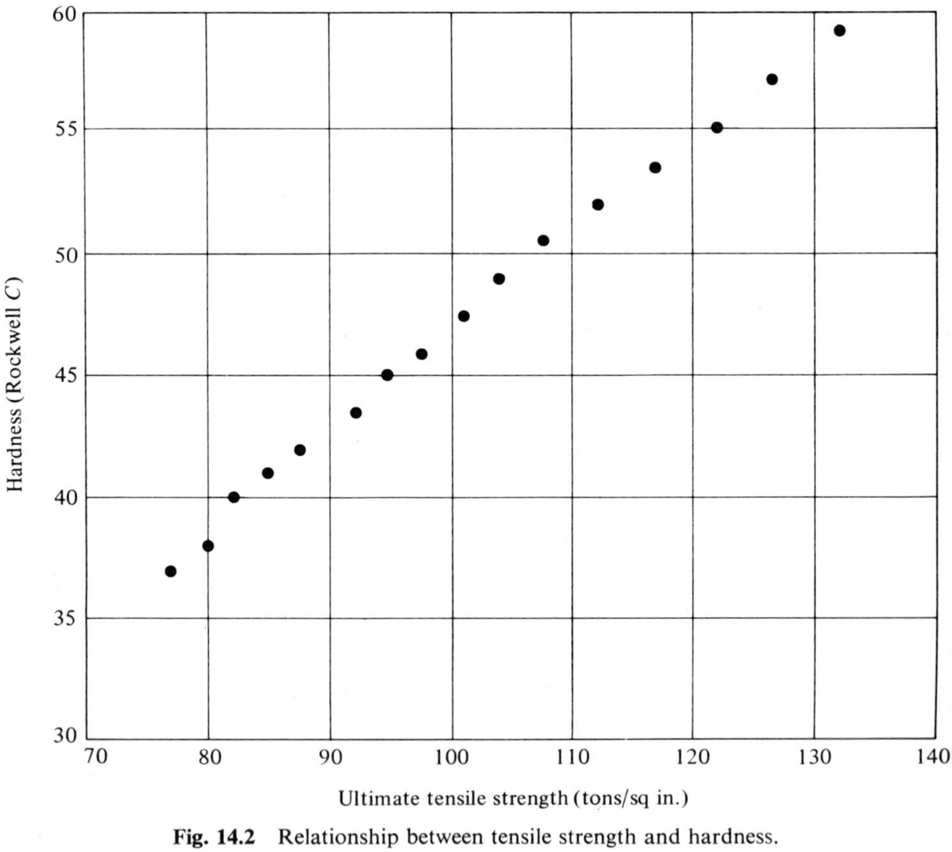

Fig. 14.2 Relationship between tensile strength and hardness.

The graphical representation shown in Fig. 14.2 exhibits a wide scattering of plotted points. As a result, "guessing" the functional from of the relationship between the two variables might be inadequate and misleading. Thus a more elaborate procedure for objectively fitting lines and curves is required. Some of these procedures are presented in the following sections.

TABLE 14.2 Physical Properties of Carbon Steels

Ultimate Tensile Strength (tons/in.^{2})	*Rockwell Hardness Number*
132	C 59
127	57
122	55
117	53.5
112	52
108	50.5
104	49
101	47.5
98	46
95	45
92	43.5
88	42
85	41
82	40
80	38
77	37

14.2 Simple Linear Regression

In simple linear regression the dependent variable depends on a single independent or controlled variable. The regression line is fitted by computing the estimates of the *intercept* and *slope* of the line. Before solving the problem of fitting a straight line to a set of observed points the following notation and assumptions are made:

14.2.1 Notation and Assumptions

At this point we will cease to preserve the distinction between random variables Y and the values they can take y. Strict consistency in this convention tends to complicate the presentation of our analysis.

The following notation will be used:

x: independent or controlled variable
y: dependent variable
$\tilde{y}$: estimated equation of regression line
A: the intercept of the regression line
a: the estimate of A
B: the slope of the regression line
b: the estimate of B
k: the number of paired observations $(x_1, y_1), \ldots, (x_k, y_k)$

The following assumptions must be satisfied:

(a) The relationship between y and x is assumed to be linear. In symbols, we write

$$E(y \mid x) = A + Bx \tag{14.1}$$

(b) x is not a random variable but some known constant associated with the random variable y. The x values are assumed to be observed without error, or are chosen.
(c) The y's are mutually independent random variables.
(d) The variance of y is constant for all x.
(e) The distribution of y_i is normal with mean $A + Bx_i$ and variance σ^2 where A, B, and σ are unknown parameters independent of x_i.

Estimates of the intercept and slope of the regression line can be derived provided that assumptions (a), (b), (c), and (d) are satisfied. It can be shown that these estimates are most efficient estimates and do not depend on any assumption about the distribution of y. However, assumption (e) is made in order to facilitate the application of standard statistical techniques to the data.

14.2.2 Least Squares Estimates of the Slope and Intercept

Suppose observations of random variable y at predetermined values of x have been plotted as shown in Fig. 14.3. Now if a straight line is fitted such that the sum of the squares of the vertical deviations of observed y_i from this line is minimum, then such a line is said to have the *least squares* property. The sum of squares $\sum_{i=1}^{k} d_i^2$ is defined as

$$\sum_{i=1}^{k} d_i^2 = \sum_{i=1}^{k} [y_i - (a + bx_i)]^2 \tag{14.2}$$

To minimize the left-hand side, we equate its partial derivatives with respect to a and b each to zero, which yields

$$\frac{\partial\left(\sum_{i=1}^{k} d_i^2\right)}{\partial a} = -2\sum_{i=1}^{k} (y_i - a - bx_i) = 0 \tag{14.3}$$

$$\frac{\partial\left(\sum_{i=1}^{k} d_i^2\right)}{\partial b} = -2\sum_{i=1}^{k} (y_i - a - bx_i)x_i = 0 \tag{14.4}$$

Upon simplifying, Eqs. (14.3) and (14.4) become

$$k\bar{y} - ka - bk\bar{x} = 0 \tag{14.5}$$

$$\sum_{i=1}^{k} x_i y_i - ka\bar{x} - b\sum_{i=1}^{k} x_i^2 = 0 \tag{14.6}$$

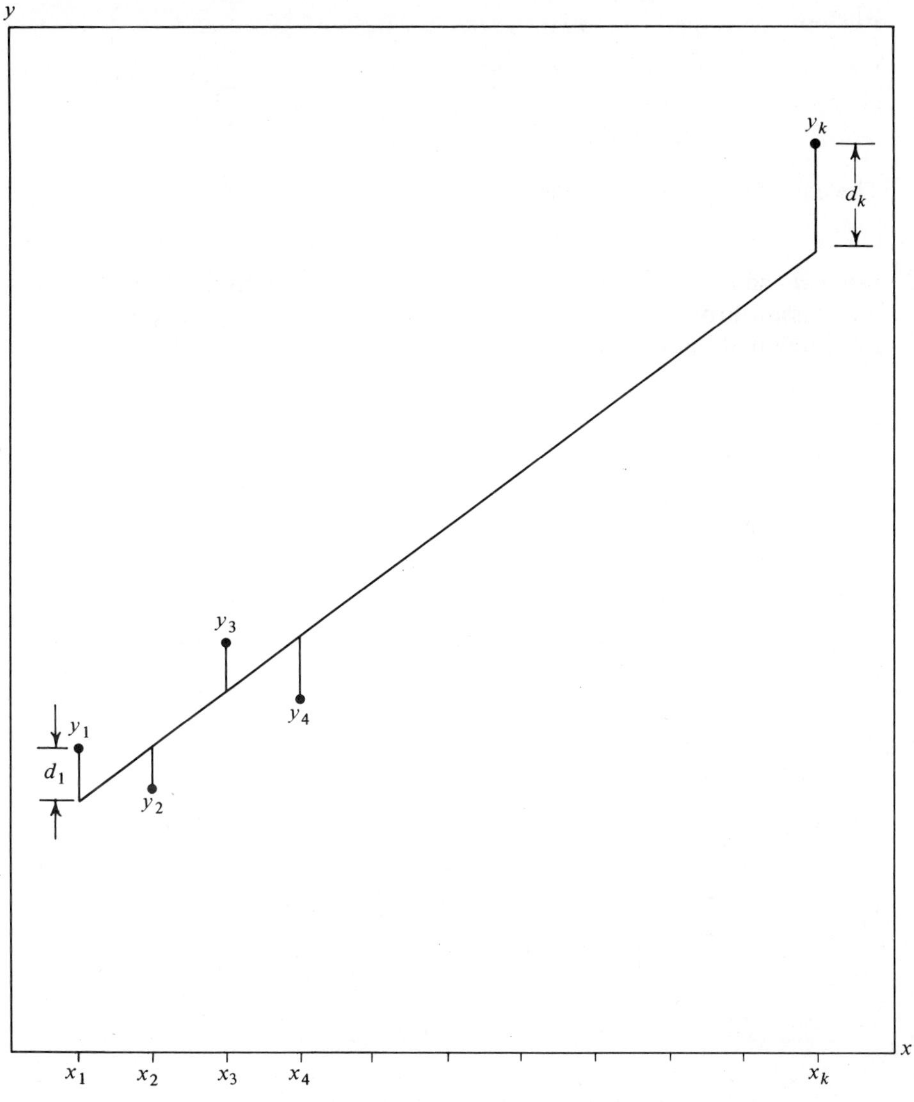

Fig. 14.3 Method of least squares.

Equations (14.5) and (14.6) are known as the *normal equations* for fitting a straight line. Equation (14.5) leads to

(14.7) $$a = \bar{y} - b\bar{x}$$

Substituting the value of a as given by Eq. (14.7) in Eq. (14.6) we have

$$\sum_{i=1}^{k} x_i y_i - k\bar{x}(\bar{y} - b\bar{x}) - b\sum_{i=1}^{k} x_i^2 = \sum_{i=1}^{k} x_i y_i - b\left(\sum_{i=1}^{k} x_i^2 - k\bar{x}^2\right) - k\bar{x}\bar{y} = 0$$

whence

$$b = \frac{\sum_{i=1}^{k} x_i y_i - k\bar{x}\bar{y}}{\sum_{i=1}^{k} (x_i - \bar{x})^2} = \frac{\sum_{i=1}^{k} (x_i - \bar{x})(y_i - \bar{y})}{\sum_{i=1}^{k} (x_i - \bar{x})^2} = \frac{\sum_{i=1}^{k} (x_i - \bar{x}) y_i}{\sum_{i=1}^{k} (x_i - \bar{x})^2} \tag{14.8}$$

The equation of the fitted line is given by

$$\tilde{y}_i = a + bx_i = \bar{y} + b(x_i - \bar{x}) \tag{14.9}$$

where a and b are given by Eqs (14.7) and (14.8), respectively. The regression line passes through $(\bar{x}, \bar{y})$, the *center of gravity* of the scatter diagram. We can show that both a and b are unbiased estimates of A and B, respectively.

Proof

$$\begin{aligned} E(b) &= \frac{1}{\sum_{i=1}^{k} (x_i - \bar{x})^2} \sum_{i=1}^{k} (x_i - \bar{x}) E(y_i) \\ &= \frac{1}{\sum_{i=1}^{k} (x_i - \bar{x})^2} \sum_{i=1}^{k} (x_i - \bar{x})(A + Bx_i) \\ &= \frac{1}{\sum_{i=1}^{k} (x_i - \bar{x})^2} \left(kA\bar{x} - kA\bar{x} + B \sum_{i=1}^{k} x_i^2 - kB\bar{x}^2 \right) \\ &= \frac{1}{\sum_{i=1}^{k} (x_i - \bar{x})^2} B \sum_{i=1}^{k} (x_i - \bar{x})^2 = B \end{aligned}$$

Similarly,

$$\begin{aligned} E(a) = E(\bar{y} - b\bar{x}) &= E\left(\frac{\sum_{i=1}^{k} y_i}{k} \right) - \bar{x}E(b) \\ &= \frac{1}{k} \sum_{i=1}^{k} E(y_i) - B\bar{x} \\ &= \frac{1}{k} (kA + kB\bar{x}) - B\bar{x} = A \end{aligned}$$

It can be shown (we omit proof) that the least squares estimates a and b are minimum variance estimates of A and B, respectively. In fact there are a variety of methods for estimating the parameters A and B. These methods lead to the same estimates given by the least squares method. We mention two of these methods: One is called *linear unbiased minimum variance* and the other is the *maximum likelihood* method. These methods will not be presented here.

A further property of the least square estimates a and b is that both estimates are *linear* combinations of the y's.

Proof

$$b = \frac{\sum_{i=1}^{k} (x_i - \bar{x})y_i}{\sum_{i=1}^{k} (x_i - \bar{x})^2}$$

Since the x_i's are assumed to be known constants,

$$\frac{(x_i - \bar{x})}{\sum_{i=1}^{k} (x_i - \bar{x})^2}$$

is equal to a constant (denoted by c_i); hence

$$b = \sum_{i=1}^{k} c_i y_i$$

is a linear combination of the y's.

Similarly, a can be expressed in terms of y_i.

Proof

$$\begin{aligned} a = \bar{y} = b\bar{x} &= \frac{1}{k}\sum_{i=1}^{k} y_i - \left(\sum_{i=1}^{k} c_i y_i\right)\bar{x} \\ &= \frac{1}{k}\sum_{i=1}^{k} y_i - \sum_{i=1}^{k} (c_i\bar{x})y_i \\ &= \sum_{i=1}^{k}\left(\frac{1}{k} - c_i\bar{x}\right)y_i \end{aligned}$$

Since $(1/k - c_i\bar{x})$ is a constant (denoted by c_i'),

$$a = \sum_{i=1}^{k} c_i' y_i$$

is a linear combination of the y's.

Now we can summarize the properties of each of the least squares estimates a and b as follows: (1) unbiased estimate, (2) minimum variance estimate, and (3) linear in y's.

14.2.3 Confidence Intervals for A and B

The variances of a and b can be obtained from Eqs. (14.7) and (14.8), respectively. Equation (14.7), $a = \bar{y} - b\bar{x} = \sum_{i=1}^{k} c_i' y_i$, shows that the estimate a is a linear combination of independent normal random variables with common variance σ^2. Thus

$$\begin{aligned} \text{Var}(a) = \sigma_a^2 = \text{Var}\left(\sum_{i=1}^{k} c_i' y_i\right) &= \sum_{i=1}^{k} c_i'^2 \,\text{Var}(y_i) \\ &= \sigma^2 \sum_{i=1}^{k} c_i'^2 \end{aligned}$$

$$= \sigma^2 \sum_{i=1}^{k} \left[\frac{1}{k} - \frac{(x_i - \bar{x})\bar{x}}{\sum_{i=1}^{k} (x_i - \bar{x})^2} \right]^2$$

$$(14.10) \qquad = \sigma^2 \left[\frac{1}{k} + \frac{\bar{x}^2}{\sum_{i=1}^{k} (x_i - \bar{x})^2} \right]$$

Thus we can state that a is normally distributed with mean A and variance σ_a^2. Similarly, the variance of b is given by

$$\text{Var}(b) = \sigma_b^2 = \text{Var}\left[\frac{\sum_{i=1}^{k} (x_i - \bar{x})y_i}{\sum_{i=1}^{k} (x_i - \bar{x})^2} \right] = \text{Var}\left[\sum_{i=1}^{k} (c_i y_i) \right]$$

$$(14.11) \qquad = \sum_{i=1}^{k} c_i^2 \sigma^2 = \frac{\sigma^2}{\sum_{i=1}^{k} (x_i - \bar{x})^2}$$

Thus b is normally distributed with mean B and variance σ_b^2.

If σ^2 is known we can assert that

$$P\left\{ a - Z_{1-\alpha/2}\sigma\sqrt{\frac{1}{k} + \frac{\bar{x}^2}{\sum_{i=1}^{k} (x_i - \bar{x})^2}} \le A \le a + Z_{1-\alpha/2}\sigma\sqrt{\frac{1}{k} + \frac{\bar{x}^2}{\sum_{i=1}^{k} (x_i - \bar{x})^2}} \right\} = 1 - \alpha$$

whence

$$(14.12) \qquad \left[a \pm Z_{1-\alpha/2}\sigma\sqrt{\frac{1}{k} + \frac{\bar{x}^2}{\sum_{i=1}^{k} (x_i - \bar{x})^2}} \right]$$

is a $100(1 - \alpha)\%$ confidence interval for A. Similarly,

$$(14.13) \qquad \left[b \pm Z_{1-\alpha/2} \frac{\sigma}{\sqrt{\sum_{i=1}^{k} (x_i - \bar{x})^2}} \right]$$

is a $100(1 - \alpha)\%$ confidence interval for B. Unfortunately, σ is usually unknown and has to be estimated. Consequently, confidence intervals must be found independent of σ. It can be proved that the unbiased estimator of σ^2 is $\sum_{i=1}^{k} (y_i - \tilde{y}_i)^2/(k - 2)$.

Proof. Consider the identity

$$\begin{aligned} (y - \bar{y}) &= (y_i - \tilde{y}_i) + (\tilde{y}_i - \bar{y}) \\ &= (y_i - \tilde{y}_i) + (a + bx_i - a - b\bar{x}) \\ &= (y_i - \tilde{y}_i) + b(x_i - \bar{x}) \end{aligned}$$

Squaring and summing over i gives

$$\sum_{i=1}^{k}(y_i - \bar{y})^2 = \sum_{i=1}^{k}(y_i - \tilde{y}_i)^2 + b^2\sum_{i=1}^{k}(x_i - \bar{x})^2$$

The cross product vanishes since $\sum_{i=1}^{k}(y_i - \tilde{y}_i)(x_i - \bar{x}) = 0$. On taking the expected values of both sides we obtain

$$E\left[\sum_{i=1}^{k}(y_i - \bar{y})^2\right] = E\left[\sum_{i=1}^{k}(y_i - \tilde{y}_i)^2 + b^2\sum_{i=1}^{k}(x_i - \bar{x})^2\right]$$

or

$$\sum_{i=1}^{k}E(y_i^2) - kE(\bar{y})^2 = E\left[\sum_{i=1}^{k}(x_i - \tilde{y}_i)^2\right] + \sum_{i=1}^{k}(x_i - \bar{x})^2E(b^2)$$

Therefore

$$\begin{aligned}E\left[\sum_{i=1}^{k}(y_i - \tilde{y}_i)^2\right] &= \sum_{i=1}^{k}E(y_i^2) - kE(\bar{y})^2 - \sum_{i=1}^{k}(x_i - \bar{x})^2E(b^2)\\ &= \sum_{i=1}^{k}[(A + Bx_i)^2 + \sigma^2] - k\left[(A + B\bar{x})^2 + \frac{\sigma^2}{k}\right]\\ &\quad -\sum_{i=1}^{k}(x_i - \bar{x})^2\left[B^2 + \frac{\sigma^2}{\sum_{i=1}^{k}(x_i - \bar{x})^2}\right]\\ &= (k - 2)\sigma^2 + \sum_{i=1}^{k}(A + Bx_i)^2 - k(A + B\bar{x})^2\\ &\quad - B^2\sum_{i=1}^{k}(x_i - \bar{x})^2\\ &= (k - 2)\sigma^2\end{aligned}$$

since

$$\sum_{i=1}^{k}(A + Bx_i)^2 - k(A + B\bar{x})^2 = \sum_{i=1}^{k}(Bx_i - B\bar{x})^2 = B^2\sum_{i=1}^{k}(x_i - \bar{x})^2$$

whence

$$\sigma^2 = \frac{E\left[\sum_{i=1}^{k}(y_i - \tilde{y}_i)^2\right]}{k - 2}$$

Therefore

$$\frac{\sum_{i=1}^{k}(y_i - \tilde{y}_i)^2}{k - 2} = \hat{\sigma}^2 = s_r^2$$

is an unbiased estimate of σ^2.

It can be shown by the partition theorem that

$$\frac{\sum_{i=1}^{k}(y_i - \tilde{y}_i)^2}{\sigma^2} = \frac{(k - 2)s_r^2}{\sigma^2}$$

is a random variable distributed as χ^2 with $k - 2$ degrees of freedom and is independent of the random variables a, b, and $\tilde{y}_i$. Since $(k-2)s_r^2/\sigma^2$ has a χ^2 distribution with $k-2$ degrees of freedom and $(a-A)/\sigma_a$ is a unit normal variate independent of $[(k-2)s_r^2]/\sigma^2$, by definition,

$$\frac{[(a-A)/\sigma_a]\sqrt{k-2}}{(s_r/\sigma)\sqrt{k-2}} = \left[(a-A)\Bigg/\sigma\sqrt{\frac{1}{k}+\frac{\bar{x}^2}{\sum_{i=1}^{k}(x_i-\bar{x})^2}}\right]\Bigg/\frac{s_r}{\sigma}$$

$$= \left[(a-A)\Bigg/s_r\sqrt{\frac{1}{k}+\frac{\bar{x}^2}{\sum_{i=1}^{k}(x_i-\bar{x})^2}}\right]$$

has the Student's t distribution with $k-2$ degrees of freedom. Hence a $100(1-\alpha)\%$ confidence interval for A is given by

(14.14)
$$\left[a \pm t_{1-\alpha/2;\,k-2}s_r\sqrt{\frac{1}{k}+\frac{\bar{x}^2}{\sum_{i=1}^{k}(x_i-\bar{x})^2}}\right]$$

Similarly, a $100(1-\alpha)\%$ confidence interval for B is given by

(14.15)
$$\left[b \pm t_{1-\alpha/2;\,k-2}s_r\sqrt{\frac{1}{\sum_{i=1}^{k}(x_i-\bar{x})^2}}\right]$$

Finally, a $100(1-\alpha)\%$ confidence interval for σ^2 is

(14.16)
$$\left[\frac{(k-2)s_r^2}{\chi^2_{1-\alpha/2;\,k-2}}, \frac{(k-2)s_r^2}{\chi^2_{\alpha/2;\,k-2}}\right]$$

14.2.4 Confidence Interval for E(y) for a Given Value of x Denoted by x_0

Here we wish to determine a $100(1-\alpha)\%$ confidence interval for $E(y\,|\,x_0) = A + Bx_0$. A point estimate for $E(y\,|\,x_0)$ is given by

$$\tilde{y}_0 = a + bx_0$$

which is a linear combination of independent normally distributed random variables. Hence, the expected value of $\tilde{y}_0$ is

(14.17)
$$E(\tilde{y}_0) = E(a) + E(b)x_0 = A + Bx_0$$

and the variance of $\tilde{y}_0$ is

$$\text{Var}(\tilde{y}_0) = \text{Var}(a + bx_0) = \text{Var}[\bar{y} + b(x_0 - \bar{x})]$$

$$= \frac{\sigma^2}{k} + (x_0 - \bar{x}^2)^2\,\text{Var}(b)$$

(14.18)
$$= \frac{\sigma^2}{k} + \frac{\sigma^2(x_0-\bar{x})^2}{\sum_{i=1}^{k}(x_i-\bar{x})^2}$$

Hence

$$\frac{\tilde{y}_0 - A - Bx_0}{\sigma_{\tilde{y}_0}}$$

is a normally distributed random variable with zero mean and unit variance and $(k - 2)s_r^2/\sigma^2$ is independently distributed as χ^2 with $k - 2$ degrees of freedom. By definition, then, the random variable

$$\frac{(\tilde{y}_0 - A - Bx_0)\sqrt{k-2}}{\sigma_{\tilde{y}_0}(s_r/\sigma)\sqrt{k-2}} = (\tilde{y}_0 - A - Bx_0)\Bigg/\left[s_r\sqrt{\frac{1}{k} + \frac{(x_0 - \bar{x})^2}{\sum_{i=1}^{k}(x_i - \bar{x})^2}}\right]$$

has the t distribution with $k - 2$ degrees of freedom. Thus, a $100(1 - \alpha)\%$ confidence interval for $A + Bx_0$ is given by

$$\left[\tilde{y}_0 \pm t_{1-\alpha/2;\,k-2}\, s_r\sqrt{\frac{1}{k} + \frac{(x_0 - \bar{x})^2}{\sum_{i=1}^{k}(x_i - \bar{x})^2}}\right] \tag{14.19}$$

14.2.5 A Joint Confidence Region for A, B

For convenience of presentation we shall assume that the origin of x has been shifted to $\bar{x}$. Then $A + Bx$ is replaced by $A^* + B^*x^*$ where

$$x^* = x - \bar{x} \quad \text{or} \quad \sum_{i=1}^{k} x_i^* = \sum_{i=1}^{k}(x_i - \bar{x}) = 0$$

$$A^* = A + B\bar{x}$$

$$B^* = B$$

It follows that $a^* = \bar{y}$ and $b^* = b$. Although this transformation does not reduce the generality of the problem, it has a desirable property: a^* and b^* become independent. The expected values and variances of a^* and b^* are as follows:

$$E(a^*) = A^* = A + B\bar{x}$$

$$E(b^*) = B^* = B$$

$$\text{Var}(a^*) = \frac{\sigma^2}{k}$$

$$\text{Var}(b^*) = \frac{\sigma^2}{\sum_{i=1}^{k}(x_i^*)^2}$$

Since a^* and b^* are normally distributed random variables,

$$\left(\frac{a^* - A^*}{\sigma/\sqrt{k}}\right)^2$$

has the chi-square distribution with one degree of freedom. Similarly,

$$\left(\frac{b^* - B^*}{\sigma\Big/\sqrt{\sum_{i=1}^{k} x_i^{*2}}}\right)^2$$

has the chi-square distribution with one degree of freedom. Since a^* and b^* are independent,

$$\left(\frac{a^* - A^*}{\sigma/\sqrt{k}}\right)^2 + \left(\frac{b^* - B^*}{\sigma\Big/\sqrt{\sum_{i=1}^{k} x_i^{*2}}}\right)^2$$

has the chi-square distribution with 2 degrees of freedom. The variate $(k - 2)s_r^2/\sigma^2$ has the chi-square distribution with $k - 2$ degrees of freedom. Thus we have two independent χ^2 distributions, and by definition the quotient

$$\left[\frac{1}{2}\left(\frac{a^* - A^*}{\sigma/\sqrt{k}}\right)^2 + \frac{1}{2}\left(\frac{b^* - B^*}{\sigma\Big/\sqrt{\sum_{i=1}^{k} x_i^{*2}}}\right)^2\right]\Bigg/\left[\frac{(k-2)s_r^2}{\sigma^2}\Big/(k-2)\right]$$

$$= \frac{k}{2}\left(\frac{a^* - A^*}{s_r}\right)^2 + \frac{\sum_{i=1}^{k} x_i^{*2}}{2}\left(\frac{b^* - B^*}{s_r}\right)^2$$

has the F distribution with 2 and $k - 2$ degrees of freedom. It follows that

$$P\left\{\frac{k}{2}\left(\frac{a^* - A^*}{s_r}\right)^2 + \frac{\sum_{i=1}^{k} x_i^{*2}}{2}\left(\frac{b^* - B^*}{s_r}\right)^2 \leq F_{1-\alpha;\,2,\,k-2}\right\} = 1 - \alpha$$

whence

$$\frac{k}{2}\left(\frac{a^* - A^*}{s_r}\right)^2 + \frac{\sum_{i=1}^{k} x_i^{*2}}{2}\left(\frac{b^* - B^*}{s_r}\right)^2 \leq F_{1-\alpha;\,2,\,k-2} \tag{14.20}$$

This means that all pairs of values of A^*, B^* that satisfy the above inequality lie inside the $100(1 - \alpha)\%$ *confidence region*. The boundary of this confidence region is specified by the equation

$$k(a^* - A^*)^2 + \sum_{i=1}^{k} x_i^{*2}(b^* - B^*)^2 - 2s_r^2 F_{1-\alpha;\,2,\,k-2} = 0 \tag{14.21}$$

which is the equation of an *ellipse*. Thus the joint confidence region of A^*, B^* is bounded by an ellipse.

Now we can obtain a confidence region for $E(y)$, which holds for all values of x. This confidence region differs from the confidence interval derived in Sec. 14.2.4. The latter gives a confidence interval for $E(y)$ at one and only one specified value of x, whereas the former gives a confidence region for $E(y)$ at all values of x. We showed that the joint confidence region for (A^*, B^*) is bounded by an ellipse. Thus all points (A^*, B^*) inside the ellipse yield a corresponding family of regression lines that lie within the confidence region for the line. The family of regression lines corresponding to the point that lie on the periphery of the ellipse would generate two *envelopes*. These two envelopes are two hyperbolas that constitute the boundary of the confidence region for the line. The equations of these hyperbolas can be derived. Points on the boundary will be determined by finding the maximum and minimum values of

$$E(y) = A^* + B^*x^* \tag{14.22}$$

for all values of x^* subject to the constraint

$$k(a^* - A^*)^2 + \sum_{i=1}^{k} x_i^{*2}(b^* - B^*)^2 - 2s_r^2 F_{1-\alpha;\,2,\,k-2} = 0$$

The extreme values of $E(y)$ can be found by the *Lagrangian multiplier technique*. We form the Lagrangian

$$L = A^* + B^*x^* + \lambda\left[k(a^* - A^*)^2 + \sum_{i=1}^{k} x_i^{*2}(b^* - B^*)^2 - 2s_r^2 F_{1-\alpha;\,2,\,k-2}\right]$$

Then

$$\frac{\partial L}{\partial A^*} = 1 - 2\lambda k(a^* - A^*) = 0 \tag{14.23}$$

$$\frac{\partial L}{\partial B^*} = x^* - 2\lambda \sum_{i=1}^{k} x_i^{*2}(b^* - B^*) = 0 \tag{14.24}$$

$$\frac{\partial L}{\partial \lambda} = k(a^* - A^*)^2 + \sum_{i=1}^{k} x_i^{*2}(b^* - B^*)^2 - 2s_r^2 F_{1-\alpha;\,2,\,k-2} = 0 \tag{14.25}$$

Solving Eqs. (14.23), (14.24), and (14.25) simultaneously we obtain

$$A^* = a^* \pm \frac{s_r}{k}\sqrt{2F_{1-\alpha;\,2,\,k-2}}\left(\frac{1}{k} + \frac{x^{*2}}{\sum_{i=1}^{k} x_i^{*2}}\right)^{-\frac{1}{2}} \tag{14.26}$$

$$B^* = b^* \pm \frac{x^* s_r}{\sum_{i=1}^{k} x_i^{*2}}\sqrt{2F_{1-\alpha;\,2,\,k-2}}\left(\frac{1}{k} + \frac{x^{*2}}{\sum_{i=1}^{k} x_i^{*2}}\right)^{-\frac{1}{2}} \tag{14.27}$$

Substituting the values of A^* and B^* in the regression equation (14.22) we obtain

$$E(y) = a^* + b^*x^* \pm s_r\sqrt{2F_{1-\alpha;\,2,\,k-2}}\left(\frac{1}{k} + \frac{x^{*2}}{\sum_{i=1}^{k} x_i^{*2}}\right)^{\frac{1}{2}}$$

$$= a + bx \pm s_r\sqrt{2F_{1-\alpha;\,2,\,k-2}}\left[\frac{1}{k} + \frac{(x - \bar{x})^2}{\sum_{i=1}^{k}(x_i - \bar{x})^2}\right]^{\frac{1}{2}} \tag{14.28}$$

The right-hand side of Eq. (14.28) will yield two values of $E(y)$ for any given x. Thus the whole sequence of x values will generate two hyperbolas that constitute the envelopes of the confidence region for $E(y)$. The confidence region is narrowest when $x = \bar{x}$ and increases as x becomes more distant from $\bar{x}$. The confidence region (shaded) for $E(y)$ is shown in Fig. 14.4.

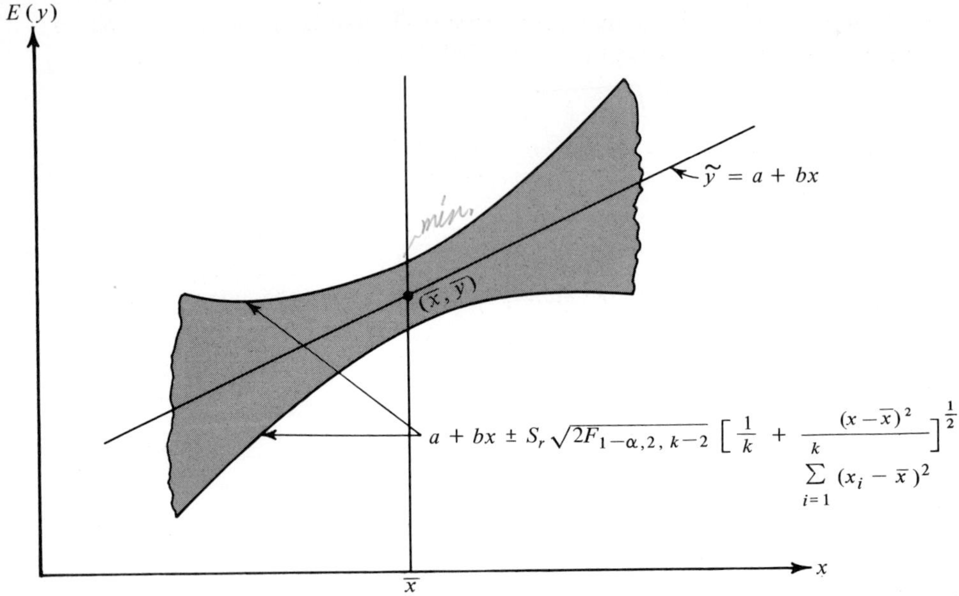

Fig. 14.4 Confidence region for $E(y) = A + Bx$.

14.2.6 Prediction Interval

Suppose that in the example cited in Sec. 14.1 the yield of the chemical process is a linear function of temperature. Furthermore, let us suppose that management is contemplating remodeling the equipment so that the process can be operated at a prescribed temperature higher than the present one, whereby a gain in yield would be achieved. Naturally management will be in a better position to make the decision on remodeling the equipment if the prospective value of the yield at the prescribed temperature can be predicted with an acceptable degree of confidence. More precisely, management desires to predict with a prescribed confidence the limits within which the prospective yield, at the specified temperature, will lie. These limits specify an interval known as the *prediction interval*. The difference between a confidence interval and a prediction interval is that the former specifies an interval for the "true" value of a parameter whereas the latter specifies an interval in which the value of a random variable may lie. In this example a confidence statement about the mean value of the yield is not useful, whereas a probability statement about the prospective value of the yield is of overriding importance. Before we proceed to derive the prediction interval we will introduce the following notation: Let y^+ = the prospective value of the dependent variable corresponding to a specified value of the controlled variable $x = x^+$, and let $\tilde{y}^+$ = the point on the fitted regression line corresponding to x^+. The probability that y^+ lies between specified limits is

$$P\{\text{lower limit} \leq y^+ \leq \text{upper limit}\} = 1 - \alpha$$

Our task, then, is to derive the limits of the specified interval. y^+ is a random variable that is normally distributed with mean

$$E(y^+ \mid x^+) = A + Bx^+$$

and variance σ^2.

$\tilde{y}^+$ is a normally distributed random variable with mean

$$E(\tilde{y}^+) = A + Bx^+$$

and variance

$$\sigma^2 \left[\frac{1}{k} + \frac{(x^+ - \bar{x})^2}{\sum_{i=1}^{k} (x_i - \bar{x})^2} \right]$$

Since y^+ and $\tilde{y}^+$ are independent, $(y^+ - \tilde{y}^+)$ is a normally distributed variate with mean

$$E(y^+ - \tilde{y}^+) = E(y^+) - E(\tilde{y}^+) = 0$$

and variance

$$\begin{aligned} \text{Var}(y^+ - \tilde{y}^+) &= \text{Var}(y^+) + \text{Var}(\tilde{y}^+) \\ &= \sigma^2 + \sigma^2 \left[\frac{1}{k} + \frac{(x^+ - \bar{x})^2}{\sum_{i=1}^{k} (x_i - \bar{x})^2} \right] \\ &= \sigma^2 \left[\frac{1 + k}{k} + \frac{(x^+ - \bar{x})^2}{\sum_{i=1}^{k} (x_i - \bar{x})^2} \right] \end{aligned}$$

It follows that

$$(y^+ - \tilde{y}^+) \Bigg/ \left[\sigma \sqrt{\frac{1 + k}{k} + \frac{(x^+ - \bar{x})^2}{\sum_{i=1}^{k} (x_i - \bar{x})^2}} \right]$$

is normally distributed with zero mean and unit variance independently of the variate $\sum_{i=1}^{k} (y_i - \tilde{y}_i)^2/\sigma^2$, which has the chi-square distribution with $k - 2$ degrees of freedom. Therefore, by definition, the quotient

$$\begin{aligned} &(y^+ - \tilde{y}^+) \Bigg/ \left[\sigma \sqrt{\frac{1 + k}{k} + \frac{(x^+ - \bar{x})^2}{\sum_{i=1}^{k} (x_i - \bar{x})^2}} \right] \div \left[\frac{\sum_{i=1}^{k} (y_i - \tilde{y}_i)^2}{\sigma^2 (k - 2)} \right]^{1/2} \\ &\qquad = (y^+ - \tilde{y}^+) \Bigg/ \left[s_r \sqrt{\frac{1 + k}{k} + \frac{(x^+ - \bar{x})^2}{\sum_{i=1}^{k} (x_i - \bar{x})^2}} \right] \end{aligned}$$

has the Student's t distribution with $k - 2$ degrees of freedom. Hence we can assert that

$$P\left\{-t_{1-\alpha/2;\,k-2} \leq (y^+ - \tilde{y}^+) \Bigg/ \left[s_r \sqrt{\frac{1+k}{k} + \frac{(x^+ - \bar{x})^2}{\sum_{i=1}^{k}(x_i - \bar{x})^2}}\right] \leq t_{1-\alpha/2;\,k-2}\right\} = 1 - \alpha$$

which is equivalent to

$$P\left\{\tilde{y}^+ - t_{1-\alpha/2;\,k-2}\, s_r \sqrt{\frac{1+k}{k} + \frac{(x^+ - \bar{x})^2}{\sum_{i=1}^{k}(x_i - \bar{x})^2}} \leq y^+ \leq \tilde{y}^+ + t_{1-\alpha/2;\,k-2}\, s_r \sqrt{\frac{1+k}{k} + \frac{(x^+ - \bar{x})^2}{\sum_{i=k}^{k}(x_i - \bar{x})^2}}\right\} = 1 - \alpha$$

whence a $100(1 - \alpha)\%$ prediction interval for y^+ is given by

$$\left[a + bx^+ \pm t_{1-\alpha/2;\,k-2}\, s_r \sqrt{\frac{1+k}{k} + \frac{(x^+ - \bar{x})^2}{\sum_{i=1}^{k}(x_i - \bar{x})^2}}\right] \tag{14.29}$$

The prediction interval given by (14.29) has the following properties:

1. The interval is narrowest when $x^+ = \bar{x}$ and increases as x^+ moves away from $\bar{x}$.
2. The prediction interval for x^+ is wider than the confidence interval for the mean $E(y)$ for any particular x.
3. The prediction interval cannot be used repeatedly since successive predictions are not independent and the error due to repeated predictions is not assessed. The probability statement on y^+ holds true only if A, B, and σ are reestimated each time a prediction on the dependent variable is made.

14.2.7 Tests of Significance

Tests of significance can be performed on the values of the parameters A, B, and σ^2. In fact, confidence intervals constructed in previous sections can be used for hypothesis testing. If the value of the *hypothesized parameter* to be tested with a given level of significance falls within the confidence interval, the null hypothesis is accepted. Otherwise the null hypothesis is rejected. An alternative approach to confidence intervals is applying the commonly used tests of significance. We will state here the rejection rule for a two-sided test concerning the parameters A, B, and σ. The rejection rule for a one-sided test is straightforward.

(1) $H_0\colon A = A_0$ against $H_1\colon A \neq A_0$

The null hypothesis H_0 is rejected at the α level of significance whenever

(14.30) $$|a - A_0| > s_r \sqrt{\frac{1}{k} + \frac{\bar{x}^2}{\sum_{i=1}^{k}(x_i - \bar{x})^2}}\, t_{1-\alpha/2;\, k-2}$$

(2) $$H_0\colon B = B_0 \quad \text{against} \quad H_1\colon B \neq B_0$$

The hypothesis $B = B_0$ is rejected at the α level of significance whenever

(14.31) $$|b - B_0| > s_r \sqrt{\frac{1}{\sum_{i=1}^{k}(x_i - \bar{x})^2}}\, t_{1-\alpha/2;\, k-2}$$

(3) $$H_0\colon \sigma^2 = \sigma_0^2 \quad \text{against} \quad H_1\colon \sigma^2 \neq \sigma_0^2$$

The null hypothesis H_0 is rejected at the α level of significance whenever

$$\frac{(k-2)s_r^2}{\sigma_0^2} < \chi^2_{\alpha/2;\, k-2}$$

or

(14.32) $$\frac{(k-2)s_r^2}{\sigma_0^2} > \chi^2_{1-\alpha/2;\, k-2}$$

14.2.8 The Regression Equation When A = 0

Frequently situations arise where the intercept of the regression line is known to be equal to zero, i.e.,

$$E(y\,|\,x) = 0 \quad \text{when} \quad x = 0$$

The equation of the regression line reduces to

$$E(y\,|\,x) = Bx$$

and the estimate of the regression line is given by

$$\tilde{y} = bx$$

By applying the least squares principle we obtain the normal equation

$$\sum_{i=1}^{k} x_i y_i - b \sum_{i=1}^{k} x_i^2 = 0$$

whence

(14.33) $$b = \frac{\sum_{i=1}^{k} x_i y_i}{\sum_{i=1}^{k} x_i^2}$$

The estimate of the slope b is a normal variate with mean B and variance

$$\frac{\sigma^2}{\sum_{i-1}^{k} x_i^2}$$

The estimate of the variability about the regression line is

$$s_r = \sqrt{\frac{\sum_{i=1}^{k} (y_i - \tilde{y}_i)^2}{k-1}}$$

Proof

$$\sum_{i=1}^{k} (y_i - \tilde{y}_i)^2 = \sum_{i=1}^{k} y_i^2 - \frac{\left(\sum_{i=1}^{k} x_i y_i\right)^2}{\sum_{i=1}^{k} x_i^2}$$

$$= \sum_{i=1}^{k} y_i^2 - b^2 \sum_{i=1}^{k} x_i^2$$

On taking the expected values of both sides we get

$$E\left[\sum_{i=1}^{k} (y_i - \tilde{y}_i)^2\right] = \sum_{i=1}^{k} (\sigma^2 + B^2 x_i^2) - \sum_{i=1}^{k} x_i^2 \left(\frac{\sigma^2}{\sum_{i=1}^{k} x_i^2} + B^2\right) = (k-1)\sigma^2$$

whence

$$\frac{\sum_{i=1}^{k} (y_i - \tilde{y}_i)^2}{k-1} = s_r^2$$

is an unbiased estimate of σ^2. The following statements can be derived in a manner similar to the preceding sections:

(a) A $100(1-\alpha)\%$ confidence interval for B is

$$\left[b \pm \frac{s_r}{\sqrt{\sum_{i=1}^{k} x_i^2}} t_{1-\alpha/2;\, k-1}\right] \tag{14.34}$$

(b) A $100(1-\alpha)\%$ confidence interval for Bx_0 is

$$\left[\tilde{y}_0 \pm t_{1-\alpha/2;\, k-1} s_r \sqrt{\frac{x_0^2}{\sum_{i=1}^{k} x_i^2}}\right] \tag{14.35}$$

(c) A $100(1-\alpha)\%$ prediction interval for y^+ is

$$\left[bx^+ \pm t_{1-\alpha/2;\, k-1} s_r \sqrt{1 + \frac{(x^+)^2}{\sum_{i=1}^{k} x_i^2}}\right] \tag{14.36}$$

where

$$s_r = \sqrt{\frac{\sum_{i=1}^{k} (y_i - \tilde{y}_i)^2}{k-1}}$$

Example 14.1 The following are data given on results of a study of the effect of carbon content in steel wires on the electrical resistance:

Carbon Content (%), x	*Resistance* (μ ohms cms at 20°C), y
.10	15
.30	18
.40	19
.55	21
.70	22.6
.80	23.8
.95	26

Suppose that for each x, values of y are normally distributed with mean $A + Bx$ and variance σ^2. Furthermore, suppose that the y's corresponding to different x's are independent random variables.

(a) Find the least squares regression line of y on x.
(b) Set up a 90% confidence interval for A.
(c) Find the point estimate of the resistance for a wire having .50% carbon content.
(d) For the above wire find the 90% confidence interval of its resistance.
(e) Find the equations of the envelopes of the 90% confidence region for the expected resistance for all values of carbon content.
(f) Test the hypothesis that $B = 13$ at a 5% level of significance.

Solution. We compute the following:

$$\sum_{i=1}^{7} x_i = 3.8 \qquad \sum_{i=1}^{7} y_i = 145.4$$

$$\sum_{i=1}^{7} x_i^2 = 2.595 \qquad \sum_{i=1}^{7} y_i^2 = 3104.2 \qquad \sum_{i=1}^{7} x_i y_i = 85.61$$

from which we derive

$$\bar{x} = \frac{\sum_{i=1}^{7} x_i}{7} = \frac{3.8}{7} = .5428$$

$$\bar{y} = \frac{\sum_{i=1}^{7} y_i}{7} = \frac{145.4}{7} = 20.7714$$

$$\sum_{i=1}^{7} (x_i - \bar{x})^2 = \sum_{i=1}^{7} x_i^2 - \frac{\left(\sum_{i=1}^{7} x_i\right)^2}{k} = 2.595 - \frac{(3.8)^2}{7} = .532$$

$$\sum_{i=1}^{7} (x_i - \bar{x})y_i = \sum_{i=1}^{7} x_i y_i - \frac{\sum_{i=1}^{7} x_i \sum y_i}{k} = 85.61 - \frac{3.8 \times 145.4}{7} = 6.687$$

$$\sum_{i=1}^{7} (y_i - \bar{y})^2 = \sum_{i=1}^{7} y_i^2 - \frac{\left(\sum_{i=1}^{7} y_i\right)^2}{k} = 3104.2 - \frac{(145.4)^2}{7} = 84.0966$$

$$\sum_{i=1}^{7} (y_i - \hat{y}_i)^2 = \sum_{i=1}^{7} (y_i - \bar{y})^2 - \frac{\left[\sum_{i=1}^{7} (x_i - \bar{x})y_i\right]^2}{\sum_{i=1}^{7} (x_i - \bar{x})^2} = .044$$

(a) The slope and the intercept of the line are

$$b = \frac{\sum_{i=1}^{7} (x_i - \bar{x})y_i}{\sum_{i=1}^{7} (x_i - \bar{x})^2} = \frac{6.687}{.532} = 12.5695$$

$$a = \bar{y} - b\bar{x} = 20.7714 - 12.5695 \times (.5428) = 13.9487$$

The least squares regression line is given by

$$\hat{y} = a + bx = 13.9487 + 12.5695x$$

(b) A 90% confidence interval for the intercept A is given by

$$a \pm t_{1-\alpha/2;\, k-2}\, s_r \sqrt{\frac{1}{k} + \frac{\bar{x}^2}{\sum_{i=1}^{7} (x_i - \bar{x})^2}}$$

$$= 13.9487 \pm t_{.95;\,5}\sqrt{\frac{.044}{5}}\sqrt{\frac{1}{7} + \frac{(.5428)^2}{.532}} = 13.9487 \pm .1575$$

$$= [13.7912,\ 14.1062]$$

(c) The point estimate of the expected resistance of a wire having .50% carbon is

$$\hat{y}_{.50} = 13.9487 + 12.5695(.5) = 20.23345$$

(d) A 90% confidence interval is given by

$$\hat{y}_{.50} \pm t_{1-\alpha/2;\, k-2}\, s_r \sqrt{\frac{1}{k} + \frac{(x_{.50} - \bar{x})^2}{\sum_{i=1}^{7} (x_i - \bar{x})^2}} = 20.23345 \pm 2.015$$

$$\times .0937\sqrt{\frac{1}{7} + \frac{(.5 - .5428)^2}{.532}} = [20.17191,\ 20.29499]$$

(e) The equations of the envelopes of a 90% confidence region of the expected resistance for all values of carbon contents (x) are

$$a + bx \pm s_r\sqrt{2F_{1-\alpha;\, 2,\, k-2}}\left[\frac{1}{k} + \frac{(x - \bar{x})^2}{\sum_{i=1}^{k} (x_i - \bar{x})^2}\right]^{1/2}$$

$$= 13.9487 + 12.5695x \pm .0937\sqrt{2 \times 3.78}\left[\frac{1}{7} + \frac{(x - .5428)^2}{.532}\right]^{1/2}$$

(f) A test for the null hypothesis that $B = 13$ is given by

$$\frac{|b - B_0|}{s_r\sqrt{\dfrac{1}{\sum_{i=1}^{k} (x_i - \bar{x})^2}}} = \frac{|12.5695 - 13|}{.0937\sqrt{\dfrac{1}{.532}}} = 3.35$$

The test statistic is distributed as Student's t with 5 degrees of freedom. Obviously the null hypothesis that $B = 13$ has to be rejected at the 5% level of significance since 3.35 is greater than $t_{.975;\,5} = 2.571$.

14.2.9 Linear Regression with n_i Observations at Each Level of x

In previous sections the analysis was focused on cases where only *one* observation on y was taken at each level of x. In this section we consider the case where n_i observations (y_{ij}, $j = 1, 2, \ldots, n_i$) are taken at x_i, $i = 1, 2, \ldots, k$. In this case *the mean of the controlled variable $\bar{x}$ is weighted in proportion to the number of observations at each level of x;* it is given by

$$\bar{x} = \frac{\sum_{i=1}^{k} n_i x_i}{\sum_{i=1}^{k} n_i} = \frac{\sum_{i=1}^{k} (n_i x_i)}{N}$$

where $N = \sum_{i=1}^{k} n_i$ = total number of observations. We define

$$\bar{y}_i = \frac{\sum_{j=1}^{n_i} y_{ij}}{n_i}$$

and

$$\bar{\bar{y}} = \frac{1}{N} \sum_{i=1}^{k} \bar{y}_i n_i = \frac{\sum_{i=1}^{k} \sum_{j=1}^{n_i} y_{ij}}{N}$$

The least squares estimates of the intercept and slope of the regression line can be obtained in a similar way using N individual observations. Thus a and b can be obtained by solving the following two normal equations:

(14.37) $$\sum_{i=1}^{k} \sum_{j=1}^{n_i} y_{ij} - a \sum_{i=1}^{k} n_i - b \sum_{i=1}^{k} n_i x_i = 0$$

(14.38) $$\sum_{i=1}^{k} n_i x_i \bar{y}_i - a \sum_{i=1}^{k} n_i x_i - b \sum_{i=1}^{k} n_i x_i^2 = 0$$

which yield

(14.39) $$a = \bar{\bar{y}} - b\bar{x}$$

and

(14.40) $$b = \frac{\sum_{i=1}^{k} n_i (x_i - \bar{x}) \bar{y}_i}{\sum_{i=1}^{k} n_i (x_i - \bar{x})^2}$$

It can be shown that

(14.41) $$\text{Var}(b) = \frac{\sigma^2}{\sum_{i=1}^{k} n_i (x_i - \bar{x})^2}, \qquad \text{Var}(a) = \sigma^2 \left[\frac{1}{N} + \frac{\bar{x}^2}{\sum_{i=1}^{k} n_i (x_i - \bar{x})^2} \right],$$

and

$$\text{Var}(\tilde{y}) = \sigma^2 \left[\frac{1}{N} + \frac{(x - \bar{x})^2}{\sum_{i=1}^{k} n_i(x_i - \bar{x})^2} \right] \tag{14.42}$$

If the relationship between the dependent and independent variables is assumed to be linear, then it can be shown that an unbiased estimate of the variability about the line is given by

$$s_r^2 = \frac{\sum_{i=1}^{k} \sum_{j=1}^{n_i} (y_{ij} - \tilde{y}_i)^2}{N - 2} \tag{14.43}$$

However, the assumption of *linearity* can now be tested. Actually we can test whether linearity prevails whenever there is more than one observation of y for some x. The analysis of variance technique can be used to test for linearity. We state here the results for testing for linearity without demonstration since it leads to mathematical consideration beyond the scope of this work. A test of the null hypothesis that linearity prevails is given by

$$F = \frac{\sum_{i=1}^{k} n_i(\bar{y}_i - \tilde{y}_i)^2/(k - 2)}{\sum_{i=1}^{k} \sum_{j=1}^{n_i} (y_{ij} - \bar{y}_i)^2/(N - k)} \tag{14.44}$$

which has the F distribution with $k - 2$ and $N - k$ degrees of freedom. Thus we reject the hypothesis of linearity at the α level of significance whenever

$$F > F_{1-\alpha;\, k-2,\, N-k} \tag{14.45}$$

Finally, the null hypothesis that $B = 0$ is rejected if

$$\frac{\sum_{i=1}^{k} n_i(\tilde{y}_i - \bar{\bar{y}})^2}{\sum_{i=1}^{k} \sum_{j=1}^{n_i} (y_{ij} - \bar{y}_i)^2/(N - k)} > F_{1-\alpha;\, 1,\, N-k} \tag{14.46}$$

Example 14.2 Assume that the data given in Example 14.1 are augmented so that corresponding to a given carbon content one or more observations on the resistance are taken. The data are given below:

Carbon Content (%), x	*Resistance* (μ ohms cms at 20°C), y
.10*	15.0
	15.3
.30*	18.0
	17.6
.40	19.0
.55*	21.0
	20.7
	20.6
.70	22.6

Carbon Content (%), x	*Resistance* (μ ohms cms at 20C°), y
.80*	23.8
	24.0
.95*	26.0
	25.9
	25.8
	25.7

*Denotes more than one observation on y.

(a) Obtain the regression line of y on x.
(b) Test the null hypothesis of linearity of this line ($\alpha = .05$).
(c) Test whether B is significant ($\alpha = .01$).
(d) Find the 90% confidence interval for $E(y \mid x = .50)$.

Solution. First we calculate the following:

$$N = \sum_{i=1}^{7} n_i = 15$$

$$\bar{\bar{y}} = \frac{\sum_{i=1}^{7} \sum_{j=1}^{n_i} y_{ij}}{15} = \frac{321}{15} = 21.4$$

$$\bar{x} = \sum_{i=1}^{7} \frac{n_i x_i}{N} = \frac{(2 \times .1 + 2 \times .3 + .4 + 3 \times .55 + .7 + 2 \times .8 + 4 \times .95}{15}$$

$$= \frac{8.95}{15} = .5966$$

To estimate B we need to calculate

$$\sum_{i=1}^{7} n_i(x_i - \bar{x})^2 = \sum_{i=1}^{7} n_i x_i^2 - \frac{\left(\sum_{i=1}^{7} n_i x_i\right)^2}{N}$$

$$= [2(.01) + 2(.09) + .16 + 3(.3025) + 1(.49) + 2(.64) + 4(.9025)] - \frac{(8.95)^2}{15}$$

$$= 1.308$$

and

$$\sum_{i=1}^{7} n_i(x_i - \bar{x})\bar{y}_i = \sum_{i=1}^{7} x_i \sum_{i=1}^{n_i} y_{ij} - \frac{\left(\sum_{i=1}^{7} \sum_{j=1}^{n_i} y_{ij}\right)\left(\sum_{i=1}^{7} n_i x_i\right)}{N}$$

$$= (.1 \times 30.3 + .3 \times 35.6 + .4 \times 19 + .55 \times 62.3 + .7 \times 22.6 + .8 \times 47.8 + .95 \times 103.4) - 21.4 \times 8.95$$

$$= 16.335$$

(a) The slope and intercept, therefore, are

$$b = \frac{\sum_{i=1}^{7} n_i(x_i - \bar{x})\bar{y}_i}{\sum_{i=1}^{7} n_i(x_i - \bar{x})^2} = \frac{16.335}{1.308} = 12.4885 \cong 12.49$$

$$a = \bar{\bar{y}} - b\bar{x} = 21.4 - 12.4885 \times .5966 = 13.9493 \cong 13.95$$

The least square regression line is given by

$$\bar{y} \cong 13.95 + 12.49x$$

(b) To test for linearity we need to calculate

$$\sum_{i=1}^{7} \sum_{j=1}^{n_i} (y_{ij} - \bar{y}_i)^2 = .0450 + .08 + .086 + .02 + .05 = .281$$

and

$$\sum_{i=1}^{7} n_i(\bar{y}_i - \tilde{y}_i)^2 = \sum_{i=1}^{7} n_i[\bar{y}_i - \bar{\bar{y}} - b(x_i - \bar{x})]^2$$

$$= \left[\sum_{i=1}^{7} \frac{\left(\sum_{j=1}^{n_i} y_{ij}\right)^2}{n_i} = \frac{\left(\sum_{i=1}^{7} \sum_{j=1}^{n_i} y_{ij}\right)^2}{N} \right]$$

$$- \frac{\left[\sum_{i=1}^{7} n_i(x_i - \bar{x})\bar{y}_i\right]^2}{\sum_{i=1}^{7} n_i(x_i - \bar{x})^2} = (7073.558 - 21.4 \times 321) - \frac{(16.335)^2}{1.308}$$

$$= .159$$

Now we obtain

$$F = \frac{\sum_{i=1}^{7} n_i(\bar{y}_i - \tilde{y}_i)^2/(k-2)}{\sum_{i=1}^{k} \sum_{j=1}^{n_i} (y_{ij} - \bar{y}_i)^2/(N-k)} = \frac{.159/5}{.281/8} = .905$$

The null hypothesis is acceptable since

$$.905 < F_{.95;\,5,8} = 3.69$$

Therefore we conclude that the relationship between x and y can be considered as linear.

(c) To test for the significance of the slope we need to calculate

$$\sum_{i=1}^{k} n_i(\tilde{y}_i - \bar{\bar{y}})^2 = \frac{\left[\sum_{i=1}^{7} n_i(x_i - \bar{x})\bar{y}_i\right]^2}{\sum_{i=1}^{7} n_i(x_i - \bar{x})^2}$$

$$= 203.999 \cong 204$$

The null hypothesis $B = 0$ is rejected whenever

$$F = \frac{\sum_{i=1}^{7} n_i(\tilde{y}_i - \bar{\bar{y}})^2}{\sum_{i=1}^{7} \sum_{j=1}^{n_i} (y_{ij} - \bar{y}_i)^2/(N-k)} = \frac{204}{.281/8} \cong 5807$$

is greater than $F_{.99,1,8} = 11.3$. Obviously we reject the null hypothesis and conclude that B is significantly different from zero.

(d) Since the hypothesis of linearity is accepted we may use

$$s_r^2 = \frac{\sum_{i=1}^{k} \sum_{j=1}^{n_i} (y_{ij} - \tilde{y}_i)^2}{N-2}$$

as an *unbiased pooled estimator* of σ^2. Hence a 90% confidence interval for $A + Bx_{.50}$ is given by

$$\left[\tilde{y}_{.50} \pm t_{1-\alpha/2;\,N-2}\, s_r \sqrt{\frac{1}{N} + \frac{(x_{.50} - \bar{x})^2}{\sum_{i=1}^{7} n_i(x_i - \bar{x})^2}}\right]$$

Since

$$s_r^2 = \frac{\sum_{i=1}^{7}\sum_{j=1}^{n_i}(y_{ij} - \tilde{y}_i)^2}{13} = \frac{.281 + .159}{13} = .03384$$

the 90% confidence interval for $A + Bx_{.50}$ is

$$13.9493 + 12.4885 \times .5 \pm t_{.95;\,13}\left[\sqrt{.03384}\sqrt{\frac{1}{15} + \frac{(.50 - .5966)^2}{1.308}}\right]$$
$$\simeq [20.11,\ 20.28]$$

This is indeed a narrow interval.

14.3 Multiple Regression

Section 14.2 dealt with the regression of a dependent variable y on a single independent (controlled) variable x. This section will be concerned with the regression of y on *several* independent (controlled) variables $(x_1, x_2, \ldots, x_r)$. However, we shall discuss only the case involving two independent variables. The model can be extended in a straightforward manner to include r independent variables. In this case y the dependent variable is assumed to be normally distributed with mean

$$E(y \mid x_1, x_2) = A + B_1x_1 + B_2x_2$$

and constant variance σ^2. The least squares estimates of A, B_1, and B_2 can easily be obtained by minimizing the sum of squares

$$\sum_{i=1}^{k}(y_i - a - b_1x_{1i} - b_2x_{2i})^2 \tag{14.47}$$

with respect to a, b_1, and b_2. Differentiating (14.47) partially with respect to a, b_1, and b_2 and setting these partial derivatives to zero, we obtain

$$\sum_{i=1}^{k} y_i - ka - b_1\sum_{i=1}^{k} x_{1i} - b_2\sum_{i=1}^{k} x_{2i} = 0 \tag{14.48}$$

$$\sum_{i=1}^{k} x_{1i}y_i - a\sum_{i=1}^{k} x_{1i} - b_1\sum_{i=1}^{k} x_{1i}^2 - b_2\sum_{i=1}^{k} x_{1i}x_{2i} = 0 \tag{14.49}$$

$$\sum_{i=1}^{k} x_{2i}y_i - a\sum_{i=1}^{k} x_{2i} - b_1\sum_{i=1}^{k} x_{1i}x_{2i} - b_2\sum_{i=1}^{k} x_{2i}^2 = 0 \tag{14.50}$$

Solution of Eqs. (14.48), (14.49), and (14.50) yields for a, b_1, and b_2

$$a = \bar{y} - b_1\bar{x}_1 - b_2\bar{x}_2$$

$$b_1 = \frac{\sum_{i=1}^{k} y_i(x_{1i} - \bar{x}_1)\sum_{i=1}^{k}(x_{2i} - \bar{x}_2)^2 - \sum_{i=1}^{k} y_i(x_{2i} - \bar{x}_2)\sum_{i=1}^{k}(x_{1i} - \bar{x}_1)(x_{2i} - \bar{x}_2)}{\sum_{i=1}^{k}(x_{1i} - \bar{x}_1)^2\sum_{i=1}^{k}(x_{2i} - \bar{x}_2)^2 - \left[\sum_{i=1}^{k}(x_{1i} - \bar{x}_1)(x_{2i} - \bar{x}_2)\right]^2}$$

(14.51)

$$b_2 = \frac{\sum_{i=1}^{k} y_i(x_{2i} - \bar{x}_2)\sum_{i=1}^{k}(x_{1i} - \bar{x}_1)^2 - \sum_{i=1}^{k} y_i(x_{1i} - \bar{x}_1)\sum_{i=1}^{k}(x_{1i} - \bar{x}_1)(x_{2i} - \bar{x}_2)}{\sum_{i=1}^{k}(x_{1i} - \bar{x}_1)^2\sum_{i=1}^{k}(x_{2i} - \bar{x}_2)^2 - \left[\sum_{i=1}^{k}(x_{1i} - \bar{x}_1)(x_{2i} - \bar{x}_2)\right]^2}$$

(14.52)

The equation of the estimated regression plane of y on x_1 and x_2 is given by

$$\tilde{y}_i = \bar{y} + b_1(x_{1i} - \bar{x}_1) + b_2(x_{2i} - \bar{x}_2) \tag{14.53}$$

where

$$\bar{x}_j = \frac{1}{k}\sum_{i=1}^{k} x_{ji} \qquad j = 1, 2$$

and

$$\bar{y} = \frac{1}{k}\sum_{i=1}^{k} y_i = \frac{1}{k}\sum_{i=1}^{k}\tilde{y}_i$$

It can be shown that a, b_1, and b_2 are normally distributed random variables with means A, B_1, and B_2, respectively, and variances

$$\begin{aligned}\text{Var}(a) = \frac{\sigma^2}{\varphi}\Bigg[\frac{\varphi}{k} &+ \sum_{i=1}^{k}(\bar{x}_{2i} - \bar{x}_2)^2 x_1^2 + \sum_{i=1}^{k}(x_{1i} - \bar{x}_1)^2\bar{x}_2^2 \\ &- 2\bar{x}_{12}\bar{x}\sum_{i=1}^{k}(x_{1i} - \bar{x}_1)(x_{2i} - \bar{x}_2)\Bigg]\end{aligned} \tag{14.54}$$

$$\text{Var}(b_1) = \frac{\sigma^2}{\varphi}\sum_{i=1}^{k}(x_{2i} - \bar{x}_2)^2 \tag{14.55}$$

$$\text{Var}(b_2) = \frac{\sigma^2}{\varphi}\sum_{i=1}^{k}(x_{1i} - \bar{x}_1)^2 \tag{14.56}$$

where

$$\varphi = \sum_{i+1}^{k}(x_{1i} - \bar{x}_1)^2\sum_{i=1}^{k}(x_{2i} - \bar{x}_2)^2 - \left[\sum_{i=1}^{k}(x_{1i} - \bar{x}_1)(x_{2i} - \bar{x}_2)\right]^2$$

It also can be shown that the estimator of $E(y \mid x_{10}, x_{20})$ is $\tilde{y}_0$ has the normal distribution with mean $A + B_1x_{10} + B_2x_{20}$ and variance

$$\begin{aligned}\text{Var}(\tilde{y}_0) = \frac{\sigma^2}{\varphi}\Bigg[\frac{\varphi}{k} &+ \sum_{i=1}^{k}(x_{2i} - \bar{x}_2)^2(x_{10} - \bar{x}_1)^2 + \sum_{i=1}^{k}(x_{1i} - \bar{x}_1)^2(x_{20} - \bar{x}_2)^2 \\ &- 2(x_{10} - \bar{x}_1)(x_{20} - \bar{x}_2)\sum_{i=1}^{k}(x_{1i} - \bar{x}_1)(x_{2i} - \bar{x}_2)\Bigg]\end{aligned} \tag{14.57}$$

Usually σ^2 is unknown and is estimated by s_r^2, which is a random variable independent of a, b_1, b_2, and $\tilde{y}_0$. It can be proven that $s_r^2(k-3)/\sigma^2$ is a random variable having the chi-square distribution with $k-3$ degrees of freedom. Its expected value is given by

$$E(s_r^2) = E\left[\frac{\sum_{i=1}^{k}(y_i - \tilde{y}_i)^2}{k-3}\right] = \sigma^2$$

Hence

$$s_r^2 = \frac{\sum_{i=1}^{k}(y_i - \tilde{y}_i)^2}{k-3} \tag{14.58}$$

is an unbiased estimator of σ^2. Confidence intervals for A, B_1, B_2, and $A + B_1x_{10} + B_2x_{20}$ can be derived in a manner similar to that of Sec. 14.2, bearing in mind that variates

$$\frac{a-A}{s_r\sqrt{\dfrac{\text{Var}(a)}{\sigma^2}}} \qquad \frac{b_1-B_1}{s_r\sqrt{\dfrac{\text{Var}(b_1)}{\sigma^2}}} \qquad \frac{b_2-B_2}{s_r\sqrt{\dfrac{\text{Var}(b_2)}{\sigma^2}}} \qquad \frac{\tilde{y}-(A+B_1x_{10}+B_2x_{20})}{s_r\sqrt{\dfrac{\text{Var}(\tilde{y}_0)}{\sigma^2}}}$$

all have the Student's t distribution with $k-3$ degrees of freedom.

14.4 Linear Correlation

14.4.1 Introduction

Linear *correlation* problems deal mainly with the degree of association between a pair of random variables, say x and y. Linear correlation provides a measure for assessing the relationship between the two variables.

The difference between correlation and regression lies in the fact that in the former we sample from a population of measurements on x and y whereas in the latter case we observe the values of y at a predetermined value of x. If the two variables x and y have *a bivariate normal* distribution, then their joint density function is given by

$$\begin{aligned} f(x, y) = \frac{1}{2\pi\sigma_x\sigma_y\sqrt{1-\rho^2}} \exp\Bigg\{-\frac{1}{2(1-\rho^2)}\Bigg[\left(\frac{x-\mu_x}{\sigma_x}\right)^2 \\ - 2\rho\left(\frac{x-\mu_x}{\sigma_x}\right)\left(\frac{y-\mu_y}{\sigma_y}\right) + \left(\frac{y-\mu_y}{\sigma_y}\right)^2\Bigg]\Bigg\} \end{aligned} \tag{14.59}$$

where ρ is defined as the population *correlation coefficient* and is equal to

$$\rho_{x,y} = \frac{\text{Cov}(x, y)}{\sigma_x\sigma_y} \tag{14.60}$$

and μ_x and σ_x^2 are the mean and variance of x and μ_y and σ_y^2 are the mean and variance of y, respectively. The coefficient of correlation can be used as a measure of the degree of the linear relationship between the two random variables x and y.

Thus the magnitude of ρ determines the degree of the relationship between the two variables x and y. When $\rho = 0$ the joint density function of x and y reduces to

$$f(x, y) = \frac{1}{2\pi\sigma_x\sigma_y} \exp\left\{-\frac{1}{2}\left[\left(\frac{x-\mu_x}{\sigma_x}\right)^2 + \left(\frac{y-\mu_y}{\sigma_y}\right)^2\right]\right\}$$

which implies that x and y are statistically independent. When $\rho = \pm 1$ we have perfect correlation; i.e., the relationship between the two variables is linear. This means that when x and y are correlated ρ must satisfy the relation $-1 < \rho < 1$. When the sign of ρ is positive, one variable tends to increase as the other variable increases; when the sign is negative, one variable tends to decrease as the other increases.

The population parameters $\sigma_x, \sigma_y, \mu_x, \mu_y$, and ρ are estimated by $s_x, s_y, \bar{x}, \bar{y}$, and r, respectively. Our task, then, is to obtain the estimates of these parameters.

14.4.2 Estimation of the Parameters of the Bivariate Normal Distribution

We use the method of maximum likelihood to obtain estimates of these parameters. The likelihood function L is given by

$$\begin{aligned} L &= \prod_{i=1}^{k} f(x, y) \\ &= \left(\frac{1}{2\pi\sigma_x\sigma_y\sqrt{1-\rho^2}}\right)^k \exp\left\{-\frac{1}{2(1-\rho^2)}\sum_{i=1}^{k}\left[\left(\frac{x_i-\mu_x}{\sigma_x}\right)^2\right.\right. \\ &\quad \left.\left. - 2\rho\left(\frac{x_i-\mu_x}{\sigma_x}\right)\left(\frac{y_i-\mu_y}{\sigma_y}\right) + \left(\frac{y_i-\mu_y}{\sigma_y}\right)^2\right]\right\} \end{aligned} \tag{14.61}$$

On taking logarithms of Eq. (14.61) we obtain

$$\begin{aligned} \ln L = -k \ln (2\pi\sigma_x\sigma_y\sqrt{1-\rho^2}) - \frac{1}{2(1-\rho^2)}\left[\frac{\sum_{i=1}^{k}(x_i-\mu_x)^2}{\sigma_x^2}\right. \\ \left. - 2\rho\frac{\sum_{i=1}^{k}(x_i-\mu_x)(y_i-\mu_y)}{\sigma_x\sigma_y} + \frac{\sum_{i=1}^{k}(y_i-\mu_y)^2}{\sigma_y^2}\right] \end{aligned} \tag{14.62}$$

Differentiating Eq. (14.62) partially with respect to σ_x and σ_y we obtain

$$\frac{\partial \ln L}{\partial \sigma_x} = \frac{-k}{\sigma_x} - \frac{1}{2(1-\rho^2)}\left[\frac{-2\sum_{i=1}^{k}(x_i-\mu_x)^2}{\sigma_x^3} + \frac{2\rho\sum_{i=1}^{k}(x_i-\mu_x)(y_i-\mu_y)}{\sigma_x^2\sigma_y}\right] \tag{14.63}$$

$$\frac{\partial \ln L}{\partial \sigma_y} = \frac{-k}{\sigma_y} - \frac{1}{2(1-\rho^2)}\left[\frac{2\rho\sum_{i=1}^{k}(x_i-\mu_x)(y_i-\mu_y)}{\sigma_x\sigma_y^2} - \frac{2\sum_{i=1}^{k}(y_i-\mu_y)^2}{\sigma_y^3}\right] \tag{14.64}$$

Equating Eq. (14.63) to zero and Eq. (14.64) to zero and rearranging we obtain

$$\text{(14.65)} \qquad \frac{\sum_{i=1}^{k}(x_i - \hat{\mu}_x)^2}{\hat{\sigma}_x^2} = \frac{\hat{\rho}\sum_{i=1}^{k}(x_i - \hat{\mu}_x)(y_i - \hat{\mu}_y)}{\hat{\sigma}_x\hat{\sigma}_y} + k(1 - \hat{\rho}^2)$$

$$\text{(14.66)} \qquad \frac{\sum_{i=1}^{k}(y_i - \hat{\mu}_y)^2}{\hat{\sigma}_y^2} = \frac{\hat{\rho}\sum_{i=1}^{k}(x_i - \hat{\mu}_x)(y_i - \hat{\mu}_y)}{\hat{\sigma}_x\hat{\sigma}_y} + k(1 - \hat{\rho}^2)$$

Adding Eqs. (14.65) and (14.66) yields

$$\frac{\sum_{i=1}^{k}(x_i - \hat{\mu}_x)^2}{\hat{\sigma}_x^2} + \frac{\sum_{i=1}^{k}(y_i - \hat{\mu}_y)^2}{\hat{\sigma}_y^2} - \frac{2\hat{\rho}\sum_{i=1}^{k}(x_i - \hat{\mu}_x)(y_i - \hat{\mu}_y)}{\hat{\sigma}_x\hat{\sigma}_y} = 2k(1 - \hat{\rho}^2)$$

(14.67)

Differentiating Eq. (14.62) partially with respect to ρ and setting the partial derivative to zero, we obtain after some rearrangement

$$k\hat{\rho} + \frac{\sum_{i=1}^{k}(x_i - \hat{\mu}_x)(y_i - \hat{\mu}_y)}{\hat{\sigma}_x\hat{\sigma}_y} - \frac{\hat{\rho}}{1 - \hat{\rho}^2}\left[\frac{\sum_{i=1}^{k}(x_i - \hat{\mu}_x)^2}{\hat{\sigma}_x^2}\right.$$

$$\text{(14.68)} \qquad \left. - \frac{2\rho\sum_{i=1}^{k}(x_i - \hat{\mu}_x)(y_i - \hat{\mu}_y)}{\hat{\sigma}_x\hat{\sigma}_y} + \frac{\sum_{i=1}^{k}(y_i - \hat{\mu}_y)^2}{\hat{\sigma}_y^2}\right] = 0$$

Substituting Eq. (14.67) in Eq. (14.68) we obtain

$$\text{(14.69)} \qquad \hat{\rho} = \frac{\sum_{i=1}^{k}(x_i - \hat{\mu}_x)(y_i - \hat{\mu}_y)}{k\hat{\sigma}_x\hat{\sigma}_y}$$

On substituting $\hat{\rho}$ in Eqs. (14.65) and (14.66) we obtain

$$\text{(14.70)} \qquad \hat{\sigma}_x^2 = s_x^2 = \frac{1}{k}\sum_{i=1}^{k}(x_i - \hat{\mu}_x)^2 \qquad \hat{\sigma}_y^2 = s_y^2 = \frac{1}{k}\sum_{i=1}^{k}(y_i - \hat{\mu}_y)^2$$

It can be proved that by differentiating Eq. (14.62) partially with respect to μ_x and μ_y and equating the partial derivatives to zero and solving, the maximum likelihood estimator for μ_x and μ_y are $\bar{x}$ and $\bar{y}$, respectively. Thus, substituting Eq. (14.70) in Eq. (14.69) and letting $\hat{\mu}_x = \bar{x}$ and $\hat{\mu}_y = \bar{y}$, $\hat{\rho}$ (which is usually denoted by $r_{x,y}$) becomes

$$r_{x,y} = \hat{\rho} = \frac{\sum_{i=1}^{k}(x_i - \bar{x})(y_i - \bar{y})}{\sqrt{\sum_{i=1}^{k}(x_i - \bar{x})^2\sum_{i=1}^{k}(y_i - \bar{y})^2}}$$

$$\text{(14.71)} \qquad = \frac{k\sum_{i=1}^{k}x_iy_i - \left(\sum_{i=1}^{k}x_i\right)\left(\sum_{i=1}^{k}y_i\right)}{\sqrt{\left[k\sum_{i=1}^{k}x_i^2 - \left(\sum_{i=1}^{k}x_i\right)^2\right]\left[k\sum_{i=1}^{k}y_i^2 - \left(\sum_{i=1}^{k}y_i\right)^2\right]}}$$

It turns out that r is a biased estimate of ρ except when $\rho = 0$. It should be noted that the value of r is merely a measure of the strength of the linear relationship between two variables. Scatter diagrams associated with various values of the sample correlation coefficient (r) are shown in Fig. 14.5. The value of $r = 0$ as shown in Fig. 14.5(a) indicates that there is no relationship between the variables x and y, whereas the value of $r = .8$, $r^2 = .64$, may be interpreted by stating that 64% of the variation in y is linearly associated with the variation in x. The rest of the variation (36%) is associated with error or other sources of variation.

An important property of the correlation coefficient is that if the data are transformed linearly, the correlation coefficient of the transformed variables remains unchanged.

Proof. Let $u = c_1x + c_2$ and $v = c_3y + c_4$, where c_1, c_2, c_3, and c_4 are constants; then by substituting for $x_i = (u_i - c_2)/c_1$, $y_i = (v_i - c_4)/c_3$, $\bar{x} = (\bar{u} - c_2)/c_1$, and $\bar{y} = (\bar{v} - c_4)/c_3$ in Eq. (14.71) we obtain

$$r_{x,y} = \frac{\sum_{i=1}^{k}\left(\frac{u_i - \bar{u}}{c_1}\right)\left(\frac{v_i - \bar{v}}{c_3}\right)}{\sqrt{\sum_{i=1}^{k}\left(\frac{u_i - \bar{u}}{c_1}\right)^2 \sum_{i=1}^{k}\left(\frac{v_i - \bar{v}}{c_3}\right)^2}} = \frac{\sum_{i=1}^{k}(u_i - \bar{u})(v_i - \bar{v})}{\sqrt{\sum_{i=1}^{k}(u_i - \bar{u})^2 \sum_{i=1}^{k}(v_i - \bar{v})^2}}$$

Hence, $r_{x,y} = r_{u,v}$.

The correlation technique has two main disadvantages. The first is that the population must have a bivariate normal distribution. The second stems from the sampling procedure that requires that neither variable be controlled.

14.4.3 Tests of Significance for ρ

Situations frequently arise where we wish to test hypotheses concerning the correlation coefficient ρ. The sampling distribution of r is very complicated for the usual values of ρ. However, when $\rho = 0$ the distribution of r can be derived. It can be shown that the quantity

$$\frac{r}{\sqrt{1 - r^2}}\sqrt{k - 2}$$

has the Student's t distribution with $k - 2$ degrees of freedom. Thus a test of the hypothesis

$$H_0: \rho = 0 \quad \text{against} \quad H_1: \rho \neq 0$$

at the $100\alpha\%$ level of significance is given by rejecting H_0 when

$$\frac{|r|\sqrt{k - 2}}{\sqrt{1 - r^2}} > t_{1-\alpha/2;\, k-2} \tag{14.72}$$

When ρ is not zero the distribution of r is not a simple function. Confidence limits of ρ have been calculated by F. N. David [3], as shown in Figs.

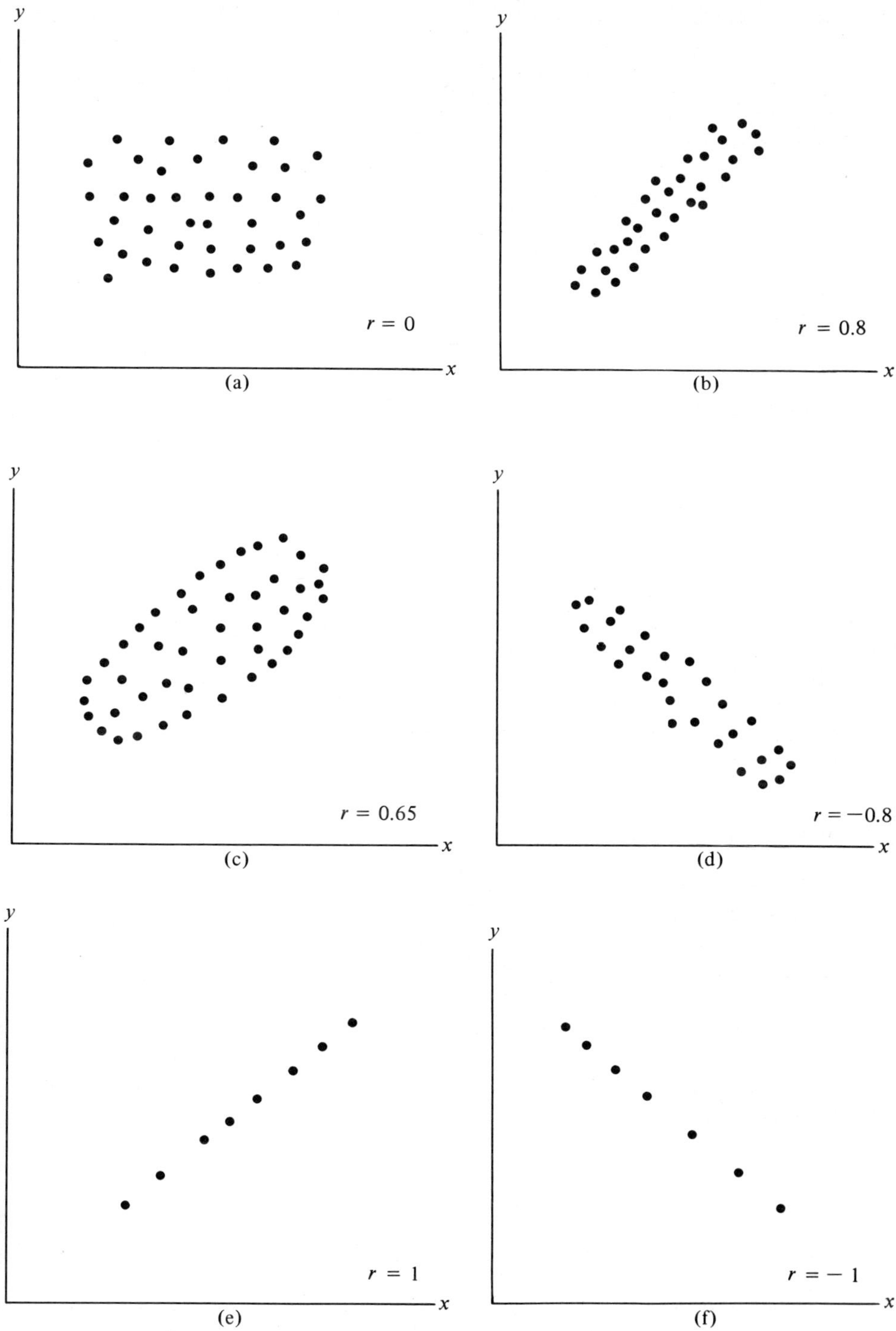

Fig. 14.5 Scatter diagrams and their associated values of r.

14.6 and 14.7. These confidence belts also can be used in testing hypotheses concerning ρ at the 5 and 1% levels of significance. If the $100(1 - \alpha)\%$ confidence interval brackets the hypothesized value of ρ we accept the null hypothesis at significance level α.

When the sample size k is large (greater than 25), Fisher [7] proved that the quantity

$$z = \frac{1}{2} \ln \frac{1 + r}{1 - r} \tag{14.73}$$

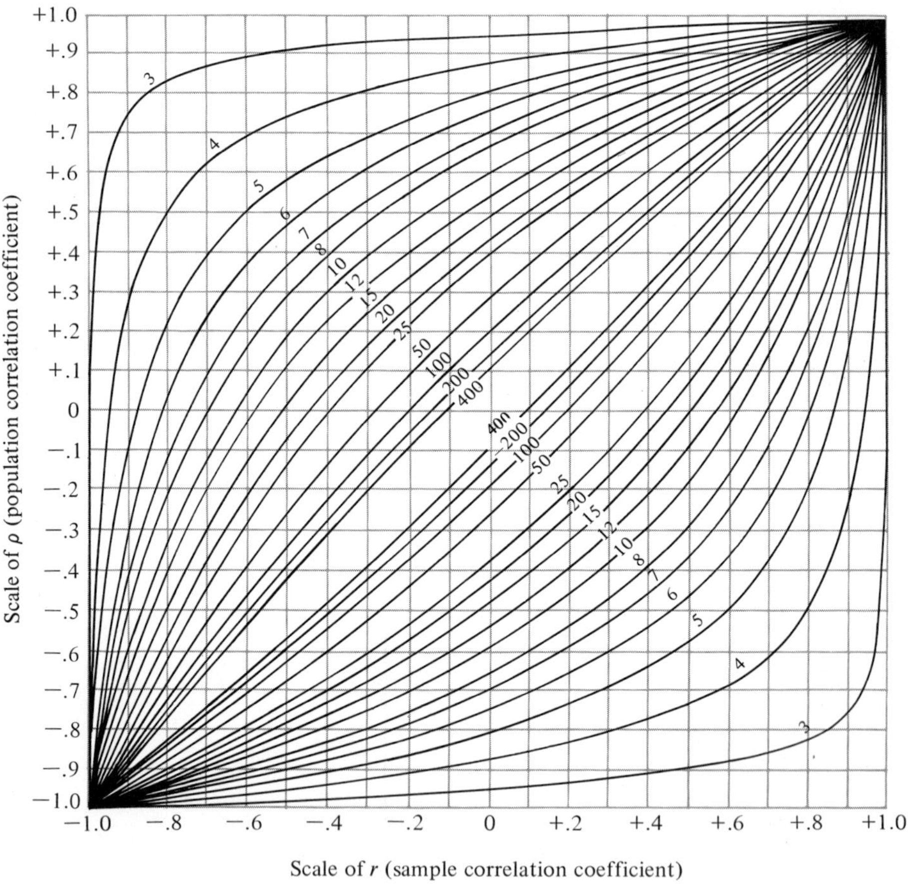

Fig. 14.6 The 95% confidence limits for the population correlation coefficient ρ given the sample correlation coefficient r. The numbers on the curves are the sample size.

[From *Biometrika Tables for Statisticians*, Vol. 1 (2nd edition), edited by E. S. Pearson and H. O. Harley, Cambridge University Press (1958); reproduced by permission of the publishers.]

is approximately normally distributed with mean

$$\frac{1}{2} \ln \frac{1+\rho}{1-\rho} \tag{14.74}$$

and variance

$$\frac{1}{k-3} \tag{14.75}$$

Thus an approximate test of significance for ρ can be performed and, as well, approximate confidence intervals for ρ can be constructed. Furthermore,

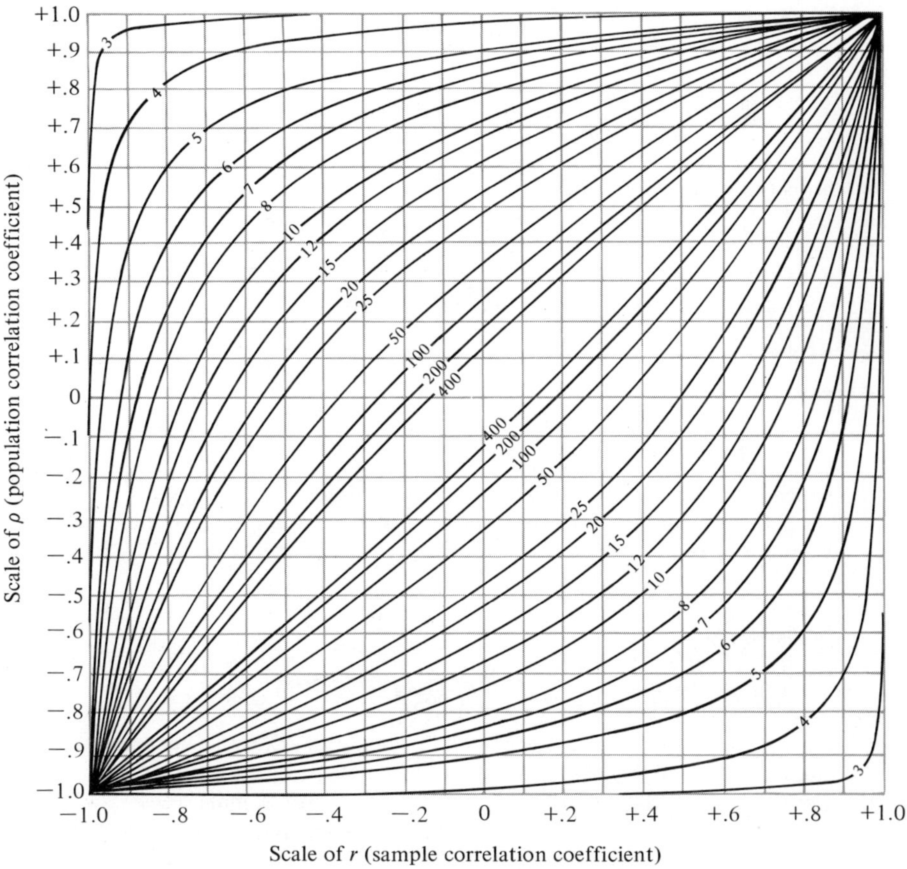

Fig. 14.7 The 99% confidence limits for the population correlation coefficient ρ given the sample correlation coefficient r. The numbers on the curves are the sample size.

the hypothesis that two independent sample correlation coefficients come from the same population ρ can be tested by the statistic

$$(z_1 - z_2)\Big/\sqrt{\frac{1}{(k_1 - 3)} + \frac{1}{(k_2 - 3)}} \tag{14.76}$$

which is approximately normally distributed with zero mean and unit variance.

Example 14.3 Using the data given in Example 14.1, find the following:
(a) Calculate the sample correlation coefficient.
(b) Is this significantly different from zero ($\alpha = .05$)?
(c) Find a 95% confidence interval for the correlation coefficient $\rho_{x,y}$.

Solution
(a) We have

$$\sum_{i=1}^{7} x_i = 3.8 \qquad \sum_{i=1}^{7} y_i = 145.4 \qquad k = 7$$

$$\sum_{i=1}^{7} x_i^2 = 2.595 \qquad \sum_{i=1}^{7} y_i^2 = 3104.2 \qquad \sum_{i=1}^{7} x_i y_i = 85.61$$

Substituting in the formula

$$r_{x,y} = \frac{k \sum_{i=1}^{k} x_i y_i - \left(\sum_{i=1}^{k} x_i\right)\left(\sum_{i=1}^{k} y_i\right)}{\sqrt{\left[k \sum_{i=1}^{k} x_i^2 - \left(\sum_{i=1}^{k} x_i\right)^2\right]\left[k \sum_{i=1}^{k} y_i^2 - \left(\sum_{i=1}^{k} y_i\right)^2\right]}}$$

we obtain

$$r_{x,y} = \frac{7 \times 85.61 - 3.8 \times 145.4}{\sqrt{[7 \times 2.595 - (3.8)^2][7 \times 3104.2 - (145.4)^2]}} = .998$$

which may be interpreted by stating that almost 100% of the variation in electrical resistance is linearly associated with variation in carbon content.

(b) We wish to test

$$H_0\colon \rho = 0 \quad \text{against} \quad H_1\colon \rho \neq 0$$

Reject H_0 if

$$\frac{|r|\sqrt{k-2}}{\sqrt{1 - r^2}} = \frac{.998\sqrt{5}}{\sqrt{1 - (.998)^2}} = 17.709 > t_{.975;\,5}$$

Obviously H_0 is rejected since $17.709 > t_{.975;\,5} = 2.571$. Therefore, we conclude that ρ is highly significant.

(c) Consulting Fig. 14.6 and entering at $r = .998$, for sample size $= 7$ we find that [.98, .999] is an approximate 95% confidence interval for ρ.

Using Fisher's approximation technique, a 95% confidence interval for $\frac{1}{2} \ln [(1 + \rho)/(1 - \rho)]$ is given by

$$\frac{1}{2} \ln \frac{1 + r}{1 - r} \pm Z_{.975} \frac{1}{\sqrt{k - 3}}$$

$$= \frac{1}{2} \ln \frac{1.998}{.002} \pm \frac{1.96}{2}$$

$$= 3.453 \pm .98 = [2.473, 4.433]$$

Equivalently,

$$2.473 < \frac{1}{2} \ln \frac{1+\rho}{1+\rho} < 4.433$$

which can be transformed to $.985 < \rho < 1$. Thus 95% limits for ρ are .985 and 1—quite close to [.98, .999] as found from Fig. 14.6.

PROBLEMS

14.1. Given the following observations of y vs x:

x:	1	2	3	4	5	6	7	8
y:	1.7	2.3	3.2	3.6	4.5	5.3	6.0	6.5

Suppose that for each x, values of y are normally distributed with mean $A + Bx$ and variance σ^2 and the y's corresponding to different x's are independent random variables.
(a) Compute the least squares regression line of y on x.
(b) If $x = 4.5$, estimate the value of y.

14.2. An experiment was run to test for a relationship between the failure time (y) of heavily overloaded resistors and resistance (x). The data are as follows:

y(min):	40.0	38.5	45.5	47.0	49.0
x(ohms):	54	55	56	57	58

(a) Estimate the relationship by the method of least squares.
(b) Estimate the value of σ^2.
(c) What value of y would you estimate for $x = 56.5$?

14.3. The following data were obtained to determine the relationship between the flow (x) and head (y) for a certain centrifugal water pump operated at constant speed:

x(gal/min):	0	1	2	3	4	6	8	9	10	12
y(ft):	100	102	102	105	108	107	116	113	115	116

(a) Determine the regression line of y on x.
(b) Estimate the value of σ^2.
(c) Find a 95% confidence limits for A.

14.4. An experiment was run to test for a relationship between the flow (x) and the efficiency (y) for a certain propellor pump. The data are as follows:

x(gal/min):	3.5	7.0	10.5	14.0	17.5	21.0
y(%):	1.6	3.0	4.6	5.1	6.4	7.3

(a) Estimate the relationship by the method of least squares.
(b) Estimate the variance (σ^2) of y.
(c) If $x = 5.0$, estimate y and find a 95% confidence interval for $E(y \mid x = 5.0)$.

14.5. The data given below were obtained in an experiment to determine the relationship between ship speed (x) and horsepower (y) for a certain type of propellor:

x(knots):	5	7	9	11	13	15
y(hp):	30	60	110	160	200	250

(a) Determine the relationship by the method of least squares.
(b) Test the hypothesis that the slope $B = 25$ at the 5% level of significance.
(c) If the true slope is equal to 22, what is the probability of accepting the hypothesis in (b)?

14.6. Consider the following data:

x:	0	1	2	3	4	5	6	7	8	9
y:	1.0	1.7	2.7	3.1	4.4	5.0	6.1	6.2	7.3	8.3

(a) Determine the least squares line.
(b) Test the hypothesis that $A = 1.0$, using $\alpha = .01$.
(c) Find a 90% confidence interval for $E(y \mid x = 4.5)$.

14.7. In Prob. 14.6 find a 90% prediction interval for a future observation at y corresponding to $x = 7.5$.

14.8. The data shown below were obtained in an experiment to determine the relationship between the cutting force (y) required for machining hard-drawn steel bars on a turret lathe and the rake angle (x) of the cutting tool under specified conditions.

x(deg):	10	15	20	25	30	35	40
y(lb):	280	250	.220	200	170	165	150

(a) Find the least squares regression of y on x.
(b) Find a 95% prediction interval for a future value of y given $x = 22$.

14.9. In Prob. 14.8 test the hypothesis that $B = -5$ at the 5% level of significance. What is the probability of accepting the hypothesis if $B = -6$?

14.10. The yield (y) of a certain chemical process is related to the operating temperature (x) with the results given below:

x(°C):	100	110	120	130	140	150	160	170	180	190
y(gal):	8	14	18	28	40	45	55	65	70	80

(a) Estimate the least squares line.
(b) Test the hypothesis that $B = 1.0$ at the 1% level of significance.
(c) Test the hypothesis thst $A = -75$ at the 1% level of significance.
(d) Find a 90% confidence interval for $E(y \mid x = 125)$.

14.11. In Prob. 14.10 find the equations of the envelopes of the 90% confidence region for the expected yield for all values of the operating temperature.

14.12. The data below were obtained in an experiment to determine the relationship between the percentage of nickel (x) and specific heat (y) for a certain alloy:

x(%):	0	0.2	0.4	0.5	0.6	0.8	0.9	1.0	1.1	1.2
y:	0.15	0.30	0.52	0.60	0.61	0.83	0.95	1.00	1.02	1.25

(a) Estimate the least squares line.
(b) Find a 90% confidence limits for B.
(c) Find a 90% confidence interval for $E(y \mid x = 0.7)$.

14.13. Consider the following data:

x:	0	2	3	5	6	7	10	11	12	14
y:	75	72	90	100	96	85	95	108	120	118

(a) Determine the regression line of y on x.
(b) Determine the regression line of x on y.
(c) Why do the two regression lines intersect? What is the point of intersection?

14.14. The following data represent the effect of annealing temperature (x) on the ductility of alpha brass, measured in terms of elongation (y):

x(°C):	300	400	500	600	700	800
y(%):	40	50	55	60	67	70

(a) Determine the regression line of y on x.
(b) Find a 90% confidence interval for $E(y \mid x = 650)$.
(c) Find a 95% confidence limits for A.

14.15. The following data represent the effect of concentration (x) of crude lactose on the wet weight (y) of mycelial inoculum:

x(%):	2	3	4	6	8
y (mg/100 ml):	2.1	2.4	1.6	2.7	3.4

(a) Determine the regression line of y on x.
(b) Find a 90% prediction interval for y corresponding to $x = 5\%$.

14.16. The cost (y) of producing a lot of a certain product depends on the lot size (x) as shown below:

y($):	100	175	285	340	460	540	590	640
x(units):	5	10	15	20	25	30	35	40

(a) Determine the regression line by the method of least squares.
(b) Find a 90% confidence limits for A.

14.17. In machining magnesium alloys, the following data were obtained to determine the effect of Brinell hardness (x) of cobalt high speed steel on the cutting speed (y).

x(BHN):	40	50	60	70	80	90	100
y(ft/min):	950	730	550	430	330	300	250

(a) Estimate the coefficients of the regression line of y on x.
(b) Find a 99% confidence interval for $E(y \mid x = 75)$.
(c) Test the hypothesis that $B = -10$ ($\alpha = .05$).

14.18. It has been determined that the relation between the tool life (t) and cutting speed (v) is given by

$$t = \left(\frac{c}{v}\right)^{1/n}$$

where c and n are unknown constants. The following data represent the tool life for machining SAE 2340 steel with single-point high speed steel at various cutting speeds:

t(min):	130.0	92.0	30.0	8.3	3.6	1.5
v(ft/min):	20	25	30	35	40	45

(a) Letting $x = \ln v$ and $y = \ln t$, determine the relationship between y and x.
(b) Estimate n by a 90% confidence interval.
(c) If $v = 32$, estimate the expected life of the cutting tool.

14.19. In a cooling experiment the temperature (t) was recorded at time (x) from the start. The data are as follows:

x(min):	0	5	10	20	25	30	40	50
t(°F):	1200	960	750	330	240	180	90	60

(a) Fit to the data a curve of the form $\ln t = A + Bx$.
(b) Estimate by a 90% prediction interval the temperature after 15 min from the start.

14.20. The weight of cement that a filling machine puts into the bag is determined by setting a dial. The following data represent the results of calibration of the dial:

Dial Setting x(lb)	25	50	100	150	200	250
	25.8	51.0	102.0	153.0	203.0	254.0
Actual weight y(lb)	25.6	50.8	102.0	152.0	204.0	253.0
	25.4	50.6	101.7	152.5	205.0	255.0

(a) Estimate the least squares line.
(b) Test for linearity at the 5% level of significance.
(c) Test the hypothesis that $B = 1$. Use $\alpha = .05$.

14.21. Consider the following data:

x	1	2	3	4	5	6	7
y	2.0	2.6	3.6	4.0	5.3	5.7	6.7
	1.8	2.8	3.8	4.4	4.9	5.5	6.8

(a) Estimate the least squares line.
(b) Test for linearity at the 5% level of significance.
(c) Find a 95% prediction interval for y corresponding to $x = 4.5$.

14.22. Suppose the regression of y on x is of the form $y = Bx$. Ten pairs of data points are found to be

x:	1	2	3	4	5	6	7	8	9	10
y:	2.1	4.0	6.0	8.3	9.7	11.9	14.7	15.2	18.3	21.1

(a) Determine the least squares line.
(b) Test the hypothesis that $B = 2$ at the 5% level of significance.
(c) Find a 90% prediction interval for y corresponding to $x = 7.5$.

14.23. An experiment was run to determine the effects of temperature (x_1) and concentration (x_2) on the yield (y) of a certain chemical process. The data are as follows:

y(gal)	x_1(°C)	x_2(%)
210	100	10
235	120	10
245	140	10
277	160	10
218	100	15
240	120	15
253	140	15
286	160	15
223	100	20
251	120	20
266	140	20
290	160	20

(a) Estimate the values of A, B_1, and B_2 for the regression plane $y = A + B_1x_1 + B_2x_2$.
(b) If $x_1 = 150$ and $x_2 = 12$, estimate the value of y.

14.24. In centrifugal casting of magnesium alloys an experiment was run to determine the effects of speed (x_1) and the diameter on which the mass is spun (x_2) on the centrifugal pressure (y). The following data were obtained:

y(lb/in.^{2}):	13	20	25	22	32	43	35	50	75
x_1(rpm):	500	600	700	500	600	700	500	600	700
x_2(in.):	12	12	12	16	16	16	20	20	20

(a) Fit an equation of the form $y = A + B_1x_1 + B_2x_2$ to the data.
(b) Estimate the variance σ^2 of y.
(c) Determine a 90% confidence interval for B_2.

14.25. In Prob. 14.4 calculate the sample correlation coefficient. Test the hypothesis that $\rho_{x,y}$ is zero. Use $\alpha = .10$.

14.26. Suppose that the calculated sample correlation turns out to be not significantly different from zero. Would you conclude that there is no relationship between the dependent and independent variables?

14.27. In a correlation problem the following sums were calculated ($k = 25$):

$$\sum x_i = 90 \qquad \sum y_i = 1200 \qquad \sum x_i y_i = 4{,}500$$
$$\sum x_i^2 = 340 \qquad \sum y_i^2 = 60{,}000$$

(a) Compute the sample correlation coefficient.
(b) Place 95% confidence limits on $\rho_{x,y}$.

14.28. From a randomly selected sample of size 30, the following sums were computed in a correlation problem:

$$\sum x_i = 4.0 \qquad \sum y_i = 130 \qquad \sum x_i y_i = 90$$
$$\sum x_i^2 = 2.6 \qquad \sum y_i^2 = 3200$$

Find a 95% confidence interval for $\rho_{x,y}$.

14.29. In Prob. 14.10 compute the sample correlation coefficient. Place 95% confidence limits on $\rho_{x,y}$.

14.30. Random variables x and y have the bivariate normal distribution with $\mu_x = \mu_y = 0$ and $\sigma_y = \sigma_x = 1$. Find the maximum likelihood estimator of $\rho_{x,y}$ based on a sample of size k.

14.31. Suppose that 50 specimens of gilding metal were tested to ascertain whether there is a relationship between yield and tensile strength. The estimate of the correlation coefficient ρ was found to be 0.85.
(a) Is there evidence of a statistically significant correlation between yield and tensile strength?
(b) Find a 95% confidence interval for ρ using Fisher's approximation.

14.32. If the assumptions underlying linear regression model are valid, show that an estimate of σ^2 is given by

$$s_r^2 = \frac{1 - r^2}{k - 2} \sum_{i=1}^{k} (y_i - \bar{y})^2$$

SUGGESTED REFERENCES

[1] Acton, F. S., *Analysis of Straight Line Data*, New York, John Wiley & Sons, Inc., 1959.

[2] Bennett, C. A., and N. L. Franklin, *Statistical Analysis in Chemistry and the Chemical Industry*, New York, John Wiley & Sons, Inc., 1954.

[3] David, F. N., *Tables of the Ordinates and Probability Integral of the Distribution of the Correlation Coefficient in Small Samples*, New York, Cambridge University Press, 1938.

[4] Dwyer, P. S., *Linear Computations*, New York, John Wiley & Sons, Inc., 1951.

[5] Ezekiel, M., and K. A. Fox, *Methods of Correlation and Regression Analysis*, 3rd ed., New York, John Wiley & Sons, Inc., 1959.

[6] Finney, D. J., *Statistical Methods in Biological Assay*, New York, Hafner Publishing Co., Inc., 1952.

[7] Fisher, R. A., "On the Probable Error of a Coefficient of Correlation from a Small Sample," *Metron*, **1**, 1921.

[8] Guest, P. G., *Numerical Methods of Curve Fitting*, New York, Cambridge University Press, 1961.

[9] Keeping, E. S., *Introduction to Statistical Inference*, Princeton, N. J., D. Van Nostrand Co., Inc., 1962.

[10] Lindley, D. V., "Estimation of a Functional Relationship," *Biometrika*, **40**, 47–49, 1953.

[11] Plackett, R. L., *Principles of Regression Analysis*, Oxford, Clarendon Press, 1960.

[12] Rushton, S., "On Least Squares Fitting by Orthogonal Polynomials Using the Choleski Method," *Journal of the Royal Statistical Society*, *Series B*, **13**, 1951.

[13] Scheffé, H., "Fitting Straight Lines When One Variable Is Controlled," *Journal of the American Statistical Association*, **53**, 106–117, 1958.

[14] Williams, E. J., *Regression Analysis*, New York, John Wiley & Sons, Inc., 1959.

15

GOODNESS-OF-FIT TESTS

15.1 Introduction

The compatibility of some observed sample distribution with a hypothesized form of a distribution is known as *goodness of fit*. The question arises whether this compatibility can be judged as satisfactory. This question can be answered if one can apply a general technique for testing the *closeness* of the observed frequencies to the theoretical frequencies. This test is known as a goodness-of-fit test.

To be specific, suppose a manufacturing process produces a certain item that can be classified as conforming to specification, oversized, or undersized with known probabilities p_1, p_2, or p_3, respectively. To test this conjecture one takes a sample of n items at random and observes the number of times conforming, oversized, and undersized items occur (denoted by n_1, n_2, and n_3, respectively). Intuitively, one would accept the postulated probabilities if the observed frequencies $n_1/n, n_2/n$, and n_3/n are close enough to p_1, p_2, and p_3, respectively. In fact, testing the closeness of a set of observed and theoretical frequencies may be viewed as a problem of testing a statistical hypothesis. In the next section we shall present a statistical test appropriate for this purpose.

15.2 The χ^2 Test

Let us consider again the example cited in the previous section. We note that n_1, n_2, and n_3 are random variables that have the multinomial distribution

$$\frac{n!}{n_1!n_2!n_3}p_1^{n_1}p_2^{n_2}p_3^{n_3} \tag{15.1}$$

where $n_1 + n_2 + n_3 = n$ and $p_1 + p_2 + p_3 = 1$. In this example we have three mutually exclusive and exhaustive outcomes, namely, conforming, oversized, and undersized. However, the problem can be formulated quite generally if we have k mutually exclusive and exhaustive outcomes (denoted by E_1, $E_2, \ldots,$ and E_k) with probabilities $p_1, p_2, \ldots,$ and p_k, respectively. Suppose n independent trials of the experiment are made and n_i denotes the number of trials producing outcome E_i. Then $n_1, n_2, \ldots, n_k$ are random variables that have the multinomial distribution

$$\frac{n!}{n_1!n_2!\cdots n_k!}p_1^{n_1}p_2^{n_2}\cdots p_k^{n_k} \tag{15.2}$$

where $\sum_{i=1}^{k} n_i = n$ and $\sum_{i=1}^{k} p_i = 1$.

We note that n_i is a binomial variable with parameter p_i, and its expected value is given by

$$E(n_i) = np_i = e_i$$

The problem, then, is to test whether the observed frequencies $n_1, n_2, \ldots,$ n_k are close enough to the hypothesized frequencies $e_1, e_2, \ldots, e_k$. This is equivalent to testing that our sample comes from a *multinomial* population with known probabilities $p_1, p_2, \ldots, p_k$. Thus the problem can be viewed as a problem of testing the hypothesis

$$H_0\colon p_i = p_{i0} \qquad i = 1, 2, \ldots, k$$

against $\quad H_1\colon H_0$ is not true

where p_{i0}'s are the postulated probabilities of the multinomial distribution. The likelihood ratio is

$$\begin{aligned}\lambda &= \frac{n!}{n_1!n_2!\cdots n_k!}p_{10}^{n_1}p_{20}^{n_2}\cdots p_{k0}^{n_k} \\ &\quad \div \frac{n!}{n_1!n_2!\cdots n_k!}\left(\frac{n_1}{n}\right)^{n_1}\left(\frac{n_2}{n}\right)^{n_2}\cdots\left(\frac{n_k}{n}\right)^{n_k} \\ &= n^n \prod_{i=1}^{k}\left(\frac{p_{i0}}{n_i}\right)^{n_i} \qquad (15.3)\end{aligned}$$

Taking logarithms,

$$\ln \lambda = n \ln n - \sum_{i=1}^{k} n_i \ln n\left(\frac{n_i}{e_i}\right)$$

$$= -\sum_{i=1}^{k} n_i \ln\left(\frac{n_i}{e_i}\right)$$

or

$$-2 \ln \lambda = 2 \sum_{i=1}^{k} n_i \ln\left(\frac{n_i}{e_i}\right) \tag{15.4}$$

It can be shown [by expanding the logarithm of (n_i/e_i)] that the right-hand side of Eq. (15.4) is approximately equal to

$$\sum_{i=1}^{k} \frac{(n_i - e_i)^2}{e_i}$$

Thus Eq. (15.4) reduces to

$$-2 \ln \lambda \cong \sum_{i=1}^{k} \frac{(n_i - e_i)^2}{e_i} \tag{15.5}$$

The random variable $-2 \ln \lambda$ has a distribution that approaches that of a χ^2 variable as n becomes infinite with $k - 1$ degrees of freedom. Thus we can assert that the distribution of the quantity $\sum_{i=1}^{k} [(n_i - e_i)^2/e_i]$ will approach that of a χ^2 with $k - 1$ degrees of freedom as n becomes infinite.

It follows that we reject the null hypothesis (H_0) at the $100\alpha\%$ level of significance if

$$\chi^2 = \sum_{i=1}^{k} \frac{(n_i - e_i)^2}{e_i} > \chi^2_{1-\alpha;\, k-1}$$

If the values of the parameters $p_1, p_2, \ldots, p_k$ are unknown, it is necessary to estimate these parameters from the data. In that case the number of degrees of freedom is reduced by the number of parameters estimated. If c parameters are estimated, then we reject the null hypothesis (H_0) at the $100\alpha\%$ level of significance if

$$\chi^2 > \chi^2_{1-\alpha;\, k-c-1}$$

The chi-square approximation is rather satisfactory provided that $e_i \geq 5$. Whenever the expected frequency of a cell is less than 5, we would combine this cell with one or more other cells until the above criterion is satisfied.

The following examples will illustrate the procedure of applying the goodness-of-fit test:

Example 15.1 In a particular hour interval on five consecutive days the number of customers who desired services at a certain restaurant were 60, 73, 48,

56, and 78. Test the null hypothesis that the expected numbers of customers per hour were the same on the five days at the 5% level of significance.

Solution. We wish to test the hypothesis

$$H_0\colon p_1 = p_2 = \cdots = p_5 = \tfrac{1}{5} \quad \text{against} \quad H_1\colon H_0 \quad \text{is not true}$$

There are a total of 315 customers; so the expected number of customers per hour per each day is

$$e_i = np_i = \tfrac{315}{5} = 63 \qquad i = 1, 2, \ldots, 5$$

The test statistic is

$$\chi^2 = \sum_{i=1}^{5} \frac{(n_i - e_i)^2}{e_i} = \frac{1}{63}[(-3)^2 + 10^2 + (-15)^2 + (-7)^2 + 15^2] = 9.65$$

The .95 point of χ^2 with 4 degrees of freedom is 9.488. Therefore, on the basis of the present evidence it is reasonable to reject the null hypothesis. Thus one would conclude that the number of customers who desired services per hour were not the same on the five days at the 5% level of significance.

Example 15.2 Suppose that a die is rolled 120 times and the following distribution of occurrences resulted:

Face	1	2	3	4	5	6
n_i	23	26	21	20	15	15

Test whether the die is fair at the 5% level of significance.

Solution. The hypothesis to be tested is

$$H_0\colon p_1 = p_2 = \cdots = p_6 = \tfrac{1}{6} \quad \text{against} \quad H_1\colon H_0 \quad \text{is not true}$$

The expected frequency is

$$e_i = np_i = \tfrac{120}{6} = 20 \qquad i = 1, 2, \ldots, 6$$

The test statistic is

$$\chi^2 = \sum_{i=1}^{6} \frac{(n_i - e_i)^2}{e_i} = \frac{1}{20}[3^2 + 6^2 + 1^2 + 0 + (-5)^2 + (-5)^2] = 4.8$$

The rejection region is

$$\sum_{i=1}^{6} \frac{(n_i - e_i)^2}{e_i} > \chi^2_{.95;\,5} = 11.07$$

Since 4.8 lies in the acceptance region, the hypothesis H_0 is accepted. Therefore, we conclude that the die is fair at the 5% level of significance.

Example 15.3 1,000 samples of aluminum foil of standard size (square foot) were chosen at random from a large lot. The number of pinholes were counted for each sample. The results are shown in Table 15.1. Is it reasonable to assume that the number of pinholes per square foot behave like a random variable having the Poisson distribution $P_X(k) = (e^{-\lambda}\lambda^k)/k!$, where λ is unknown? Assume $\alpha = .05$.

TABLE 15.1

Number of Pinholes per Sample (k_i)	*Observed Sample Frequency* (n_i)
0	120
1	266
2	260
3	169
4	96
5	35
6	17
7	3
8	2
9	1

Solution. Since λ is unknown it must be estimated. The maximum likelihood estimator of λ is

$$\begin{aligned}\hat{\lambda}_L &= \sum_{i=0}^{\infty} \frac{n_i k_i}{1{,}000} \\ &= \frac{1}{1{,}000}[(120 \times 0) + (266 \times 1) + (260 \times 2) + (169 \times 3) + (96 \times 4) \\ &\quad + (35 \times 5) + (17 \times 6) + (3 \times 7) + (2 \times 8) + (1 \times 9)] \\ &= 2\end{aligned}$$

The probability that a random area of a square foot will have a particular value of k pinholes is

$$P_X(k) \cong \frac{e^{-2} \times 2^k}{k!}$$

The expected number of samples with particular values of k is

$$1{,}000 P_X(k) \cong 1{,}000\left(\frac{e^{-2} \times 2^k}{k!}\right) \qquad k = 0, 1, 2, \ldots$$

For example,

$$1{,}000 P_X(0) \cong 1{,}000(e^{-2}) = 135.3$$

Similarly, the expected number of samples with different values of k can be evaluated as shown in Table 15.2. We note that the last three expected frequencies in column 3 of Table 15.2 are less than 5. Since the χ^2 approximation of the multinomial distribution is not very satisfactory for $e_i < 5$, the cells for $k_i \geq 6$ are combined into a single cell. The results are shown in Table 15.3. The table shows that we have seven classes. Hence the number of degrees of freedom for this χ^2 is $k - c - 1 = 7 - 1 - 1 = 5$ (using the fact that $c = 1$, since we have estimated the single parameter λ). Since the upper 5% significance point of $\chi^2_{.95;\,5}$ is 11.07, $[\sum (n_i - e_i)^2]/e_i = 5.92$ falls in the acceptance region. Thus we conclude that the number of pinholes per square foot behaved like a random variable having the Poisson distribution at the 5% level of significance.

TABLE 15.2

Number of Pinholes per Sample (k_i)	Observed Sample Frequency (n_i)	Expected Number of Sample Frequency (e_i)
0	120	135.3
1	266	270.7
2	260	270.7
3	169	180.4
4	96	90.2
5	35	36.1
6	17	12.0
7	3	3.4
8	2	.9
9	1	.2

TABLE 15.3

Number of Pinholes per Sample (k_i)	Observed Sample Frequency (n_i)	Expected Number of Sample Frequency (e_i)	$(n_i - e_i)^2$	$\frac{(n_i - e_i)^2}{e_i}$
0	120	135.3	234.09	1.730
1	266	270.7	22.09	.081
2	260	270.7	114.49	.423
3	169	180.4	129.96	.720
4	96	90.2	33.64	.373
5	35	36.1	1.21	.033
≥ 6	23	16.5	42.25	2.560
				5.920

15.3 Contingency Tables (Two-by-Two Case)

Suppose the elements of a population can be classified according to two characteristics. Furthermore, suppose that each characteristic is classified into r and s groups, where $r = 2$ and $s = 2$. For example, consider a manufacturing process that produces a certain plastic tubing. Items produced by this process are classified according to two characteristics: the length of the tubing L and the outside diameter O.D. Each characteristic is further classified into two classes, acceptable and unacceptable. Now suppose n independent items are inspected and that n_{11} items have acceptable length and outside diameter, n_{12} have unacceptable length but acceptable outside diameter, n_{21} have acceptable length but unacceptable outside diameter, and n_{22} have unacceptable length and outside diameter. The elements of the sample may be represented in the form of Table 15.4, where the *marginal* totals $n_{1.}, n_{2.}, n_{.1}$, and $n_{.2}$ are defined as follows:

$$n_{1.} = n_{11} + n_{12} \qquad n_{2.} = n_{21} + n_{22}$$

$$n_{.1} = n_{11} + n_{21} \qquad n_{.2} = n_{12} + n_{22}$$

Table 15.4. is called a 2×2 contingency table. It is obvious that the following relationships are satisfied:

TABLE 15.4

O.D. \ L	*Acceptable*	*Unacceptable*	*Total*
Acceptable	n_{11}	n_{12}	$n_{1.}$
Unacceptable	n_{21}	n_{22}	$n_{2.}$
Total	$n_{.1}$	$n_{.2}$	n

$$n = \sum_{i=1}^{2} \sum_{j=1}^{2} n_{ij} = n_{.1} + n_{.2} = n_{1.} + n_{2.}$$

Table 15.4 serves the purpose of studying the relationship between the two characteristics. Suppose that the design engineer desires to know whether the length and outside diameter of the tubing are independent. Before considering how to test this conjecture we introduce the following notation:

α_1 = the probability that an item selected at random has acceptable length

α_2 = the probability that an item selected at random has unacceptable length

β_1 = the probability that an item selected at random has acceptable outside diameter

β_2 = the probability that an item selected at random has unacceptable outside diameter

where $\alpha_1 + \alpha_2 = \beta_1 + \beta_2 = 1$. Under the assumption that the two characteristics (length and outside diameter) are independent, the joint probability of an item selected at random having, say, acceptable length and outside diameter therefore is $\alpha_1\beta_1$. In a straightforward manner the joint distribution

TABLE 15.5

O.D. \ L	*Acceptable*	*Unacceptable*	*Total*
Acceptable	$\alpha_1\beta_1$	$\alpha_2\beta_1$	β_1
Unacceptable	$\alpha_1\beta_2$	$\alpha_2\beta_2$	β_2
Total	α_1	α_2	1

can be arranged as shown in Table 15.5. The joint distribution of the observed frequencies n_{11}, n_{12}, n_{21}, and n_{22} is multinomial, namely,

$$\frac{n!}{n_{11}!n_{12}!n_{21}!n_{22}!}(\alpha_1\beta_1)^{n_{11}}(\alpha_2\beta_1)^{n_{12}}(\alpha_1\beta_2)^{n_{21}}(\alpha_2\beta_2)^{n_{22}} \tag{15.6}$$

If the true marginal probabilities α_j and β_i are known, the quantity

$$\chi^2 = \sum_{i=1}^{2}\sum_{j=1}^{2}\left[\frac{(n_{ij} - n\alpha_j\beta_i)^2}{n\alpha_j\beta_i}\right] \tag{15.7}$$

would have approximately the χ^2 distribution with 3 degrees of freedom. However, usually we do not know the true values of the marginal probabilities and have to estimate them from the data. It can be shown that the maximum likelihood estimators of α_i and β_j are

$$\begin{cases} \hat{\alpha}_1 = \dfrac{n_{.1}}{n} \\ \hat{\alpha}_2 = \dfrac{n_{.2}}{n} \\ \hat{\beta}_1 = \dfrac{n_{1.}}{n} \\ \hat{\beta}_2 = \dfrac{n_{2.}}{n} \end{cases} \tag{15.8}$$

where $\hat{\alpha}_1 + \hat{\alpha}_2 = 1$ and $\hat{\beta}_1 + \hat{\beta}_2 = 1$. Thus the maximum likelihood estimators of the joint probabilities (under the assumption that the two characteristics are independent) are

$$\hat{\alpha}_j\hat{\beta}_i = \frac{n_{.j}n_{i.}}{n^2} \qquad i = 1, 2 \qquad j = 1, 2 \tag{15.9}$$

Now if we replace $\alpha_j\beta_i$ in Eq. (15.7) by their maximum likelihood estimators we obtain

$$\chi^2 = \sum_{i=1}^{2}\sum_{j=1}^{2}\left[\left(n_{ij} - \frac{n_{.j}n_{i.}}{n}\right)^2 \Big/ \frac{n_{.j}n_{i.}}{n}\right] \tag{15.10}$$

The degrees of freedom of this χ^2 is $k - c - 1$, where c is the number of parameters estimated. Here $k = 4$ and $c = 2$ since we have estimated only one α and only one β (note $\hat{\alpha}_1 + \hat{\alpha}_2 = 1$, $\hat{\beta}_1 + \hat{\beta}_2 = 1$). The degrees of freedom for the χ^2 in Eq. (15.10) are thus $4 - 2 - 1 = 1$. It follows that for large n the random variable χ^2 defined by Eq. (15.10) has approximately the χ^2 distribution with 1 degree of freedom. Thus the statistic given by Eq. (15.10) can be used to test the hypothesis that the two characteristics are independent. In symbols, the hypothesis of independence is rejected at the $100\alpha\%$ level of significance if

$$\sum_{i=1}^{2}\sum_{j=1}^{2}\left[\left(n_{ij} - \frac{n_{.j}n_{i.}}{n}\right)^2 \Big/ \frac{n_{.j}n_{i.}}{n}\right] > \chi^2_{1-\alpha;\,1} \tag{15.11}$$

The following example will illustrate the procedure of testing the independence of two characteristics:

Example 15.4 The industrial engineer of the XYZ Company wishes to test whether the length and the outside diameter of a certain plastic tubing are related. To test this conjecture a sample of 600 items was chosen at random. The length and outside diameter of each tubing were inspected and classified as acceptable and unacceptable, as shown in Table 15.6. Is it reasonable to assume that the length and outside diameter are independent ($\alpha = .05$)?

TABLE 15.6

Outside Diameter \ *Length*	*Acceptable*	*Unacceptable*	*Total*
Acceptable	380	20	400
Unacceptable	160	40	200
Total	540	60	600

Solution. We compute the estimates of the expected frequencies:

$$\frac{n_{.1}n_{1.}}{n} = \frac{540 \times 400}{600} = 360$$

$$\frac{n_{.2}n_{1.}}{n} = \frac{60 \times 400}{600} = 40$$

$$\frac{n_{.1}n_{2.}}{n} = \frac{540 \times 200}{600} = 180$$

$$\frac{n_{.2}n_{2.}}{n} = \frac{60 \times 200}{600} = 20$$

Applying Eq. (15.10) we obtain

$$\chi^2 = \frac{(380 - 360)^2}{360} + \frac{(20 - 40)^2}{40} + \frac{(160 - 180)^2}{180} + \frac{(40 - 20)^2}{20}$$
$$= 33.33$$

We reject the hypothesis of independence since

$$33.33 > \chi^2_{.95;\,1} = 3.84$$

Thus we conclude that the two dimensions are indeed dependent at the 5% level of significance. The χ^2 test applied to a contingency table indicates only whether the two characteristics are independent. It does not show in what specific way these dimensions differ. However, an inspection of the data shows a relatively high number of items with unacceptable outside diameter. It also shows that the dimension of the outside diameter appears to have an affect on the dimension of the length.

15.4 Contingency Tables ($r \times s$ Case)

The two-by-two *contingency table* is a special case of the $r \times s$ contingency table. In this case we have two characteristics classified into r and s groups, where r and s are not necessarily equal to 2. In a straightforward manner Eq. (15.7) becomes

$$\chi^2 = \sum_{i=1}^{r} \sum_{j=1}^{s} \left[\frac{(n_{ij} - n\alpha_j\beta_i)^2}{n\alpha_j\beta_i}\right] \tag{15.12}$$

The right-hand side of Eq. (15.12) is asymptotically distributed as χ^2 with $rs - 1$ degrees of freedom. However, in practice the true values of α_j and β_i are usually unknown and have to be estimated from the data. If these estimates are inserted in Eq. (15.12) we obtain

$$\chi^2 = \sum_{i=1}^{r} \sum_{j=1}^{s} \left[\frac{(n_{ij} - n_{.j}n_{i.}/n)^2}{n_{.j}n_{i.}/n}\right] \tag{15.13}$$

Since $\sum_{j=1}^{s} \alpha_j = 1$, only $(s - 1)\alpha$'s have to be estimated. Similarly, $(r - 1)\beta$'s have to be estimated. The degrees of freedom for the χ^2 in Eq. (15.13) thus are

$$rs - [(r - 1) + (s - 1)] - 1 = (r - 1)(s - 1) \tag{15.14}$$

Thus the hypothesis of independence is rejected at the $100\alpha\%$ level of significance if

$$\sum_{i=1}^{r} \sum_{j=1}^{s} \left[\frac{(n_{ij} - n_{.j}n_{i.}/n)^2}{n_{.j}n_{i.}/n}\right] > \chi^2_{1-\alpha;\,(r-1)(s-1)} \tag{15.15}$$

Range \ *Dispersion*	*Low*	*High*
Short	470	27
Medium	350	50
Long	380	73

Example 15.5 The following data show the characteristics of dispersion and ranges of a certain caliber of small arms ammunition: Test the hypothesis that dispersion and range are independent at the 5% level of significance.

Solution. We compute the estimates of the expected frequencies

$$\frac{n_{.1}n_{1.}}{n} = 1200 \times \frac{497}{1350} = 441.77$$

$$\frac{n_{.2}n_{1.}}{n} = 150 \times \frac{497}{1350} = 55.22$$

$$\frac{n_{.1}n_{2.}}{n} = 1200 \times \frac{400}{1350} = 355.55$$

$$\frac{n_{.2}n_{2.}}{n} = 150 \times \frac{400}{1350} = 44.44$$

$$\frac{n_{.1}n_{3.}}{n} = 1200 \times \frac{453}{1350} = 402.66$$

$$\frac{n_{.2}n_{3.}}{n} = 150 \times \frac{453}{1350} = 50.33$$

The test statistic is

$$\chi^2 = \frac{(470 - 441.77)^2}{441.77} + \frac{(27 - 55.22)^2}{55.22} + \frac{(350 - 355.55)^2}{355.55} + \frac{(50 - 44.44)^2}{44.44} + \frac{(380 - 402.66)^2}{402.66} + \frac{(73 - 50.33)^2}{50.33}$$

$$= 28.5$$

Since $28.5 > \chi^2_{.95;\,2} = 5.99$, we reject the hypothesis of independence and conclude that there is strong evidence of a highly significant degree of dependence between dispersion and range at the 5% level of significance.

15.5 The Kolmogorov-Smirnov Tests

The Kolmogorov-Smirnov tests are nonparametric tests for differences between two continuous cumulative distribution functions. The *two-sample* Kolmogorov-Smirnov *test* is concerned with the agreement between two cumulative distributions; it tests the hypothesis that two independent samples are drawn from identical continuous distributions. The distribution of the absolute difference between the values of the two observed cumulative distributions is used to establish the critical region for the test.

The *one-sample test* is concerned with the agreement between an observed and hypothesized cumulative distributions; it tests the hypothesis that a random sample comes from a population with a specified continuous cumulative distribution function. Therefore, it is a test of goodness of fit.

The Kolmogorov-Smirnov one-sample test possesses the advantage of being an exact method, whereas the chi-square test for goodness of fit is valid only for comparatively large samples. We recall that the chi-square test requires all cell frequencies to exceed 5. This restriction is no longer required because there is no need to classify the observations in carrying out the present test. It should be noted, however, that the present test is *not* valid if the theoretical distribution function is not continuous.

The one-sample test is based on the statistic

$$D_n = \max |F(x) - S_n(x)| \tag{15.16}$$

where $S_n(x)$ is the empirical cumulative distribution function of a sample of size n drawn from a population in which random variable X has a continuous cumulative distribution function F(x). The null hypothesis is *rejected* if the observed value of the statistic D_n falls in the critical region at the desired level of significance. The distribution of D_n can be derived numerically for any particular value of n by using combinatorial methods that are too laborious and complicated to be presented here. Certain critical values of the distribution of D_n, however, are given in Table 15.7.

TABLE 15.7 Critical Values of D_n in the Kolmogorov-Smirnov One-Sample Test†

n	$D_{.10}$	$D_{.05}$	$D_{.01}$	n	$D_{.10}$	$D_{.05}$	$D_{.01}$
3	.642	.708	.828	12	.338	.375	.450
5	.510	.565	.669	15	.304	.338	.404
6	.470	.521	.618	20	.264	.294	.356
7	.438	.486	.577	25	.240	.270	.320
8	.411	.457	.543	30	.220	.240	.290
9	.388	.422	.514	40	.190	.200	.240
10	.368	.410	.490	large values	$\frac{1.220}{\sqrt{n}}$	$\frac{1.360}{\sqrt{n}}$	$\frac{1.630}{\sqrt{n}}$

†Adapted from F. J. Massey, Jr., "The Kolmogorov-Smirnov test for goodness of fit," *Journal of the American Statistical Association*, **46** (1951), p. 70, with permission.

We shall show how to apply the preceding technique by a numerical illustration.

Example 15.6 The following are the repair times (in hours) of a certain machine: 6.10, 1.30, 11.05, 4.60, .50, 1.75, 8.20, 10.50, 15.00, 20.50. Use the Kolmogorov-Smirnov test with $\alpha = .05$ to determine whether these data can be looked upon as a sample from an exponential population with parameter $\lambda = .20$.

Solution. Under the null hypothesis that repair times are exponentially distributed, the theoretical cumulative distribution is given by

$$F(x) = \begin{cases} \int_0^x .2e^{-.2t}\,dt & \\ 1 - e^{-.2x} & x \geq 0 \\ 0 & \text{otherwise} \end{cases}$$

In order to compare the observed and theoretical cumulative distributions we construct the following table:

x	$S(x)$	$F(x)$	D
.50	.1	.0952	.0048
1.30	.2	.2290	.0710
1.75	.3	.2953	.0047
4.60	.4	.6015	.2015
6.10	.5	.7048	.2048
8.20	.6	.8061	.2061
10.50	.7	.8776	.1776
11.05	.8	.8903	.0903
15.00	.9	.9503	.0503
20.50	1.0	.9835	.0165

Hence, the maximum difference is .2061. To determine whether this difference is significant we find the critical value of D in Table 15.7. For $n = 10$ and $\alpha = .05$ the critical value is .410 which is greater than .2061. Hence the null hypothesis (that repair times are exponentially distributed with $\lambda = .20$) cannot be rejected at the 5% level of significance.

PROBLEMS

15.1. Suppose a coin is tossed 800 times with the result that 432 heads are obtained. Apply the χ^2 test to see whether the coin is unbiased at the 5% level of significance.

15.2. Suppose that a die is rolled 150 times and the number of times each face comes up is recorded. The following results are obtained:

Face	1	2	3	4	5	6
Observed frequency	29	19	19	27	26	30

Are these results consistent with the hypothesis that the die is fair at the 1% level of significance?

15.3. According to past records the proportion of items produced by a certain manufacturing process is as follows: 5% scrapped, 3% reworked, and 92% acceptable. Inspection of 500 items showed that 40 items must be scrapped and 28 items can be reworked. Are these results consistent with past records? Assume $\alpha = .05$.

15.4. According to the genetic model the proportion of individuals possessing the four blood types O, A, B, and AB should be in the proportions

$$q^2 : p^2 + 2pq : r^2 + 2qr : 2pr$$

respectively, where $p + q + r = 1$. 1,000 individuals were tested, and their blood types were found to have the proportions 18%: 51%: 19%: 12%. Are these results compatible with the model if $p = q = .4$? Assume $\alpha = .05$.

15.5. The record of industrial accidents in a certain plant is as follows:

Month	*Frequency*	*Month*	*Frequency*
January	18	July	16
February	6	August	24
March	8	September	13
April	12	October	9
May	21	November	8
June	7	December	14

Test the null hypothesis that there is no seasonal variation over the month of the year in which accidents had occurred. Assume $\alpha = .01$.

15.6. In a particular hour interval on seven consecutive days of the week the number of express trains arriving at a certain subway station were 15, 9, 18, 12, 13, 10, and 14. Test the null hypothesis that the expected numbers of arrivals per hour were the same on the seven days of the week. Assume $\alpha = .05$.

15.7. The number of oil tankers arriving at a certain refinery on any day was recorded for 1,000 days. The results are shown in the following table:

Number of Tankers Arriving per Day	*Observed Frequency*
0	372
1	360
2	191
3	57
4	16
5	2
6	1
7	1
	1,000

Is it reasonable to assume on the basis of the above data that the number of oil tankers arriving on any day is a random variable having the Poisson distribution? Assume $\alpha = .05$.

15.8. The following table gives the frequency distribution of demand for a certain commodity per day based on the records of 1,000 days.

Demand per Day	*Observed Frequency*
0	626
1	274
2	80
3	15
4	4
5	1

Fit a Poisson distribution to the above data and test for goodness of fit at the 5% level of significance.

15.9. The time study engineer of the XYZ Company obtained the following distribution that shows the cycle time in minutes for the assembly of a certain electronic product:

Cycle Time (min)	*Frequency*	*Cycle Time* (min)	*Frequency*	*Cycle Time* (min)	*Frequency*
1.32	2	1.38	8	1.44	7
1.33	1	1.39	7	1.45	12
1.34	2	1.40	15	1.46	3
1.35	5	1.41	8	1.47	5
1.36	4	1.42	22	1.48	1
1.37	19	1.43	14		

The time study engineer concludes that these data behave like a sample from a normal population at the 5% level of significance. Is he right? Justify your answer.

15.10. 49 items were placed on life test, and the test was run until all items failed. The observed failure times $t_1, \ldots, t_{49}$ were (from *Statistical Techniques in Life Testing*, Technical Report No. 5):

1.2	13.7	38.9	72.4	102.8	151.6	203.0
2.2	15.1	47.9	73.6	108.5	152.6	204.3
4.9	15.2	48.4	76.8	128.7	164.2	229.5
5.0	23.9	49.3	83.8	133.6	166.8	253.1
6.8	24.3	53.2	95.1	144.1	178.6	304.1
7.0	25.1	55.6	97.9	147.6	185.2	341.7
2.1	35.8	62.7	99.6	150.6	187.1	354.4

Use the χ^2 test to test the hypothesis that the underlying distribution of life is exponential. Assume $\alpha = .05$.

15.11. A thousand individuals were classified according to whether they gamble and according to whether they smoke as follows:

	Smoker	*Nonsmoker*
Gambler	120	30
Nongambler	479	371

Test the hypothesis that the two criteria of classification are independent at the 1% level of significance.

15.12. An experiment was performed to study the effect of a lubricant used in a certain drawing operation on the quality of items produced. The following results were obtained:

	Defective	*Nondefective*
Without lubricant	22	238
With lubricant	4	126

Would you conclude from these data that the use of lubricant reduced the proportion of defectives? Assume $\alpha = .05$.

15.13. The following data on the color of eyes and hair for 6,800 individuals were compiled (from Kendall):

Eye \ Hair	Fair	Brown	Black	Red
Blue	1768	807	189	47
Hazel or green	946	1387	746	43
Brown	115	438	288	16

Test the hypothesis that hair color and eye color are independently distributed. Assume $x = .01$.

15.14. Grades in a probability course and a statistical quality control course taken simultaneously at a certain school for a sample of students is shown in the following table:

	Statistical Quality Control Grade			
Probability Grade	A	B	C	D
A	38	24	8	2
B	16	12	8	8
C	10	19	35	23
D	3	6	7	13

Test the hypothesis at the 1% level of significance that grades in both courses are independently distributed.

15.15. Photographs are taken and printed according to two methods, A and B, and are rated as either satisfactorily exposed, underexposed, or overexposed. Test the hypothesis that the type of exposure is distributed independently of the method used at the 5% level of significance.

Method \ *Exposed*	*Satisfactory*	*Underexposed*	*Overexposed*
A	115	15	10
B	120	12	8

SUGGESTED REFERENCES

[1] Barnard, G. A., "Significance Test for 2 × 2 Tables," *Biometrika*, **34**, 1947.

[2] Cochran, W. G., "Some Methods for Strengthening the Common χ^2 Tests," *Biometrics*, **10**, 1954.

[3] Cochran, W. G., "The χ^2 Test of Goodness of Fit," *Annals of Mathematical Statistics*, **23**, 1952.

[4] Cramér, H., *Mathematical Methods of Statistics*, Princeton, N.J., Princeton University Press, 1946.

[5] Goodman, L. A., "Simple Methods for Analyzing Three-Factor Interaction in Contingency Tables," *Journal of the American Statistical Association*, **59**, 1954.

[6] Goodman, L. A., and W. H. Kruskal, "Measures of Association for Cross Classification," *Journal of the American Statistical Association*, **49**, 1954.

[7] Gumbel, E. J., "On the Reliability of the Classical Chi-Square Test," *Annals of Mathematical Statistics*, **14**, 1943.

[8] Guttman, I., S. S. Wilks, and J. S. Hunter, *Introductory Engineering Statistics*, 2nd ed., New York, John Wiley & Sons, Inc., 1971.

[9] Hoel, P. G., *Introduction to Mathematical Statistics*, 3rd ed., New York, John Wiley & Sons, Inc., 1965.

[10] Lehmann, E. L., *Testing of Statistical Hypotheses*, New York, John Wiley & Sons, Inc., 1959.

[11] Mood, A. M., and F. A. Graybill, *Introduction to the Theory of Statistics*, New York, McGraw-Hill Book Company, 1963.

[12] Pearson, E. S., "The Choice of Statistical Tests Illustrated on the Interpretation of Data Classified in a 2 × 2 Table," *Biometrika*, **34**, 1947.

[13] Smirnov, N., "Tables for Estimating the Goodness of Fit of Empirical Distributions," *Annals of Mathematical Statistics*, **19**, 1948.

16

CONTROL CHARTS

16.1 Statistical Stability

Let us consider a simple productive operation of mass producing an electrical resistor. Suppose that the design specification for the measurable characteristic (resistance) is $10^{\pm .10}$ ohms. If samples were taken in order of production and the resistance of each sample was measured to hundreths of an ohm, we would find that the measured resistance of each sample varies from one sample to the next. This *inherent variability* in the measured resistance may be due to variation in material used, manufacturing process, labor skill, measurement, etc.

Littauer in his article "Social Aspects of Scientific Method in Industrial Production" [10] states:

> If exact design, production and measurement were possible the implication would be that there was no variation in the measurement or in the making of things and that within a few operative steps one could tell just how to adjust the productive process so that it would continuously produce exactly uniform resistors. The fact of the matter is that neither measurement nor manufacture can be performed without variation, and the important problem to resolve in any experimental inquiry is distinguishing between inherent and removable variation.

Thus any measurable characteristic is always subject to two sources of variation: *random variation* and *nonrandom variation.* The random variation is usually known as variation due to *chance causes.* If these random variations exhibit a stable pattern, then we say the process is operating under a stable system of chance causes. A process that exhibits a stable pattern of chance causes is said to be in a *state of statistical stability* or in a *state of statistical control* or simply in *control.* Variations that are not within the stable pattern of chance causes are known as *assignable causes of variation.* A process subject to the occurrence of assignable causes of variation is said to be *out* of control.

Obviously it is not economically feasible to measure the resistance of every resistor produced to verify whether it conforms to specification ($10^{\pm .10}$ ohms). One thus must resort to an evaluative procedure that enables one to judge whether the process is in a state of statistical stability and accordingly whether the process capabilities meet the specification requirements. Littauer [10] states:

> Shewhart developed his now famous control chart procedure . . . a profound but practically simple technique to apply . . . for detecting the presence of removable causes, so that these causes can be discovered and eliminated. As a result of eliminating these causes the productive operation becomes more uniform . . . its variation is reduced and stays for reasonable periods of time within computable limits. The cycle of specification, production and inspection evaluation employing the Shewhart statistical control procedure provides a continuing self correcting cycle of operation or experimental inference by means of which, conceptually at least, the error of operation or of inference approaches zero. That is, theoretically we make all resistors exactly alike and exactly in accord with the specification. The enactment of the cycle establishes a state of control in the making of resistors . . . a state in which, on the basis of past observation one can predict within limits future performance, with reasonable assurance of verifying one's predictions.

The application of control charts technique to maintain the desired state of statistical stability will be considered in detail in the following sections.

CONTROL CHARTS FOR VARIABLES

16.2 Control Chart for the Mean: $\bar{X}$ Chart

In this chapter we will adhere to the standard notation used in the literature of statistical quality control. The prime (′) symbol is used to identify a known parameter of a universe. For example, $\bar{X}'$ and σ' denote the mean and standard deviation of the normal distribution, respectively, whereas $\bar{X}$ and σ

denote the estimate of the mean and the estimate of the standard deviation of the normal distribution, respectively. Let us assume that the measurable characteristic under study is normally distributed with known mean $\bar{X}'$ and known standard deviation σ'. If successive samples of size n are taken in order of production, the mean of each sample is

$$\bar{X}_j = \frac{1}{n}\sum_{i=1}^{n} x_i \qquad j = 1, 2, \ldots \tag{16.1}$$

which is normally distributed with mean $\bar{X}'$ and standard deviation $\sigma'/\sqrt{n}$. Now if the process is in a state of control with respect to the characteristic measured—i.e., the sequence of observations $x_1, x_2, \ldots, x_n$ behaves like identically distributed random variables with mean $\bar{X}'$ and standard deviation σ'—we expect that the sample means $\bar{X}_1, \bar{X}_2, \ldots$ will lie within the interval $[\bar{X}' - (3\sigma'/\sqrt{n}), \bar{X}' + (3\sigma'/\sqrt{n})]$ with probability .9973. In symbols, we write

$$P\left\{\bar{X}' - \frac{3\sigma'}{\sqrt{n}} \leq \bar{X}_j \leq \bar{X}' + \frac{3\sigma'}{\sqrt{n}}\right\} = .9973 \tag{16.2}$$

That is, on the average the means of only 27 subgroups out of 10,000 will fall outside the interval $[\bar{X}' \pm (3\sigma'/\sqrt{n})]$. If the means of these subgroups are plotted as shown in Fig. 16.1, we expect that on the average only 27 in 10,000 sample averages will fall outside the control limits $[\bar{X}' \pm (3\sigma'/\sqrt{n})]$. Figure 16.1 is known as a 3-sigma control chart. The center solid line is known as the

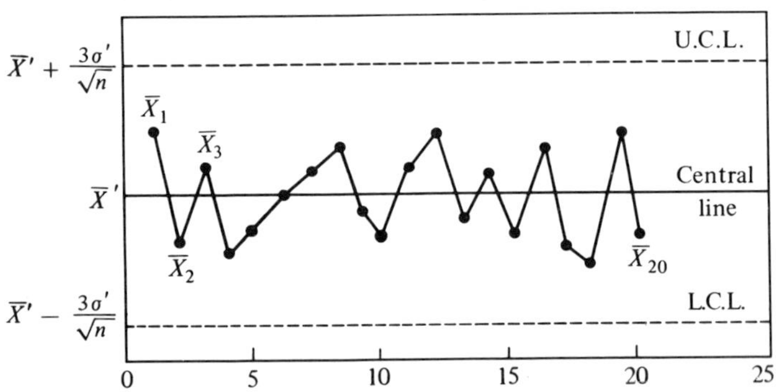

Fig. 16.1 3-sigma control chart ($\bar{X}$ chart).

center control line, whereas the dotted lines $\bar{X}' + (3\sigma'/\sqrt{n})$ and $\bar{X}' - (3\sigma'/\sqrt{n})$ are referred to as the *upper control limit* (U.C.L.) and *lower control limit* (L.C.L.), respectively.

If the plotted sample averages $\bar{X}_1, \bar{X}_2, \ldots$ fall between the control limits, we say that *the process is in control* and take no further action. If a sample average $(\bar{X}_j)$ falls outside the control limits, we say *the process is out of control* and we investigate the process to discover and remove the assignable causes of variation. Strictly speaking, when the control chart signals the presence of assignable causes of variation, this signal can be interpreted as either a *false* or a *real* alarm. False alarm means that a sample average ($\bar{X}_j$) falls outside the control limits (indicating lack of control) when actually the process is in a state of control. The chance of this happening on the average is 27 in 10,000 times, which is indeed very insignificant and, consequently, one can hardly expect this happening to be very realistic. What happens if we decide to decrease the spread of the control limits, say, to $\bar{X}' \pm (2.5\sigma'/\sqrt{n})$? We can immediately compute the probability that a sample average $\bar{X}_j$ will fall outside the control limits. This probability is

$$1 - P\left\{\bar{X}' - \frac{2.5\sigma'}{\sqrt{n}} \leq \bar{X}_j \leq \bar{X}' + \frac{2.5\sigma'}{\sqrt{n}}\right\} = .0124 \tag{16.3}$$

That is, on the average only 124 in 10,000 sample averages will fall outside the 2.5-sigma limits.

To summarize, the probability of a false alarm for 3-sigma control limits is .0027, whereas the probability of a false alarm for 2.5-sigma limits is .0124. The probability of a false alarm is known as the probability of committing a type I error, which is the level of significance in testing hypotheses. Thus we conclude that the probability of committing a type I error increases as the multiple of $\sigma'/\sqrt{n}$ decreases, and vice versa. Now what happens if the process goes out of control due to the occurrence of an assignable cause of variation which, say, takes the form of a shift in the mean $\bar{X}'$ to a new level $\bar{X}' + \delta\sigma'$ (σ' remains stable)? $\bar{X}_j$ now behaves like a random variable having the normal distribution with mean $\bar{X}' + \delta\sigma'$ and standard deviation $\sigma'/\sqrt{n}$, as shown in Fig. 16.2. The probability that a sample average $\bar{X}_j$ will fall between the control limits is

$$\int_{\bar{X}'-(k\sigma'/\sqrt{n})}^{\bar{X}'+(k\sigma'/\sqrt{n})} \frac{\sqrt{n}}{\sigma'\sqrt{2\pi}} \exp\left[-\frac{1}{2}\left(\frac{\bar{X} - \bar{X}' - \delta\sigma'}{\sigma'/\sqrt{n}}\right)^2\right] d\bar{X} = \Phi(k - \delta\sqrt{n}) - \Phi(-k - \delta\sqrt{n}) \tag{16.4}$$

where k = the factor that determines the spread of the control limits. Let

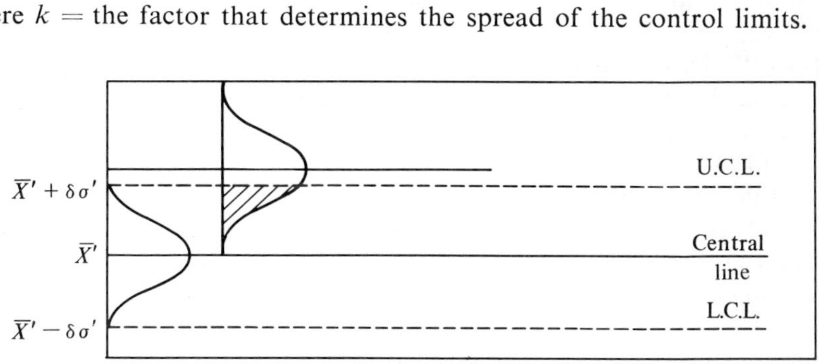

Fig. 16.2 Effects of change in process mean.

us suppose that $\delta = 2$, $k = 3$ (3-sigma limits), and $n = 4$. Then the probability that $\bar{X}_j$ will fall within the control limits becomes

$$\Phi(3 - 4) - \Phi(-3 - 4) \cong .158 \tag{16.5}$$

This means that if the first sample average ($\bar{X}_j$) selected falls between the control limits we may conclude erroneously that the process is in control when actually the mean has shifted by $2\sigma'$. Thus we have committed a type II error; the probability of this error is .158. The probability of committing a type II error if $k = 2.5$ (2.5-sigma limits) is

$$\Phi(2.5 - 4) - \Phi(-2.5 - 4) \cong .066 \tag{16.6}$$

From Eqs. (16.5) and (16.6) it is evident that the probability of committing a type II error decreases as k (multiple of $\sigma'/\sqrt{n}$) decreases, and vice versa. In the light of this analysis it can readily be seen that the errors of the first and second kind are related; i.e., an increase in one results in a decrease in the other, and vice versa. It follows that a type I error can be made as small as desirable at the expense of increasing the type II error. The converse is true; i.e., a type II error can be made as small as possible at the expense of increasing the type I error. Obviously one can determine the sample size (n) and the factor that determines the spread of the control limits (k) that control both errors.

The conventional 3-sigma control limits are extensively used in the United States, whereas 2.5-sigma control limits are used widely in the United Kingdom. It should be noted that neither the 3-sigma nor the 2.5-sigma control limits are validated on an economical or statistical basis but rather on their effectiveness validated by industrial experience. However, the 3-sigma control limits are preferred if the process is subject to the occurrence of large shifts in the process mean whereas the 2.5-sigma control limits are more sensitive in detecting small shifts in the process mean. Shewhart recommended using 3-sigma control limits and sample $n = 4$ or 5, leaving the interval between samples to be determined by the quality control engineer. In recent years there have been a few efforts at establishing the optimal parameters of $\bar{X}$ charts, namely, the sample size, the factor that determines the spread of the control limits, and the intersample interval that minimizes the overall cost of monitoring, charting, loss due to defective items, etc. The explication of these modern statistical quality control techniques is beyond the scope of this text.

To summarize, the statistical control chart is a useful device for maintaining the statistical stability of a repetitive process. Actually the control chart is a dynamic technique that is fed information on the process as it is in operation and that provides operational criteria to distinguish between random and nonrandom variation and thereby provides a basis for action to eliminate removable causes of variation. In practice there are two major kinds of uses of the control chart. The first is the use of a control chart to *maintain the existing state of control*. The second major use of a control chart is to *bring a process under control* by establishing operational criteria for control from the observed data. The former has already been discussed and plotted in Fig. 16.1. The latter will be discussed as we proceed.

Example 16.1 The resistance of a certain resistor is known to be normally distributed with mean = 10 ohms and standard deviation = .05 ohm. The following are the $\bar{X}$ values for 20 subgroups of 5 readings:

Subgroup number	1	2	3	4	5	6	7	8	9	10
$\bar{X}$	10.04	10.03	9.96	9.98	10.01	10.01	10.04	9.94	10.02	9.99
Subgroup number	11	12	13	14	15	16	17	18	19	20
$\bar{X}$	9.98	9.96	9.96	10.03	10.05	10.04	9.95	10.01	10.02	9.97

(a) Is the process probably in control?
(b) If the specification limits for the resistor are 9.90 to 10.10 ohms, what percent defective items will be produced?
(c) If it has been decided to reduce the sample size to four per subgroup, what should be the new control limits?

Solution

(a) The upper control limit is

$$\bar{X}' + \frac{3\sigma'}{\sqrt{n}} = 10 + \frac{3 \times .05}{\sqrt{5}} = 10.067$$

The lower control limit is

$$\bar{X}' - \frac{3\sigma'}{\sqrt{n}} = 10 - \frac{3 \times .05}{\sqrt{5}} = 9.933$$

Since all sample averages lie within the control limits [9.933, 10.067] the process is assumed to be in statistical control.

(b) If the process will be statistically controlled, the percent defective items that will be produced is

$$1 - [\Phi(2) - \Phi(-2)] = .0456$$

(c) The new control limits are

$$\text{Upper control limit} = 10 + \frac{3 \times .05}{2} = 10.075$$

$$\text{Lower control limit} = 10 - \frac{3 \times .05}{2} = 9.925$$

We note that the control limits widen as the sample size decreases, and consequently the type II error increases. Note that the type I error remains unchanged and is equal to .0027 since 3-sigma limits are used for both sample sizes.

In most practical applications the values of $\bar{X}'$ and σ' are usually unknown and have to be estimated from the observed data. There is no rule of thumb as to the number of observations required for estimating $\bar{X}'$ and σ'; however, on the basis of wide experience it is desirable that these estimates be based on at least 100 observations or at least 25 subgroups of $n = 4$ or 5 observations. Increasing the number of subgroups increases the accuracy of the estimates provided that the process is in a state of control. Increasing n may increase the accuracy of estimating σ' but also may increase the chance of variation within a subgroup. As a result, the 3-sigma control charts become insensitive to changes in the process mean.

The unbiased estimate of X' is given by

$$\bar{\bar{X}} = \frac{1}{N}\sum_{j=1}^{N} \bar{X}_j \tag{16.7}$$

where $N =$ number of subgroups. It can be shown that the unbiased estimate of σ' when X is a normal variate is given by

$$\bar{\sigma}/c_2 \tag{16.8}$$

where
$$\bar{\sigma} = \frac{1}{N}\sum_{j=1}^{N} \sigma_i = \frac{1}{N}\sum_{j=1}^{N}\sum_{i=1}^{n}\sqrt{\frac{(X_i - \bar{X}_j)^2}{n}}$$

and
$$c_2 = \sqrt{\frac{2}{n}}\left(\frac{n-2}{2}\right)! \div \left(\frac{n-3}{2}\right)!$$

The factor c_2 depends on n and is tabulated in Table L in the Appendix. The formulas for 3-sigma control limits then become

$$\begin{cases} \text{U.C.L.}_{\bar{X}} = \bar{\bar{X}} + 3\dfrac{\bar{\sigma}}{c_2\sqrt{n}} \\ \text{L.C.L.}_{\bar{X}} = \bar{\bar{X}} - 3\dfrac{\bar{\sigma}}{c_2\sqrt{n}} \end{cases} \tag{16.9}$$

By letting $A_1 = 3/c_2\sqrt{n}$, Eq. (16.9) becomes

$$\begin{cases} \text{U.C.L.}_{\bar{X}} = \bar{\bar{X}} + A_1\bar{\sigma} \\ \text{L.C.L.}_{\bar{X}} = \bar{\bar{X}} - A_1\bar{\sigma} \end{cases} \tag{16.10}$$

where A_1 depends on n and is tabulated in Table L. It should be noted that the computation of $\bar{\sigma}$ is rather tedious and laborious for most practical applications. An alternative approach that involves simpler computations for estimating σ' is as follows: Let R_j denote the *range* of a subgroup j based on n observations, and let $\bar{R}$ denote the average of the ranges of N subgroups, where

$$\bar{R} = \frac{1}{N}\sum_{j=1}^{N} R_j \tag{16.11}$$

By integration methods it can be shown that $\bar{R}/d_2$ is an unbiased estimate of σ', where d_2 depends on n and is tabulated in Table L. The formulas for 3-sigma control limits then become

$$(16.12)\qquad \begin{cases} \text{U.C.L.}_{\bar{X}} = \bar{\bar{X}} + \dfrac{3\bar{R}}{d_2\sqrt{n}} \\ \text{L.C.L.}_{\bar{X}} = \bar{\bar{X}} - \dfrac{3\bar{R}}{d_2\sqrt{n}} \end{cases}$$

By letting $3/d_2\sqrt{n} = A_2$, Eq. (16.12) becomes

$$(16.13)\qquad \begin{cases} \text{U.C.L.}_{\bar{X}} = \bar{\bar{X}} + A_2\bar{R} \\ \text{L.C.L.}_{\bar{X}} = \bar{\bar{X}} - A_2\bar{R} \end{cases}$$

where A_2 depends on n and is tabulated in Table L. Although both $\bar{\sigma}/c_2$ and $\bar{R}/d_2$ are unbiased estimates of σ', $\bar{\sigma}/c_2$ is preferred since it is more efficient. The control limits given by Eqs. (16.10) and (16.13) are referred to as *trial control limits*. Control charts having trial control limits are sometimes called *empirical control charts*. The procedure of starting the control charts will be taken up in Sec. 16.5.

16.3 Control Chart for the Range: *R* Chart

If the standard deviation of the measured characteristic σ' increases, say, to $3\sigma'$, whereas the mean $\bar{X}'$ remains stable, the probability of a sample average $\bar{X}_j$ falling outside the 3-sigma control limits is

$$(16.14)\qquad 2\left[1 - \Phi\left(\frac{3\sigma'/\sqrt{n}}{3\sigma'/\sqrt{n}}\right)\right] = .3174$$

That is, on the average of 3,174 in 10,000 times an $\bar{X}$ chart will show lack of control whereas if σ' were stable the chance of this happening is 27 in 10,000. To obviate this ambiguity we have to keep a record of the variability of observations within each subgroup by plotting the *range* of each sample on a control chart known as the *range chart* or simply *R chart*. It is a useful device for detecting changes in the variability of the quality characteristic. The distribution of the range R from samples from a normal universe is approximately related to the chi-square distribution. Statistical theory gives the values of the mean and standard deviation of the range as

$$E(R) = d_2\sigma' \quad \text{and} \quad \sigma_R = d_3\sigma'$$

where d_2 and d_3 depend on n and are tabulated in Table L. Thus 3-sigma control limits of the R chart are

$$(16.15)\qquad \begin{cases} \text{U.C.L.}_R = d_2\sigma' + 3d_3\sigma' = (d_2 + 3d_3)\sigma' \\ \text{L.C.L.}_R = d_2\sigma' - 3d_3\sigma' = (d_2 - 3d_3)\sigma' \end{cases}$$

or

$$(16.16) \qquad \begin{cases} \text{U.C.L.}_R = D_2\sigma' \\ \text{L.C.L.}_R = D_1\sigma' \end{cases}$$

where $D_2 = d_2 + 3d_3$ and $D_1 = d_2 - 3d_3$ can be obtained from Table L.

When σ' is unknown, it is estimated from $\bar{R}/d_2$. On substituting $\bar{R}/d_2$ for σ' in Eq. (16.15) we obtain

$$(16.17) \qquad \begin{cases} \text{U.C.L.}_R = \left(1 + \dfrac{3d_3}{d_2}\right)\bar{R} \\ \text{L.C.L.}_R = \left(1 - \dfrac{3d_3}{d_2}\right)\bar{R} \end{cases}$$

By letting $D_4 = 1 + (3d_3/d_2)$ and $D_3 = 1 - (3d_3/d_2)$, Eq. (16.17) becomes

$$(16.18) \qquad \begin{cases} \text{U.C.L.}_R = D_4\bar{R} \\ \text{L.C.L.}_R = D_3\bar{R} \end{cases}$$

where D_3 and D_4 depend on n and are tabulated in Table L. Note that control limits computed from the tabulated values of D_3 and D_4 are the .5 and 99.5 percentiles of the sampling distribution of the statistic R for samples from a normal universe. The R chart has the same appearance as the $\bar{X}$ chart and is plotted in Fig. 16.3.

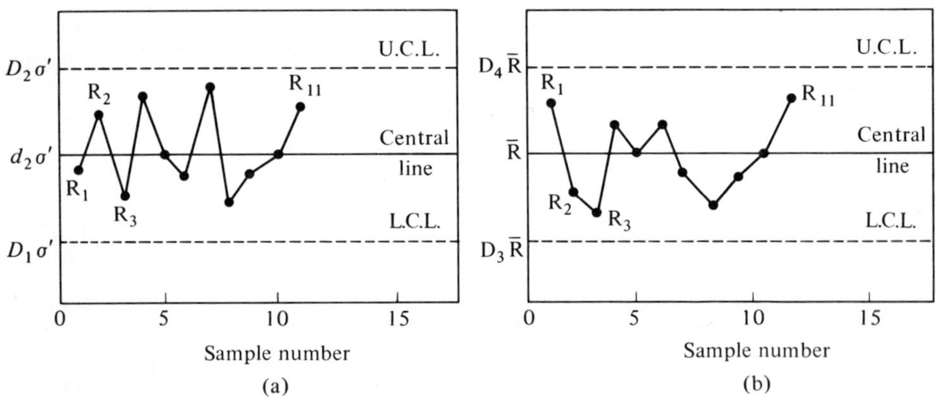

Fig. 16.3 *R* chart (a) known σ' (b) unknown σ'.

16.4 Control Chart for the Standard Deviation: σ Chart

The *σ chart* is an alternative device for detecting shifts in the variability of the quality characteristic. Both σ and R charts serve the same purpose.

The latter, however, is more widely used because of the ease with which the range can be computed, whereas the former is more sensitive to changes in variability when the sample size is comparatively large ($n > 20$). The distribution of σ for a normal universe is related to the chi-square distribution. It can be shown that the expected value and the standard deviation of σ (denoted by σ_σ) are

$$E(\sigma) = c_2\sigma' \quad \text{and} \quad \sigma_\sigma = [2(n-1) - 2nc_2^2]^{1/2}\frac{\sigma'}{\sqrt{2n}}$$

The control limits then are

$$\text{(16.19)} \qquad \begin{cases} \text{U.C.L.}_\sigma = c_2\sigma' + 3\sigma_\sigma = B_2\sigma' \\ \text{L.C.L.}_\sigma = c_2\sigma' - 3\sigma_\sigma = B_1\sigma' \end{cases}$$

When σ' is unknown, we use its unbiased estimate $\bar{\sigma}/c_2$. Thus the estimated control limits become

$$\text{(16.20)} \qquad \begin{cases} \text{U.C.L.}_\sigma = \bar{\sigma} + 3[2(n-1) - 2nc_2^2]^{1/2}\dfrac{\bar{\sigma}}{c_2\sqrt{2n}} = B_4\bar{\sigma} \\ \\ \text{L.C.L.}_\sigma = \bar{\sigma} - 3[2(n-1) - 2nc_2^2]^{1/2}\dfrac{\bar{\sigma}}{c_2\sqrt{2n}} = B_3\bar{\sigma} \end{cases}$$

The factors B_1, B_2, B_3, and B_4 are also given in Table L.

16.5 Starting the Control Charts

The first step in starting the control charts is to decide on the subgroup size and frequency of sampling. The basic concept of the control charts is *to select subgroups in a way that makes each subgroup as homogeneous as possible and that gives maximum opportunity for variation from one subgroup to another*. Shewhart recommends four or five as the ideal subgroup size. The frequency of sampling is determined by the quality control engineer on the basis of experience. $\bar{X}_j$ and R_j are computed for each subgroup and plotted on the $\bar{X}$ and R charts, respectively. If $\bar{X}'$ and σ' are known, the control limits for the $\bar{X}$ and R charts are computed from $[\bar{X}' \pm (3\sigma'/\sqrt{n})]$ and $[D_2\sigma', D_1\sigma']$, respectively. If all points fall within the control limits we conclude that the process is probably in control. However, there is no positive assurance that the process is really in control. Our conclusion merely means that for practical purposes it pays to act as if no assignable causes of variation are present.

When $\bar{X}'$ and σ' are unknown, trial control limits are computed from at least 25 subgroups. The trial control limits for $\bar{X}$ and R charts are computed from $[\bar{\bar{X}} \pm A_2\bar{R}]$ and $[D_4\bar{R}, D_3\bar{R}]$, respectively. If all points fall within control limits, the trial control limits should be extended to apply for future production; however, they should be revised from time to time as additional data are accumulated. If the control charts indicate lack of control by points falling outside the trial control limits, we conclude that the process is not in a state of control and, accordingly, that the observed data do not come from a stable distribution. Hence these trial control limits cannot be applied for maintaining control of future operations unless they are revised. The *revised* limits should be based upon observations drawn from a *stable* distribution. The procedure of revising the trial control limits is as follows:

1. When the R chart indicates a possible lack of control by points falling outside the trial control limits it is desirable to estimate the value of $\bar{R}$ and σ' that might be attained if the variability were brought into control. This can be achieved by eliminating the values of R falling outside the trial control limits and recomputing the values of the control limits based on the remaining observations. If the revised control limits throw more points outside the control limits the procedure is repeated until all points fall within the control limits. The revised $\bar{R}$ may be used for computing new control limits for the $\bar{X}$ chart.
2. If the R chart exhibits a state of control whereas the $\bar{X}$ chart shows lack of control, the spread of the trial control limits on the $\bar{X}$ chart should continue to be $2A_2\bar{R}$. The location of the central line is revised either by centering the process average at an aimed at level, designated at $\bar{X}'$, or by eliminating those points falling outside the trial control limits and recomputing new trial control limits (based on revised $\bar{\bar{X}}$) until all points fall within the control limits.
3. If both R and $\bar{X}$ charts exhibit lack of control, revised control limits are computed as described in 1 and 2. The following example will illustrate this procedure:

Example 16.2 Samples of four items are taken from a process at regular intervals. The following are the $\bar{X}$ and R values for 25 subgroups:

Subgroup (j):	1	2	3	4	5	6	7	8	
$\bar{X}_j$:	2.060	2.071	2.073	2.051	2.050	1.975	1.971	1.960	
R_j:	.091	.093	.081	.086	.129	.186	.083	.075	
Subgroup (j):	9	10	11	12	13	14	15	16	
$\bar{X}_j$:	1.968	2.066	2.051	2.006	1.965	2.156	1.966	1.992	
R_j:	.030	.230	.082	.097	.038	.088	.145	.053	
Subgroup (j):	17	18	19	20	21	22	23	24	25
$\bar{X}_j$:	1.973	2.046	2.168	2.075	1.961	1.963	1.936	1.800	2.057
R_j:	.041	.026	.237	.099	.071	.060	.102	.059	.148

(a) Is the process probably in control?
(b) If not, determine the control limits for future production.

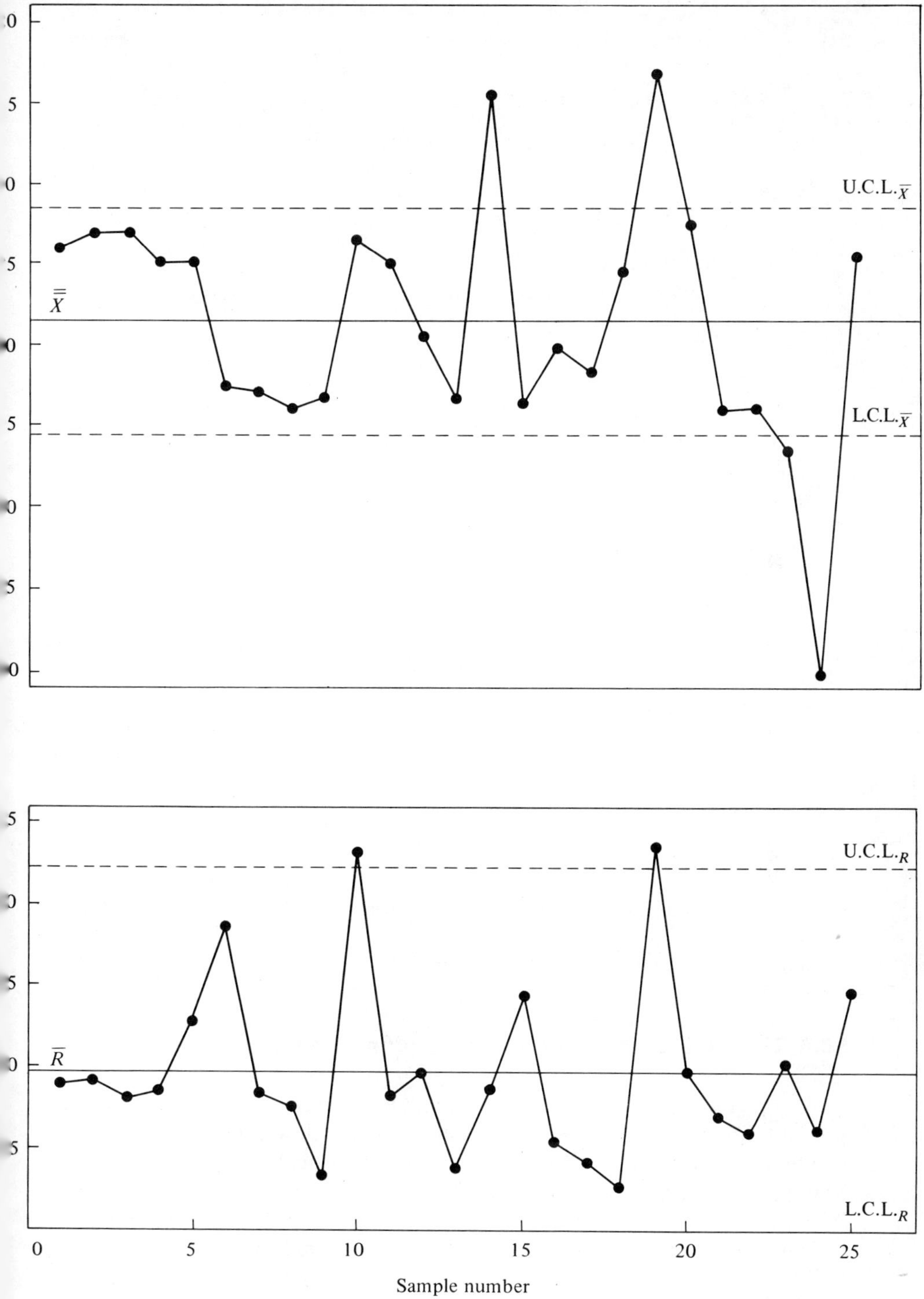

Fig. 16.4 Control charts for $\bar{X}$ and R.

Solution

(a) We compute

$$\bar{\bar{X}} = \frac{1}{25}\sum_{j=1}^{25}\bar{X}_j = \frac{50.36}{25} = 2.0144$$

$$\bar{R} = \frac{1}{25}\sum_{j=1}^{25}R_j = \frac{2.430}{25} = .0972$$

The trial control limits then are

$$\text{U.C.L.}_{\bar{X}} = \bar{\bar{X}} + A_2\bar{R} = 2.0144 + (.73)(.0972) = 2.0853$$

$$\text{L.C.L.}_{\bar{X}} = \bar{\bar{X}} - A_2\bar{R} = 2.0144 - (.73)(.0972) = 1.9435$$

$$\text{U.C.L.}_R = D_4\bar{R} = (2.28)(.0972) = .2216$$

$$\text{L.C.L.}_R = D_3\bar{R} = (0)(.0972) = 0$$

The control charts for $\bar{X}$ and R are shown in Fig. 16.4. We note that $\bar{X}_{14}$, $\bar{X}_{19}$, $\bar{X}_{23}$, $\bar{X}_{24}$, R_{10}, and R_{19} fall outside the trial control limits. Thus we conclude the process is not in a state of control.

(b) Eliminating $\bar{X}_{14}$, $\bar{X}_{19}$, $\bar{X}_{23}$, $\bar{X}_{24}$, R_{10}, and R_{19} and recomputing $\bar{\bar{X}}$ and $\bar{R}$ we obtain

$$\bar{\bar{X}} = \frac{1}{21}\sum \bar{X}_j = \frac{42.30}{21} = 2.0143$$

and

$$\bar{R} = \frac{1}{23}\sum R_j = \frac{1.963}{23} = .085$$

The revised control limits then are

$$\text{U.C.L.}_{\bar{X}} = 2.0143 + (.73)(.085) = 2.0763$$

$$\text{L.C.L.}_{\bar{X}} = 2.0143 - (.73)(.085) = 1.9523$$

$$\text{U.C.L.}_R = (2.28)(.085) = .1938$$

$$\text{L.C.L.}_R = (0)(.085) = 0$$

Since none of the remaining points fall outside the revised control limits, the process is assumed to be in control at this level. The revised control limits can be used for controlling future production; however, they should be revised as additional data are accumulated.

16.6 The Role of Normality in Sampling Theory

In previous sections it was assumed that the measured characteristic is normally distributed with mean $\bar{X}'$ and standard deviation σ'. Under this assumption 3-sigma control limits as well as the probability of a false alarm were established for the $\bar{X}$, R, and σ charts. The question arises of whether we can establish a control chart for a measurable characteristic that is not normally distributed. To answer this question let us first investigate the distribution of the sample mean ($\bar{X}$) drawn from nonnormal universes. According to the *central limit theorem* the mean of n identically distributed independent variates will be approximately normally distributed regardless of the distribution of the individual variables. Obviously the larger the sample size n the closer will be the frequency distribution of $\bar{X}$ to the normal curve. Shewhart [12] showed that the distribution of $\bar{X}$ values from samples of four drawn from rectangular and triangular universes is approximately normal. In practice, the observed distributions of the majority of industrial quality characteristics exhibit normal or nearly normal behavior. The distribution of $\bar{X}$ will be very close to normal, the closer the underlying distribution is to a normal curve.

Finally, we wish to examine the distributions of $\bar{R}$ and $\bar{\sigma}$. According to the central limit theorem the distributions of R and σ tend to be normal as the number of subgroups increases. Summarizing, we can assume that the distributions of $\bar{X}$, $\bar{R}$, and $\bar{\sigma}$ tend to be close to normal even though their underlying distributions are not normal. Now we can answer the question previously raised by saying that we can establish control charts for quality characteristics whose underlying distributions are not necessarily normal. It should be noted that the magnitude of type I error based on 3-sigma limits and normality is not applicable. However, in practice the magnitude of type I error for nonnormal universes tends to be quite small.

16.7 Natural Tolerance Limits

Design engineers specify on their blueprints the dimensions of the product to be manufactured to meet established work standards. More precisely, they specify the *basic* value for each dimension and its permissible variation. For example, if the diameter of a shaft is specified as 2 in.$^{\pm.001}$, then 2 in. is referred to as the "basic size" and $\pm.001$ as the "tolerance." When a tolerance distribution is both plus and minus of a basic size it is referred to as a *bilateral tolerance*, whereas if the distribution is on one side only it is referred to as a *unilateral tolerance*. In the example cited, the diameter of the shaft must be at least 1.999 in. but not larger than 2.001 in. The former is known as the *lower specification limit* (L), whereas the latter is known as the *upper specification limit* (U). These limits represent the extreme dimensions beyond which acceptable product cannot extend.

The values of the tolerance are determined by the functional fitness of the product. This functional fitness can be attained if parts produced are screened by rejecting all parts nonconforming to the specification. Let us suppose that the diameter of this shaft is normally distributed with mean $\bar{X}' = 2.00$ and $\sigma' = .003$. Furthermore, suppose that the manufacturing process is under the surveillance of $\bar{X}$ and R charts. If the process is in a state of control, then the variability of the diameter can be described by a stable normal distribution function; i.e., the individual measurements behave like identically distributed random variables with mean $\bar{X}' = 2.00$ and $\sigma' = .003$. Thus we expect on the average 99.73% of the parts produced will fall between $\bar{X}' \pm 3\sigma' = 2 \pm .009$ provided the process remains in control.

The spread $6\sigma' = .018$ is known as the *natural tolerance spread*, and $\bar{X}' \pm 3\sigma' = [1.991, 2.009]$ are referred to as the *natural tolerance limits* that include 99.73% of the population. These natural tolerance limits are then compared with the *specification limits*. If the natural tolerance spread ($6\sigma'$) is less than or equal to the difference between the specification limits ($U - L$), the manufactured product will meet the specification requirements as long as the process remains in a state of control. If ($6\sigma'$) exceeds ($U - L$) the manufactured product will not meet the specification despite the fact that the process is in control. In this case we attempt either to reduce the variability of the process or to revise the specification limits.

In practice, the values of $\bar{X}'$ and σ' are usually unknown and have to be estimated from the observed data. These estimates should be based upon data from a controlled process with an adequate sample size and a relatively large number of subgroups. The same procedure of comparing natural tolerance and specification limits can be applied using these estimates. However, we may need to adjust the centering of the process average so that the natural tolerance limits fall well within the specification limits. If the underlying distribution is known but not normal, the natural tolerance limits can be computed with specified odds. If the underlying distribution is completely unknown, the technique of distribution-free tolerance limits described in Chapter 7 may be applied.

Example 16.3 Suppose in Example 16.2 that the specification limits for this product characteristic are $2 \pm .020$. Will the process be able to meet this specification?

Solution. The estimates of $\bar{X}'$ and σ' are

$$\bar{\bar{X}} = 2.0143 \qquad \sigma = \frac{\bar{R}}{d_2} = \frac{.085}{2.059} = .0412$$

The estimated natural tolerance spread that includes 99.73% of the population then is $6\sigma = 6(.0412) = .2472$. Since $.2472 > .040$, the process definitely will not be able to meet these specification limits.

CONTROL CHARTS FOR ATTRIBUTES

16.8 Control Chart for Fraction Defective: *p* Chart

The *p chart* can be applied to quality characteristics that are observed as *attributes*. In practice, measurable characteristics can be observed as attributes since items can be inspected by "go–not go" gauges and classified as good or bad. The p chart is an alternative device for maintaining control of a repetitive process; however, it is not as sensitive as the $\bar{X}$ and R charts in detecting assignable causes of variation. It is less informative than the $\bar{X}$ and R charts in the sense that it does not reveal the behavior of the operation that can be used as an aid for the diagnosis of the causes of trouble. The p chart can be applied to one or many quality characteristics, whereas separate $\bar{X}$ and R charts are required for each measurable characteristic. The monitoring cost of a p chart is much less than the monitoring cost of an $\bar{X}$ or R chart, since the cost of inspection, charting, and computing are much less for attributes than for variables.

Let p' denote the probability of an item being defective, and let d_i be the number of defectives in sample i of size n. Then $d_i/n = p_i$ is a random variable having the binomial distribution. Its expected value and standard deviation are

$$\begin{cases} E(p_i) = p' \\ \sigma_{p_i} = \sqrt{\dfrac{p'(1-p')}{n}} \end{cases} \tag{16.21}$$

The control limits are thus established at

$$\begin{cases} \text{U.C.L.}_p = p' + 3\sqrt{\dfrac{p'(1-p')}{n}} \\ \text{L.C.L.}_p = p' - 3\sqrt{\dfrac{p'(1-p')}{n}} \end{cases} \tag{16.22}$$

If p' is unknown, trial control limits are computed using $\bar{p}$ as the estimate of p', where $\bar{p}$ is the ratio of the total number of defectives actually found in all samples to the total number of items inspected. $\bar{p}$ should be computed from observations drawn from a controlled process and should be revised as additional data are accumulated. The trial control limits then are given by

$$\begin{cases} \text{U.C.L.}_p = \bar{p} + 3\sqrt{\dfrac{\bar{p}(1-\bar{p})}{n}} \\ \text{L.C.L.}_p = \bar{p} - 3\sqrt{\dfrac{\bar{p}(1-\bar{p})}{n}} \end{cases} \tag{16.23}$$

Although it is customary to use 3-sigma limits for the p charts, the probability of a false alarm is not .0027 as for the $\bar{X}$ chart. The exact probability of a false alarm can be computed from the expression

$$1 - p\{np' - 3\sqrt{p'(1-p')} \leq d_i \leq np' + 3\sqrt{p'(1-p')}\} \tag{16.24}$$

and can be obtained from the tables of binomial probability distribution (Table J in the Appendix). The normal approximation can be used when the sample size is relatively large.

It should be noted that when subgroup size is variable, limits should be computed separately for each subgroup.

Example 16.4 A sample of 200 bolts is drawn at regular intervals from the production line, and each bolt is checked by "go–not go" thread gauge. The number of defective bolts in 20 successive samples are given in the following table:

Subgroup	*d*	*Subgroup*	*d*
1	3	11	2
2	3	12	3
3	1	13	2
4	3	14	1
5	2	15	1
6	3	16	3
7	2	17	3
8	2	18	3
9	3	19	2
10	3	20	3

(a) Draw a p control chart.
(b) Is the process in control?

Solution. We compute $\bar{p} = 48/4{,}000 = .012$. Thus the trial control limits are

$$\text{U.C.L.}_p = .012 + 3\sqrt{\frac{(.012)(.988)}{200}} = .035$$

$$\text{L.C.L.}_p = .012 - 3\sqrt{\frac{(.012)(.988)}{200}} = 0$$

Since no points fall outside these limits we conclude that the process is probably in controi. These trial limits can be applied for controlling future production. The p chart is shown in Fig. 16.5.

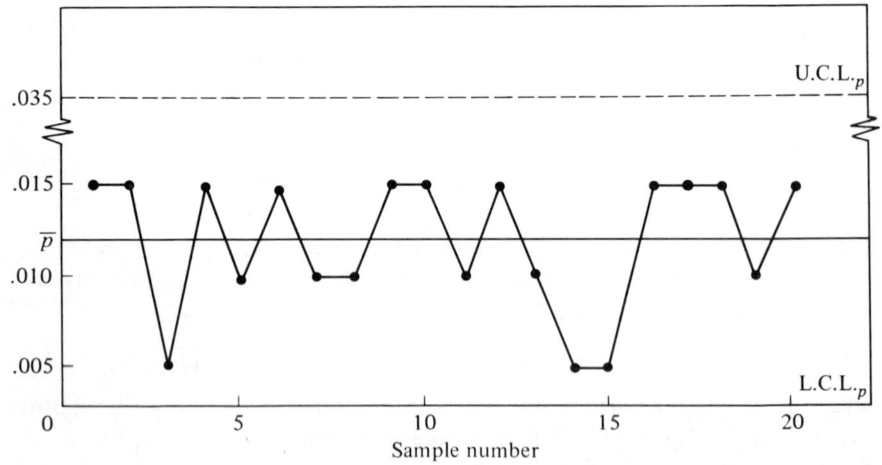

Fig. 16.5 p chart.

16.9 Control Chart for the Number of Defects in a Sample: *np* Chart

If the subgroup size is constant, the *np chart* may be used. It is more convenient than the *p* chart, simply because it saves the calculation of p_i for each subgroup. The control limits for the *np* chart are

$$\text{(16.25)} \qquad \begin{cases} \text{U.C.L.}_{np} = np' + 3\sqrt{np'(1-p')} \\ \text{L.C.L.}_{np} = np' - 3\sqrt{np'(1-p')} \end{cases}$$

When p' is unknown its estimate $\bar{p}$ is used for computing the trial control limits.

16.10 Control Chart for Defects: *c* Chart

The *c chart* is used for the control of the number of defects per unit. The *c* chart takes into account the number of defects found in a *single* item, group of items, part of an item, etc. For example, we may be interested in the control of the number of pinholes in aluminum foil of a given size, the number of defective rivets on an airplane wing, the number of defective items in a radio assembly, or the number of blemishes on a given surface. The degree of defectiveness of such items is characterized by the number of defects found in each item. The number of defects c per unit is assumed to follow the Poisson distribution. Hence,

$$\text{(16.26)} \qquad p(c) = \frac{e^{-c'}c'^{c}}{c!} \qquad c = 0, 1, 2, \ldots$$

where c' is the mean and $\sqrt{c'}$ is the standard deviation of the Poisson distribution. Here we are assuming that the normal distribution is a good approximation to the Poisson. This approximation is found satisfactory if c' is greater than ten and the number of units inspected is large. Under these assumptions, and using 3-sigma limits, the control limits for the c chart are

$$\text{(16.27)} \qquad \begin{cases} \text{U.C.L.}_{c} = c' + 3\sqrt{c'} \\ \text{L.C.L.}_{c} = c' - 3\sqrt{c'} \end{cases}$$

when c' is unknown, trial control limits are computed by using $\bar{c}$, where $\bar{c}$ is the estimate of c' and is defined as the ratio of the total number of defects found in all units or samples to the total number of units or samples inspected. Thus the trial control limits are given by

$$\text{(16.28)} \qquad \begin{cases} \text{U.C.L.}_{c} = \bar{c} + 3\sqrt{\bar{c}} \\ \text{L.C.L.}_{c} = \bar{c} - 3\sqrt{\bar{c}} \end{cases}$$

Note that the exact probability limits are computed by using the Poisson distribution.

Example 16.5 In the inspection of vinyl sheets a sample of 30 yd was taken at regular intervals and the number of defects were recorded in the table below.
(a) Draw a control chart for c.
(b) Is the process in control?

Sample Number	*Number of Defects*	*Sample Number*	*Number of Defects*
1	3	11	4
2	4	12	9
3	8	13	3
4	5	14	10
5	6	15	6
6	9	16	6
7	8	17	7
8	3	18	4
9	12	19	13
10	6	20	2

Solution. We compute

$$\bar{c} = \frac{\text{total number of defects}}{\text{total number of items inspected}} = \frac{128}{20} = 6.4$$

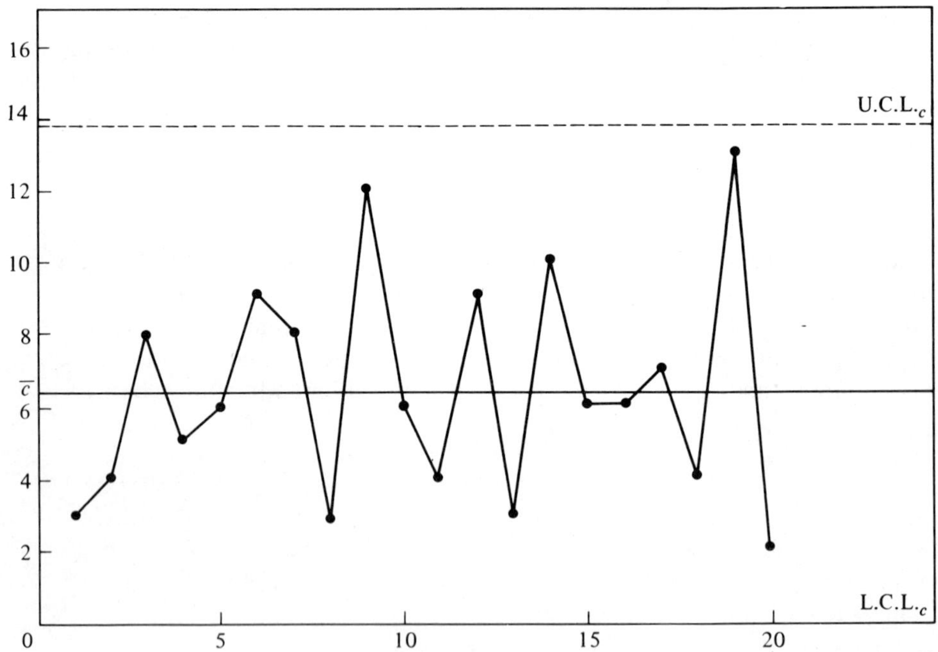

Fig. 16.6 c chart.

The trial control limits are established at

$$\text{U.C.L.}_{\cdot c} = \bar{c} + 3\sqrt{\bar{c}} = 6.4 + 3\sqrt{6.4} = 13.98$$
$$\text{L.C.L.}_{\cdot c} = \bar{c} - 3\sqrt{\bar{c}} = 6.4 - 3\sqrt{6.4} = 0$$

Since none of the points fall outside the trial control limits, we conclude that the process is in control. These trial control limits can be applied for control of future production; however, they should be revised as additional data are accumulated. The c chart is plotted in Fig. 16.6.

PROBLEMS

16.1. Define "chance cause." What is a nonchance or discoverable cause?

16.2. What constitutes a homogeneous subgroup? What merit does the homogeneous subgroup have in maintaining statistical stability?

16.3. Show by the aid of a diagram that an increase in the process variability with no change in process average throws the $\bar{X}$ chart out of control as well as the R chart.

16.4. What errors can be made in applying a statistical stability criterion such as the plus and minus 3-sigma limits for sample averages? How do these errors relate to the Neyman–Pearson errors of the first and second kind?

16.5. In studying each of the following cases indicate which type of control charts—namely, the X, R, p, np, and c charts—is the most appropriate one:
(a) The life of incandescent lamps.
(b) The number of cracked glass vessels per case.
(c) The rate of production of electronic tubes.
(d) The number of defective fuses per gross.
(e) The number of blemishes per roll of cloth.
(f) Detection of slow fluctuations.
(g) The gross weight of cement bags.
(h) The out of straightness of seamless tubing.
(i) The number of accidents per month in a certain plant.
(j) Detection of fast fluctuations.
(k) The demand for a certain commodity per week.
(l) The number of machine failures per month in a manufacturing plant.
(m) The blowing time of electric fuses.
(n) The shear strength of a spot weld.
(o) The percentage of alcohol in successive batches of wine.

16.6. A normally distributed measurable characteristic is under the surveillance of an $\bar{X}$ chart. Samples of four items are taken at regular intervals. Compute the control limits such that the probability of a false alarm is .05.

16.7. Suppose that 3-sigma control limits of an $\bar{X}$ chart have been set at 28.6 (L.C.L.) and 31.8 (U.C.L.) on a certain operation in a foundry for samples of nine items taken at regular intervals.
(a) Compute the estimate of σ'.
(b) If the process is in control and the specification limits for the item are 27.70–35.20 inclusive, what percent defective will be produced? Assume normality.

(c) If it has been decided to reduce the sample size to four, what should the new control limits be?

16.8. Given the following table of 20 samples of 4 observations each on the outcomes of a process:

Sample Number	1	2	3	4	5	6	7	8	9	10
$\bar{X}$	.173	.175	.178	.177	.170	.171	.169	.172	.177	.176
R	.011	.013	.007	.006	.013	.008	.006	.013	.012	.009
Sample Number	11	12	13	14	15	16	17	18	19	20
$\bar{X}$	.166	.168	.171	.173	.167	.167	.170	.175	.175	.174
R	.012	.015	.007	.008	.008	.004	.016	.010	.008	.014

(a) Is the process probably in control?

(b) If the specification of this product characteristic is $.175^{\pm.005}$ in., what percentage of the product might be expected to conform to specification provided that this process remains in control? Assume normality.

(c) What do you suggest for reducing the percentage defective?

16.9. Control charts are maintained for $\bar{X}$ and σ on the diameter of a motor shaft. The subgroup size is five. After 30 subgroups, $\sum \bar{X} = 60.88$ and $\sum \sigma = .018$. Compute the values of the 3-sigma control limits for the $\bar{X}$ and σ charts.

16.10. In Prob. 16.9 assume that $\bar{X}'$ and σ' are equal to 2.00 and .007, respectively. What should the control limits be?

16.11. A process is under the surveillance of an $\bar{X}$ chart by taking samples of four items at regular intervals. The standard deviation is stable throughout the process and is equal to .06. The mean is equal to 2.15; however, it occasionally sustains large shifts.

(a) Compute 2.5-sigma limits for the $\bar{X}$ chart.

(b) If the mean shifts by $2\sigma'$, what is the probability of catching the shift on the first sample taken after the shift has occurred? Assume normality.

16.12. The specification of a hole punched in a brass disk is $.250^{+.005}$ in. Control charts for $\bar{X}$ and R are maintained by taking samples of five disks at hourly intervals. After 30 subgroups, $\sum \bar{X} = 7.6173$ and $\sum R = .067$.

(a) Compute the control limits for the $\bar{X}$ and R charts.

(b) If oversized disks are scrapped, what percentage might be expected to be scrapped, provided that the process remains in control? Assume normality.

(c) What do you suggest to make this scrap exactly .1%?

16.13. A normally distributed quality characteristic is under the surveillance of $\bar{X}$ and R charts. The process is subject to the occurrence of an assignable cause of variation that shifts $\bar{X}'$ to a new level $\bar{X}' + 2\sigma'$ but does not change σ'. If the subgroup size is four, compute the following:

(a) The probability that a sample average will fall outside the control limits after the shift has occurred.

(b) The probability that a sample range will fall outside the control limits after the shift has occurred.

16.14. The quality characteristic of a certain product appears to be in statistical control with an $\bar{X}'$ of 20 and a σ' of 5. The standard deviation is stable throughout the process, whereas the mean occasionally sustains a sudden shift. The magnitude of this shift is $-1.5\sigma'$. It is required to design a control chart to meet the following requirements:

(a) The probability of a false alarm is .005.

(b) The probability of catching the shift at the first sample after it has occurred is .88. Compute the control limits.

16.15. The following table gives the results of daily inspection of vacuum cleaners. Draw a p chart and state your conclusions.

Date (February)	*Number Inspected*	*Number of Defectives*	*Date (February)*	*Number Inspected*	*Number of Defectives*
3	200	18	15	100	7
4	200	16	16	100	6
5	200	13	17	100	8
6	200	10	18	250	18
7	200	21	19	250	16
9	300	24	21	250	22
10	300	21	22	300	27
11	200	14	23	300	18
12	250	20	24	300	12
13	200	12	25	300	27

16.16. A manufacturer purchases small, standard plastic gears in kegs that usually contain several thousand gears. Each shipment consists of 20 kegs. As part of the acceptance procedure for these gears, 500 are selected at random from each keg and inspected for several possible sources of defectiveness. In a certain shipment the respective percentages of defectives in the samples were found to be: 1.6, 2.2, 1.0, .8, 0, 1.2, 2.4, 2.6, 2.0, 1.8, 0, 1.2, 2.8, 0, .8, .6, 1.2, 1.4, 2.2, and 2.0.
(a) Draw an appropriate control chart for these data.
(b) What do you conclude?

16.17. The following data represent the number of defective carbon film resistors produced for a daily production of 5,000 units:

Day:	1	2	3	4	5	6	7	8	9	10	11	12
Defectives:	108	107	76	132	118	79	101	89	74	79	109	98

(a) Compute the trial control limits for a p chart.
(b) What do you recommend as the value of p' for future production?

16.18. A sample of 100 vacuum tubes is drawn from the production line at regular intervals. The number of defective tubes found in 20 successive samples was as follows:

Sample number:	1	2	3	4	5	6	7	8	9	10
Defectives:	1	4	2	3	0	2	3	5	2	0
Sample number:	11	12	13	14	15	16	17	18	19	20
Defectives:	3	2	3	3	0	5	2	3	0	2

(a) Is the process probably in control?
(b) Can the manufacturer guarantee to supply 2% or less defective lots provided that the process remains in control?

16.19. The p chart for a certain product has shown excellent statistical control with respect to the quality characteristics examined. The process average fraction defective $100p'$ is 1.8.
(a) If the subgroup size is 100, compute 3-sigma control limits.
(b) Use the Poisson approximation to find the probability of a false alarm.

16.20. The average fraction defective of a controlled process has been shown to be .03. Your control chart calls for taking daily samples of 400 items. What is the probability that if the process average should suddenly shift to .06 you would catch the shift
(a) On the first sample?
(b) On or before the second sample taken after the shift has occurred?

16.21. Draw an np chart for the data given in Prob. 16.16, and state your conclusions.

16.22. For the data given in Prob. 16.18, determine the following:
(a) Compute the trial control limits for an *np* chart.
(b) If the estimate of p' remains unchanged, what is the probability that the twenty-first sample will contain exactly four defectives?

16.23. In the inspection of vinyl tablecloth material a sample of 20 yd was taken at regular intervals. The following table gives the results of inspection:

Sample number:	1	2	3	4	5	6	7	8	9	10
Number of defects:	8	4	6	2	3	2	4	2	8	5
Sample number:	11	12	13	14	15	16	17	18	19	20
Number of defects:	2	1	0	6	8	3	4	2	5	10

(a) Draw a control chart for c.
(b) What value of c' would you suggest for future periods?

16.24. Find .995 and .005 probability limits for a c chart when $c' = 8.50$.

16.25. A control chart on the number of flaws per airplane wing showed excellent control with $c' = 12.0$
(a) Compute 3-sigma control limits.
(b) What is the probability of a point falling outside the control limits?

SUGGESTED REFERENCES

[1] Burr, I. W., *Engineering Statistics and Quality Control*, New York, McGraw-Hill Book Company, 1953.

[2] Burr, I. W., "The Effects of Non-Normality on Constants for $\bar{X}$ and R Charts," *Industrial Quality Control*, **23**(11), 1967.

[3] Cowden, D. J., *Statistical Methods in Quality Control*, Englewood Cliffs, N.J., Prentice-Hall, Inc., 1957.

[4] Duncan, A. J., *Quality Control and Industrial Statistics*, 3rd ed., Homewood, Ill., Richard D. Irwin, Inc., 1965.

[5] Grant, E. L., and R. Leavenworth, *Statistical Quality Control*, 4th ed., New York, McGraw-Hill Book Company, 1972.

[6] Hald, A., *Statistical Theory with Engineering Applications*, New York, John Wiley & Sons, Inc., 1952.

[7] Hansen, B. L., *Quality Control: Theory and Applications*, Englewood Cliffs, N.J. Prentice-Hall, Inc., 1963.

[8] Johnson, N. L., and F. C. Leone, "Cumulative Sum Control Charts," *Industrial Quality Control*, **18,** 19, 1962.

[9] Juran, J. M. (ed.), *Quality Control Handbook*, 2nd ed., New York, McGraw-Hill Book Company, 1962.

[10] Littauer, S. B., "Social Aspects of Scientific Method in Industrial Production," *Philosophy of Science*, **21**, No. 2, 1954.

[11] Littauer, S. B., "Technological Stability," *Transactions of New York Academy of Science*, 1950.

[12] Shewhart, W. A., *Economic Control of Quality of Manufactured Product*, Princeton, N.J., D. Van Nostrand Co., Inc., 1931.

[13] Shewhart, W. A., *Statistical Method from the Viewpoint of Quality Control*, Washington, D.C., The Graduate School, Department of Agriculture, 1939.

[14] Western Electric Company, *Statistical Quality Control Handbook*, 2nd ed., Western Electric Company, Inc., New York, 1958.

17

ACCEPTANCE SAMPLING

17.1 Introduction

Most likely the manufacturer who buys his product (parts, subassemblies, material, etc.) in lots of considerable size, from one or more suppliers, desires to know whether the quality characteristic within each lot conforms to his specification. Obviously the manufacturer would like to accept submitted lots if their percent defective does not exceed the specified acceptable quality level. Therefore, each lot must be inspected to determine whether it is acceptable. More precisely, if a lot of size N items is submitted, every item in the lot will be inspected and classified as defective or satisfactory. The lot will be accepted if the number of defective items in the lot is less than or equal to an allowable number; otherwise it will be rejected. When the lot size N is exceptionally large, 100% inspection will be costly and time-consuming. Moreover, 100% inspection may not be feasible or advisable on the following grounds:

1. The loss incurred due to a defective item is very low. In some cases no inspection at all is the most economical course of action.
2. 100% inspection is impossible when inspection is destructive. For instance, a lot of small caliber ammunition is accepted as satisfactory if 99% of the shots fall within a specified distance from the center of a target at a given range. Hence the decision to accept or reject the lot will be reached after destroying the entire lot.

3. 100% inspection is not 100% perfect since manual or mechanical inspection is subject to some margin of error.

In light of the previous discussion, can a receiver use better inspection to assure the quality of product or work submitted by a producer? The answer is to use *acceptance sampling plans*. That is, the decision to accept or reject a lot will be based on a series of samples drawn at random from the submitted lot. Sampling plans are not only economical but also are as effective as 100% inspection. In many instances a well-designed sampling plan may produce better results than 100% inspection.

It should be borne in mind that an acceptance sampling plan may accept occasional lots with a much higher fraction of defective items than the consumer is willing to tolerate. However, if submitted lots differ in quality, the sampling plan will accept the good lots more frequently than the bad lots, and as a result a long-range average quality level, consistent with the quality specified, can be maintained.

Sampling plans may be based on two different kinds of measurements. Inspection may be performed by grading the product as defective or non-defective or as good or bad, e.g., checking the size of cylindrical male parts by go and not-go ring gages. Inspection also may be performed by measuring the product to verify whether it conforms to specification, e.g., measuring the pitch diameter of a screw with a thread micrometer. When related to sampling inspection, the first is known as sampling by *attributes* whereas the second is known as sampling by *variables*. In general, inspection by attributes is less expensive than by variables. However, inspection by variables is more informative than attributes and requires smaller sample size to provide the same protection against accepting lots of poor quality.

SAMPLING BY ATTRIBUTES

17.2 Single Sampling Plans

A sampling plan based on one sample is called a *single sampling plan*. It is characterized by two numbers (n, c), where n is the sample size and c is the acceptance number.

A sample of size n is drawn from the lot and inspected by attributes. The lot is accepted if the number of defectives (d) in the sample does not exceed the acceptance number (c). That is, accept the lot if $d \leq c$ and reject the lot if $d > c$. Now on what basis can one determine the values of n and c? Obviously the optimal selection of n and c should be based on economic considerations. However, the formulation of an economic model which includes relevant cost parameters is complicated. Therefore the values of n and c are determined so that the sampling plan will discriminate between good and bad lots with specified odds for any level of fraction defective in the submitted lots.

17.2.1 The Operating Characteristic (OC) Curve—Lot Quality

Let p denote the fraction defective in a submitted lot. Suppose that the consumer will accept a submitted lot if its fraction defective is less than or equal to 1% and invariably will reject a lot of poorer quality. A plan that would discriminate perfectly between lots with $p \leq 1\%$ and lots with $p > 1\%$ would have the *operating characteristic* (*OC*) curve shown in Fig. 17.1.

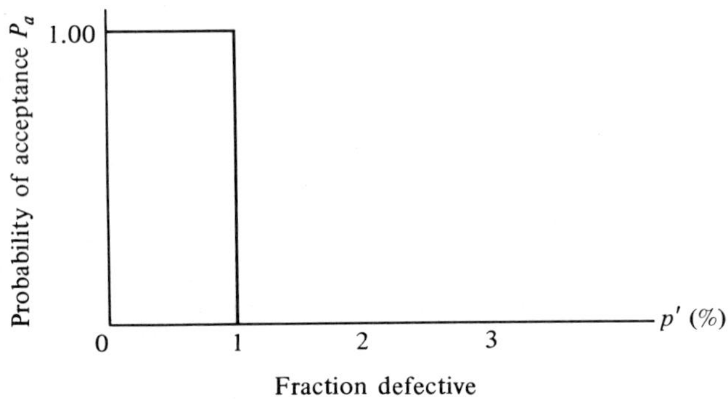

Fig. 17.1 Ideal *OC* curve for a sampling plan.

This ideal *OC* curve can be achieved only with 100% inspection, provided that 100% inspection is infallible. Unfortunately no sampling plan will have an ideal *OC* curve as such. A well-designed sampling plan, however, can approach such a curve. Now if the consumer will reject a submitted lot whenever its fraction defective exceeds 1% (using 100% inspection), the producer will have to screen the rejected lot to eliminate defectives. This means that both the consumer and producer will sustain excessive inspection cost. Consequently, it seems necessary to seek a more realistic approach to this problem, an approach by which it would be feasible to reduce the prohibitive cost of inspection. This dilemma has been solved by instituting acceptance sampling plans.

In practice, the producer and consumer reach an agreement on a sampling plan that is fair to both. Obviously the consumer wants to protect himself against accepting a poor quality lot having a sizable fraction of defectives. He must define the risk he is willing to take in having a poor quality lot accepted by the sampling plan. In other words, the consumer specifies the probability of the sampling plan accepting a lot that has a fraction defective p_2. This probability is usually denoted by β. Similarly, the producer specifies the probability of the sampling plan rejecting a lot that has a fraction defective p_1. This probability is usually denoted by α. Once the consumer and producer have come to agreement on the values of α, β, p_1, and p_2, a sampling

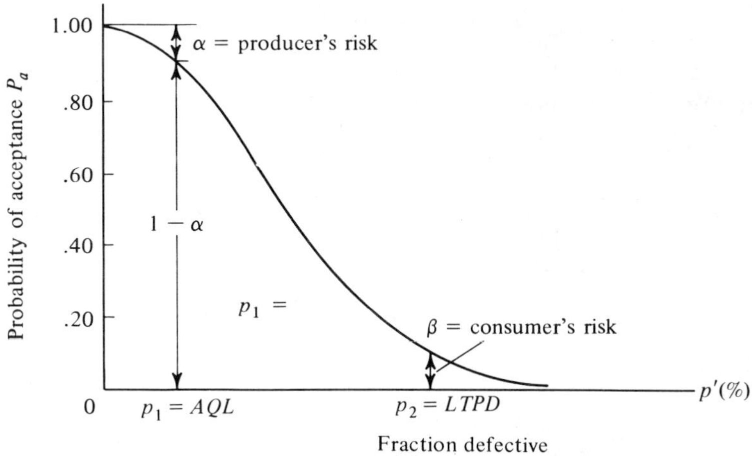

Fig. 17.2 *OC* curve for a single sampling plan.

plan is determined. The *OC* of this sampling plan *should* pass through the two points (p_1, α) and (p_2, β), as shown in Fig. 17.2. The following nomenclature will be adopted for these points:

α = producer's risk
β = consumer's risk
p_1 = acceptable quality level (denoted by AQL)
p_2 = lot tolerance percentage defective (denoted by LTPD) or sometimes called rejectable quality level (denoted by RQL)

The area between the AQL and LTPD is known as the *indifference zone*. From Fig. 17.2 it can be seen that if the quality of the submitted lot is better than p_1, the lot will be accepted with probability greater than $(1 - \alpha)$; if worse than p_2, the lot will be accepted with probability less than β. Thus the *OC* curve for any sampling plan will give the probability with which the plan will discriminate between good and bad or acceptable and unacceptable lots for any level of fraction defective.

Let us now derive the probability of accepting a lot submitted with fraction defective p'. If the incoming lot size is N and we are sampling without replacement, then the probability distribution of the number of defectives (k) in a sample of size n is hypergeometric. In symbols,

$$P_X(k) = \frac{\binom{Np'}{k}\binom{N - Np'}{n - k}}{\binom{N}{n}} \tag{17.1}$$

In practice, the lot size N runs into hundreds, thousands, or even larger. In Sec. 6.19.4 we proved that the hypergeometric distribution with parameters n, p', and N approaches the binomial distribution with parameters n and p. Consequently, Eq. (17.1) can be written

$$P_X(k) \cong \binom{n}{k} p'^k(1 - p')^{n-k} \tag{17.2}$$

We proved in Sec. 6.2 that if $n \longrightarrow \infty$ and $p' \longrightarrow 0$ so that $np' = \lambda$, the limiting distribution of the binomial is a Poisson. Thus Eq. (17.2) becomes

$$P_X(k) \cong \frac{e^{-\lambda}\lambda^k}{k!} \tag{17.3}$$

where $\lambda = np'$.

Now if the sampling plan is specified by (n, c) and the lot quality is p', then the probability of accepting the lot is

$$P_a \cong \sum_{k=0}^{k=c} \frac{e^{-\lambda}\lambda^k}{k!} = \sum_{k=0}^{k=c} \frac{e^{-np'}(np')^k}{k!} \tag{17.4}$$

Equation (17.4) can be evaluated by using Table I in the Appendix. The product $np' = \lambda$ is used to enter Table I; in the column headed $k = c$, to find the P_a value. The following example will illustrate the use of Table I:

Example 17.1 A single sampling plan uses a sample size of 40 and an acceptance number of 1. The lot size is large in comparison with sample size. Use Table I to compute the probabilities of acceptance of lots .5, 1, 2, 3, 4, 5, 6, 7, 8, and 10% defective. Plot the *OC* curve for the sampling plan.

Solution. Here we have a sampling plan with $n = 40$, $c = 1$. That is, a sample of 40 items is drawn from the lot and inspected. The lot is accepted if the sample contains not more than one defective. If $p' = .5\%$, then $\lambda = .20$. The probability of accepting .5% defective lot is

$$P_a(p' = .5\%) \cong \sum_{k=0}^{k=1} \frac{e^{-.20}(.20)^k}{k!} = .982$$

(Note that .982 is read out of Table I with entries $\lambda = .2$ and $k = 1$.) Similarly, we can compute the values of P_a for the specified values of p', which are tabulated below:

p'	.5%	1%	2%	3%	4%	5%	6%	7%	8%	10%
P_a	.982	.938	.809	.663	.525	.406	.308	.231	.171	.092

The *OC* curve for the sampling plan is shown in Fig. 17.3.

Note that the probability of accepting a lot given a specified lot quality (p') depends solely on the sample size (n) and the acceptance number (c). Thus the two numbers n and c completely determine the *OC* curve. Let us now study the effect of n and c on the shape of the *OC* curve. Suppose in

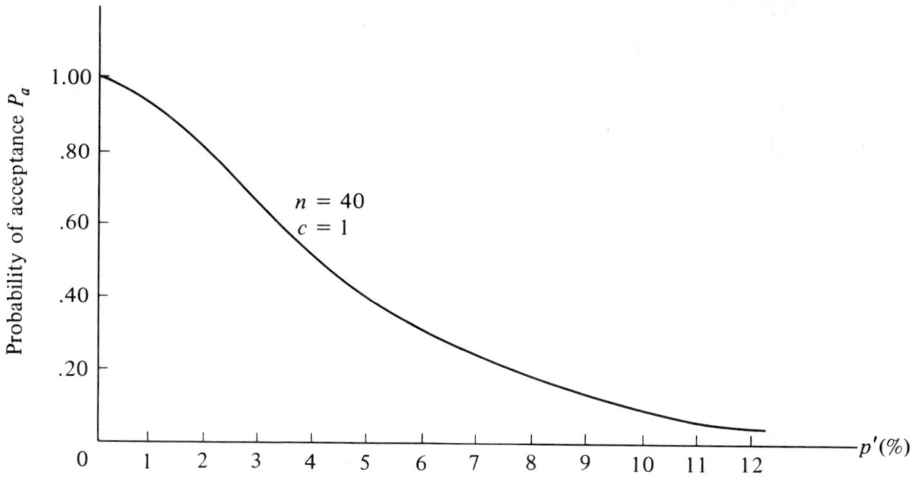

Fig. 17.3 *OC* curve for Example 17.1.

Example 17.1 that we keep the sample size ($n = 40$) constant and change the acceptance number c. The *OC* curves resulting from changing the acceptance number alone are shown in Fig. 17.4.

If we keep the acceptance number ($c = 1$) constant and change the sample size n we obtain the *OC* curves shown in Fig. 17.5. From Figs. 17.4 and 17.5 we draw the following conclusions:

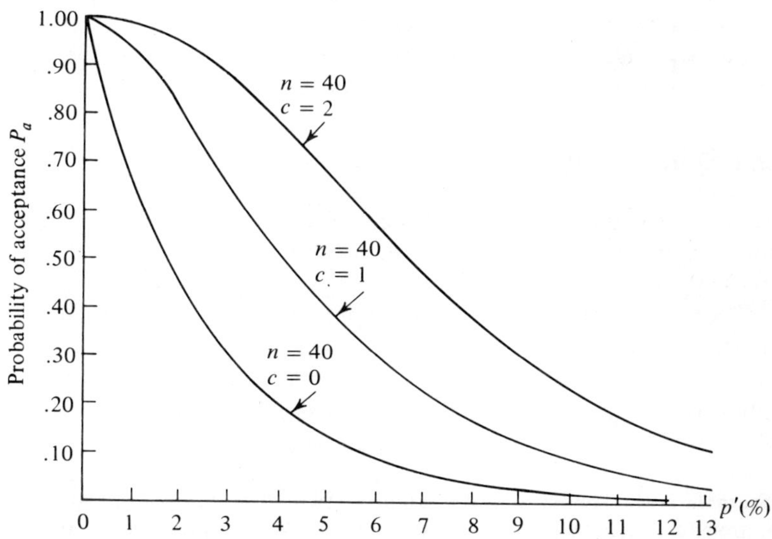

Fig. 17.4 Comparison of *OC* curves with different acceptance numbers.

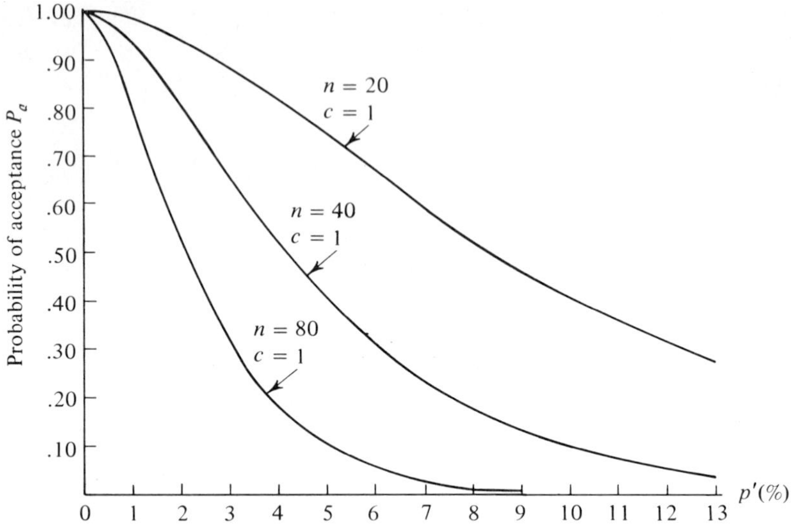

Fig. 17.5 Comparison of *OC* curves with different sample sizes.

1. For the same sample size, increasing the c value serves to move the curve farther from the origin. Thus for a fixed sample size the sampling plan would give better discrimination among lots of different quality if the acceptance number is reduced. To illustrate, suppose a 4% defective lot is submitted for inspection; the lot would be accepted 20% of the time if $c = 0$, whereas the same lot would be accepted 78% of the time if $c = 2$.
2. For the same acceptance number, increasing the sample size causes the slope of the *OC* to become steeper. The steeper the curve, the better the protection against accepting lots of poorer quality.

17.2.2 Determination of Sampling Plan

Assume that the producer and the consumer have agreed to use a single sampling plan for attributes that will protect specified values of AQL and LTPD with specified values of α and β, respectively. What is now needed is an *OC* curve that will pass through the points (AQL, $1 - \alpha$) and (LTPD, β). This *OC* curve is uniquely determined by the numbers n and c. It should be noted that since n and c can take on integral values only, it is usually not possible to find an *OC* curve that will pass through these points exactly; however, it is possible to find a curve that will closely approach these points.

Example 17.2 Devise a single sampling plan that will provide the following protection: $\alpha = .05$, AQL $= p_1 = .02$ and $\beta = .05$, LTPD $= p_2 = .08$.

Solution. To find the sample size n and the acceptance number c, we first assume that $c = 0$ and then find the ratio p_2/p_1. If this ratio is equal to $.08/.02 = 4$, then the acceptance number of the sampling plan is 0 and the corresponding sample size can be determined. If $p_2/p_1 > 4$ for $c = 0$, we try $c = 1$. We continue in this way until we find the value of c that yields a value for p_2/p_1 equal or closest to 4. To illustrate, if $c = 0$, then

$$P_a(p_1) = e^{-np_1} = .95 \tag{17.5}$$

whence $np_1 = .05$. Similarly, $P_a(p_2) \equiv e^{-np_2} = .05$, whence $np_2 = 3.0$. Proceeding in this way we obtain the following results:

c	$np_1(1 - \alpha = .95)$	$np_2(\beta = .05)$	$\frac{p_2}{p_1}$
0	.05	3.0	60.0
1	.35	4.8	13.7
2	.80	6.3	7.9
3	1.36	7.8	5.7
4	1.95	9.17	4.7
5	2.61	10.5	4.02
6	3.20	11.9	3.71

Therefore, $c = 5$ and $n = 2.61/.02 \approx 131$ appear to correspond closely with the requirements. Actually, we obtain the following protection with $n = 131$ and $c = 5$

$$\text{AQL} = .02 \qquad \text{LTPD} = .08$$
$$\alpha = .0494 \qquad \beta = .05$$

which is close indeed to the requirements.

The same result can be obtained by using the Peach–Littauer [14] method. The following table was developed by P. Peach and S. B. Littauer for designing single sampling schemes ($\alpha = \beta = .05$):

R_0	c	np_1
58.	0	.05
12.	1	.36
7.5	2	.82
5.7	3	1.37
4.6	4	1.97
4.0	5	2.61
3.6	6	3.29
3.3	7	3.98
3.1	8	4.70
2.7	10	6.17
2.37	14	9.25
2.03	21	14.89
1.81	30	22.44
1.61	47	37.20
1.51	63	51.43
1.335	129	111.83
1.251	215	192.41

Directions for use of table:

1. Calculate $R_0 = \frac{P_2}{P_1}$.
2. Find R_0 in table. If it does not appear, use the next larger value shown.
3. Read directly the acceptance number c.
4. Davide np_1 by p_1 to get n, the sample size.

The authors [14] proved that if the number of defectives in a sample of n follows a Poisson law, then

$$\frac{\chi^2_{\alpha;\,2(c+1)}}{\chi^2_{1-\beta;\,2(c+1)}} = \frac{2np_1}{2np_2} = \frac{p_1}{p_2} \tag{17.6}$$

Applying Eq. (17.6) to Example 17.2 we obtain

$$\frac{\chi^2_{.05,\,2(c+1)}}{\chi^2_{.95,\,2(c+1)}} = \frac{1}{4} \tag{17.7}$$

From Table G in the Appendix we find that Eq. (17.7) is satisfied when $2(c+1) = 12$; i.e., $c = 5$ and the corresponding sample size can be obtained from the equations

$$\chi^2_{\alpha;\,2(c+1)} = 2np_1 \quad \text{or} \quad \chi^2_{1-\beta;\,2(c+1)} = 2np_2 \tag{17.8}$$

On substitution, Eq. (17.8) becomes

$$\chi^2_{.05;\,12} = 5.226 = 2n(.02) \quad \text{or} \quad \chi^2_{.95;\,12} = 21.026 = 2n(.08)$$

whence $n \cong 131$, which agrees with our previous finding.

17.3 Average Outgoing Quality

Sampling plans also may be specified according to the *quality level* of lots that leave the inspection point. Suppose that lots of size N are being subjected to a single sampling plan specified by n and c. Furthermore, suppose that lots of but one quality level p' are submitted for inspection. If inspection is nondestructive and the lot size is very large compared to the sample size, then the sampling plan will reject p'% defective lot with probability

$$1 - \sum_{k=0}^{c} \frac{e^{-np'}(np')^k}{k!}$$

Now if rejected lots are 100% inspected and the defectives are removed and replaced by nondefectives, none of these lots will be rejected by the sampling plan. These lots are called *rectified lots* and the inspection scheme is known as *rectifying inspection*. Thus lots accepted by the sampling plan will contain either (1) approximately the percent defective submitted (p') although they will be slightly improved by the replacement of any defectives found in the samples by nondefectives whenever $c \geq 1$ or (2) no defectives if rejected lots are rectified. This means that the average quality of the lots leaving the inspection point is a combination of 1 and 2. More precisely, the *average outgoing quality* (denoted by AOQ) leaving the inspection point is

$$\text{(17.9)} \qquad \text{AOQ} = (N - n)p' \sum_{k=0}^{c} \frac{e^{-np'}(np')^k}{k!} \Big/ N = p'P_a(p')\left(1 - \frac{n}{N}\right)$$

If for large N and small n the ratio n/N can be neglected, then Eq. (17.9) becomes

$$\text{(17.10)} \qquad \text{AOQ} \simeq p'P_a(p')$$

Suppose that the quality level of submitted lots is not stable, i.e., p' is not constant for all lots. Then, as the quality of lots subjected to the sampling plan improves, the probability of acceptance increases, and as a result less sorting and screening will be called for as more lots are accepted. On the other hand, as the quality of submitted lots becomes poorer, more lots are rejected and, as a result, more lots are screened. Consequently, the average outgoing quality will improve since rejected lots are rectified. Let us illustrate this situation graphically by considering Example 17.1. The values of average outgoing quality for various values of p' are as follows:

p'(%)	P_a	AOQ(%) $\simeq p'P_a(p')$
.2	.997	.199
.5	.982	.491
1	.938	.938
2	.809	1.618
3	.663	1.989
4	.525	2.100
5	.406	2.030
6	.308	1.848
7	.231	1.617
8	.171	1.368
9	.126	1.134
10	.092	.920
12	.048	.576
14	.024	.336

Figure 17.6 illustrates the variation of AOQ(%) with incoming quality p'(%). The maximum value of AOQ represents the *worst* expected quality that will leave the inspection point and is called the *average outgoing quality limit* (AOQL). From Fig. 17.6 the AOQL $\simeq$ 2.1%. Consequently, the average quality level of all lots subjected to the sampling plan ($n = 40$, $c = 1$) will not exceed 2.1%. It should be noted that the value of AOQL may be exceeded in a particular instance; however, in the long run the worst AOQ will not exceed the AOQL. Referring to Fig. 17.6 we find that the AOQL occurs at

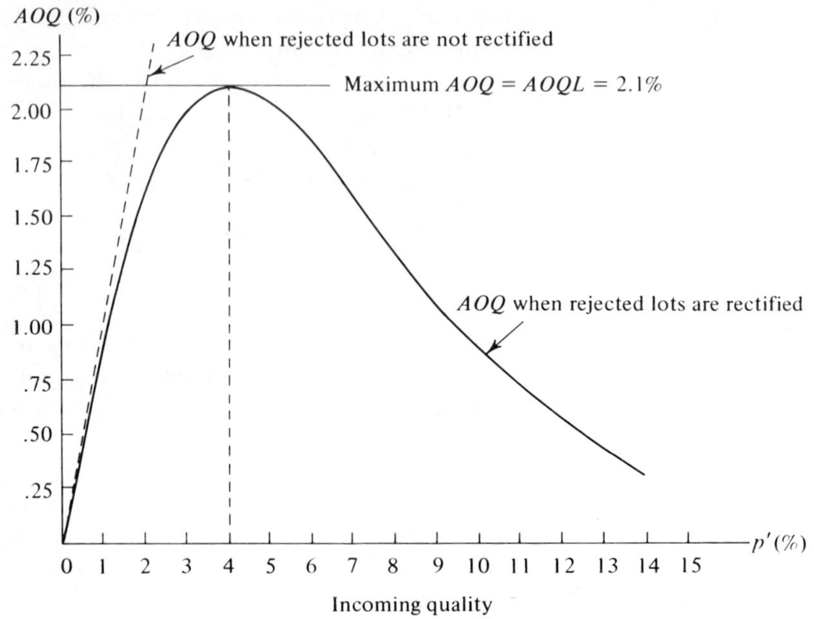

Fig. 17.6 Average outgoing quality for Example 17.1.

$p' = 4\%$. This means if the quality of submitted lots were 4%, the AOQ is equivalent to the worst average outgoing quality if the quality of submitted lots is higher or lower than 4%. If rejected lots are not rectified the accepted lots would leave the inspection point with the same quality level as submitted. A graph of this relationship would be a straight line, as shown in Fig. 17.6.

17.4 Double Sampling Plans

The *double sampling plan* is characterized by the sample sizes of two samples and their acceptance number $(n_1c_1; n_2c_2)$. A first sample of size n_1 is drawn from a lot of size N, and the items are inspected. If the number of defectives (d_1) found in the first sample is less than or equal to the first acceptance number c_1, the lot is accepted. If the number of defective items (d_1) in the first sample is greater than the second acceptance number c_2, the lot is rejected. If the number of defective items (d_1) in the first sample exceeds the first acceptance number c_1 but is less than or equal to the second acceptance number c_2, a second sample of size n_2 is drawn and the items are inspected. If the number of defectives $(d_1 + d_2)$ in the two samples combined is less than or equal to the second acceptance number c_2, the lot is accepted.

If the number of defectives $(d_1 + d_2)$ in the two samples combined exceeds the second acceptance number c_2, the lot is rejected. In symbols,

$$\text{First stage} \begin{cases} \text{accept the lot if } d_1 \leq c_1 \mid n_1 \\ \text{reject the lot if } d_1 > c_2 \mid n_1 \quad c_1 < c_2 \end{cases}$$

$$\text{draw second sample if } c_1 < d_1 \leq c_2 \mid n_1$$

$$\text{Second stage} \begin{cases} \text{accept the lot if } d_1 + d_2 \leq c_2 \mid n_1 + n_2 \\ \text{reject the lot if } d_1 + d_2 > c_2 \mid n_1 + n_2 \end{cases}$$

In general, double sampling requires less average inspection than single sampling. If the unit inspection cost is relatively high, considerable saving can be achieved with double plans. This is particularly true when the fraction defective of the submitted lots deviates too much from the specified AQL and the sample size is relatively large. Double sampling possesses psychological advantages over single sampling. It appears more convincing to reject a lot on the basis of the outcomes of more than one sample. Usually the producer objects to rejecting a lot on the basis of a single defective item found in the sample. Double sampling plans will not reject a lot when a single defective item has been found. However, single sampling plans are much easier to administer than double sampling. They provide better information on the process average of the lots being submitted. This is due to the fact that single sampling requires larger sample size than double sampling to provide the same protection.

Example 17.3 In a double sampling plan, $N = 5{,}000$, $n_1 = 20$, $c_1 = 0$, $n_2 = 40$, and $c_2 = 2$.

(a) Plot the OC curves for the sampling plan.
(b) Suppose that rejected lots will be rectified. What will be the AOQ if the submitted product is 1% defective?
(c) Considering both the inspection of samples and the inspection of rejected lots, what will be the average number of items inspected per lot if the submitted product is 5% defective?

Solution

(a) Suppose that the incoming lot is p'% defective. The probability of accepting the lot on the first sample is

$$P_{a_1}(p') = P\{d_1 \leq 0 \mid p', n_1\}$$

The probability of accepting the lot after n_2 items have been inspected is

$$P_{a_2}(p') = P\{d_1 = 1 \mid p', n_1\}P\{d_2 \leq 1 \mid p', n_2\} + P\{d_1 = 2 \mid p', n_1\} \times P\{d_2 = 0 \mid p', n_2\}$$

The probability of acceptance for the combined sample thus is

$$P_a(p') = P\{d_1 \leq 0 \mid p', n_1\} + P\{d_1 = 1 \mid p', n_1\}P\{d_2 \leq 1 \mid p', n_2\} + P\{d_1 = 2 \mid p', n_1\}P\{d_2 = 0 \mid p', n_2\}$$

For example, if $p' = .05$, then

$$P_a(.05) = P_{a_1}(.05) + P_{a_2}(.05) = e^{-1} + e^{-1}\sum_{k=0}^{1}\frac{e^{-2}2^k}{k!} + \frac{e^{-1}}{2}e^{-2}$$
$$= .368 + .368(.406) + .184(.135) = .542$$

Proceeding in the same way, the probabilities of acceptance for various values of p' are as follows:

p'	.01	.03	.05	.07	.09	.12
$P_{a_1}(p')$	.819	.549	.368	.247	.165	.091
$P_{a_2}(p')$	.164	.248	.174	.093	.044	.012
$P_a(p')$	.983	.797	.542	.340	.209	.103
$P\{d_1 > 2 \mid n_1, p'\}$	.001	.023	.080	.167	.269	.430

The last row of the above table gives the probabilities of rejection for different values of p' after n_1 items have been inspected. The *OC* curves for the given double sampling plan are shown in Fig. 17.7.

(b) In the case of rectifiable inspection, the average outgoing quality is given by

$$\text{AOQ}(p') = \frac{p'}{N}[(N - n_1)P_{a_1}(p') + (N - n_1 - n_2)P_{a_2}(p')]$$
$$= \frac{p'}{N}[(N - n_1)P_a(p') - n_2 P_{a_2}(p')]$$
$$\cong p'P_a(p')$$

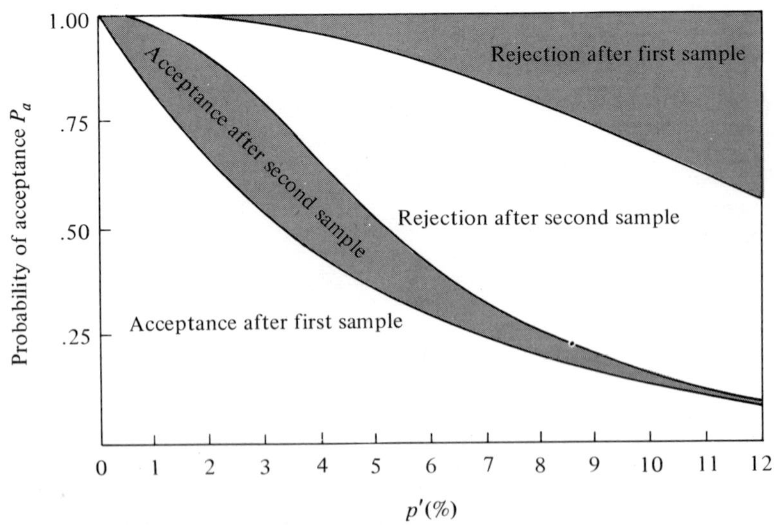

Fig. 17.7 *OC* curve for a double sampling plan for Example 17.3.

since for large lot size and small samples the ratios n_1/N and n_2/N could be neglected. Consequently,

$$\text{AOQ}(p' = .01) \cong .01(.983) = .00983$$

(c) The average total inspection (ATI) in a lot of size N when rejected lots are rectified would be

$$\text{ATI}(p') = n_1 + n_2 P\{c_1 < d_1 \leq c_2 \mid n_1, p'\} + (N - n_1)P\{d_1 > c_2 \mid n_1, p'\} + (N - n_1 - n_2)[1 - P_a(p') - P\{d_1 > c_2 \mid n_1, p'\}]$$

On substituting we obtain

$$\text{ATI}(.05) = 20 + 40 \sum_{k=1}^{2} \frac{e^{-1}}{k!} + (5000 - 20)(.08) + (5000 - 60)(1 - .542 - .08)$$
$$= 20 + 40(.552) + 398.4 + 1867.32 \cong 2308$$

The average total inspection increases as poorer lots are submitted for inspection.

The derivation of a double sampling plan with specified AQL, LTPD, α, and β is an arbitrary procedure. It is customary to determine the *OC* curve that provides the desired protection for a single sampling plan and then arbitrarily fix the ratio n_1/n_2 so that by trial and error c_1 and c_2 can be determined to satisfy the specified risks. It should be noted that on deciding on the ratio of n_1 to n_2, the first sample should provide appropriate discrimination for the fractions defective that will be presented for inspection. The first sample would not be effective if only a few lots with fractions defective in the range of discrimination provided by this sample are submitted for inspection.

Multiple sampling plans are extensions of double sampling plans. They are characterized by a series of samples $n_1, n_2, \ldots, n_k$, and their respective *acceptance* and *rejection* numbers $(c_1, r_1), (c_2, r_2), \ldots, (c_k, r_k)$. Note that $c_1 < c_2 < \cdots < c_k$ and $c_i < r_i$ $i = 1, \ldots, k$. The size of each sample in multiple sampling is smaller than that of single or double sampling plans. The following multiple sampling plan will be used to illustrate how the decision on acceptance or rejection is reached:

Sample Number	*Individual Sample Size*	*Combined Sample Size*	*Acceptance Number*	*Rejection Number*
1	n_1	n_1	c_1	r_1
2	n_2	$n_1 + n_2$	c_2	r_2
3	n_3	$n_1 + n_2 + n_3$	c_3	r_3
4	n_4	$n_1 + n_2 + n_3 + n_4$	c_4	r_4

A first sample of n_1 items is drawn; the submitted lot is accepted if the number of defective items (d_1) in the sample does not exceed c_1. The lot is rejected if r_1 or more defectives are found. If $c_1 < d_1 < r_1$, a second sample of n_2 items is drawn; the lot is accepted if the number of defective items $(d_1 + d_2)$ in the combined sample of $n_1 + n_2$ does not exceed c_2. The lot is rejected if r_2 or more defectives are found in the combined sample of $n_1 + n_2$. If $c_2 < d_1 + d_2 < r_2$, a third sample of n_3 items is drawn. The procedure is continued until a decision acceptance or rejection is reached. Usually a final decision is forced with the last sample (if the lot is neither accepted nor rejected by the end of $k - 1$ samples). Multiple sampling plans usually require better trained inspectors and adequate storage facilities to facilitate sampling. The administration of the sampling program for multiple plans usually involves higher costs than single or double sampling plans.

The main advantage of multiple plans is their low inspection cost. This is due to the fact that the average amount of inspection using multiple plans is smaller than for single or double plans providing the same protection.

17.5 Sequential Sampling Plans

Sequential sampling is similar to multiple sampling. However, in sequential sampling, after each unit has been inspected, *one of three* decisions must be made: accept, reject, or continue sampling. There is no specified limit on the total number of units to be inspected. Theoretically inspection could continue indefinitely. The sequential sampling plan is completely determined by p_1(AQL), p_2(LTPD), α, and β. Actually one requires to devise a statistical test to discriminate between two simple hypotheses

$$H_0 : p = p_1 \quad \text{against} \quad H_1 : p = p_2 > p_1$$

This discrimination can be accomplished by using the likelihood ratio (assuming that the distribution of the number of defective items is binomial)

$$\lambda_n = \frac{p_2^{d_n}(1 - p_2)^{n-d_n}}{p_1^{d_n}(1 - p_1)^{n-d_n}} \tag{17.11}$$

where n = the size of cumulative sample and d_n = the cumulative number of defective units. Now consider a testing procedure defined by the following rules:

1. If $\lambda_n \geq B$, stop sampling and accept H_1 (reject H_0).
2. If $\lambda_n \leq A$, stop sampling and accept H_0 (reject H_1).
3. If $A < \lambda_n < B$, draw $(n + 1)$st observation.

The constants A and B are chosen to make the probability of types I and II errors equal to α and β, respectively. Exact values of A and B are difficult to determine. However, the following approximate formulas are found to be satisfactory:

$$A \cong \frac{\beta}{1 - \alpha} \tag{17.12}$$

$$B \cong \frac{1 - \beta}{\alpha} \tag{17.13}$$

Equations (17.12) and (17.13) provide good approximations when both α and β are less than one-half. Accordingly, sampling continues as long as

$$\frac{\beta}{1-\alpha} < \lambda_n < \frac{1-\beta}{\alpha}$$

i.e., as long as

$$\frac{\beta}{1-\alpha} < \left(\frac{p_2}{p_1}\right)^{d_n}\left(\frac{1-p_2}{1-p_1}\right)^{n-d_n} < \frac{1-\beta}{\alpha} \tag{17.14}$$

On taking logarithms, (17.14) becomes

$$\ln\frac{\beta}{1-\alpha} < d_n \ln\frac{p_2}{p_1} + (n-d_n)\ln\frac{1-p_2}{1-p_1} < \ln\frac{1-\beta}{\alpha} \tag{17.15}$$

If we plot in the (n, d_n) plane the lines

$$d_n \ln\frac{p_2}{p_1} + (n-d_n)\ln\frac{1-p_2}{1-p_1} = \ln\frac{\beta}{1-\alpha} \tag{17.16}$$

and

$$d_n \ln\frac{p_2}{p_1} + (n-d_n)\ln\frac{1-p_2}{1-p_1} = \ln\frac{1-\beta}{\alpha} \tag{17.17}$$

[Equation (17.16) is known as the equation of the *acceptance line*, and Eq. (17.17) is known as the equation of the *rejection line*.] then as long as the sequence of sample points (n, d_n), $n = 1, 2, \ldots$ fall between the acceptance and rejection lines sampling is continued. As soon as a sample point (n, d_n) falls above the rejection line sampling is discontinued and H_0 is rejected (H_1 accepted). As soon as a sample point (n, d_n) falls below the rejection line sampling is discontinued and H_0 is accepted (H_1 rejected). The following example will illustrate this procedure:

Example 17.4

(a) Devise an item-by-item sequential plan that will give the following protection:

$$\alpha = .05 \qquad p_1(\text{AQL}) = .10$$
$$\beta = .20 \qquad p_2(\text{LTPD}) = .20$$

(b) Determine whether you would accept or reject a lot on the basis of the following observations:

ggbgg bgggb gbggb gggbb

(*b* denotes a defective item and *g* denotes a good item.)

Solution

(a) First we compute the following:

$$\ln\frac{p_2}{p_1} = \ln 2 = .693$$

$$\ln\frac{\beta}{1-\alpha} = \ln\frac{.20}{.95} = -1.558$$

$$\ln \frac{1-\beta}{\alpha} = \ln \frac{.80}{.05} = 2.772$$

$$\ln \frac{1-p_2}{1-p_1} = \ln \frac{.80}{.90} = -.118$$

This gives the following equations:

Acceptance line: $.811d_n - .118n + 1.558 = 0$

Rejection line: $.811d_n - .118n - 2.772 = 0$

These lines can be converted into an item-by-item table of acceptance and rejection numbers as shown below:

n	*Acceptance Number*	*Rejection Number*	n	*Acceptance Number*	*Rejection Number*
1	†	‡	15	0	6
2	†	‡	16	0	6
3	†	‡	17	0	6
4	†	‡	18	0	7
5	†	5	19	0	7
6	†	5	20	0	7
7	†	5	21	1	7
8	†	5	22	1	7
9	†	5	23	1	7
10	†	5	24	1	7
11	†	6	25	1	8
12	†	6	26	1	8
13	†	6	27	2	8
14	0	6	...	...	...

†Acceptance requires a sample of at least 14 items.
‡Rejection requires a sample of at least 5 items.

Note that the acceptance number is the next integer below d_n whereas the rejection number is the next integer above d_n.

(b) The lot would be rejected since the cumulative number of defective units is equal to the rejection number for $n = 20$. In Fig. 17.8, d_n is plotted against n and the plan is represented by the two parallel lines. The result of each observation is recorded by a line one unit to the right if the item is good and a line drawn one unit up if the item is defective. Figure 17.8 shows that sampling is stopped at the twentieth observation and that the lot would be rejected.

An operating characteristic curve for this plan can be constructed from five points. Two of these are points used in designing the plan [(AQL, $1 - \alpha$), (LTPD, β)]. Two additional points are ($P_a = 1$, $p' = 0\%$ and $P_a = 0$, $p' = 100\%$). A fifth point can be obtained from the relationship that when p' = slope of acceptance or rejection line. The slope is given by

$$\frac{\ln [(1-p_1)/(1-p_2)]}{\ln [p_2(1-p_1)/p_1(1-p_2)]} = \frac{.118}{.811} = .145$$

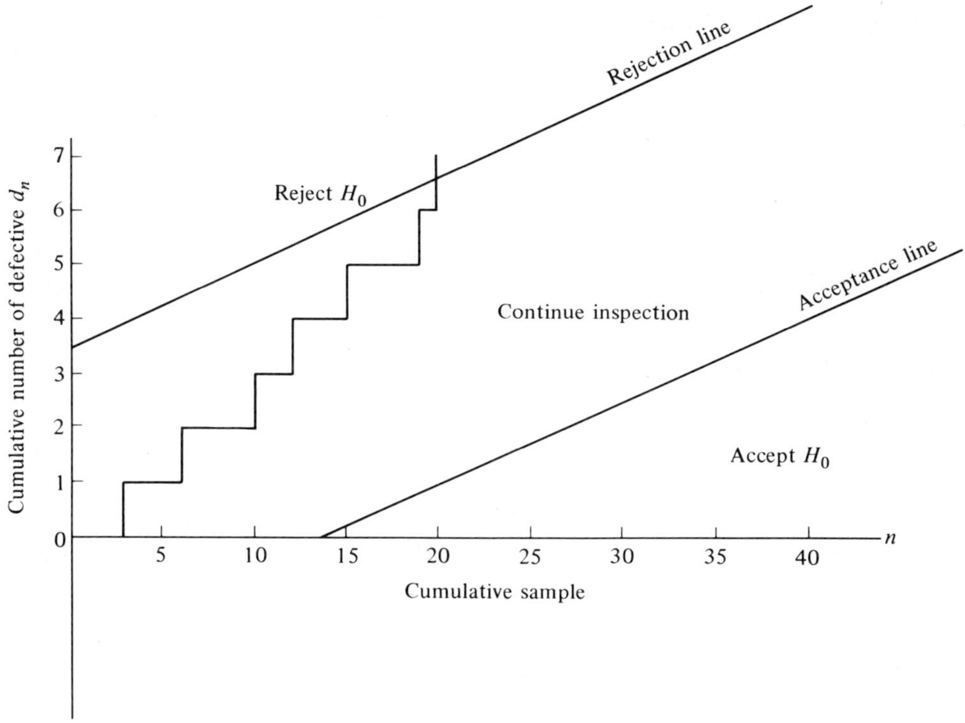

Fig. 17.8 Sequential sampling plan for Example 17.4.

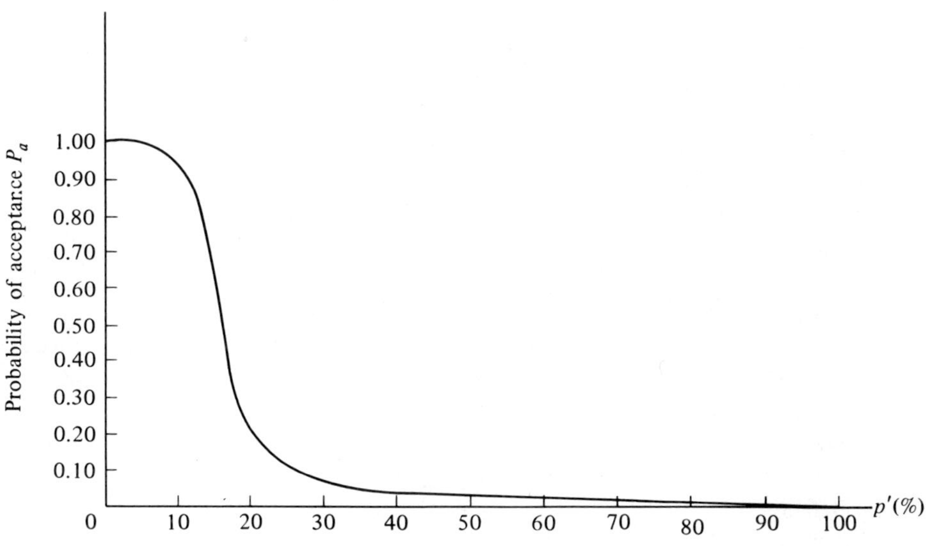

Fig. 17.9 *OC* curve for a sequential sampling plan for Example 17.4.

The probability of acceptance, then, is

$$P_a(.145) = \ln \frac{1-\beta}{\alpha} \bigg/ \left[\ln \frac{1-\beta}{\alpha} + \ln \left(\frac{1-\alpha}{\beta} \right) \right] = \frac{2.772}{2.772 + 1.558} = .64$$

The *OC* curve for this plan is shown in Fig. 17.9.

Sequential sampling is usually adopted in order to reduce inspection cost. The inspection cost tends to be lower for sequential sampling since it requires less average amount of inspection than multiple sampling, giving about the same quality protection.

17.6 Other Types of Attributes Plans

Volumes of journals have been filled with various notions on the attributes sampling plans. Obviously it is not a feasible task to embrace this body of knowledge here. However, we shall briefly describe some of the most important sampling plans that have achieved popularity in many industries.

17.6.1 Military Standard 105D Sampling Plans

MIL-STD-105D[12], lot-by-lot sampling inspection by attributes, was adopted by the U.S. Department of Defense in 1963. This standard has an international character. It is designated as ABC-STD-105, having been prepared under the Tripartite Alliance (the standardization agreement between the United States, the United Kingdom, and Canada). The "ABC" designation is derived from the more commonly used "American–British–Canadian" terminology. The ABC standard is based on the AQL concept. If the AQL value to be maintained has been specified and the lot size is known, the ABC standard provides the appropriate sampling plan to be used. The ABC standard has the following features:

1. The relationship between lot size and sample size is based on empirical grounds rather than statistical grounds.
2. It provides data for single, double, and multiple sampling plans.
3. It calls for shifting from normal inspection to tightened or reduced inspection as evidence has been established that poorer or better quality lots were submitted.
4. An attempt was made to match *OC* curves as closely as practicable for single, double, and multiple sampling plans.
5. Defects are classified as critical, major, and minor.
6. It protects the producer against the rejection of submitted lots with AQL or better. The consumer's protection, though undefined, can be obtained from the *OC* curve of the sampling plan.

17.6.2 Dodge–Romig Sampling Plans

H. F. Dodge and H. G. Romig [5] devised sampling plans to protect the consumer against acceptance of poor quality lots at a minimum total inspection cost. If the incoming process average and lot size are known, the sampling plan associated with it will minimize total inspection provided that rejected lots will be rectified. If the incoming process average is unknown the rightmost columns of the tables are used until information on the process average accumulates and consequently a more economical plan can be chosen. Dodge–Romig tables contain single and double sampling plans for two acceptance criteria—LTPD and AOQL. The single and double sampling LTPD tables provide sampling plans for the following values of LTPD (.5, 1.0, 2.0, 3.0, 4.0, 5.0, 7.0, and 10.0%), with β set at 10.0%.

17.6.3 Continuous Production Sampling Plans

A sampling plan for individual units manufactured in quantity by an essentially continuous process was devised by H. F. Dodge [4]. His CSP-1 plan is characterized by two numbers, f and i. The procedure is as follows:

1. At the outset, inspect 100% of the units produced and continue inspection until i units in succession are found without defects.
2. When i units have been found free of defects, discontinue 100% inspection and inspect only a fraction f of the units by selecting them one at a time from the flow of product.
3. If a sample unit is found defective, revert to 100% inspection and continue until i units in succession are found free of defects.
4. Correct and replace with good units all defective units found.

The selection of a plan for any desired AOQL can be obtained from a graph in f, i, and p_2 (LTPD). Two modifications of CSP-1 were developed later by Dodge and Torrey [6] and are referred to as CSP-2 and CSP-3.

Other types of continuous sampling plans were developed by Lieberman and Solomon [11], Wald and Wolfowitz [19], Girshick [9], and Derman, Littauer, and Solomon [3].

ACCEPTANCE SAMPLING BY VARIABLES

17.7 Introduction

Sampling plans by *variables* serve the same purpose as sampling plans by attributes. However, in variables plans the quality characteristic is actually measured on a variable scale and the decision of acceptance or rejection of lots submitted for inspection is based on whether the measured characteristic satisfies certain criteria. The following are some advantages and limitation of acceptance sampling by variables:

1. It provides more information about the quality characteristic under study. This information can be utilized for quality improvement.
2. It provides protection equivalent to attributes plans with a smaller sample size.
3. Acceptance criteria must be applied separately to each quality characteristic. Consequently, the cost of acceptance inspection tends to be higher than sampling by attributes.
4. The cost of inspection by item is usually higher than by attributes.
5. The underlying distribution of the measured characteristic is assumed to be normal. The appropriateness of this assumption should be ascertained prior to applying the acceptance or rejection criteria and calculating the *OC* curve of the sampling plan.

There are many different plans involving acceptance sampling by variables. The following two classifications are usually used:

1. Sampling plans based on the mean when the population standard deviation is known and constant. These plans are referred to as "known-sigma plans" or "plans with known variability."
2. Sampling plans based on the mean where the population standard deviation is unknown. Such plans are referred to as "unknown-sigma plans" or "plans with unknown variability."

In the following sections we shall derive variables sampling plans with known variability. The derivation of other plans is beyond the scope of this book but may be found in the references given at the end of this chapter.

17.8 Single Sampling Plans by Variables—Sigma Known and Constant

Case 1. Single Specification Limit

Suppose that in a given situation the lower limit on shear strength for cotter pins to be acceptable is 47,500 psi. Furthermore, suppose that the standard deviation is known to be 2,500 psi and measurements are normally distributed. If the acceptance criterion is the fraction defective in submitted lots—i.e., a specimen is considered defective if its shear strength is below 47,500 psi—and the sampling plan has to provide the following protection:

$$\alpha = .05 \qquad \text{AQL} = p_1 = .02$$
$$\beta = .10 \qquad \text{LTPD} = p_2 = .13$$

then a sample of size $n = 41$ specimens and acceptance number $c = 2$ would provide the required protection. However, we can develop a more sensitive plan if we use the actual values of the shear strength of the specimens rather than the number of specimens with shear strength less than 47,500 psi. Now let us prove this assertion.

Since the lower limit in shear strength for cotter pins to be acceptable is 47,500 psi, the acceptable mean shear strength μ_A is obtained from the equation

$$.02 = \Phi(-2.055) = \Phi\left(\frac{47{,}500 - \mu_A}{2{,}500}\right)$$

whence $\mu_A = 52{,}638$ psi. Similarly, the rejectable mean shear strength μ_R is obtained from the equation

$$.13 = \Phi(-1.126) = \Phi\left(\frac{47{,}500 - \mu_R}{2{,}500}\right)$$

whence $\mu_R = 50{,}315$ psi. However, it is required that lots from a population whose mean shear strength is 50,315 psi should only be accepted 10% of the time whereas lots whose mean shear strength is 52,638 psi should be accepted 95% of the time. If a sample of size n is drawn from each lot and the sample mean $\bar{x}$ is computed, then the lot will be accepted with probability .10 if $\bar{x}$ is found to be greater than a prescribed constant k. Consequently, we must have

$$P\{\bar{x} > k \mid \mu_R\} = P\left\{\frac{\bar{x} - 50{,}315}{2{,}500/\sqrt{n}} > \frac{k - 50{,}315}{2{,}500/\sqrt{n}}\right\} = .10$$

and, hence

$$\frac{k - 50{,}315}{2{,}500/\sqrt{n}} = 1.282 \tag{17.18}$$

Similarly, a lot with mean shear strength 52,638 psi will be accepted with probability .95 if $\bar{x} > k$. Thus

$$P\{\bar{x} > k \mid \mu_A\} = P\left\{\frac{\bar{x} - 52{,}638}{2{,}500/\sqrt{n}} > \frac{k - 52{,}638}{2{,}500/\sqrt{n}}\right\} = .95$$

and, hence

$$\frac{k - 52{,}638}{2{,}500/\sqrt{n}} = -1.645 \tag{17.19}$$

The solution of Eqs. (17.18) and (17.19) yields the following values of n and k: $n = 10$ and $k = 51{,}333$ psi. Hence, our sampling plan reads as follows: Choose 10 specimens at random, and obtain the shear strength of each. Find the mean shear strength. If it is at least 51,333 psi, accept the lot; otherwise reject it. Thus we proved that attribute sampling requires larger sample size than variables sampling to provide the same reliability regarding the acceptance or rejection of a lot.

The *OC* curve for this plan can be computed from the equation

$$P_a(\mu) = P\{\bar{x} \geq 51{,}333 \mid \mu\} = 1 - \Phi\left(\frac{51{,}333 - \mu}{2{,}500/\sqrt{10}}\right) \tag{17.20}$$

For example, the probability of accepting a lot whose mean shear strength is 51,500 psi is

$$P_a(51{,}500) = 1 - \Phi\left(\frac{51{,}333 - 51{,}500}{2{,}500/\sqrt{10}}\right) = .583$$

Case 2. Double Specification Limit—Sigma Known and Constant

In many situations the measurable characteristic is considered acceptable if its value lies between upper and lower specification limits denoted by U and L, respectively. Similar to case 1, we can devise a sampling plan that will accept lots with acceptable means with high probability but also accept lots with rejectable means with low probability. The desired sampling plan is characterized by two numbers, n and k. That is, choose n specimens and record the value of the measurable characteristic for each. If $\bar{x}$ for the n measurements satisfies the inequality

$$\mu_A - k \leq \bar{x} \leq \mu_A + k \tag{17.21}$$

where $\mu_A = (U + L)/2$, the lot is accepted; otherwise it is rejected.

Since the probability of accepting a lot with mean μ_A is $1 - \alpha$, we must have

$$P_a(\mu_A) = P\{\mu_A - k \leq \bar{x} \leq \mu_A + k \mid \mu_A\} = 1 - \alpha$$
$$= \Phi\left(\frac{k\sqrt{n}}{\sigma}\right) - \Phi\left(\frac{-k\sqrt{n}}{\sigma}\right) = 1 - 2\Phi\left(\frac{-k\sqrt{n}}{\sigma}\right)$$

whence

$$\frac{k\sqrt{n}}{\sigma} = Z_{1-\alpha/2} \tag{17.22}$$

Now if the rejectable mean μ_R deviates from μ_A by $\pm\Delta$, then the probability of accepting a lot with mean μ_R is

$$P_a(\mu_A + \Delta) = P\{\mu_A - k \leq \bar{x} \leq \mu_A + k \mid \mu_A + \Delta\}$$
$$= \Phi\left(\frac{k - \Delta}{\sigma/\sqrt{n}}\right) - \Phi\left(\frac{-k - \Delta}{\sigma/\sqrt{n}}\right)$$
$$= \Phi\left(\frac{k + \Delta}{\sigma/\sqrt{n}}\right) - \Phi\left(\frac{\Delta - k}{\sigma/\sqrt{n}}\right) = \beta \tag{17.23}$$

and

$$P_a(\mu_A - \Delta) = P\{\mu_A - k \leq \bar{x} \leq \mu_A + k \mid \mu_A - \Delta\}$$
$$= \Phi\left(\frac{k + \Delta}{\sigma/\sqrt{n}}\right) - \Phi\left(\frac{\Delta - k}{\sigma/\sqrt{n}}\right) = \beta \tag{17.24}$$

Equations (17.23) and (17.24) are identical because of the symmetry of the distribution of the standard normal variate. In practice, $[(k + \Delta)/(\sigma/\sqrt{n})]$ will be large enough so that $\Phi[(k + \Delta)/(\sigma/\sqrt{n})]$ is approximately equal to 1. Hence Eq. (17.23) or (17.24) can be written, with a high degree of approximation, as

$$\beta \cong 1 - \Phi\left(\frac{\Delta - k}{\sigma/\sqrt{n}}\right) = \Phi\left(\frac{k - \Delta}{\sigma/\sqrt{n}}\right)$$

whence

$$Z_\beta \cong \frac{k - \Delta}{\sigma/\sqrt{n}} \tag{17.25}$$

The solution of Eqs. (17.22) and (17.25) yields the following values of n and k:

$$n \cong \frac{(Z_{1-\beta} + Z_{1-\alpha/2})^2\sigma^2}{\Delta^2} \tag{17.26}$$

$$k \cong \frac{\Delta Z_{1-\alpha/2}}{Z_{1-\beta} + Z_{1-\alpha/2}} \tag{17.27}$$

Example 17.5 Specifications on a certain type of resistor are .750 ± .005 ohm. Measurements are assumed to be normally distributed with standard deviation of .002 ohm. Design a variables sampling plan that will accept a lot whose mean resistance is .750 ohm 95% of the time and will reject a lot whose mean resistance is .751 or .749 ohm 90% of the time. Compute the probability of accepting a lot whose mean resistance is .7505 ohm.

Solution. Here we have

$$\alpha = .05 \qquad \beta = .10 \qquad \Delta = .001$$

Hence

$$n \cong \frac{(Z_{.90} + Z_{.975})^2(.002)^2}{(.001)^2} = 42$$

$$k \cong \frac{(.001)(1.96)}{3.242} = .000604$$

Thus our sampling plan reads as follows: Choose 42 resistors at random and record the resistance of each. Compute the mean resistance $\bar{x}$. Accept the lot if $.749396 \leq \bar{x} \leq .750604$; reject it otherwise. The probability of accepting a lot whose mean resistance is .7505 ohm is

$$P_a(.7505) = P\{.749396 \leq \bar{x} \leq .750604 \mid \mu = .7505\} = .631$$

17.9 Sequential Sampling Plans by Variables—Sigma Known and Constant

The test procedure is similar to the test of proportions developed in Sec. 17.5. We wish to develop a test of the hypothesis

$$H_0 = \mu = \mu_0 \quad \text{against} \quad H_1: \mu = \mu_1 > \mu_0$$

where μ is the mean of a normal population with known standard deviation σ. As before, sampling will continue as long as

$$\frac{\beta}{1-\alpha} < \lambda_n < \frac{1-\beta}{\alpha} \tag{17.28}$$

where

$$\lambda_n = \frac{L_1(x_1, x_2, \ldots, x_n)}{L_0(x_1, x_2, \ldots, x_n)} = \frac{\left(\frac{1}{\sqrt{2\pi}\sigma}\right)^n \exp\left[\frac{-1}{2\sigma^2}\sum_{i=1}^{n}(x_i-\mu_1)^2\right]}{\left(\frac{1}{\sqrt{2\pi}\sigma}\right)^n \exp\left[\frac{-1}{2\sigma^2}\sum_{i=1}^{n}(x_i-\mu_0)^2\right]}$$

$$= \exp\left[\frac{-1}{2\sigma^2}\left(\sum_{i=1}^{n} x_i^2 + n\mu_1^2 - 2\mu_1\sum_{i=1}^{n} x_i - \sum_{i=1}^{n} x_i^2 - n\mu_0^2 + 2\mu_0\sum_{i=1}^{n} x_i\right)\right]$$

$$= \exp\left[\frac{(\mu_1-\mu_0)\sum_{i=1}^{n} x_i}{\sigma^2} + \frac{n(\mu_0^2-\mu_1^2)}{2\sigma^2}\right]$$

On substituting the value of λ_n in (17.28) we obtain

$$\frac{\beta}{1-\alpha} < \exp\left[\frac{(\mu_1-\mu_0)\sum_{i=1}^{n} x_i}{\sigma^2} + \frac{n(\mu_0^2-\mu_1^2)}{2\sigma^2}\right] < \frac{1-\beta}{\alpha}$$

By taking natural logarithms the above inequality becomes

$$\ln\frac{\beta}{1-\alpha} < \frac{(\mu_1-\mu_0)\sum_{i=1}^{n} x_i}{\sigma^2} + \frac{n(\mu_0^2-\mu_1^2)}{2\sigma^2} < \ln\frac{1-\beta}{\alpha} \tag{17.29}$$

Hence, the equation of the acceptance line is

$$\frac{(\mu_1-\mu_0)\sum_{i=1}^{n} x_i}{\sigma^2} + \frac{n(\mu_0^2-\mu_1^2)}{2\sigma^2} = \ln\frac{\beta}{1-\alpha} \tag{17.30}$$

and the equation of the rejection line is

$$\frac{(\mu_1-\mu_0)\sum_{i=1}^{n} x_i}{\sigma^2} + \frac{n(\mu_0^2-\mu_1^2)}{2\sigma^2} = \ln\frac{1-\beta}{\alpha} \tag{17.31}$$

To illustrate this procedure, consider testing the hypothesis

$$H_0\colon \mu = \mu_0 = 100 \quad \text{against} \quad H_1\colon \mu = \mu_1 = 110$$

with $\alpha = \beta = .01$ and $\sigma = 10$.
Substituting these values in Eqs. (17.30) and (17.31) we obtain

$$.1\sum_{i=1}^{n} x_i - \frac{n}{200}(2{,}100) = \ln\left(\frac{.01}{.99}\right)$$

$$.1\sum_{i=1}^{n} x_i - \frac{n}{200}(2{,}100) = \ln\left(\frac{.99}{.01}\right)$$

and, reducing,

$$\sum_{i=1}^{n} x_i = 105n - 45.95 \quad \text{(acceptance line)} \tag{17.32}$$

$$\sum_{i=1}^{n} x_i = 105n + 45.95 \quad \text{(rejection line)} \tag{17.33}$$

Sampling is continued as long as the sequence of sample points $(n, \sum_{i=1}^{n} x_i)$, $n = 1, 2, \ldots$ falls inside the band formed by the acceptance and rejection lines (see Fig. 17.10). As soon as a sample point $(n, \sum_{i=1}^{n} x_i)$ falls above the rejection line, sampling is discontinued and H_0 is rejected. H_0 is accepted and sampling is discontinued as soon as a sample point $(n, \sum_{i=1}^{n} x_i)$ falls below the acceptance line.

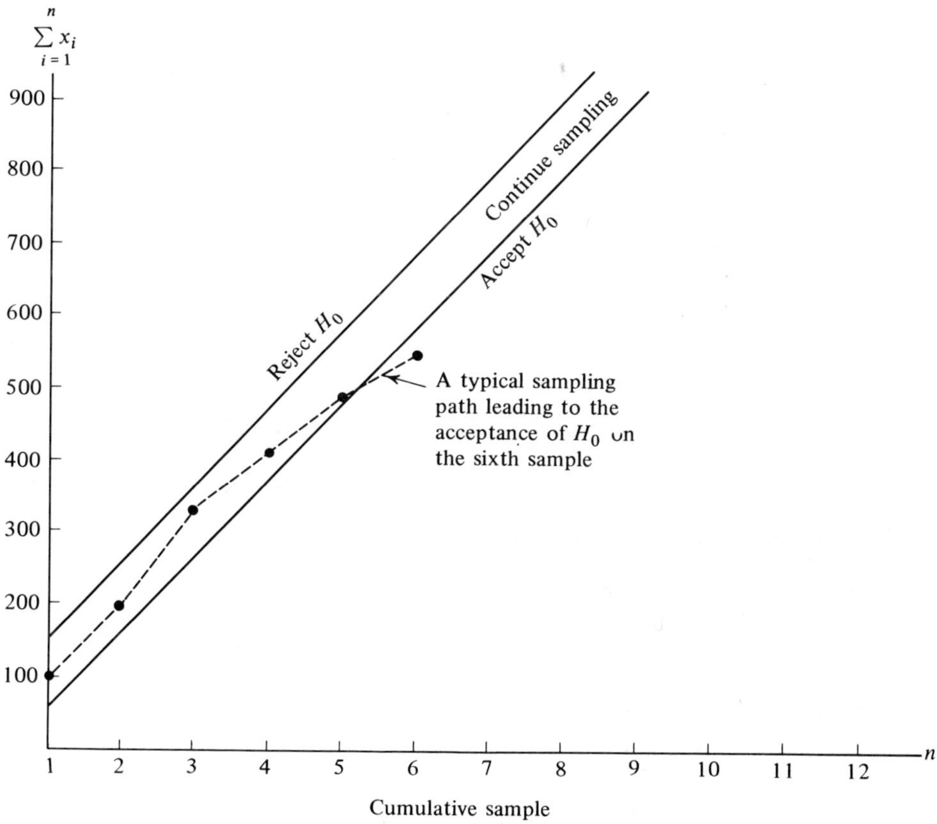

Fig. 17.10 Sequential test for means.

PROBLEMS

17.1. What is the probability of accepting a lot whose incoming quality is 10% defective using a sample of size four and an acceptance number of zero?

17.2. A buyer accepts a large lot submitted for inspection by a vendor if none or one defective is found in a sample of size 20. Use the Poisson approximation to find the probability of accepting a lot whose incoming quality is 3% defective.

17.3. A company purchases lots of a certain product by using the sampling plan $n = 10$, $c = 1$. Assuming the lot is very large compared to the sample size, determine the probability of acceptance if the submitted quality is 2% defective. Use the Poisson approximation.

17.4. A product in lots is purchased by using the sampling plan $n = 40$, $c = 2$. Graph the *OC* curve for this plan by using the Poisson approximation. Estimate the LTPD if the consumer's risk is .125.

17.5. The following single sampling plan is used in the acceptance inspection of a certain product: $n = 100$, $c = 4$. Graph the *OC* curve for this plan by using the Poisson approximation. Estimate the AQL if the producer's risk is .04.

17.6. Graph the AOQ curve for the plan in Prob. 17.5. Estimate the AOQL for the plan.

17.7. A wholesaler uses the following single sampling plan for purchasing large lots of a certain product: $n = 50$, $c = 3$.
(a) Graph the *OC* curve for this plan by using the Poisson approximation.
(b) Graph the AOQ for this plan and estimate the AOQL.
(c) Estimate the LTPD if the consumer's risk is .105.

17.8. A contractor uses the following double sampling plan for purchasing large lots of a certain product: $n_1 = 50$, $c_1 = 0$ and $n_2 = 50$, $c_2 = 2$.
(a) Find the probability of accepting a lot whose incoming quality is 4% defective. Use the Poisson approximation.
(b) Estimate the AQL if the producer's risk is .07.

17.9. Calculate the average number of items that will be inspected per sample under the plan in Prob. 17.8.

17.10. The following double sampling plan is used in the acceptance inspection of a product: $n_1 = 20$, $c_1 = 0$ and $n_2 = 20$, $c_2 = 1$.
(a) Find the probability of accepting a lot whose incoming quality is 5% defective. Use the Poisson approximation.
(b) Graph the AOQ for this plan.
(c) Calculate the average sampling number for this plan.
(d) Estimate the LTPD if the consumer's risk is .10.

17.11. If the lot size is 2,000 and the process average is 5%, what is the average total inspection under the plan of Prob. 17.10, assuming the rejected lots are inspected 100%?

17.12. Devise a single sampling plan that will give the following protection: AQL = .4%, $\alpha = .09$ and LTPD = 3.8%, $\beta = .05$.

17.13. Devise a single sampling plan that will meet the following requirements: AQL = .01, $\alpha = .08$ and LTPD = .05, $\beta = .125$.

17.14. Graph the AOQ curve for the plan in Prob. 17.13.

17.15. Devise a double sampling plan which will give approximately the same protection as that afforded by the plan of Prob. 17.12. Assume $n_1 = n_2$.

17.16. If the lot size is 10,000 and the process average is 2%, what is average total inspection under the plan in Prob. 17.15, assuming the rejected lots are inspected 100%?

17.17. Calculate the average total inspection under the plan in Prob. 17.13, assuming the rejected lots are inspected 100%. Assume that the lot size is 10,000 and the process average is 5%.

17.18. A firm is considering two possible acceptance sampling plans to form the basis of a rectifying inspection scheme.

$$\text{Plan I}: n = 50,\ c = 2$$
$$\text{Plan II}: n = 100,\ c = 4$$
$$\text{Lot size} = 1{,}000$$

The incoming fraction defective (p') has the following probability function:

$$P(p') = \begin{cases} .40 & p' = .03 \\ .60 & p' = .04 \end{cases}$$

If the cost of inspection of an item is \$1 and the penalty incurred per defective item is \$30, which plan is more economical?

17.19. XYZ Company is considering two possible acceptance sampling plans by attributes to form the basis of a receiving inspection scheme.

$$\text{Plan I}: \ n = 100,\ c = 4$$
$$\text{Plan II}: n_1 = 50,\ c_1 = 2 \text{ and } n_2 = 50,\ c_2 = 3$$
$$\text{Lot size} = 2{,}000$$

The cost of inspection is \$.50 per unit, and the penalty incurred per defective item is \$20. Which plan is more economical if the incoming fraction defective is 4%?

17.20. The specification for a certain quality characteristic is 100 ± 10.85 mm. The process is under the surveillance of an $\bar{X}$ chart and is in control with a process mean of 100 and a standard deviation of 5. Compute the probability of a lot being rejected by the consumer if he is using the single sampling plan $n = 50$, $c = 2$.

17.21. Determine the equations of the acceptance and rejection lines for an item-by-item sequential plan that will provide the following protection: AQL $= .10$, $\alpha = .05$ and LTPD $= .30$, $\beta = .20$.

17.22. The following data were drawn one observation at a time in the order recorded: *ggggg*, *ggggg*, *gbggb*, and *ggbbg* (*b* denotes a defective item, and *g* denotes a good item). The experiment was performed to test the following hypothesis:

$$H_0: p = p_1 = .05 \quad \text{against} \quad H_1: p = p_2 = .15$$

where p denotes the proportion defective items in the population. It is desired to reject H_0 when it is true with probability .10 and to accept H_0 when H_1 is true with probability .20. Using sequential analysis, would you accept or reject H_0 before you had used the entire sample?

17.23. Devise an item-by-item sequential plan to test the hypothesis

$$H_0: p = p_1 = .10 \quad \text{against} \quad H_1: p = p_2 = .20$$

at the 1% level of significance. It is desired to reject H_0 when H_1 is true with probability .95.

17.24. Specifications require that a certain quality characteristic of a manufactured product have a minimum value of 1,000 units. The past practice regarding acceptance inspection of large lots by attributes has been to test 155 items from each lot. If more than six items failed, the lot was rejected; otherwise it was accepted. The AQL and LTPD are 1.5 and 6.75%, respectively. The suggestion is made that acceptance decision be based on a variables sampling plan that will provide the same protection as that afforded by the attributes plan at the AQL and LTPD levels. Past experience indicates that the quality characteristic is normally distributed with a stable standard deviation of 100 units.
(a) Derive the acceptance criteria for the variables sampling plan.
(b) Which plan provides better protection if the lot is 5% defective?

17.25. An aircraft company considers an airscrew as defective if the pitch diameter (D) is not within $1.995 \leq D \leq 2.005$. Past experience indicates that measurements are normally distributed with $\sigma = .0031$. Devise a variables sampling plan with an AQL = 2.000, $\alpha = .05$, and $\beta = .10$.

17.26. In a chemical process the maximum acidity of a certain solution is specified as 7.6 pH. Suppose that measurements for determining pH are normally distributed with a known standard deviation of .2. Derive a variables sampling plan with AQL = 3%, $\alpha = .05$, LTPD = 7%, and $\beta = .10$. Find an attributes plan that satisfies the same requirements, and compare the sample size with that of the variables plan.

17.27. The minimum tensile strength of a certain type of steel is specified as 180,000 psi. Assume measurements are normally distributed with a known standard deviation of 2,500 psi. Draft a variables sampling plan with an AQL = 1%, $\alpha = .05$ and LTPD = 3% and $\beta = .10$. Find an attributes plan that satisfies the same requirements, and compare the sample size with that of the variables plan.

17.28. Suppose that a measurable characteristic X is normally distributed with unknown mean μ and known standard deviation $\sigma = 10$. Devise an item-by-item sequential plan to test the hypothesis

$$H_0\colon \mu = \mu_0 = 200 \quad \text{against} \quad H_1\colon \mu = \mu_1 = 220$$

with $\alpha = \beta = .01$.

17.29. A manufacturer claims that the average breaking strength of his yarn is 200 psi. Past experience indicates that the breaking strength is normally distributed with $\sigma = 5$ psi. Devise an item-by-item sequential plan so that the following requirements are satisfied: (1) If the mean breaking strength was at the normal value of 200 psi, there should be a 99% chance of acceptance. (2) If the mean breaking strength was as low as 180 psi, there should be a 95% chance of rejection.

17.30. A producer claims that the diameter of ball bearings he manufactures have a standard deviation of .001 in. Past experience indicates that measurements of the diameter are normally distributed with a mean of 1 in. Devise an item-by-item sequential plan so that the following requirements are satisfied: (1) If the standard deviation was at the normal value .001 in., there should be a 99% chance of acceptance. (2) If the standard deviation was as high as .002 in., there should be a 99% chance of rejection.

SUGGESTED REFERENCES

[1] Bowker, A. H., and H. P. Goode, *Sampling Inspection by Variables*, New York, McGraw-Hill Book Company, 1952.

[2] Burr, I. W., *Engineering Statistics and Quality Control*, New York, McGraw-Hill Book Company, 1953.

[3] Derman, C., S. B. Littauer, and H. Solomon, "Tightened Multi-Level Continuous Sampling Plans," *Annals of Mathematical Statistics*, **28**, 1957.

[4] Dodge, H. F., "Sampling Plans for Continuous Production," *Industrial Quality Control*, **4**(3), 1947.

[5] Dodge, H. F., and H. G. Romig, *Sampling Inspection Tables—Single and Double Sampling*, New York, John Wiley & Sons, Inc., 1959.

[6] Dodge, H. F., and M. N. Torrey, "Additional Continuous Sampling Plans," *Industrial Quality Control*, **7**(5), 1951.

[7] Duncan, A. J., "*Quality Control and Industrial Statistics*, 3rd ed., Homewood, Ill., Richard D. Irwin, Inc., 1965.

[8] Freeman, H. A., M. Friedman, F. Mosteller, and W. A. Wallis (eds.), *Sampling Inspection*, New York, McGraw-Hill Book Company, 1948.

[9] Girshick, M. A., "A Sequential Inspection Plan for Quality Control," *Applied Mathematics and Statistics Laboratory*, Technical Rept. No. 16: Stanford, Stanford University Press, 1954.

[10] Grant, E. L., and R. Leavenworth, *Statistical Quality Control*, 4th ed., New York, McGraw-Hill Book Company, 1972.

[11] Lieberman, G. J., and H. Solomon, "Multi-Level Continuous Sampling Plans," *Annals of Mathematical Statistics*, **26**, 1955.

[12] *Military Standard* 105*D—Sampling Procedures and Tables for Inspection by Attributes*, Washington, D.C., Superintendent of Documents, U.S. Government Printing Office, 1963 (the ABC standard).

[13] *Military Standard* 414*—Sampling Procedures and Tables for Inspection by Variables for Percent Defective*, Washington, D.C., Superintendent of Documents, U.S. Government Printing Office, 1957.

[14] Peach, P., and S. B. Littauer, "A Note on Sampling Inspection," *Annals of Mathematical Statistics*, **17**(1), 1946.

[15] Resnikoff, G. J., "Some Modifications of the Lieberman–Solomon Multi-Level Continuous Sampling Plans," *Applied Mathematics and Statistics Laboratory*, Technical Rept. No. 26, Stanford, Stanford University Press, 1958.

[16] Statistical Research Group, Columbia University, *Sequential Analysis of Statistical Data: Applications*, New York, Columbia University Press, 1945.

[17] Statistical Research Group, Columbia University, *Sampling Inspection*, New York, McGraw-Hill Book Company, 1948.

[18] Wald, A., *Sequential Analysis*, New York, John Wiley & Sons, Inc., 1947.

[19] Wald, A., and J. Wolfowitz, "Sampling Plans for Continuous Production which Insure a Prescribed Limit on the Outgoing Quality," *Annals of Mathematical Statistics*, March 1945.

18

SOME ASPECTS OF QUEUEING THEORY

18.1 Introduction

Queueing theory deals with the mathematical study of *queues*, or *waiting lines*. Probability theory plays an important role in the analysis of queueing systems.

Theoretical research into the properties of queues began in connection with problems of telephone operation. In 1905 a Danish telephone engineer, A. K. Erlang, studied the problems of telephone traffic congestion with the objective of finding a solution to the demand of services provided by the Copenhagen Telephone Systems. Erlang's contribution provided the structure of what is known nowadays as queueing theory. Until about two decades ago most of the work on queueing theory involved telephone problems. In recent years queueing theory has been applied to a wide field of operational problems. Some of the disciplines in which queueing theory is used are operations research, maintenance theory, production systems, inventory theory, road traffic theory, and management sciences.

Queueing theory deals with the problem of providing *service facilities* to meet the demand for service created by the arrival of a stream of *things*. In queueing problems we often refer to things as *customers*, such as patients waiting for medical treatment, machines waiting to be repaired, and planes waiting to land, and to service facilities such as physicians' clinics, repairmen in shops, runways in air fields, toll gates, cafeterias, and ticket offices.

If the customers arrive too infrequently, the service facilities remain idle until additional customers arrive. An example of this situation would be a fleet of taxicabs waiting for passengers or ambulances waiting for emergency calls.

Thus a waiting line or queue is created either by customers or by idle facilities. In the former the current demand for service exceeds the capacity of facilities, whereas in the latter existing facilities exceed customers' demand.

Queues are usually studied with the objective of improving the efficiency of the system as a whole. This may involve organizing congestion situations so that the total cost associated with providing the facilities of service and waiting for that service is minimized. In this connection we wish to emphasize that queueing theory is not directly concerned with optimization. It merely seeks a solution by predicting the behavior of the waiting line and thereby provides vital information required for such an optimization. The following sections present some elementary queueing processes and their basic characteristics.

18.2 Basic Elements of Queueing Systems

In order to study a queueing system the following basic elements must be specified:

18.2.1 The Pattern of Arrivals

To simplify the analysis we shall assume that the population of distinct potential customers is infinite. The analysis of the finite case is more complicated because the number of customers in the queueing system affects the number of potential calling customers outside the system at any point in time.

In practice, we encounter various *patterns of arrival*. Some of these are described below:

1. *Distributed arrivals*. Under this pattern of arrival the interarrival times are mutually independent random variables with known probability distribution other than the exponential.
2. *Scheduled arrivals*. Customers arrive according to scheduled times. An example of this pattern would be patients who adhere to their schedule.
3. *Batch arrivals*. Under this pattern the demand for service does not occur individually but in groups. For example, people waiting for buses after a show.

The most interesting and certainly the most frequently encountered pattern of arrivals is the *completely random pattern*. Under this pattern arrivals occur at random and the probability of an arrival at any time remains constant regardless of what happened previously and the length of the waiting line. Under these conditions it can be shown that arrivals obey the Poisson probability law. Thus the probability of n arrivals in any finite interval of time t is given by

$$P_t(n) = \frac{(\lambda t)^n e^{-\lambda t}}{n!} \qquad n = 0, 1, \ldots \qquad \lambda > 0 \tag{18.1}$$

If we substitute $n = 0$ in Eq. (18.1), we obtain the probability of no arrivals in a period of duration t:

$$P_t(0) = e^{-\lambda t}$$

It follows that the probability of one or more arrivals in a period of duration t is

$$1 - P_t(0) = 1 - e^{-\lambda t} = F(t) \tag{18.2}$$

Differentiating Eq. (18.2) with respect to t yields

$$f(t) = \lambda e^{-\lambda t} \qquad t \geq 0 \qquad \lambda > 0 \tag{18.3}$$

the probability density function of the time between consecutive arrivals. Consequently, under the Poisson arrival pattern, the interarrival time obeys the exponential distribution. This interesting result is a direct consequence of the assumption of completely random arrivals.

18.2.2 The Service Mechanism

This criterion refers to the number of service facilities available, the capacity of each service facility, and the distribution of service time. The *service mechanism* may consist of one or more service facilities, each of which may contain one or more *parallel service channels* (servers). The arrangement of facilities and the number of servers at each one must be specified for a meaningful study of the system.

The distribution of service time also must be specified. The service time can be a constant for all customers or can vary as a function of queue length or can obey a known probability distribution. In specific situations the exponentially distributed service time is rather justified. For example, statistical investigations indicate that the distributions of repair time, fault-finding time, and length of a telephone conversation are very close to an exponential distribution.

18.2.3 Queue Discipline

This criterion refers to the order in which customers are selected for service. Many rules have been established by which customers waiting in line are served. One rule is known as random assignment. This means that the assignment of a priority to any specific customer is made by a random choice among the set of possible priorities. A special type of priority assignment is known as preemptive priorities. Under such a rule the service mechanism, upon the arrival of a customer with a higher priority, discharges the customer being serviced and services the newly arrived customer.

Another rule is a time-honored one, i.e., first come, first served. This rule is implicitly assumed by queueing models unless stated otherwise.

18.2.4 Elementary Queueing Systems

The behavior of many different types of queueing systems has been discussed in the literature; only few elementary systems will be discussed here. In our analysis we shall assume independent exponential distributions for service and interarrival times. Furthermore, we shall assume that customers waiting in line are served according to the rule of first come, first served. Our rule also assumes that no customer in the system will depart before he receives service. Under these assumptions we shall study the following types of service systems:

1. *Loss system.* Rejects calling customers who arrive at a time when all service channels are busy. A typical example is a parking lot, with each parking space representing a service channel; cars arriving when no parking space is available are simply turned away and not given service. Note in this situation that parking time is the service time. Figure 18.1 displays this system.

2. *Delay system.* Allows calling customers who arrive at a time when all service channels are busy to join a queue and wait until service becomes

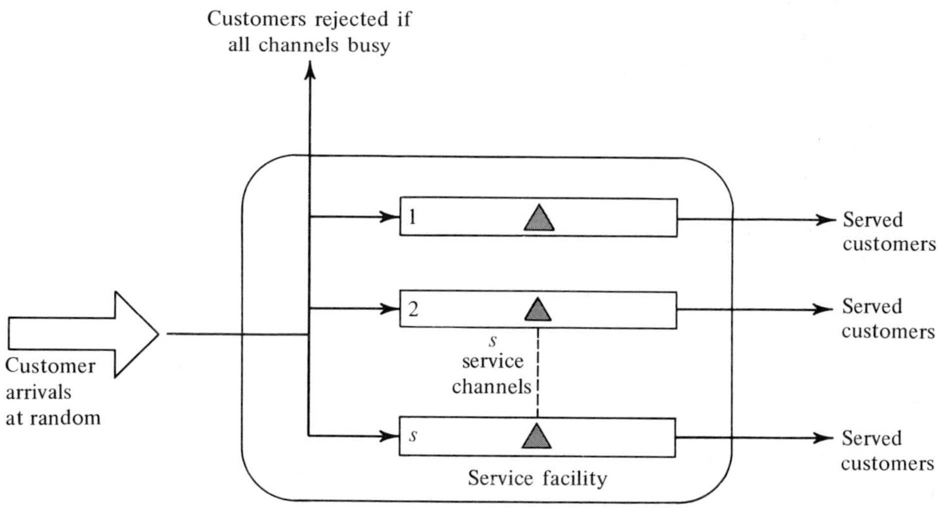

Fig. 18.1 Loss system.

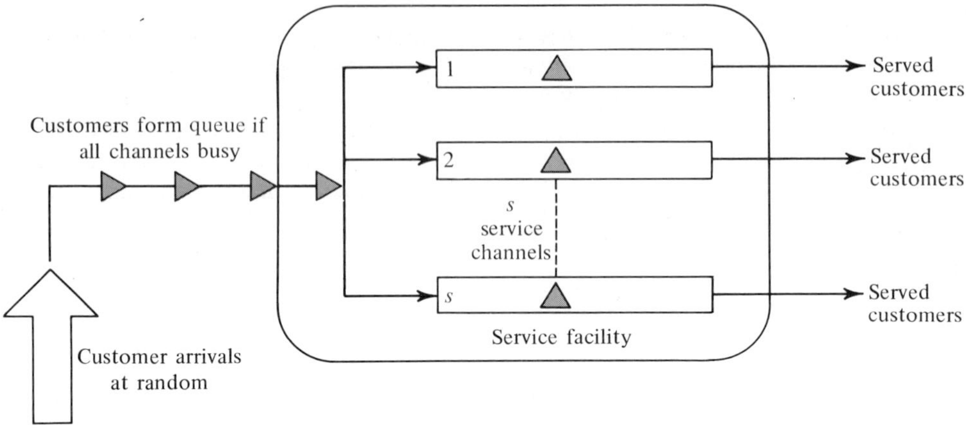

Fig. 18.2 Delay system.

available. Machine breakdowns and bus stations are typical cases. This system is shown graphically in Fig. 18.2.

3. *Combined system.* Allows calling customers who arrive when all channels are busy to join a queue up to a certain limit but rejects any further customers. A typical example is a gasoline station with limited space where cars arriving when the station is full are not permitted to wait on the street.

18.3 The State of the System

A fundamental concept in the analysis of queueing processes is the *state* of the system. We shall illustrate this concept by a numerical example. Suppose that the probability of demand per week for a certain commodity is given by the following:

Demand	*Probability*
0	.2
1	.4
2	.4

We further suppose that there is 1-week lag in delivery. Orders are placed at the start of the week and are delivered in time for the start of the following week. Suppose our ordering policy is as follows:

Stock Level	*Order*
0	2
1	1
2	0

We shall assume that our inventory level at the start of the first week (initial state) is one unit. Then, according to our policy, we will order one unit and it will be delivered in time for the start of the second week. Therefore, we will have either one or two units at the start of the second week. More precisely, we will have one unit if there was a demand for at least one unit during the week. This probability is .8. Otherwise, with probability .2 we will have two units at the start of the second week. In symbols, we write

$$\begin{aligned} P_1\{0\} &= 0 & P_2\{0\} &= 0 \\ P_1\{1\} &= 1 & P_2\{1\} &= .8 \\ P_1\{2\} &= 0 & P_2\{2\} &= .2 \end{aligned}$$

where $P_n\{k\}$ denotes the probability of having k units at the start of the nth week. Likewise, we will have at the start of the third week no stock if we started the second week with two units and there was a demand for two units. This probability is

$$P_3\{0\} = P_2\{2\}P\{D = 2\} = .2 \times .4 = .08$$

where D denotes demand. We will have at the start of the third week one unit in stock if we started the second week with one unit and there was a demand for at least one unit or if we started with two units and there was a demand for one unit. This probability is

$$P_3\{1\} = P_2\{1\}P\{D \geq 1\} + P_2\{2\}P\{D = 1\} = (.8 \times .8) + (.2 \times .4) = .72$$

Otherwise we will have at the start of the third week two units in stock if we started with no stock or if we started with one unit and there was no demand or if we started with two units and there was no demand. This probability is

$$\begin{aligned} P_3\{2\} &= P_2\{0\} + P_2\{1\}P\{D = 0\} + P_2\{2\}P\{D = 0\} \\ &= 0 + (.8 \times .2) + (.2 \times .2) = .2 \end{aligned}$$

Generalizing the foregoing analysis we obtain

$$\begin{aligned} P_{n+1}\{0\} &= .4P_n\{2\} \qquad n \geq 1 \\ P_{n+1}\{1\} &= .8P_n\{1\} + .4P_n\{2\} \qquad n \geq 1 \\ P_{n+1}\{2\} &= P_n\{0\} + .2P_n\{1\} + .2P_n\{2\} \qquad n \geq 1 \end{aligned}$$

By the above equations we can immediately compute the probabilities for $n = 1, 2, \ldots$. Table 18.1 shows the results for $n = 1, 2, \ldots, 12$. It is readily seen from Table 18.1 that the *state probabilities* settle down to steady values as n becomes large and, in fact, a rigorous proof that they do has been established. The steady-state probabilities sometimes known as *equilibrium probabilities* have the following interpretations: In the long run, independent of the state of the first week (initial state), 11.76% of the weeks will start with zero stock, 58.87% of the weeks with one unit, and 29.37% of the weeks with two units. These state probabilities reach steady state in the long run. However,

TABLE 18.1 State Probabilities

n	1	2	3	4	5	6	7	8	9	10	11	12
$P_{n+1}\{0\}$	0	.08	.080	.1056	.1056	.1138	.1138	.1164	.1164	.1172	.1172	.1176
$P_{n+1}\{1\}$	.8	.72	.656	.6304	.6099	.6017	.5952	.5926	.5905	.5897	.5889	.5887
$P_{n+1}\{2\}$	.2	.20	.264	.2640	.2845	.2845	.2910	.2910	.2931	.2931	.2939	.2937

"long run" here often is not so long. They describe the long-run behavior of the system and hence can be used to predict the long-run profits and costs. In most queueing models steady-state probabilities exist independently of the initial state of the queue. In the cited example state probabilities could only change once a week, whereas in queueing models they could change at any point of time (since arrivals and departures can occur at any point of time).

18.4 Terminology and Notation

The following standard terminology and notation will be used:

Line length = number of customers in the system
Queue length = number of customers waiting for service
= line length minus number of units being served
$P_n\{t\}$ = probability that exactly n customers are in the system at time t
λ = mean arrival rate
μ = mean service rate
s = number of servers (parallel service channels) in the system
$\rho = \lambda/\mu$ = utilization factor
N = maximum permissible length of the line
Δt = sufficiently small interval of time
$o(\Delta t)$ = function of Δt that is of a smaller order of magnitude than Δt

The following notation pertains to a system that has reached a steady-state condition:

P_n = probability that exactly n customers are in the system
L = expected line length
L_q = expected queue length
W = expected waiting time in the system (includes service time)
W_q = expected waiting time in the queue (excludes service time)

18.5 Single-Channel, Infinite, Poisson Arrival, Exponential Service Queues

In this section we shall derive the solution of a simple delay system, in which the arrivals are Poisson with parameter λt (λ = mean arrival rate) and there is single server with a service time having an exponential distribution with parameter μ (mean service rate). We shall assume that μ is greater than λ since otherwise the queue would grow indefinitely. Under the foregoing assumptions we shall derive the following characteristics:

18.5.1 The Probability of Exactly n Units in the System $\{P_n\}$

Consider the diagrammatic representation of completely random arrivals in Fig. 18.3 where the $\times$'s represent calling customers and the distance

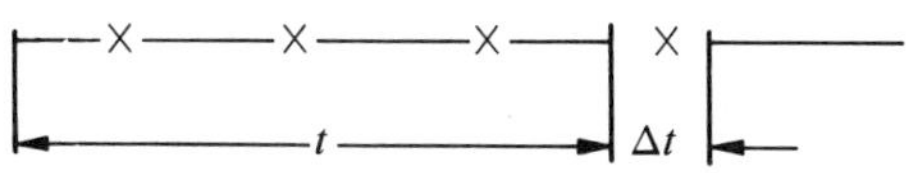

Fig. 18.3 Representation of completely random arrivals.

between any two $\times$'s represents the interarrival time that is exponentially distributed with parameter λ. Let Δt represent a sufficiently short duration. Since the exponential distribution is endowed with a complete lack of memory (see Sec. 6.11) the probability of arrivals during the time interval $(t, t + \Delta t)$ is independent of t and depends only on time Δt. Hence we can assert that the probability of no arrivals during time Δt is

$$e^{-\lambda \Delta t} = 1 - \lambda \Delta t + \frac{\lambda^2(\Delta t)^2}{2!} - \cdots = 1 - \lambda \Delta t + o(\Delta t) \tag{18.4}$$

The probability that exactly one arrival occurs during the time Δt is

$$\lambda \Delta t e^{-\lambda \Delta t} = \lambda \Delta t + o(\Delta t) \tag{18.5}$$

It follows that the probability of two or more arrivals during the time Δt is $o(\Delta t)$ and, consequently, will be considered negligible. Likewise, the probability that a service is completed in the time interval $(t, t + \Delta t)$ is $\mu \Delta t + o(\Delta t)$. When the number of customers in the system at time t is greater than or equal to one ($n \geq 1$), the probability that there will be n customers in the system at the time $(t + \Delta t)$ may be expressed as the sum of four independent compound probabilities as follows:

(18.6)
$$\begin{cases} \text{(a) } P\{n \text{ customers in the system at the time } t\}P\{\text{no arrivals during the time interval } (t, t+\Delta t)\}P\{\text{no customers are serviced during the time interval } (t, t+\Delta t)\} \\ \text{(b) } P\{(n+1) \text{ customers in the system at the time } t\}P\{\text{no customers arrive during the time interval } (t, t+\Delta t)\}P\{\text{one customer is serviced during the time interval } (t, t+\Delta t)\} \\ \text{(c) } P\{(n-1) \text{ customers in the system at the time } t\}P\{\text{one customer arrives during the time interval } (t, t+\Delta t)\}P\{\text{no customers are serviced during the time interval } (t, t+\Delta t)\} \\ \text{(d) } P\{n \text{ customers in the system at the time } t\}P\{\text{one customer arrives during the time interval } (t, t+\Delta t)\}P\{\text{one customer is serviced during the time interval } (t, t+\Delta t)\} \end{cases}$$

Since these probabilities are assumed to be independent we can assert that

$$\begin{aligned} P_n\{t+\Delta t\} = P_n\{t\}[1-\lambda\Delta t + o(\Delta t)][1-\mu\Delta t + o(\Delta t)] + P_{n+1}\{t\} \\ [1-\lambda\Delta t + o(\Delta t)][\mu\Delta t + o(\Delta t)] + P_{n-1}\{t\}[\lambda\Delta t + o(\Delta t)] \\ [1-\mu\Delta t + o(\Delta t)] + P_n\{t\}[\lambda\Delta t + o(\Delta t)][\mu\Delta t + o(\Delta t)] \end{aligned} \tag{18.7}$$

Equation (18.7) can be arranged to give

$$\begin{aligned} \frac{P_n\{t+\Delta t\} - P_n\{t\}}{\Delta t} = \lambda P_{n-1}\{t\} - (\lambda+\mu)P_n\{t\} + \mu P_{n+1}\{t\} \\ + \frac{o(\Delta t)}{\Delta t} \end{aligned} \tag{18.8}$$

Since Eq. (18.8) holds for all positive values of Δt, the

$$\begin{aligned} \lim_{\Delta t\to 0}\left[\frac{P_n\{t+\Delta t\} - P_n\{t\}}{\Delta t}\right] = \frac{dP_n\{t\}}{dt} = \lambda P_{n-1}\{t\} - (\lambda+\mu)P_n\{t\} \\ + \mu P_{n+1}\{t\} \qquad n \geq 1 \end{aligned} \tag{18.9}$$

The probability of no customers in the system ($n = 0$) at the time $(t + \Delta t)$ is the sum of two independent probabilities as follows:

(a) $P\{$no customers in the system at the time $t\}P\{$no customers arrive during the time interval $(t, t + \Delta t)\}$

(b) $P\{$one customer in the system during the time $t\}P\{$one customer is serviced during the time interval $(t, t + \Delta t)\}P\{$no customers arrive during the time interval $(t, t + \Delta t)\}$

As before, we obtain

$$\begin{aligned} P_0\{t+\Delta t\} = P_0\{t\}[1-\lambda\Delta t + o(\Delta t)] \\ + P_1\{t\}[\mu\Delta t + o(\Delta t)][1-\lambda\Delta t + o(\Delta t)] \end{aligned} \tag{18.10}$$

Equation (18.10) can be arranged to give

$$\frac{P_0(t+\Delta t) - P_0\{t\}}{\Delta t} = \mu P_1\{t\} - \lambda P_0\{t\} + \frac{o(\Delta t)}{\Delta t} \tag{18.11}$$

Furthermore,

$$\lim_{\Delta t \to 0} \left[\frac{P_0\{t + \Delta t\} - P_0\{t\}}{\Delta t} \right] = \frac{dP_0\{t\}}{dt} = \mu P_1\{t\} - \lambda P_0\{t\} \qquad n = 0 \tag{18.12}$$

We now assume that the steady state is reached so that $P_n\{t\}$ is independent of t. Consequently,

$$\lim_{t \to \infty} P_n\{t\} = P_n \quad \text{and} \quad \lim_{t \to \infty} \left[\frac{dP_n\{t\}}{dt} \right] = 0$$

Hence, when $t \longrightarrow \infty$, Eqs. (18.9) and (18.12) become

$$\lambda P_{n-1} - (\lambda + \mu)P_n + \mu P_{n+1} = 0 \qquad n \geq 1 \tag{18.13}$$

$$\mu P_1 - \lambda P_0 = 0 \qquad n = 0 \tag{18.14}$$

From Eq. (18.14) we have

$$P_1 = \frac{\lambda}{\mu} P_0 \tag{18.15}$$

From Eq. (18.13), with $n = 1$, we obtain

$$\lambda P_0 - (\lambda + \mu)P_1 + \mu P_2 = 0$$

whence

$$P_2 = \left(1 + \frac{\lambda}{\mu}\right) P_1 - \frac{\lambda}{\mu} P_0 \tag{18.16}$$

Substituting the value of P_1 as given by Eq. (18.15) into Eq. (18.16) we obtain

$$P_2 = \left(1 + \frac{\lambda}{\mu}\right) \frac{\lambda}{\mu} P_0 - \frac{\lambda}{\mu} P_0 = \left(\frac{\lambda}{\mu}\right)^2 P_0 \tag{18.17}$$

Likewise,

$$P_n = \left(\frac{\lambda}{\mu}\right)^n P_0 \qquad n = 1, 2, \ldots \tag{18.18}$$

On taking summation we obtain

$$\sum_{n=0}^{\infty} P_n = P_0 \sum_{n=0}^{\infty} \left(\frac{\lambda}{\mu}\right)^n \tag{18.19}$$

However, it is apparent that

$$\sum_{n=0}^{\infty} P_n = 1 \tag{18.20}$$

Hence, from Eqs. (18.19) and (18.20) we have

$$P_0 = 1 / \sum_{n=0}^{\infty} \left(\frac{\lambda}{\mu}\right)^n = \frac{1}{1 + (\lambda/\mu) + (\lambda/\mu)^2 + \cdots} \tag{18.21}$$

Since we have assumed that $\lambda/\mu < 1$ the summation of the infinite geometric series $[1 + (\lambda/\mu) + (\lambda/\mu)^2 + \cdots]$ is $1/[1 - (\lambda/\mu)]$. Therefore, Eq. (18.21) becomes

$$P_0 = 1 - \frac{\lambda}{\mu} \tag{18.22}$$

Substituting this value into Eq. (18.18) we get

$$P_n = \left(\frac{\lambda}{\mu}\right)^n \left(1 - \frac{\lambda}{\mu}\right) = \rho^n(1 - \rho) \qquad \begin{array}{l} 0 \leq \rho < 1 \\ n = 0, 1, 2, \ldots \end{array} \tag{18.23}$$

where $\rho = \lambda/\mu$ is known as the *utilization factor* or *traffic intensity*. It measures the proportion of time that the service facility (server) is kept busy or the probability that a random arrival has to wait.

18.5.2 The Mean Number of Customers in the System (L)

The expected value of the line length is given by

$$\begin{aligned} L &= \sum_{n=0}^{\infty} n\rho^n(1 - \rho) \\ &= (1 - \rho)(\rho + 2\rho^2 + 3\rho^3 + \cdots) \\ &= (1 - \rho)\rho(1 + 2\rho + 3\rho^2 + \cdots) \\ &= (1 - \rho)\rho \frac{d}{d\rho}\left(\sum_{n=0}^{\infty} \rho^n\right) \\ &= (1 - \rho)\rho \frac{d}{d\rho}\left(\frac{1}{1 - \rho}\right) \\ &= \frac{\rho}{1 - \rho} = \frac{\lambda}{\mu - \lambda} \qquad 0 \leq \rho < 1 \end{aligned} \tag{18.24}$$

It should be noted that as ρ approaches 1 the expected value of the line length tends to infinity.

18.5.3 The Mean Number of Customers in the Queue (L_q)

The expected value of the queue length is given by

$$\begin{aligned} L_q &= \sum_{n=1}^{\infty} (n - 1)\rho^n(1 - \rho) \\ &= \sum_{n=1}^{\infty} n\rho^n(1 - \rho) - \sum_{n=1}^{\infty} \rho^n(1 - \rho) \\ &= L - (1 - P_0) = \frac{\rho^2}{1 - \rho} = \frac{\lambda^2}{\mu(\mu - \lambda)} \qquad 0 \leq \rho < 1 \end{aligned} \tag{18.25}$$

18.5.4 The Mean Waiting Time in the Queue (W_q)

If the service facility is not busy when the customer arrives, then no queue is formed and the service is provided immediately. If one customer is ahead of him, then the probability of waiting in the queue longer than τ is $e^{-\mu\tau}$. If two customers are ahead of the arriving customer, then the total waiting time in the queue will be for the completion of two services—that of the preceding two customers. We recall that the sum of k independent and identically distributed service times has the Erlang distribution (see Sec. 6.13). Likewise, if n customers are ahead of the arriving customer the probability that the customer will still be in the queue after time τ is

$$P\{T > \tau\} = \int_{\tau}^{\infty} \frac{\mu^n a^{n-1} e^{-\mu a}}{(n-1)!}\, da \qquad n \geq 1$$

However, at any random instant the probability that the system contains n customers is $P_n = \rho^n(1-\rho)$. Hence the expected probability that the arriving customer has to wait in the queue longer than τ is

$$\begin{aligned}
P\{T > \tau\} = 1 - F_T(\tau) &= \sum_{n=1}^{\infty} P_n \int_{\tau}^{\infty} \frac{\mu^n a^{n-1} e^{-\mu a}}{(n-1)!}\, da \\
&= \sum_{n=1}^{\infty} \rho^n(1-\rho) \sum_{j=0}^{n-1} \frac{e^{-\mu\tau}(\mu\tau)^j}{j!} \\
&= (1-\rho)e^{-\mu\tau}\left\{\rho + \rho^2(1+\mu\tau) + \rho^3\left[1 + \mu\tau + \frac{(\mu\tau)^2}{2!}\right]\right. \\
&\quad \left. + \rho^4\left[1 + \mu\tau + \frac{(\mu\tau)^2}{2!} + \frac{(\mu\tau)^3}{3!}\right] + \cdots\right\} \\
&= (1-\rho)e^{-\mu\tau}\rho\left\{1 + \rho(1+\mu\tau) + \rho^2\left[1 + \mu\tau\right.\right. \\
&\quad \left.\left. + \frac{(\mu\tau)^2}{2!}\right] + \cdots\right\} \\
&= (1-\rho)e^{-\mu\tau}\rho\left[(1 + \rho + \rho^2 + \cdots)\left(1 + \rho\mu\tau\right.\right. \\
&\quad \left.\left. + \rho^2\frac{(\mu\tau)^2}{2!} + \cdots\right)\right] \\
&= \rho e^{-\mu\tau}(e^{\rho\mu\tau}) = \rho e^{(\lambda-\mu)\tau} \qquad \tau > 0
\end{aligned}$$

The case for $\tau = 0$ corresponds to the occasion when $n = 0$; thus there is a discontinuity at $\tau = 0$. Hence, the probability density function of T is

$$\begin{aligned}
f_T(\tau) = \frac{d}{d\tau}F_T(\tau) &= \rho(\mu-\lambda)e^{-(\mu-\lambda)\tau} \qquad \tau > 0 \\
&= 1 - \rho \qquad \tau = 0
\end{aligned}$$

Therefore

$$\begin{aligned}
W_q = E(T) &= \rho\int_0^{\infty} \tau(\mu-\lambda)e^{-(\mu-\lambda)\tau}\, d\tau = \frac{\rho}{\mu-\lambda} \\
&= \frac{\lambda}{\mu(\mu-\lambda)} \qquad 0 \leq \rho < 1 \qquad (18.26)
\end{aligned}$$

18.5.5 The Mean Waiting Time in the System (W)

It is obvious that the mean waiting time in the system is the sum of the mean waiting time in the queue plus the mean service time for each customer. Therefore,

$$W = W_q + \frac{1}{\mu} = \frac{\lambda}{\mu(\mu - \lambda)} + \frac{1}{\mu} = \frac{1}{\mu - \lambda}$$

$$= \frac{1}{\mu(1 - \rho)} \qquad 0 \leq \rho < 1 \tag{18.27}$$

Incidentally, it has been proven by Little [4] that for infinite queues, $L = \lambda W$ and $L_q = \lambda W_q$ under essentially general conditions. This useful relation between L and W holds for a wide variety of service and arrival distributions for one or for several channels. We shall make use of this relationship in our forthcoming analysis. The following examples will illustrate the foregoing results.

Example 18.1 Arrivals at an information desk that has one clerk are considered to be Poisson with mean rate of 20 per hr. The time that customers spend in seeking information of one kind or another is assumed to be exponentially distributed with a mean of 2 min. Find the characteristics of this queueing process.

Solution

(a) The average number of arrivals and the average number of servicings, respectively, are $\lambda = \frac{1}{3}$ arrival per min and $\mu = \frac{1}{2}$ servicing per min.

(b) The expected fraction of time the clerk is busy is $\rho = \lambda/\mu = \frac{2}{3}$, or 66.66% of the time.

(c) The probability of forming a queue of length n with $n = 0, 1, 2, \ldots$ is given by

$$P_0 = 1 - \rho = \tfrac{1}{3}$$
$$P_1 = \rho(1 - \rho) = \tfrac{2}{9}$$
$$P_2 = \rho^2(1 - \rho) = \tfrac{4}{27}$$
$$P_3 = \rho^3(1 - \rho) = \tfrac{8}{81}$$
$$\ldots$$

Thus, when a customer arrives at the information desk he has a probability of $\frac{1}{3}$ of securing the required information immediately, a probability of $\frac{2}{9}$ of standing in the queue alone, a probability of $\frac{4}{27}$ of being the second customer in the waiting line, and so on.

(d) The mean number of customers in the system is

$$L = \frac{\rho}{1 - \rho} = \frac{2/3}{1/3} = 2 \text{ persons}$$

(e) The mean number of customers in the queue is

$$L_q = L - (1 - P_0) = 2 - \tfrac{2}{3} = 1\tfrac{1}{3} \text{ persons}$$

(f) The mean waiting time in the queue is

$$W_q = \frac{\rho}{\mu - \lambda} = \frac{2/3}{1/6} = 4 \text{ min}$$

(g) The mean waiting time in the system is

$$W = W_q + \frac{1}{\mu} = 4 + 2 = 6 \text{ min}$$

Example 18.2 In an assembly plant, pneumatic presses break down according to the Poisson distribution with a mean rate of .5 breakdown per hr, and the cost of downtime of each press waiting for repair and being repaired is estimated at \$20 per hr. Repairs are made by a single repairman. Suppose that repair time is exponentially distributed with a mean of 80 min. If the facility cost of repairing one press at the present service rate is \$20, compute the expected total daily cost. What is the optimal service rate?

Solution. The expected total daily cost is the sum of the expected downtime cost plus the expected repair cost per day. We note that $\lambda = .5$ and $\mu = 60/80 = .75$. Thus the expected number of presses in the system, both those being repaired and those in the queue, is

$$L = \frac{\lambda}{\mu - \lambda} = \frac{.5}{.75 - .5} = 2$$

The expected downtime in an 8-hr day is (2)(8) = 16. The expected total daily cost, then, is

$$(16)(20) + (8)(20)(.75) = \$440$$

To find the optimal service rate μ_0, we have to formulate the expected total daily cost as a function of μ. This is

$$\text{E.T.C. per day} = (8)\left(\frac{.5}{\mu - .5}\right)(20) + (8)(20)\mu$$

Differentiating the above equations with respect to μ and setting the result equal to zero we obtain

$$\frac{d\text{E.T.C.}}{d\mu} = \frac{-80}{(\mu - .5)^2} + 160 = 0$$

whence

$$\mu_0 = \tfrac{1}{2} + \sqrt{\tfrac{1}{2}} = 1.207 \text{ presses per day}$$

To generalize the foregoing result, we write

$$\mu_0 = \lambda + \sqrt{\lambda R}$$

where $R = \dfrac{\text{waiting cost per unit per period}}{\text{cost of servicing one unit}}$

18.6 Single-Channel, Infinite, Poisson Arrival, Nonexponential Service Queues

In Sec. 18.5 we postulated that the number of arrivals per period obeys the Poisson probability law and that service times are exponentially distributed. The assumption that arrivals occur according to a Poisson process has a sound practical basis. Although the Poisson assumption is not always likely to be valid, much evidence exists to indicate that this assumption is frequently justified. Evidence in support of the assumption that service times are exponentially distributed is not as strong. Often this assumption is made to simplify the analysis. In this section we shall violate this assumption and merely assume that service times are independent, identically distributed random variables with mean $1/\mu$ and variance σ^2. Other assumptions made in Sec. 18.5 remain unchanged. Queueing models with arbitrary service time distribution have been studied by Kendall [2], Pollaczek, and others. The mathematical analysis of these models is much more difficult; however, we shall give the results (without proof) for a single-channel model under the foregoing assumptions.

In the steady state the following characteristics are verified when $\rho = \lambda/\mu < 1$:

$$P_0 = 1 - \frac{\lambda}{\mu} \tag{18.28}$$

$$L_q = \frac{\lambda^2\sigma^2 + (\lambda/\mu)^2}{2[1 - (\lambda/\mu)]} \tag{18.29}$$

$$L = L_q + \frac{\lambda}{\mu} \tag{18.30}$$

$$W_q = \frac{L_q}{\lambda} \tag{18.31}$$

$$W = W_q + \frac{1}{\mu} \tag{18.32}$$

Note carefully, for a constant service time, the variance $\sigma^2 = 0$. As a result, the values of L_q, L, W_q, and W are minimum. Obviously, these values increase indefinitely by increasing the variance of the service time. It is apparent, if we wish to reduce the expected length of the queue, that we should attempt to reduce the variance σ^2, provided that λ is constant and $1/\mu$ cannot be changed.

When the service time is exponential, $\sigma^2 = 1/\mu^2$, the foregoing results, Eqs. (18.29)–(18.32), will reduce to those given in the preceding section. For example, Eq. (18.29) becomes

$$L_q = \frac{\lambda^2/\mu^2 + (\lambda/\mu)^2}{2[1 - (\lambda/\mu)]} = \frac{\rho^2}{1 - \rho}$$

which is the same as Eq. (18.25).

18.7 Single-Channel, Finite, Poisson Arrival, Exponential Service Queues

This model is identical with the model given in Sec. 18.5, except that the maximum permissible length of the line is N rather than infinity. In essence, this is a combined system. Here we have a finite number of states for values of

n from zero to N. It is evident that Eqs. (18.13) and (18.14) hold for $0 \leq n \leq N - 1$. We shall now consider the case of $n = N$.

18.7.1 The Probability of Exactly N Units in the System $\{P_N\}$

The probability that there will be N customers in the system at the time $(t + \Delta t)$ may be expressed as the sum of two independent compound probabilities as follows:

(a) $P\{N$ customers in the system at the time $t\}P\{$no customers are serviced during the time interval $(t, t + \Delta t)\}$

(b) $P\{N - 1$ customers in the system at the time $t\}P\{$one customer arrives during the time interval $(t, t + \Delta t)\}P\{$no customers are serviced during the time interval $(t, t + \Delta t)\}$

Thus

$$P_N\{t + \Delta t\} = P_N\{t\}[1 - \mu\Delta t + o(\Delta t)] + P_{N-1}\{t\} [\lambda\Delta t + o(\Delta t)][1 - \mu\Delta t + o(\Delta t)] \tag{18.33}$$

which after some rearrangement gives

$$\frac{P_N\{t + \Delta t\} - P_N\{t\}}{\Delta t} = \lambda P_{N-1}\{t\} - \mu P_N\{t\} + \frac{o(\Delta t)}{\Delta t} \tag{18.34}$$

As before,

$$\lim_{\Delta t \to 0} \left[\frac{P_N\{t - \Delta t\} - P_N\{t\}}{\Delta t}\right] = \frac{dP_N\{t\}}{dt} = \lambda P_{N-1}\{t\} - \mu P_N\{t\} \tag{18.35}$$

Now if we assume that a steady state is reached so that $(dP_N\{t\})/dt = 0$, we obtain

$$P_{N-1} = \frac{\mu}{\lambda} P_N \tag{18.36}$$

However, for $n < N$, P_{N-1} can be obtained from Eq. (18.18); therefore $P_{N-1} = (\lambda/\mu)^{N-1}P_0$. Hence, substituting this value in Eq. (18.36) we obtain

$$P_N = \frac{\lambda}{\mu}\left(\frac{\lambda}{\mu}\right)^{N-1} P_0 = \left(\frac{\lambda}{\mu}\right)^N P_0 \tag{18.37}$$

Therefore, we can assert that

$$P_n = \left(\frac{\lambda}{\mu}\right)^n P_0 \qquad 0 \leq n \leq N \tag{18.38}$$

On taking summation we obtain

$$\sum_{n=0}^{N} P_n = P_0 \sum_{n=0}^{N} \left(\frac{\lambda}{\mu}\right)^n = 1$$

whence

$$P_0 = \frac{1}{\sum_{n=0}^{N} (\lambda/\mu)^n} = \frac{1}{\sum_{n=0}^{N} \rho^n} = \frac{1}{1 + \rho + \rho^2 + \cdots + \rho^N}$$

The denominator is the sum of a finite geometric series; hence,

$$P_0 = \frac{1 - \rho}{1 - \rho^{N+1}} \tag{18.39}$$

Substituting this value of P_0 into Eq. (18.38) we obtain

$$P_n = \rho^n\left(\frac{1 - \rho}{1 - \rho^{N+1}}\right) \qquad 0 \leq n \leq N \tag{18.40}$$

18.7.2 The Mean Number of Customers in the System (*L*)

As before,

$$\begin{aligned} L &= \sum_{n=0}^{N} nP_n \\ &= \left(\frac{1 - \rho}{1 - \rho^{N+1}}\right) \sum_{n=0}^{N} n\rho^n \\ &= \left(\frac{1 - \rho}{1 - \rho^{N+1}}\right) \rho(1 + 2\rho + 3\rho^2 + \cdots + N\rho^{N-1}) \\ &= \left(\frac{1 - \rho}{1 - \rho^{N+1}}\right) \rho \frac{d}{d\rho}\left(\sum_{n=0}^{N} \rho^n\right) \\ &= \left(\frac{1 - \rho}{1 - \rho^{N+1}}\right) \rho \frac{d}{d\rho}\left(\frac{1 - \rho^{N+1}}{1 - \rho}\right) \\ &= \left(\frac{1 - \rho}{1 - \rho^{N+1}}\right) \rho \left[\frac{1 - \rho^{N+1}}{(1 - \rho)^2} - \frac{(N + 1)\rho^N}{1 - \rho}\right] \\ &= \rho \frac{1 - (N + 1)\rho^N + N\rho^{N+1}}{(1 - \rho)(1 - \rho^{N+1})} \end{aligned} \tag{18.41}$$

Notice that in this case it is not necessary to assume that $\rho < 1$.

18.7.3 The Mean Number of Customers in the Queue (L_q)

$$\begin{aligned} L_q &= \sum_{n=1}^{N} (n - 1)P_n \\ &= \sum_{n=1}^{N} nP_n - \sum_{n=1}^{N} P_n \\ &= L - (1 - P_0) \end{aligned}$$

$$= \rho\frac{1-(N+1)\rho^N + N\rho^{N+1}}{(1-\rho)(1-\rho^{N+1})} - \left(1 - \frac{1-\rho}{1-\rho^{N+1}}\right)$$

$$= \rho^2\frac{1 - N\rho^{N-1} + (N-1)\rho^N}{(1-\rho)(1-\rho^N)} \tag{18.42}$$

18.7.4 The Mean Waiting Time in the Queue (W_q)

We shall follow the same line of argument as that of Sec. 18.5.4. Here we assume that the maximum permissible length of the line does not exceed N.

Since the waiting time in the queue of the arriving customer is a random variable T having the Erlang distribution, his expected waiting time in the queue if n customers are ahead of him is

$$W_q = \sum_{n=0}^{N} P_n E(T)$$

$$= \sum_{n=0}^{N} P_n \int_0^\infty \frac{\tau\mu^n\tau^{n-1}e^{-\mu\tau}}{(n-1)!}\,d\tau$$

$$= \sum_{n=0}^{N} \frac{n}{\mu} P_n = \frac{1}{\mu}\sum_{n=1}^{N} nP_n$$

$$= \frac{L}{\mu} = \frac{\rho}{\mu}\left[\frac{1-(N+1)\rho^N + N\rho^{N+1}}{(1-\rho)(1-\rho^{N+1})}\right] \tag{18.43}$$

The reader may verify that Eq. (18.43) converges to Eq. (18.26) as $N \longrightarrow \infty$.

18.7.5 The Mean Waiting Time in the System (W)

$$W = W_q + \frac{1}{\mu}$$

$$= \frac{1 + \rho^{N+2}(1+N) - \rho^{N+1}(2+N)}{\mu(1-\rho)(1-\rho^{N+1})} \tag{18.44}$$

Note that when $N = 1$ the system reduces to a pure loss system, which can be in one of two possible states: busy with one customer or idle. The probability that the system is busy with one customer can be obtained by substituting $N = 1$ and $n = 1$ in Eq. (18.40). This gives

$$P\{\text{the system is busy}\} = \rho\left(\frac{1-\rho}{1-\rho^2}\right) = \frac{\rho}{1+\rho} \tag{18.45}$$

The system will be idle when $n = 0$; so

$$P\{\text{the system is idle}\} = \frac{1-\rho}{1-\rho^2} = \frac{1}{1+\rho} \tag{18.46}$$

It is evident that as $\rho \longrightarrow 0$, Eq. (18.45) tends to zero and Eq. (18.46) tends to one, i.e., the server will be idle most of the time. Conversely, as $\rho \longrightarrow \infty$ the server will be busy most of the time.

18.8 Multichannel, Finite, Poisson Arrival, Exponential Service Queues

If more than one channel is available to provide services, then we say that the service facility is multichannel with $s \geq 2$. This model is essentially the same as that discussed in Sec. 18.7, except that service is provided by s channels arranged in parallel. We shall assume that the service facility has s equal exponential channels, each with a mean service rate μ. The number of arrivals is again assumed to follow the Poisson probability law with mean arrival rate λ. Arriving customers join a single queue if all channels are busy until they come to the front of the line, where they enter the first unoccupied channel. For this model we postulate that the line length is not permitted to exceed some specified number N. Examples of this sort of combined system are restaurants with counter seats and repair facilities.

With s parallel channels and n customers we shall investigate the following three cases:

$$\text{(a) } n = 0 \qquad (b)\ 0 < n < s \qquad \text{(c) } s \leq n \leq N$$

We shall begin with

Case (a)

When there are no customers in the system the same arguments leading to Eq. (18.14) apply unchanged. That is,

$$P_1 = \frac{\lambda}{\mu} P_0 = \rho P_0 \qquad n = 0 \tag{18.47}$$

Case (b)

When $0 < n < s$ there will be no queue since the number of the customers is less than the number of channels.

There are four possible independent events leading to n channels being occupied at time $(t + \Delta t)$. These events are given by (18.6). Consider now the first event given by [18.6(a)]. If there are n customers in the system at time t, then the probability that no customers are serviced during the time interval $(t, t + \Delta t)$ is given by the binomial probability law:

$$\binom{n}{0}[\mu\Delta t + o(\Delta t)]^0[1 - \mu\Delta t + o(\Delta t)]^n$$

$$= 1 - \binom{n}{1}[\mu\Delta t + o(\Delta t)] + \binom{n}{2}[\mu\Delta t + o(\Delta t)]^2 - \cdots$$

$$= 1 - n\mu\Delta t + o(\Delta t)$$

Hence, [18.6(a)] becomes

$$P_n\{t\}[1 - \lambda\Delta t + o(\Delta t)][1 - n\mu\Delta t + o(\Delta t)]$$

Likewise, Eq. (18.7) becomes

$$\begin{aligned} P_n\{t + \Delta t\} = {} & P_n\{t\}[1 - \lambda\Delta t + o(\Delta t)][1 - n\mu\Delta t + o(\Delta t)] \\ & + P_{n+1}\{t\}[1 - \lambda\Delta t + o(\Delta t)][(n+1)\mu\Delta t + o(\Delta t)] \\ & + P_{n-1}\{t\}[\lambda\Delta t + o(\Delta t)][1 - (n-1)\mu\Delta t + o(\Delta t)] \\ & + P_n\{t\}[\lambda\Delta t + o(\Delta t)][n\mu\Delta t + o(\Delta t)] \end{aligned} \tag{18.48}$$

By the usual procedure as in Eqs. (18.8), (18.9), and (18.13) we find the steady-state equation

$$\lambda P_{n-1} - (\lambda + n\mu)P_n + (n+1)\mu P_{n+1} = 0 \qquad 0 < n < s \tag{18.49}$$

From Eq. (18.47) and by repeated application of the recurrent equation (18.49) we find

$$P_n = \frac{\rho^n}{n!}P_0 \qquad 0 < n < s \tag{18.50}$$

Now we shall examine the case when $n = s$ under the assumption that a loss system exists. The probability that there are s customers in the system at $(t + \Delta t)$ is

$$\begin{aligned} P_s\{t + \Delta t\} = {} & P_s\{t\}[1 - s\mu\Delta t + o(\Delta t)] \\ & + P_{s-1}\{t\}[\lambda\Delta t + o(\Delta t)][1 - (s-1)\mu\Delta t + o(\Delta t)] \end{aligned}$$

This yields, for $t \longrightarrow \infty$,

$$-s\mu P_s + \lambda P_{s-1} = 0 \tag{18.51}$$

Note that when $n = 0$, Eq. (18.50) holds true. It also holds true when $n = s$, since by Eqs. (18.51) and (18.50)

$$P_s = \frac{\lambda}{s\mu}P_{s-1} = \frac{\rho}{s}P_{s-1} = \frac{\rho}{s}\frac{\rho^{s-1}}{(s-1)!}P_0 = \frac{\rho^s}{s!}P_0$$

Thus, Eq. (18.50) can be written

$$P_n = \frac{\rho^n}{n!}P_0 \qquad 0 \leq n \leq s \tag{18.52}$$

From the condition

$$\sum_{n=0}^{s} P_n = P_0 \sum_{n=0}^{s} \frac{\rho^n}{n!} = 1$$

we find P_0 and, on substituting its value in Eq. (18.52), we obtain

$$P_n = \frac{\rho^n/n!}{\sum_{n=0}^{s} (\rho^n/n!)} \qquad 0 \leq n \leq s \tag{18.53}$$

which is the probability that exactly n channels will be busy after the service complex has reached the steady-state condition. Equation (18.53) is known as the *first Erlang formula*. In particular, the probability that all s channels are busy is

$$P_s = \frac{\rho^s/s!}{\sum_{n=0}^{s} (\rho^n/n!)} \tag{18.54}$$

This is, in fact, the probability of losing customers—under the assumption that we have a loss system from which customers depart when they find all channels busy.

Case (c)

Here a queue is formed when all channels are busy, so that the events given by (18.6) apply without change. However, in this case the service facility contains s channels instead of one. Bearing this in mind the event specified by [18.6(a)], then, is

$$P_n\{t\}[1 - \lambda\Delta t + o(\Delta t)]\binom{s}{0}[\mu\Delta t + o(\Delta t)]^0[1 - \mu\Delta t + o(\Delta t)]^s$$
$$= P_n\{t\}[1 - \lambda\Delta t + o(\Delta t)][1 - s\mu\Delta t + o(\Delta t)]$$

Likewise, Eq. (18.7) becomes

$$\begin{aligned} P_n\{t + \Delta t\} = P_n\{t\}&[1 - \lambda\Delta t + o(\Delta t)][1 - s\mu\Delta t + o(\Delta t)] \\ &+ P_{n+1}\{t\}[1 - \lambda\Delta t + o(\Delta t)][s\mu\Delta t + o(\Delta t)] \\ &+ P_{n-1}\{t\}[\lambda\Delta t + o(\Delta t)][1 - s\mu\Delta t + o(\Delta t)] \\ &+ P_n\{t\}[\lambda\Delta t + o(\Delta t)][s\mu\Delta t + o(\Delta t)] \end{aligned} \tag{18.55}$$

By the same procedure as in Eqs. (18.8), (18.9), and (18.13) we find the steady-state equation

$$-(\lambda + s\mu)P_n + \lambda P_{n-1} + s\mu P_{n+1} = 0 \qquad s < n < N \tag{18.56}$$

If $n = s$, we find after the usual manipulation the steady-state equation

$$-(\lambda + s\mu)P_n + \lambda P_{n-1} + s\mu P_{n+1} = 0 \qquad n = s \tag{18.57}$$

which is identical with Eq. (18.56). Thus Eqs. (18.56) and (18.57) can be combined to give

$$-(\lambda + s\mu)P_n + \lambda P_{n-1} + s\mu P_{n+1} = 0 \qquad s \le n \le N \tag{18.58}$$

Finally, when $n = N$, we find

$$\begin{aligned} P_N\{t + \Delta t\} = P_N\{t\}&[1 - s\mu\Delta t + o(\Delta t)] \\ &+ P_{N-1}\{t\}[\lambda\Delta t + o(\Delta t)][1 - s\mu\Delta t + o(\Delta t)] \end{aligned} \tag{18.59}$$

This yields, for $t \longrightarrow \infty$,

$$\lambda P_{N-1} - s\mu P_N = 0 \qquad n = N \tag{18.60}$$

Now, letting $n = s$ in Eq. (18.58), we obtain

$$P_{s+1} = \frac{1}{s\mu}[(\lambda + s\mu)P_s - \lambda P_{s-1}]$$

However, substituting in this the values of P_s and P_{s-1} as given by Eq. (18.52) we get

$$P_{s+1} = \frac{1}{s\mu}\left[(\lambda + s\mu)\frac{\rho^s}{s!}P_0 - \lambda\frac{\rho^{s-1}}{(s-1)!}P_0\right]$$

$$= P_0\frac{\rho^s}{s!}\left(1 + \frac{\rho}{s} - 1\right) = \frac{\rho^{s+1}}{s!s}P_0 \tag{18.61}$$

Likewise, we find that

$$P_n = \frac{\rho^n}{s!s^{n-s}}P_0 \qquad s \le n < N \tag{18.62}$$

However, from Eqs. (18.60) and (18.62) we have

$$P_N = \frac{\lambda}{s\mu}P_{N-1} = \left(\frac{\rho}{s}\right)\frac{\rho^{n-1}}{s!s^{n-s-1}}P_0 = \frac{\rho^n}{s!s^{n-s}}P_0$$

which is identical with Eq. (18.62). Hence, we can assert that

$$P_n = \begin{cases} \dfrac{\rho^n}{n!}P_0 & 0 \le n \le s \\ \dfrac{\rho^n}{s!s^{n-s}}P_0 & s \le n \le N \end{cases} \tag{18.63}$$

Eliminating P_0 by using the condition

$$\sum_{n=0}^{N} P_n = 1 \tag{18.64}$$

we find that

$$P_0 = \left(\sum_{n=0}^{s-1}\frac{\rho^n}{n!} + \sum_{n=s}^{N}\frac{\rho^n}{s!s^{n-s}}\right)^{-1} \tag{18.65}$$

The second term on the right-hand side is the sum of a finite geometric series; on summing Eq. (18.65) becomes

$$P_0 = \left\{\sum_{n=0}^{s-1}\frac{\rho^n}{n!} + \left[\frac{\rho^s}{s!}\frac{1-(\rho/s)^{N-s+1}}{1-(\rho/s)}\right]\right\}^{-1} \tag{18.66}$$

Hence, for a combined loss-delay system we have

$$P_n = \begin{cases} \dfrac{\rho^n}{n!}\left\{\displaystyle\sum_{n=0}^{s-1}\frac{\rho^n}{n!} + \left[\frac{\rho^s}{s!}\frac{1-(\rho/s)^{N-s+1}}{1-(\rho/s)}\right]\right\}^{-1} & 0 \le n \le s \\ \dfrac{\rho^n}{s!s^{n-s}}\left\{\displaystyle\sum_{n=0}^{s-1}\frac{\rho^n}{n!} + \left[\frac{\rho^s}{s!}\frac{1-(\rho/s)^{N-s+1}}{1-(\rho/s)}\right]\right\}^{-1} & s \le n \le N \end{cases} \tag{18.67}$$

Now we can proceed to find the remaining characteristics of this queueing system.

If $0 \le n \le s$, it follows that

$$L_q = 0 \quad \text{and} \quad L = \sum_{n=0}^{s} nP_n \tag{18.68}$$

If $s \leq n \leq N$, it follows that

$$L_q = \sum_{n=s}^{N} (n-s)P_n = \sum_{j=0}^{N-s} jP_{s+j} = \sum_{j=0}^{N-s} j\frac{\rho^{s+j}}{s!s^j}P_0$$

$$= \frac{P_0}{s!}\sum_{j=0}^{N-s}\frac{j\rho^{s+j}}{s^j} = \frac{P_0}{s!}\frac{\rho^{s+1}}{s}\left[1 + 2\frac{\rho}{s} + \cdots + \frac{(N-s)\rho^{N-s-1}}{s^{N-s-1}}\right]$$

$$= \frac{P_0}{s!}\frac{\rho^{s+1}}{s}\frac{d}{d(\rho/s)}\left[\sum_{j=0}^{N-s}\left(\frac{\rho}{s}\right)^j\right]$$

$$= \frac{P_0\rho^{s+1}}{s![1-(\rho/s)]^2 s}\left[1 - \left(\frac{\rho}{s}\right)^{N-s} - (N-s)\left(\frac{\rho}{s}\right)^{N-s}\left(1-\frac{\rho}{s}\right)\right] \tag{18.69}$$

and

$$L = \sum_{n-0}^{N} nP_n = \sum_{n=0}^{s-1} nP_n + \sum_{n=s}^{N}(n-s+s)P_n$$

$$= \sum_{n=0}^{s-1} nP_n + \sum_{n=s}^{N}(n-s)P_n + s\sum_{n=s}^{N}P_n$$

$$= \sum_{n=0}^{s-1} nP_n + L_q + s\left(1 - \sum_{n=0}^{s-1}P_n\right) \tag{18.70}$$

18.9 Multichannel, Infinite, Poisson Arrival, Exponential Service Queues

This model is identical with that of Sec. 18.8, except that no restriction is imposed on the possible length of the line. Hence the system becomes a pure delay system. However, it is necessary to assume that $\rho/s < 1$, since otherwise the queue would grow indefinitely. Letting $N \longrightarrow \infty$, Eq. (18.66) becomes

$$P_0 = \left\{\sum_{n=0}^{s-1}\frac{\rho^n}{n!} + \frac{\rho^s}{s![1-(\rho/s)]}\right\}^{-1} \tag{18.71}$$

and Eq. (18.67) becomes

$$P_n = \begin{cases} \dfrac{\rho^n/n!}{\sum_{n=0}^{s-1}(\rho^n/n!) + \rho^s/s![1-(\rho/s)]} & 0 \leq n \leq s \\[2ex] \dfrac{\rho^n/s!s^{n-s}}{\sum_{n=0}^{s-1}(\rho^n/n!) + \rho^s/s![1-(\rho/s)]} & n \geq s \end{cases} \tag{18.72}$$

Equation (18.72) is sometimes referred to as the *second Erlang formula*. At first glance these equations seem cumbersome for numerical calculation. However, computation can be simplified if they can be expressed in terms of the Poisson probability function. We shall take this point up in our next example.

From Eq. (18.72) the probability that a calling customer will have to wait in line is

$$P\{n \geq s\} = \sum_{n=s}^{\infty} \frac{\rho^n}{s!s^{n-s}} P_0 = \frac{P_0}{s!}\left[\frac{\rho^s}{1-(\rho/s)}\right] \tag{18.73}$$

Now we evaluate

$$L_q = \sum_{n=s}^{\infty} (n-s)P_n = \sum_{j=0}^{\infty} jP_{s+j}$$

Similar to the manipulation that led to Eq. (18.69), we obtain

$$L_q = \frac{\lambda\mu\rho^s}{(s-1)!(s\mu-\lambda)^2} P_0 \tag{18.74}$$

$$L = L_q + \frac{\lambda}{\mu} = L_q + \rho \tag{18.75}$$

$$W_q = \frac{L_q}{\lambda} \tag{18.76}$$

and

$$W = W_q + \frac{1}{\mu} \tag{18.77}$$

where P_0 is given by Eq. (18.71).

Example 18.3 A firm that produces aluminum foil has a large number of rolling machines. Studies indicate that the number of breakdowns follow the Poisson probability law with a mean rate of three per hr. The firm has two service stations where service times for each station follow the exponential distribution with a mean of 30 min. The operating cost of each service station is $20 per hr, and the downtime of each machine costs $25 per hr in lost profits. The machines are operated 8 hr a day.

(a) What is the probability that both service stations are idle at any instant?
(b) What is the probability that a machine will have to wait for service?
(c) What is the expected daily cost of this system?
(d) What is the optimal number of service stations?

Solution

(a) This system is obviously a pure delay system with $\lambda = 3$, $\mu = 2$, and $s = 2$ service channels. The utilization factor per service station is $\rho = 1.5$. Hence, the probability that both service stations are idle at any instant is

$$P_0 = \left[\sum_{n=0}^{1} \frac{\rho^n}{n!} + \frac{\rho^2}{2![1-(\rho/2)]}\right]^{-1} = \left[1 + 1.5 + \frac{(1.5)^2}{2!(1-.75)}\right]^{-1} = \frac{1}{7}$$

(b) The probability that a machine arrival will have to wait for service is, from Eq. (18.73),

$$P\{n \geq s\} = \frac{P_0}{s!}\frac{\rho^s}{1-(\rho/s)} = \frac{(1.5)^2}{(7)(2!)(1-.75)} = .643$$

(c) By Eq. (18.77) the average time a machine spends in the system is

$$W = \frac{1}{\mu} = W_q = \frac{1}{\mu} + \frac{\mu\rho^s}{(s-1)!(s\mu-\lambda)^2}P_0$$

$$= .5 + \frac{(2)(1.5)^2}{7} = 1.143$$

The expected daily cost of downtime is

$$(\lambda W)(25)(8) = (3)(1.143)(25)(8) = \$685.8$$

The operating cost per day is

$$(s)(20)(8) = (2)(20)(8) = \$320$$

The expected daily cost of this system, then, is

$$685.8 + 320 = \$1005.8$$

(d) The optimal number of channels can be obtained by successive computation for $s = 1, 2, \ldots$. For $s = 1$, the expected daily cost tends to infinity under the assumption that the population of machines is infinite. For $s = 3$, we compute

$$W = \frac{1}{\mu} + \frac{\mu\rho^s}{(s-1)!(s\mu-\lambda)^2}\left[\sum_{n=0}^{s-1}\frac{\rho^n}{n!} + \frac{\rho^s}{s![1-(\rho/s)]}\right]^{-1}$$

$$= \frac{1}{\mu} + \frac{\mu s e^{\rho}}{(s\mu-\lambda)^2}\left(\frac{e^{-\rho}\rho^s}{s!}\right)\left[e^{\rho}\sum_{n=0}^{s-1}\frac{e^{-\rho}\rho^n}{n!} + \frac{e^{\rho}(e^{-\rho}\rho^s/s!)}{1-(\rho/s)}\right]^{-1}$$

$$= \frac{1}{\mu} + [\mu(s-\rho)]^{-1}\left[\frac{[1-(\rho/s)]F(s-1,\rho)}{P(s,\rho)} + 1\right]^{-1}$$

$$= .5 + [(2)(1.5)]^{-1}\left[\frac{(.5)(.809)}{.125} + 1\right]^{-1} = .579$$

where $F(s-1,\rho) = \sum_{n=0}^{s-1}\frac{e^{-\rho}\rho^n}{n!}$ (cumulative Poisson distribution)

and $P(s,\rho) = \frac{e^{-\rho}\rho^s}{s!}$ (Poisson probability function)

are computed from the table of the Poisson distribution function (Table I in the Appendix). The expected daily cost of the system with three service stations therefore, is

$$(\lambda W)(200) + 480 = (3)(.579)(200) + 480 = \$827.4$$

For $s = 4$,

$$W = .5 + [(2)(2.5)]^{-1}\left[\left(\frac{5}{8}\right)\left(\frac{.934}{.047}\right) + 1\right]^{-1} = .5147$$

The expected daily cost of the system with four service stations, then, is

$$(3)(.5147)(200) + 640 = \$948.82$$

Thus we see that the optimal number of service stations is three. The reader may verify that when $s \geq 5$, the reduction in downtime cost will not offset the increase in the operating cost.

18.10 Simulation of Queues

In previous sections we developed analytical solutions of some elementary queueing models, under the assumptions that arrivals follow the Poisson distribution and that service times are exponentially distributed. Here we must emphasize that these assumptions simplified the mathematical analysis to a great extent. When models become more complex and realistic, the mathematical analysis becomes extremely complicated. Although analytical solutions of many realistic models have been established, there still remain many potentially important queueing models that defy analytical solution. In such situations we resort to *simulation* technique, which is discussed in Sec. 2.6.

Usually most of the queueing problems are concerned with finding the number of service channels necessary to achieve high efficiency without unduly impairing the quality of service. The optimal number of facilities can be found by simulating many different numbers of facilities. However, in practice not "too many" simulation runs are required because very often it is possible to hedge the optimal number of facilities with acceptable confidence. With the aid of high-speed computers it is possible to solve complex queueing problems by simulation. Usually the computer running time for simulated experiments is very short. Moreover, the task of writing a simulation program is extremely simplified by the recent development of "simulation languages" as SIMSCRIPT, DYNAMO, GPSS 360, etc. These programs provide a rapid way of converting a simulation model into a computer program.

Among the significant merits of simulation is that it can be applied to theoretical distributions as well as empirical distributions. Simulation is a very useful tool in studying the transient behavior of queueing processes. This is particularly true when a service facility is provided for only a short period of time, and as a result the steady-state condition is not likely to be reached. Very often it is difficult to study the transient conditions of complex queueing problems by straightforward analytical techniques whereas it is quite simple to do so by the use of simulation.

PROBLEMS

18.1. A retailer of TV sets places an order every time he makes a sale so that the sum of sets on hand plus the unfilled orders is a constant, say N. Demand per week follows the

Poisson distribution with parameter λ. The distribution of times between order and delivery is exponential with parameter μ per week. Prospective customers are lost whenever the retailer is out of stock. Demonstrate that this inventory problem is a queueing system by describing its components.

18.2. Consider a single-server queueing system with a Poisson input and exponential service times. If $\lambda = 10$ per hr and $\mu = 15$ per hr, find the probability that a calling customer will have to wait in the system more than 15 min.

18.3. The mean arrival rate of customers at a repair shop is 10 per hr. The shop has one service station where service time follows the exponential distribution with a mean of 5 min.
(a) What is the utilization factor of this system?
(b) What is the probability of forming a waiting line that contains three customers?
(c) What is the average number of the customers in the queue?

18.4. The number of jobs that arrive at a certain production facility is Poisson with $\lambda = 8$ per min. Processing time is exactly .10 min.
(a) What is the proportion of time that the production facility remains idle?
(b) What is the expected number of jobs in the system?
(c) What is the average time a job spends in the system?

18.5. Work out the preceding problem if processing time is a random variable with a mean of .1 and a variance of .0025 min.

18.6. The mean arrival rate at a box office ticket window is 25 per hr. The window is manned by a single clerk whose service time follows the exponential distribution with parameter $\mu = 40$ per hr.
(a) What is the utilization factor of this system?
(b) What is the average number of customers in the queue?
(c) What is the average time a customer spends in the queue?

18.7. Arrivals at an information desk that has one clerk are considered to be Poisson with $\lambda = 15$ per hr. The service time is normally distributed with a mean of 2 min and a standard deviation of $\sigma = .25$ min.
(a) What is the probability that a person arriving at the desk will have to wait?
(b) What is the average number of persons in the queue?

18.8. A gas station has one multiple-purpose pump. Arrivals are Poisson with parameter $\lambda = 12$ per hr, and service times follow the exponential distribution with parameter $\mu = 15$ per hr. Suppose that the available space in the station will only accommodate three cars.
(a) What is the probability of losing potential customers?
(b) What is the average percentage of idle time of the pump?

18.9. A parking lot has four parking spaces. Arrivals are Poisson with $\lambda = 2$ per hr, and parking times are exponentially distributed with a mean of 60 min. What is the expected number of cars in the lot at any time?

18.10. In Prob. 18.1 assume that $\lambda = 6$ and $\mu = \frac{1}{2}$ and $N = 15$. Find the probability of losing potential customers.

18.11. Consider a barber shop that has only one barber chair and two chairs for waiting customers. If there are no vacant chairs, customers will not enter the shop. Customers arrive according to a Poisson input process with a mean rate of 8 per hr. The time for a haircut follows the exponential distribution with a mean time of 15 min. (Note that for purposes of illustration we assume exponential distribution.)

(a) What is the average percentage idle time of the barber?
(b) What is the probability that a calling customer will find a seat (either in a barber chair or a waiting chair)?
(c) If the barber charges \$2 per haircut, what is his expected revenue per hr?

18.12. Consider a parking lot with space for four cars. Arrivals are Poisson with parameter $\lambda = 4$ per hr, and the time spent parked is exponentially distributed with a mean of 30 min. City regulations allow only two cars to stay in the street while waiting for a space to be vacated. Suppose that the lot charges \$3 per hr for parking and the operating expenses are estimated at \$3 per hr. Compute the expected hourly profit from the lot.

18.13. Suppose that radios are brought to a repair shop manned with a single repairman. Arrivals follow the Poisson distribution with parameter λ. Service time is exponentially distributed with parameter μ when there is only one radio to be repaired. When more radios are awaiting service the repairman tends to work faster. The repair time is found to be exponentially distributed with parameter $= n^2\mu$, where n is the number of radios in the shop.
(a) Show that $P_n = \frac{(\lambda/\mu)^n}{(n!)^2} P_0 \qquad n = 0, 1, \ldots.$
(b) What is the probability that more than two radios are waiting for repair in the shop?

18.14. Consider a queueing system with Poisson input λ. When there is only one unit in the system it is processed by a single clerk with an expected service time of $1/\mu$ min. When there is more than one unit in the system the clerk is joined by an assistant who helps him to process the same unit, thereby reducing the mean service time to $1/m$ min. In both cases assume that service times are exponentially distributed.
(a) Show that $P_n = (\lambda/m)^{n-1}(\lambda/\mu)P_0$.
(b) If $\lambda/\mu = .75$ and $\lambda/m = \frac{1}{2}$, find the expected idle time of the clerk and the assistant in an 8-hr day.
(c) What is the probability that an arriving unit will have to wait for service?

18.15. Suppose that the number of arrivals per period is Poisson with $\lambda = 4$ units per hr. The cost of waiting per unit per hour is \$10, and the facility cost for servicing one unit at the persent service rate is \$15. Service time follows the exponential distribution with parameter μ. Find the value of μ so that the total expected cost of the system is minimized.

18.16. Consider a single-server queueing system with a Poisson input and exponential service time. If $\lambda = 5$ units per hr and the expected number of units in the system must not exceed ten, what is the minimum service rate? What will be the mean waiting time in the queue?

18.17. Arrivals at a check-out counter of a certain supermarket are Poisson with a mean rate of 18 customers per hr. Service time is exponentially distributed with a mean of 2.50 min.
(a) What is the probability that a customer will not have to wait?
(b) What is the probability that there will be a waiting line?
(c) What is the average time a customer spends in the system?

18.18. An ambulance service has two ambulances in operation. Accidents and other emergency calls follow the Poisson distribution with parameter $\lambda = \frac{1}{3}$ call per hr. The time necessary to reach and deliver the victim to the hospital and return to readiness is exponentially distributed with a mean of .30 hr. Find the probability that an emergency call will find no ambulance available.

18.19. Consider a self-service discount store with two cash registers. Assume that customers tend to choose a check-out lane at random and to remain there until served. Arrival to the

store follows the Poisson distribution with parameter $\lambda = 50$ customers per hr. Service times are exponential with $\mu = 30$ customers per hr.
(a) What is the average number of customers waiting for service?
(b) What is the average time a customer spends in the system?

18.20. The queueing process in the preceding problem was studied by an operations research analyst. He recommended that customers stay in a single queue until they come to the front of the line and then that they proceed to the first free cashier. Assuming that λ and μ remain unchanged, would you support the analyst's recommendation?

18.21. Consider a parking lot with ten parking spaces. Arrivals are Poisson with parameter $\lambda = 8$ cars per hr. Parking time is exponentially distributed with a mean of 75 min.
(a) What is the probability of having five cars in the lot?
(b) What is the probability that no parking space is available?

18.22. In the preceding problem show that as the number of parking spaces tends to infinity, the number of cars in the parking lot follows the Poisson distribution.

18.23. Arrivals at a bakery shop are Poisson with a mean rate of 240 customers per day. Customers are served in the order of arrival. The shop has two clerks, each having a mean service rate of 150 customers per day.
(a) What is the probability that both clerks are idle? that one is idle?
(b) What is the probability that a customer has to wait for service?
(c) What is the mean length of the queue?
(d) What is the mean waiting time of a customer in the queue?
(e) What is the average time a customer spends in the system?

18.24. A factory has two service stations where service times are exponentially distributed with a mean of 30 min. The time between calls for service is exponentially distributed with a mean of 20 min. Suppose a maximum of five machines are allowed to remain waiting for repair. Arrivals in excess of five are diverted to an outside contractor.
(a) What is the probability that both service stations are idle?
(b) What is the expected number of machines in the queue?
(c) What is the expected number of machines repaired by the contractor in an 8-hr day?

18.25. Trucks arrive at a loading dock according to a Poisson input process with a mean rate of 8 trucks per day. Each truck is loaded by a team of three-men on a first-come, first-served basis. Loading time follows the exponential distribution with a mean rate of 5 trucks per day. The cost of waiting per truck per day is $40, while the loading cost per team is $75 per day. How many teams should be employed in order to minimize expected total cost per 8-hr day?

SUGGESTED REFERENCES

[1] Cox, D. R., and W. Smith, *Queues*, London, Methuen & Co. Ltd., 1961.

[2] Kendall, D. G., "Some Problems in the Theory of Queues," *Journal of the Royal Statistical Society*, *B*, **13**, 1951.

[3] Kendall, D. G., "Stochastic Processes in the Theory of Queues and Their Analysis by the Method of the Imbedded Markov Chains," *Annals of Mathematical Statistics*, **24**, 1953.

[4] Little, J. D. C., "A Proof for the Queueing Formula: $L = \lambda W$," *Journal of Operations Research*, **9** (3), 1961.

[5] Morse, P. M., *Queues, Inventories, and Maintenance*, New York, John Wiley & Sons, Inc., 1958.

[6] Prabhu, N. U., *Queues and Inventories*, New York, John Wiley & Sons, Inc., 1965.

[7] Riordan, J., *Stochastic Service Systems*, New York, John Wiley & Sons, Inc., 1962.

[8] Saaty, T. L., *Elements of Queueing Theory*, New York, McGraw-Hill Book Company, 1961.

[9] Takács, L., *Introduction to the Theory of Queues*, London, Oxford University Press, 1962.

APPENDIX TABLES

A RANDOM DIGITS
B POWERS AND ROOTS
C VALUES OF e^{-x}
D THE CUMULATIVE STANDARDIZED NORMAL DISTRIBUTION FUNCTION
E ORDINATES OF THE NORMAL DENSITY FUNCTION
F THE t-DISTRIBUTION
G THE χ^2-DISTRIBUTION
H THE F-DISTRIBUTION
I THE POISSON DISTRIBUTION FUNCTION
J THE BINOMIAL DISTRIBUTION FUNCTION
K DISTRIBUTION OF THE STUDENTIZED RANGE
L FACTORS FOR COMPUTING CONTROL CHART LINES

TABLE A Random Digits†

(1)	(2)	(3)	(4)	(5)	(6)	(7)	(8)	(9)	(10)
1306	1189	5731	3968	5606	5084	8947	3897	1636	7810
0422	2431	0649	8085	5053	4722	6598	5044	9040	5121
6597	2022	6168	5060	8656	6733	6364	7649	1871	4328
7965	6541	5645	6243	7658	6903	9911	5740	7824	8520
7695	6937	0406	8894	0441	8135	9797	7285	5905	9539
5160	7851	8464	6789	3938	4197	6511	0407	9239	2232
2961	0551	0539	8288	7478	7565	5581	5771	5442	8761
1428	4183	4312	5445	4854	9157	9158	5218	1464	3634
3666	5642	4539	1561	7849	7520	2547	0756	1206	2033
6543	6799	7454	9052	6689	1946	2574	9386	0304	7945
9975	6080	7423	3175	9377	6951	6519	8287	8994	5532
4866	0956	7545	7723	8085	4948	2228	9583	4415	7065
8239	7068	6694	5168	3117	1586	0237	6160	9585	1133
8722	9191	3386	3443	0434	4586	4150	1224	6204	0937
1330	9120	8785	8382	2929	7089	3109	6742	2468	7025
2296	2952	4764	9070	6356	9192	4012	0618	2219	1109
3582	7052	3132	4519	9250	2486	0830	8472	2160	7046
5872	9207	7222	6494	8973	3545	6967	8490	5264	9821
1134	6324	6201	3792	5651	0538	4676	2064	0584	7996
1403	4497	7390	8503	8239	4236	8022	2914	4368	4529
3393	7025	3381	3553	2128	1021	8353	6413	5161	8583
1137	7896	3602	0060	7850	7626	0854	6565	4260	6220
7437	5198	8772	6927	8527	6851	2709	5992	7383	1071
8414	8820	3917	7238	9821	6073	6658	1280	9643	7761
8398	5224	2749	7311	5740	9771	7826	9533	3800	4553
0995	8935	2939	3092	2496	0359	0318	4697	7181	4035
6657	0755	9685	4017	6581	7292	5643	5064	1142	1297
8875	8369	7868	0190	9278	1709	4253	9346	4335	3769
8399	6702	0586	6428	7985	2979	4513	1970	1989	3105
6703	1024	2064	0393	6815	8502	1375	4171	6970	1201
4730	1653	9032	9855	0957	7366	0325	5178	7959	5371
8400	6834	3187	8688	1079	1480	6776	9888	7585	9998
3647	8002	6726	0877	4552	3238	7542	7804	3933	9475
6789	5197	8037	2354	9262	5497	0005	3986	1767	7981
2630	2721	2810	2185	6323	5679	4931	8336	6662	3566
1374	8625	1644	3342	1587	0762	6057	8011	2666	3759
1572	7625	9110	4409	0239	7059	3415	5537	2250	7292
9678	2877	7579	4935	0449	8119	6969	5383	1717	6719
0882	6781	3538	4090	3092	2365	6001	3446	9985	6007
0006	4205	2389	4365	1981	8158	7784	6256	3842	5603
4611	9861	7916	9305	2074	9462	0254	4827	9198	3974
1093	3784	4190	6332	1175	8599	9735	8584	6581	7194
3374	3545	6865	8819	3342	1676	2264	6014	5012	2458
3650	9676	1436	4374	4716	5548	8276	6235	6742	2154
7292	5749	7977	7602	9205	3599	3880	9537	4423	2330
2353	8319	2850	4026	3027	1708	3518	7034	7132	6903
1094	2009	8919	5676	7283	4982	9642	7235	8167	3366
0568	4002	0587	7165	1094	2006	7471	0940	4366	9554
5606	4070	5233	4339	6543	6695	5799	5821	3953	9458
8285	7537	1181	2300	5294	6892	1627	3372	1952	3028

† From Donald B. Owen, *Handbook of Statistical Tables:* Reading, Mass., Addison-Wesley Publishing Co., 1962. Used by permission of the publisher and the courtesy of the U. S. Atomic Energy Commission.

TABLE A—*Cont.*

(1)	(2)	(3)	(4)	(5)	(6)	(7)	(8)	(9)	(10)
2444	9039	4803	8568	1590	2420	2547	2470	8179	4617
5748	7767	2700	6289	2814	8381	1549	9519	3341	1192
7761	8583	0852	5619	6864	8506	9643	7763	9611	1289
6838	9280	2654	0812	3988	2146	5095	0150	8043	9079
6440	2631	3033	9167	4998	7036	0133	7428	9702	1376
8829	0094	2887	3802	5497	0318	5168	6377	9216	2802
9845	4796	2951	4449	1999	2691	5328	7674	7004	6212
5072	9000	3887	5739	7920	6074	4715	3681	2721	2701
9035	0553	1272	2600	3828	8197	8852	9092	8027	6144
5562	1080	2222	0336	1411	0303	7424	3713	9278	1918
2757	2650	8727	3953	9579	2442	8041	9869	2887	3933
6397	1848	1476	0787	4990	4666	1208	2769	3922	1158
9208	7641	3575	4279	1282	1840	5999	1806	7809	5885
2418	9289	6120	8141	3908	5577	3590	2317	8975	4593
7300	9006	5659	8258	3662	0332	5369	3640	0563	7939
6870	2535	8916	3245	2256	4350	6064	2438	2002	1272
2914	7309	4045	7513	3195	4166	0878	5184	6680	2655
0868	8657	8118	6340	9452	7460	3291	5778	1167	0312
7994	6579	6461	2292	9554	8309	5036	0974	9517	8293
8587	0764	6687	9150	1642	2050	4934	0027	1376	5040
8016	8345	2257	5084	8004	7949	3205	3972	7640	3478
5581	5775	7517	9076	4699	8313	8401	7147	9416	7184
2015	3364	6688	2631	2152	2220	1637	8333	4838	5699
7327	8987	5741	0102	1173	7350	7080	7420	1847	0741
3589	1991	1764	8355	9684	9423	7101	1063	4151	4875
2188	6454	7319	1215	0473	6589	2355	9579	7004	6209
2924	0472	9878	7966	2491	5662	5635	2789	2564	1249
1961	1669	2219	1113	9175	0260	4046	8142	4432	2664
2393	9637	0410	7536	0972	5153	0708	1935	1143	1704
7585	4424	2648	6728	2233	3518	7267	1732	1926	3833
0197	4021	9207	7327	9212	7017	8060	6216	1942	6817
9719	5336	5532	8537	2980	8252	4971	0110	6209	1556
8866	4785	6007	8006	9043	4109	5570	9249	9905	2152
5744	3957	8786	9023	1472	7275	1014	1104	0832	7680
7149	5721	1389	6581	7196	7072	6360	3084	7009	0239
7710	8479	9345	7773	9086	1202	8845	3163	7937	6163
5246	5651	0432	8644	6341	9661	2361	8377	8673	6098
3576	0013	7381	0124	8559	9813	9080	6984	0926	2169
3026	1464	2671	4691	0353	5289	8754	2442	7799	8983
6591	4365	8717	2365	5686	8377	8675	9798	7745	6360
0402	3257	0480	5038	1998	2935	1306	1190	2406	2596
7105	7654	4745	4482	8471	1424	2031	7803	4367	6816
7181	4140	1046	0885	1264	7755	1653	8924	5822	4401
3655	3282	2178	8134	3291	7262	8229	2866	7065	4806
5121	6717	3117	1901	5184	6467	8954	3884	0279	8635
3618	3098	9208	7429	1578	1917	7927	2696	3704	0833
0166	3638	4947	1414	4799	9189	2459	5056	5982	6154
6187	9653	3658	4730	1652	8096	8288	9368	5531	7788
1234	1448	0276	7290	1667	2823	3755	5642	4854	8844
8949	8731	4875	5724	2962	1182	2930	7539	4526	7252

TABLE A—Cont.

(1)	(2)	(3)	(4)	(5)	(6)	(7)	(8)	(9)	(10)
4357	4146	8353	9952	8004	7945	1530	5207	4730	1967
5339	7325	6862	7584	8634	3485	2278	5832	0612	8118
6583	8433	0717	0606	9284	2719	1888	2889	0285	2765
6564	3526	2171	3809	3428	5523	9078	0648	7768	3326
4811	1933	3763	6265	8931	0649	8085	6177	4450	2139
6931	7236	1230	0441	4013	1352	6563	1499	7332	3068
8755	3390	6120	7825	9005	7012	1643	9934	4044	7022
6742	2260	3443	0190	9278	1816	7697	7933	0067	2906
6655	3930	9014	6032	7574	1685	5258	3100	5358	1929
8514	4806	4124	9286	0449	5051	4772	4651	0038	1580
8135	5004	7299	8981	4689	1950	2271	2201	8344	3852
4414	6855	0127	5489	5157	6386	7492	3736	7164	0498
3727	7959	5056	5983	8021	0204	7616	4325	7454	5039
5434	7342	0314	7252	0067	2800	6292	4706	3454	6881
7195	8828	9869	2785	3186	8375	7414	7232	0401	2483
2705	8245	6251	9611	1077	0641	0195	7024	6202	3899
1547	8981	4972	1280	4286	5678	0338	8096	8284	7010
3424	1435	1354	7631	7260	7361	0151	8903	9056	8684
8969	7551	3695	4915	7921	2913	3840	9031	9747	9735
5225	8720	8898	2478	3342	9200	8836	7269	2992	6284
6432	9861	1516	2849	2539	2208	4595	8616	6170	5865
3085	5903	8319	2744	0814	7318	8619	7614	3265	5999
0264	1246	3687	9759	6995	6565	3949	1012	0179	0059
8710	2419	6065	0036	9650	2027	6042	5467	1839	5577
5736	9001	3132	4521	9973	5070	8078	4150	2276	5059
7529	1339	4802	5751	3785	7125	4922	8877	9530	6499
5133	7995	8030	7408	2186	0725	5554	5664	6791	9677
3170	9915	6960	2621	6718	4059	9919	1007	6469	5410
3024	0680	1127	8088	0200	5868	0084	6362	6808	3727
4398	3121	7749	8191	2087	8270	5233	3980	6774	8522
0082	5419	7659	2061	2506	7573	1157	3979	2309	0811
4351	6516	6814	5898	3973	8103	3616	2049	7843	0568
3268	0086	7580	1337	3884	5679	4830	4509	9587	2184
4391	8487	4884	1488	2249	6661	5774	7205	2717	0730
7328	0705	0652	9424	7082	8579	5647	5571	9667	5855
3835	2938	2671	4691	0559	8382	2825	4928	5379	8635
8731	4980	8674	4506	7262	8127	2022	2178	7463	4842
2995	7868	0683	3768	0625	9887	7060	0514	0034	8600
5597	9028	5660	5006	8325	9677	2169	3196	0357	7811
3081	5876	8150	1360	1868	9265	3277	8465	7502	6458
7406	4439	5683	6877	2920	9588	3002	2869	3746	3690
5969	9442	7696	7510	1620	4973	1911	1288	6160	9797
4765	9647	4364	1037	4975	1998	1359	1346	6125	5078
3129	2532	7577	2815	8696	9248	9410	9282	6572	3940
6906	8859	5044	8826	6218	3206	9034	0843	9832	2703
7993	3141	0103	4528	7988	4635	8478	9094	9077	5306
2549	3737	7686	0723	4505	6841	1379	6460	1869	5700
3672	7033	4844	0149	7412	6370	1884	0717	5740	8477
2217	0293	3978	5933	1032	5192	1732	2137	9357	5941
3162	9968	6369	1258	0416	4326	7840	6525	2608	5255

TABLE A—*Cont.*

(1)	(2)	(3)	(4)	(5)	(6)	(7)	(8)	(9)	(10)
1758	1489	2774	6033	9813	1052	1816	7484	1699	7350
6430	8803	0478	4157	5626	1603	1339	4666	1207	2135
4893	8857	1717	1533	6572	8408	2173	4754	0272	1305
1516	2733	7326	8674	9233	1799	5281	0797	0885	0947
4950	3171	5756	3036	9047	8719	8498	1312	7124	4787
0549	6775	9360	6639	0990	0037	7309	4702	0812	4195
1018	7027	7569	7549	2539	2315	8030	7663	3881	8264
2241	9965	9729	7092	4891	9239	0738	1804	3025	1030
1602	0708	2201	9848	6241	1084	8142	8555	7291	5016
5840	8381	1549	9902	6935	3681	6420	0214	8489	5911
1676	0367	7484	1595	5693	3008	9816	7311	6162	1024
6048	4175	8940	9029	8306	8892	4127	1709	4043	6591
5549	9621	2563	0515	0560	9021	0632	4309	4044	7010
5317	4584	9418	4600	0640	9668	6379	6515	6310	7916
2532	7784	6469	4793	5957	4123	6555	3237	6915	6960
2300	5412	3106	4877	6936	4109	8060	1896	6881	7028
1499	8699	4534	5367	7557	2701	2587	2521	2159	6991
6201	3791	2946	2863	5684	5517	7448	2227	8991	7505
6839	9736	8312	8068	7339	5395	9559	3416	6169	5484
0092	5537	1933	3186	8482	6680	2656	1864	4535	2193
1862	3253	6515	6299	2929	2219	9145	7511	2146	4962
9886	6744	3097	8894	0446	3494	8211	1723	6138	3181
5289	9071	1231	0651	9109	7448	2228	9700	0224	4595
2685	7104	7193	5506	2993	7028	4830	3866	8698	0277
6055	7092	4786	6847	4543	7448	2017	4114	8385	3625
4092	4995	0280	9371	3375	3503	4496	8642	5388	1831
5951	4937	3670	5797	5030	6524	2265	7748	7875	6976
7687	3849	4821	2373	1157	4208	3623	9399	7349	6663
8665	3463	2055	4872	2702	1807	9056	8576	4237	8757
3193	3011	8899	6721	0086	2623	7977	7578	4024	1997
9181	7365	9135	1669	2007	7784	6363	6913	6017	6588
9459	2175	5728	4933	0111	6703	1234	2410	1620	4859
9874	5278	2849	3163	6372	2600	9887	7060	3919	1111
7729	2099	7513	2774	6030	8260	9023	1368	9513	6122
4699	8102	3001	7947	1659	6571	1969	7152	7356	1062
1872	7244	3954	7422	2688	8649	0156	1965	5012	2461
9636	0123	2438	1757	4204	1650	2486	0002	4724	7412
6403	9054	7632	7469	8973	3332	0294	5062	0303	7315
4433	3293	2314	7431	2389	4094	5062	0118	0046	6070
2361	3933	8026	0431	8012	3214	8927	7355	0585	7638
4077	8463	6580	6983	0181	3327	6812	1755	3387	4569
6678	0006	3686	8478	9187	2291	9032	9852	1450	7940
6499	2582	5207	4627	0456	4245	3583	8996	1006	5839
6663	9021	0319	7908	1241	9977	7042	6923	2539	2103
9999	4503	6105	5525	1068	6272	7036	0200	6291	2841
9048	6982	3845	6865	9029	8700	0349	3416	8236	1129
5136	9653	3654	2863	5565	8923	5596	8389	9927	9092
9906	1070	5693	3012	1218	7309	4361	3041	4327	8423
4198	7035	8182	6270	6461	2079	2998	5507	9605	6734
2030	5878	8989	6789	4359	1820	5063	9199	7751	6337

TABLE B Powers and Roots

n	n^2	$\sqrt{n}$	$\sqrt{10n}$	n^3	$\sqrt[3]{n}$	$\sqrt[3]{10n}$	$\sqrt[3]{100n}$
1	1	1.000 000	3.162 278	1	1.000 000	2.154 435	4.641 589
2	4	1.414 214	4.472 136	8	1.259 921	2.714 418	5.848 035
3	9	1.732 051	5.477 226	27	1.442 250	3.107 233	6.694 330
4	16	2.000 000	6.324 555	64	1.587 401	3.419 952	7.368 063
5	25	2.236 068	7.071 068	125	1.709 976	3.684 031	7.937 005
6	36	2.449 490	7.745 967	216	1.817 121	3.914 868	8.434 327
7	49	2.645 751	8.366 600	343	1.912 931	4.121 285	8.879 040
8	64	2.828 427	8.944 272	512	2.000 000	4.308 869	9.283 178
9	81	3.000 000	9.486 833	729	2.080 084	4.481 405	9.654 894
10	100	3.162 278	10.00000	1 000	2.154 435	4.641 589	10.00000
11	121	3.316 625	10.48809	1 331	2.223 980	4.791 420	10.32280
12	144	3.464 102	10.95445	1 728	2.289 428	4.932 424	10.62659
13	169	3.605 551	11.40175	2 197	2.351 335	5.065 797	10.91393
14	196	3.741 657	11.83216	2 744	2.410 142	5.192 494	11.18689
15	225	3.872 983	12.24745	3 375	2.466 212	5.313 293	11.44714
16	256	4.000 000	12.64911	4 096	2.519 842	5.428 835	11.69607
17	289	4.123 106	13.03840	4 913	2.571 282	5.539 658	11.93483
18	324	4.242 641	13.41641	5 832	2.620 741	5.646 216	12.16440
19	361	4.358 899	13.78405	6 859	2.668 402	5.748 897	12.38562
20	400	4.472 136	14.14214	8 000	2.714 418	5.848 035	12.59921
21	441	4.582 576	14.49138	9 261	2.758 924	5.943 922	12.80579
22	484	4.690 416	14.83240	10 648	2.802 039	6.036 811	13.00591
23	529	4.795 832	15.16575	12 167	2.843 867	6.126 926	13.20006
24	576	4.898 979	15.49193	13 824	2.884 499	6.214 465	13.38866
25	625	5.000 000	15.81139	15 625	2.924 018	6.299 605	13.57209
26	676	5.099 020	16.12452	17 576	2.962 496	6.382 504	13.75069
27	729	5.196 152	16.43168	19 683	3.000 000	6.463 304	13.92477
28	784	5.291 503	16.73320	21 952	3.036 589	6.542 133	14.09460
29	841	5.385 165	17.02939	24 389	3.072 317	6.619 106	14.26043
30	900	5.477 226	17.32051	27 000	3.107 233	6.694 330	14.42250
31	961	5.567 764	17.60682	29 791	3.141 381	6.767 899	14.58100
32	1 024	5.656 854	17.88854	32 768	3.174 802	6.839 904	14.73613
33	1 089	5.744 563	18.16590	35 937	3.207 534	6.910 423	14.88806
34	1 156	5.830 952	18.43909	39 304	3.239 612	6.979 532	15.03695
35	1 225	5.916 080	18.70829	42 875	3.271 066	7.047 299	15.18294
36	1 296	6.000 000	18.97367	46 656	3.301 927	7.113 787	15.32619
37	1 369	6.082 763	19.23538	50 653	3.332 222	7.179 054	15.46680
38	1 444	6.164 414	19.49359	54 872	3.361 975	7.243 156	15.60491
39	1 521	6.244 998	19.74842	59 319	3.391 211	7.306 144	15.74061
40	1 600	6.324 555	20.00000	64 000	3.419 952	7.368 063	15.87401
41	1 681	6.403 124	20.24846	68 921	3.448 217	7.428 959	16.00521
42	1 764	6.480 741	20.49390	74 088	3.476 027	7.488 872	16.13429
43	1 849	6.557 439	20.73644	79 507	3.503 398	7.547 842	16.26133
44	1 936	6.633 250	20.97618	85 184	3.530 348	7.605 905	16.38643
45	2 025	6.708 204	21.21320	91 125	3.556 893	7.663 094	16.50964
46	2 116	6.782 330	21.44761	97 336	3.583 048	7.719 443	16.63103
47	2 209	6.855 655	21.67948	103 823	3.608 826	7.774 980	16.75069
43	2 304	6.928 203	21.90890	110 592	3.634 241	7.829 735	16.86865
49	2 401	7.000 000	22.13594	117 649	3.659 306	7.883 735	16.98499
50	2 500	7.071 068	22.36068	125 000	3.684 031	7.937 005	17.09976

TABLE B—Cont.

n	n^2	$\sqrt{n}$	$\sqrt{10n}$	n^3	$\sqrt[3]{n}$	$\sqrt[3]{10n}$	$\sqrt[3]{100n}$
50	2 500	7.071 068	22.36068	125 000	3.684 031	7.937 005	17.09976
51	2 601	7.141 428	22.58318	132 651	3.708 430	7.989 570	17.21301
52	2 704	7.211 103	22.80351	140 608	3.732 511	8.041 452	17.32478
53	2 809	7.280 110	23.02173	148 877	3.756 286	8.092 672	17.43513
54	2 916	7.348 469	23.23790	157 464	3.779 763	8.143 253	17.54411
55	3 025	7.416 198	23.45208	166 375	3.802 952	8.193 213	17.65174
56	3 136	7.483 315	23.66432	175 616	3.825 862	8.242 571	17.75808
57	3 249	7.549 834	23.87467	185 193	3.848 501	8.291 344	17.86316
58	3 364	7.615 773	24.08319	195 112	3.870 877	8.339 551	17.96702
59	3 481	7.681 146	24.28992	205 379	3.892 996	8.387 207	18.06969
60	3 600	7.745 967	24.49490	216 000	3.914 868	8.434 327	18.17121
61	3 721	7.810 250	24.69818	226 981	3.936 497	8.480 926	18.27160
62	3 844	7.874 008	24.89980	238 328	3.957 892	8.527 019	18.37091
63	3 969	7.937 254	25.09980	250 047	3.979 057	8.572 619	18.46915
64	4 096	8.000 000	25.29822	262 144	4.000 000	8.617 739	18.56636
65	4 225	8.062 258	25.49510	274 625	4.020 726	8.662 391	18.66256
66	4 356	8.124 038	25.69047	287 496	4.041 240	8.706 588	18.75777
67	4 489	8.185 353	25.88436	300 763	4.061 548	8.750 340	18.85204
68	4 624	8.246 211	26.07681	314 432	4.081 655	8.793 659	18.94536
69	4 761	8.306 624	26.26785	328 509	4.101 566	8.836 556	19.03778
70	4 900	8.366 600	26.45751	343 000	4.121 285	8.879 040	19.12931
71	5 041	8.426 150	26.64583	357 911	4.140 818	8.921 121	19.21997
72	5 184	8.485 281	26.83282	373 248	4.160 168	8.962 809	19.30979
73	5 329	8.544 004	27.01851	389 017	4.179 339	9.004 113	19.39877
74	5 476	8.602 325	27.20294	405 224	4.198 336	9.045 042	19.48695
75	5 625	8.660 254	27.38613	421 875	4.217 163	9.085 603	19.57434
76	5 776	8.717 798	27.56810	438 976	4.235 824	9.125 805	19.66095
77	5 929	8.774 964	27.74887	456 533	4.254 321	9.165 656	19.74681
78	6 084	8.831 761	27.92848	474 552	4.272 659	9.205 164	19.83192
79	6 241	8.888 194	28.10694	493 039	4.290 840	9.244 335	19.91632
80	6 400	8.944 272	28.28427	152 000	4.308 869	9.283 178	20.00000
81	6 561	9.000 000	28.46050	531 441	4.326 749	9.321 698	20.08299
82	6 724	9.055 385	28.63564	551 368	4.344 481	9.359 902	20.16530
83	6 889	9.110 434	28.80972	571 787	4.362 071	9.397 796	20.24694
84	7 056	9.165 151	28.98275	592 704	4.379 519	9.435 388	20.32793
85	7 225	9.219 544	29.15476	614 125	4.396 830	9.472 682	20.40828
86	7 396	9.273 618	29.32576	636 056	4.414 005	9.509 685	20.48800
87	7 569	9.327 379	29.49576	658 503	4.431 048	9.546 403	20.56710
88	7 744	9.380 832	29.66479	681 472	4.447 960	9.582 840	20.64560
89	7 921	9.433 981	29.83287	704 969	4.464 745	9.619 002	20.72351
90	8 100	9.486 833	30.00000	729 000	4.481 405	9.654 894	20.80084
91	8 281	9.539 392	30.16621	753 571	4.497 941	9.690 521	20.87759
92	8 464	9.591 663	30.33150	778 688	4.514 357	9.725 888	20.95379
93	8 649	9.643 651	30.49590	804 357	4.530 655	9.761 000	21.02944
94	8 836	9.965 360	30.65942	830 584	4.546 836	9.795 861	21.10454
95	9 025	9.746 794	30.82207	857 375	4.562 903	9.830 476	21.17912
96	9 216	9.797 959	30.98387	884 736	4.578 857	9.864 848	21.25317
97	9 409	9.848 858	31.14482	912 673	4.594 701	9.898 983	21.32671
98	9 604	9.899 495	31.30495	941 192	4.610 436	9.932 884	21.39975
99	9 801	9.949 874	31.46427	970 299	4.626 065	9.966 555	21.47229
100	10 000	10.000 000	31.62278	1 000 000	4.641 589	10.000 000	21.54435

TABLE C Values of e^{-x}†

x	e^{-x} Value	x	e^{-x} Value	x	e^{-x} Value
0.00	1.00000	0.45	.63763	0.90	.40657
0.01	.99005	0.46	.63128	0.91	.40252
0.02	.98020	0.47	.62500	0.92	.39852
0.03	.97045	0.48	.61878	0.93	.39455
0.04	.96079	0.49	.61263	0.94	.39063
0.05	.95123	0.50	.60653	0.95	.38674
0.06	.94176	0.51	.60050	0.96	.38289
0.07	.93239	0.52	.59452	0.97	.37908
0.08	.92312	0.53	.58860	0.98	.37531
0.09	.91393	0.54	.58275	0.99	.37158
0.10	.90484	0.55	.57965	1.00	.36788
0.11	.89583	0.56	.57121	1.01	.36422
0.12	.88692	0.57	.56553	1.02	.36060
0.13	.87809	0.58	.55990	1.03	.35701
0.14	.86936	0.59	.55433	1.04	.35345
0.15	.86071	0.60	.54881	1.05	.34994
0.16	.85214	0.61	.54335	1.06	.34646
0.17	.84366	0.62	.53794	1.07	.34301
0.18	.83527	0.63	.53259	1.08	.33960
0.19	.82696	0.64	.52729	1.09	.33622
0.20	.81873	0.65	.52205	1.10	.33287
0.21	.81058	0.66	.51685	1.11	.32956
0.22	.80252	0.67	.51171	1.12	.32628
0.23	.79453	0.68	.50662	1.13	.32303
0.24	.78663	0.69	.50158	1.14	.31982
0.25	.77880	0.70	.49659	1.15	.31664
0.26	.77105	0.71	.49164	1.16	.31349
0.27	.76338	0.72	.48675	1.17	.31037
0.28	.75578	0.73	.48191	1.18	.30728
0.29	.74826	0.74	.47711	1.19	.30422
0.30	.74082	0.75	.47237	1.20	.30119
0.31	.73345	0.76	.46767	1.21	.28920
0.32	.72615	0.77	.46301	1.22	.29523
0.33	.71892	0.78	.45841	1.23	.29229
0.34	.71177	0.79	.45384	1.24	.28938
0.35	.70469	0.80	.44933	1.25	.28650
0.36	.69768	0.81	.44486	1.26	.28365
0.37	.69073	0.82	.44043	1.27	.28083
0.38	.68386	0.83	.43605	1.28	.27804
0.39	.67706	0.84	.43171	1.29	.27527
0.40	.67032	0.85	.42741	1.30	.27253
0.41	.66365	0.86	.42316	1.31	.26982
0.42	.65705	0.87	.41895	1.32	.26714
0.43	.65051	0.88	.41478	1.33	.26448
0.44	.64404	0.89	.41066	1.34	.26185

† From R. C. Burington, *Handbook of Mathematical Tables and Formulas,* 4th ed.: New York, McGraw-Hill Book Company, 1965. Used by permission.

TABLE C—*Cont.*

x	e^{-x} Value	x	e^{-x} Value	x	e^{-x} Value
1.35	.25924	1.80	.16530	2.25	.10540
1.36	.25666	1.81	.16365	2.26	.10435
1.37	.25411	1.82	.16203	2.27	.10331
1.38	.25158	1.83	.16041	2.28	.10228
1.39	.24908	1.84	.15882	2.29	.10127
1.40	.24660	1.85	.15742	2.30	.10026
1.41	.24414	1.86	.15567	2.31	.09926
1.42	.24171	1.87	.15412	2.32	.09827
1.43	.23931	1.88	.15259	2.33	.09730
1.44	.23693	1.89	.15107	2.34	.09633
1.45	.23457	1.90	.14957	2.35	.09537
1.46	.23224	1.91	.14808	2.36	.09442
1.47	.22993	1.92	.14661	2.37	.09348
1.48	.22764	1.93	.14515	2.38	.09255
1.49	.22537	1.94	.14370	2.39	.09163
1.50	.22313	1.95	.14227	2.40	.09072
1.51	.22091	1.96	.14086	2.41	.08982
1.52	.21871	1.97	.13946	2.42	.08892
1.53	.21654	1.98	.13807	2.43	.08804
1.54	.21438	1.99	.13670	2.44	.08716
1.55	.21225	2.00	.13534	2.45	.08629
1.56	.21014	2.01	.13399	2.46	.08543
1.57	.20805	2.02	.13266	2.47	.08458
1.58	.20598	2.03	.13134	2.48	.08374
1.59	.20393	2.04	.13003	2.49	.08291
1.60	.20190	2.05	.12873	2.50	.08208
1.61	.19989	2.06	.12745	2.51	.08127
1.62	.19790	2.07	.12619	2.52	.08046
1.63	.19593	2.08	.12493	2.53	.07966
1.64	.19398	2.09	.12369	2.54	.07887
1.65	.19205	2.10	.12246	2.55	.07808
1.66	.19014	2.11	.12124	2.56	.07730
1.67	.18825	2.12	.12003	2.57	.07654
1.68	.18637	2.13	.11884	2.58	.07577
1.69	.18452	2.14	.11765	2.59	.07502
1.70	.18268	2.15	.11648	2.60	.07427
1.71	.18087	2.16	.11533	2.61	.07353
1.72	.17907	2.17	.11418	2.62	.07280
1.73	.17728	2.18	.11304	2.63	.07208
1.74	.17552	2.19	.11192	2.64	.07136
1.75	.17377	2.20	.11080	2.65	.07065
1.76	.17204	2.21	.10970	2.66	.06995
1.77	.17033	2.22	.10861	2.67	.06925
1.78	.16864	2.23	.10753	2.68	.06856
1.79	.16696	2.24	.10646	2.69	.06788

TABLE C—Cont.

x	e^{-x} Value	x	e^{-x} Value	x	e^{-x} Value
2.70	.06721	2.95	.05234	4.00	.01832
2.71	.06654	2.96	.05182	4.10	.01657
2.72	.06587	2.97	.05130	4.20	.01500
2.73	.06522	2.98	.05079	4.30	.01357
2.74	.06457	2.99	.05029	4.40	.01227
2.75	.06393	3.00	.04979	4.50	.01111
2.76	.06329	3.05	.04736	4.60	.01005
2.77	.06266	3.10	.04505	4.70	.00910
2.78	.06204	3.15	.04285	4.80	.00823
2 79	.06142	3.20	.04076	4.90	.00745
2.80	.06081	3.25	.03877	5.00	.00674
2.81	.06020	3.30	.03688	5.10	.00610
2.82	.05961	3.35	.03508	5.20	.00552
2.83	.05901	3.40	.03337	5.30	.00499
2.84	.05843	3.45	.03175	5.40	.00452
2.85	.05784	3.50	.03020	5.50	.00409
2.86	.05727	3.55	.02872	5.60	.00370
2.87	.05670	3.60	.02732	5.70	.00335
2.88	.05613	3.65	.02599	5.80	.00303
2.89	.05558	3.70	.02472	5.90	.00274
2.90	.05502	3.75	.02352	6.00	.00248
2.91	.05448	3.80	.02237	6.25	.00193
2.92	.05393	3.85	.02128	6.50	.00150
2.93	.05340	3.90	.02024	6.75	.00117
2.94	.05287	3.95	.01925	7.00	.00091
				7.50	.00055
				8.00	.00034
				8.50	.00020
				9.00	.00012
				9.50	.00007
				10.00	.00005

TABLE D *The Cumulative Standardized Normal Distribution Function*†

(Note: $.0^{2}1350 = .001350$)

Entry $= P\{Z < Z_\alpha\} = \alpha$

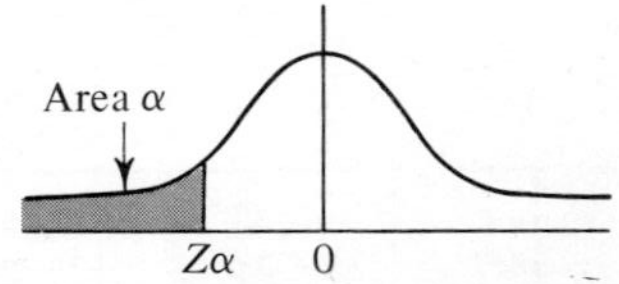

Z_α	.00	.01	.02	.03	.04	.05	.06	.07	.08	.09
− .0	.5000	.4960	.4920	.4880	.4840	.4801	.4761	.4721	.4681	.4641
− .1	.4602	.4562	.4522	.4483	.4443	.4404	.4364	.4325	.4286	.4247
− .2	.4207	.4168	.4219	.4090	.4052	.4013	.3974	.3936	.3897	.3859
− .3	.3821	.3783	.3745	.3707	.3669	.3632	.3594	.3557	.3520	.3483
− .4	.3446	.3409	.3372	.3336	.3300	.3264	.3228	.3192	.3156	.3121
− .5	.3085	.3050	.3015	.2981	.2946	.2912	.2877	.2843	.2810	.2776
− .6	.2743	.2709	.2676	.2643	.2611	.2578	.2546	.2514	.2483	.2451
− .7	.2420	.2389	.2358	.2327	.2297	.2266	.2236	.2206	.2177	.2148
− .8	.2119	.2090	.2061	.2033	.2005	.1977	.1949	.1922	.1894	.1867
− .9	.1841	.1814	.1788	.1762	.1736	.1711	.1685	.1660	.1635	.1611
−1.0	.1587	.1562	.1539	.1515	.1492	.1469	.1446	.1423	.1401	.1379
−1.1	.1357	.1335	.1314	.1292	.1271	.1251	.1230	.1210	.1190	.1170
−1.2	.1151	.1131	.1112	.1093	.1075	.1056	.1038	.1020	.1003	.09853
−1.3	.09680	.09510	.09342	.09176	.09012	.08851	.08691	.08534	.08379	.08226
−1.4	.08076	.07927	.07780	.07636	.07493	.07353	.07215	.07078	.06944	.06811
−1.5	.06681	.06552	.06426	.06301	.06178	.06057	.05938	.05821	.05705	.05592
−1.6	.05480	.05370	.05262	.05155	.05050	.04947	.04846	.04746	.04648	.04551
−1.7	.04457	.04363	.04272	.04182	.04093	.04006	.03920	.03836	.03754	.03673
−1.8	.03593	.03515	.03438	.03362	.03288	.03216	.03144	.03074	.03005	.02938
−1.9	.02872	.02807	.02743	.02680	.02619	.02559	.02500	.02442	.02385	.02330
−2.0	.02275	.02222	.02169	.02118	.02068	.02018	.01970	.01923	.01876	.01831
−2.1	.01786	.01743	.01700	.01659	.01618	.01578	.01539	.01500	.01463	.01426
−2.2	.01390	.01355	.01321	.01287	.01255	.01222	.01191	.01160	.01130	.01101
−2.3	.01072	.01044	.01017	$.0^{2}9903$	$.0^{2}9642$	$.0^{2}9387$	$.0^{2}9137$	$.0^{2}8894$	$.0^{2}8656$	$.0^{2}8424$
−2.4	$.0^{2}8198$	$.0^{2}7976$	$.0^{2}7760$	$.0^{2}7549$	$.0^{2}7344$	$.0^{2}7143$	$.0^{2}6947$	$.0^{2}6756$	$.0^{2}6569$	$.0^{2}6387$
−2.5	$.0^{2}6210$	$.0^{2}6037$	$.0^{2}5868$	$.0^{2}5703$	$.0^{2}5543$	$.0^{2}5386$	$.0^{2}5234$	$.0^{2}5085$	$.0^{2}4940$	$.0^{2}4799$
−2.6	$.0^{2}4661$	$.0^{2}4527$	$.0^{2}4396$	$.0^{2}4269$	$.0^{2}4145$	$.0^{2}4025$	$.0^{2}3907$	$.0^{2}3793$	$.0^{2}3681$	$.0^{2}3573$
−2.7	$.0^{2}3467$	$.0^{2}3364$	$.0^{2}3264$	$.0^{2}3167$	$.0^{2}3072$	$.0^{2}2980$	$.0^{2}2890$	$.0^{2}2803$	$.0^{2}2718$	$.0^{2}2635$
−2.8	$.0^{2}2555$	$.0^{2}2477$	$.0^{2}2401$	$.0^{2}2327$	$0^{2}2256$	$.0^{2}2186$	$.0^{2}2118$	$.0^{2}2052$	$.0^{2}1988$	$.0^{2}1926$
−2.9	$.0^{2}1866$	$.0^{2}1807$	$.0^{2}1750$	$.0^{2}1695$	$.0^{2}1641$	$.0^{2}1589$	$.0^{2}1538$	$.0^{2}1489$	$.0^{2}1441$	$.0^{2}1395$
−3.0	$.0^{2}1350$	$.0^{2}1306$	$.0^{2}1264$	$.0^{2}1223$	$.0^{2}1183$	$.0^{2}1144$	$.0^{2}1107$	$.0^{2}1070$	$.0^{2}1035$	$.0^{2}1001$

† From W. C. Guenther, *Analysis of Variance:* Englewood Cliffs, N. J., Prentice-Hall, Inc. Used by permission.

TABLE D—Cont.

(Note: $.9^{2}8650 = .998650$)

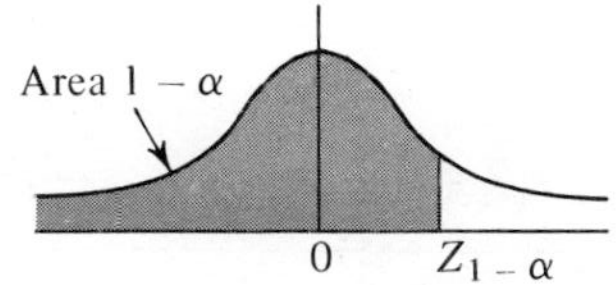

Entry = $P\{Z < Z_{1-\alpha}\} = 1 - \alpha$

$Z_{1-\alpha}$	.00	.01	.02	.03	.04	.05	.06	.07	.08	.09
.0	.5000	.5040	.5080	.5120	.5160	.5199	.5239	.5279	.5319	.5359
.1	.5398	.5438	.5478	.5517	.5557	.5596	.5636	.5675	.5714	.5753
.2	.5793	.5832	.5871	.5910	.5948	.5987	.6026	.6064	.6103	.6141
.3	.6179	.6217	.6255	.6293	.6331	.6368	.6406	.6443	.6480	.6517
.4	.6554	.6591	.6628	.6664	.6700	.6736	.6772	.6808	.6844	.6879
.5	.6915	.6950	.6985	.7019	.7054	.7088	.7123	.7157	.7190	.7224
.6	.7257	.7291	.7324	.7357	.7389	.7422	.7454	.7486	.7517	.7549
.7	.7580	.7611	.7642	.7673	.7703	.7734	.7764	.7794	.7823	.7852
.8	.7881	.7910	.7939	.7967	.7995	.8023	.8051	.8078	.8106	.8133
.9	.8159	.8186	.8212	.8238	.8264	.8289	.8315	.8340	.8365	.8389
1.0	.8413	.8438	.8461	.8485	.8508	.8531	.8554	.8577	.8599	.8661
1.1	.8643	.8665	.8686	.8708	.8729	.8749	.8770	.8790	.8810	.8830
1.2	.8849	.8869	.8888	.8907	.8925	.8944	.8962	.8980	.8997	.90147
1.3	.90320	.90490	.90658	.90824	.90988	.91149	.91309	.91466	.91621	.91774
1.4	.91924	.92073	.92220	.92364	.92507	.92647	.92785	.92922	.93056	.93189
1.5	.93319	.93448	.93574	.93699	.93822	.93943	.94062	.94179	.94295	.94408
1.6	.94520	.94630	.94738	.94845	.94950	.95053	.95154	.95254	.95352	.95449
1.7	.95543	.95637	.95728	.95818	.95907	.95994	.96080	.96164	.96246	.96327
1.8	.96407	.96485	.96562	.96638	.96712	.96784	.96856	.96926	.96995	.97062
1.9	.97128	.97193	.97257	.97320	.97381	.97441	.97500	.97558	.97615	.97670
2.0	.97725	.97778	.97831	.97882	.97932	.97982	.97030	.98077	.98124	.98169
2.1	.98214	.98257	.98300	.98341	.98382	.98422	.98461	.98500	.98537	.98574
2.2	.98610	.98645	.98679	.98713	.98745	.98778	.98809	.98840	.98870	.98899
2.3	.98928	.98956	.98983	$.9^{2}0097$	$.9^{2}0358$	$.9^{2}0613$	$.9^{2}0863$	$.9^{2}1106$	$.9^{2}1344$	$.9^{2}1576$
2.4	$.9^{2}1802$	$.9^{2}2024$	$.9^{2}2240$	$.9^{2}2451$	$.9^{2}2656$	$.9^{2}2857$	$.9^{2}3053$	$.9^{2}3244$	$.9^{2}3431$	$.9^{2}3613$
2.5	$.9^{2}3790$	$.9^{2}3963$	$.9^{2}4132$	$.9^{2}4297$	$.9^{2}4457$	$.9^{2}4614$	$.9^{2}4766$	$.9^{2}4915$	$.9^{2}5060$	$.9^{2}5201$
2.6	$.9^{2}5339$	$.9^{2}5473$	$.9^{2}5604$	$.9^{2}5731$	$.9^{2}5855$	$.9^{2}5975$	$.9^{2}6093$	$.9^{2}6207$	$.9^{2}6319$	$.9^{2}6427$
2.7	$.9^{2}6533$	$.9^{2}6636$	$.9^{2}6736$	$.9^{2}6833$	$.9^{2}6928$	$.9^{2}7020$	$.9^{2}7110$	$.9^{2}7197$	$.9^{2}7282$	$.9^{2}7365$
2.8	$.9^{2}7445$	$.9^{2}7523$	$.9^{2}7599$	$.9^{2}7673$	$.9^{2}7744$	$.9^{2}7814$	$.9^{2}7882$	$.9^{2}7948$	$.9^{2}8012$	$.9^{2}8074$
2.9	$.9^{2}8134$	$.9^{2}8193$	$.9^{2}8250$	$.9^{2}8305$	$.9^{2}8359$	$.9^{2}8411$	$.9^{2}8462$	$.9^{2}8511$	$.9^{2}8559$	$.9^{2}8605$
3.0	$.9^{2}8650$	$.9^{2}8684$	$.9^{2}8736$	$.9^{2}8777$	$.9^{2}8817$	$.9^{2}8856$	$.9^{2}8893$	$.9^{2}8930$	$.9^{2}8965$	$.9^{2}8999$

TABLE E Ordinates of the Normal Density Function

$$f(x) = \frac{1}{\sqrt{2\pi}} e^{-x^2/2}$$

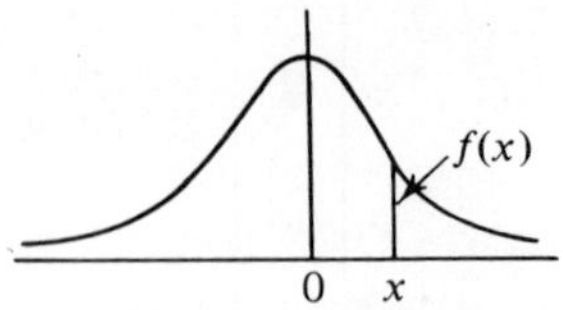

x	.00	.01	.02	.03	.04	.05	.06	.07	.08	.09
.0	.3989	3989	.3989	.3988	.3986	.3984	.3982	.3980	.3977	.3973
.1	.3970	.3965	.3961	.3956	.3951	.3945	.3939	.3932	.3925	.3918
.2	.3910	.3902	.3894	.3885	.3876	.3867	.3857	.3847	.3836	.3825
.3	.3814	.3802	.3790	.3778	.3765	.3752	.3739	.3725	.3712	.3697
.4	.3683	.3668	.3653	.3637	.3621	3605	.3589	.3572	.3555	.3538
.5	.3521	.3503	.3485	.3467	.3448	.3429	.3410	.3391	.3372	.3352
.6	.3332	.3312	.3292	.3271	.3251	.3230	.3209	.3187	.3166	.3144
.7	.3123	.3101	.3079	.3056	.3034	.3011	.2989	.2966	.2943	.2920
.8	.2897	.2874	.2850	.2827	.2803	.2780	.2756	.2732	.2709	.2685
.9	.2661	.2637	.2613	.2589	.2565	.2541	.2516	.2492	.2468	.2444
1.0	.2420	.2396	.2371	.2347	.2323	.2299	.2275	.2251	.2227	.2203
1.1	.2179	.2155	.2131	.2107	.2083	.2059	.2036	.2012	.1989	.1965
1.2	.1942	.1919	.1895	.1872	.1849	.1826	.1804	.1781	.1758	.1736
1.3	.1714	.1691	.1669	.1647	.1626	.1604	.1582	.1561	.1539	.1518
1.4	.1497	.1476	.1456	.1435	.1415	.1394	.1374	.1354	.1334	.1315
1.5	.1295	.1276	.1257	.1238	.1219	.1200	.1182	.1163	.1145	.1127
1.6	.1109	.1092	.1074	.1057	.1040	.1023	.1006	.0989	.0973	.0957
1.7	.0940	.0925	.0909	.0893	.0878	.0863	.0848	.0833	.0818	.0804
1.8	.0790	.0775	.0761	.0748	.0734	.0721	.0707	.0694	.0681	.0669
1.9	.0656	.0644	.0632	.0620	.0608	.0596	.0584	.0573	.0562	.0551
2.0	.0540	.0529	.0519	.0508	.0498	.0488	.0478	.0468	.0459	.0449
2.1	.0440	.0431	.0422	.0413	.0404	.0396	.0387	.0379	.0371	.0363
2.2	.0355	.0347	.0339	.0332	.0325	.0317	.0310	.0303	.0297	.0290
2.3	.0283	.0277	.0270	.0264	.0258	.0252	.0246	.0241	.0235	.0229
2.4	.0224	.0219	.0213	.0208	.0203	.0198	.0194	.0189	.0184	.0180
2.5	.0175	.0171	.0167	.0163	.0158	.0154	.0151	.0147	.0143	.0139
2.6	.0136	.0132	.0129	.0126	.0122	.0119	.0116	.0113	.0110	.0107
2.7	.0104	.0101	.0099	.0096	.0093	.0091	.0088	.0086	.0084	.0081
2.8	.0079	.0077	.0075	.0073	.0071	.0069	.0067	.0065	.0063	.0061
2.9	.0060	.0058	.0056	.0055	.0053	.0051	.0050	.0048	.0047	.0046
3.0	.0044	.0043	.0042	.0040	.0039	.0038	.0037	.0036	.0035	.0034
3.1	.0033	.0032	.0031	.0030	.0029	.0028	.0027	.0026	.0025	.0025
3.2	.0024	.0023	.0022	.0022	.0021	.0020	.0020	.0019	.0018	.0018
3.3	.0017	.0017	.0016	.0016	.0015	.0015	.0014	.0014	.0013	.0013
3.4	.0012	.0012	.0012	.0011	.0011	.0100	.0010	.0010	.0009	.0009
3.5	.0009	.0008	.0008	.0008	.0008	.0007	.0007	.0007	.0007	.0006
3.6	.0006	.0006	.0006	.0005	.0005	.0005	.0005	.0005	.0005	.0004
3.7	.0004	.0004	.0004	.0004	.0004	.0004	.0003	.0003	.0003	.0003
3.8	.0003	.0003	.0003	.0003	.0003	.0002	.0002	.0002	.0002	.0002
3.9	.0002	.0002	.0002	.0002	.0002	.0002	.0002	.0002	.0001	.0001

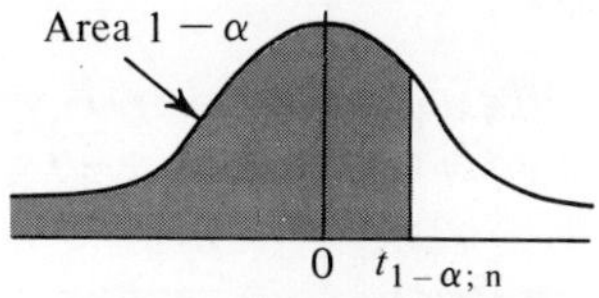

TABLE F The t-Distribution†

Entry $= t_{1-\alpha;\,n}$ **where** $P\{t_n < t_{1-\alpha;\,n}\} = 1 - \alpha$

$t_{1-\alpha;\,n} = -t_{\alpha;\,n}$

n \ $1-\alpha$	.60	.70	.80	.90	.95	.975	.990	.995	.999	.9995
1	.325	.727	1.376	3.078	6.314	12.71	31.82	63.66	318.3	636.6
2	.289	.617	1.061	1.886	2.920	4.303	6.965	9.925	22.33	31.60
3	.277	.584	.978	1.638	2.353	3.182	4.541	5.841	10.22	12.94
4	.271	.569	.941	1.533	2.132	2.776	3.747	4.604	7.173	8.610
5	.267	.559	.920	1.476	2.015	2.571	3.365	4.032	5.893	6.859
6	.265	.553	.906	1.440	1.943	2.447	3.143	3.707	5.208	5.959
7	.263	.549	.896	1.415	1.895	2.365	2.998	3.499	4.785	5.405
8	.262	.546	.889	1.397	1.860	2.306	2.896	3.355	4.501	5.041
9	.261	.543	.883	1.383	1.833	2.262	2.821	3.250	4.297	4.781
10	.260	.542	.879	1.372	1.812	2.228	2.764	3.169	4.144	4.587
11	.260	.540	.876	1.363	1.796	2.201	2.718	3.106	4.025	4.437
12	.259	.539	.873	1.356	1.782	2.179	2.681	3.055	3.930	4.318
13	.259	.538	.870	1.350	1.771	2.160	2.650	3.012	3.852	4.221
14	.258	.537	.868	1.345	1.761	2.145	2.624	2.977	3.787	4.140
15	.258	.536	.866	1.341	1.753	2.131	2.602	2.947	3.733	4.073
16	.258	.535	.865	1.337	1.746	2.120	2.583	2.921	3.686	4.015
17	.257	.534	.863	1.333	1.740	2.110	2.567	2.898	3.646	3.965
18	.257	.534	.862	1.330	1.734	2.101	2.552	2.878	3.611	3.922
19	.257	.533	.861	1.328	1.729	2.093	2.539	2.861	3.579	3.883
20	.257	.533	.860	1.325	1.725	2.086	2.528	2.845	3.552	3.850
21	.257	.532	.859	1.323	1.721	2.080	2.518	2.831	3.527	3.819
22	.256	.532	.858	1.321	1.717	2.074	2.508	2.819	3.505	3.792
23	.256	.532	.858	1.319	1.714	2.069	2.500	2.807	3.485	3.767
24	.256	.531	.857	1.318	1.711	2.064	2.492	2.797	3.467	3.745
25	.256	.531	.856	1.316	1.708	2.060	2.485	2.787	3.450	3.725
26	.256	.531	.856	1.315	1.706	2.056	2.479	2.779	3.435	3.707
27	.256	.531	.855	1.314	1.703	2.052	2.473	2.771	3.421	3.690
28	.256	.530	.855	1.313	1.701	2.048	2.467	2.763	3.408	3.674
29	.256	.530	.854	1 311	1.699	2.045	2.462	2.756	3.396	3.659
30	.256	.530	.854	1.310	1.697	2.042	2.457	2.750	3.385	3.646
40	.255	.529	.851	1.303	1.684	2.021	2.423	2.704	3.307	3.551
50	.255	.528	.849	1.298	1.676	2.009	2.403	2.678	3.262	3.495
60	.254	.527	.848	1.296	1.671	2.000	2.390	2.660	3.232	3.460
80	.254	.527	.846	1.292	1.664	1.990	2.374	2.639	3.195	3.415
100	.254	.526	.845	1.290	1.660	1.984	2.365	2.626	3.174	3.389
200	.254	.525	.843	1.286	1.653	1.972	2.345	2.601	3.131	3.339
500	.253	.525	.842	1.283	1.648	1.965	2.334	2.586	3.106	3.310
∞	.253	.524	.842	1.282	1.645	1.960	2.326	2.576	3.090	3.291

† From W. C. Guenther, *Analysis of Variance:* Englewood Cliffs, N. J., Prentice-Hall, Inc. Used by permission.

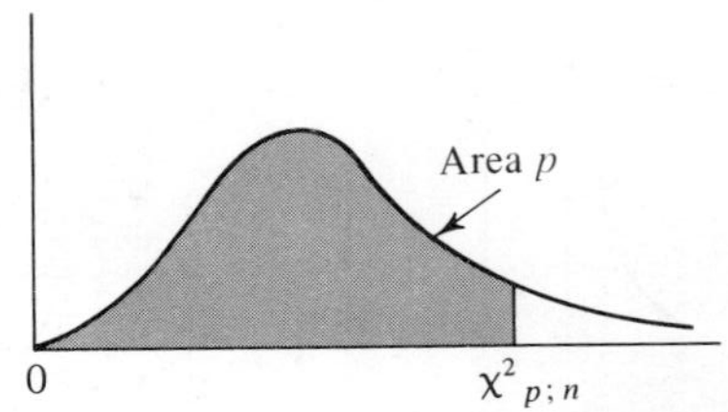

TABLE G The χ^2-Distribution†

Entry $= \chi^2_{p;\,n}$ where $P\{\chi^2_n < \chi^2_{p;\,n}\} = p$

n \ p	0.005	0.010	0.025	0.050	0.100	0.250	0.500	0.750	0.900	0.950	0.975	0.990	0.995	0.999
1	0.00004	0.00016	0.00098	0.00393	0.01579	0.1015	0.4549	1.323	2.706	3.841	5.024	6.635	7.879	10.83
2	0.0100	0.0201	0.0506	0.1026	0.2107	0.5754	1.386	2.773	4.605	5.991	7.378	9.210	10.60	13.82
3	0.0717	0.1148	0.2158	0.3518	0.5844	1.213	2.366	4.108	6.251	7.815	9.348	11.34	12.84	16.27
4	0.2070	0.2971	0.4844	0.7107	1.064	1.923	3.357	5.385	7.779	9.488	11.14	13.28	14.86	18.47
5	0.4117	0.5543	0.8312	1.145	1.610	2.675	4.351	6.626	9.236	11.07	12.83	15.09	16.75	20.52
6	0.6757	0.8721	1.2373	1.635	2.204	3.455	5.348	7.841	10.64	12.59	14.45	16.81	18.55	22.46
7	0.9893	1.239	1.690	2.167	2.833	4.255	6.346	9.037	12.02	14.07	16.01	18.48	20.28	24.32
8	1.344	1.646	2.180	2.733	3.490	5.071	7.344	10.22	13.36	15.51	17.53	20.09	21.96	26.12
9	1.735	2.088	2.700	3.325	4.168	5.899	8.343	11.39	14.68	16.92	19.02	21.67	23.59	27.88
10	2.156	2.558	3.247	3.940	4.865	6.737	9.342	12.55	15.99	18.31	20.48	23.21	25.19	29.59
11	2.603	3.053	3.816	4.575	5.578	7.584	10.34	13.70	17.28	19.68	21.92	24.72	26.76	31.26
12	3.074	3.571	4.404	5.226	6.304	8.438	11.34	14.85	18.55	21.03	23.34	26.22	28.30	32.91
13	3.565	4.107	5.009	5.892	7.041	9.299	12.34	15.98	19.81	22.36	24.74	27.69	29.82	34.53
14	4.075	4.660	5.629	6.571	7.790	10.17	13.34	17.12	21.06	23.68	26.12	29.14	31.32	36.12
15	4.601	5.229	6.262	7.261	8.547	11.04	14.34	18.25	22.31	25.00	27.49	30.58	32.80	37.70
16	5.142	5.812	6.908	7.962	9.312	11.91	15.34	19.37	23.54	26.30	28.85	32.00	34.27	39.25
17	5.697	6.408	7.564	8.672	10.09	12.79	16.34	20.49	24.77	27.59	30.19	33.41	35.72	40.79
18	6.265	7.015	8.231	9.390	10.86	13.68	17.34	21.60	25.99	28.87	31.53	34.81	37.16	42.31
19	6.844	7.633	8.907	10.12	11.65	14.56	18.34	22.72	27.20	30.14	32.85	36.19	38.58	43.82
20	7.434	8.260	9.591	10.85	12.44	15.45	19.34	23.83	28.41	31.41	34.17	37.57	40.00	45.32
21	8.034	8.897	10.28	11.59	13.24	16.34	20.34	24.93	29.62	32.67	35.48	38.93	41.40	46.80
22	8.643	9.542	10.98	12.34	14.04	17.24	21.34	26.04	30.81	33.92	36.78	40.29	42.80	48.27
23	9.260	10.20	11.69	13.09	14.85	18.14	22.34	27.14	32.01	35.17	38.08	41.64	44.18	49.73
24	9.886	10.86	12.40	13.85	15.66	19.04	23.34	28.24	33.20	36.42	39.36	42.98	45.56	51.18
25	10.52	11.52	13.12	14.61	16.47	19.94	24.34	29.34	34.38	37.65	40.65	44.31	46.93	52.62
26	11.16	12.20	13.84	15.38	17.29	20.84	25.34	30.43	35.56	38.89	41.92	45.64	48.29	54.05
27	11.81	12.88	14.57	16.15	18.11	21.75	26.34	31.53	36.74	40.11	43.19	46.96	49.64	55.48
28	12.46	13.56	15.31	16.93	18.94	22.66	27.34	32.62	37.92	41.34	44.46	48.28	50.99	56.89
29	13.12	14.26	16.05	17.71	19.77	23.57	28.34	33.71	39.09	42.56	45.72	49.59	52.34	58.30

TABLE G—Cont.

n \ p	0.005	0.010	0.025	0.050	0.100	0.250	0.500	0.750	0.900	0.950	0.975	0.990	0.995	0.999
30	13.79	14.95	16.79	18.49	20.60	24.48	29.34	34.80	40.26	43.77	46.98	50.89	53.67	59.70
40	20.71	22.16	24.43	26.51	29.05	33.66	39.34	45.62	51.80	55.76	59.34	63.69	66.77	73.40
50	27.99	29.71	32.36	34.76	37.69	42.94	49.33	56.33	63.17	67.50	71.42	76.15	79.49	86.66
60	35.53	37.48	40.48	43.19	46.46	52.29	59.33	66.98	74.40	79.08	83.30	88.38	91.95	99.61
70	43.28	45.44	48.76	51.74	55.33	61.70	69.33	77.58	85.53	90.53	95.02	100.4	104.2	112.3
80	51.17	53.54	57.15	60.39	64.28	71.14	79.33	88.13	96.58	101.9	106.6	112.3	116.3	124.8
90	59.20	61.75	65.65	69.13	73.29	80.62	89.33	98.65	107.6	113.1	118.1	124.1	128.3	137.2
100	67.33	70.06	74.22	77.93	82.36	90.13	99.33	109.1	118.5	124.3	129.6	135.8	140.2	149.4

†Reproduced with permission from E. S. Pearson and H. O. Hartley, *Biometrika Tables for Statisticians,* 2nd ed., Vol. 1, pp. 130–131: New York, Cambridge University Press, 1958.

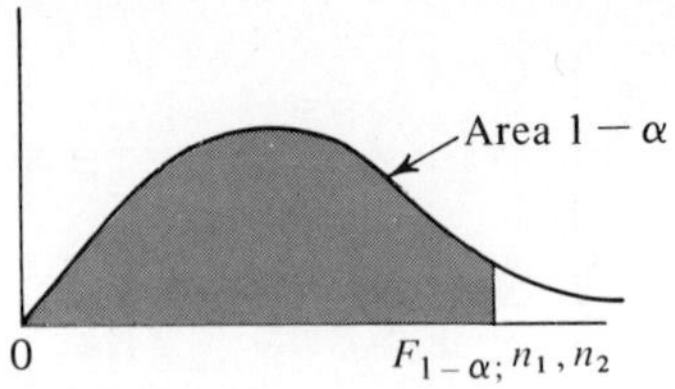

TABLE H The F-Distribution†

Entry $= F_{1-\alpha;\, n_1, n_2}$
where $P\{F_{n_1, n_2} < F_{1-\alpha;\, n_1, n_2}\} = 1 - \alpha$

df for denominator (n_2)	$1-\alpha$	df for numerator (n_1)											
		1	2	3	4	5	6	7	8	9	10	11	12
1	.75	5.83	7.50	8.20	8.58	8.82	8.98	9.10	9.19	9.26	9.32	9.36	9.41
	.90	39.9	49.5	53.6	55.8	57.2	58.2	58.9	59.4	59.9	60.2	60.5	60.7
	.95	161	200	216	225	230	234	237	239	241	242	243	244
2	.75	2.57	3.00	3.15	3.23	3.28	3.31	3.34	3.35	3.37	3.38	3.39	3.39
	.90	8.53	9.00	9.16	9.24	9 29	9.33	9.35	9.37	9.38	9.39	9.40	9.41
	.95	18.5	19.0	19.2	19.2	19.3	19.3	19.4	19.4	19.4	19.4	19.4	19.4
	.99	98.5	99.0	99.2	99.2	99.3	99.3	99.4	99.4	99.4	99.4	99.4	99.4
3	.75	2.02	2.28	2.36	2.39	2.41	2.42	2.43	2.44	2.44	2.44	2.45	2.45
	.90	5.54	5.46	5.39	5.34	5.31	5.28	5.27	5.25	5.24	5.23	5.22	5.22
	.95	10.1	9.55	9.28	9.12	9.10	8.94	8.89	8.85	8.81	8.79	8.76	8.74
	.99	34.1	30.8	29.5	28.7	28.2	27.9	27.7	27.5	27.3	27.2	27.1	27.1
4	.75	1.81	2.00	2.05	2.06	2.07	2.08	2.08	2.08	2.08	2.08	2.08	2.08
	.90	4.54	4.32	4.19	4.11	4 05	4.01	3.98	3.95	3.94	3.92	3.91	3.90
	.95	7.71	6.94	6.59	6.39	6.26	6.16	6.09	6.04	6.00	5.96	5.94	5.91
	.99	21.2	18.0	16.7	16.0	15.5	15.2	15.0	14.8	14.7	14.5	14.4	14.4
5	.75	1.69	1.85	1.88	1.89	1.89	1.89	1.89	1.89	1.89	1.89	1.89	1.89
	.90	4.06	3.78	3.62	3.52	3.45	3.40	3.37	3.34	3.32	3.30	3.28	3.27
	.95	6.61	5.79	5.41	5.19	5.05	4.95	4.88	4.82	4.77	4.74	4.71	4.68
	.99	16.3	13.3	12.1	11.4	11.0	10.7	10.5	10.3	10.2	10.1	9.96	9.89
6	.75	1.62	1.76	1.78	1.79	1.79	1.78	1.78	1.77	1.77	1.77	1.77	1.77
	.90	3.78	3.46	3.29	3.18	3.11	3.05	3.01	2.98	2.96	2.94	2.92	2.90
	.95	5.99	5.14	4.76	4.53	4.39	4.28	4.21	4.15	4.10	4.06	4.03	4.00
	.99	13.7	10.9	9.78	9.15	8.75	8.47	8.26	8.10	7.98	7.87	7.79	7.72
7	.75	1.57	1.70	1.72	1.72	1.71	1.71	1.70	1.70	1.69	1.69	1.69	1.68
	.90	3.59	3.26	3.07	2.96	2.88	2.83	2.78	2.75	2.72	2.70	2.68	2.67
	.95	5.59	4.74	4.35	4.12	3.97	3.87	3.79	3.73	3.68	3.64	3.60	3.57
	.99	12.2	9.55	8.45	7.85	7.46	7.19	6.99	6.84	6.72	6.62	6.54	6.47
8	.75	1.54	1.66	1.67	1.66	1.66	1.65	1.64	1.64	1.64	1.63	1.63	1.62
	.90	3.46	3.11	2.92	2.81	2.73	2.67	2.62	2.59	2.56	2.54	2.52	2.50
	.95	5.32	4.46	4.07	3.84	3.69	3.58	3.50	3.44	3.39	3.35	3.31	3.28
	.99	11.3	8.65	7.59	7.01	6.63	6.37	6.18	6.03	5.91	5.81	5.73	5.67
9	.75	1.51	1.62	1.63	1.63	1.62	1.61	1.60	1.60	1.59	1.59	1.58	1.58
	.90	3.36	3.01	2.81	2.69	2.61	2.55	2.51	2.47	2.44	2.42	2.40	2.38
	.95	5.12	4.26	3.86	3.63	3.48	3.37	3.29	3.23	3.18	3.14	3.10	3.07
	.99	10.6	8 02	6.99	6.42	6.06	5.80	5.61	5.47	5.35	5.26	5.18	5.11
10	.75	1.49	1.60	1.60	1.59	1.59	1.58	1.57	1.56	1.56	1.55	1.55	1.54
	.90	3.28	2.92	2.73	2.61	2.52	2.46	2.41	2.38	2.35	2.32	2.30	2.28
	.95	4.96	4.10	3.71	3.48	3.33	3.22	3.14	3.07	3.02	2.98	2.94	2.91
	.99	10.0	7.56	6.55	5.99	5.64	5.39	5.20	5.06	4.94	4.85	4.77	4.71
11	.75	1.47	1.58	1.58	1.57	1.56	1.55	1.54	1.53	1.53	1.52	1.52	1,51
	.90	3.23	2.86	2.66	2.54	2.45	2.39	2.34	2.30	2.27	2.25	2.23	2.21
	.95	4 84	3.98	3.59	3.36	3.20	3.09	3.01	2.95	2.90	2.85	2.82	2.79
	.99	9.65	7.21	6.22	5.67	5.32	5.07	4.89	4.74	4.63	4.54	4.46	4.40
12	.75	1.46	1.56	1.56	1.55	1.54	1.53	1.52	1.51	1.51	1.50	1.50	1.49
	.90	3.18	2.81	2.61	2.48	2.39	2.33	2.28	2.24	2.21	2.19	2.17	2.15
	.95	4.75	3.89	3.49	3.26	3.11	3.00	2.91	2.85	2.80	2.75	2.72	2.69
	.99	9.33	6.93	5.95	5.41	5.06	4.82	4.64	4.50	4.39	4.30	4.22	4.16

† This table is abridged from Table 18 in E. S. Pearson and H. O. Hartley, *Biometrika Tables for Statisticians,* 2nd ed., Vol. 1: New York: Cambridge University Press, 1958. Used by permission.

TABLE H—Cont.

df for numerator (n_1)												$1-\alpha$	df for denominator (n_2)
15	20	24	30	40	50	60	100	120	200	500	∞		
9.49	9.58	9.63	9.67	9.71	9.74	9.76	9.78	9.80	9.82	9.84	9.85	.75	1
61.2	61.7	62.0	62.3	62.5	62.7	62.8	63.0	63.1	63.2	63.3	63.3	.90	
246	248	249	250	251	252	252	253	253	254	254	254	.95	
3.41	3.43	3.43	3.44	3.45	3.45	3.46	3.47	3.47	3.48	3.48	3.48	.75	2
9.42	9.44	9.45	9.46	9.47	9.47	9.47	9.48	9.48	9.49	9.49	9.49	.90	
19.4	19.4	19.5	19.5	19.5	19.5	19.5	19.5	19.5	19.5	19.5	19.5	.95	
99.4	99.4	99.5	99.5	99.5	99.5	99.5	99.5	99.5	99.5	99.5	99.5	.99	
2.46	2.46	2.46	2.47	2.47	2.47	2.47	2.47	2.47	2.47	2.47	2.47	.75	3
5.20	5.18	5.18	5.17	5.16	5.15	5.15	5.14	5.14	5.14	5.14	5.13	.90	
8.70	8.66	8.64	8.62	8.59	8.58	8.57	8.55	8.55	8.54	8.53	8.53	.95	
26.9	26.7	26.6	26.5	26.4	26.4	26.3	26.2	26.2	26.2	26.1	26.1	.99	
2.08	2.08	2.08	2.08	2.08	2.08	2.08	2.08	2.08	2.08	2.08	2.08	.75	4
3.87	3.84	3.83	3.82	3.80	3.80	3.79	3.78	3.78	3.77	3.76	3.76	.90	
5.86	5.80	5.77	5.75	5.72	5.70	5.69	5.66	5.66	5.65	5.64	5.63	.95	
14.2	14.0	13.9	13.8	13.7	13.7	13.7	13.6	13.6	13.5	13.5	13.5	.99	
1.89	1.88	1.88	1.88	1.88	1.88	1.87	1.87	1.87	1.87	1.87	1.87	.75	5
3.24	3.21	3.19	3.17	3.16	3.15	3.14	3.13	3.12	3.12	3.11	3.10	.90	
4.62	4.56	4.53	4.50	4.46	4.44	4.43	4.41	4.40	4.39	4.37	4.36	.95	
9.72	9.55	9.47	9.38	9.29	9.24	9.20	9.13	9.11	9.08	9.04	9.02	.99	
1.76	1.76	1.75	1.75	1.75	1.75	1.74	1.74	1.74	1.74	1.74	1.74	.75	6
2.87	2.84	2.82	2.80	2.78	2.77	2.76	2.75	2.74	2.73	2.73	2.72	.90	
3.94	3.87	3.84	3.81	3.77	3.75	3.74	3.71	3.70	3.69	3.68	3.67	.95	
7.56	7.40	7.31	7.23	7.14	7.09	7.06	6.99	6.97	6.93	6.90	6.88	.99	
1.68	1.67	1.67	1.66	1.66	1.66	1.65	1.65	1.65	1.65	1.65	1.65	.75	7
2.63	2.59	2.58	2.56	2.54	2.52	2.51	2.50	2.49	2.48	2.48	2.47	.90	
3.51	3.44	3.41	3.38	3.34	3.32	3.30	3.27	3.27	3.25	3.24	3.23	.95	
6.31	6.16	6.07	5.99	5.91	5.86	5.82	5.75	5.74	5.70	5.67	5.65	.99	
1.62	1.61	1.60	1.60	1.59	1.59	1.59	1.58	1.58	1.58	1.58	1.58	.75	8
2.46	2.42	2.40	2.38	2.36	2.35	2.34	2.32	2.32	2.31	2.30	2.29	.90	
3.22	3.15	3.12	3.08	3.04	3.02	3.01	2.97	2.97	2.95	2.94	2.93	.95	
5.52	5.36	5.28	5.20	5.12	5.07	5.03	4.96	4.95	4.91	4,88	4.86	.99	
1.57	1.56	1.56	1.55	1.55	1.54	1.54	1.53	1.53	1.53	1.53	1.53	.75	9
2.34	2.30	2.28	2.25	2.23	2.22	2.21	2.19	2.18	2.17	2.17	2.16	.90	
3.01	2.94	2.90	2.86	2.83	2.80	2.79	2.76	2.75	2.73	2.72	2.71	.95	
4.96	4.81	4.73	4.65	4.57	4.52	4.48	4.42	4.40	4.36	4.33	4.31	.99	
1.53	1.52	1.52	1.51	1.51	1.50	1.50	1.49	1.49	1.49	1.48	1.48	.75	10
2.24	2.20	2.18	2.16	2.13	2.12	2.11	2.09	2.08	2.07	2.06	2.06	.90	
2.85	2.77	2.74	2.70	2.66	2.64	2.62	2.59	2.58	2.56	2.55	2.54	.95	
4.56	4.41	4.33	4.25	4.17	4.12	4.08	4.01	4.00	3.96	3.93	3.91	.99	
1.50	1.49	1.49	1.48	1.47	1.47	1.47	1.46	1.46	1.46	1.45	1.45	.75	11
2.17	2.12	2.10	2.08	2.05	2.04	2.03	2.00	2.00	1.99	1.98	1.97	.90	
2.72	2.65	2.61	2.57	2.53	2.51	2.49	2.46	2.45	2.43	2.42	2.40	.95	
4.25	4.10	4.02	3.94	3.86	3.81	3.78	3.71	3.69	3.66	3.62	3.60	.99	
1.48	1.47	1.46	1.45	1.45	1.44	1.44	1.43	1.43	1.43	1.42	1.42	.75	12
2.10	2.06	2.04	2.01	1.99	1.97	1.96	1.94	1.93	1.92	1.91	1.90	.90	
2.62	2.54	2.51	2.47	2.43	2.40	2.38	2.35	2.34	2.32	2.31	2.30	.95	
4.01	3.86	3.78	3.70	3.62	3.57	3.54	3.47	3.45	3.41	3.38	3.36	.99	

TABLE H—Cont.

df for denominator (n_2)	$1-\alpha$	df for numerator (n_1)											
		1	2	3	4	5	6	7	8	9	10	11	12
13	.75	1.45	1 54	1.54	1.53	1.52	1.51	1.50	1.49	1.49	1.48	1.47	1.47
	.90	3.14	2.76	2.56	2.43	2.35	2.28	2.23	2.20	2.16	2.14	2.12	2.10
	.95	4.67	3.81	3.41	3.18	3.03	2.92	2.83	2.77	2.71	2.67	2.63	2.60
	.99	9.07	6.70	5.74	5.21	4.86	4.62	4.44	4.30	4.19	4 10	4 02	3.96
14	.75	1.44	1.53	1.53	1.52	1.51	1.50	1.48	1.48	1.47	1.46	1.46	1.45
	.90	3.10	2.73	2.52	2.39	2.31	2.24	2.19	2.15	2.12	2.10	2.08	2.05
	.95	4.60	3.74	3.34	3.11	2.96	2.85	2.76	2.70	2.65	2 60	2.57	2.53
	.99	8.86	6.51	5.56	5.04	4.69	4.46	4.28	4.14	4.03	3.94	3.86	3.80
15	.75	1.43	1.52	1.52	1.51	1.49	1.48	1.47	1.46	1.46	1.45	1.44	1.44
	.90	3.07	2.70	2.49	2.36	2.27	2.21	2.16	2.12	2.09	2.06	2.04	2.02
	.95	4.54	3.68	3.29	3.06	2.90	2.79	2.71	2.64	2.59	2.54	2.51	2.48
	.99	8.68	6.36	5.42	4.89	4.56	4.32	4.14	4.00	3.89	3.80	3.73	3.67
16	.75	1.42	1.51	1.51	1.50	1.48	1.48	1.47	1.46	1.45	1.45	1.44	1,44
	.90	3.05	2.67	2.46	2.33	2.24	2.18	2.13	2.09	2.06	2.03	2.01	1.99
	.95	4.49	3.63	3.24	3.01	2.85	2.74	2.66	2.59	2.54	2.49	2.46	2.42
	.99	8.53	6.23	5.29	4.77	4.44	4.20	4.03	3.89	3.78	3.69	3.62	3.55
17	.75	1.42	1.51	1.50	1.49	1.47	1.46	1.45	1.44	1.43	1.43	1.42	1.41
	.90	3.03	2.64	2.44	2.31	2.22	2.15	2.10	2.06	2.03	2.00	1.98	1.96
	.95	4.45	3.59	3.20	2.96	2.81	2.70	2.61	2.55	2.49	2.45	2.41	2.38
	.99	8.40	6.11	5.18	4.67	4.34	4.10	3.93	3.79	3.68	3.59	3.52	3.46
18	.75	1.41	1.50	1.49	1.48	1.46	1.45	1.44	1.43	1.42	1.42	1.41	1.40
	.90	3.01	2.62	2.42	2.29	2.20	2.13	2.08	2.04	2.00	1.98	1.96	1.93
	.95	4.41	3.55	3.16	2.93	2.77	2.66	2.58	2.51	2.46	2.41	2.37	2.34
	.99	8.29	6.01	5.09	4.58	4.25	4.01	3.84	3.71	3.60	3.51	3.43	3.37
19	.75	1.41	1.49	1.49	1.47	1.46	1.44	1.43	1.42	1.41	1.41	1.40	1.40
	.90	2.99	2.61	2.40	2.27	2.18	2.11	2.06	2.02	1.98	1.96	1.94	1.91
	.95	4.38	3.52	3.13	2.90	2.74	2.63	2.54	2.48	2.42	2.38	2.34	2.31
	.99	8.18	5.93	5.01	4.50	4.17	3.94	3.77	3.63	3.52	3.43	3.36	3.30
20	.75	1.40	1.49	1.48	1.46	1.45	1.44	1.42	1.42	1.41	1.40	1.39	1.39
	.90	2.97	2.59	2.38	2.25	2.16	2.09	2.04	2.00	1.96	1.94	1.92	1.89
	.95	4.35	3.49	3.10	2.87	2.71	2.60	2.51	2.45	2.39	2.35	2.31	2.28
	.99	8.10	5.85	4.94	4.43	4.10	3.87	3.70	3.56	3.46	3.37	3.29	3.23
22	.75	1.40	1.48	1.47	1.45	1.44	1.42	1.41	1.40	1.39	1.39	1.38	1.37
	.90	2.95	2.56·	2.35	2.22	2.13	2.06	2.01	1.97	1.93	1.90	1.88	1.86
	.95	4.30	3.44	3.05	2.82	2.66	2.55	2.46	2.40	2.34	2.30	2.26	2.23
	.99	7.95	5.72	4.82	4.31	3.99	3.76	3.59	3.45	3.35	3.26	3.18	3.12
24	.75	1.39	1.47	1.46	1.44	1.43	1.41	1.40	1.39	1.38	1.38	1.37	1.36
	.90	2.93	2.54	2.33	2.19	2.10	2.04	1.98	1.94	1.91	1.88	1.85	1.83
	.95	4.26	3.40	3.01	2.78	2.62	2.51	2.42	2.36	2.30	2.25	2.21	2.18
	.99	7.82	5.61	4.72	4.22	3.90	3.67	3.50	3.36	3.26	3.17	3.09	3.03
26	.75	1.38	1.46	1.45	1.44	1.42	1.41	1.40	1.39	1.37	1.37	1.36	1.35
	.90	2.91	2.52	2.31	2.17	2.08	2.01	1.96	1.92	1.88	1.86	1.84	1.81
	.95	4.23	3.37	2.98	2.74	2.59	2.47	2.39	2.32	2.27	2.22	2.18	2.15
	.99	7.72	5.53	4.64	4.14	3.82	3.59	3.42	3.29	3.18	3.09	3.02	2.96
28	.75	1.38	1.46	1.45	1.43	1.41	1.40	1.39	1.38	1.37	1.36	1.35	1.34
	.90	2.89	2.50	2.29	2.16	2.06	2.00	1.94	1.90	1.87	1.84	1.81	1.79
	.99	4.20	3.34	2.95	2.71	2.56	2.45	2.36	2.29	2.24	2.19	2.15	2.12
	.99	7.64	5.45	4.57	4.07	3.75	3.53	3.36	3.23	3.12	3.03	2.96	2.90

TABLE H—Cont.

df for numerator (n_1)												$1-\alpha$	df for denominator (n_2)
15	20	24	30	40	50	60	100	120	200	500	∞		
1.46	1.45	1.44	1.43	1.42	1.42	1.42	1.41	1.41	1.40	1.40	1.40	.75	13
2.05	2.01	1.98	1.96	1.93	1.92	1.90	1.88	1.88	1.86	1.85	1.85	.90	
2.53	2.46	2.42	2.38	2.34	2.31	2.30	2.26	2.25	2.23	2.22	2.21	.95	
3.82	3.66	3.59	3.51	3.43	3.38	3.34	3.27	3.25	3.22	3.19	3.17	.99	
1.44	1.43	1.42	1.41	1.41	1.40	1.40	1.39	1.39	1.39	1.38	1.38	.75	14
2.01	1.96	1.94	1.91	1.89	1.87	1.86	1.83	1.83	1.82	1.80	1.80	.90	
2.46	2.39	2.35	2.31	2.27	2.24	2.22	2.19	2.18	2.16	2.14	2.13	.95	
3.66	3.51	3.43	3.35	3.27	3.22	3.18	3.11	3.09	3.06	3.03	3.00	.99	
1.43	1.41	1.41	1.40	1.39	1.39	1.38	1.38	1.37	1.37	1.36	1.36	.75	15
1.97	1.92	1.90	1.87	1.85	1.83	1.82	1.79	1.79	1.77	1.76	1.76	.90	
2.40	2.33	2.29	2.25	2.20	2.18	2.16	2.12	2.11	2.10	2.08	2.07	.95	
3.52	3.37	3.29	3.21	3.13	3.08	3.05	2.98	2.96	2.92	2.89	2.87	.99	
1.41	1.40	1.39	1.38	1.37	1.37	1.36	1.36	1.35	1.35	1.34	1.34	.75	16
1.94	1.89	1.87	1.84	1.81	1.79	1.78	1.76	1.75	1.74	1.73	1.72	.90	
2.35	2.28	2.24	2.19	2.15	2.12	2.11	2.07	2.06	2.04	2.02	2.01	.95	
3.41	3.26	3.18	3.10	3.02	2.97	2.93	2.86	2.84	2.81	2.78	2.75	.99	
1.40	1.39	1.38	1.37	1.36	1.35	1.35	1.34	1.34	1.34	1.33	1.33	.75	17
1.91	1.86	1.84	1.81	1.78	1.76	1.75	1.73	1.72	1.71	1.69	1.69	.90	
2.31	2.23	2.19	2.15	2.10	2.08	2.06	2.02	2.01	1.99	1.97	1.96	.95	
3.31	3.16	3.08	3.00	2.92	2.87	2.83	2.76	2.75	2.71	2.68	2.65	.99	
1.39	1.38	1.37	1.36	1.35	1.34	1.34	1.33	1.33	1.32	1.32	1.32	.75	18
1.89	1.84	1.81	1.78	1.75	1.74	1.72	1.70	1.69	1.68	1.67	1.66	.90	
2.27	2.19	2.15	2.11	2.06	2.04	2.02	1.98	1.97	1.95	1.93	1.92	.95	
3.23	3.08	3.00	2.92	2.84	2.78	2.75	2.68	2.66	2.62	2.59	2.57	.99	
1.38	1.37	1.36	1.35	1.34	1.33	1.33	1.32	1.32	1.31	1.31	1.30	.75	19
1.86	1.81	1.79	1.76	1.73	1.71	1.70	1.67	1.67	1.65	1.64	1.63	.90	
2.23	2.16	2.11	2.07	2.03	2.00	1.98	1.94	1.93	1.91	1.89	1.88	.95	
3.15	3.00	2.92	2.84	2.76	2.71	2.67	2.60	2.58	2.55	2.51	2.49	.99	
1.37	1.36	1.35	1.34	1.33	1.33	1.32	1.31	1.31	1.30	1.30	1.29	.75	20
1.84	1.79	1.77	1.74	1.71	1.69	1.68	1.65	1.64	1.63	1.62	1.61	.90	
2.20	2.12	2.08	2.04	1.99	1.97	1.95	1.91	1.90	1.88	1.86	1.84	.95	
3.09	2.94	2.86	2.78	2.69	2.64	2.61	2.54	2.52	2.48	2.44	2.42	.99	
1.36	1.34	1.33	1.32	1.31	1.31	1.30	1.30	1.30	1.29	1.29	1.28	.75	22
1.81	1.76	1.73	1.70	1.67	1.65	1.64	1.61	1.60	1.59	1.58	1.57	.90	
2.15	2.07	2.03	1.98	1.94	1.91	1.89	1.85	1.84	1.82	1.80	1.78	.95	
2.98	2.83	2.75	2.67	2.58	2.53	2.50	2.42	2.40	2.36	2.33	2.31	.99	
1.35	1.33	1.32	1.31	1.30	1.29	1.29	1.28	1.28	1.27	1.27	1.26	.75	24
1.78	1.73	1.70	1.67	1.64	1.62	1.61	1.58	1.57	1.56	1.54	1.53	.90	
2.11	2.03	1.98	1.94	1.89	1.86	1.84	1.80	1.79	1.77	1.75	1.73	.95	
2.89	2.74	2.66	2.58	2.49	2.44	2.40	2.33	2.31	2.27	2.24	2.21	.99	
1.34	1.32	1.31	1.30	1.29	1.28	1.28	1.26	1.26	1.26	1.25	1.25	.75	26
1.76	1.71	1.68	1.65	1.61	1.59	1.58	1.55	1.54	1.53	1.51	1.50	.90	
2.07	1.99	1.95	1.90	1.85	1.82	1.80	1.76	1.75	1.73	1.71	1.69	.95	
2.81	2.66	2.58	2.50	2.42	2.36	2.33	2.25	2.23	2.19	2.16	2.13	.99	
1.33	1.31	1.30	1.29	1.28	1.27	1.27	1.26	1.25	1.25	1.24	1.24	.75	28
1.74	1.69	1.66	1.63	1.59	1.57	1.56	1.53	1.52	1.50	1.49	1.48	.90	
2.04	1.96	1.91	1.87	1.82	1.79	1.77	1.73	1.71	1.69	1.67	1.65	.95	
2.75	2.60	2.52	2.44	2.35	2.30	2.26	2.19	2.17	2.13	2.09	2.06	.99	

TABLE H—Cont.

df for denominator (n_2)	$1-\alpha$	df for numerator (n_1)											
		1	2	3	4	5	6	7	8	9	10	11	12
30	.75	1.38	1.45	1.44	1.42	1.41	1.39	1.38	1.37	1.36	1.35	1.35	1.34
	.90	2.88	2.49	2.28	2.14	2.05	1.98	1.93	1.88	1.85	1.82	1.79	1.77
	.95	4.17	3.32	2.92	2.69	2.53	2.42	2.33	2.27	2.21	2.16	2.13	2.09
	.99	7.56	5.39	4.51	4.02	3.70	3.47	3.30	3.17	3.07	2.98	2.91	2.84
40	.75	1.36	1.44	1.42	1.40	1.39	1.37	1.36	1.35	1.34	1.33	1.32	1.31
	.90	2.84	2.44	2.23	2.09	2.00	1.93	1.87	1.83	1.79	1.76	1.73	1.71
	.95	4.08	3.23	2.84	2.61	2.45	2.34	2.25	2.18	2.12	2.08	2.04	2.00
	.99	7.31	5.18	4.31	3.83	3.51	3.29	3.12	2.99	2.89	2.80	2.73	2.66
60	.75	1.35	1.42	1.41	1.38	1.37	1.35	1.33	1.32	1.31	1.30	1.29	1.29
	.90	2.79	2.39	2.18	2.04	1.95	1.87	1.82	1.77	1.74	1.71	1.68	1.66
	.95	4.00	3.15	2.76	2.53	2.37	2.25	2.17	2.10	2.04	1.99	1.95	1.92
	.99	7.08	4.98	4.13	3.65	3.34	3.12	2.95	2.82	2.72	2.63	2.56	2.50
120	.75	1.34	1.40	1.39	1.37	1.35	1.33	1.31	1.30	1.29	1.28	1.27	1.26
	.90	2.75	2.35	2.13	1.99	1.90	1.82	1.77	1.72	1.68	1.65	1.62	1.60
	.95	3.92	3.07	2.68	2.45	2.29	2.17	2.09	2.02	1.96	1.91	1.87	1.83
	.99	6.85	4.79	3.95	3.48	3.17	2.96	2.79	2.66	2.56	2.47	2.40	2.34
200	.75	1.33	1.39	1.38	1.36	1.34	1.32	1.31	1.29	1.28	1.27	1.26	1.25
	.90	2.73	2.33	2.11	1.97	1.88	1.80	1.75	1.70	1.66	1.63	1.60	1.57
	.95	3.89	3.04	2.65	2.42	2.26	2.14	2.06	1.98	1.93	1.88	1.84	1.80
	.99	6.76	4 71	3.88	3.41	3.11	2.89	2.73	2.60	2.50	2.41	2.34	2.27
∞	.75	1.32	1.39	1.37	1.35	1.33	1.31	1.29	1.28	1.27	1.25	1.24	1.24
	.90	2.71	2.30	2.08	1.94	1.85	1.77	1.72	1.67	1.63	1.60	1.57	1.55
	.95	3.84	3.00	2.60	2.37	3.21	2.10	2.01	1.94	1.88	1.83	1.79	1.75
	.99	6.63	4.61	3.78	3.32	3.02	2.80	2.64	2.51	2.41	2.32	2.25	2.18

TABLE H—Cont.

df for numerator (n_1)												$1-\alpha$	df for denominator (n_2)
15	20	24	30	40	50	60	100	120	200	500	∞		
1.32	1.30	1.29	1.28	1.27	1.26	1.26	1.25	1.24	1.24	1.23	1.23	.75	30
1.72	1.67	1.64	1.61	1.57	1.55	1.54	1.51	1.50	1.48	1.47	1.46	.90	
2.01	1.93	1.89	1.84	1.79	1.76	1.74	1.70	1.68	1.66	1.64	1.62	.95	
2.70	2.55	2.47	2.39	2.30	2.25	2.21	2.13	2.11	2.07	2.03	2.01	.99	
1.30	1.28	1.26	1.25	1.24	1.23	1.22	1.21	1.21	1.20	1.19	1.19	.75	40
1.66	1.61	1.57	1.54	1.51	1.48	1.47	1.43	1.42	1.41	1.39	1.38	.90	
1.92	1.84	1.79	1.74	1.69	1.66	1.64	1.59	1.58	1.55	1.53	1.51	.95	
2.52	2.37	2.29	2.20	2.11	2.06	2.02	1.94	1.92	1.87	1.83	1.80	.99	
1.27	1.25	1.24	1.22	1.21	1.20	1.19	1.17	1.17	1.16	1.15	1.15	.75	60
1.60	1.54	1.51	1.48	1.44	1.41	1.40	1.36	1.35	1.33	1.31	1.29	.90	
1.84	1.75	1.70	1.65	1.59	1.56	1.53	1.48	1.47	1.44	1.41	1.39	.95	
2.35	2 20	2.12	2.03	1.94	1.88	1.84	1.75	1.73	1.68	1.63	1.60	.99	
1.24	1.22	1.21	1.19	1.18	1.17	1.16	1.14	1.13	1.12	1.11	1.10	.75	120
1.55	1.48	1.45	1.41	1.37	1.34	1.32	1.27	1.26	1.24	1.21	1.19	.90	
1.75	1.66	1.61	1.55	1.50	1.46	1.43	1.37	1.35	1.32	1.28	1.25	.95	
2.19	2.03	1.95	1.86	1.76	1.70	1.66	1.56	1.53	1.48	1.42	1.38	.99	
1.23	1.21	1.20	1.18	1.16	1.14	1.12	1.11	1.10	1.09	1.08	1.06	.75	200
1.52	1.46	1.42	1.38	1.34	1.31	1.28	1.24	1.22	1.20	1.17	1.14	.90	
1.72	1.62	1.57	1.52	1.46	1.41	1.39	1.32	1.29	1.26	1.22	1.19	.95	
2.13	1.97	1.89	1.79	1.69	1.63	1.58	1.48	1.44	1.38	1.33	1.28	.99	
1.22	1.19	1.18	1.16	1.14	1.13	1.12	1.09	1.08	1.07	1.04	1.00	.75	∞
1.49	1.42	1.38	1.34	1.30	1.26	1.24	1.18	1.17	1.13	1.08	1.00	.90	
1.67	1.57	1.52	1.46	1.39	1.35	1.32	1.24	1.22	1.17	1.11	1.00	.95	
2.04	1.88	1.79	1.70	1.59	1.52	1.47	1.36	1.32	1.25	1.15	1.00	.99	

TABLE I The Poisson Distribution Function†

$$F_X(k) = \sum_{s=0}^{k} \frac{e^{-\lambda}\lambda^s}{s!}$$

λ \ k	0	1	2	3	4	5	6	7	8	9
0.02	980	1,000								
0.04	961	999	1,000							
0.06	942	998	1,000							
0.08	923	997	1,000							
0.10	905	995	1,000							
0.15	861	990	999	1,000						
0.20	819	982	999	1,000						
0.25	779	974	998	1,000						
0.30	741	963	996	1,000						
0.35	705	951	994	1,000						
0.40	670	938	992	999	1,000					
0.45	638	925	989	999	1,000					
0.50	607	910	986	998	1,000					
0.55	577	894	982	998	1,000					
0.60	549	878	977	997	1,000					
0.65	522	861	972	996	999	1,000				
0.70	497	844	966	994	999	1,000				
0.75	472	827	959	993	999	1,000				
0.80	449	809	953	991	999	1 000				
0.85	427	791	945	989	998	1 000				
0.90	407	772	937	987	998	1,000				
0.95	387	754	929	984	997	1,000				
1.00	368	736	920	981	996	999	1,000			
1.1	333	699	900	974	995	999	1,000			
1.2	301	663	879	966	992	998	1,000			
1.3	273	627	857	957	989	998	1,000			
1.4	247	592	833	946	986	997	999	1,000		
1.5	223	558	809	934	981	996	999	1,000		
1.6	202	525	783	921	976	994	999	1,000		
1.7	183	493	757	907	970	992	998	1,000		
1.8	165	463	731	891	964	990	997	999	1,000	
1.9	150	434	704	875	956	987	997	999	1,000	
2.0	135	406	677	857	947	983	995	999	1,000	

† From *Poisson's Exponential Binomial Limit* by E. C. Molina, copyright © 1942 by Litton Educational Publishing, Inc., by permission of Van Nostrand Reinhold Company.

TABLE I—Cont.

λ \ k	0	1	2	3	4	5	6	7	8	9
2.2	111	355	623	819	928	975	993	998	1,000	
2.4	091	308	570	779	904	964	988	997	999	1,000
2.6	074	267	518	736	877	951	983	995	999	1,000
2.8	061	231	469	692	848	935	976	992	998	999
3.0	050	199	423	647	815	916	966	988	996	999
3.2	041	171	380	603	781	895	955	983	994	998
3.4	033	147	340	558	744	871	942	977	992	997
3.6	027	126	303	515	706	844	927	969	988	996
3.8	022	107	269	473	668	816	909	960	984	994
4.0	018	092	238	433	629	785	889	949	979	992
4.2	015	078	210	395	590	753	867	936	972	989
4.4	012	066	185	359	551	720	844	921	964	985
4.6	010	056	163	326	513	686	818	905	955	980
4.8	008	048	143	294	476	651	791	887	944	975
5.0	007	040	125	265	440	616	762	867	932	968
5.2	006	034	109	238	406	581	732	845	918	960
5.4	005	029	095	213	373	546	702	822	903	951
5.6	004	024	082	191	342	512	670	797	886	941
5.8	003	021	072	170	313	478	638	771	867	929
6.0	002	017	062	151	285	446	606	744	847	916

	10	11	12	13	14	15	16
2.8	1,000						
3.0	1,000						
3.2	1,000						
3.4	999	1,000					
3.6	999	1,000					
3.8	998	999	1,000				
4.0	997	999	1,000				
4.2	996	999	1,000				
4.4	994	998	999	1,000			
4.6	992	997	999	1,000			
4.8	990	996	999	1,000			
5.0	986	995	998	999	1,000		
5.2	982	993	997	999	1,000		
5.4	977	990	996	999	1,000		
5.6	972	988	995	998	999	1,000	
5.8	965	984	993	997	999	1,000	
6.0	957	980	991	996	999	999	1,000

TABLE I—Cont.

λ \ k	0	1	2	3	4	5	6	7	8	9
6.2	002	015	054	134	259	414	574	716	826	902
6.4	002	012	046	119	235	384	542	687	803	886
6.6	001	010	040	105	213	355	511	658	780	869
6.8	001	009	034	093	192	327	480	628	755	850
7.0	001	007	030	082	173	301	450	599	729	830
7.2	001	006	025	072	156	276	420	569	703	810
7.4	001	005	022	063	140	253	392	539	676	788
7.6	001	004	019	055	125	231	365	510	648	765
7.8	000	004	016	048	112	210	338	481	620	741
8.0	000	003	014	042	100	191	313	453	593	717
8.5	000	002	009	030	074	150	256	386	523	653
9.0	000	001	006	021	055	116	207	324	456	587
9.5	000	001	004	015	040	089	165	269	392	522
10.0	000	000	003	010	029	067	130	220	333	458

λ \ k	10	11	12	13	14	15	16	17	18	19
6.2	949	975	989	995	998	999	1,000			
6.4	939	969	986	994	997	999	1,000			
6.6	927	963	982	992	997	999	999	1,000		
6.8	915	955	978	990	996	998	999	1,000		
7.0	901	947	973	987	994	998	999	1,000		
7.2	887	937	967	984	993	997	999	999	1,000	
7.4	871	926	961	980	991	996	998	999	1,000	
7.6	854	915	954	976	989	995	998	999	1,000	
7.8	835	902	945	971	986	993	997	999	1,000	
8.0	816	888	936	966	983	992	996	998	999	1,000
8.5	763	849	909	949	973	986	993	997	999	999
9.0	706	803	876	926	959	978	989	995	998	999
9.5	645	752	836	898	940	967	982	991	996	998
10.0	583	697	792	864	917	951	973	986	993	997

λ \ k	20	21	22
8.5	1,000		
9.0	1,000		
9.5	999	1,000	
10.0	998	999	1,000

TABLE I—Cont.

λ \ k	0	1	2	3	4	5	6	7	8	9
10.5	000	000	002	007	021	050	102	179	279	397
11.0	000	000	001	005	015	038	079	143	232	341
11.5	000	000	001	003	011	028	060	114	191	289
12.0	000	000	001	002	008	020	046	090	155	242
12.5	000	000	000	002	005	015	035	070	125	201
13.0	000	000	000	001	004	011	026	054	100	166
13.5	000	000	000	001	003	008	019	041	079	135
14.0	000	000	000	000	002	006	014	032	062	109
14.5	000	000	000	000	001	004	010	024	048	088
15.0	000	000	000	000	001	003	008	018	037	070

λ \ k	10	11	12	13	14	15	16	17	18	19
10.5	521	639	742	825	888	932	960	978	988	994
11.0	460	579	689	781	854	907	944	968	982	991
11.5	402	520	633	733	815	878	924	954	974	986
12.0	347	462	576	682	772	844	899	937	963	979
12.5	297	406	519	628	725	806	869	916	948	969
13.0	252	353	463	573	675	764	835	890	930	957
13.5	211	304	409	518	623	718	798	861	908	942
14.0	176	260	358	464	570	669	756	827	883	923
14.5	145	220	311	413	518	619	711	790	853	901
15.0	118	185	268	363	466	568	664	749	819	875

λ \ k	20	21	22	23	24	25	26	27	28	29
10.5	997	999	999	1,000						
11.0	995	998	999	1,000						
11.5	992	996	998	999	1,000					
12.0	988	994	997	999	999	1,000				
12.5	983	991	995	998	999	999	1,000			
13.0	975	986	992	996	998	999	1,000			
13.5	965	980	989	994	997	998	999	1,000		
14.0	952	971	983	991	995	997	999	999	1,000	
14.5	936	960	976	986	992	996	998	999	999	1,000
15.0	917	947	967	981	989	994	997	998	999	1,000

TABLE I—Cont.

λ \ k	4	5	6	7	8	9	10	11	12	13
16	000	001	004	010	022	043	077	127	193	275
17	000	001	002	005	013	026	049	085	135	201
18	000	000	001	003	007	015	030	055	092	143
19	000	000	001	002	004	009	018	035	061	098
20	000	000	000	001	002	005	011	021	039	066
21	000	000	000	000	001	003	006	013	025	043
22	000	000	000	000	001	002	004	008	015	028
23	000	000	000	000	000	001	002	004	009	017
24	000	000	000	000	000	000	001	003	005	011
25	000	000	000	000	000	000	001	001	003	006
	14	15	16	17	18	19	20	21	22	23
16	368	467	566	659	742	812	868	911	942	963
17	281	371	468	564	655	736	805	861	905	937
18	208	287	375	469	562	651	731	799	855	899
19	150	215	292	378	469	561	647	725	793	849
20	105	157	221	297	381	470	559	644	721	787
21	072	111	163	227	302	384	471	558	640	716
22	048	077	117	169	232	306	387	472	556	637
23	031	052	082	123	175	238	310	389	472	555
24	020	034	056	087	128	180	243	314	392	473
25	012	022	038	060	092	134	185	247	318	394
	24	25	26	27	28	29	30	31	32	33
16	978	987	993	996	998	999	999	1,000		
17	959	975	985	991	995	997	999	999	1,000	
18	932	955	972	983	990	994	997	998	999	1,000
19	893	927	951	969	980	988	993	996	998	999
20	843	888	922	948	966	978	987	992	995	997
21	782	838	883	917	944	963	976	985	991	994
22	712	777	832	877	913	940	959	973	983	989
23	635	708	772	827	873	908	936	956	971	981
24	554	632	704	768	823	868	904	932	953	969
25	473	553	629	700	763	818	863	900	929	950
	34	35	36	37	38	39	40	41	42	43
19	999	1,000								
20	999	999	1,000							
21	997	998	999	999	1,000					
22	994	996	998	999	999	1,000				
23	988	993	996	997	999	999	1,000			
24	979	987	992	995	997	998	999	999	1,000	
25	966	978	985	991	994	997	998	999	999	1,000

TABLE J The Binomial Distribution Function†

$$F_X(x) = \sum_{k=0}^{x} \binom{n}{k} p^k (1-p)^{n-k}$$

n	x	p = 0.05	0.10	0.15	0.20	0.25	0.30	0.35	0.40	0.45	0.50
2	0	0.9025	0.8100	0.7225	0.6400	0.5625	0.4900	0.4225	0.3600	0.3025	0.2500
	1	0.9975	0.9900	0.9775	0.9600	0.9375	0.9100	0.8775	0.8400	0.7975	0.7500
3	0	0.8574	0.7290	0.6141	0.5120	0.4219	0.3430	0.2746	0.2160	0.1664	0.1250
	1	0.9928	0.9720	0.9392	0.8960	0.8438	0.7840	0.7182	0.6480	0.5748	0.5000
	2	0.9999	0.9990	0.9966	0.9920	0.9844	0.9730	0.9571	0.9360	0.9089	0.8750
4	0	0.8145	0.6561	0.5220	0.4096	0.3164	0.2401	0.1785	0.1296	0.0915	0.0625
	1	0.9860	0.9477	0.8905	0.8192	0.7383	0.6517	0.5630	0.4752	0.3910	0.3125
	2	0.9995	0.9963	0.9880	0.9728	0.9492	0.9163	0.8735	0.8208	0.7585	0.6875
	3	1,0000	0.9999	0.9995	0.9984	0.9961	0.9919	0.9850	0.9744	0.9590	0.9375
5	0	0.7738	0.5905	0.4437	0.3277	0.2373	0.1681	0.1160	0.0778	0.0503	0.0312
	1	0.9774	0.9185	0.8352	0.7373	0.6328	0.5282	0.4284	0.3370	0.2562	0.1875
	2	0.9988	0.9914	0.9734	0.9421	0.8965	0.8369	0.7648	0.6826	0.5931	0.5000
	3	1.0000	0.9995	0.9978	0.9933	0.9844	0.9692	0.9460	0.9130	0.8688	0.8125
	4	1.0000	1.0000	0.9999	0.9997	0.9990	0.9976	0.9947	0.9898	0.9815	0.9688
6	0	0.7351	0.5314	0.3771	0.2621	0.1780	0.1176	0.0754	0.0467	0.0277	0.0156
	1	0.9672	0.8857	0.7765	0.6554	0.5339	0.4202	0.3191	0.2333	0.1636	0.1094
	2	0.9978	0.9842	0.9527	0.9011	0.8306	0.7443	0.6471	0.5443	0.4415	0.3438
	3	0.9999	0.9987	0.9941	0.9830	0.9624	0.9295	0.8826	0.8208	0.7447	0.6562
	4	1.0000	0.9999	0.9996	0.9984	0.9954	0.9891	0.9777	0.9590	0.9308	0.8906
	5	1.0000	1.0000	1.0000	0.9999	0.9998	0.9993	0.9982	0.9959	0.9917	0.9844
7	0	0.6983	0.4783	0.3206	0.2097	0.1335	0.0824	0.0490	0.0280	0.0152	0.0078
	1	0.9556	0.8503	0.7166	0.5767	0.4449	0.3294	0.2338	0.1586	0.1024	0.0625
	2	0.9962	0.9743	0.9262	0.8520	0.7564	0.6471	0.5323	0.4199	0.3164	0.2266
	3	0.9998	0.9973	0.9879	0.9667	0.9294	0.8740	0.8002	0.7102	0.6083	0.5000
	4	1.0000	0.9998	0.9988	0.9953	0.9871	0.9712	0.9444	0.9037	0.8471	0.7734
	5	1.0000	1.0000	0.9999	0.9996	0.9987	0.9962	0.9910	0.9812	0.9643	0.9375
	6	1.0000	1.0000	1.0000	1.0000	0.9999	0.9998	0.9994	0.9984	0.9963	0.9922
8	0	0.6634	0.4305	0.2725	0.1678	0.1001	0.0576	0.0319	0.0168	0.0084	0.0039
	1	0.9428	0.8131	0.6572	0.5033	0.3671	0.2553	0.1691	0.1064	0.0632	0.0352
	2	0.9942	0.9619	0.8948	0.7969	0.6785	0.5518	0.4278	0.3154	0.2201	0.1445
	3	0.9996	0.9950	0.9786	0.9437	0.8862	0.8059	0.7064	0.5941	0.4770	0.3633
	4	1.0000	0.9996	0.9971	0.9896	0.9727	0.9420	0.8939	0.8263	0.7396	0.6367
	5	1.0000	1.0000	0.9998	0.9988	0.9958	0.9887	0.9747	0.9502	0.9115	0.8555
	6	1.0000	1.0000	1.0000	0.9999	0.9996	0.9987	0.9964	0.9915	0.9819	0.9648
	7	1.0000	1.0000	1.0000	1.0000	1.0000	0.9999	0.9998	0.9993	0.9983	0.9961
9	0	0.6302	0.3874	0.2316	0.1342	0.0751	0.0404	0.0207	0.0101	0.0046	0.0020
	1	0.9288	0.7748	0.5995	0.4362	0.3003	0.1960	0.1211	0.0705	0.0385	0.0195
	2	0.9916	0.9470	0.8591	0.7382	0.6007	0.4628	0.3373	0.2318	0.1495	0.0898
	3	0.9994	0.9917	0.9661	0.9144	0.8343	0.7297	0.6089	0.4826	0.3614	0.2539
	4	1.0000	0.9991	0.9944	0.9804	0.9511	0.9012	0.8283	0.7334	0.6214	0.5000
	5	1.0000	0.9999	0.9994	0.9969	0.9900	0.9747	0.9464	0.9006	0.8342	0.7461
	6	1.0000	1.0000	1.0000	0.9997	0.9987	0.9957	0.9888	0.9750	0.9502	0.9102
	7	1.0000	1.0000	1.0000	1.0000	0.9999	0.9996	0.9986	0.9962	0.9909	0.9805
	8	1.0000	1.0000	1.0000	1.0000	1.0000	1.0000	0.9999	0.9997	0.9992	0.9980

† Source: This table is extracted from *Tables of the Binomial Probability Distribution:* Washington, D.C., National Bureau of Standards, Applied Mathematics Series, No. 6, U.S. Department of Commerce, 1952.

TABLE J—Cont.

n	x	0.05	0.10	0.15	0.20	0.25	0.30	0.35	0.40	0.45	0.50
						p					
10	0	0.5987	0.3487	0.1969	0.1074	0.0563	0.0282	0.0135	0.0060	0.0025	0.0010
	1	0.9139	0.7361	0.5443	0.3758	0.2440	0.1493	0.0860	0.0464	0.0232	0.0107
	2	0.9885	0.9298	0.8202	0.6778	0.5256	0.3828	0.2616	0.1673	0.0996	0.0547
	3	0.9990	0.9872	0.9500	0.8791	0.7759	0.6496	0.5138	0.3823	0.2660	0.1719
	4	0.9999	0.9984	0.9901	0.9672	0.9219	0.8497	0.7515	0.6331	0.5044	0.3770
	5	1.0000	0.9999	0.9986	0.9936	0.9803	0.9527	0.9051	0.8338	0.7384	0.6230
	6	1.0000	1.0000	0.9999	0.9991	0.9965	0.9894	0.9740	0.9452	0.8980	0.8281
	7	1.0000	1.0000	1.0000	0.9999	0.9996	0.9984	0.9952	0.9877	0.9726	0.9453
	8	1.0000	1.0000	1.0000	1.0000	1.0000	0.9999	0.9995	0.9983	0.9955	0.9893
	9	1.0000	1.0000	1.0000	1.0000	1.0000	1.0000	1.0000	0.9999	0.9997	0.9990
11	0	0.5688	0.3138	0.1673	0.0859	0.0422	0.0198	0.0088	0.0036	0.0014	0.0005
	1	0.8981	0.6974	0.4922	0.3221	0.1971	0.1130	0.0606	0.0302	0.0139	0.0059
	2	0.9848	0.9104	0.7788	0.6174	0.4552	0.3127	0.2001	0.1189	0 0652	0.0327
	3	0.9984	0.9815	0.9306	0.8389	0.7133	0 5696	0.4256	0.2963	0.1911	0.1133
	4	0.9999	0.9972	0.9841	0.9496	0.8854	0.7897	0.6683	0.5328	0.3971	0.2744
	5	1.0000	0.9997	0.9973	0.9883	0.9657	0.9218	0.8513	0.7535	0.6331	0.5000
	6	1.0000	1.0000	0.9997	0.9980	0.9924	0.9784	0.9499	0.9006	0.8262	0.7256
	7	1.0000	1.0000	1.0000	0.9998	0.9988	0.9957	0.9878	0.9707	0.9390	0.8867
	8	1.0000	1.0000	1.0000	1.0000	0.9999	0.9994	0.9980	0.9941	0.9852	0.9673
	9	1.0000	1.0000	1.0000	1.0000	1.0000	1.0000	0.9998	0.9993	0.9978	0.9941
	10	1.0000	1.0000	1.0000	1.0000	1.0000	1.0000	1.0000	1.0000	0.9998	0.9995
12	0	0.5404	0.2824	0.1422	0.0687	0.0317	0.0138	0.0057	0.0022	0.0008	0.0002
	1	0.8816	0.6590	0 4435	0.2749	0.1584	0.0850	0.0424	0.0196	0.0083	0.0032
	2	0.9804	0.8891	0.7358	0.5583	0.3907	0.2528	0.1513	0.0834	0.0421	0.0193
	3	0.9978	0.9744	0.9078	0.7946	0.6488	0.4925	0.3467	0.2253	0.1345	0.0730
	4	0.9998	0.9957	0.9761	0.9274	0.8424	0.7237	0.5833	0.4382	0.3044	0.1938
	5	1.0000	0.9995	0.9954	0.9806	0.9456	0.8822	0.7873	0.6652	0.5269	0.3872
	6	1.0000	0.9999	0.9993	0.9961	0.9857	0.9614	0.9154	0.8418	0.7393	0.6128
	7	1.0000	1.0000	0.9999	0.9994	0.9972	0.9905	0.9745	0.9427	0.8883	0.8062
	8	1.0000	1.0000	1.0000	0.9999	0.9996	0.9983	0.9944	0.9847	0.9644	0.9270
	9	1.0000	1.0000	1.0000	1.0000	1.0000	0.9998	0.9992	0.9972	0.9921	0.9807
	10	1.0000	1.0000	1.0000	1.0000	1.0000	1.0000	0.9999	0.9997	0.9989	0.9968
	11	1.0000	1.0000	1.0000	1.0000	1.0000	1.0000	1.0000	1.0000	0.9999	0.9998
13	0	0.5133	0.2542	0.1209	0.0550	0.0238	0.0097	0.0037	0.0013	0.0004	0.0001
	1	0.8646	0.6213	0.3983	0.2336	0.1267	0.0637	0.0296	0.0126	0.0049	0.0017
	2	0.9755	0.8661	0.6920	0.5017	0.3326	0.2025	0.1132	0.0579	0.0269	0.0112
	3	0.9969	0.9658	0.8820	0.7473	0.5843	0.4206	0.2783	0.1686	0.0929	0.0461
	4	0.9997	0.9935	0.9658	0.9009	0.7940	0.6543	0.5005	0.3530	0.2279	0.1334
	5	1.0000	0.9991	0.9925	0.9700	0.9198	0.8346	0 7159	0.5744	0.4268	0.2905
	6	1.0000	0.9999	0.9987	0.9930	0.9757	0.9376	0.8705	0.7712	0.6437	0.5000
	7	1.0000	1.0000	0.9998	0.9988	0.9944	0.9818	0.9538	0.9023	0.8212	0.7095
	8	1.0000	1.0000	1.0000	0.9998	0.9990	0.9960	0.9874	0.9679	0.9302	0.8666
	9	1.0000	10.000	1.0000	1.0000	0.9999	0.9993	0.9975	0.9922	0.9797	0.9539
	10	1.0000	1.0000	1.0000	1.0000	1.0000	0.9999	0.9997	0.9987	0.9959	0.9888
	11	1.0000	1.0000	1.0000	1.0000	1.0000	1.0000	1.0000	0.9999	0.9995	0.9983
	12	1.0000	1.0000	1.0000	1.0000	1.0000	1.0000	1.0000	1.0000	1.0000	0.9999
14	0	0.4877	0.2288	0.1028	0.0440	0.0178	0.0068	0.0024	0.0008	0.0002	0.0001
	1	0.8470	0.5846	0.3567	0.1979	0.1010	0.0475	0.0205	0.0081	0.0029	0.0009

TABLE J—Cont.

n	x	p = 0.05	0.10	0.15	0.20	0.25	0.30	0.35	0.40	0.45	0.50
14	2	0.9699	0.8416	0.6479	0.4481	0.2811	0.1608	0.0839	0.0398	0.0170	0.0065
	3	0.9958	0.9559	0.8535	0.6982	0.5213	0.3552	0.2205	0.1243	0.0632	0.0287
	4	0.9996	0.9908	0.9533	0.8702	0.7415	0.5842	0.4227	0.2793	0.1672	0.0898
	5	1.0000	0.9985	0.9885	0.9561	0.8883	0.7805	0.6405	0.4859	0.3373	0.2120
	6	1.0000	0.9998	0.9978	0.9884	0.9617	0.9067	0.8164	0.6925	0.5461	0.3953
	7	1.0000	1.0000	0.9997	0.9976	0.9897	0.9685	0.9247	0.8499	0.7414	0.6047
	8	1.0000	1.0000	1.0000	0.9996	0.9978	0.9917	0.9757	0.9417	0.8811	0.7880
	9	1.0000	1.0000	1.0000	1.0000	0.9997	0.9983	0.9940	0.9825	0·9574	0.9102
	10	1.0000	1.0000	1.0000	1.0000	1.0000	0.9998	0.9989	0.9961	0 9886	0.9713
	11	1.0000	1.0000	1.0000	1.0000	1.0000	1.0000	0.9999	0.9994	0.9978	0.9935
	12	1.0000	1.0000	1.0000	1.0000	1.0000	1.0000	1.0000	0.9999	0.9997	0.9991
	13	1.0000	1.0000	1.0000	1.0000	1.0000	1.0000	1.0000	1.0000	1.0000	0.9999
15	0	0.4633	0.2059	0.0874	0.0352	0.0134	0.0047	0.0016	0.0005	0.0001	0.0000
	1	0.8290	0.5490	0.3186	0.1671	0.0802	0.0353	0.0142	0.0052	0.0017	0.0005
	2	0.9638	0.8159	0.6042	0.3980	0.2361	0.1268	0.0617	0.0271	0.0107	0.0037
	3	0.9945	0.9444	0.8227	0.6482	0.4613	0.2969	0.1727	0.0905	0.0424	0.0176
	4	0.9994	0.9873	0.9383	0.8358	0.6865	0.5155	0.3519	0.2173	0.1204	0.0592
	5	0.9999	0.9978	0.9832	0.9389	0.8516	0.7216	0.5643	0.4032	0.2608	0.1509
	6	1.0000	0.9997	0.9964	0.9819	0.9434	0.8689	0.7548	0.6098	0.4522	0.3036
	7	1.0000	1.0000	0.9996	0.9958	0.9827	0.9500	0.8868	0.7869	0.6535	0.5000
	8	1.0000	1.0000	0.9999	0.9992	0.9958	0.9849	0.9578	0.9050	0.8182	0.6964
	9	1.0000	1.0000	1.0000	0.9999	0.9992	0.9963	0.9876	0.9662	0.9231	0.8491
	10	1.0000	1.0000	1.0000	1.0000	0.9999	0.9993	0.9972	0.9907	0.9745	0.9408
	11	1.0000	1.0000	1.0000	1.0000	1.0000	0.9999	0.9995	0.9981	0.9937	0.9824
	12	1.0000	1.0000	1.0000	1.0000	1.0000	1.0000	0.9999	0.9997	0.9989	0.9963
	13	1.0000	1.0000	1.0000	1.0000	1.0000	1.0000	1.0000	1.0000	0.9999	0.9995
	14	1.0000	1.0000	1.0000	1.0000	1.0000	1.0000	1.0000	1.0000	1.0000	.10000
16	0	0.4401	0.1853	0.0743	0.0281	0.0100	0.0033	0.0010	0.0003	0.0001	0.0000
	1	0.8108	0.5147	0.2839	0.1407	0.0635	0.0261	0.0098	0.0033	0.0010	0.0003
	2	0.9571	0.7892	0.5614	0.3518	0.1971	0.0994	0.0451	0.0183	0.0066	0.0021
	3	0.9930	0.9316	0.7899	0.5981	0.4050	0.2459	0.1339	0.0651	0.0281	0.0106
	4	0.9991	0.9830	0.9209	0.7982	0.6302	0.4499	0.2892	0.1666	0.0853	0.0384
	5	0.9999	0.9967	0.9765	0.9183	0.8103	0.6598	0.4900	0.3288	0.1976	0.1051
	6	1.0000	0.9995	0.9944	0.9733	0.9204	0.8247	0.6881	0.5272	0.3660	0.2272
	7	1.0000	0.9999	0.9989	0.9930	0.9729	0.9256	0.8406	0.7161	0.5629	0.4018
	8	1.0000	1.0000	0.9998	0.9985	0.9925	0.9743	0.9329	0.8577	0.7441	0.5982
	9	1.0000	1.0000	1.0000	0.9998	0.9984	0.9929	0.9771	0.9417	0.8759	0.7728
	10	1.0000	1.0000	1.0000	1.0000	0.9997	0.9984	0.9938	0.9809	0.9514	0.8949
	11	1.0000	1.0000	1.0000	1.0000	1.0000	0.9997	0.9987	0.9951	0.9851	0.9616
	12	1.0000	1.0000	1.0000	1.0000	1.0000	1.0000	0.9998	0.9991	0.9965	0.9894
	13	1.0000	1.0000	1.0000	1.0000	1.0000	1.0000	1.0000	0.9999	0.9994	0.9979
	14	1.0000	1.0000	1.0000	1.0000	1.0000	1.0000	1.0000	1.0000	1.0000	0.9997
	15	1.0000	1.0000	1.0000	1.0000	1.0000	1.0000	1.0000	1.0000	1.0000	1.0000
17	0	0.4181	0.1668	0.0631	0.0225	0.0075	0.0023	0.0007	0.0002	0.0000	0.0000
	1	0.7922	0.4818	0.2525	0.1182	0.0501	0.0193	0.0067	0.0021	0.0006	0.0001
	2	0.9497	0.7618	0.5198	0.3096	0.1637	0.0774	0.0327	0.0123	0.0041	0.0012
	3	0.9912	0.9174	0.7556	0.5489	0.3530	0.2019	0.1028	0.0464	0.0184	0.0064
	4	0.9988	0.9779	0.9013	0.7582	0.5739	0.3887	0.2348	0.1260	0.0596	0.0245

TABLE J—Cont.

n	x	0.05	0.10	0.15	0.20	0.25	0.30	0.35	0.40	0.45	0.50
		p									
17	5	0.9999	0.9953	0.9681	0.8943	0.7653	0.5968	0.4197	0.2639	0.1471	0.0717
	6	1.0000	0.9992	0.9917	0.9623	0.8929	0.7752	0.6188	0.4478	0.2902	0.1662
	7	1.0000	0.9999	0.9983	0.9891	0.9598	0.8954	0.7872	0.6405	0.4743	0.3145
	8	1.0000	1.0000	0.9997	0.9974	0.9876	0.9597	0.9006	0.8011	0.6626	0.5000
	9	1.0000	1.0000	1.0000	0.9995	0.9969	0.9873	0.9617	0.9081	0.8166	0.6855
	10	1.0000	1.0000	1.0000	0.9999	0.9994	0.9968	0.9880	0.9652	0.9174	0.8338
	11	1.0000	1.0000	1.0000	1.0000	0.9999	0.9993	0.9970	0.9894	0.9699	0.9283
	12	1.0000	1.0000	1.0000	1.0000	1.0000	0.9999	0.9994	0.9975	0.9914	0.9755
	13	1.0000	1.0000	1.0000	1.0000	1.0000	1.0000	0.9999	0.9995	0.9981	0.9936
	14	1.0000	1.0000	1.0000	1.0000	1.0000	1.0000	1.0000	0.9999	0.9997	0.9988
	15	1.0000	1.0000	1.0000	1.0000	1.0000	1.0000	1.0000	1.0000	1.0000	0.9999
	16	1.0000	1.0000	1.0000	1.0000	1.0000	1.0000	1.0000	1.0000	1.0000	1.0000
18	0	0.3972	0.1501	0.0536	0.0180	0.0056	0.0016	0.0004	0.0001	0.0000	0.0000
	1	0.7735	0.4503	0.2241	0.0991	0.0395	0.0142	0.0046	0.0013	0.0003	0.0001
	2	0.9419	0.7338	0.4797	0.2713	0.1353	0.0600	0.0236	0.0082	0.0025	0.0007
	3	0.9891	0.9018	0.7202	0.5010	0.3057	0.1646	0.0783	0.0328	0.0120	0.0038
	4	0.9985	0.9718	0.8794	0.7164	0.5187	0.3327	0.1886	0.0942	0.0411	0.0154
	5	0.9998	0.9936	0.9581	0.8671	0.7175	0.5344	0.3550	0.2088	0.1077	0.0481
	6	1.0000	0.9988	0.9882	0.9487	0.8610	0.7217	0.5491	0.3743	0.2258	0.1189
	7	1.0000	0.9998	0.9973	0.9837	0.9431	0.8593	0.7283	0.5634	0.3915	0.2403
	8	1.0000	1.0000	0.9995	0.9957	0.9807	0.9404	0.8609	0.7368	0.5778	0.4073
	9	1.0000	1.0000	0.9999	0.9991	0.9946	0.9790	0.9403	0.8653	0.7473	0.5927
	10	1.0000	1.0000	1.0000	0.9998	0.9988	0.9939	0.9788	0.9424	0.8720	0.7597
	11	1.0000	1.0000	1.0000	1.0000	0.9998	0.9986	0.9938	0.9797	0.9463	0.8811
	12	1.0000	1.0000	1.0000	1.0000	1.0000	0.9997	0.9986	0.9942	0.9817	0.9519
	13	1.0000	1.0000	1.0000	1.0000	1.0000	1.0000	0.9997	0.9987	0.9951	0.9846
	14	1.0000	1.0000	1.0000	1.0000	1.0000	1.0000	1.0000	0.9998	0.9990	0.9962
	15	1.0000	1.0000	1.0000	1.0000	1.0000	1.0000	1.0000	1.0000	0.9999	0.9993
	16	1.0000	1.0000	1.0000	1.0000	1.0000	1.0000	1.0000	1.0000	1.0000	0.9999
19	0	0.3774	0.1351	0.0456	0.0144	0.0042	0.0011	0.0003	0.0001	0.0000	0.0000
	1	0.7547	0.4203	0.1985	0.0829	0.0310	0.0104	0.0031	0.0008	0.0002	0.0000
	2	0.9335	0.7054	0.4413	0.2369	0.1113	0.0462	0.0170	0.0055	0.0015	0.0004
	3	0.9868	0.8850	0.6841	0.4551	0.2630	0.1332	0.0591	0.0230	0.0077	0.0022
	4	0.9980	0.9648	0.8556	0.6733	0.4654	0.2822	0.1500	0.0696	0.0280	0.0096
	5	0.9998	0.9914	0.9463	0.8369	0.6678	0.4739	0.2968	0.1629	0.0777	0.0318
	6	1.0000	0.9983	0.9837	0.9324	0.8251	0.6655	0.4812	0.3081	0.1727	0.0835
	7	1.0000	0.9997	0.9959	0.9767	0.9225	0.8180	0.6656	0.4878	0.3169	0.1796
	8	1.0000	1.0000	0.9992	0.9933	0.9713	0.9161	0.8145	0.6675	0.4940	0.3238
	9	1.0000	1.0000	0.9999	0.9984	0.9911	0.9674	0.9125	0.8139	0.6710	0.5000
	10	1.0000	1.0000	1.0000	0.9997	0.9977	0.9895	0.9653	0.9115	0.8159	0.6762
	11	1.0000	1.0000	1.0000	1.0000	0.9995	0.9972	0.9886	0.9648	0.9129	0.8204
	12	1.0000	1.0000	1.0000	1.0000	0.9999	0.9994	0.9969	0.9884	0.9658	0.9165
	13	1.0000	1.0000	1.0000	1.0000	1.0000	0.9999	0.9993	0.9969	0.9891	0.9682
	14	1.0000	1.0000	1.0000	1.0000	1.0000	1.0000	0.9999	0.9994	0.9972	0.9904
	15	1.0000	1.0000	1.0000	1.0000	1.0000	1.0000	1.0000	0.9999	0.9995	0.9978
	16	1.0000	1.0000	1.0000	1.0000	1.0000	1.0000	1.0000	1.0000	0.9999	0.9996
	17	1.0000	1.0000	1.0000	1.0000	1.0000	1.0000	1.0000	1.0000	1.0000	1.0000

TABLE J—Cont.

n	x	p: 0.05	0.10	0.15	0.20	0.25	0.30	0.35	0.40	0.45	0.50
20	0	0.3585	0.1216	0.0388	0.0115	0.0032	0.0008	0.0002	0.0000	0.0000	0.0000
	1	0.7358	0.3917	0.1756	0.0692	0.0243	0.0076	0.0021	0.0005	0.0001	0.0000
	2	0.9245	0.6769	0.4049	0.2061	0.0913	0.0355	0.0121	0.0036	0.0009	0.0002
	3	0.9841	0.8670	0.6477	0.4114	0.2252	0.1071	0.0444	0.0160	0.0049	0.0013
	4	0.9974	0.9568	0.8298	0.6296	0.4148	0.2375	0.1182	0.0510	0.0189	0.0059
	5	0.9997	0.9887	0.9327	0.8042	0.6172	0.4164	0.2454	0.1256	0.0553	0.0207
	6	1.0000	0.9976	0.9781	0.9133	0.7858	0.6080	0.4166	0.2500	0.1299	0.0577
	7	1.0000	0.9996	0.9941	0.9679	0.8982	0.7723	0.6010	0.4159	0.2520	0.1316
	8	1.0000	0.9999	0.9987	0.9900	0.9591	0.8867	0.7624	0.5956	0.4143	0.2517
	9	1.0000	1.0000	0.9998	0.9974	0.9861	0.9520	0.8782	0.7553	0.5914	0.4119
	10	1.0000	1.0000	1.0000	0.9994	0.9961	0.9829	0.9468	0.8725	0.7507	0.5881
	11	1.0000	1.0000	1.0000	0.9999	0.9991	0.9949	0.9804	0.9435	0.8692	0.7483
	12	1.0000	1.0000	1.0000	1.0000	0.9998	0.9987	0.9940	0.9790	0.9420	0.8684
	13	1.0000	1.0000	1.0000	1.0000	1.0000	0.9997	0.9985	0.9935	0.9786	0.9423
	14	1.0000	1.0000	1.0000	1.0000	1.0000	1.0000	0.9997	0.9984	0.9936	0.9793
	15	1.0000	1.0000	1.0000	1.0000	1.0000	1.0000	1.0000	0.9997	0.9985	0.9941
	16	1.0000	1.0000	1.0000	1.0000	1.0000	1.0000	1.0000	1.0000	0.9997	0.9987
	17	1.0000	1.0000	1.0000	1.0000	1.0000	1.0000	1.0000	1.0000	1.0000	0.9998
	18	1.0000	1.0000	1.0000	1.0000	1.0000	1.0000	1.0000	1.0000	1.0000	1.0000

TABLE K *Distribution of the Studentized Range*†

$1 - \alpha = .95$

Entry $= q_{1-\alpha;\, \nu_1, \nu_2}$ where $P\{q_{\nu_1, \nu_2} < q_{1-\alpha;\, \nu_1, \nu_2}\} = 1 - \alpha$ ν_1 = number of treatments, ν_2 = *d. f.* associated with estimates of variance

ν_2 \ ν_1	2	3	4	5	6	7	8	9	10	11	12	13	14	15	16	17	18	19	20
1	18.0	27.0	32.8	37.1	40.4	43.1	45.4	47.4	49.1	50.6	52.0	53.2	54.3	55.4	56.3	57.2	58.0	58.8	59.6
2	6.09	8.3	9.8	10.9	11.7	12.4	13.0	13.5	14.0	14.4	14.7	15.1	15.4	15.7	15.9	16.1	16.4	16.6	16.8
3	4.50	5.91	6.82	7.50	8.04	8.48	8.85	9.18	9.46	9.72	9.95	10.15	10.35	10.52	10.69	10.84	10.98	11.11	11.24
4	3.93	5.04	5.76	6.29	6.71	7.05	7.35	7.60	7.83	8.03	8.21	8.37	8.52	8.66	8.79	8.91	9.03	9.13	9.23
5	3.64	4.60	5.22	5.67	6.03	6.33	6.58	6.80	6.99	7.17	7.32	7.47	7.60	7.72	7.83	7.93	8.03	8.12	8.21
6	3.46	4.34	4.90	5.31	5.63	5.89	6.12	6.32	6.49	6.65	6.79	6.92	7.03	7.14	7.24	7.34	7.43	7.51	7.59
7	3.34	4.16	4.68	5.06	5.36	5.61	5.82	6.00	6.16	6.30	6.43	6.55	6.66	6.76	6.85	6.94	7.02	7.09	7.17
8	3.26	4.04	4.53	4.89	5.17	5.40	5.60	5.77	5.92	6.05	6.18	6.29	6.39	6.48	6.57	6.65	6.73	6.80	6.87
9	3.20	3.95	4.42	4.76	5.02	5.24	5.43	5.60	5.74	5.87	5.98	6.09	6.19	6.28	6.36	6.44	6.51	6.58	6.64
10	3.15	3.88	4.33	4.65	4.91	5.12	5.30	5.46	5.60	5.72	5.83	5.93	6.03	6.11	6.20	6.27	6.34	6.40	6.47
11	3.11	3.82	4.26	4.57	4.82	5.03	5.20	5.35	5.49	5.61	5.71	5.81	5.90	5.99	6.06	6.14	6.20	6.26	6.33
12	3.08	3.77	4.20	4.51	4.75	4.95	5.12	5.27	5.40	5.51	5.62	5.71	5.80	5.88	5.95	6.03	6.09	6.15	6.21
13	3.06	3.73	4.15	4.45	4.69	4.88	5.05	5.19	5.32	5.43	5.53	5.63	5.71	5.79	5.86	5.93	6.00	6.05	6.11
14	3.03	3.70	4.11	4.41	4.64	4.83	4.99	5.13	5.25	5.36	5.46	5.55	5.64	5.72	5.79	5.85	5.92	5.97	6.03
15	3.01	3.67	4.08	4.37	4.60	4.78	4.94	5.08	5.20	5.31	5.40	5.49	5.58	5.65	5.72	5.79	5.85	5.90	5.96
16	3.00	3.65	4.05	4.33	4.56	4.74	4.90	5.03	5.15	5.26	5.35	5.44	5.52	5.59	5.66	5.72	5.79	5.84	5.90
17	2.98	3.63	4.02	4.30	4.52	4.71	4.86	4.99	5.11	5.21	5.31	5.39	5.47	5.55	5.61	5.68	5.74	5.79	5.84
18	2.97	3.61	4.00	4.28	4.49	4.67	4.82	4.96	5.07	5.17	5.27	5.35	5.43	5.50	5.57	5.63	5.69	5.74	5.79
19	2.96	3.59	3.98	4.25	4.47	4.65	4.79	4.92	5.04	5.14	5.23	5.32	5.39	5.46	5.53	5.59	5.65	5.70	5.75
20	2.95	3.58	3.96	4.23	4.45	4.62	4.77	4.90	5.01	5.11	5.20	5.28	5.36	5.43	5.49	5.55	5.61	5.66	5.71
24	2.92	3.53	3.90	4.17	4.37	4.54	4.68	4.81	4.92	5.01	5.10	5.18	5.25	5.32	5.38	5.44	5.50	5.54	5.59
30	2.89	3.49	3.84	4.10	4.30	4.46	4.60	4.72	4.83	4.92	5.00	5.08	5.15	5.21	5.27	5.33	5.38	5.43	5.48
40	2.86	3.44	3.79	4.04	4.23	4.39	4.52	4.63	4.74	4.82	4.91	4.98	5.05	5.11	5.16	5.22	5.27	5.31	5.36
60	2.83	3.40	3.74	3.98	4.16	4.31	4.44	4.55	4.65	4.73	4.81	4.88	4.94	5.00	5.06	5.11	5.16	5.20	5.24
120	2.80	3.36	3.69	3.92	4.10	4.24	4.36	4.48	4.56	4.64	4.72	4.78	4.84	4.90	4.95	5.00	5.05	5.09	5.13
∞	2.77	3.31	3.63	3.86	4.03	4.17	4.29	4.39	4.47	4.55	4.62	4.68	4.74	4.80	4.85	4.89	4.93	4.97	5.01

† Reprinted with permission from E. S. Pearson and H. O. Hartley, *Biometrika Tables for Statisticians*, 2nd ed., New York, Cambridge University Press, 1954.

TABLE K—*Cont.*

$1 - \alpha = .99$

ν_2 \ ν_1	2	3	4	5	6	7	8	9	10	11	12	13	14	15	16	17	18	19	20
1	90.0	135	164	186	202	216	227	237	246	253	260	266	272	277	282	286	290	294	298
2	14.0	19.0	22.3	24.7	26.6	28.2	29.5	30.7	31.7	32.6	33.4	34.1	34.8	35.4	36.0	36.5	37.0	37.5	37.9
3	8.26	10.6	12.2	13.3	14.2	15.0	15.6	16.2	16.7	17.1	17.5	17.9	18.2	18.5	18.8	19.1	19.3	19.5	19.8
4	5.61	8.12	9.17	9.96	10.6	11.1	11.5	11.9	12.3	12.6	12.8	13.1	13.3	13.5	13.7	13.9	14.1	14.2	14.4
5	5.70	6.97	7.80	8.42	8.91	9.32	9.67	9.97	10.24	10.48	10.70	10.89	11.08	11.24	11.40	11.55	11.68	11.81	11.93
6	5.24	6.33	7.03	7.56	7.97	8.32	8.61	8.87	9.10	9.30	9.49	9.65	9.81	9.95	10.08	10.21	10.32	10.43	10.54
7	4.95	5.92	6.54	7.01	7.37	7.68	7.94	8.17	8.37	8.55	8.71	8.86	9.00	9.12	9.24	9.35	9.46	9.55	9.65
8	4.74	5.63	6.20	6.63	6.96	7.24	7.47	7.68	7.87	8.03	8.18	8.31	8.44	8.55	8.66	8.76	8.85	8.94	9.03
9	4.60	5.43	5.96	6.35	6.66	6.91	7.13	7.32	6.49	7.65	7.78	7.91	8.03	8.13	8.25	8.32	8.41	8.49	8.57
10	4.48	5.27	5.77	6.14	6.43	6.67	6.87	7.05	7.21	7.36	7.48	7.60	7.71	7.81	7.91	7.99	9.07	8.15	8.22
11	4.39	5.14	5.62	5.97	6.25	6.48	6.67	6.84	6.99	7.13	7.25	7.36	7.46	7.56	7.65	7.73	7.81	7.88	7.95
12	4.32	5.04	5.50	5.84	6.10	6.32	6.51	6.67	6.81	6.94	7.06	7.17	7.26	7.36	7.44	7.52	7.59	7.66	7.73
13	4.26	4.96	5.40	5.73	5.98	6.19	6.37	6.53	6.67	6.79	6.90	7.01	7.10	7.19	7.27	7.34	7.42	7.48	7.55
14	4.21	4.89	5.32	5.63	5.88	6.08	6.26	6.41	6.54	6.66	6.77	6.87	6.96	7.05	7.12	7.20	7.27	7.33	7.39
15	4.17	4.83	5.25	5.56	5.80	5.99	6.16	6.31	6.44	6.55	6.66	6.76	6.84	6.93	7.00	7.07	7.14	7.20	7.26
16	4.13	4.78	5.19	5.49	5.72	5.92	6.08	6.22	6.35	6.46	6.56	6.66	6.74	6.82	6.90	6.97	7.03	7.09	7.15
17	4.10	4.74	5.14	5.43	5.66	5.85	6.01	6.15	6.27	6.38	6.48	6.57	6.66	6.73	6.80	6.87	6.94	7.00	7.05
18	4.07	4.70	5.09	5.38	5.60	5.79	5.94	6.08	6.20	6.31	6.41	6.50	6.58	6.65	6.72	6.79	6.85	6.91	6.96
19	4.05	4.67	5.05	5.33	5.55	5.73	5.89	6.02	6.14	6.25	6.34	6.43	6.51	6.58	6.65	6.72	6.78	6.84	6.89
20	4.02	4.64	5.02	5.29	5.51	5.69	5.84	5.97	6.09	6.19	6.29	6.37	6.45	6.52	6.59	6.65	6.71	6.76	6.82
24	3.96	4.54	4.91	5.17	5.37	5.54	5.69	5.81	5.92	6.02	6.11	6.19	6.26	6.33	6.39	6.45	6.51	6.56	6.61
30	3.89	4.45	4.80	5.05	5.24	5.40	5.54	5.65	5.76	5.85	5.93	6.01	6.08	6.14	6.20	6.26	6.31	6.36	6.41
40	3.82	4.37	4.70	4.93	5.11	5.27	5.39	5.50	5.60	5.69	5.77	5.84	5.90	5.96	6.02	6.07	6.12	6.17	6.21
60	3.76	4.28	4.60	4.82	4.99	5.13	5.25	5.36	5.45	5.53	5.60	5.67	5.73	5.79	5.84	5.89	5.93	5.98	6.02
120	3.70	4.20	4.50	4.71	4.87	5.01	5.12	5.21	5.30	5.38	5.44	5.51	5.56	5.61	5.66	5.71	5.75	5.79	5.83
∞	3.64	4.11	4.40	4.60	4.76	4.88	4.99	5.08	5.16	5.23	5.29	5.35	5.40	5.45	5.49	5.54	5.57	5.61	5.65

TABLE L *Factors for Computing Control Chart Lines*†

Number of Observations in Sample, n	Chart for Averages: Factors for Control Limits			Chart for Standard Deviations: Factors for Central Line		Chart for Standard Deviations: Factors for Control Limits				Chart for Ranges: Factors for Central Line		Chart for Ranges: Factors for Control Limits					
	A	A_1	A_2	c_2	$1/c_2$	B_1	B_2	B_3	B_4	d_2	$1/d_2$	d_3	D_1	D_2	D_3	D_4
2	2.121	3.760	1.880	0.5642	1.7725	0	1.843	0	3.267	1.128	0.8865	0.853	0	3.686	0	3.267
3	1.732	2.394	1.023	0.7236	1.3820	0	1.858	0	2.568	1.693	0.5907	0.888	0	4.358	0	2.575
4	1.500	1.880	0.729	0.7979	1.2533	0	1.808	0	2.266	2.059	0.4857	0.880	0	4.698	0	2.282
5	1.342	1.596	0.577	0.8407	1.1894	0	1.756	0	2.089	2.326	0.4299	0.864	0	4.918	0	2.115
6	1.225	1.410	0.483	0.8686	1.1512	0.026	1.711	0.030	1.970	2.534	0.3946	0.848	0	5.078	0	2.004
7	1.134	1.277	0.419	0.8882	1.1259	0.105	1.672	0.118	1.882	2.704	0.3698	0.833	0.205	5.203	0.076	1.924
8	1.061	1.175	0.373	0.9027	1.1078	0.167	1.638	0.185	1.815	2.847	0.3512	0.820	0.387	5.307	0.136	1.864
9	1.000	1.094	0.337	0.9139	1.0942	0.219	1.609	0.239	1.761	2.970	0.3367	0.808	0.546	5.394	0.184	1.816
10	0.949	1.028	0.308	0.9227	1.0837	0.262	1.584	0.284	1.716	3.078	0.3249	0.797	0.687	5.469	0.223	1.777
11	0.905	0.973	0.285	0.9300	1.0753	0.299	1.561	0.321	1.679	3.173	0.3152	0.787	0.812	5.534	0.256	1.744
12	0.866	0.925	0.266	0.9359	1.0684	0.331	1.541	0.354	1.646	3.258	0.3069	0.778	0.924	5.592	0.284	1.716
13	0.832	0.884	0.249	0.9410	1.0627	0.359	1.523	0.382	1.618	3.336	0.2998	0.770	1.026	5.646	0.308	1.692
14	0.802	0.848	0.235	0.9453	1.0579	0.384	1.507	0.406	1.594	3.407	0.2935	0.762	1.121	5.693	0.329	1.671
15	0.775	0.816	0.223	0.9490	1.0537	0.406	1.492	0.428	1.572	3.472	0.2880	0.755	1.207	5.737	0.348	1.652
16	0.750	0.788	0.212	0.9523	1.0501	0.427	1.478	0.448	1.552	3.532	0.2831	0.749	1.285	5.779	0.364	1.636
17	0.728	0.762	0.203	0.9551	1.0470	0.445	1.465	0.466	1.534	3.588	0.2787	0.743	1.359	5.817	0.379	1.621
18	0.707	0.738	0.194	0.9576	1.0442	0.461	1.454	0.482	1.518	3.640	0.2747	0.738	1.426	5.854	0.392	1.608
19	0.688	0.717	0.187	0.9599	1.0418	0.477	1.443	0.497	1.503	3.689	0.2711	0.733	1.490	5.888	0.404	1.596
20	0.671	0.697	0.180	0.9619	1.0396	0.491	1.433	0.510	1.490	3.735	0.2677	0.729	1.548	5.922	0.414	1.586
21	0.655	0.679	0.173	0.9638	1.0376	0.504	1.424	0.523	1.477	3.778	0.2647	0.724	1.606	5.950	0.425	1.575
22	0.640	0.662	0.167	0.9655	1.0358	0.516	1.415	0.534	1.466	3.819	0.2618	0.720	1.659	5.979	0.434	1.566
23	0.626	0.647	0.162	0.9670	1.0342	0.527	1.407	0.545	1.455	3.858	0.2592	0.716	1.710	6.006	0.443	1.557
24	0.612	0.632	0.157	0.9684	1.0327	0.538	1.399	0.555	1.445	3.895	0.2567	0.712	1.759	6.031	0.452	1.548
25	0.600	0.619	0.153	0.9696	1.0313	0.548	1.392	0.565	1.435	3.931	0.2544	0.709	1.804	6.058	0.459	1.541
Over 25	$\frac{3}{\sqrt{n}}$	$\frac{3}{\sqrt{n}}$	...	...	...	*	**	*	**	...	...	...	...	...	...	...

*$1 - \frac{3}{\sqrt{2n}}$ **$1 + \frac{3}{\sqrt{2n}}$

† Reproduced by permission from *ASTM Manual on Quality Control of Materials:* Philadelphia, Pa., American Society for Testing Materials, 1951

ANSWERS TO SELECTED EXERCISES

Chapter 1

1.3. (a) $A' = \{4, 5, 6, 7\}$; $B' = \{1, 3, 5, 7\}$; $C' = \{2, 4, 6\}$ (b) $\{1, 2, 3, 4, 6\}$ (c) $\{2\}$ (d) $\varnothing$

1.10. $\frac{63}{64}$

1.12. 0.30

1.17. (a) 0.495 (b) 0.36 (c) 0.115 (d) 0.74

1.22. 0.75

1.26. 0.625

1.31. $\frac{4}{17}$

1.35. 120

1.39. 41

1.40. 0.0799

Chapter 2

2.2. \$5500

2.6. $Z(\omega) = \omega \qquad \omega = 3, 4, \ldots, 18$

2.10. (a) $1 - e^{-.01a} \qquad 0 \le a \le \infty$ (b) 0.301 (c) 0.081

2.14. $\frac{1}{24}$

2.18. 30

2.19. (a) 1 (b) $1 - e^{-t^2/2} \qquad 0 \le t \le \infty$ (c) 0.368

2.24. (a) 0.8 (b) $1 - (0.2)^k \qquad k = 1, 2, \ldots$ (c) 0.008

Chapter 3

3.2. (a) $f_{X_1,X_2}(t_1, t_2) = \begin{cases} 1 & 0 < t_1 < 1 \quad 0 < t_2 < 1 \\ 0 & \text{otherwise} \end{cases}$ (b) 0.25

3.4. (a) $f_{X_1,X_2}(t_1, t_2) = \begin{cases} t_1 t_2 e^{-(1/2)(t_1^2 + t_2^2)} & t_1 \ge 0 \quad t_2 \ge 0 \\ 0 & \text{otherwise} \end{cases}$ (b) 0.3402

3.7. (a)

k:	2	3	4	5	6
$P_Y(k)$:	0.04	0.16	0.32	0.32	0.16

(b)

k:	0	1	2
$P_Y(k)$:	0.36	0.48	0.16

3.10. (a) $\frac{1}{2}$ (b) $F_{X_1,X_2}(t_1,t_2) = \begin{cases} \frac{1}{2}(t_1-1)(t_2-1) & 1 \le t_1 \le 3 \quad 1 \le t_2 \le 2 \\ t_2 - 1 & 3 \le t_1 \quad 1 \le t_2 \le 2 \\ \frac{1}{2}(t_1-1) & 1 \le t_1 \le 3 \quad 2 \le t_2 \\ 1 & t_1 \ge 3 \quad t_2 \ge 2 \end{cases}$

(c) $\frac{3}{4}$

3.13. (a) $P_{X_A}(x_A)P_{X_B}(x_B)$ for all X_A and X_B (b) 0.15 (c) 0.375 (d) 0.0085

3.15. (a) x_1, x_2, x_3: 0, 0, 1 0, 1, 0 1, 0, 0

$P_{X_1,X_2,X_3}(x_1,x_2,x_3)$: $\frac{1}{3}$; $\frac{1}{3}$; $\frac{1}{3}$

(b) $P_{X_1}(x_1) = \begin{cases} \frac{2}{3} & x_1 = 0 \\ \frac{1}{3} & x_1 = 1 \end{cases}$

3.17. (a) $\frac{1}{19}(5 + 3a_1)$ $a_1 = 1, 2$ (b) $\frac{1}{19}(3 + 2a_2^2)$ $a_2 = 0, 1, 2$

(c) $\frac{a_1 + a_2^2}{3 + 2a_2^2}$ $a_1 = 1, 2$ $a_2 = 0, 1, 2$

3.21. (a) $f_{X_1}(a_1) = 3a_1^2$ $0 < a_1 < 1$ $f_{X_2}(a_2) = \frac{3}{2}(1 - a_2^2)$ $0 < a_2 < 1$

(b) $\frac{1}{a_1}$ $0 < a_2 < a_1$ (c) X_1 and X_2 are not independent

3.24. $\frac{11}{16}$

3.28. (a) 32 (b) $\frac{5}{48}$

3.30. (a) $f_{X_1}(a_1) = 4a_1(1 - a_1^2)$ $0 < a_1 < 1$ $f_{X_2}(a_2) = 4a_2^3$ $0 < a_2 < 1$
(b) X_1 and X_2 are not independent (c) $\frac{2a_1}{a_2^2}$ $0 < a_1 < a_2$ (d) 0.25 (e) 0.40

Chapter 4

4.3. $f_Y(a) = \begin{cases} \frac{1}{a} & 1 < a < e \\ 0 & \text{otherwise} \end{cases}$

4.5. $f_Y(a) = \begin{cases} \frac{e^{-\sqrt{a/3}}}{2\sqrt{3a}} & a \ge 0 \\ 0 & \text{otherwise} \end{cases}$

4.8. $P_G(a) = \begin{cases} \frac{2}{3} & a = 1 \\ \frac{1}{3} & a = 2 \end{cases}$

4.11. $P_Y(a) = \begin{cases} 0.1 & a = 0 \\ 0.6 & a = 3 \\ 0.3 & a = 6 \end{cases}$

4.15. $f_Y(a) = \begin{cases} \frac{4}{3}a & 0 \le a \le 1 \\ \frac{2}{3}a & 1 < a \le \sqrt{2} \end{cases}$

4.20. $f_Y(a) = \frac{3}{2}\left(\frac{3}{2} + a\right)^{-2}$ $a \ge 0$

4.23. $f_Y(a) = e^{-a}\frac{a^2}{2} \qquad a \geq 0$

4.28. $f_{U,V}(u, v) = \begin{cases} \frac{1}{8}e^{-(1/2)u} & u \geq 0 \qquad -u \leq v \leq u \\ 0 & \text{otherwise} \end{cases}$

Chapter 5

5.2. (a) 1 (b) 1 (c) $\frac{13}{12}$

5.4. $E(X) = 0$; $\text{Var}(X) = \frac{1}{2}$

5.8. $E(X) = \sqrt{\frac{\pi}{2}}$; $\text{Var}(X) \cong 2.429$

5.13. (a) 4.5 (b) $\frac{1}{3}$ (c) $e - 1$

5.15. 2

5.17. (a) 1 (b) 2 (c) 1

5.21. (a) $-\frac{1}{144}$ (b) $-\frac{5}{43}$

5.24. 2001

5.26. $\frac{3}{8}$

5.31. (a) $\frac{0.01}{0.01 - \theta}e^{10\theta}$ (b) 110; 10,000

5.34. $\frac{0.10}{\theta}(e^{11\theta} - e^{\theta})$

5.39. 0.057

5.40. 0.841

Chapter 6

6.2. (a) 0.441 (b) 0.0308 (c) 0.4694

6.6. (a) 0.9786 (b) 15

6.8. (a) 0.87 (b) 0.029 (c) 0.333 (d) 0.725

6.11. 23

6.15. $\frac{5}{1296}$

6.18. 0.2525

6.21. 0.376

6.26. 0.8125

6.30. 0.4987

6.32. 0.695

6.36. 78.12

6.39. (a) $2981 (b) 0.058

6.42. (a) 0.368 (b) 0.129 (c) 0.67

6.48. 0.20

6.52. (a) 0.01 (b) 0.60 (c) 0.199

6.54. (a) 0.25 (b) 0.90

6.58. $\frac{1}{2} + \frac{1}{\pi}\arctan(t) \qquad -\infty < t < \infty$

6.61. (a) 0.1056 (b) 0.6826

6.64. (a) 0.816 (b) 0.894

Chapter 7

7.1. (a) 0.47 (b) 2.75

7.5. (a) $\frac{1}{1024}$ (b) $\simeq 1$ (c) 0.944

7.7. (a) 0.52 (b) 1.25

7.11. $g(v) = \begin{cases} \frac{3}{16}v^3(4 - v^2) & 0 \le v \le 1 \\ \frac{3}{32}v(4 - v) & 1 < v \le 4 \end{cases}$

7.14. $f_R(r) = \frac{3}{2}r(1 - r)(5 - 2r) \quad 0 \le r \le 1$

7.16. 0.472

7.18. $n \simeq 38$

Chapter 9

9.1. 0.518

9.4. (a) \$20 (b) \$17.2 (c) \$18.32

9.6. $\sum t_i > \frac{\chi^2_{1-\alpha;\,4n}}{2\lambda_0}$

9.9. $\sum \left(\frac{x_i - \mu_0}{\sigma_0}\right)^2 \le \chi^2_{\alpha;\,n}$

9.12. The company's claim is not true; $\pi = 0.954$

9.14. Reject

9.18. The variance of the challenger does not differ from one

9.21. (a) Reject the producer's claim (b) 0.115 (c) $n \simeq 11$

9.24. 36

9.28. (a) 0.2401 (b) $\simeq .518$

9.30. (a) Accept H_0 (b) 15

9.36. 12

9.40. (a) Percentage defective is greater than 5% (b) 0.984

Chapter 10

10.1. 87

10.4. $n_X = 15$; $n_Y = 40$

10.7. (a) Reject H_0 (b) 0.921

10.9. (a) Reject H_0 (b) 0.24

10.13. $n = m = 8$

10.17. (a) Accept equality (b) 0.19

10.21. Variances are equal; $n = m \simeq 41$

10.27. Accept H_0

10.30. $n = m \simeq 15$

10.35. 18

10.40. Accept H_0.

10.42. Vaccine *B* is superior

10.45. Variances are equal

Chapter 11

11.2. Accept H_0.
11.4. Methods are the same.
11.8. Accept equality.
11.10. The hypothesis of randomness is rejected.
11.16. The distributions are not the same, since $7.687 > \chi^2_{.95;\,2} = 5.991$
11.20. The hypothesis of independence is rejected.

Chapter 12

12.2. $\hat{\theta}_2$ is preferred
12.3. $\hat{\theta} = \frac{30}{77}X_1 + \frac{20}{77}X_2 + \frac{15}{77}X_3 + \frac{12}{77}X_4$
12.8. 90%
12.9. $\frac{\sigma^2}{2n}$
12.13. $\hat{\lambda} = \frac{2}{\bar{t}}$
12.14. Largest observation (X_n)
12.17. $\hat{q} = \frac{\sum k_i}{n + \sum k_i}$
12.21. [3.87; 4.10]
12.25. [0.456; 2.6]
12.28. [4.8; 15.2]
12.30. [0.033; 0.218]

Chapter 13

13.3. (a)

SOV	*SS*	*df*	*MS*	*F-ratio*
Batteries	615.6	2	307.8	17.07
Within	216.4	12	18.03	
Total	832	14		

(b) Since $F_{.95,2,12} = 3.89$, the batteries are significantly different
13.6. Lubricants (A and D), (A and C), and (B and D) are different; [1.606, 1.734]
13.8. 0.98
13.9. Scheme (a) is better
13.12. A and D differ
13.13. (a)

SOV	*SS*	*df*	*MS*	*EMS*
Transformers	10.44	2	5.22	$\sigma_e^2 + 3\sigma_I^2 + 15\sigma_{\text{Tr.}}^2$
Testers	5.45	4	1.36	$\sigma_e^2 + 3\sigma_I^2 + 9\sigma_{\text{Test}}^2$
Interaction	20.23	8	2.52	$\sigma_e^2 + 3\sigma_I^2$
Within	43.2	30	1.44	σ_e^2
Total	79.32	44		

(b) Since $\frac{5.22}{2.52} = 2.07 < F_{.95;\,2,\,8}$, conclude that the transformers are homogeneous.
(c) 0.25
13.17. (a)

SOV	*SS*	*df*	*MS*	*F-ratio*
Temperature	0.732	4	0.183	4.094

Within	0.447	10	0.0447
Total	1.179	14	

Temperature has an effect on yield since $4.094 > F_{.95;\,4,\,10} = 3.48$ (b) Linear response is significant (c) 0.0149

13.23. (a)

SOV	*SS*	*df*	*MS*	*F-ratio*
Temperature	11.50	3	3.83	0.707
Concentration	44.33	2	22.16	4.09
Interaction	27.00	6	4.50	0.83
Within	65.00	12	5.41	
Total	147.83	23		

(b) Concentration effects are significant, whereas temperature and interaction effects are not significant

13.29. (a)

SOV	*SS*	*df*	*MS*	*F-ratio*
Formulas	127.5	2	63.75	7.75
Ovens	6.6	2	3.3	0.192
Interaction	32.9	4	8.225	0.478
Within	621.6	36	17.266	
Total	788.6	44		

(b) Interaction and Oven effects are not significant. Formula effects are barely significant (c) Formulas A and B are different.

13.32. (a)

SOV	*SS*	*df*	*MS*	*F-ratio*
Engines	102.77	2	51.38	3.41
Dynamometers	38.11	2	19.05	1.26
Interaction	60.22	4	15.05	1.92
Within	70.50	9	7.83	
Total	271.60	17		

(b) All effects are not significant (c) $n = 4$

Chapter 14

14.1. (a) $\tilde{y} = 0.9605 + 0.706x$ (b) 4.1375

14.4. (a) $\tilde{y} = 0.746 + 0.32x$ (b) 0.0925 (c) [1.805; 2.887]

14.7. [6.5135; 7.4465]

14.11. $a + bx \pm 5.381\left[0.1 + \dfrac{(x - 145)^2}{8250}\right]^{1/2}$

14.14. (a) $\tilde{y} = 24.6285 + 0.05886x$ (b) [61.016; 64.755] (c) [17.5334; 31.7037]

14.17. (a) $\tilde{y} = 1300.711 - 11.357x$ (b) [322.188; 575.684] (c) The hypothesis that $B = -10$ is not rejected

14.21. (a) $\tilde{y} = 1.1714 + 0.7767x$ (b) $F = 1.419 < F_{.95;\,5,\,7} = 3.97$. Hence we conclude that linearity prevails (c) [4.27; 5.06]

14.25. $r_{x,y} = 0.992$. Since $\dfrac{|r|\sqrt{4}}{\sqrt{1 - r^2}} = 15.72 > t_{.95;\,4} = 2.132$, ρ is highly significant

14.29. $r_{x,y} = 0.9967$; [0.989, 0.999]

14.31. (a) Since $\dfrac{|0.85|\sqrt{48}}{\sqrt{1 - (0.85)^2}} = 11.17 > t_{.995;\,48} = 2.678$ there is a statistically significant correlation between yield and tensile strength (b) [0.745; 0.914]

Chapter 15

15.3. $\chi^2 = 21.97 > \chi^2_{.95;\,2} = 5.991$. Hence frequencies differ at the 5% level

15.6. H_0 is accepted since $\chi^2 = 4.3 < \chi^2_{.95;\,6} = 12.59$
15.11. H_0 is rejected since $\chi^2 = 29.687 > \chi^2_{.95;\,1} = 6.635$
15.13. H_0 is rejected since $\chi^2 = 1075.2 > \chi^2_{.99;\,6} = 16.812$
15.15. H_0 is accepted since $\chi^2 = 0.661 < \chi^2_{.95;\,2} = 5.991$

Chapter 16

16.6. $\bar{X}' \pm 0.98\sigma'$
16.8. (a) The process is in control (b) 62.4% (c) Revise the specification tolerance or reduce the process variability if possible
16.11. (a) $\text{U.C.L.}_{\bar{X}} = 2.225$; $\text{L.C.L.}_{\bar{X}} = 2.075$ (b) 0.9332
16.13. (a) 0.8413 (b) 0.01
16.16. $\bar{p} = 0.0139$; $\text{U.C.L.}_p = 0.0296$; $\text{L.C.L.}_p = 0$. The shipment appears to be in control.
16.20. (a) 0.608 (b) 0.846
16.24. $\text{U.C.L.}_c = 16.50$; $\text{L.C.L.}_c = 1.50$

Chapter 17

17.3. 0.982
17.6. AOQL $\simeq$ 2.54%
17.11. 1006.76
17.13. $n = 100$; $c = 2$
17.17. 8762.5
17.20. 0.191
17.23. Acceptance line: $0.145n - 3.68$; Rejection line: $0.145n + 5.61$
17.28. Acceptance line: $\sum x_i = 210n - 22.975$; Rejection line: $\sum x_i = 210n + 22.975$

Chapter 18

18.2. 0.286
18.5. (a) $P_0 = 0.2$ (b) 2.8 (c) 0.35
18.10. 0.085
18.12. \$2.867
18.16. 5.5; $1\frac{9}{11}$
18.18. 4.7×10^{-3}
18.21. (a) 0.065 (b) 0.214
18.24. (a) 0.179 (b) 0.63 (c) 2.04

INDEX

D

E

F